MW00836832

FLIES

FLIES

The Natural History & Diversity of Diptera

Stephen A. Marshall

Firefly Books

A FIREFLY BOOK

Published by Firefly Books Ltd. 2012

First printing

Publisher Cataloging-in-Publication Data (U.S.)
Marshall, Stephen A.
Flies : the natural history and diversity of diptera / Stephen A. Marshall.
[616] p. : col. photos. ; cm.
Includes bibliographical references and index.
Summary: History and diversity of flies, with scientific updates.
ISBN-13: 978-1-77085-100-9
1. Flies. 2. Diptera. I. Title.
595.77 dc23 QL533.M26 2012

Library and Archives Canada Cataloguing in Publication
Marshall, S. A. (Stephen Archer)
Flies : the natural history and diversity of Diptera / Stephen A. Marshall.
Includes bibliographical references and index.
ISBN-13: 978-1-77085-100-9
1. Diptera. 2. Flies. I. Title.
QL531.M37 2012 595.77 C2012-901352-8

Published in the United States by
Firefly Books (U.S.) Inc.
P.O. Box 1338, Ellicott Station
Buffalo, New York 14205

Published in Canada by
Firefly Books Ltd.
66 Leek Crescent
Richmond Hill, Ontario L4B 1H1

Project management: Hartley Millson
Cover and interior design: Hartley Millson
Additional formatting: Jolie Dobson
Editing and index: Gillian Watts

Printed in China

The publisher gratefully acknowledges the financial support for our publishing program by the Government of Canada through the Canada Book Fund as administered by the Department of Canadian Heritage.

Image page 2: *Rhingia nasica*, Syrphidae.

Dedication

This book is dedicated to the Diptera taxonomists on whom we depend for the specimens, names and systematic context needed to explore the world of flies. Those experts — the relatively few individuals with the accumulated expertise to identify, organize, describe, key and classify "their" special parts of the fly tree of life — have made it possible to summarize and interpret dipteran diversity here. They need to be recognized not only for their individual generosity in sharing unique knowledge, but also for their collective contribution to our knowledge of flies through publication of the manuals, catalogs, reviews and revisions that form the foundation of dipterology.

INTRODUCTION

The Dominance of Diptera

Most species of animals belong to one of four large orders: the Coleoptera (beetles), Lepidoptera (moths and butterflies), Hymenoptera (ants, wasps, bees and sawflies) or Diptera (flies). Of these megadiverse groups, the flies are arguably the most important, if only because they kill millions of people a year by transmitting our most devastating diseases. But this enormous impact is due to only a few dozen species, and dipteran damage to forests, crops and stored products is similarly due to mere dozens of different fly genera. Vastly more — thousands of species — are beneficial, contributing to the pollination of plants, biological control of pest insects, and disposal of the dung, carrion and other organic matter that would otherwise quickly carpet the planet. Despite all this, a relatively large proportion of the species in the order Diptera remains undiscovered, unnamed or unidentifiable. This is perhaps in part because of the natural attraction of insect enthusiasts to shining beetles and colorful moths, and perhaps in part because so many groups of flies are relatively small and soft-bodied, and thus more difficult to preserve and study. Up until very recently the study of most groups of flies was also rendered more challenging by a dearth of accessible literature.

The 160,000 or so species of flies so far discovered and described represent just over 10% of named animal species, but it is anybody's guess how many species remain to be formally named. The number of named species in the order is currently expanding by about 1 percent per year; at that rate it will be a while before we know the real number, but we probably share the planet with between 400,000 and 800,000 fly species. Of course an indeterminate number of species still await discovery in the other megadiverse orders as well, especially among the parasitic Hymenoptera, but it seems likely that at least 15–20% of all animal species are flies. And in most terrestrial and aquatic habitats the proportion of individuals is much higher, with adult flies generally comprising 35–75% of the insect specimens taken by sampling devices ranging from aerial plankton nets down to interception, Malaise and pan traps at ground level. You can get a sense of this by sweeping a net around a wetland or forest and examining the entire catch, or by scrutinizing the splatters on your car windshield. See for yourself — flies rule!

An adult fly looks like a fly mostly because of the way its muscle-packed thorax is dominated by a bulging middle segment (mesothorax) that supports its single pair of wings, which are the source of the name *diptera*, coined by the Greek philosopher Aristotle more than 2,000 years ago — *di*

OPPOSITE PAGE
Gall midges massed on a leaf in Costa Rica.

Figure 1.1 ***Basic Fly Anatomy***

The insect thorax is divided into three leg-bearing segments, with the second and third normally supporting wings as well as legs. Only in the Diptera is the middle thoracic segment greatly enlarged to support a single pair of wings. The third thoracic segment is reduced and bears only a small pair of halteres in place of the second pair of wings. The "palpus" in Diptera is the maxillary palpus. Each labial palpus is reduced to a fleshy labellum, and these in turn fuse into a characteristic cuplike lobe (also often called a labellum) that this tachinid fly is using to sponge up honeydew from a leaf surface.

means "two" and *ptera* means "wings." With a few distinctive exceptions (including some mayflies, male scales, male strepsipterans, a few barklice, a few grasshoppers, the odd minute parasitic wasp and the occasional planthopper), other winged insects have four wings. Flies, however, differ from other insects in having their second pair of wings reduced to a pair of characteristically stalked and clublike halteres on a dramatically atrophied third thoracic segment, the metathorax. Each halter, which looks like a knob on a stick between the thorax and abdomen, wobbles up and down on a base full of sense organs, serving as a gyroscopic flight indicator that helps maintain stability during flight. Adult flies conspicuously rule the air with their marathon flights and astonishing acrobatics. Larval flies, in contrast, are generally pallid denizens of mucky and hidden underworlds such as moist decomposing materials or the insides of plant or animal tissues.

Flies have traditionally been divided into two suborders: Nematocera (loosely translated as "thread horn") and Brachycera ("short horn"). The suborder Nematocera is a heterogeneous assemblage of generally delicate flies, such as mosquitoes and crane flies, with long appendages. Brachycera, in contrast, are generally more compact flies — House Flies, horse flies and fruit flies, for instance — with relatively short antennae. Flies are usually easily recognizable as either Nematocera or Brachycera, but the flies traditionally treated as Nematocera make up an artificial group characterized by primitive or default (plesiomorphic) characteristics similar to those found in closely related orders (such as Mecoptera). The suborder Brachycera, on the other hand, is a natural, or *monophyletic* group, defined by characteristics (including the shortened antennae and palpi) that originated in a common ancestor not shared with other groups.

This sort of traditional classification of organisms into groups defined by the presence of new characteristics (*apomorphies*, such as shortened antennae) versus other groups recognized by the absence of those derived characteristics is intuitive, but modern classification systems strive to use only natural groups characterized by new (apomorphic) characteristics. Groups defined on the basis of plesiomorphies are being replaced as classifications are changed to approximate strict reflections of phylogeny. Such classifications name only monophyletic, or natural, groups (a monophyletic group, or *clade*, includes an ancestor and all of its descendants).

Thus, artificial groups such as the Nematocera (that is, flies that lack the special characteristics of Brachycera) are disappearing from formal classifications even though their names remain in common use. The informal terms "lower Diptera" and "nematoceran flies" are used in this book to refer to the old suborder Nematocera in a way that reflects this change in classification. Similar problems arise at various levels throughout the order. For example, the division of the Cyclorrhapha into Aschiza and Schizophora recognizes one natural group — the Schizophora — defined by a remarkable ability to open up the front of the head to pump out an airbag-like structure, leaving a scar-like ptilinal fissure, or "schism," and another assemblage of families — the Aschiza — defined by the absence of a ptilinal fissure (*aschiza* means "without schism"). Predictably, the latter assemblage is now widely considered to be artificial.

Such artificial groups often comprise the basal lineages of a larger group. The "Aschiza" can be envisaged as the basal lineages of the Cyclorrhapha, and nematoceran flies can be seen as the basal lineages of the order Diptera. Think of a tree with just a few low branches coming off the trunk, well below the multi-branched leafy main canopy. Those branches could be considered the "basal lineages" of the entire tree, and isolated twigs near the base of a main branch would be basal lineages of that branch. The higher classification of Diptera is more fully discussed in Part Two and summarized in Figure 1.2.

There is a voluminous literature on the biology and diversity of flies, much of it encapsulated in the Diptera catalogs and manuals that provide an overview of the fly faunas of various regions. The Diptera catalogs, with chapters by leading experts on various families, list all known fly species and their distributions in the Nearctic (Stone et al. 1965), Neotropical (Papavero 1966–78), Afrotropical (Crosskey 1980), Palaearctic (Soos and Papp, 1984–93), Oriental (Delfinado and Hardy 1973–77) and Pacific (Evenhuis 1989) regions. World catalogs have also appeared for several families of flies such as the Tipulidae, Stratiomyidae, Bombyliidae, Sarcophagidae, Ephydridae, Empididae, Dolichopodidae, Sepsidae and Sphaeroceridae, and updated information about fly names can be found in the database of world Diptera at http://www.diptera.org (Pape and Thompson 2010).

The Diptera manuals, covering the Nearctic (McAlpine et al. 1981, 1987, 1989), Palaearctic (Papp and Darvas 1997, 1998, 2000), Neotropical (Brown et al. 2009, 2010) and Afrotropical (Kirk-Spriggs et al., in preparation) regions, are also collaborative efforts, with chapters written by fly taxonomists from around the world (often the same experts who wrote the catalog chapters and not infrequently the same experts I turned to for help with this book). The regional Diptera manuals include exhaustive treatments of adult and larval morphology as well as keys to the families and genera that occur in their respective regions; they are essential reference works for serious students of Diptera. The manuals are strongly influenced by earlier works, of which arguably the most important are the books and papers of Willi Hennig, especially his 1973 textbook treatment of the order. Fewer works interpret dipterology for the general reader or even for non-specialists; the classic accessible work on Diptera biology is Harold Oldroyd's 1964 *The Natural History of Flies*. Oldroyd not only made dipterology interesting for a wide range of students, naturalists and general readers but also conveyed the excitement of the field in a way that influenced many of today's fly specialists to choose their research paths.

The present book is meant to follow in the footsteps of Oldroyd in making the world of flies accessible to a wide readership, but with an updated, more comprehensive and more detailed coverage than that of Oldroyd's slim volume. Part One presents an overview of the biology of flies and considers their interactions with plants and animals. Part Two provides a brief introduction to the evolution and distribution of flies, followed by a family-by-family overview of the world's fly families and subfamilies. Part Three includes an illustrated key to the world's families of flies, less technical and hopefully a bit more user-friendly than the more comprehensive regional keys found in the manuals.

Figure 1.2 *The Main Subgroups of the Order Diptera*

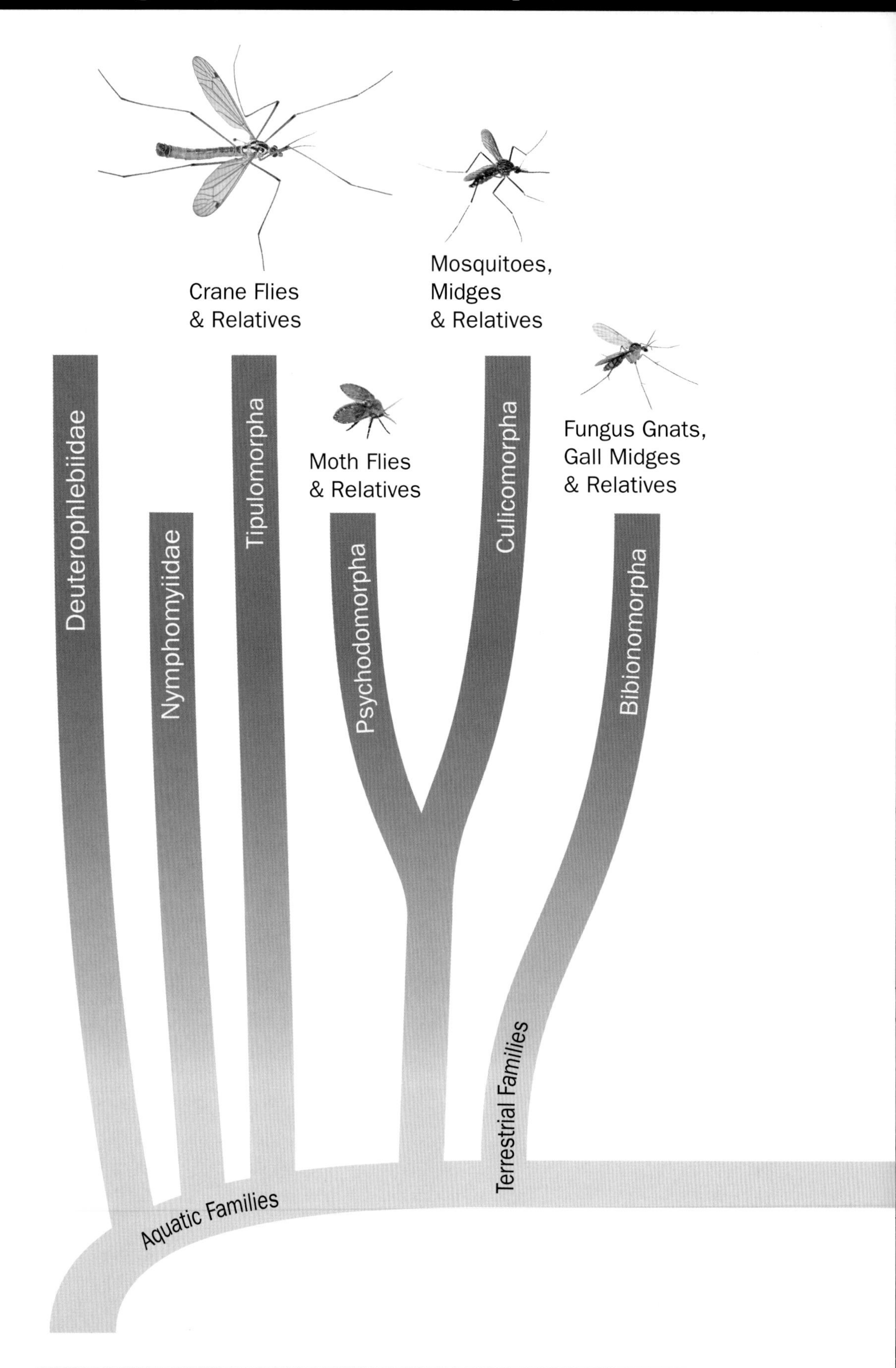

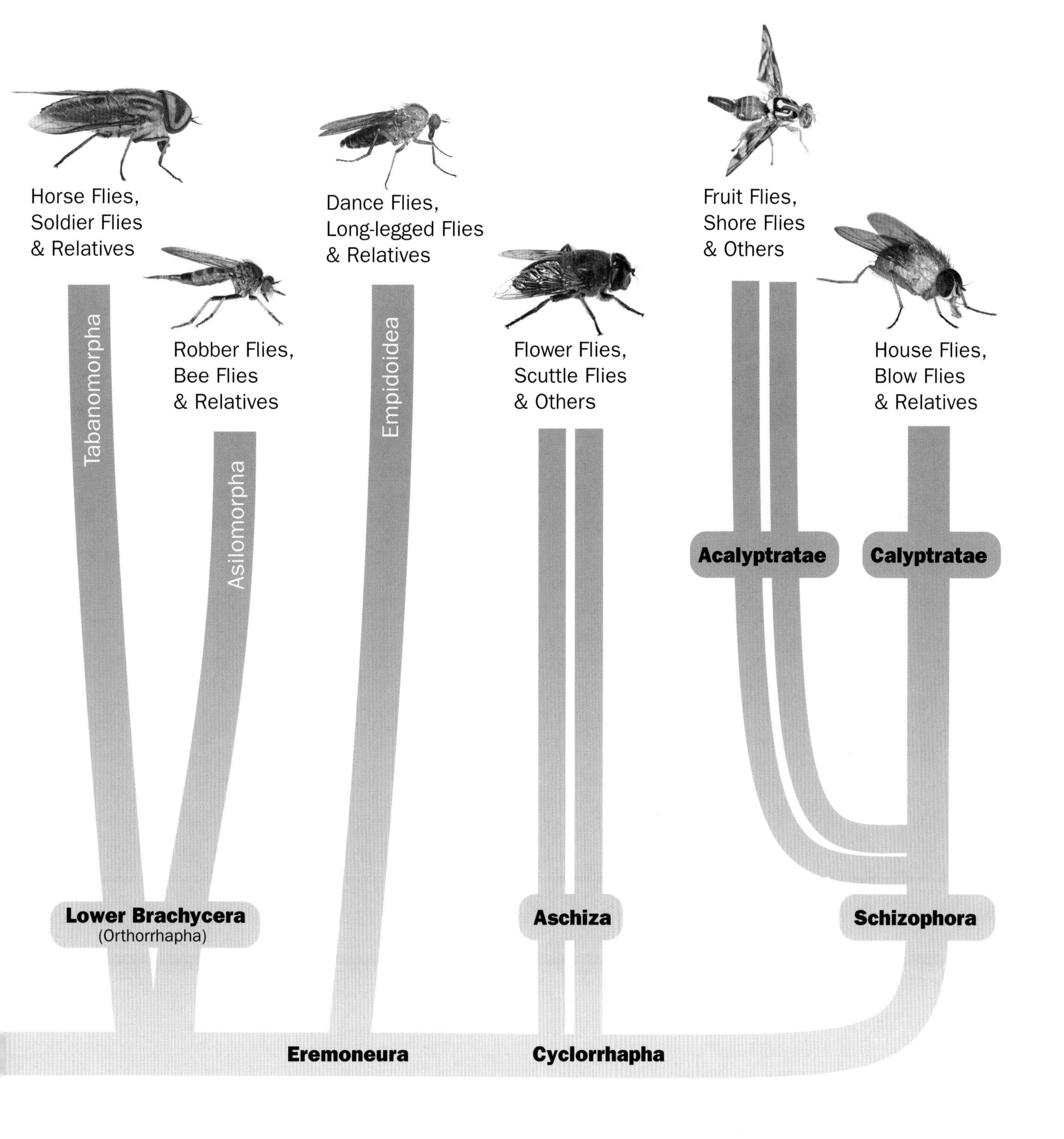

Brachycera (Higher Diptera)

PART 1

CHAPTER 1

CHAPTER 2

Life Histories, Habits and Habitats of Flies

CHAPTER 3

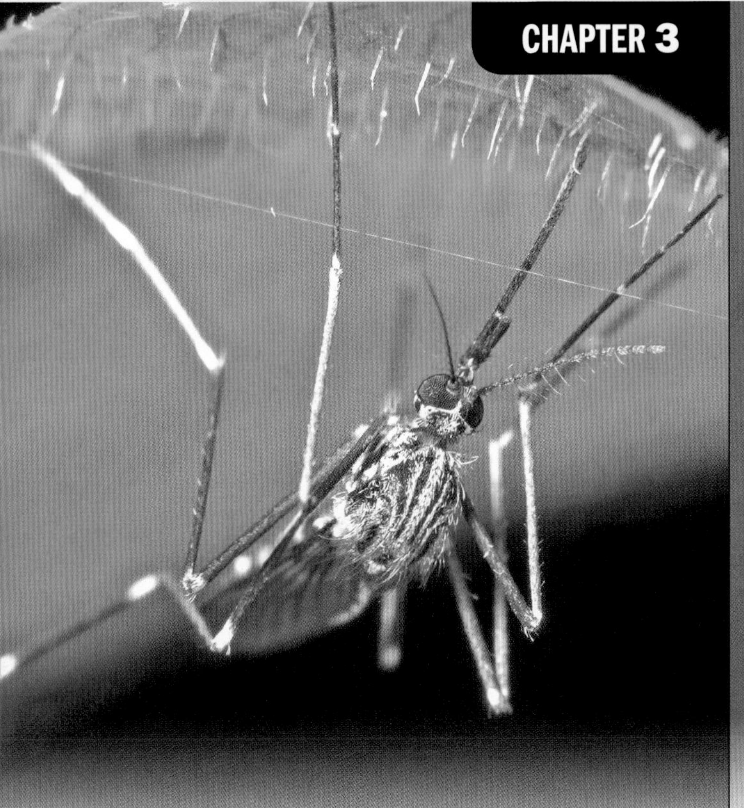

CHAPTER 4

1

Life Histories of Flies

FLIES, ARGUABLY THE MOST UBIQUITOUS OF ALL insects, are richly represented in almost all terrestrial and freshwater habitats, and a few even make inroads into the otherwise almost insect-free marine environment. How has this single lineage dominated the planet so effectively that any aerial insect sample invariably includes a diversity of two-winged adult flies, and almost any adequately moist organic material abounds in their relatively obscure legless larvae? One reason is certainly the effective division of habitat between the flightless larvae and flying adults, but that fails to differentiate Diptera from other holometabolous orders that, together with the flies, make up the majority of all animal species. Why is the order Diptera so speciose and why are flies so abundant relative to the other megadiverse orders of animals — beetles, wasps and moths? What are the key features of different life stages that might explain the exuberant success and overwhelming importance of this group? To answer these questions we need to look at the diversity and roles of each dipteran life stage: egg, larva, pupa and adult.

Larvae, adults and pupae of flies differ widely among the major grades and clades of the Diptera, and so the discussions of these stages below are divided according to the traditional two suborders, Nematocera (lower Diptera) and Brachycera (higher Diptera). Within the suborder Brachycera, the "lower Brachycera" and Empidoidea are treated separately from the higher Brachycera (or Cyclorrhapha). These are relatively familiar and easily recognized groups and thus convenient for a general discussion of life histories, even though the "Nematocera" and "lower Brachycera" are arguably artificial groups. Higher classification of the Diptera is summarized in Figure 2 and further discussed in Section Two.

Fly Eggs
Exceptionally Diverse but Essentially Aquatic

Flies, like other insects with complete metamorphosis, generally have a life cycle that includes egg, larval, pupal and adult stages, although the egg stage of some kinds of flies is very short, and a few lineages bypass the egg stage altogether. Flies that do lay eggs (oviparous and ovoviviparous lineages) produce what seems at first glance to be an infinite variety of egg types delivered onto, into or near appropriate hosts or habitats

OPPOSITE PAGE
All Diptera larvae lack true legs, and larvae of higher Brachycera (like this aphid-eating flower fly) lack a distinct head capsule.

This deer fly is attaching a clutch of eggs to a marsh plant; newly hatched larvae will drop into the marsh, where they will feed mostly on other aquatic organisms.

using a dazzling array of oviposition strategies. For example, some parasitoids produce thousands of tiny eggs and scatter them on foliage for ingestion by host caterpillars, while others produce fewer eggs and place them directly on or in their hosts. A few parasitic flies have harpoon-like eggs, and many others have simple eggs that hatch immediately into tough, mobile maggots able to hunt hidden hosts. Fly parasites of vertebrates often use surprising strategies such as squirting streams of eggs or newly hatched larvae up an animal's nose or, stranger yet, attaching eggs to another fly (such as a mosquito) that is likely to obligingly deliver the parasite's eggs to a vertebrate victim. The diversity of oviposition strategies among predaceous, phytophagous and saprophagous fly species is no less impressive. Aquatic species lay eggs in, on, beside or above aquatic or potentially aquatic habitats, and terrestrial species oviposit in or on various materials, including plants, fungi, soil, dung, carrion and rotting wood.

Flies lack the specialized appendicular ovipositor (a specialized egg-laying tube derived from abdominal appendages) found in most other insect orders. Instead they usually deliver their eggs using a simple, variably telescoping abdomen. In most flies the female abdomen is soft and flexible but in some groups the tip of the abdomen is hardened or otherwise specialized for injecting eggs into the right host or substrate. Derived lineages with specialized oviposition strategies include several parasitoid groups that use abdominal clamps to hold potential hosts, parasitic and predaceous groups that release sticky eggs into an abdominal "sand chamber" before dropping the sand-coated eggs into dry environments, and many phytophagous species that use stiffened abdominal tips to penetrate plant hosts. The oldest fly lineages, however, are aquatic, and the range of ovipositional strategies and egg structures in the Diptera still reflects those watery beginnings.

Mosquito eggs floating on the surface of a puddle, a mass of House Fly eggs at the bottom of a garbage can, fruit fly eggs on a rotting banana, a cluster of horse fly eggs suspended on a leaf over a pond — indeed, virtually all fly eggs — show similarities that reflect an aquatic heritage. The egg's outer shell, or chorion, is covered with ridges or an ornately sculptured meshwork penetrated by a couple of different kinds of openings, one of which, the micropyle, allows for the entry of sperm during fertilization. The other small openings are the aeropyles or respiratory pores. Almost all fly eggs, including the eggs of terrestrial species, are at least potentially or periodically "aquatic" dwellers — in aqueous microhabitats such as rotting materials, rainwater or host fluids.

The eggs' conspicuous sculpturing or ridging serves to separate the respiratory pores from the fluid surroundings, so it is not merely decoration. It is, rather, a plastron that retains a permanent layer of air — a physical gill through which oxygen can diffuse inward and carbon dioxide can diffuse outward. Some fly eggs have tubular or flattened projections (respiratory horns) that project the plastron above the surface of the material in which the eggs are embedded.

The plastron effectively equips the egg stage to survive periodic immersion, but fly eggs face a wide variety of other environmental challenges that are variously and ingeniously dealt with by different lineages. Eggs of parasitoid species laid on the body surfaces of active hosts, for example, must be very tough, at least on the exposed upper surfaces. Eggs of aquatic species often adhere to the substrate in gelatinous strings, float on the surface in rafts or are laid as single eggs equipped with floatation chambers. Eggs that undergo long periods of dormancy before hatching, such as those deposited by some mosquitoes in dry depressions due for inundation many months later, must be resistant to desiccation. Eggs deposited directly in ephemeral habitats such as dung or carrion must hatch very quickly, before the resource is gone or consumed by competitors.

The pressure to speed through the egg stage and get on with the business of eating and growing has driven some flies to lay eggs that hatch immediately, just after or even just before leaving the female's reproductive tract (this is called ovoviviparity). Other flies carry this trend to extremes by hatching their eggs internally and depositing active young larvae (viviparity or larviparity), or by hatching a single egg at a time internally and "nursing" the larva inside the uterus before depositing it as a mature larva that pupates immediately (pseudo-placental viviparity, or "pupiparity"). The egg stage is bypassed in at least 22 families of Diptera, with ovoviviparity and viviparity most often appearing among flies that are either parasitoids or dependent on ephemeral habitats such as dung or small carrion, and pupiparity popping up in a couple of lineages where adult flies imbibe enough nutrients from vertebrate blood to bypass a feeding larval stage. Viviparity can take the form of unilarviparity (one larva at a time), multilarviparity (many larvae at a time) or, more rarely, oligolarviparity, in which only a few larvae develop in the uterus at one time.

Fly Larvae

Legless Larvae and Life in the Muck

Although some groups shorten or entirely bypass the larval stage, this is the main feeding stage and the longest part of the life cycle of most flies. It is also the stage entomologists use to categorize fly lineages by biology (aquatic, terrestrial, saprophagous, phytophagous, predaceous, parasitic, etc.), even though the immature stages of Diptera remain surprisingly poorly known. The larvae of most fly species, many fly genera and even a few fly families have yet to be discovered, but the immature stages we do know about occupy almost every conceivable kind of living and dead organic material, including vertebrate and invertebrate hosts, various parts of living fungi and plants, and all kinds of decaying matter.

Larval flies all lack true legs, and most are soft, pale creatures that dwell inside moist matter. Few fly larvae munch on exposed foliage as do caterpillars or sawflies, and only a few fly lineages have larvae that live as exposed hunters after the fashion of so many beetle larvae. Instead, flies typically develop within nutrient-rich media. The first Diptera larvae probably immersed themselves in aquatic or semiaquatic slurries of algae or similar food, much as the oldest lineages of flies continue to do today.

Larvae of the Lower Diptera ("Nematocera")

The earliest fly lineages extend back some 250 million years. As might be expected in such old and diverse lineages, larvae of the major nematocerous (lower Diptera) superfamilies have successfully explored a variety of feeding strategies, such as mining in leaf tissue, predation and even parasitism. These strategies, however, invariably

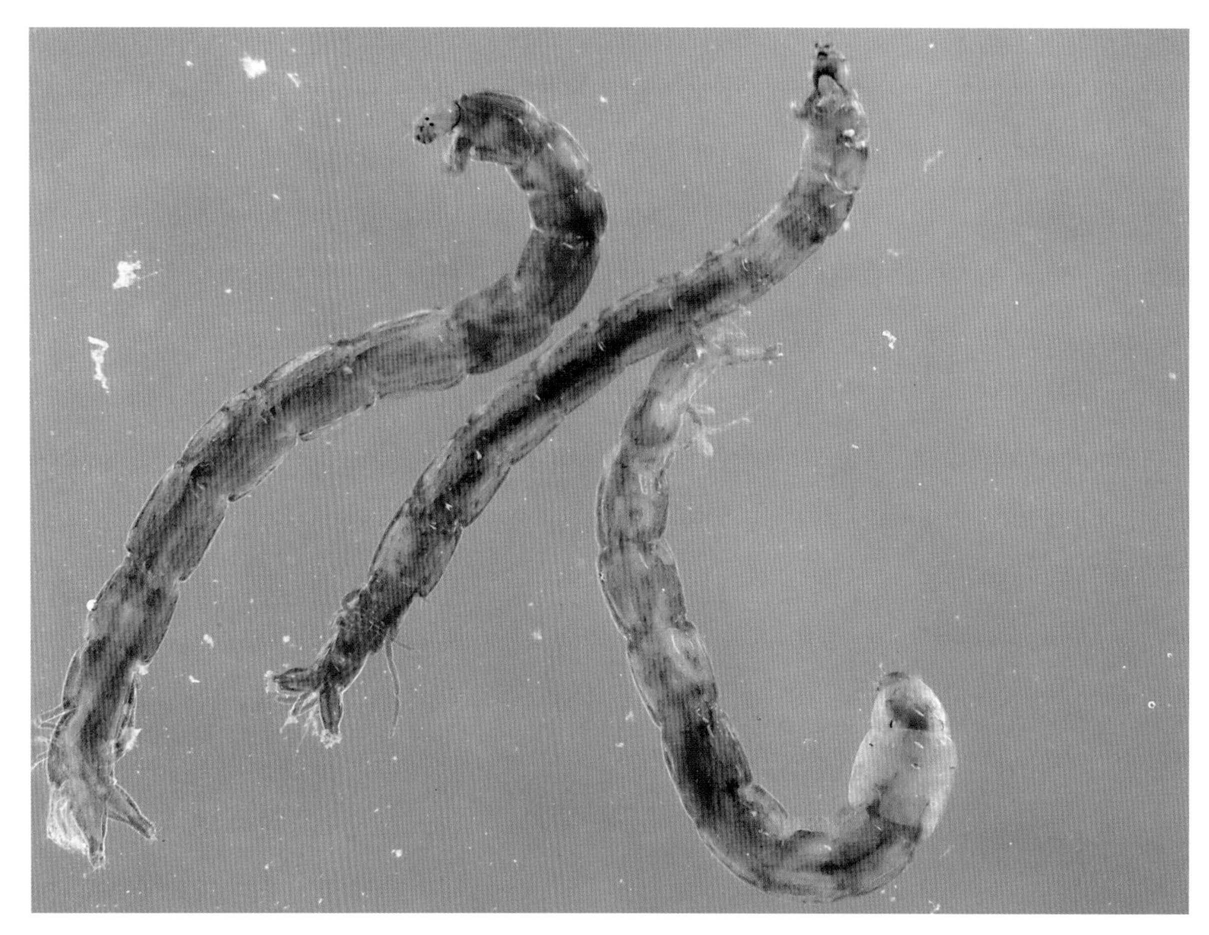

In some aquatic habitats the most abundant insects are larval midges, like these haemoglobin-rich Chironomidae extracted from an oxygen-poor ditch. Chironomid midges have no open spiracles; they respire through their body wall.

developed from the earlier and still more general strategy of gathering or filtering organic material from aquatic or subaquatic surroundings. Larvae of phantom crane flies, for example, can live immersed in the fine muck of springs and seeps, breathing through a snorkel-like tail that can be extended to the water surface. Larvae in the lower Diptera family Axymyiidae, found deep in waterlogged wood, have a similar snorkel-like tail, but with a hardened wood-drilling tip. Like other lower Diptera larvae, Axymyiidae larvae are soft-bodied, except for a hard head capsule supporting chewing mandibles that work from side to side to condition and ingest organic material.

Members of the huge family Chironomidae (midges) are relatively diverse both in habits and habitats; larvae are found in most aquatic and semiaquatic environments. Midge species ingest a range of detritus, algae and fungi by filtering, shredding, scraping or gathering material from their surroundings. A few have leaf-mining, parasitic or predaceous larvae, but these are relatively recently evolved lineages (most of the predaceous midge species eat other midge larvae). The success of this family derives from the abundance of species that follow the basic fly strategy of ingesting particulate organic material from their watery surroundings. The Chironomidae is the most diverse and abundant family of aquatic insects, so this seems to be a successful strategy.

Some midges have special combs and other structures used to concentrate the good bits from the organic soups in which they live, a strategy taken to extremes by larvae of the closely related mosquito and black fly families (Culicidae and Simuliidae). Both groups are made up almost entirely of specialized aquatic filter-feeders: black fly larvae use labral fans to filter particles from running water, and mosquito larvae usually use a complex network of fans and brushes to filter particles out of standing water or from surface films (although a few species have become predators). Meniscus midges (Dixidae) also filter particulate food from surface films, while larvae of a related family

Larval *Eucorethra* (Chaoboridae) lie just below the water surface as they await aquatic and terrestrial prey, breathing through open posterior spiracles surrounded by water-repellent hairs (seen here from above the water).

(Thaumaleidae, or seepage midges) are specialized grazers of diatoms in shallow films of flowing water.

Like some mosquitoes, the phantom midges (Chaoboridae) and many biting midges (Ceratopogonidae) have abandoned the ancestral mode of submerged particle feeding and instead develop as predators of other invertebrates. Larval phantom midges in the genus *Chaoborus* are abundant planktonic predators that grasp zooplankton (planktonic crustaceans) by using grossly enlarged antennae; those in the genus *Eucorethra* ambush other insects in or on temporary woodland pools. Larvae of some biting midges, which look like tiny aquatic snakes, penetrate the body walls of their larger prey before consuming them from the inside; others have abandoned the aquatic environment entirely and seek a variety of food in semiaquatic or terrestrial habitats. Predaceous larvae appear in a few other lineages of lower dipterans, such as those fungus gnats that spin webs to capture other insects.

A few predaceous fungus gnats, such as the famous glowworms (*Arachnocampa* spp.) of New Zealand and eastern Australia and the "dismalites" (*Orfelia fultoni*) of the southeastern United States, have independently converged on bioluminescence as a strategy for luring other flies into their silken snares. Larvae of *Arachnocampa* use a single blue-green bioluminescent organ (derived from the Malpighian tubules) to attract prey to a dangling sticky thread; larvae of *Orfelia* use dual anterior and posterior blue bioluminescent organs (derived from secretory cells) to lure insects into a sticky web. Related fungus gnats in the genus *Keroplatus* sometimes have luminescent fat cells that impart a larval glow. All light-producing Diptera are in the same fungus gnat family, Keroplatidae. Glowworms and dismalites are predators, but other Keroplatidae include fungus feeders as well as some odd groups, such as the Tasmanian species that develops as a parasitoid inside land planarians.

There are thousands of species of parasitoids amongst the higher Diptera, but very few among the lower Diptera. One of these exceptional parasitoid lineages is a group of Cecidomyiidae

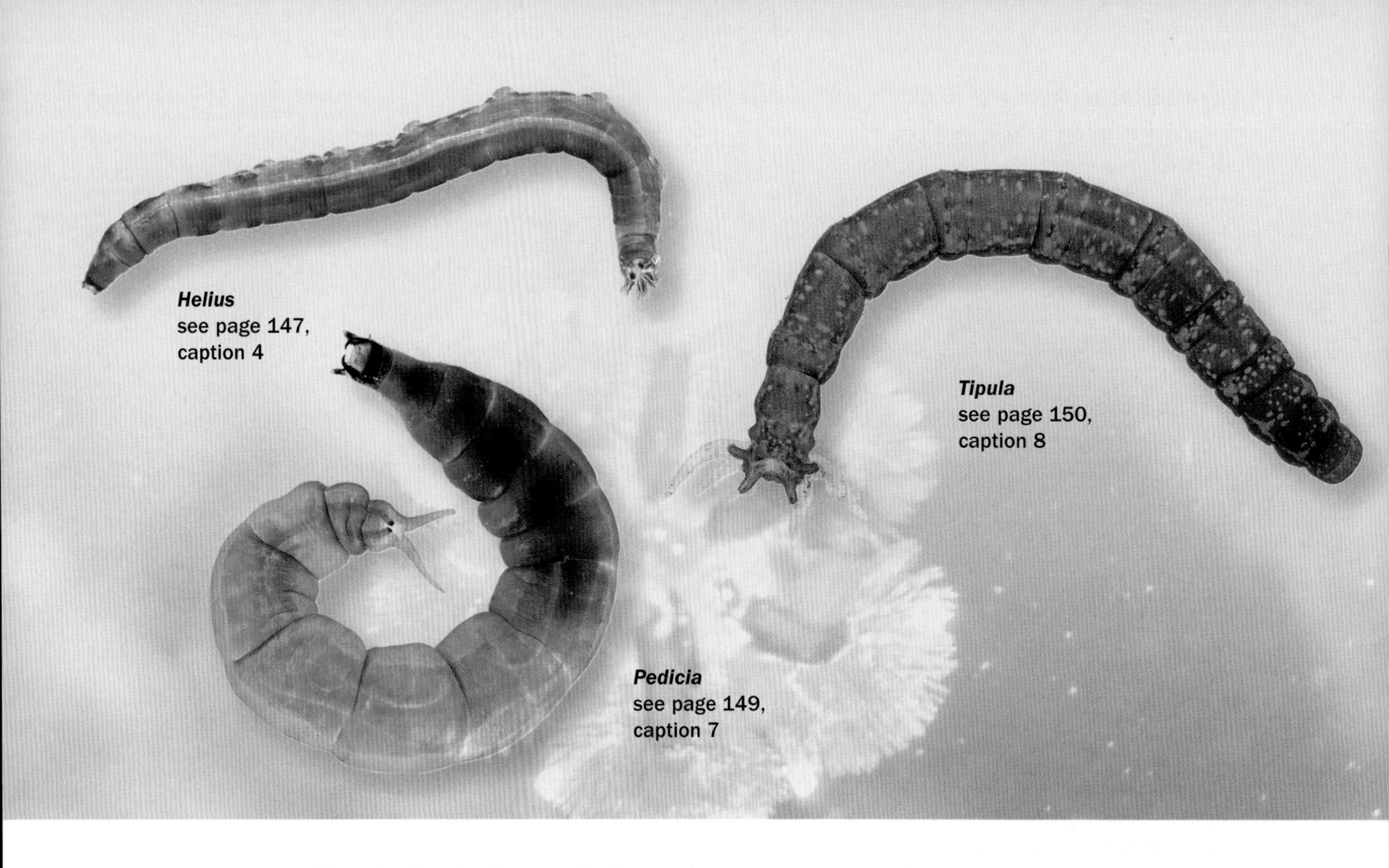

(the gall midge family) that attack plant-sucking bugs. Although the common name "gall midges" is used for the whole family, Cecidomyiidae larvae have a wide variety of predatory and plant-feeding habits, including their eponymous habit of inducing host plants to form custom galls in which the larvae develop. Directing plants to make galls is a life-history strategy that shows up in several fly families, but most gall-forming Diptera (and most gall-forming insects) are delicate little nematocerans in the huge family Cecidomyiidae.

Other lower dipteran families also have larvae that diverge in interesting ways from the immersed particle-feeding general form. The bizarre larvae of Deuterophlebiidae and Blephariceridae, for example, are adapted to torrential waters, where their extravagant body forms enable them to stick to the surfaces of water-buffeted rocks while they feed as scrapers of diatoms and other particulate food. Some lower dipterans have largely abandoned the ancestral aquatic habitats in favor of fungi, plant tissue and other terrestrial turf. For example, larvae of Psychodidae (moth flies) and Scatopsidae (black scavenger flies) occur in a wide variety of wet to moist habitats, where they typically graze fungal mycelia or microorganisms. The drain flies living in the biofilm (slimy sludge) in your bathroom pipes and the black scavenger flies grazing in your compost pile are typical of this group but, as with any major lineage, there are also specialists. Some psychodids cling to rocks in swift streams and some abound in the trickling filters of sewage plants; some scatopsids inhabit the wet layer under bark of recently fallen trees, while others live deep in the sand of rolling dunes.

Larvae of crane flies and their relatives (Tipulomorpha, the largest and one of the oldest lineages of lower Diptera) exhibit a spectrum of life-history strategies that corresponds to the size of the group. Most are particle feeders in a range of freshwater aquatic or semiaquatic environments, but a few are predaceous or phytophagous and many occur in other habitats such as deserts, fungi, soils, animal nests, wood and plant tissues. Most crane fly larvae breathe through a pair of posterior spiracles surrounded by tufts of water-repellent hairs and integrated with mesh-like structures that serve as physical gills (plastrons), but some species that live in well-oxygenated water lack spiracles and breathe directly through the body wall.

Despite the diversity of larval lower Diptera, common themes prevail. Almost all nematoceran larvae are soft bodied except for a hard head capsule that supports a pair of horizontally or obliquely swinging mandibles, and most have some sort of filter mechanism used to concentrate particulate matter from the moist media in which they live. Almost all have four larval instars (more in black flies, less in gall midges) and, although some atypical gall midges reproduce as larvae, most lower Diptera molt from the fourth instar larva into a simple pupal stage.

Aquatic soldier fly larvae on the surface of a pond. Their tails are tipped with a circle of water-repellent hairs that surround the posterior spiracles.

Larvae of the Suborder Brachycera

The Lower Brachycera

The suborder Brachycera — "short-horned flies" — forms a natural group defined both by adult structures such as the short antennae (the "short horns") and larval features such as mandibles rotated to work vertically, like the fangs of a snake. Lower Diptera, in contrast, normally have mandibles that work horizontally, meeting in the middle like clapping hands. A few have partially rotated mandibles, and the mandibles of some species of some lower Diptera are rotated to the point where they are almost as vertical as the "fangs" of larval Brachycera. The larval mouthparts of lower Brachycera, however, differ from those of lower Diptera in being subdivided into basal inverted U-shaped parts and a more conspicuous pair of curved and pointed sclerites. These "fangs" (distal mandibular hooks) articulate with the basal sclerite and work parallel to one another to rasp, rake or impale food.

Larvae in the oldest lineages of Brachycera (the lower Brachycera) are predators that use their fang-like distal mandibular hooks to strike, penetrate and consume the contents of other invertebrates encountered in aquatic, semiaquatic and terrestrial substrates. The filter-feeding structures found in so many lower Diptera are absent from predaceous Brachycera, which imbibe their prey's high-protein contents without needing any kind of filter to concentrate nutrients. Larval lower Brachycera are often general predators of other invertebrates, but some seek only particular prey, such as wood-boring or root-feeding beetle larvae, grasshopper eggs or even ants. Larval Vermileonidae (wormlions) make conical pitfall traps in loose, dusty soil or fine sand, much like the pits made by common antlions (larval Myrmeleontidae, order Neuroptera). Like antlions, wormlions vigorously strike at ants and other insects that fall into their pits. Larval Tabanidae normally feed on other insects in wet or submerged soils, but some *Tabanus* (horse flies) will strike at small vertebrates from underfoot, and some *Chrysops* (deer flies) seem to be facultative predators, able to develop at least partly on plant material.

While predation is the general rule in lower Brachycera, some lineages no longer consume multiple prey items and instead spend their entire larval lives feeding in or on a single host. Larvae that undergo their entire development at the expense of a single doomed victim are called parasitoids (in contrast with predators, which consume multiple prey items, and parasites, which live on a single host without killing it). The parasitoid lineages of lower Brachycera — small-headed flies, tangle-veined flies and bee flies — have active first-instar larvae that seek out and penetrate or attach to their hosts before transforming into immobile parasitoids.

Larval lower Brachycera usually consume other invertebrates, but a few are saprophagous or phytophagous. Stratiomyidae (soldier flies) and some related lineages, for example, have apparently reverted to a filter-feeding strategy not unlike that of many lower Diptera, and soldier fly larvae often occur in the same liquid or semi-liquid environments as crane flies, moth flies and other Nematocera. Analogous pharyngeal filters also appear in the ground plan of the Cyclorrhapha (higher Brachycera) but, if our current understanding of fly phylogeny is correct, these superficially similar filtering mechanisms must have evolved independently, because the Stratiomyidae and the saprophagous Cyclorrhapha are descended from different predaceous ancestors that lacked filtering structures.

Maggots: Larvae of the Higher Brachycera

Most of us have a good idea of what a maggot looks like, often from first-hand experience with flyblown garbage, maggoty foodstuffs or perhaps a corpse undulating with masses of maggots. Those soft, tapered, usually cream-colored denizens of decay that we recognize as maggots are larvae of higher Brachycera (Cyclorrhapha). The pointy end of a maggot is what is left of its head, and it is generally translucent enough to reveal the hardened bits and pieces that define the maggot equivalent of a skeleton. The black hooks that protrude through the mouth are called mouthhooks; in at least the last two of the usual three larval stages in this group these are fang-like sclerites apparently derived from the two-piece mandibles of their lower brachyceran ancestors.

Larvae of higher Brachycera — maggots — are usually relatively featureless cream-colored insects, but these Lesser House Flies (*Fannia canicularis*) are distinctively ornamented.

Although Cyclorrhapha do just about anything and live almost anywhere, most maggots macerate various kinds of muck with their mouthhooks, drawing in mixtures of microbes and fluid that they concentrate with the help of a unique filter system. Most maggots concentrate their food by pushing the mass of fluid and food down against a filter on the floor of the pharynx (the chamber just inside the mouth), concentrating a bolus of microorganisms and other solids and ejecting the excess fluid. Think of maggots as small-scale analogs of whales: whales filter feed on krill, using their fringe-like baleen to concentrate food from a sea of salt water, while maggots filter and concentrate bacteria from little oceans of decomposition by using their pharyngeal filters. And don't forget that there are billions and billions of maggots for every whale.

Filter feeding is a common strategy among maggots and is possibly the ancestral habit for the whole Cyclorrhapha, although even the basal Cyclorrhapha lineages have now diversified to include phytophagous, predaceous and parasitic forms. Scuttle flies (Phoridae) and flower flies (Syrphidae), for example, include common filter-feeding species found in wet muck or decomposing material, but both families also include large and diverse subgroups with larvae that feed on or in other invertebrates. Most of the familiar bright black and yellow flower flies have leech-like predaceous larvae that feed on aphids, and some of the most important scuttle fly lineages are parasitoids that develop inside bees and ants. The scuttle fly family (Phoridae) is not only the most diverse group of Aschiza but is arguably the most biologically diverse family of animals, with larvae that develop in or on a wide range of invertebrate hosts and a variety of living and dead organic material.

Some phorid larvae are even true parasites of other invertebrates, living inside Neotropical passalid beetles without killing them. Others, such as those whose larvae wrap themselves around the necks of host ants to become food-stealing necklaces, are kleptoparasites.

Most species of higher Brachycera (and probably most species of Diptera) belong to the single derived and diverse lineage called Schizophora. Larvae of most families of Schizophora are typical maggots: cream-colored and soft, tapering to an apparently headless, pointed front end with a pair of prominent sickle-like mouthhooks visible at the tip. Maggots are important inhabitants of wet decomposing material, where they use a pharyngeal filter to concentrate microorganisms for ingestion. The speed with which flies are able to locate and populate decomposing material, and the role of maggots in the rapid breakdown of dung, carrion and other rotting material, renders this group the earth's most significant cleanup crew. Filter-feeding maggots are found in an astonishing range of microbe-rich microhabitats, including nests of vertebrates, colonies of social insects, foodstuffs stored by other insects, insectivorous plants, festering wounds, soil, mud and bits of decay around such odd places as the excretory glands of land crabs.

Although decaying matter is the most common habitat of larval Schizophora, many maggots are neither filter feeders nor microbial grazers; they instead develop as leaf miners, stem borers, pollen feeders, fungus feeders, predators, parasites or parasitoids. Thousands of fly species develop as parasitoids that kill their invertebrate hosts; true parasites (which do not kill their hosts) occur in relatively few lineages and are mostly restricted to vertebrate hosts. The most important lineages with vertebrate-parasitizing maggots are in the families Calliphoridae, Sarcophagidae and Oestridae, but similar habits appear occasionally in other lineages, including Chloropidae, Muscidae and Piophilidae. Larvae of most parasitic Calliphoridae are flesh-feeding screwworms; exceptions include the Congo Floor Maggot (*Auchmeromyia senegalensis*), which feeds on the blood of sleeping humans and other animals, and the *Protocalliphora* species that attack nestling birds. Bird blood-feeding maggots are also found in the Piophilidae (*Neottiophilum*) and Muscidae (*Passeromyia* and *Philornis*). Other higher Brachycera larvae that live on but do not kill host animals include the Australian *Batrachomyia* (Chloropidae), which develop under the skin on the backs of frogs, and the many species of *Cladochaeta* (Drosophilidae), found only in the spittle masses of New World *Clastoptera* spittlebug nymphs.

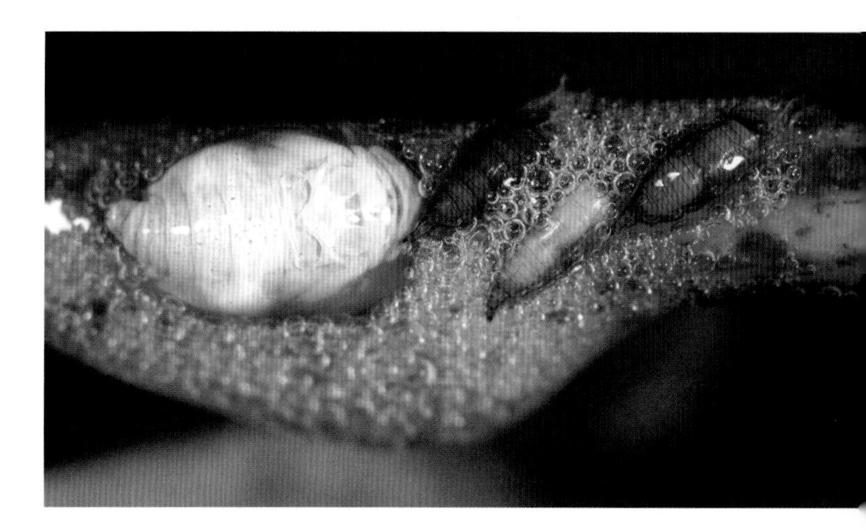

These three small brown objects found in a spittlebug's frothy domicile are puparia of Drosophilidae that underwent their larval development attached to the spittlebug nymph. Larvae of the large genus *Cladochaeta* are found only in spittle masses.

Parasitoid maggots, which kill their hosts, appear independently in no fewer than fifteen Schizophora families, including the thick-headed flies (Conopidae), which dive-bomb adult bees; the snail-killing flies (Sciomyzidae), which attack a wide variety of mollusks; and the Pyrgotidae, which come out at dusk to lay eggs on flying adult scarab beetles. Most calyptrate families include at least a few parasitoid species, and all 10,000 or so species in the largest family of higher Diptera, the calyptrate family Tachinidae, are parasitoids that gain access to invertebrate hosts in a variety of devious ways. Some deposit eggs directly in or on the host, others lay eggs near the host and leave the larvae to finish the job, and still others leave huge numbers of tiny eggs on the host's food.

Predaceous maggots occur in many families of Schizophora, including the Drosophilidae, a family best known for yeast-feeding larvae such as the iconic laboratory fruit fly, and the Chloropidae, a family best known for grass-feeding species. Some species in both families have larvae that are significant predators of pest

The adult wing, legs and antenna are visible on this mosquito pupa, which is breathing at the water surface using a pair of funnel-like respiratory "trumpets" on top of the thorax.

homopterans such as aphids and scale insects. The families Muscidae and Scathophagidae, generally thought of as dung flies, include many species with larvae that are either obligate or facultative predators of other insects in dung. Even the Dryomyzidae, an acalyptrate family with common saprophagous species on fungi and dung, includes a predaceous lineage — and a really odd one at that. *Oedoparena glauca*, a dryomyzid fly common on the Pacific coast of North America, is a predator of barnacles. Each larva, hatching from an egg deposited on a closed barnacle at low tide, enters an opening barnacle as the tide comes in, later eats the contents of the closed and submerged barnacle, and then moves on to another barnacle at a later low tide.

Many mature maggots, near the end of their third and final instar, are capable of prodigious leaps, primarily to propel them away from the generally putrescent material in which they developed prior to pupation, but probably also as a defense mechanism. A mature maggot is able to arch over to grab its tail end with its mouthhooks before contracting or compressing itself into a sort of living spring. When it releases its mouthhooks and straightens out its body, it is forcefully catapulted into the air with an audible snap, often ending up an impressive distance from the starting point. I have opened up rotting fruit to find writhing masses of maggots one moment and leaping larvae in my open camera case, an arm's length away, the next. Larval leaping is best known among familiar pests of stored products in the Piophilidae (such as the descriptively named Cheese Skippers) but is widespread among acalyptrates that abruptly vacate the larval pabulum prior to pupariation.

Pupae and Puparia

The Perilous Pre-adult Period

Like all holometabolous insects (insects with complete metamorphosis), flies require two metamorphic molts. The first is a larval–pupal molt to evert the wings, which are hidden internally during the larval stage, and the second is the pupal–adult molt, which allows the transformation from a pupa with undeveloped wings to an adult with fully developed wings. The pupa is often described as a resting stage between larva and adult, but the pupae of many flies are mobile and may be capable of defending themselves by impaling potential predators in mobile spiked traps (the "gin traps" of some pupal Ceratopogonidae) or by actively moving out of harm's way. Pupal mosquitoes, called tumblers, breathe at the water surface with trumpet-like anterior spiracles, but they swiftly swim to the bottom with a tumbling motion if disturbed. Some other aquatic fly pupae are equally mobile, but in general, fly pupae are relatively inactive forms with their wing buds and other incipient adult appendages visible but non-functional.

Several lower Diptera, especially in the family Mycetophilidae and related fungus gnat families, pupate attached to some sort of silken structure such as a cocoon, a web or even just a thread spun from the ceiling of a cave; some higher Diptera also use silk to construct a shelter or cocoon in which to pupate. Some significant groups of flies have the unusual habit of pupating inside a cocoon-like shelter formed from the inflated hardened skin of the last larval stage, called a puparium. This habit originated independently in a few lower Diptera (some gall midges), the lower brachyceran family Stratiomyidae (soldier flies) and the enormous lineage known as the Cyclorrhapha, or higher Brachycera. The name *cyclorrhapha* literally means "circular-seamed," and refers to one of the remarkable adaptations by which adult flies of this group escape from their hard pupal shelters. The head end of the puparium has a circular line of weakness, or "seam," that allows the adult fly to pop off the cap of its shelter before emerging. Cyclorrhaphan puparia are seed-like structures easily recognized by this cap and seam.

Most Diptera other than the Cyclorrhapha have pupae that are not confined in a puparium, and often use pupal mobility in combination with specialized armature to make their way out of the soil or other substrate prior to adult emergence. Bee flies that parasitize mason bees, for example, would be trapped inside their host's masonry nest if the pupa could not drill its way through the nest wall before emerging as an adult. Pupae and the pupal exuviae of lower Diptera (Nematocera) and lower Brachycera are often seen with the anterior (front) half of the body projecting out of the soil, tree trunks and other habitats where the larvae developed. If adults have already emerged from the pupae, the pupal exuviae will have a straight or T-shaped opening on the back through which the adult fly escaped.

The puparia of Cyclorrhapha are completely immobile, so newly emerged adults of Cyclorrhapha that pupate deep in the soil or in other confined areas need some way to dig or cut their way to the open air. Since adult flies lack the mandibles that would serve this function in other groups of insects, most Cyclorrhapha use a large, spiny balloon-like structure called a ptilinum instead. The ptilinum is pumped out of the front of the new adult's head to pop the cap off the puparium and to punch a path through the soil or other substrate to the surface. Once the fly escapes its confines, it withdraws the ptilinum, leaving a characteristic scar — the ptilinal fissure — arcing above the antennae.

Two-Winged Adults
The Ultimate Aerial Invasion

Enumerate the insects caught in a net swept through almost any sort of vegetation, count the insects visiting flowers, scan the swarming invertebrates over water, look at the smaller creatures attracted to your porch lights, glance at the unwelcome visitors sucking your blood, inspect the insects infesting your fruit or just try to focus on the ever-present but generally ignored "aerial plankton" on a warm summer evening. We are, as these exercises suggest, surrounded by flies. In above-ground terrestrial habitats, adult flies often outnumber all other adult insects combined, and

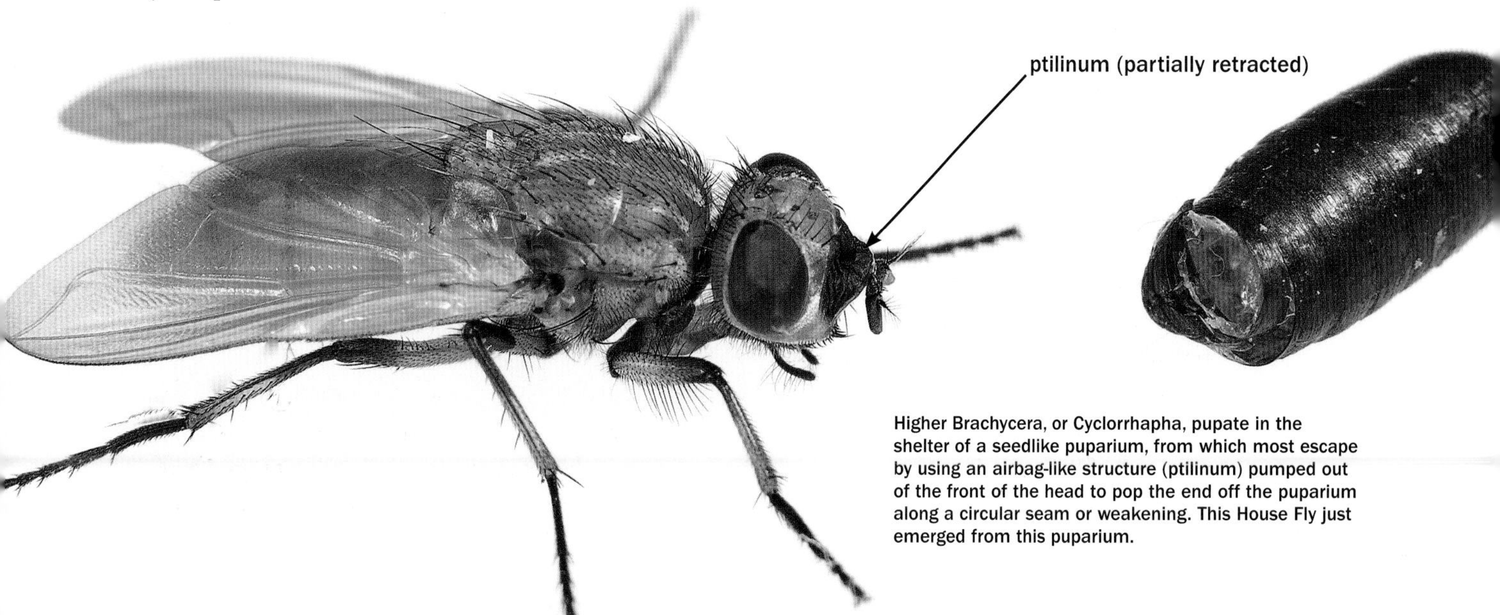

Higher Brachycera, or Cyclorrhapha, pupate in the shelter of a seedlike puparium, from which most escape by using an airbag-like structure (ptilinum) pumped out of the front of the head to pop the end off the puparium along a circular seam or weakening. This House Fly just emerged from this puparium.

in many environments you can find more species of flies than species of any other group of animals. Much of that dipteran dominance derives from the success of the legless larvae of flies in a wide range of habitats, but the two-winged adults also contribute, through dispersal, reproduction and ingestion of liquid or liquefied food.

Adults of the Lower Diptera ("Nematocera")

The first adult flies were probably long-legged, fragile-looking insects with long antennae, perhaps a bit like scorpionflies in the family Nannochoristidae but for the single pair of wings, and probably not much different from today's crane flies (Tipulidae). Crane flies, mosquitoes and midges are familiar lower dipteran families with the relatively delicate appearance and thread-like antennae that give today's "Nematocera" adults a common gestalt, unlike the more compact Brachycera. But the similarities that appear to unite the lower Diptera into a group traditionally known as Nematocera are really just a lack of the special (derived) attributes that define the suborder Brachycera.

Adult lower Diptera, like most flies, are fluid feeders; many kinds can be seen feeding on nectar or honeydew, the sugar-rich waste product of plant-sucking bugs. The spongy lower lip (labium tipped with a broad labellum) of most flies allows them to liquefy and imbibe honeydew that gets spattered on leaves by plant-sucking bugs (homopterans); flies taste this ubiquitous sweet material with their feet, and their evolution as high-energy flying machines was probably fueled by honeydew long before flowers were available. Males of many lower Diptera lineages do not feed at all, and females of several groups load up on the protein needed for egg production by piercing the blood vessels of vertebrates or the bodies of other insects. The most significant families of blood-feeding flies are found in a single nematoceran lineage, the infraorder Culicomorpha. Here we find the punkies, or biting midges (Ceratopogonidae), a large family that includes a range of bloodthirsty females, from predators and parasites that puncture other insects through to the minute crepuscular biting flies known to some human victims as "no-see-ums." The black flies (Simuliidae) and mosquitoes (Culicidae) are also in this group, and are of tremendous importance because of their biting females and the diseases they transmit. One culicomorph family (Corethrellidae) is made up almost entirely of species that suck blood from male frogs, which the bloodsucking female flies locate by homing in on the frogs' serenades. But Culicomorpha are not all blood feeders, and in fact the largest family in the group (Chironomidae) is known as the "non-biting midges." Blood feeding appears elsewhere in the lower Diptera; a sucking subgroup of the moth fly family (Psychodidae) also includes biting females that transmit serious diseases of humans and animals.

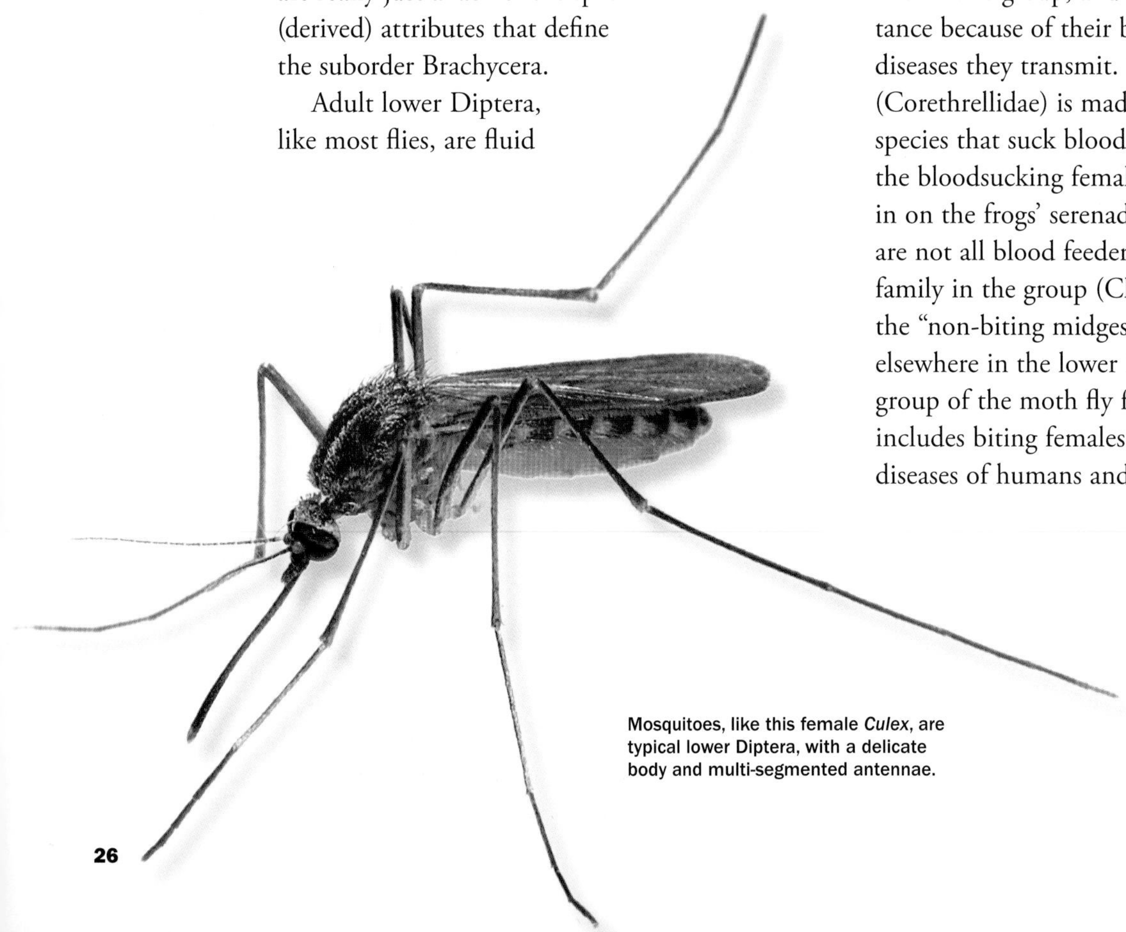

Mosquitoes, like this female *Culex*, are typical lower Diptera, with a delicate body and multi-segmented antennae.

Most of the lower Diptera adults that feed on body fluids of other insects are found among the Ceratopogonidae, but females of some species of torrent midges (Blephariceridae) are active predators that hang from vegetation as they imbibe the contents of recently captured midges, crane flies or similar small insects. An unusual form of insect-eating occurs among spider-associated fly species in the gall midge family (Cecidomyiidae). Many gall midges (as well as many crane flies) hang out in spider webs, often dangling by their front legs with the tarsi placed between adhesive droplets, presumably because this habit protects them from other insect predators. Adults of at least one spider-associated gall midge species are kleptoparasites that sip body fluids from the prey of their spider hosts; similar habits are found in some biting midges (Ceratopogonidae). Some ceratopogonid kleptoparasites seem to be highly specific: the only known hosts of one recently discovered Neotropical species are termites freshly captured by a tiny spider, which in turn seems to hunt only one species of termite. The spider suspends its prey in silken sacs hung by threads from termite nests in trees, and the female flies swiftly home in on the immobilized termites. Although more associations like this undoubtedly remain to be discovered, kleptoparasitism is relatively uncommon in the lower Diptera; it is much more frequently encountered among the higher Diptera.

Mating in the lower Diptera often takes place in a swarm made up of multiple males maneuvering in a tight-flying cloud as they await the arrival of fecund females. Like other swarming insects, swarming male Diptera almost always have enlarged eyes, which are used to maintain their position in the swarm, and like all flies they have a pair of club-like appendages called halteres that assist their aerobatics by serving as the insect equivalent of gyroscopic flight stabilizers. The halteres, derived from the vestigial hind wings, provide fine control of pitch and yaw — essential for mating success in a competitive swarm. Mating pairs usually leave the swarm promptly, but often stay *in copula* for extended periods. Some lower Diptera (some species of the cosmopolitan genus *Plecia*, family Bibionidae) are known as "Lovebugs" because of their exceptionally prolonged copulations (several days). Squished pairs

Female punkies, or biting midges, are sometimes predators and sometimes parasites that puncture other insects. This one is taking a blood meal from the thorax of another lower dipteran (a crane fly). All adult flies lack chewing mandibles, although blood-feeding and haemolymph-feeding female flies in the lower Diptera and blood-feeding flies in the lower Brachycera have piercing or cutting mandibles.

of Lovebugs — big-eyed males and small-eyed females — are familiar windshield decorations for anyone who has driven in rural southern United States in late spring.

As is to be expected in a heterogeneous group that includes tens of thousands of species, there are many exceptional life histories among the lower Diptera. Wing loss in one or both sexes is frequent here, as it is throughout the order, and wingless nematocerans are especially common under cold conditions. Wingless crane flies (Tipulidae) in the Holarctic genus *Chionea*, for example, are commonly seen walking on the snow, and other wingless crane flies occur on islands and at high latitudes and high altitudes. Winged male marine midges in the genus *Clunio* frenetically search the surfaces of wave-washed rocks to find wingless larviform females, often cutting into the female pupae to mate with the pharate adults even before they can emerge from the pupal skin. Stranger still are the Nymphomyiidae, minute aquatic flies that start their adult lives with narrow, fringed wings that they later snap off along basal shear points, much as winged termites break off their wings after their mating flights. After shedding their wings, the short-lived, non-feeding nymphomyiid males and females remain *in copula* as they enter swift, cold streams to deposit masses of eggs, which remain attached to the soon-drowned female. And perhaps strangest of all are the species of dark-winged fungus gnats (Sciaridae) in which males and females pupate in a common cocoon; the adults mate and lay eggs there, never leaving the cocoon.

Adults of the Suborder Brachycera

The Lower Brachycera and Empidoidea

Horse flies, robber flies, bee flies and soldier flies are the most familiar groups of lower Brachycera. These groups of common, usually conspicuous flies conveniently span the normal range of adult feeding behaviors among the basal brachyceran lineages. The blood-feeding habits of many adult female horse flies and their cleg and deer fly relatives (Tabanidae) are known to everyone, although few of us take the time to fully appreciate the elegant structures of these robust flies. If you look beyond their brilliantly colored eyes and examine the biting bits of a blood-feeding female, you might see that she is anchoring her mouthparts into your skin with piercing maxillae while slicing through skin and capillaries with scissors-like mandibles before imbibing the resultant pool of blood with her sponging labium. Some tabanid species have spectacular beaks that greatly exceed the rest of the body in length. The long-beaked species are most likely to be seen hovering over a correspondingly long-tubed flower, but many also take blood meals, penetrating hosts by using a long "syntrophium" made up of the mandibles and maxillae along with the food canal (labrum-epipharynx) and salivary canal (hypopharynx). When these piercing parts are sheathed in a long labium, it normally bends out of the way during biting. Females of some Tabanidae are strictly nectar feeders, and males (as in other lower Brachycera and lower Diptera) are often nectar feeders but never blood feeders. The males, which have strikingly enlarged and modified eyes that meet at the top of the head (female eyes are separate), are less commonly seen than females, but some hover conspicuously at eye level and many can be seen on flowers shortly after emergence. Several other families of lower Brachycera also routinely visit flowers to feed on pollen and nectar.

Bee flies (Bombyliidae) are frequent enough on flowers to serve as significant pollinators, especially in open areas such as grasslands, where their generally subterranean larval hosts (such as bee larvae, wasp larvae and grasshopper egg pods) are common. Although bee flies and the related tangle-veined flies (Nemestrinidae) are non-biting flies, adults of both groups often have strikingly long beaks, used to drink nectar from long-necked flowers. Soldier flies (Stratiomyidae) make up another common group of non-biting adult flower visitors, especially near the wet areas where most larvae of these slow-moving, often colorful flies live. However, not all lower Brachycera are

Many flies, like this long-beaked bee fly, are important pollinators.

attracted to flowers. Nectar feeding is unknown in the large family Asilidae (robber flies), a group composed entirely of predaceous adults with prominent mouthparts used to pierce other insects, injecting them with neurotoxic, protein-dissolving saliva before ingesting their liquefied body contents. Other predaceous Brachycera (Empidoidea, the dance flies and long-legged flies) are occasional flower visitors and some are specialized nectar and pollen feeders. Among some predaceous dance flies (subfamily Empidinae) the females have lost their ability to hunt their own prey, leaving those that require protein meals dependent on prey captured by males and transferred to their mates as nuptial gifts.

Mating in the lower Brachycera, especially those with big-eyed males, is often expedited by swarming or hovering. Some kinds of soldier flies, for example, form large male swarms more or less at eye level, while the big-eyed males of some Tabanidae form aggregations of hovering males at treetops and hilltops. The general phenomenon of "hilltopping," which brings numbers of otherwise rarely seen flies to prominent hilltops in search of mates, is important for various groups of Diptera, including both lower and higher Diptera. Flies that normally occur at low population densities — such as host-specific parasitoids and phytophagous species dependent on uncommon host plants — are especially common at hilltops, where some swarm above rocks, shrubs or other prominent objects, at or near the highest points.

Most fly swarms are made up of males intent on intercepting incoming females, but the tables are turned in some species of dance flies (Empidinae) with females that rely on males to bring prey as nuptial gifts. Some species of the enormous empidine genus *Rhamphomyia* form conspicuous crepuscular swarms of females that compete for gift-bearing males by trying to appear as large and fecund as possible, a kind of competition that has led to a proliferation of female-only ornaments such as fringed legs and inflated abdominal sacs. Related *Rhamphomyia* and *Empis* are often seen hanging from twigs or vegetation in copulating pairs, with the male hanging on

Flower flies (Syrphidae), like this *Eristalis flavipes*, are the best-known, and usually the most important, fly pollinators.

while the female feeds on his nuptial gift. Male dance flies in another group (*Hilara* and related genera) have silk glands in their swollen front tarsi; they produce silk to wrap the nuptial gift into a balloon-like package that is used to attract a mate and ensure a lengthy copulation. Other predaceous Diptera, including robber flies and long-legged flies, often exhibit elaborate mating behaviors: males advertise their sexual desirability through energetic flipping, hovering and other moves, as well as occasionally outrageous ornaments on various parts of the head, body or appendages.

The Cyclorrhapha, or Higher Brachycera

Adult Cyclorrhapha can be thought of as "maggots grown up," as they develop from a unique kind of larva that warrants the strictest definition of the term *maggot*. Maggots pupate inside a cocoon-like puparium made from the skin of the last larval stage. The adult fly escapes from the puparium by knocking off a cap along a predetermined line of weakness (the "circular seam"), and most Cyclorrhapha adults are forever scarred as a result of the remarkable tools they use in this process. In the main lineage of Cyclorrhapha, the ready-to-emerge adult is able to pump a large, spiky sac (the ptilinum) out the front of its head. The ptilinum is used not only to knock off the puparial cap but also to punch a path out of the substrate in which pupation took place (such as soil or dung). The ptilinum is withdrawn into the head early in the life of the adult, leaving a visible scar (the ptilinal fissure) arcing above the antennae, marking the area where the head was at one time opened up to release the balloon-like structure. The ptilinal fissure serves as a distinguishing feature that is shared by most but not all Cyclorrhapha; flies with this fissure (or "schism") comprise a distinctive lineage — the Schizophora.

A few lineages of Cyclorrhapha predate the origin of the ptilinal fissure, giving us the familiar pattern of a group divided into a "higher" subgroup, characterized by a new feature (in this case the ptilinal fissure), and a "lower" group that is a heterogeneous assemblage of basal lineages lacking that derived feature. The "lower" Cyclorrhapha are called the Aschiza (literally, "without schism"), a term still routinely used, even though the lower Cyclorrhapha (like the lower Diptera) probably do not form a natural (monophyletic) group.

The best-known group of Aschiza, the Syrphidae (flower flies), is arguably the single most important family of insect pollinators. These often brightly colored flies frequent flowers to feed on nectar and pollen, although some species seek their sugars and proteins from leaf surfaces spattered by honeydew or sprinkled with wind-blown pollen. Several syrphids, especially species that sit exposed on flowers, are convincing mimics of stinging insects. Bee and wasp mimics are found in several other families of flies (for example, Asilidae, Stratiomyidae, Xylomyidae, Conopidae and Micropezidae), but most of the species of insects with striking similarities to wasps and bees are flower flies. The best known of these are the adults of rat-tailed maggots, called Drone Flies because of their uncanny resemblance to Honey Bees (especially the non-stinging males, or drones), but apparent bee and wasp mimics are found across the family Syrphidae.

Syrphids and the closely related Pipunculidae (big-headed flies) are the most accomplished hovering flies, with the Pipunculidae being especially adept at hovering in small areas among foliage (or even within the confines of an insect net). Hovering in female Pipunculidae facilitates their usual strategy of seeking active leafhopper hosts among dense vegetation, but hovering in other flies is often a way for males to attract and interact with females. Males of some flower flies, for example, hover in shafts of sunlight, and male smoke flies (*Microsania*, family Platypezidae) swarm in plumes of smoke over campfires. Other Aschiza perform mating displays or form aggregations on substrates such as leaf surfaces (some Phoridae, some Platypezidae), but mating behavior remains unknown for most flies. Some, including most populations of the most common and widespread species of the Aschiza family Lonchopteridae, have no mating behavior because they have no males. Parthenogenesis is more unusual in the Cyclorrhapha than in the lower Diptera but it does occur in several independent lineages, including the acalyptrate families Sphaeroceridae, Chamaemyiidae, Drosophilidae and Agromyzidae as well as the Lonchopteridae. A very few fly lineages take parthenogenesis to its extreme, sometimes reproducing without bothering with the adult stage. This sort of "larva-to-larva" reproduction, called paedogenesis, is best known in the lower dipteran genus *Miastor* (Cecidomyiidae) but has also been reported to occur in the Cyclorrhapha (at least occasionally in the Drone Fly, *Eristalis tenax*, family Syrphidae).

Mating behaviors in the Cyclorrhapha are often elaborate and may involve displays, vigorous defense of mating territories on land and in the air, and a wide range of sexual dimorphisms. Swarming is uncommon in the higher Cyclorrhapha (Schizophora), but shafts of sunlight are sometimes enlivened by male swarms of Milichiidae, Lonchaeidae, Fanniidae and some Muscidae. More common are the male aggregations known as leks, which occur on leaves, rocks, tree trunks and other surfaces distinct from and often distant from larval habitats. Lekking males often hold their prime positions in the mating arena through ritualized displays and impressive body ornaments such as broadened heads, antler-like processes, greatly enlarged bristles and other structures. Intrasexual interactions can involve fights, shoving matches or mutual measurement (the best-endowed male usually gets the female); intersexual interactions may involve flashy displays, eversible scent glands or other perfume-wafting structures, and "kissing." This oral exchange of fluids between sexes occurs in Drosophilidae, Tephritidae, Platystomatidae, Ephydridae and Micropezidae; it can be compared to the presentation of high-protein nuptial gifts to females by predaceous male dance flies.

The adults of only relatively few species of Cyclorrhapha actively hunt other invertebrates, the main exceptions being a few acalyptrates (including some Ephydridae and one subfamily of Micropezidae); some lineages of Muscoidea (some very common Muscidae, a few Anthomyiidae and many Scathophagidae); and the Afrotropical and Oriental genus *Bengalia* (Calliphoridae). Adults of several higher flies are kleptoparasites of other predaceous arthropods, sharing prey captured by larger predators that digest their prey extra-orally. Web-spinning spiders, robber flies, rove beetles and predaceous Hemiptera are all frequently attended by small kleptoparasitic acalyptrates in the families Milichiidae, Lonchaeidae and Chloropidae, some of which not only use the larger predator as a source of fluid food but also as a mating arena. Other cyclorrhaphan kleptoparasites are mostly associated with social insects. The wingless phoretic adult bee lice (Braulidae), for example, take liquids from the mouths of their host adult bees. Some adult Calliphoridae and Milichiidae are associated with ants, feeding on regurgitated food, anal drops or brood. The Aschiza family Phoridae includes many termite-associated and ant-associated species, some of which are kleptoparasitic as adults. The most remarkable of these is the Malaysian phorid *Vestigipoda myrmolarvoidea*, adult females of which are legless and wingless flies that look almost identical to ant larvae. They stay in ant nests, where they are tended by the host worker ants along with the ant brood.

Among adult Cyclorrhapha, scavenging on dead insects is more common than predation; adult flies of various families, including Lonchopteridae, Sciomyzidae, Somatiidae, Pallopteridae, Psilidae and Lauxaniidae, are known to probe dead insects. Most other adult Cyclorrhapha use a broad, sponging labellum to take up liquid or liquefied matter from dung, carrion, flowers, plant and animal exudates or honeydew. Decomposing material, especially dung, is utilized by a wide variety of adult flies, including groups that breed elsewhere as well as those that feed, lay eggs and develop on the same smelly substrates. While there are many widespread generalists among the saprophagous Cyclorrhapha (*Musca domestica* springs to mind), adults of many groups can be found only on particular types of decaying material. Some wombat flies (*Borboroides*, family Heleomyzidae), for example, are found almost exclusively on wombat dung; the wingless adults of the one and only known species in the family Mormotomyiidae (*Mormotomyia hirsuta*) have been found only on bat dung from a single cave in Kenya; and the lone extant species of Mystacinobiidae (*Mystacinobia zelandica*) lives only with New Zealand short-tailed bats. Adults of some dung-breeding Sphaeroceridae are phoretic on dung-storing scarab beetles or dung-producing millipede hosts; some other sphaerocerids and some drosophilids are phoretic on certain land crabs and lay eggs in the microbe-rich waste around the crabs' excretory glands. Other adult flies are associated with decomposing fungi, various kinds and ages of carrion and particular types of decomposing vegetation. Seaweed fly adults (Coelopidae, Heterocheilidae and Helcomyzidae as well as some specialized subgroups of Sphaeroceridae, Sepsidae, Anthomyiidae and Scathophagidae) are found in impressive numbers on the piles of marine seaweed in which their larvae develop, but they rarely occur elsewhere. Many adult flies are known only in association with fermenting plant exudates, especially bleeding tree wounds.

Adults of only a few groups of Cyclorrhapha attack vertebrates, but some of the lineages that have secondarily adopted blood feeding are of considerable importance. Both male and female adults of all blood-feeding Cyclorrhapha take blood meals; this contrasts with the biting species of lower Diptera and lower Brachycera, in which only females bite. The most familiar blood-feeding higher Brachycera include the biting Muscidae (such as stable flies, horn flies and moose flies), which lay eggs and develop in dung or similar microbially rich, moist material. Tstetse flies (*Glossina* spp., family Glossinidae) do all their feeding as adults, taking blood meals from vertebrates and in so doing transmitting major diseases of humans (sleeping sickness) and livestock (nagana). Glossinidae, like the other entirely blood-feeding family of Cyclorrhapha (Hippoboscidae: bat flies and louse flies), are referred to as pupiparous because one larva at a time is nourished by "milk glands" in the uterus of the female fly and retained in the female's abdomen until mature. The mature larva is deposited and pupariates immediately, without feeding on anything outside its mother's body. Many bat flies are wingless and specialized ectoparasites as adults,

Bot fly adults like this *Cuterebra* don't feed, but their larvae feed and develop in the flesh of vertebrate hosts.

with that specialization reaching its extreme in females of *Ascodipteron*, which are reduced to featureless sacs of developing young embedded in the host bat's skin. This is in marked contrast to another lineage of Schizophora, the Oestridae, in which it is the adults that don't feed and the parasitic larvae that live in the flesh, nasal cavities or stomachs of vertebrates.

Bot and warble flies are bee mimics, but otherwise mimicry of bees and wasps is much less common in the Schizophora than the Aschiza. Ant mimicry, on the other hand, is frequent and appears in several families, including the Syringogastridae, Richardiidae, Ulidiidae, Sepsidae, Sphaeroceridae and Micropezidae. The latter two families include fully winged species that use wing pigmentation to imitate ant shapes, as well as entirely wingless species that resemble ants right down to the fake petiole. Some, like a Western Australian micropezid that develops inside the endangered Albany Pitcher Plant, are also ant-like in behavior; this entirely wingless ant mimic, *Badisis ambulans*, even jumps like its ant models. Winglessness is rare in the Micropezidae but very common in the Sphaeroceridae, where it has probably been independently derived more than a hundred times. A few of the many wingless Sphaeroceridae are astonishingly accurate ant mimics, and some fully winged species are apparent beetle mimics that press their wings against the body to create an elytra-like appearance. Similar apparent beetle mimicry is found in other higher Brachycera families, including the Lauxaniidae and Drosophilidae; the ultimate beetle mimicry is found in the beetle fly family Celyphidae, in which the scutellum is expanded to form a beetle-like shell.

Impressive though the variety of known fly life-history strategies seems to be, those known strategies can be only a small fraction of what actually occurs in nature. With less than half of the world's fly species even named, and surely no more than 5 percent of those species recognized as larvae, much remains to be discovered.

2

Flies, Plants and Fungi

Pollination, Phytophagy, Fungivory and Decomposition

The lives of most plants, ranging from the crops that provide food and fibers through to the exotic and beautiful species that support us in more subtle ways, are tied into the lives of flies. Flowers — the sexual organs of flowering plants — have been visited by adult flies since the origin of angiosperms, and the leaves, stems, fruits and roots of living plants are all home to a variety of larval flies. As well, the recycling of plant tissue into nutrients available to the next generation of plants is mediated by a huge variety of dipteran decomposers. Each of these roles — pollination, phytophagy and saprophagy — is played by both specialists and generalists from throughout the order Diptera.

Flies and Flowers
Pollination and More

Adult flies are generally fast and frenetic insects, able to use their muscle-packed mesothorax and a single pair of powerful wings to take flight instantaneously and to effectively flee threats or swiftly search for ephemeral or scattered resources. Such energy-expensive activities demand concentrated fuels, which Diptera frequently find in the form of liquid or easily liquefied sugars. Fluid sugar sources such as flowing sap are uncommon now and were probably even scarcer when the order was diversifying into its main lineages, some 200 million years ago, tens of millions of years before nectar-rich flowering plants appeared on the scene. The first flies probably obtained the necessary carbohydrates for flight by picking up honeydew from leaf surfaces. Scales, aphids and other honeydew-producing homopterans were diverse and abundant long before flowering plants, and the sweet waste they egest was probably easily lapped up to serve as the original fly fuel. Honeydew feeding remains critical to fly life histories today, although many flies now fuel their flight with sugar obtained from nectar instead of (or in addition to) honeydew.

Flies taste with sugar-sensitive feet, and most respond to the taste of sugar by extending a specialized lower lip tipped with a fleshy labellum made up of a pair of sponge-like lobes at the tip of the labium. The labellum of most flies can be spread out to form a broad surface; in the Brachycera the labellar surface bears radiating gutter-like grooves, or pseudotracheae. Watch a House Fly stroll from a tabletop onto a doughnut: its labium pushes down

OPPOSITE PAGE Diptera, like this flower fly (Syrphidae), play an important role as pollinators.

The common *Alstroemeria* lilies of central Chile are frequented by several long-tongued species of Acroceridae (small-headed flies) in the genus *Lasia*.

and the labellum spreads out on the sugary surface as soon as the fly's feet touch the doughnut. The fly regurgitates onto a sweet surface, liquefying the sugars so they can flow by capillary action up the trough-like pseudotracheae towards the mouth. The pseudotracheae are reduced or absent in non-feeding flies, some bloodsucking species and species that have developed long beaks for sucking nectar from deep flowers, but they remain the rule in most groups of higher flies. When you see a fly dancing or darting about on foliage, it is probably using its feet to search for thin films of sweet homopteran waste to liquefy and imbibe through the pseudotracheae of its honeydew-adapted labellum. I routinely take advantage of this behavior by spraying a mixture of honey and water onto leaf surfaces; this artificial honeydew often lures great gatherings of flies to where they can be easily collected or photographed. Honey spraying is a particularly effective strategy after a heavy rain, when natural honeydew is in short supply.

Flies were already diverse and widely dependent on sugar sources for their energy-expensive aerial lifestyle by the time flowering plants first appeared, so they were pre-adapted to partner with flowers. Animal-mediated pollination offers plants many advantages and may in fact be the reason for the great diversity of angiosperms today. Using a mobile intermediary for pollination allows flowering plants to get the male gamete (pollen) to the female parts of a distant plant with minimal waste and maximal outcrossing compared to wind-pollinated plants. Unlike the latter, animal-pollinated plants can continue to cross-fertilize at very low densities, especially if their relationship with the animal pollinator is a very specific one.

The first flower–pollinator associations might have been triggered by beetles or flies visiting flowers in search of pollen as food, in much the same way as midges in the family Ceratopogonidae commonly pierce pollen grains in water lily flowers today. But perhaps the first flower-associated flies were attracted to primitive plants by other rewards. Fossil evidence shows that Middle Jurassic to Early Cretaceous insect communities included lower brachyceran flies with long beaks that might have been used to sip ovulate fluids from gymnosperm plants, probably performing a bit of incidental pollination along the way. Once flowers started encouraging these kinds of beneficial associations by providing nectar rewards to visiting insects, they would have attracted a great variety of generalist flies already dependent on sugary fuels such as honeydew. By then the evolution of specialized flower-feeding flies was inevitable, and the explosive diversification of flowering plants in the Late Cretaceous was quickly followed by the appearance of numerous nectar-feeding lineages among the flies, wasps and moths. Bees, by the way, were relatively late arrivals at the floral party, first crashing the scene millions of years after the origin of flowering plants.

Flies and flowers today have a myriad of complicated interactions other than the obvious association between generalized sugar-seeking species and flowers with an easily accessible nectar reward. Lots of flies visit flowers to seek nectar to fuel flight; flower-visiting flies range from opportunistic generalists to highly specialized species that visit only a particular species or type of flower. Of these, certainly the most spectacular are the long-tongued flies that hover conspicuously over correspondingly long-necked flowers. Exceptionally long mouthparts with reduced labella appear in the same groups of lower Brachycera (Nemestrinidae, Tabanidae, Acroceridae) in widely separated parts of the world with Mediterranean climates (especially Chile, California, South Africa and Western Australia), with the most spectacular specialists occurring in South Africa. One species of South African tangle-veined fly (*Moegistorhynchus longirostris*, family Nemestrinidae) has a beak sometimes exceeding 8 centimeters in length; it occurs in the same communities as some similarly well-hung horse flies (Tabanidae) and small-headed flies (Acroceridae). As discussed below, the plants visited by these long-tongued flies are often entirely dependent on one or a few species of similarly endowed dipterans for pollination.

Although flies most frequently visit flowers in search of food, adult flies also arrive at flowers in search of other things, such as sex, shelter, warmth, prey or places to lay eggs. Males of certain fruit flies (Tephritidae), for example, visit particular kinds of flowers to pick up perfumes from the surfaces of petals and sepals. These volatile oils are subsequently stored in the rectal glands and released to attract females. For some flies, the flowers themselves are the attractant for sexually receptive individuals that cue into specific blooms as preferred places for mate location and copulation. Most flower-visiting flies are associated with flowers only as adults, but there are many examples of flies that develop in the same flowers they pollinate. *Elachiptera formosa* (Chloropidae), for example, is the main pollinator of North American *Peltandra virginica* (Green Arrow Arum), which attracts flies by releasing volatile scents that advertise flowers of the right stage and age. Once an appropriate plant is located, the flies feed on pollen, mate, oviposit and complete larval development in the same kinds of flowers.

Globe Flower (*Trollius europaeus*), a familiar yellow blossom in the buttercup family (Ranunculaceae), is also pollinated by flies that lay eggs in the same flowers they pollinate. Larvae of several species of the anthomyiid genus *Chiastocheta* develop in the Globe Flower seeds, forming a flower–pollinator relationship comparable to the better-known mutualism between *Yucca* and prodoxid moths that pollinate *Yucca* flowers and develop in the seeds. Similar mutualistic relationships are found between various other plants and their fly pollinators. Cempedak, a Southeast Asian tree fruit (*Artocarpus integer*, family Moraceae), is pollinated by gall midges (*Contarinia* spp., family Cecidomyiidae) that develop on fungal mycelia on male inflorescences, while other gall midges develop on the decomposing remains of the same flowers they pollinate as adults. Similar relationships exist between species in the aroid genus *Alocasia* and its acalyptrate pollinators — *Colocasiomyia* species (Drosophilidae) in Japan and *Neurochaeta inversa* (Neurochaetidae) in Australia — and between flowers of *Aristolochia* and flies in the genus *Megaselia* (Phoridae) in Central and South America.

Aristolochia is an enormous genus of more than 500 species, including the familiar vines known as pipevine or Dutchman's Pipe. Flowers of different *Aristolochia* species attract flies of several families, including Anthomyiidae, Calliphoridae, Chloropidae, Milichiidae, Phoridae, Syrphidae and Sarcophagidae, but the association is not always

Many flowers create carrion-like or dung-like scents to attract saprophagous Diptera, like this flesh fly (Sarcophagidae) attracted to a foul-smelling *Aristolochia* bloom.

a mutualistic one. Some species attract flies with strong odors resembling fungi, carrion or other smelly stuff irresistible to female flies in search of oviposition sites; these are generally false lures — the flies are being "used" by the plant for pollination. Several *Aristolochia* species (and other flowers too) not only deceive flies with false olfactory cues, they also trap them inside for a period of time before the fooled fly is released to carry off its cargo of pollen. In some cases the false promise of dung, fungi or carrion conveyed by these floral scents leads the flies to waste their eggs in flowers where they will hatch and promptly die for lack of food.

Duped dipteran pollinators can be found at the deviously scented flowers of many stinking plants, ranging from the famously enormous *Rafflesia arnoldii* (Stinking Corpse Lily) flowers of Southeast Asia through to a range of less exotic woodland, house and garden plants. Red Trillium or Stinking Benjamin (*Trillium erectum*), Jack-in-the-Pulpit (*Arisaema triphyllum*) and skunk cabbage (species of *Symplocarpus* and *Lysichiton*) are among the common North American woodland plants pollinated by flies; Jack-in-the-Pulpit attracts fungus flies such as Mycetophilidae and the others attract carrion flies such as Calliphoridae. Stinking Corpse Lily belongs to an odd family of parasitic plants called the Rafflesiaceae, while Jack-in-the-Pulpit and skunk cabbage join several other fly-pollinated plants, including Corpse Flower (*Amorphophallus titanum)*, in the arum family (Araceae). Corpse Flower is an Indonesian giant with flowers clustered in typical aroid fashion along a central penis-like spadix rising from a vase-shaped spathe. With a spathe as wide as 3.5 meters (12 feet) across and a heat-generating spadix up to 2.5 meters (8 feet) tall, these enormous blossoms put out one very powerful stink.

Some arums trap their fly pollinators to ensure pollination. The Mediterranean Dead-Horse Arum (*Helicodiceros muscivorus*), for example, lures carrion flies (mostly Calliphoridae) down its stinking spadix and deep into its vase-like spathe, where they are detained by stiff hairs reminiscent of those rows of one-way spikes used to prevent cars from escaping parking garages without paying (analogous minnow trap–like mechanisms or one-way routes down slippery or slimy chutes show up in several fly-trapping plants). The hairs wilt and allow the flies to escape once the male flowers above begin to produce pollen, forcing the flies to pick up a load of pollen on their way out. Similar strategies of temporarily trapping particular groups of flies are employed by other aroids, such as the European Wild Arum or Cuckoo Pint (*Arum maculatum*), which detains moth flies (Psychodidae) for pollination, and by some members of the milkweed family. Some of these, in the Old World genus *Ceropegia* (condom flower or bushman's pipe), have flower tubes lined with downward-pointing hairs that form a trap for small flies attracted by the flower's odor. The insects cannot escape until the hairs wither; once released, they are often burdened with *Ceropegia* pollinia (pollen packets). The same sorts of fly-trapping pollination systems occur in superficially similar trap flowers of members of the pipevine or birthwort family (Aristolochiaceae).

Purposely putrid scents are also characteristic of many fly-pollinated members of the milkweed family (Asclepiadaceae or Apocynaceae), including the carrion flowers (*Stapelia gigantea*, *S. grandiflora* and related plants) that originate in arid parts of southern Africa. Similar stinking inflorescences are found in the soursop family (Annonaceae), such as the eastern North American pawpaw (*Asimima* spp.) trees pollinated by carrion flies attracted to their meat-colored, downward-facing, foul-smelling flowers. *Stapelia* species are now common as potted plants the world over, but blooming carrion plants have such an unpleasant smell that they are best kept out of doors, where the flowers' foul scent (and perhaps their general resemblance to dead furry animals) will attract shiny green blow flies. The visiting blow flies pick up pollen and leave behind batches of (doomed) eggs, usually visible as white masses deep inside the blossom.

Orchids too can expedite pollination by duping dipterans, sometimes by attracting fungus flies to a mushroom-like or yeasty odor, and sometimes by inducing carrion flies to visit by producing putrescent odors and visual patterns that resemble maggot-riddled carrion or flies already clustered on something rotting. Some orchids use aphid-like structures on the lip to lure predaceous Syrphidae that deposit their eggs among the fake aphids, where of course the syrphid larvae cannot survive. A number of terrestrial orchids are pollinated by flies in the families Sciaridae, Cecidomyiidae, Mycetophilidae and Tachinidae; some males in the latter two families are lured in by orchids that use sexual deception, seducing them with the false promise of a receptive female. As if it were not devious enough to use false advertising to fool flies, many orchids also force the arriving insects to play the pollinator role by using astonishingly complex tricks to steer the visitors into contact with pollinia. Some orchids in the widely cultivated Asian genus *Bulbophyllum*, for example, lure flies to an attractive ramp-like lip. Once the fly has walked up the ramp and beyond a tipping point, it is violently pitched into the flower column. Another orchid, *Epipactis thunbergii*, was recently studied by the Japanese scientist Naoto Sugiura, who found that the plant is entirely dependent on only four species for pollination, all flies in the family Syrphidae. He describes a part of the orchid flower as "throwing the syrphid fly onto the stigmatic surface," where it is slapped with a sticky pollinium.

Pollination of orchids by flies and other insects has intrigued biologists for a very long time, as evidenced by Sprengel's oft-quoted 18th-century description of insect pollination as "one of the most admirable arrangements of nature." This "admirable arrangement" was catapulted into the forefront of biology with the publication of Charles Darwin's 1862 book *On the Various Contrivances by Which British and Foreign Orchids Are Fertilised by Insects*, in which Darwin reported his insights on insects interacting with orchids from all over the world, describing for the first time the extent and complexity of relationships between orchid flowers and insects. He realized that the extreme diversity of orchids was driven by the very specific relationships between orchids and particular animal pollinators, including moths, bees, birds, bats and (of course) flies. Some of his most famous remarks pertained to the long floral tubes of many orchids, which he noted were often closely matched with the pollinator's proboscis. Every biology student is familiar with the story of Darwin's prediction that a Malagasy orchid with a floral tube 30 centimeters (1 foot) long had to be pollinated by a moth with an equally long proboscis. His contemporaries apparently laughed at this ridiculous prediction, which was not borne out until scientists recognized an appropriately endowed sphingid moth (*Xanthopan morganii*) in Madagascar more than 40 years later. How much more Darwin's detractors would have guffawed had he suggested that such long-tubed orchids

Anthophilous Diptera

Syrphidae

Calliphoridae

Anthomyiidae

Tachinidae

Stratiomyidae

Muscidae

must be matched with correspondingly long-tongued flies! In fact some of the most specific orchid–pollinator partnerships are those that have resulted from coevolution of orchids and long-tongued flies, especially the Nemestrinidae and Tabanidae able to hover like hummingbirds over long-tubed flowers while threading elongate mouthparts down into the nectar reward deep in the flower.

The evolution of long-tubed flowers, with their corollas so elongated that only flies with correspondingly long tongues can reach the nectar therein, has led to specific fly–plant relationships and thus to efficient pollination. The more specialized the relationship, the less likely it is that pollen will be wasted by the fly's visiting other flower species, and the more likely it is that a fit between the length of the beak and the depth of the flower will result in efficient transfer of pollen from flower to fly and from fly to flower. A fly, of course, doesn't care about pollination and does not need to precisely match its proboscis length to flower depth; it can get as much or more nectar if its tongue is slightly longer than the flower (albeit at the cost of developing and wielding that long tongue). An extra-long tongue can be used to reach nectar in a range of depths without the fly being forced to touch its face or body to the flower, where it would pick up pollen, defeating (from the flower's perspective) the purpose of the nectar reward. The appearance of longer-tongued flies would thus lead to selection for longer tubed flowers that would be more likely to transfer pollen to the longer-tongued flies, which would in turn reward even longer fly tongues, and so on and so on — an evolutionary arms race that has selected for long-tubed flowers and flies with improbably long tongues that seem to defy the laws of leverage (try moving around while holding a straw five times your body length!).

The specialized partnerships between flies and flowers that have resulted from this coevolution are now reflected in very efficient pollination, but also in extreme interdependency for the long-tubed flowers and the long-tongued flies. The best-studied groups of flowers associated with long-tongued flies are those made up of "guilds" of similar (but not necessarily closely related) South African flowers that are collectively dependent on corresponding guilds of long-tongued flies for pollination. One such guild, reviewed by Manning and Goldblatt (1996), is made up of 28 species of the plant families Iridaceae and Geraniaceae pollinated by two closely related species of long-tongued tangle-veined flies (Nemestrinidae) whose mouthparts are 20 to 50 millimeters (up to 2 inches) long. Similarly, South African orchids in the *Disa draconis* complex depend on a guild of specialized Tabanidae and Nemestrinidae that have even longer mouthparts; one of the flies in this group, the nemestrinid *Moegistorhynchus longirostris*, holds the world record for the longest fly proboscis and for the longest insect mouthparts in relation to body length. Long-tongued flies in the families Acroceridae, Nemestrinidae and Tabanidae visit similar groups of flowers in Chile. For example, the common *Alstroemeria* lilies of central Chile are frequented by several long-tongued species of small-headed flies in the genus *Lasia*, as well as a spectacularly long-tongued species of horse fly (*Mycteromyia conica*). Other *Lasia* species join another genus of small-headed flies (*Eulonchus*) to form a similar pollinator guild in the southwestern United States.

Some of the long-tongued flies that effectively pollinate orchids and other long-tubed flowers clearly did not develop their elongate mouthparts for picking up nectar, as they belong to lineages long committed to blood feeding or predation. Pollen-dusted individuals of some long-beaked predators in the dance fly family (Empididae), for example, are common on flowers, and *Empis* species were among the few flies named as orchid pollinators in Darwin's famous book on orchid pollination. Mosquitoes are also common flower visitors and are the main pollinators of some small orchid flowers, such as those of some of the boreal forest bog orchids in the genus *Habenaria*; the flowers present an appropriately dimensioned spur full of nectar accessible by the same proboscis used to suck blood. Male mosquitoes, which do not suck blood, are also common visitors to a wide range of flowers, particularly at high latitudes. Some of the flower-visiting horse flies (Tabanidae) that sport strikingly long tongues are able to

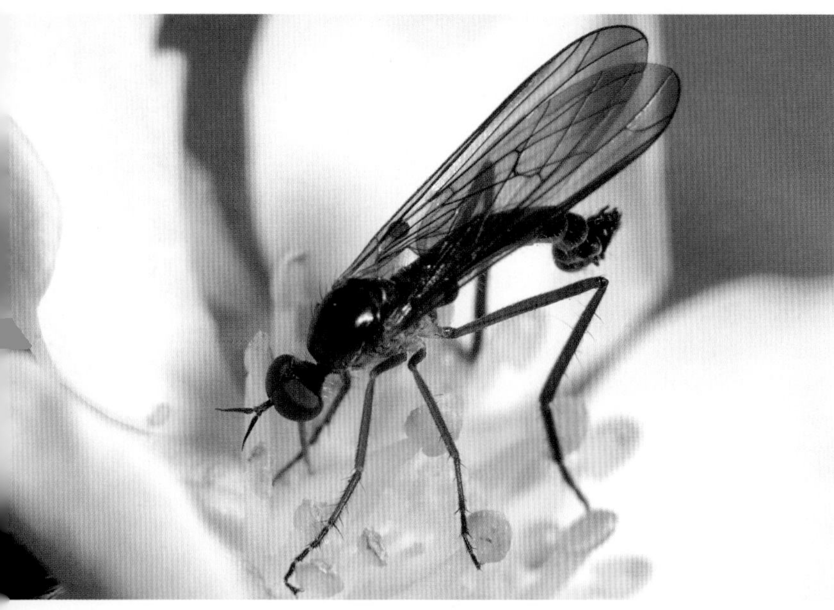

blood feed too; long-beaked Neotropical and Australian *Scione* species can deliver an impressively deep jab (see page 239, captions 7 and 8). Several nectar-feeding horse flies in the large Old World genus *Philoliche* seem even less prepared for blood-feeding as they wield a proboscis up to three times their body length. But bite they do, apparently bending the long labium to the side while the mandibles and maxillae do the subcutaneous dirty work. Members of another biting fly family, the Ceratopogonidae, frequently imbibe nectar, and are common visitors to many plants, including Cacao (*Theobroma cacao*). Species of the large and widespread biting midge genus *Forcipomyia* are major pollinators of Cacao flowers, and thus critical to the production of chocolate; related *Theobroma* species are also pollinated by superficially similar but only distantly related little nematocerous flies in the gall midge family.

Although the spectacular long-tongued flies of South Africa and similar Mediterranean climates have received much of the attention paid to flower–pollinator coevolution, long-tongued flies occur in almost every part of the world and appear in various families across the Diptera. Several nematocerous flies have conspicuously long mouthparts and are frequent flower visitors, although relatively little is known of their floral relationships. The North American *Eugnoriste* (Sciaridae) and *Gnoriste* (Mycetophilidae), for example, both use strikingly long tubular mouthparts to nectar feed, as do the many common limoniine crane flies (subfamily Limoniinae) with mosquito-like beaks. One western North American *Gnoriste* species has a long (about 6.5 millimeters/0.25 inch) proboscis that it uses specifically to probe for nectar in the long floral tubes of Piggyback Plant (*Tolmiea menziesii*), for which it is the primary pollinator.

Long mouthparts are found in many flower-visiting fly groups in addition to those already mentioned, including the Empidoidea, Vermileonidae, Bombyliidae, Tachinidae, Milichiidae, Conopidae, Stylogasteridae and Syrphidae. The Syrphidae (flower flies), widely considered to be the most important group of pollinating flies, usually have short, sponging mouthparts, but a few genera (especially *Rhingia*, *Tuberculanostoma* and *Lycastrirhyncha*) have long noses that conceal elongate mouthparts. The bee fly family (Bombyliidae) includes the largest number of long-tongued species of any family, although the group includes both long- and short-tongued species. Bee flies are most diverse in dry environments, but some long-tongued species are conspicuous flower visitors in early spring. The common Holarctic species *Bombylius major*, for example, can almost invariably be seen hovering over the patches of Coltsfoot (*Tussilago farfara*) that appear before most other spring flowers across Canada and northern Europe; this conspicuous fly is also one of the most important pollinators of native spring wildflowers.

The long mouthparts of bee flies and many other flower-visiting flies are essentially straws used to reach exclusive sources of carbohydrate-rich nectar, but many flower-visiting flies also feed on protein-rich pollen. Since a long straw is not the ideal tool for picking up pollen, bee flies often have other structures to help them gather pollen and get it to the end of their elongate mouthparts. The forelegs of some species of long-tongued bee flies are raked across the flower's pollen-bearing surfaces, picking up pollen that is then transferred to the tip of the proboscis, where it is mixed with fluids and sucked up the straw. Other flies, including flower flies, employ leg structures to assist in pollen feeding; many require the proteins in pollen for egg development. Even wind-pollinated plants are visited by pollen-hungry flies, and females of some species of flower flies are common on grass flowers.

The myriad linkages between flies and flowers are fascinating but also critically important because of the dependence of so many crops on insect pollinators. Most economically valuable plants will opportunistically (promiscuously!) use a variety of insects to get their pollen to the opposite sexual parts of other flowers, but how do flies stack up as pollinators of crops? This is an important question, given that the worldwide economic value of pollination is enormous, with estimated values running into the hundreds of billions of dollars per year for food crops alone. At the same time, our current overdependence on an originally European bee (the Honey Bee) for crop pollination worldwide is a matter of concern as numbers of commercial Honey Bee colonies precipitously decline because of colony collapse disorder, pest mites and other problems. Wild pollinators, especially native bees, usually fare poorly in agroecosystems that have traditionally been managed without regard for the variety of nearby microhabitats needed for maintenance of pollinator diversity.

There is now an overdue interest in the role of wild pollinators, including flies, and a growing appreciation that the pollination services we all depend on are provided by more than just the familiar social bees. Many plants, including some of economic significance, are entirely dependent on a few specialized flies for pollination, but vastly more are at least partly dependent on pollination by a broad range of relatively opportunistic flower-visiting Diptera. Just how dependent they are is influenced by a spectrum of variables such as plant species, season, climate, latitude and altitude. Things are further complicated by the difficulty of distinguishing between flies that are merely visitors to flowers and those that are significant pollinators. Still, some generalizations are possible, one of which is "flies rule where it's cool." At high altitudes and high latitudes, Diptera are the most abundant insects and the most important pollinators, and the same is probably true under other relatively cool and moist conditions, such as those that prevail early in the day, early in the season and in relatively humid environments like cloud forests.

Even though most individual animals (and the vast majority of species) visiting flowers worldwide are undoubtedly flies, they are often considered to be less valuable as pollinators than bees, because bees are equipped to gather and transport pollen back to a nest to feed their pollen-dependent larvae. Fly adults, in contrast, visit flowers to meet their own needs rather than those of their larvae, so they don't need structures for transporting pollen to a nest (structures that would also facilitate transfer of pollen from flower to flower). Despite that, many flower-visiting flies are covered with conspicuous hairs that pick up piles of pollen, and more than a few have special structures for squeezing or raking pollen from

OPPOSITE PAGE Some long-beaked flower-visiting flies in the families Mycetophilidae **(TOP)**, Empididae **(MIDDLE)** and Bombyliidae **(BOTTOM)**. The mycetophilid (*Gnoriste megarrhina*) is specifically associated with one particular kind of long-tubed flower (*Tolmiea menziesii*); the other flies shown here visit a wide range of flowers.

anthers. The role of flies in pollination is probably generally underestimated, in part because of the common assumption that bees are more important. The lack of attention to flies in pollination studies might also reflect the relative difficulty of studying speedy, skittish flies compared to the generally larger and more easily observed, captured and identified bees. Several groups of flies remain essentially unidentifiable because the lack of adequate keys or identification guides for most regions impedes studies of dipteran pollinators. Work on economically important dipteran pollinators has focused on the Syrphidae and Calliphoridae, although other groups (especially the Muscidae, Anthomyiidae, Tachinidae, Sarcophagidae, Bombyliidae and Stratiomyidae) are diverse and abundant flower visitors on a wide range of plants in a wide range of habitats. The name of one of these groups —Anthomyiidae — speaks explicitly to an association with flowers (the Greek for flower is *anthos* and *myia* refers to a fly), but all, especially the calyptrate families, are common on flowers.

Most fly families include flower-visiting species, and members of many families are abundant on flowers under particular circumstances. I have seen, for example, flowers of the dominant Patagonian shrub Calafate (*Berberis buxifolia*) mobbed by March flies and dance flies (Bibionidae and Empididae); flowers of spring ephemerals such as Trout Lily (*Erythronium americanum*) in Canada's boreal zone crowded with *Anthalia* species (Empididae); and flowers of Bogbean (*Menyanthes trifoliata*) in Canadian peatlands loaded with tiny black scavenger flies (*Swammerdamella*, family Scatopsidae). Many authors have commented on the general dominance of flies on northern flowers, and flies that breed in large numbers in particular microhabitats often abound on nearby flowers, as in the case of the seaweed flies (Coelopidae) reported as major pollinators along the Finnish and South African coasts. Most such interesting fly–flower associations have yet to be investigated; this is a potentially rich field as more and more flies become identifiable following new taxonomic and faunal studies.

The Syrphidae, or flower flies, are the best known and probably the most beneficial of the flower-visiting fly families. Most of the 6,000 or so species visit flowers, and many are efficient pollinators. Most (80 percent in one European study) flowering plants are visited by syrphids, often in large numbers, yet studies of the value of syrphids

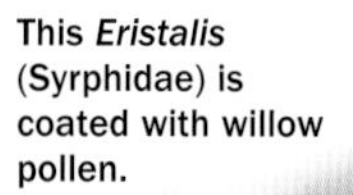

This *Eristalis* (Syrphidae) is coated with willow pollen.

as pollinators are rare, and mostly limited to demonstrations that some widespread syrphid species are indeed valuable pollinators. Recent studies, for example, show that the common syrphid *Episyrphus balteatus* significantly improves seed set and yield in Canola (*Brassica napus*) and that the ubiquitous flower flies known as Drone Flies (*Eristalis tenax*) are good pollinators of greenhouse sweet pepper (*Capsicum annuum*) flowers. Drone Flies, long recognized as important pollinators of fruit crops ranging from apples to strawberries, develop in foul ditches and similar habitats, where their aquatic larvae (rat-tailed maggots) feed on microorganisms. *Episyrphus balteatus* larvae, on the other hand, are aphid predators, so this species is valuable both as a pollinator and as a biological control agent on the same crops. *Eristalis tenax* and *Episyrphus balteatus* are both originally from Europe but are now widespread and common in disturbed areas, including agroecosystems, where the total syrphid diversity is relatively low and dominated by widespread species. Total flower fly diversity is dramatically greater in natural areas, where a variety of native fly species undoubtedly play a major role in pollination of native plants.

The Calliphoridae (blow flies and cluster flies) are less diverse than the Syrphidae but can nonetheless be very common on flowers. Only a few of the 1,100 or so species of Calliphoridae routinely visit flowers other than those that deliberately lure carrion flies with putrescent odors, but those few species often abound on flowers. Cluster flies (*Pollenia* spp.), especially some secondarily widespread species that develop as parasitoids of originally European earthworms, can be conspicuously abundant in disturbed habitats. Cluster flies overwintering in houses often reach high enough densities to qualify as pests, but these same abundant adult flies play a beneficial role as common pollinators in spring and summer. Other calliphorids, including the common carrion-breeding bluebottle and greenbottle flies, can be important natural pollinators and have also been cost-effectively reared in large numbers for commercial pollination. One recent Canadian study showed that *Calliphora vicina* was an efficient pollinator and that it could be used to pollinate leeks at about one-fifth of the cost of bringing in the Honey Bees required for the same task.

Cluster flies (*Pollenia* species) are often abundant flower visitors and important pollinators.

Parasites of Plants
Phytophagy in the Diptera

At first glance, flies are unlikely plant pests. Adult flies lack the equipment to chew on foliage, seeds or stems, and for some reason they have never moved into the plant-sucking niche so prominently occupied by the aphids, leafhoppers and other Hemiptera (although some fly larvae in the family Cecidomyiidae do suck up plant juices). Many flies are attracted to exposed sap and a few adult Agromyzidae feed on sap that female flies deliberately liberate from leaves using their sharp ovipositors, but adult flies rarely damage foliage or penetrate plants with their mouthparts. If their feeding damages plants at all, the damage is generally restricted to pollen, and the flies that feed on pollen probably do more good as pollinators than they do harm as pollen eaters. Fly adults, then, are more likely to be plant partners than plant pests. Their larvae are another story.

If you see a fly larva exposed on a leaf or stem, it is almost certainly a predator, probably a

Most leaf-mining flies are in the family Agromyzidae. The *Phytomyza* larvae that made the columbine leaf mines (LEFT) have pupariated and emerged, but the yellow *Liriomyza* larva (RIGHT) is visible within its mine.

syrphid or cecidomyiid larva feeding on aphids. With very few and rarely seen exceptions (such as some cylindrotomine crane flies), fly larvae never live exposed lives munching on leaves after the fashion of caterpillars and sawflies. Different dipteran lineages with plant-feeding larvae have been independently derived from saprophagous (microbe-grazing) ancestors on multiple occasions, but in almost every case the shift to phytophagy has been a logical transition from one kind of immersion feeding to another. There is, for example, a fine line between living in a microbe-rich soup of rotting plant material and feeding within fresh or living plant tissue. It is easy to envisage a gradual transition from feeding on decomposing material in muck (like many basal Diptera lineages) to feeding on roots or mycorrhiza in wet soil (many Tipulidae and Sciaridae) and even moving up into the roots or stems to feed within plant tissue (many lineages). And simply feeding within plant tissue (essentially an aquatic environment) is but a step away from more specialized forms of phytophagy, such as leaf-mining and galling.

Larvae living ("mining") in the thin tissue layer between the upper and lower surfaces of leaves create the characteristic blotchy patches or squiggly signatures seen on the foliage of many, if not most, kinds of plants. Similar leaf-mining habits have developed independently in several orders of insects: many mines are made by sawflies and a few groups of beetles, but most are the work of either micro moths or flies in the acalyptrate family Agromyzidae. Miners from these two groups can sometimes be distinguished from one another by the pattern of frass (feces) left in the mine; micro moths often leave a dense single line of frass, while many agromyzid flies leave a sparse double line of alternating rows along the edges of the mines. Most leaf mines tell a story with their expanding serpentine shape, starting small where the larva hatches from the egg and gradually expanding as it moves and grows, ending at the widest point, where the larva either pupariates in the leaf or drops from the leaf to pupariate on the ground (the usual strategy). Some mines, especially those made by flies other than Agromyzidae, are blotch-like rather than serpentine and may be shared by several larvae. Large blotch mines are commonly the product of Anthomyiidae, but some Drosophilidae and Tephritidae make similar mines. A few miners are also found in the Syrphidae (some *Cheilosia*) and Dolichopodidae (*Thrypticus*), as well as the families Sciaridae, Chironomidae, Anthomyzidae, Scathophagidae, Ephydridae and Psilidae.

The family Agromyzidae is the most important group of leaf-mining flies and arguably the most economically significant group of leaf miners, although only a few of the 2,000 or so named species in this family damage crops. Some economically important agromyzids are highly host-specific, but the most important pests are species that feed on a wide range of ornamental and vegetable crops. Chief among these polyphagous pests are species in the large genus *Liriomyza*. The Pea Leafminer (*Liriomyza huidobrensis*) is an interesting example, since this originally South American fly now has an almost worldwide distribution and has recently transformed from a sporadic and easily controlled seasonal pest into a major problem. This polyphagous species attacks more than a dozen families of plants and now damages many vegetable and flower crops throughout the growing season. The current global problems with *Liriomyza huidobrensis* are likely due to a recently developed resistance to pesticides, but assessment is complicated because of the size and taxonomic difficulty of this genus of almost 400 species of tiny acalyptrate flies.

Leaf-mining flies are often easiest to identify on the basis of host plant and mine topology; adults of many species appear almost identical until the male genitalia are examined. Even then some species are difficult to tell apart: the Pea Leafminer is part of a complex of species that includes a North American species (*L. langei*) that was previously confused with the South American *L. huidobrensis*. Other important members of this genus include the widespread polyphagous Vegetable Leafminer (*L. sativae*) and the American Serpentine Leafminer (*L. trifolii*). Despite its common name "American," the characteristic tightly coiled mines of *L. trifolii* can be found on the foliage of an impressive range of hosts in Africa, Asia and Europe as well as in North America, where it is the most common agromyzid pest. Many common non-pest *Liriomyza* can also be found on weeds, where they probably serve as useful reservoirs for parasitoids that help regulate populations of pest species.

Although most agromyzids are leaf miners, some attack roots, shoots or flowers and others feed at least partly on decomposing material. Some larvae introduce fungi into their plant hosts as they feed, and then they feed on the fungus inside the plant tissue. Similar habits are found in the family Cecidomyiidae, some of which induce plants to form thickenings around the mines and some of which force their host plants into more elaborate growths that form a feeding shelter for the developing larva. Such insect-specific plant growths are called galls.

Galls are remarkable structures, abnormal but symmetrical growths that plants are "forced" to make by specialized mites, nematodes or insects (somewhat similar growths can be caused by bacteria, viruses, nematodes and fungi). Insect-induced galls are remarkably complex, species-specific structures found on the roots, stems, leaves, buds, flowers or fruits of a wide range of plants. Dozens of insect families have

The large genus *Liriomyza* includes many economically damaging leaf-mining species. This bubbling fly has a parasitic larva of a prostigmatan mite attached to its abdomen.

Most gall-inducing fly larvae, like this *Harmandiola* larva exposed in a cut-open poplar leaf gall, belong to the huge family Cecidomyiidae.

independently hit upon gall induction as a strategy for forcing plants to essentially build them custom houses, complete with kitchens continuously supplied with high-quality food. This strategy suits the default dipteran larval strategy of living immersed in wet food, so it is not surprising that several families of flies include gall makers. Some of the most familiar galls are the large ones made by acalyptrate flies in the family Tephritidae (such as on thistles and goldenrod); some smaller galls are induced by gall flies in the acalyptrate families Chloropidae and Agromyzidae. Gall flies in the small (one genus) acalyptrate family Fergusoninidae inoculate plants in the family Myrtaceae with symbiotic nematodes as they lay their eggs in plant tissue. The nematodes induce the plant to form large galls, each providing shelter and food for a single fly larva, which in turn serves as a host for several nematodes. Although originally Australasian and Southeast Asian, flies in the family Fergusoniidae are now under consideration for release in the southeastern United States for biological control of the invasive shrub *Melaleuca quinquenervia* (Broad-leaved Paperbark).

Most gall flies, however, are in the lower dipteran family Cecidomyiidae, an enormous group of almost 6,000 described species and probably tens of thousands of undescribed species. Despite the common name "gall midges," the basal lineages of the family are fungus feeders and the group includes many predators, as well as a diversity of plant-feeding species that don't make galls. All of the predaceous and phytophagous species belong to the Cecidomyiinae — only one of the four subfamilies of gall midges — and it too includes fungus feeders. Even the gall midges that do make galls are not always plant feeders. Some species (probably more than is widely appreciated) feed on fungi in the gall. The Holly Berry Gall Midge (*Asphondylia ilicicola*), for example, inserts eggs into holly flower ovaries in spring. The midge-infested berries don't turn red (as they otherwise would) the following winter, and if you open one of those abnormally green berries you will find a non-feeding gall midge larva and a patch of white fungus. The fungus starts to grow in spring, at which time the larva feeds and develops, not on plant tissue but on the symbiotic fungus. Several other gall midges have similar stories, often inoculating the plant with a symbiotic fungus when they lay eggs, but most gall midges have moved from feeding on fungi to feeding on plant tissue.

Larvae of phytophagous Cecidomyiidae have a remarkable way of attacking their plant hosts with energetic "kissing" — pressing the front of the conical head against the host plant's tissue and sucking up of a slurry of tissue mixed with saliva — as the larva grasps, pierces or abrades the plant tissue using vertically articulated mandibles. The feeding behavior of economically important species has been studied with sensitive recording equipment, revealing an audible slurping as the larva's pharyngeal dilator muscles create the suction necessary to draw material up into its digestive system. Gall midge larvae also have a unique bib in the form of a spatula-shaped plate under the front of the head, but the bib probably has nothing to do with their method of sucking food. Gall-forming species that pupate outside the gall usually use the larval spatula like a shovel, digging their way into the soil before pupating in relative safety underground. Those that pupate in the gall may use the spatula to make a tunnel through the plant tissue almost to the gall surface before pupating, so the pupa can later easily push its way through the surface. Other species pupate deep within the safety of the gall; they use elaborate pupal appendages to drill out of the gall before emerging as a delicate adult. Species that let the pupal stage do the drilling or digging often lack a larval spatula.

Galls induced by Cecidomyiidae are usually fairly host-specific and come in an extraordinary range of species–specific shapes. One group of Indian gall midges in the large genus *Contarinia*, for example, induces host *Acacia* plants to form remarkable leaflet galls called "cylinder-piston" galls. Some of the hundreds of species of *Asphondylia* midges produce swollen white galls that look like cotton balls on saltbush branches in the American southwest; like the superficially similar tephritid galls that abound on *Baccharis* shrubs in southern South America, these wooly *Asphondylia* galls are more solid than they look. Other American cecid galls can look like pimples, sea urchins, nipples, nozzles, folds, little red peppers, bonbons, ping-pong balls, sponges or swellings of various other sorts on stems, leaves, flowers, fruits and shoots.

Many gall midges, including several notorious pest species, feed on plant tissue without inducing galls. The infamous Hessian Fly (*Mayetiola destructor*), one of many originally European insects accidentally introduced to North America, lays eggs on the upper sides of wheat, triticale or barley leaves. Hatching larvae move down the leaf to find an enclosed, high-humidity feeding site between the leaf and the stem, where they scrape and suck without causing a gall but often causing weakening or breaking of the stem. Hessian Flies pupate inside the skin of the last larval stage, which forms a seed-like shelter similar to the puparium of higher Diptera. Other *Mayetiola* species are similar cereal pests in other parts of the world, and other non-galling gall midges are serious pests of other crops. The now widely distributed Sorghum Midge (*Contarinia sorghicola*), for example, feeds on sorghum flower spikelets, destroying the seeds. The variety of plants attacked by gall midges is reflected in the pantheon of common names for the pest species: Alfalfa Gall Midge, Grape Blossom Midge, Apple Blossom Midge, Pea Midge, Pear Midge, Coffee Flower Midge, Rose Midge and so on.

Flies that attack fruits are the most infamous of phytophagous Diptera, but the true fruit fly family (Tephritidae) also includes species with larvae that mine in leaves, species that induce host plants to form galls and others that live in seeds or stems. The 70 or so serious agricultural pests in the Tephritidae, like the 4,000 or so non-pest species in the family, use a stiffened abdominal tip to insert their eggs into plant tissue. The most infamous of these is the Mediterranean Fruit Fly (*Ceratitis capitata*), or "Medfly," a tephritid described by the California Department of Food

The Mediterranean Fruit Fly (*Ceratitis capitata*), or "Medfly," attacks a wide variety of fruit and is considered to be one of the most important agricultural pests in the world.

and Agriculture's website as the "most important agricultural pest in the world." Medfly females lay eggs under the skin of fruits of all sorts (the host range of Medflies includes some 300 kinds of cultivated and wild fruits), where the resulting maggots macerate the fruit with their mouthhooks, ingesting the pulp while introducing decay-causing microbes and rendering the fruit unfit for consumption. Although this originally African species has already spread around much of the world, some fruit-growing areas of suitable climate (such as Chile, California and New Zealand) still remain Medfly-free, thanks to expensive quarantine, agricultural inspection and eradication programs. Hundreds of millions of dollars have been spent to keep Medflies at bay in California alone, where residents are so aware of this species that it was even at the heart of a threatened "ecoterrorist" attack on the use of insecticides. An extremist group (called the "Breeders") claimed to be breeding and releasing Medflies in California to protest against environmentally hazardous chemical eradication programs, with the rationale that eradication programs are useless once the pest is firmly established. Indeed, the Medfly has been established in Hawaii since 1910, and no eradication program is in place there. It is unlikely that Medflies (other than sterile males), were ever deliberately released in California, but the state has successfully eradicated several accidental introductions since the "Breeders" were active in the 1980s.

Medflies are highly polyphagous, damaging a dangerously wide range of fruits (and some vegetables), but many Tephritidae feed only on single hosts. The Olive Fruit Fly (*Bactrocera oleae*), for example, attacks only olives, but it does so with a vengeance, potentially rendering an entire crop valueless if not controlled. Many other members of the large genus *Bactrocera*, such as the Queensland Fruit Fly (*B. tryoni*) and the Oriental Fruit Fly (*B. dorsalis*), feed on a wide variety of fruit crops and have the potential to devastate a regional fruit industry, not only by damaging the crop but also by making it impossible to export to areas not yet home to these pests. The threat of accidental movement of *Bactrocera* species is illustrated by the story of the Asian Fruit Fly (*Bactrocera invadens*), a polyphagous pest described as new to science only in 2005: this fly recently invaded Africa and is now rapidly spreading through the central part of the continent, where it attacks a wide variety of fruits. Some of the 200 or so species in the New World genus *Anastrepha* have the potential to generate similar horror stories, with the Mexican Fruit Fly (*A. ludens*) being the potentially invasive species perhaps most feared among fruit growers in the southeastern United States, right alongside the Caribbean Fruit Fly (*A. suspensa*), the South American Fruit Fly (*A. fraterculus*) and the Inga Fruit Fly (*A. distincta*). Nor are pest fruit flies restricted to the tropics and subtropics. Apple Maggots (*Rhagoletis pomonella*), cherry fruit flies (*R. cingulata* in North America, *R. cerasi* in Europe) and Blueberry Maggots (*R. mendax*) are among the major fruit pests of north temperate countries.

The largest subfamily of Tephritidae, the Tephritinae, includes about 2,000 species that develop in the flower heads, stems or roots of Asteraceae, plants that are themselves sometimes weedy pests and potentially subject to control using introduced Tephritinae. For example, the tephritines *Urophora quadrifasciata* and *U. affinis* were imported from Europe to North America to help control the invasive Spotted Knapweed. Another dozen species of tephritines have been moved around the world in attempts at biological control of other pest Asteraceae.

The Tephritidae, or true fruit flies, include most of the fruit-eating fly species, but maggots of other families are also found in fruit. The clouds of tiny *Drosophila* that invariably appear above fruit left out a bit too long, for example, derive from maggots that develop rapidly in yeasty overripe or rotting fruit, accelerating its decay in the process. Drosophilidae, sometimes also called fruit flies but more properly called vinegar flies or just drosophilids, are ubiquitous on decomposing fruit. They form an enormously diverse family that includes lineages that have abandoned microbial grazing in favor of other strategies, including predation and parasitism (see Chapter 8). Most drosophilids are typical maggots that live in moist masses from which they filter microorganisms, but a few cross the line from feeding on yeasts (like the famous

The Asian Fruit Fly, *Bactrocera invadens*, **(RIGHT)** and Oriental Fruit Fly, *Bactrocera dorsalis*, **(LEFT)** are two of the many species of Tephritidae that have spread beyond their native ranges to become invasive pests.

Laboratory Fruit Fly, *Drosophila melanogaster*) to inflicting Tephritidae-like damage to commercial fruits. The pretty little striped *Zaprionus indianus*, for example, is an originally African fly that now threatens figs in parts of South America, and the originally Oriental *Drosophila suzukii* has recently appeared as a pest of cherries, raspberries and other soft-skinned fruits in the United States and Canada. Another acalyptrate species, *Neosilba pendula* (Lonchaeidae), has also been reported as a pest of coffee and Barbados Cherry in South America. Seeds too are home to flies, such as the cone maggots (anthomyiids in the genus *Strobilomyia* and lonchaeids in the genus *Earomyia*) that are pests of conifers.

The default maggot strategy seems to be to filter microorganisms from within moist masses of organic material, so of course the underground parts of plants are especially susceptible to invasion by dipterous lineages. Some Syrphidae, including the Bulb Fly (*Merodon equestris*) and the "lesser bulb flies" in the genus *Eumerus* damage belowground parts of narcissus, amaryllis, tulips and various *Allium* (onion and garlic) species, feeding on the damaged and decaying bulbs. Similar damage is inflicted on European garlic bulbs by an acalyptrate, the Garlic Fly (*Suillia univittata*, family Heleomyzidae). Much worse damage is inflicted on onion crops by the Onion Maggot (*Delia antiqua*, family Anthomyiidae), a serious pest of onions and related *Allium* species in much of the northern hemisphere. Like other root maggots in the genus *Delia*, larvae invade the bulbs, causing decay and feeding to some extent on microbes as well as plant tissue; there seems to be a continuum between root maggot species that feed only on decaying tissue and more strictly phytophagous species. Other *Delia* pests include the Cabbage Maggot (*D. radicum*), Seed Corn Maggot (*D. platura*) and many others. Root maggots in the genera *Delia* and *Hylemya* (closely related Anthomyiidae) are certainly the most important root-feeding flies, but other families of Diptera sometimes damage the roots or bulbs of economically important plants. The cosmopolitan Carrot Rust Fly (*Psila rosae*, family Psilidae), for example, develops in fleshy taproots, causing serious damage to carrots as well as parsley, parsnips and celery.

Turf is often damaged by "leatherjackets" (the larvae of some terrestrial Tipulidae), such as the European (recently introduced to North America) crane fly species *Tipula paludosa* and *T. oleraceae*. Indoors, the larvae of other lower dipterans, especially the tiny dark-winged fungus gnats in the genus *Bradysia* (Sciaridae), can be serious pests on the roots of potted plants.

Flies and Fungi
Fungivorous Diptera

The spores, mycelia and fruiting bodies of fungi are consumed by a wide range of fly families, most of which show relatively little specificity to particular kinds of fungi and most of which feed on fungi that have at least begun to decay. They thus straddle the fuzzy line between being saprophagous and being truly fungivorous or mycophagous (an even fuzzier line than that between saprophagy and phytophagy). Those clearly on the mycophagous side of the line — the true fungus-eating flies such as the Platypezidae and some Sciaroidea — tend to be much more host specific. Within the Platypezidae (flat-footed flies), for example, the Holarctic genus *Polyporivora* develops only on polypore fungi, and larvae of the Nearctic genus *Melanderomyia* feed only on stinkhorns (Phallaceae); most other platypezid species develop in only a few related species or genera. The family Platypezidae is entirely mycophagous, in contrast with many other fungus-associated families that include species with a spectrum of other habits. Even the mainly fungivorous family Mycetophilidae (fungus gnats) also includes species that eat other insects, rotten wood or other materials, and almost all fungus-associated families include at least some saprophagous species. Larvae of the many fungus-associated species of Phoridae, for example, are often saprophages, parasitoids or predators of other insects in the fungi rather than strict fungivores, although the fungivorous phorid *Megaselia halterata* is a significant pest of commercial mushroom production, and several other phorids feed in a wide variety of mushrooms. Drosophilidae are almost invariably the most common acalyptrate flies on soft fungi in nature, but even fungus-associated drosophilid genera such as *Hirtodrosophila* and *Mycodrosophila* are probably obligate consumers of microorganisms in decaying fungi, and are thus more saprophages than fungivores.

There are flies associated with most groups of soft fungi and some hard fungi, but the Boletaceae, or boletes — the fleshy pore mushrooms so popular with mushroom foragers — are especially good places to look for fungus-associated flies. Fresh boletes in Europe and North America are home to several species of Mycetophilidae, Anthomyiidae (*Pegomya*

species) and fungus-feeding syrphids (*Cheilosia*); as they decay they come to abound in Sciaridae and fungus-associated but saprophagous Sphaeroceridae, Phoridae and Drosophilidae. *Cheilosia*, sciarids, phorids and *Suillia* species (Heleomyzidae) have also been associated with European truffles, those coveted underground mushrooms of culinary fame and considerable value. Truffle hunters apparently lie on the ground to spot hovering swarms of *Suillia*, which mark good spots to dig down a foot or two in search of the mushrooms. Related heleomyzid flies in the genus *Tapeigaster* are among the most conspicuous fungus-breeding flies in Australia, but Heleomyzidae are uncommon on tropical fungi.

Truffle flies are considered beneficial insects by truffle hunters, but some other Diptera, especially Phoridae and Sciaridae, are serious pests of edible mushrooms. Sciaridae (dark-winged fungus gnats) are probably the worst pests of commercial mushrooms: one widespread species, *Lycoriella ingenua*, causes significant damage as it feeds on mycelia and new button mushrooms. Cecid (Cecidomyiidae) pests in mushroom houses are also mycelial feeders, as are many cecids, sciarids and mycetophilids that feed in rotting wood. In fact, many flies that live on decomposing material feed at least in part on fungi, so species usually considered saprophagous are often really fungivores, even though the terms *fungivorous* and

Stinkhorn fungi (Phallaceae) expose their spores in a stinky gelatinous coating that attracts a variety of flies, such as these Black Blow Flies. Dung mosses (Splachnaceae) have a similar strategy of attracting flies on which they depend for spore dispersal.

mycophagous are most generally applied only to flies that feed on mushrooms and other conspicuous fungus fruiting bodies.

Diptera and Decomposing Plant Material

Wrack, Wood, Nests and Dung

Every living thing, including you and me, will eventually die and decompose, and the odds are good that our reintegration with the primordial ooze will be sped along by microbe-munching maggots. Most kinds of terrestrial decomposition, including that of organisms eaten and egested (as feces) by other organisms, are catalyzed by the larvae of Diptera. There is an enormous range of host specificity among the decomposer Diptera, not only with regard to the type of material (dung, carrion, vegetation, fungi) but also with regard to aspect and age. Relatively long-lived pockets of decay, such as a human body or a fallen tree, will support waves of different decomposers; even ephemeral resources such as the fruiting bodies of fungi show a continuum from species that feed on fresh fungi right through to those that feed on the foci of microorganisms left when the fungus is almost entirely rotted. One of the most specialized of all saprophagous fly communities is that associated with the piles of seaweed known as wrack.

Seaweed Flies

Wrack is the term usually used for the conspicuous windrows of kelp seen along marine shores the world over, especially on colder coastlines. Once tossed up by the monthly high tide, these piles remain exposed to terrestrial decomposition until sufficiently high tides come along a month or so later to wash what remains back into the ocean, replacing it with new heaps of kelp to start the "wrack cycle" anew. Wrack piles are essentially the world's largest compost heaps, extending in discontinuous linear strands from the Arctic Ocean to Tierra del Fuego. Like compost heaps, they are loaded with flies, but the flies in decomposing seaweed form an impressively specialized community of wrack-restricted species, almost all of which differ widely from those found in compost or other kinds of decay. Windrows of wrack invariably writhe with maggots; the wrack cycle pivots on the relationship between maggots, microbes and decaying seaweed. The players — the key species — in the wrack cycle make up an astoundingly specialized community.

Unlike the flies found in compost heaps, most of which are species that occur in a wide variety of decomposing materials across two or more continents, most of the flies found in seaweed occur only in seaweed. Furthermore, most of them belong to major groups of species — genera, subfamilies or even entire families — found only in these coastal piles of seaweed. Almost no other restricted habitat hosts such a unique fauna. Wrack faunas the world over include the wrack-restricted genera *Thoracochaeta* (Sphaeroceridae) and *Fucellia* (Anthomyiidae), and wrack faunas everywhere but South America include the wrack-restricted family Coelopidae and wrack-restricted species in the shoreline genus *Chersodromia* (Empididae); otherwise, every region supports its own remarkably special wrack community. European and northeastern North American wrack, for example, supports Holarctic species in the wrack-only genera *Orygma* (Sepsidae), *Coelopa* (Coelopidae), *Fucellia* and *Thoracochaeta*, along with more narrowly distributed species in the latter three genera and a couple of wrack-restricted species in the diverse family Scathophagidae. Western North America, on the other hand, is home to an almost entirely different set of *Fucellia*, *Thoracochaeta* and *Coelopa* species, along with members of the seaweed fly family Helcomyzidae and one of the two species of the wrack-restricted family Heterocheilidae (the other species is known only from seaweed-strewn European seacoasts). Chilean wrack supports yet a different community of specialized Sphaeroceridae and Helcomyzidae, and Australian wrack is home to some remarkable groups of Coelopidae and Sphaeroceridae. Studies of wrack in Europe, Namibia, California and New Zealand have all uncovered further species of wrack specialists.

Coelopidae | *Coelopa*

Helcomyzidae | *Paractora*

Heterocheilidae | *Heterocheila*

Anthomyiidae | *Fucellia*

Sepsidae | *Orygma*

Deposits of seaweed, or wrack, found along rocky marine shores the world over are home to several specialized groups of flies. The families Coelopidae, Helcomyzidae and Heterocheilidae **(UPPER LEFT, UPPER RIGHT AND MIDDLE LEFT)** are wrack-restricted, as are the genera *Fucellia* **(MIDDLE RIGHT)** and *Orygma* **(BOTTOM)** in the families Anthomyiidae and Sepsidae.

The remarkable wrack community is important to us for a number of reasons, above and beyond the interesting observation that untold millions of maggots inhabit rotting kelp from southern New Zealand to the Canadian Arctic. The specialized interaction between most wrack flies and micro-organisms allows for rapid and efficient decomposition of wrack in the short period available between high tides, transforming the immense primary productivity of offshore kelp beds into soluble decay products that are washed back into the ocean at each high tide. Since kelp beds are the most productive plant communities on the planet and there is almost no herbivory on living kelp, wrack maggots play a pivotal role in marine food chains. And those wrack maggots show an astounding degree of specificity at the species, genus, subfamily and even family level, something that might initially be unexpected in such a spatially restricted community made up mostly of microbial grazers. In fact, specialization, at least at the species or genus level, is the rule rather than the exception among the flies that live in many kinds of decomposing material.

Deadwood Diptera

Xylophilous Flies

A diversity of Diptera develop in wood, with specialists found in the bark or wood of different tree species and in habitats of differing size and moisture levels. Most of the 50 or so North American fly families with tree-associated larvae depend on decomposing wood rather than living plant material, and many thrive in the sticky fermenting tissue under the bark of recently fallen trees. Fermenting sap, either under bark or flowing from tree wounds, supports a wide variety of specialized microbial grazers along with an associated community of predators. Subcortical (under-bark) species occur throughout the order, and a rich community of flies can be found in other woody habitats ranging from moist heartwood to wet tree holes, with some larger families occupying more or less the entire spectrum. Crane flies (Tipulidae), for example, are common under bark, where larvae of the distinctive little black *Gnophomyia* crane flies (subfamily Limoniinae) often abound, but different species also occur in wood of various ages and densities. Even solid heartwood is home to some crane flies, especially the stout-jawed larvae of the big, colorful *Ctenophora* species (subfamily Tipulinae). Certain other crane fly species are found only in partially submerged wood.

Among the lower Diptera, the Tipulidae are probably the most conspicuous xylophilous ("wood-loving") larvae, but several species of fungus gnats can also be abundant under bark or in fungus-riddled deadwood. Most feed on fungi but some, especially in the family Keroplatidae, are predators. The glistening mucous tubes sometimes conspicuous on the surface of rotting wood or fungi are made of keroplatid larval salivary secretions, serving as slick subway lines and shelters for these slimy larvae. Other lower Diptera larvae commonly found under bark include a few Ceratopogonidae, certain exceptional Cecidomyiidae and some Scatopsidae. Subcortical Cecidomyiidae in the genus *Miastor* can rapidly build up numbers by skipping sex and bypassing the egg and adult stages. Larvae simply develop inside parent larvae, effectively devouring their mothers from inside. This type of life cycle (called paedogenesis) continues until conditions deteriorate as the food (fungus) is used up or the bark dries out; then a "normal" life cycle with pupae and adults is resumed until the winged adults locate a new habitat.

Although they are most common under bark, lower Diptera are also found in more unusual sorts of deadwood habitats. Larvae in the odd little family Axymyiidae, for example, burrow deep within waterlogged deciduous logs that are partially (but not entirely) submerged in small woodland streams. Axymyiid larvae have robust mandibles to mine deep into fairly solid wood, but they maintain contact with the exposed wood surface by using a long, snakelike tail that ends in a pair of spiracles (they have another pair of spiracles just behind the head). Species now placed in the family Axymyiidae used to be treated as part of another odd little wood-eating lower dipteran family, the Pachyneuridae. Known pachyneurid larvae are associated with decaying wood: the one

North American species lives in rotting twigs and branches of alder shrubs on the west coast.

Lower brachyceran larvae are no less diverse in woody habitats than lower Diptera. The flattened, sandpaper-skinned larvae of soldier flies (Stratiomyidae, especially subfamily Pachygastrinae) often occur in dense aggregations in the sappy subcortical parts of dead or damaged trunks, with different genera feeding on fungi, sap and microorganisms under the bark of different kinds of trees. A couple of lower brachyceran families are so strongly associated with wood that their formal names include the prefix *xylo-*. The relatively large white-skinned and long-headed larval Xylophagidae are common subcortical predators, and larvae of the small family Xylomyidae sometimes have much the same habits as pachygastrine Stratiomyidae.

Certainly the most spectacular of the xylophagous (wood-eating) lower Brachycera are the giant Pantophthalmidae of Central and South America, which look like enormous horse flies. Pantophthalmids bore in dying or recently dead trees to feed on a mixture of wood and microorganisms; a single tree of a preferred species (often trees with latex-loaded or mucilaginous sap) can be peppered with cylindrical holes or studded with pantophthalmid pupae hanging partway out of the holes. Larval Empidoidea (dance flies and long-legged flies) and lower brachyceran predators in the families Therevidae (stiletto flies), and Asilidae (robber flies) can also occur in wood, although most species in these families are soil inhabitants. Predaceous larvae of xylophilous species often hunt other insects in beetle burrows or channels deep within rotting wood. Among the best known of these are the little dolichopodids in the genus *Medetera* (Empidoidea, family Dolichopodidae). Adult *Medetera* ("woodpecker flies") can almost always be found perched at about a 45-degree angle on trunks of standing dead or damaged trees, and their predaceous larvae presumably abound in beetle burrows penetrating the same trees.

Xylophilous maggots (wood-associated Cyclorrhapha larvae) are diverse and occur in each of the major subgroups of higher Brachycera. A few families of Aschiza occur in wood or under bark, although the Platypezidae found under bark are fungivorous and only secondarily associated with wood, and relatively few species in the huge family Phoridae live in wood. The hundreds of species of Syrphidae associated with wood, on the other hand, comprise the most diverse group of xylophilous Diptera. Most members of the largest syrphid subfamily, Eristalinae, develop in or on wood in various states of decay, where they filter feed on microorganisms from wet wood, in tree holes, under bark, in fermenting sap or even in fairly solid wood of standing trees.

Many species of calyptrate and acalyptrate maggots occur in wood, but the most common ones belong either to one genus (*Lonchaea*) in the acalyptrate family Lonchaeidae or to one genus (*Phaonia*) of the calyptrate family Muscidae. Both are probably at least facultatively predaceous on larvae of wood-boring

The spectacular antler flies in the genus *Phytalmia* (Tephritidae) are associated with fallen trees in New Guinea and Queensland, Australia.

beetles, and both can be very common under the bark of fallen trees at the right stage of decay. Subcortical species also occur in several other acalyptrate families, including Clusiidae, Ulidiidae and Psilidae. Other higher Brachycera have a wider range of wood associations. Bleeding tree wounds, especially those that blend with associated fungi to create an oozing "slime flux," are hotspots for dipteran diversity. Specialized slime flux associates among the Acalyptratae include the otherwise rarely encountered families Odiniidae, Periscelididae and some Aulacigastridae, as well as some members of the diverse family Drosophilidae.

Perhaps the most spectacular of all xylophagous Diptera are specialized members of the primarily fruit-associated acalyptrate family Tephritidae, of which several tropical Pacific species in the genus *Phytalmia* are known as "antler flies" because the males have massive, antler-like head projections. The "antlers" are used to defend territories over rotting sapwood on fallen trees of selected species, where females mate with the guarding males before they lay their eggs. Some Clusiidae also have ornamented or outlandishly widened heads that they use in ritual head-butting or head-measuring battles, usually on fallen trees near optimal oviposition sites. Similarly equipped males are found in a few other families, such as the Diopsidae and Ulidiidae, in which some species form male aggregations, or leks.

Diptera and Dung

Coprophagous Flies

Shit happens — reliably and copiously. In fact, fecal material is among the most abundant sources of insect food and shelter on the planet, creating a ubiquitous microhabitat to which both maggots and adult flies are admirably adapted. The average horse egests almost 20 kilograms (44 pounds) of manure a day, cows create about 26 kilograms (57 pounds) of meadow muffins a day, a chicken-sized bird generates guano to the tune of 1 kilogram (2.2 pounds) per month, 7 billion people on the planet are each pumping out around a kilogram of poo per day, and millions of other animal species (mostly invertebrates) are cranking out awesome amounts of fecal material by the minute. All this nutritious dung is a hotly contested resource among insects, and the winners in the poop sweepstakes are almost invariably Diptera, probably because fly larvae are ideally adapted for living inside substrates that are rich in moisture and microorganisms (a typical turd is largely water and bacteria). Adult flies are also well adapted for feeding on dung (or at least the moist surfaces of fresh deposits) and many species (including those that breed elsewhere) regularly or irregularly visit droppings to sponge a snack. Many other species associated with feces are predators, either as adult flies that visit the dung in search of prey or as maggots that feed mostly on larvae of other dung Diptera during at least part of their development.

Diptera that develop in the droppings of domestic animals have been studied most extensively, partly because some such species are serious pests of humans and animals, and partly because the droppings themselves would degrade pasture and cause other problems if not removed by coprophagous insects. The most serious coprophagous pests are flies in the family Muscidae, including bloodsucking muscids such as the Stable Fly (*Stomoxys calcitrans*) and Horn Fly (*Haematobia irritans*), as well as non-biting pest species such as the ubiquitous House Fly (*Musca domestica*) and the infamous Australian Bush Fly (*Musca vetustissima*). While there are many different native dung fly species in different parts of the world, the fauna of domestic animal dung is remarkably similar worldwide, at least at the family level and often at the generic and species level as well. Cow dung almost anywhere, for example, will be routinely visited by 50 to 100 fly species, mostly secondarily widespread coprophagous species in the families Muscidae, Sarcophagidae, Calliphoridae, Sphaeroceridae and Sepsidae but usually including Fanniidae, Anthomyiidae, Scathophagidae, Tipulidae, Stratiomyidae, Sciaridae and Phoridae, and several other families as well.

About 30 fly families include members that breed in dung, but no significant insect family (that is, no family with more than one or two species) is completely restricted to dung. Even the families thought of as dung specialists include members that develop in other wet, rotting, bacteria-rich substrates. The family Sepsidae (the ant-like scavenger flies) is probably the closest to an entirely coprophagous family, but even there the basal lineages live in rotting seaweed, and several species are also known from dead invertebrates and other kinds of decomposing material. The family Sphaeroceridae (small dung flies) undoubtedly includes more species of dung-associated Diptera than any other family, but even in this group most species develop as microbial grazers in other wet, bacteria-rich material such as decomposing plant material that hasn't been processed through an animal's stomach.

The importance of dung-breeding flies in the spread of disease is enormous (see Chapter 3), and it is not without reason that the House Fly (*Musca domestica*) is sometimes described as the most dangerous animal on the planet. Less widely recognized, however, is the important role that flies play in removing waste, both naturally and as a result of manipulation for control or conditioning of dung. Even the human dung processed through sewage treatment plants is partly broken down by millions of Psychodidae larvae that keep trickle filters trickling (tens of millions of these little lower dipterans are discharged every day in the effluent from some sewage plants). The speed with which maggots can turn a wet mass of feces into an aerated pile of dry compost is impressive, and flies are likely to play an ever-increasing role in the commercial processing of both human and animal waste in the future. At the moment only a few species are employed in this fashion, but the order offers a vast pool of coprophagous species to pick from. The fly species most often deliberately introduced to manure-composting systems at the moment is a member of the Stratiomyidae, the Black Soldier Fly (*Hermetia illucens*), which contributes to the management of poultry and pig waste by speeding up the composting process and by consuming competing maggots of House Flies. House Flies themselves, although generally considered pests because of their habit of flying from filth to food, are admirably adapted to the manure environment and can greatly speed up the composting process. Properly composted manure is a high-value product, and the fly puparia can be harvested as a valuable animal feed (in aquaculture, for example). Wingless or weak-flying species, such as many coprophagous Sphaeroceridae, have great potential as partners in the commercial processing of dung.

Considerations of coprophagous flies as cleanup crews, nuisances or disease carriers invariably concentrate on the ubiquitous species associated with humankind and domestic animals, often overlooking the astonishing diversity of flies associated with dung and other waste products of different species in different microhabitats. Some Australian wombat flies (*Borboroides*, family Heleomyzidae), for example, occur only on wombat dung, and the monotypic wingless fly families Mormotomyiidae and Mystacinobiidae live only on bat dung, the former in a single cave in Kenya and the latter only in colonies of New Zealand short-tailed bats. The nests, burrows and dens of mammals are occupied by such a diverse fauna of mostly dung-developing Diptera that Finnish dipterist Walter Hackman was able to find 240 species of flies in vole burrows alone.

More specialized coprophagous flies are found at the excretory extremes of inconspicuous invertebrate waste and ponderous piles of pachyderm poop. Many species (in fact, several genera) of Sphaeroceridae (small dung flies) are known only from elephant dung, and one species in the same family prefers millipede frass to the point that gravid female flies cling to millipedes while awaiting an opportunity to oviposit on a fresh diplopod deposit. A few other species in both New World and Old World genera of Sphaeroceridae have a similar habit of clinging to invertebrates while awaiting an opportunity to oviposit in dung — but not necessarily dung produced by their invertebrate hosts. Several such species (in the genera *Ceroptera* and *Norrbomia*) are kleptoparasitic on dung-rolling scarab beetles, riding on their hosts in order to lay eggs in their dung balls just prior to burial. Yet another sphaerocerid is found only in and around the green (excretory) glands of land crabs. Similar habits have evolved repeatedly in a number of lineages of Drosophilidae, a family that also includes species that feed on insect frass.

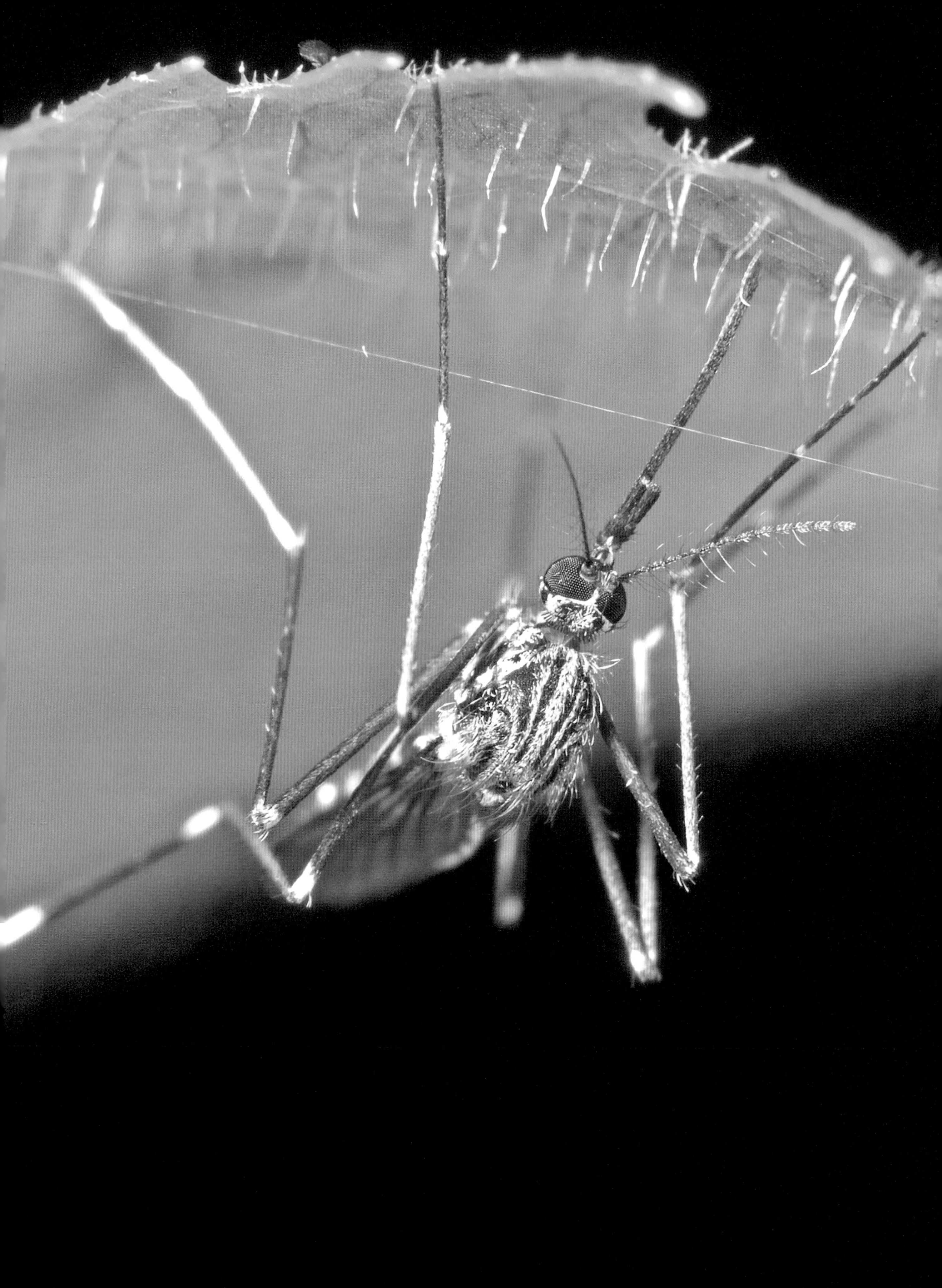

3

Flies and Vertebrates

Blood Feeding, Disease, Myiasis, Dead Bodies and Forensic Dipterology

MANY FLIES HAVE AN INTIMATE ASSOCIATION with mankind and other vertebrates, imbibing their blood, consuming their tissues or developing in their remains. Females of some groups of lower Diptera and lower Brachycera acquire protein for egg production by biting vertebrates, a habit best known in the mosquitoes (Culicidae), black flies (Simuliidae), punkies (Ceratopogonidae), horse flies (Tabanidae) and sand flies (Psychodidae) but also found in the frog midges (Corethrellidae), some snipe flies (Rhagionidae) and a few athericid flies (Athericidae). Only a few lineages of higher Brachycera (Cyclorrhapha) have blood-feeding adults, but in these groups (including the entire Hippoboscoidea and several genera of Muscidae), both sexes bite. In some Muscidae, Piophilidae and Calliphoridae the larvae do the blood feeding, usually attacking nestling birds, although some African calliphorid maggots imbibe mammal blood (including that of humans).

Blood-feeding flies are important disease carriers, but diseases of humans and animals are also carried by non-biting species. Eyes and other moist body surfaces attract flies, such as some Chloropidae and Drosophilidae, that can transmit conjunctivitis. Many other flies, especially Muscidae and Calliphoridae, move freely from filth to food and carry along a variety of microorganisms responsible for enteric (gastrointestinal) diseases. When flies have an even more intimate association with living vertebrates, developing under the skin or in the digestive system, it is referred to as myiasis. All species of Oestridae, some Calliphoridae and a few Sarcophagidae are obligate myiasis-causing flies whose larvae live as parasites inside vertebrates. Other families of flies have a more opportunistic approach to myiasis, only occasionally developing in or on vertebrates (including people). For example, larvae of many kinds of flies, including Stratiomyidae, Syrphidae, Phoridae, Piophilidae and Drosophilidae, can occur in materials ingested by people and are thus occasionally implicated in facultative enteric myiasis.

Even normally innocuous fly groups that neither bite nor cause myiasis can affect vertebrates, especially when fly population densities trigger respiratory problems or allergic reactions. Midges (Chironomidae), for example, often reach such densities, as do phantom midges (Chaoboridae). Midge species that use larval haemoglobin as a respiratory pigment can cause serious allergy problems when particles from decomposing adult flies become airborne.

OPPOSITE PAGE Mosquitoes, like this *Aedes* resting under a leaf, are the most important biting Diptera. The blood-feeding females of these delicate flies are familiar nuisances and fearsome carriers of some of humankind's most serious diseases.

Although *Anopheles* is arguably the most dangerous animal genus, the distinctively patterned biting females are also beautiful insects. This South American *An. malefactor* is resting on a cabin wall in a typical head-down posture.

Mosquitoes and Disease

In the time it takes you to read this sentence, someone somewhere will be killed by a fly-borne disease and hundreds of others will become sick because of fly-associated microorganisms. Some may be defeated by dehydration due to enteric diseases carried from feces to food by House Flies or other synanthropic flies, while others might be devastated by flesh-eating protozoa carried by sand flies, stricken with lethargy caused by trypanosomes injected by tsetse flies, or blinded by filarial worms delivered by black flies. The strongest odds, however, favor death or disability following injection of a virus, worm or protozoan by a female mosquito. Mosquito-borne viruses such as those that cause yellow fever, dengue, encephalitis, West Nile disease and chikungunya are serious threats, and mosquito-borne filarial worms can cause the horrifying symptoms known as elephantiasis. But the most infamous of insect-borne diseases are the malarias carried by a select few mosquito species.

Malaria

Many types of malaria can infect animals and birds, but a mere four species of *Plasmodium* cause the malarias that kill more than a million people a year and sicken hundreds of million more. About half the people in the world live in regions where the bite of an *Anopheles* mosquito may be accompanied by injection of saliva containing tiny *Plasmodium* sporozoites. The sporozoites change form and multiply, first in the liver and then in the blood cells of their human hosts, causing intermittent fevers and other problems of a severity that depends partly on the species of *Plasmodium* involved (deaths are usually due to *P. falciparum*). Some species of *Plasmodium* (*P. vivax*, *P. ovale*) are able to hide in the liver, enabling the disease to reappear years or even decades after its first appearance. The human body is a dead end for the malaria-causing protists if a female of the right species of *Anopheles* does not come along to pick up *Plasmodium* gametocytes along with a blood meal. However, once inside a mosquito's gut, the gametocytes enter a sexual part of their life cycle that can take place only inside an appropriate *Anopheles*. There they transform into male or female gametes that fuse into yet another form (the ookinete) that settles into the mosquito's gut wall before rupturing to release a batch of sporozoites that travel to the salivary glands, from whence they get injected into another host to start the cycle anew. Just over a million of those human hosts (mostly children in Africa) are killed by malaria every year. Millions more suffer but survive the chills and fevers of the disease.

It seems remarkable that the world's most devastating disease is completely dependent on one sex of just some species of one genus of fly. Only a few dozen of the hundreds of *Anopheles* species transmit malaria, and much of the global effort to break the malaria cycle is focused on that handful of delicate little mosquito species. Understanding the taxonomy, distribution and larval habits of those species is a prerequisite to selective and effective control, and an understanding of adult habits can also be critical. Since *Anopheles* mosquitoes usually bite at night (or dawn or dusk), bed nets treated with safe insecticides are probably the most cost-effective weapons in the fight against malaria in parts of the world where dwellings are not normally air-conditioned (or even screened). Regional, and perhaps ultimately global, eradication of malaria is likely to follow the elimination of poverty and the concomitant elevation of living standards (remember that much of Europe and North America were once malarious zones), but in the meantime an increasingly sophisticated technological arsenal is being deployed against the mosquitoes that transmit malaria and other diseases from person to person. Aside from variations on the main strategy of killing mosquito adults and larvae with biological and chemical weapons, some of the up-and-coming approaches to targeting *Anopheles* and other dangerous mosquitoes include mini-lasers linked to cameras that can distinguish them from other insects and selectively zap them, microorganisms that infect adult mosquitoes and shorten their lives to the point where they don't transmit disease, and genetically modified mosquitoes that are immune to *Plasmodium* invasion.

Dengue

Malaria is justifiably recognized as the world's most significant insect-borne disease, but mosquitoes and other flies continue to threaten us through their association with a variety of other malevolent microorganisms. Those of us in the developed world are at minimal risk, but mosquito-borne viruses, including those responsible for dengue, West Nile disease and encephalitis, are of increasing concern even in North America and Europe. Dengue, or breakbone fever, is much in the headlines right now because of increasingly frequent appearances of locally acquired cases in temperate countries. Travelers routinely return from the tropics unknowingly infected with mosquito-borne diseases, but locally acquired cases of mosquito-vectored diseases have been rare in North America and Europe for most of the past century. That seems about to change as dengue — a normally flu-like but sometimes fatal disease, long a serious problem in warmer parts of the world — responds to climate change and to changing mosquito distributions.

The four related viruses that cause different forms of dengue are normally carried by *Aedes aegypti* and other *Aedes* mosquitoes, including species such as *Aedes japonicus* and *Aedes albopictus* with seemingly ever-expanding ranges. Many of the mosquitoes that carry dengue are container breeders that thrive in plastic garbage, old tires and the like in urban areas, perhaps one of the reasons why dengue is doing so well today. And recent figures from the World Health Organization show that dengue is on the march: Brazil, for example, saw almost 600 dengue deaths in 2010, almost double the previous year, and Venezuela's 100,000 diagnosed dengue cases in 2010 more than doubled those reported in 2009. The rise of dengue in some Asian countries (where the disease first appeared in the 1950s) is even more alarming, and

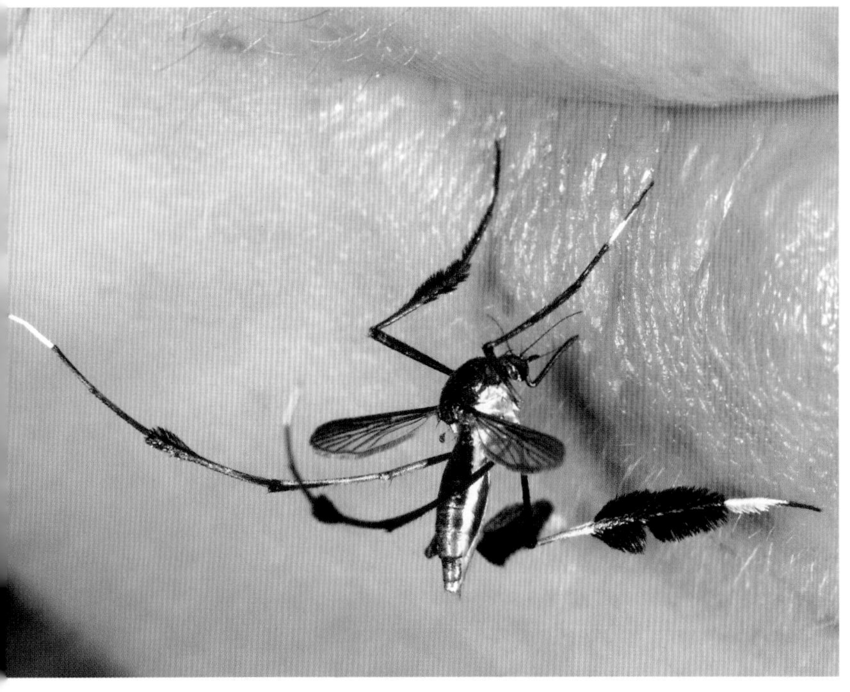

recent World Health Organization figures suggest that total annual infections may be around 50 million worldwide. So, dengue threatens billions, infects tens of millions, seriously affects hundreds of thousands and kills tens of thousands every year. Most of the mortality results from a particularly virulent form known as dengue hemorrhagic fever. There is currently no vaccine or effective treatment, so understanding and controlling (or avoiding!) the mosquito vectors is the key to dealing with dengue.

Yellow Fever

Although dengue is currently the most important mosquito-borne virus, it is only so because yellow fever has for many years been kept at bay with effective vaccines (vaccines for dengue and malaria remain elusive). Just over a century ago the yellow fever virus was one of the great scourges of mankind, and it still stands as one of the most dangerous diseases of all times. Accounts of famous yellow fever epidemics, such as the 1793 epidemic in Philadelphia, include bone-chilling descriptions of jaundiced victims bleeding from the mouth and eyes while also vomiting blood. Although the overall fatality rate for the disease is relatively low (around 3 percent), that particular epidemic killed almost 10 percent of the city's population. The United States has not had a yellow fever epidemic since 1905 and an effective vaccine exists, but, remarkably, the disease is still with us and (according to the World Health Organization) currently causes 200,000 cases and 30,000 deaths per year, mostly in unvaccinated populations in Africa.

The virus is often transmitted by the same *Aedes* species that are major vectors for dengue, with *Aedes aegypti* (the aptly named Yellow Fever Mosquito) being almost entirely responsible for the important urban outbreaks still occurring in Africa. In rural areas the disease is more often transmitted by other mosquitoes, including Afrotropical *Aedes* species as well as the forest-loving Neotropical genera *Haemagogus* and *Sabethes*, which can carry the disease between humans and other primates. Yellow fever is originally an African disease and the Yellow Fever Mosquito is originally an African fly, although it is now "circumtropical," found in warm areas worldwide. Both the disease and the fly were virtually eliminated from the New World by the mid-1900s, thanks to vaccination and vector control, but experts now fear a resurgence, as *Aedes aegypti* has again become common from the southeastern United States southward.

This *Culex territans* is full of blood drawn from a frog in a Canadian pond. The blood will nourish her eggs, soon to be laid in a raft on the water surface.

OPPOSITE PAGE This Yellow Fever Mosquito (*Aedes aegypti*, also known as the Dengue Mosquito; **TOP**) is just emerging from its floating pupa. Larvae and pupae of this dangerous mosquito can be common in artificial habitats such as water tanks, flower vases and discarded tires or plastic containers. The **BOTTOM** image is a South American *Sabethes (Sabethes) tarsopus*.

Other Mosquito-Borne Viruses

Aedes aegypti and related species are also responsible for the spread of several emerging viral diseases, of which one of the most worrisome is chikungunya, a dengue-like disease of increasing importance in the Old World tropics. In addition to major occurrences in Africa and India, a 2006 outbreak on the small Indian Ocean island of Réunion infected almost a quarter of a million people and killed more than 200. Chikungunya is not yet reported from the New World or the Pacific, but the major vectors for this disease are *Aedes aegypti* and *A. albopictus*, both of which are now firmly established in much of the world, including Hawaii and North America. There was an outbreak of chikungunya in Italy in 2007, and it seems just a matter of time before this *Aedes*-associated virus becomes more widespread.

Mosquitoes harbor hundreds of different viruses, of which about a hundred can develop in both mosquito and human cells. The *Aedes*-borne dengue, yellow fever and chikungunya viruses are currently the most important of these, but other significant viruses are carried by other mosquitoes. Some can cause a brain inflammation known as encephalitis or encephalomyelitis. Unlike dengue and yellow fever, which are strictly diseases of humans and other primates, the most important mosquito-borne encephalitis viruses are normally found in birds; for them humans can be thought of as abnormal dead ends. That is small comfort, however, for those bitten by a mosquito carrying eastern equine encephalomyelitis, a severe disease that kills as many as half the people it infects. Mosquitoes of several species carry this disease, either between birds or from birds to mammals (often horses; less often, people). Eastern equine encephalomyelitis occurs throughout much of the New World, including northeastern North America, where annually about half a dozen unlucky people get the disease.

Western equine encephalomyelitis is a related disease that is widespread in the New World, also cycled between birds and mammals. West Nile virus, an originally Old World encephalitis virus recently arrived in North America, is similarly maintained in a bird reservoir and transmitted to humans by bird-biting *Culex* mosquitoes. West Nile first appeared in North America in 1999, killing seven people in New York in the first year. It has since spread throughout much of North America, infecting about 1,000 Americans and killing 39 in 2010 — offering a cautionary tale to those who think the "Developed World" is safe from mosquito-borne diseases. Other significant mosquito-borne viruses include Ross River virus in Australia, Sindbis virus in the Old World and Australia, and Rift Valley fever in eastern Africa.

Filariasis

Viruses and protists are not the only mosquito associates that have the potential to cause us misery, a point unforgettably made by a photograph I once saw of a man with such a grotesquely swollen scrotum that he had to use a wheelbarrow to move around. The poor fellow was suffering from elephantiasis caused by filarial worms (*Wuchereria bancrofti*) that had migrated to his lymphatic vessels and lymph nodes, where they developed into adult worms. Although I have seen that particular unforgettable example of mosquito-borne misery only in photographs, I have encountered people in the Caribbean and Africa with grossly swollen elephantine stumps that seemed to have replaced their legs — another common manifestation of chronic lymphatic filariasis.

Three different kinds of worms, carried by a variety of mosquitoes in much of the tropical world, cause similar unpleasant chronic symptoms in about half of the over 100 million people currently infected with these strictly mosquito-delivered filarial nematodes, while others suffer from episodic fever, swelling and pain. Several different mosquitoes, including *Aedes* and *Anopheles* species, are involved, but *Culex quinquefasciatus* seems to be the most important vector in urban areas, in the New World and in parts of Africa and India. Filariasis is not a serious threat to people who live in temperate parts of the world, but animals are not so lucky. Dog heartworm, for example, is a filarial nematode delivered to dogs by the bites of several kinds of mosquitoes. Adult nematodes lodge in the heart and pulmonary arteries, sometimes killing the animal. Other filarial nematodes are associated with other vertebrate hosts and mosquito vectors throughout the world.

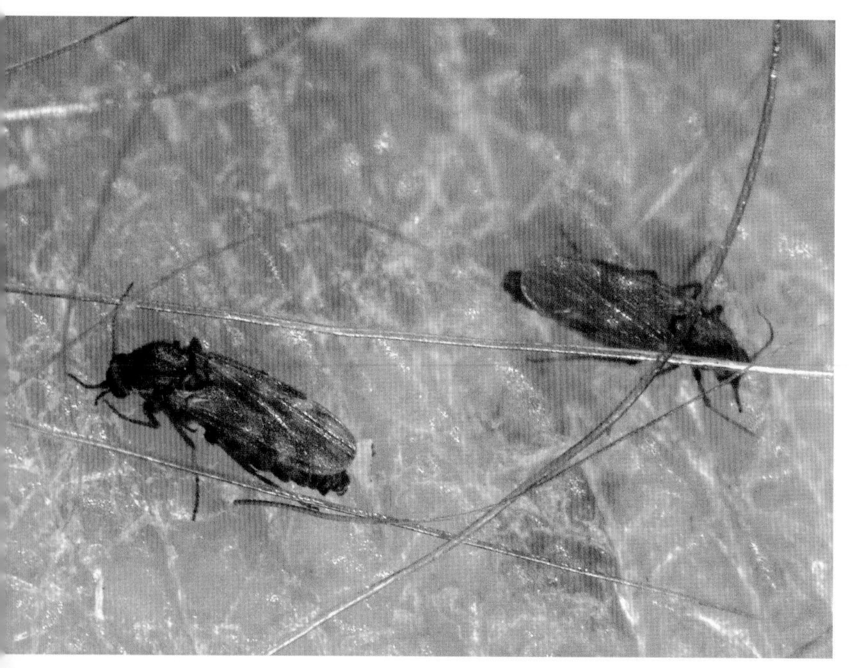

Biting midges in the genus *Culicoides* are sometimes called "no-see-ums" in reference to their minute size.

Other Biting Flies and Disease

Mosquitoes monopolize the delivery of most of the major arthropod-borne diseases, but other biting flies bring their share of misery in the form of associated viruses, protists and nematodes. Lower Diptera ("Nematocera") with blood-feeding females occur in the Corethrellidae, Ceratopogonidae, Simuliidae and Psychodidae. Disease-causing organisms are carried by members of all of these groups, although the parasitic protists carried by Corethrellidae affect only the male frogs that are their sole hosts.

Ceratopogonidae

The bites of certain Ceratopogonidae (biting midges, or punkies) can cause a burning itch out of all proportion to their minute size, but the pain passes quickly and these tiny flies are less likely than mosquitoes to cause more lasting problems. They do carry a few viruses and nematodes that affect humans, including Oropouche fever virus in the neotropics (especially the Brazilian Amazon), but the main impact of biting midges is on animals. Sheep, horses, cattle, poultry and a variety of wild animals suffer from ceratopogonid-carried protists, nematodes and viruses, of which the most important (at least at the moment) is bluetongue disease of sheep.

Bluetongue is of special interest now because it has been on the move from its original warm temperate range (the disease was first recorded in South Africa) to historically bluetongue-free areas, including northern Europe and the United

Kingdom. The spread of bluetongue might be related to climate change, although the disease is now transmitted by widespread biting midge (*Culicoides*) species as well as species restricted to the original range of the disease. The blurring of lines between areas affected by bluetongue and bluetongue-free areas is significant, because one way this serious livestock disease was contained in the past was through regulation of stock movement out of affected regions. Animals with bluetongue suffer various lesions and other symptoms that include a cyanotic (blue) tongue, fevers, difficult breathing and sometimes death.

Related *Culicoides*-carried viral diseases include African horse sickness in Africa and the Arabian Peninsula and epizootic hemorrhagic disease in New World deer and other ruminants. Biting midges can also be vectors of protist diseases of poultry and can transmit the nematode *Onchocerca cervicalis*, which causes equine onchocerciasis. But *Onchocerca cervicalis* is much less important than another *Onchocerca* species, carried by black flies (Simuliidae).

Female black flies, like this Andean *Simulium bicoloratum*, can be abundant blood-feeding pests near the running waters in which their larvae develop.

Simuliidae

Black flies are well known to most Canadians (including me) for the ravenous hordes of small, stout flies that make early spring fishing trips a disturbingly desanguinating activity. These persistent flies bite only outdoors, only during the day, and usually only for a fairly short season. They are annoying, but most black fly victims in temperate countries suffer no more than the loss of a bit of blood and some local swelling and itching. Elsewhere, especially in sub-Saharan Africa, the story is different: bites of certain *Simulium* species can deliver larvae of the filarial nematode *Onchocerca volvulus*, an agent of the notorious disease known as onchocerciasis, or river blindness.

These little larvae develop into thread-like adult worms that concentrate in nodules, where they mate and produce huge numbers of tiny progeny called microfilariae, which in turn migrate to the skin (where they can be picked up by biting black flies), eyes and elsewhere in the host. It is the myriad of microfilariae that can make infected people miserable by causing unbearable itching and other issues, including lesions and rashes. When migrating microfilariae make it to the host's eyes, they cause a variety of visual problems, sometimes including blindness ("river blindness" refers to this and to the river habitat of the black fly larvae). Many of the severe ocular impacts follow from an immune response to particular bacteria (*Wolbachia*) associated with the nematodes; together the worms and their endosymbiotic bacteria are responsible for about a quarter of a million current cases of blindness, almost all in West Africa. Millions more people (mostly in Africa, but some in the neotropics) are infected but show less severe symptoms.

Another filarial nematode, *Mansonella ozzardi*, is carried by black flies in South America, but the symptoms of mansonellosis are relatively mild. Black flies also transmit a variety of diseases of wildlife and livestock, including other *Onchocerca* nematodes that infect cattle and several protists (in the genus *Leucocytozoon*) that cause malaria-like diseases of poultry. The latter are of diminishing economic importance now that poultry are largely raised indoors, safe from the bites of black fly vectors.

Lutzomyia is a large genus of tiny "sand flies," including the carriers of leishmaniasis and other diseases in the New World.

Psychodidae

The moth fly family (Psychodidae) is best known for innocuous non-biting species such as the common drain flies that routinely emerge from bathroom sinks everywhere, but two subfamilies of Psychodidae have blood-feeding females that are able to transmit microorganisms to their vertebrate hosts. One of these subfamilies (Sycoracinae) is a small group only known to bite frogs, but the other (Phlebotominae, or sand flies) is an important lineage of biting flies with several tiny but exceedingly dangerous members. A few dozen species of sand flies in two genera, *Phlebotomus* in the Old World and *Lutzomyia* in the New World, are responsible for the transmission of a number of viruses, bacteria and protists that cause disease in both humans and domestic animals.

The most frightening of these diseases is leishmaniasis, which is caused by about 20 species of subtropical or tropical parasites in the protist genus *Leishmania*. Unlike malaria and onchocerciasis, which depend on a pool of infected people from which blood-feeding female flies can pick up parasites to pass on to other people, most kinds of leishmaniasis lurk in non-human reservoirs, including a wide variety of rodents, monkeys, opossums, horses and other mammals. The disease-causing *Leishmania* species are different in the New World and the Old World, with several species in a range of habitats and climates ranging from South American forests to the war-torn deserts of Afghanistan. The disease occurs in both cutaneous and visceral forms, with the more common cutaneous form (known as oriental sore, tropical sore, Baghdad boil or Delhi boil) manifesting itself as nasty open skin lesions, sometimes around the nose or mouth (mucocutaneous leishmaniasis). Visceral leishmaniasis (kala-azar, or dumdum fever) is even worse, with systemic effects that often lead to death of the infected host. According to the World Health Organization, about 350 million people in 88 countries are at risk from leishmaniasis, and around 12 million are currently infected. Treatment remains difficult and expensive — prohibitively so for many Third World sufferers.

Other diseases carried by the same two genera of biting sand fly include viruses such as the vesicular stomatitis virus that affects livestock and occasionally humans in the New World; the Chandipura virus, which is increasingly affecting children in India; and the sand fly fever viruses that now affect humans in various parts of the world. Carrion's disease, a bacterial disease of humans caused by *Bartonella bacilliformis*, is transmitted by Andean sand flies in the genus *Lutzomyia*. Infection can manifest itself as an ugly but rarely life-threatening chronic skin disease called Peruvian wart (verruga peruana) or as an acute, often fatal form called Oroya fever. Carrion's disease was named after a Peruvian medical student, Daniel Carrión, who died from Oroya fever in 1885, after establishing that it was caused by the same agent as Peruvian wart.

Biting Brachycera

In comparison to the numerous nasty nematocerans discussed above, biting lower Brachycera are of negligible importance as disease vectors. Horse flies and deer flies (Tabanidae) rarely have the sort of intimate association with disease organisms that characterize the major diseases carried by lower flies, in which the pathogen is dependent on the

Both males and females in the stable fly genus *Stomoxys* **(LEFT)** penetrate hosts using a stiff, bayonet-like labium tipped with sharp, toothlike processes. This is typical for biting higher Brachycera and contrasts with deer flies and other biting lower Brachycera **(RIGHT)**, which bite only as females and slice into skin using scissors-like mandibles.

fly for part of its development as well as transmission. Instead, horse flies and deer flies perform more like "dirty needles" to which pathogens sometimes stick as they move quickly from host to host with still-bloody mouthparts (entomologists call this "mechanical transmission"). The most significant exceptions to this generalization are some filarial nematodes, especially the African eyeworm, *Loa loa*. This worm, currently restricted to equatorial West Africa, needs to go through part of its life cycle inside deer fly (*Chrysops*) hosts. Biting deer flies pick up the tiny microfilariae from one human host and later deliver larval nematodes to another human host (a related worm is found in monkeys). The larval *Loa loa* nematodes develop into adults under the host's skin, sometimes showing up as visible worms moving across the eye. They ultimately produce lots of little microfilariae, available to be picked up by further deer flies, in which they undergo larval development.

Other biting lower Brachycera (some Athericidae, Rhagionidae) are unimportant as disease vectors, but the biting higher Brachycera (blood-feeding Cyclorrhapha) include a few species of medical or veterinary significance. Both males and females of the blood-feeding calyptrate families Muscidae, Glossinidae and Hippoboscidae bite, unlike the lower Brachycera and lower Diptera, in which only the females take a blood meal. The most notorious biting muscid is the ubiquitous Stable Fly, *Stomoxys calcitrans*, a vicious and persistent biter common not only in the vicinity of stables but also along freshwater shorelines, where the larvae develop in decomposing vegetation. No pathogens are known to develop in Stable Flies, but they do have the potential to mechanically transfer diseases from host to host. The same is generally true for other biting muscids, although some nematodes that infect cattle can develop in Horn Fly (*Haematobia irritans*) adults. Non biting Muscidae such as House Flies are much more important in disease transmission than are biting species.

Tsetse flies (*Glossina* species, family Glossinidae) are infamous for their role in the life cycle and transmission of the protists (*Trypanosoma brucei*) that cause African sleeping sickness (African trypanosomiasis) and the related livestock disease nagana. African sleeping sickness is a devastating disease that continues to kill thousands of people in tropical Africa annually, although the number of new cases per year appears to be declining (the World Health Organization cites fewer than 10,000 new reported cases in 2009, the lowest in 60 years; this had further declined to around 7,000 cases in 2010). Both male and female tsetse flies derive all their nourishment and water from blood meals. Ingested trypanosomes develop and multiply in the fly's midgut before transforming into a stage able to leave the gut and migrate through the body to the salivary glands. There they transform into another stage that multiplies and transforms yet again into "metatrypanosomes," which are injected into the next host along with the fly's saliva.

Only half a dozen of the 31 species of tsetse are significant in spreading sleeping sickness, and these vary by distribution, habitat and even the severity of the disease they carry. Some species (*Glossina fuscipes*, *G. palpalis*, *G. tachinoides*) are mostly associated with forests in West Africa and transmit only a chronic form of the disease characterized by the lethargy that gave rise to the name "sleeping sickness." Others (*Glossina morsitans*, *G. pallidipes*, *G. swynnertoni*) occupy savannas in East Africa, where they transmit a more acute illness that kills its victims quickly. Most cases today take the chronic form (*Trypanosoma brucei gambiense*), as the disease is confined largely to West and Central Africa (especially the Congo); according to the World Health Organization the eastern, acute form of sleeping sickness (*Trypanosoma brucei rhodesiense*) accounts for less than 5 percent of human infections. Animal infections by *T. brucei brucei* and related *Trypanosoma* species that cause the livestock disease nagana still represent a serious problem and have been credited with blocking practical animal agriculture in large areas of Africa. Mortality of imported domestic livestock in tsetse areas is especially high, and furthermore, infected animals that don't die can serve as a reservoir for both human and animal trypanosomiasis.

The Hippoboscidae (louse flies and bat flies) are ectoparasitic flies that suck blood from vertebrate hosts, but since none normally bite people, they are of negligible importance to human health. At least one paper raises the frightening possibility that bat flies could transmit Ebola from bats to humans, but unless you handle tropical bats on a regular basis, I would not worry about being bitten by any bat fly, let alone one infected with Ebola. Hippoboscids do, however, transmit pathogens between their hosts (normally birds or bats). One common louse fly is suspected of moving West Nile virus from bird to bird, and several other hippoboscids carry nematodes and protists pathogenic to their bird and mammal hosts.

Non-biting Flies & Disease

Filth Flies, Eye Flies and Others

A few kinds of small acalyptrate flies are irritatingly attracted to moist membranes, and some have the annoying habit of spitting on and slurping around your eyeballs as they mop up fluids with their spongy lower lips. The filthy habits of these particular flies (Drosophilidae and Chloropidae, especially chloropids in the genus *Liohippelates*) render them efficient mechanical carriers of various pathogens that cause pinkeye (conjunctivitis, which is associated with several different bacteria) and worse (yaws). Yaws is a skin, bone and joint infection caused by a spirochete bacterium closely related to syphilis bacterium. *Liohippelates* flies can pick up the bacteria while feeding at oozing sores. Adults of other higher Brachycera are attracted to similarly unappetizing substrates, often flitting from filth to food in a way that lends itself to the spread of a frightening variety of pathogenic microorganisms. The higher flies as a whole are pre-adapted to this role by their general association with microorganisms (since their larvae often develop in rotting material) and by their usual practice of spitting or vomiting on food before mopping up the liquefied mass with a spongy lower lip.

The House Fly, *Musca domestica*, and other *Musca* species are among the flies most commonly seen moving back and forth between foodstuffs and foul substrates such as feces, garbage and carrion. Since the larvae (maggots) of these flies feed partly by filtering microorganisms from rotting material, it is unsurprising that the adult flies are like a living inoculum, their hairy bodies and sticky tarsal pads teeming with potential pathogens. More than a hundred species of pathogenic microbes have been isolated from their digestive tracts, a

Musca is one of the most dangerous animal genera, with several species implicated in the transmission of disease-causing bacteria, viruses and protists. Other filth-associated calyptrate flies, especially in the families Muscidae and Calliphoridae, pose similar threats to health.

disturbing statistic considering that these flies routinely regurgitate on their food (and our food!).

Both the cosmopolitan House Fly and the closely related Bazaar Fly (*Musca sorbens*, originally from the Old World tropics and introduced to Hawaii) are well known for contaminating food with the bacteria, viruses and protists responsible for the enteric diseases (diarrhea and dysentery) that remain among the world's most serious health issues. In fact, according to the World Health Organization, enteric diseases kill and sicken more children worldwide than any other infectious disease. Recent research has confirmed the role of House Flies in the spread of cholera and demonstrated their potential to pick up infective bacteria from people with anthrax and other diseases. Eye pathogens, such as the *Chlamydia* that causes trachoma eye disease, are often introduced by *Musca* species feeding on or around people's eyes. All in all, *Musca* probably vies with *Anopheles* as the world's most dangerous genus.

Non-biting flies also play a significant role in the transmission of animal diseases, involving many of the same species and the same mechanical transmission mechanisms as for human diseases, although some flies are strictly livestock pests. Face Flies (*Musca autumnalis*), for example, represent a serious livestock problem because of their habit of swarming around the heads of cattle and horses, settling and feeding around the eyes. Not surprisingly, they are mechanical vectors of eye diseases, including bacterially induced pinkeye.

Myiasis

Dead bodies of all sorts swiftly attract ovipositing or larvipositing female flies, and prompt penetration of vertebrate carcasses by maggots is the norm. Invasion of living vertebrates by fly larvae, however, is a bit more remarkable and warrants the special descriptive term *myiasis*, which is used to describe many different relationships between maggots and vertebrates. Myiasis can range from obligate (caused by specialized parasitic larvae that develop only in living vertebrate hosts) to facultative (larvae able to develop in dead or live vertebrate hosts) or even accidental (larvae develop opportunistically in living vertebrate hosts). Accidental myiasis can be caused by several kinds of flies that occasionally end up in a vertebrate's stomach, where they sometimes survive or even thrive. The scuttle fly larvae that suddenly appear in your day-old sandwich, the Cheese Skippers living in that piece of Cheddar from yesterday's lunch, and those larval soldier flies hidden in your juicy ripe mango belong to three of the 50 or so fly families that occasionally cause accidental intestinal myiasis. Cheese Skippers (*Piophila casei*, family Piophilidae) and scuttle flies (Phoridae, especially *Megaselia scalaris*) have both been recorded as surviving the trip through the human digestive system, sometimes even pupating and emerging as adults before ending up in the toilet. *Megaselia scalaris* has also been recorded as causing urogenital myiasis, as well as facultative myiasis under the skin of frogs, in turtle eggs and in wound tissue of various vertebrates, including *Homo sapiens*.

Having a maggot burrowing around under your skin may sound unappealing, but it is not always a bad thing. In fact, it has long been appreciated that untreated battlefield wounds infested by maggots are far less likely to become seriously infected than those without maggots. This ancient bit of field wisdom has been updated with the deliberate medicinal use of selected cultures of sterile blow fly (*Lucilia sericata*, family Calliphoridae) maggots to treat difficult infections, and maggot therapy is now a recognized and useful form of applied myiasis. Medicinal maggots consume only the necrotic tissue and associated microorganisms, egesting antimicrobial compounds as they go. Other blow fly maggots, including some strains of *Lucilia sericata*, sometimes fail to respect the line between necrotic and healthy tissue, becoming "secondary screwworms" that opportunistically consume living flesh.

A few kinds of blow flies have gone a step further to become specialists on living flesh, often invading hosts through a nick or cut and going on to become "primary screwworms" that can cause great damage or even death. The main primary screwworms in the New World are *Cochliomyia* species; those in the Old World are *Chrysomya* species, but some *Chrysomya* species are now secondarily widespread. The American Screwworm (*Cochliomyia hominivorax*) used to be a serious livestock pest throughout the New World, homing in on even tiny scratches on cattle and horses to lay eggs that develop into flesh-eating larvae. This species was successfully eradicated from North America by an intensive control program that involved the release of massive numbers of sterilized male flies, but American Screwworms are still important pests in the neotropics. *Chrysomya bezziana* is a similar primary screwworm in the Old World tropics, but other species in both genera are able to develop on carrion as well as living flesh. Many blow fly species normally associated with carrion are facultative flesh eaters, with maggots that routinely attack living livestock through small wounds or soiled hair around orifices. The resulting myiasis, called "strike," is a serious problem in Australia and New Zealand, where sheep strike caused by one or two species of *Lucilia* (especially *L. cuprina*) causes more than $100 million worth of damage per year.

Nestling birds, naked and exposed, look susceptible to invasion by maggots, and such is indeed the case. Maggots that parasitize nestlings are found in the Piophilidae (European Nest Skipper flies, *Neottiophilum praeustum*), Muscidae (*Philornis* in the neotropics, *Passeromyia* in the Old World tropics) and Calliphoridae (bird blow flies, *Protocalliphora* and *Trypocalliphora*). The bird blow flies are closely related north temperate (Holarctic) genera often treated together as the single genus *Protocalliphora*, although maggots of *Trypocalliphora* burrow right under the nestlings' skin while those of *Protocalliphora* merely hang out in the nest, visiting the nestlings periodically to suck a bit of blood. There are bloodsucking maggots elsewhere in the Calliphoridae; the Congo Floor Maggot (*Auchmeromyia senegalensis*) sucks human blood, but most bloodsucking maggots attack only nestling birds. Bird maggots in the Muscidae and Piophilidae have habits similar to *Protocalliphora*, and larvae of the remarkable acalyptrates in the genus *Batrachomyia* (Chloropidae) develop under the skin of Australian frogs.

Screwworms are flies that develop in living flesh. The New World screwworm genus *Cochliomyia* includes primary screwworms that cause serious livestock damage, as well as secondary screwworms (like this species) that are more often associated with dead tissue.

There is a fine line between being a blood-sucking maggot or a flesh-eating screwworm and becoming an obligate internal parasite, and several flies cross (or at least straddle) that line. Some of the straddlers are, unsurprisingly, in the blow fly family Calliphoridae, in which some species spend their larval lives in cyst-like pockets, or "warbles," under the host's skin. Most of the calliphorids known to develop in this way attack only exotic wild hosts such as elephants or water buffalo, but one African species (the Tumbu Fly, *Cordylobia anthropophaga*) includes *Homo sapiens* in its host range. Tumbu Flies normally lay their eggs on feces or dirt but often oviposit in soiled clothing such as diapers, from which the parasitic larvae can launch their attacks on human flesh. The family Sarcophagidae (flesh flies) includes a few species with similarly parasitic larvae, but in contrast to calliphorids, they skip the external egg stage and drop their larvae directly around orifices and wounds. The sarcophagid genus *Wohlfahrtia* includes saprophagous members with habits much like most blow flies, but larvae of some European and North American *Wohlfahrtia* species aggregate in cyst-like swellings under the skin of living vertebrates. *Wohlfahrtia* species rarely attack adult humans, although the European species *W. magnifica* sometimes shows up under the tender skin of infants.

Bot Flies

One small family of higher Brachycera, the calyptrate family Oestridae, comprises about 150 species of mostly robust, beelike bot flies with parasitic larvae that develop under the skin, in the sinuses or in the digestive tracts of their vertebrate hosts. The Oestridae are discussed more fully in the family section in Chapter 8, but the main modes of bot fly development are illustrated by the following examples from the four subfamilies Gasterophilinae, Oestrinae, Cuterebrinae and Hypodermatinae.

Stomach bots (subfamily Gasterophilinae) are found in rhinos, horses and elephants, but horse stomach bots in the genus *Gasterophilus* are the most important and best known. *Gasterophilus* species differ in where they deposit their eggs; the most common species (the cosmopolitan *G. intestinalis*) usually lays eggs on the forelegs, where they get licked off and hatch on the horse's tongue or oral tissues. Second instar larvae later move to the gut and attach themselves to the stomach lining, where they feed until mature; then they release and get pooped out to pupariate in the soil. Nose bots, or "snot bots" (subfamily Oestrinae), are more diverse, with a number of well-known species that attack horses, sheep, deer, elephants and related mammals. The Sheep Nose Bot (*Oestrus ovis*) is typical of the group in that the beelike adults attack their hosts (sheep and goats) on the fly, zooming in to shoot living larvae at the nostrils. The larvae develop in the sheep's sinuses, ultimately getting sneezed or snorted out to pupariate in the ground.

New World skin bots (subfamily Cuterebrinae) develop under the host's skin in localized warbles like the parasitic sarcophagid and calliphorid maggots discussed above, and usually with attack strategies not unlike those of the Tumbu Flies, which deposit eggs away from the host. Most New World skin bots are *Cuterebra* species that lay eggs along runs or other areas frequented by rabbits and other rodents. Larvae latch on to host fur and make their way inside through a convenient orifice. These are relatively big flies, and their warbles look excruciatingly large for their hosts. Mature larvae penetrate the host's skin to escape to the soil prior to pupariation. The Human Bot Fly (*Dermatobia hominis*), which uses other insects to deliver its eggs to humans and other hosts, is an unusual member of the Cuterebrinae.

Old World skin bot flies (subfamily Hypodermatinae) are so called because all but the Holarctic Reindeer Skin Bot (*Hypoderma tarandi*) and a few secondarily widespread species occur only in the Old World. Hypodermatines lay eggs directly on the host, in contrast with their New World counterparts in the Cuterebrinae. Unlike the more or less sedentary cuterebrine maggots, hypodermatine maggots can be distressingly mobile, migrating painfully about the host's body. Although members of this group also develop in deer and rodents, the best-known species are warble flies of cattle. The Common Cattle Grub (*Hypoderma lineatum*), for example, is a Holarctic species that hatches from eggs laid on cattle hair. Larvae burrow (with the help of protein-dissolving enzymes in their saliva) directly through the host's hide at the base of the hair. After hatching in early summer, maggots migrate through the host's tissue to the esophageal submucosa, where they remain until spring; then they move to big, boil-like swellings in the animal's back. Each swelling, or warble, has a breathing hole through which you can see the maggot; that hole is enlarged by the mature maggot so it can pop out and drop to the ground to pupariate. The holes ruin the hides, and cattle often injure themselves when they react to the ovipositing flies by totally freaking out (a phenomenon more formally known as "gadding").

Dead Bodies and Flies

Sarcosaprophagous Diptera and Forensic Entomology

The importance of blow flies in decomposition has long been recognized and is neatly reflected in an oft-quoted line penned by Carolus Linnaeus about 250 years ago: "Three flies consume the corpse of a horse as quickly as a lion." I'm not too sure about a horse, but I do know that a road-killed possum tossed into the bush in the heat of an Australian summer is reduced to little more than a writhing mass of maggots in three or four days, and the progeny of a single female blow fly could theoretically account for the whole lot. Of course the maggots don't consume the corpse on their

Gasterophilus intestinalis

Cuterebra fontinella

Dermatobia hominis

Bot flies (Oestridae) develop as parasites inside vertebrates. Horse Stomach Bots **(UPPER LEFT)** lay eggs on the horse's hair, where they are licked off and hatch into larvae that develop attached to the lining of the stomach or intestinal tract. Rodent bots **(UPPER RIGHT AND BOTTOM)** lay eggs in areas frequented by their hosts, where larvae are likely to get the opportunity to penetrate their hosts and develop under the skin. The Human Bot Fly (**MIDDLE**, showing a newly emerged adult) attaches its eggs to other flies such as mosquitoes and House Flies; the eggs hatch when the other flies land on a warm body.

Cuterebra fontinella **larva**

Cuterebra fontinella **adult**

own but instead work in concert with millions of bacteria, and an exposed corpse will swiftly attract many more than three flies. Carrion represents a hotly contested resource for an impressively diverse community of flies and associated insects almost from the moment it appears. The resulting community of sarcosaprophagous flies normally exhibits fairly predictable changes in composition as the dead animal is transformed from a pile of putrefying meat to mere skin and bones.

The species involved differ according to locality, season and microhabitat, but the succession normally starts with blow flies, especially bluebottle flies and their relatives in the genus *Calliphora*, which drop their loads of eggs around wounds and moist membranes almost before the body hits the ground. Groups of blow fly maggots start to appear in the juicy parts within a day of death, working as a mass to break down tissue and together producing alkaline feces that help "condition" the carcass for maggot development. Greenbottle flies (*Lucilia* species) usually arrive a bit later, along with some muscids and flesh flies (Sarcophagidae). Flesh flies larviposit (lay living larvae) rather than oviposit (lay eggs), so their larvae hit the ground running (or at least creeping) and develop quickly. Muscidae (especially House Flies, *Musca domestica*) usually arrive well before the body starts to dry out following the rapid development of the initial invaders. They are soon followed by shining black Cheese Skippers (Piophilidae) and a diversity of other Diptera, including several species of small flies in the large family Sphaeroceridae and at least one genus of Heleomyzidae. Some Fanniidae, Sepsidae, Drosophilidae and often the odd black scavenger fly (Stratiomyidae) are also likely to be found on the still-stinking corpse; Phoridae, including the infamous "coffin flies," are more common on older cadavers, as are some moth flies (Psychodidae) and some species of Muscidae. In addition to the many sarcosaprophagous flies that develop in carrion, a wide variety of adult flies will occasionally visit decomposing flesh for a snack, even though they breed elsewhere.

The buzz surrounding bodies of dead animals is not only of interest to naturalists with an eye for dipteran diversity, it also intrigues those concerned with how, where and when those animals came to be dead in the first place — especially when the animals in question are *Homo sapiens* who met their demise under questionable circumstances. Residents always reflect the histories of their homes at some level, and so it is with flies. I was, for example, once asked what the flies had to say about a murder victim found in an open field. Soil samples taken around the corpse yielded large numbers of empty blow fly puparia, which told us something important about when the body was dumped in the field. Blow fly larvae go through three stages, or instars, before migrating away from the decomposing food mass in which they developed in order to pupariate in the soil. Pupae (inside puparia) undergo a period of development before turning into adult flies that move on, leaving behind an empty puparium with its cap popped off along a neat circular seam. The empty puparia in this case suggested that the body had been in place at least long enough for the flies to develop from mature larvae to adults, disputing a critical claim that the victim had been dumped there just before the police found her body.

This sort of application of dipterology to legal investigations has blossomed into a high-profile discipline in its own right, called forensic entomology. Forensic entomologists now have enough background data on the development times of different fly species under different conditions to come up with fairly sophisticated estimates of postmortem intervals, based on careful identification of fly adults, larvae, puparia and sometimes even eggs. The main species used in this sort of calculation are blow flies, and most forensic entomology is focused on a few well-known species of Calliphoridae. Several labs around the world are now working to improve data on blow fly development times by studying calliphorid life cycles on dead pigs under different conditions, meant to match potential human murder victims (obviously this kind of study cannot use real human bodies). The possible permutations and combinations of microhabitat, weather and location are endless, as are the condition of the corpses and their clothing, but the data have nonetheless been accepted as adequate to establish the guilt or innocence of many an accused murderer.

Most forensic entomology investigations involve identification and measurement of fly larvae, carefully considered against weather records (temperature, humidity, rainfall), other variables and known development times for each species. But sometimes a simple species identification is enough to tell a story, as when the presence of species outside their normal distributions or with special habitat preferences suggests that a body has been moved. Most of the forensic dipterology case histories in the literature dwell on successful convictions, but flies have also facilitated successful exonerations — as in the "case of the erroneously condemned Hungarian ferry skipper" recounted by Nuorteva (1977). A ferry skipper had been given a life sentence for the knife murder of a man found on his ferry one September evening, some hours after the skipper had arrived for work at 6 p.m. The original autopsy report recorded fly eggs and larvae on the body. Forensic entomology did not play a role in the original trial, but when the case was reopened several years later, an entomologist testified that no sarcosaprophagous flies were active after 6 p.m. in Hungary at that time of year, and that blow fly eggs would have needed more than 10 hours to hatch at the prevailing temperature. This meant that the eggs had been deposited the previous day, contrary to the claims of the prosecution. The ferry skipper was exonerated and released.

A more recent case involved the macabre discovery of some decomposed bodies stuffed into suitcases and dumped in an Australian forest. The person accused of this crime had been out of the country for a while when the bodies were found, and his conviction depended in part on the assumption that the bodies had been in the forest for months prior to their discovery. Investigators had diligently collected entomological evidence when the bodies were first discovered and had reported the presence of a relatively rare genus and species of Sphaeroceridae. When a forensic entomologist reinvestigated the case at a later date, he asked my opinion on whether the Sphaeroceridae reported from the original investigation were consistent with such a long period of decomposition.

Blow flies (Calliphoridae) are usually the first insects to arrive at a dead body. This female is using her soft, telescoping abdominal tip to deposit eggs on some exposed tissue.

As part of his reinvestigation, he had already gathered a year's worth of fly samples at the crime scene, using pigs in suitcases to emulate the decomposing victims. The samples were full of flies, and those from the pig carcasses exposed for several months (but sampled in the same season as the human victims were discovered) contained numerous larvae, adults and puparia of an interesting and previously poorly known sphaerocerid fly, the cosmopolitan *Phthitia empirica*. There were, however, no specimens of the sphaerocerid genus listed in the original forensic report. So, at that point it seemed that the flies reported in the original investigation were not consistent with the species found after a long period of decomposition. But when I examined the samples from the original forensic investigation, I found that all the sphaerocerids in those samples were in fact *Phthitia empirica*. They had simply been misidentified as another genus by the original investigator. Thus, the dipterological evidence suggested that the bodies had indeed been in the forest for several months.

As the above examples show, forensic dipterology is about more than just measuring maggots. It often involves a combination of good taxonomy, common sense and an understanding of the natural history of the flies involved.

4

Flies and Invertebrates

Predators, Parasitoids, Parasites and Thieves

ALTHOUGH THE ANCESTRAL FLY PROBABLY developed as an aquatic, saprophagous larva, the past 250 million years of fly evolution have seen repeated shifts from saprophagy to other strategies. Numerous lineages of flies have independently moved from saprophagy to blood feeding, plant feeding, predation or parasitism, with the latter strategies leading to a remarkable complexity of interrelationships between flies and other invertebrates.

Predaceous Larvae

Predation, perhaps the simplest of the interrelationships between flies and other invertebrates, appears over and over in the Diptera. Groups with predaceous larvae have apparently been derived from dozens of different groups with saprophagous larvae; lineages characterized by predaceous adults are much less frequent but also appear across the order.

Each of the major lineages of lower Diptera includes some predaceous larvae. The mostly saprophagous crane fly family (Tipulidae), for example, has a subfamily (Pediciinae) made up almost entirely of robust aquatic or semiaquatic species with larvae that prey on other aquatic or semiaquatic invertebrates. The moth flies and their relatives (Psychodomorpha) and the related mosquito/midge lineage (Culicomorpha) are also generally saprophagous, but larval predation appears in most families of Culicomorpha and characterizes the entire culicomorph families Chaoboridae (phantom midges) and Corethrellidae (frog midges). Both Chaoboridae and Corethrellidae have aquatic larvae that resemble mosquito larvae, but with remarkable prehensile antennae used to grasp other aquatic invertebrates. Mosquito (Culicidae) larvae themselves are mostly filter feeders that gather microorganisms and organic matter with their elaborate mouth brushes, but some (such as the giant treehole mosquitoes in the genus *Toxorhynchites*) instead use massive mandibles to munch other mosquito larvae. Predaceous larvae in the midge family (Chironomidae) also feed on family members, commonly consuming their saprophagous relatives. Predators also appear within the biting midge family Ceratopogonidae. One derived biting midge lineage (in the subfamily Ceratopogoninae) is characterized by wiry, active aquatic larvae that consume other aquatic

OPPOSITE PAGE Like all robber flies (family Asilidae), this *Mallophora* is a predator that pierces prey with a prominent beak, immobilizing and liquefying its victims with injections of neurotoxic, proteolytic saliva. Larval robber flies, like most lower Brachycera, are also predators.

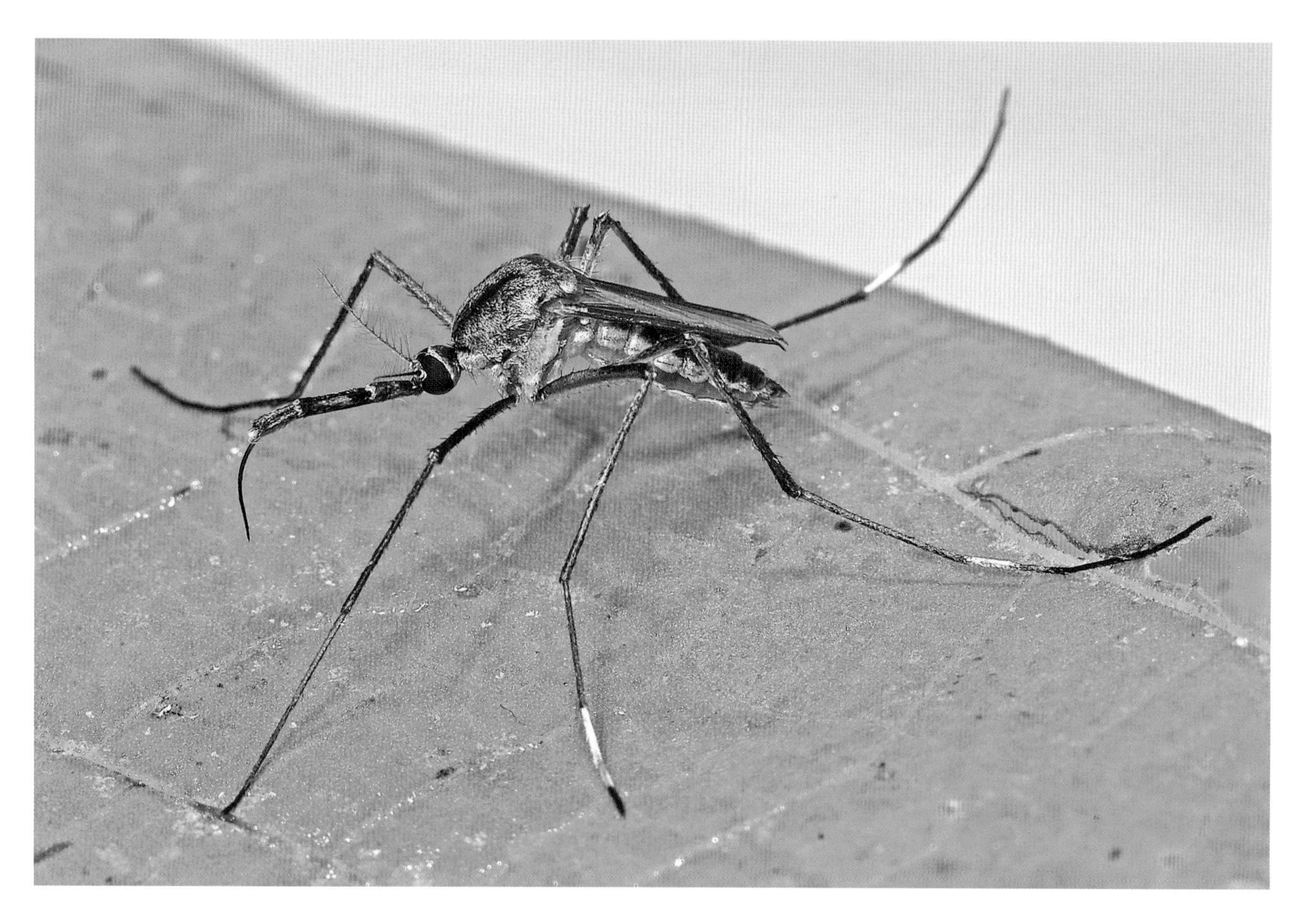

Not all mosquitoes bite. Members of the widespread genus *Toxorhynchites* (elephant mosquitoes or giant treehole mosquitoes) are predators as larvae and feed only on nectar as adults.

organisms, sometimes penetrating much larger invertebrates and consuming them from the inside.

Most terrestrial lower Diptera belong to a single lineage, the Bibionomorpha, a cluster of mostly saprophagous families closely related to the higher Diptera. Larval predation pops up a few times in the Bibionomorpha and is sometimes remarkable, as in the glowworms (*Arachnocampa* spp.) that live in silken tubes and lure prey with bioluminescence before trapping them on sticky silk lines. Several other groups of predaceous Keroplatidae spin webs to capture small prey, but only a few are bioluminescent. Members of the largest bibionomorph family, the Cecidomyiidae, are known as gall midges because of the conspicuous success of phytophagous members of one derived subfamily, the Cecidomyiinae. The same subfamily, however, includes predaceous larvae such as the bright orange *Aphidoletes* commonly seen munching their way through aphid colonies. Larval predation is common in the subfamily Cecidomyiinae, but other gall midge subfamilies, including the more "basal" or "primitive" groups, are saprophagous or fungivorous. The same is true for the basal lineages of the whole infraorder Bibionomorpha, and thus probably for the common ancestor of the Bibionomorpha and the Brachycera, even though most lower Brachycera are predaceous as larvae.

The rotated mandibles and other larval structures that characterize the suborder Brachycera probably originated as part of a suite of adaptations for predation in the first lower Brachycera. Larvae of lower Brachycera, such as horse flies, snipe flies, robber flies and stiletto flies, are thus predators by default, although, as discussed further below, a few lineages have moved from predation to parasitism and larvae of one superfamily (Stratiomyoidea) are saprophagous. Lower brachyceran larvae are usually roving general predators in soil, wood, mud or water, but some are highly specialized. Of these, perhaps the most remarkable are the larvae of Vermileonidae (wormlions), which make pits much like those of the better-known antlions (Neuroptera). Wormlion larvae lie across the bottoms of their pits and swiftly strike out at invertebrates unfortunate enough to tumble in, reaching around the victim's body to envelop and impale their prey with a pair of fang-like mandibles.

The higher Brachycera (Cyclorrhapha) are probably descended from a predaceous ancestor held in common with the dance flies and long-legged flies (Empidoidea), almost all of which are predators as larvae. The actual sister group (the most closely related lineage) to the Cyclorrhapha, however, is now thought to be a southwestern American species unknown as a larva and only inferred to be predaceous (probably on other invertebrates in mammal burrows). That species, the sole representative of the family Apystomyiidae, was only recently identified as the possible closest relative to the Cyclorrhapha. Since the origin of the Cyclorrhapha probably pivoted on an evolutionary shift from predation to a new kind of saprophagy, it would be interesting to know the larval biology of the apystomyiid "missing link."

Within the Cyclorrhapha the basal lineages are known as the Aschiza because they lack the ptilinal fissure ("schism") of the main lineage, the Schizophora. Most of those basal lineages are saprophagous or fungivorous, but the Phoridae, the biggest family of Aschiza, is a staggeringly diverse group that includes a few predators and many parasitoids. Some of the most common predaceous phorid larvae can be found in grasshopper and spider egg masses. The family Syrphidae also includes lineages that have moved from saprophagy to predation, and the large subfamily Syrphinae is characterized by conspicuous predaceous larvae that graze among aphid colonies like voracious land leeches. The exposed, often colorful larvae of syrphine syrphids are unusual in an order generally characterized by inconspicuous larvae that literally immerse themselves in their food or in wet microhabitats, although similarly conspicuous aphid-eating larvae are also found in the gall midge family (Cecidomyiidae).

Larvae of the Schizophora, the main lineage of higher Brachycera, are largely microbial grazers that rasp away inside various moist materials with fang-like mouthhooks, ingesting matter and microbes to be concentrated by a pharyngeal filter into an ingestible bolus. Many Schizophora, however, have shifted partly or entirely away from ingesting and filtering microbes to ingesting their neighbors, and are thus either facultative predators that retain their microbial filters (many muscids, for example), or obligate predators in which the pharyngeal filter is lost. Some of the obligate predators belong to host-specific predaceous (or predaceous and parasitic) families, such as the mollusk-eating Sciomyzidae and the aphid fly family Chamaemyiidae, while others are general predators in particular microhabitats such as bark beetle burrows or plant stems. Many higher brachyceran predators, such as the barnacle-eating larvae of the genus *Oedoparena* (Dryomyzidae), belong to specialized genera in otherwise saprophagous families. Equally peculiar predators pop up within large, mostly saprophagous acalyptrate genera such as *Drosophila* (Drosophilidae). Although best known for the yeast-eating larvae of the common vinegar flies or Laboratory Fruit Flies, different members of this genus prey on black fly and midge larvae,

Adults and larvae of the genus *Oedoparena* (Dryomyzidae) occur in the intertidal zones of western North America and Japan, where the larvae develop as specialized predators of barnacles. This *O. glauca* is laying eggs in a barnacle on the Pacific coast of the United States.

devour dragonfly eggs and consume living embryos in the egg masses of Neotropical frogs.

Drosophila also includes lineages that have taken the seemingly small evolutionary step from developing as predators of several prey organisms (eggs, larvae or adults) to developing at the expense of a single victim. Larvae that consume and kill only a single individual organism are parasitoids rather than predators. It does not seem like much of a leap from preying on a mass of eggs or aphids to feeding inside a single egg, aphid or other host, but relatively few arthropod lineages outside the Diptera have made that jump from predator to parasitoid. Most of the independent origins of endoparasitism have taken place within the order Diptera, perhaps because fly larvae are pre-adapted for feeding within a moist mass of food, whether mud, decomposing organic matter or a host's body.

Parasitoid Larvae

Flies with parasitoid larvae are often referred to as "parasitic Diptera," even though true parasitism of other invertebrates (in which the host survives) is very rare. *Parasitoid* is the better term for insects that develop in or on a single host, killing it in the process. The relationship between parasitoid and host is often very specific, beginning with the strategies used to get at hidden or recalcitrant hosts and continuing on through the mechanisms that allow parasitoid fly larvae to develop in synchrony with the host. Some have mobile, armored maggots that take on the job of seeking hosts, some lay eggs on their hosts, some inject eggs into their hosts and others place tiny eggs on materials likely to be ingested by hosts prior to hatching into parasitoid larvae. Let's look briefly at a few examples that illustrate these strategies and show how parasitoid lineages have independently cropped up over and over in the Diptera. The dozens of independent origins of parasitoidism in the Diptera exceed the number in all other insects combined, and the overall abundance and importance of parasitoid Diptera can be compared only to the parasitoid wasps. In contrast to the many parasitic fly lineages, however, parasitic Hymenoptera all belong to one lineage with a single parasitoid ancestor.

Despite their great age, very few lower Diptera have evolved parasitoid lineages, and most of these are in the gall midge family (Cecidomyiidae). Larvae of parasitoid gall midges in the genera *Endaphis* and *Endopsylla* bore into their host bugs (aphids, lace bugs or jumping plant lice), where they develop as internal parasitoids that kill the host before leaving to pupate in the soil. One recently described *Endaphis*, *E. fugitiva*, has the remarkable ability to detect when its host (the Banana Aphid) is under attack and is thus likely to be consumed by a predator. Although this fly would normally leave the host (by way of the anus) only upon maturity, Muratori et al. (2010) showed that if the host is under attack the larva swiftly pops out to pupate prematurely. Aphids and their relatives are normal prey for related gall midges (like the well-known *Aphidoletes*), so it is not too surprising to find that a couple of cecidomyiid lineages have shifted from aphid predation to aphid parasitoidism.

A more remarkable choice of hosts is found in the fungus gnat (Keroplatidae) genus *Planarivora*, which develops as an internal parasitoid of land planarians. *Planarivora* was first discovered in Tasmania in 1965 and then found in South America almost 20 years later. The majority of the terrestrial lower Diptera remain unknown as larvae, so we can expect more such discoveries in the years to come. True parasitoids are unknown among the aquatic lower Nematocera, although a few midges (Chironomidae) skirt the edges of parasitoidism by developing inside sponges or under the mantles of aquatic snails, or by feeding externally on the body fluids of other aquatic invertebrates such as stonefly or mayfly nymphs.

Parasitoid lineages are most often derived from predaceous ancestors, so you might expect to see this strategy appearing repeatedly throughout the predominantly predaceous lower Brachycera. But instead, parasitoidism in the lower Brachycera is restricted to two highly specialized and entirely parasitic families — small-headed flies and tangle-veined flies — and a diversity of lineages

Several members of the diverse family Phoridae are parasitoids, and some are known as "ant-decapitating flies." This one is darting in to oviposit on a carpenter ant; its larva will develop inside the ant, ultimately decapitating it prior to pupariating in the detached head capsule. Members of the same phorid genus (*Apocephalus*) also parasitize Honey Bees and bumble bees.

among the bee flies (Bombyliidae). The family Bombyliidae includes both predators and parasitoids, with the parasitoids probably derived from predaceous ancestors that specialized on subterranean hosts such as bee larvae. Most Bombyliidae larvae latch on to a single host and consume it from the outside (as ectoparasitoids), but a few are endoparasitoids that develop inside a single host. Eggs are usually deposited near (but not on) hosts, leaving the newly hatched larva with the task of crawling into a nest or burrow where it can latch on to a host.

The parasitoid families Nemestrinidae (tangle-veined flies) and Acroceridae (small-headed flies) also leave the task of host finding to active first-instar larvae. Most larvae of tangle-veined flies seek out, penetrate and parasitize nymphs and adults of grasshoppers and their relatives, but some genera are parasitoids of subterranean larval and pupal beetles. The active first-stage larvae (planidia) of small-headed flies are even more specific in their choice of hosts: all known species in the family apparently develop only as endoparasitoids of spiders. Despite being legless (like all fly larvae), the planidial first-instar larvae are able to loop, jump or crawl along in search of hosts; some can creep along the silken strands of host webs before penetrating the spider, typically through the thin membrane at a leg joint. Some species attach to the host's book lungs and remain quiescent for a long period before undergoing a burst of activity, eating the host from the inside and emerging to pupate in a silken hammock obligingly produced by the dying spider.

Parasitoid larvae are common in the higher Brachycera (Cyclorrhapha), appearing in at least 10 separate lineages of Schizophora and in a couple of basal lineages of Aschiza. The scuttle flies (Phoridae), the biggest family of Aschiza and one of the most diverse families of all Diptera, appear particularly predisposed to parasitoidism. They have been reared from a wide variety of invertebrates, including mollusks, myriapods, worms, arachnids, beetles, scale insects, termites and lower Diptera. Among the most important and interesting parasitoids in the group are the ant-decapitating flies and the bee-killing flies. Ant-decapitating flies can be found as internal parasitoids of ants as far north as Canada's boreal region. The common name of this group refers to the habit of some species (especially in the genus *Apocephalus*, but also other genera) of severing the host's head prior to pupariation, which sometimes takes place in the ant's detached head capsule. Some *Apocephalus* species also attack bees, but most bee-killing flies are in the large Neotropical genus *Melaloncha*. Several phorid parasitoids specialize in attacking injured or dying insects, blurring the boundary between parasitoidism and saprophagy and suggesting that some parasitoids in this group might be descended from scavenging ancestors. All known Aschiza parasitoids outside the Phoridae are in the entirely parasitic family Pipunculidae, or big-headed flies, which are usually parasitoids of Auchenorrhyncha (leafhoppers, spittlebugs and their relatives). Females use their enormous eyes to hunt down hoppers that they scoop up, inject with an egg and then drop. The only pipunculids known to have different habits are the *Nephrocerus* species recently (2007) found to attack adult crane flies (Tipulidae), injecting a single egg into each host.

The Schizophora, an enormous group and one of the main branches of the order Diptera, is usually (albeit artificially) divided into the acalyptrates and calyptrates. The 75 or so widely recognized acalyptrate families include relatively few endoparasitoid lineages, concentrated in half a dozen mostly small parasitic families. Conopidae, or thick-headed flies, make up the largest strictly parasitic acalyptrate family, with all 800 or so described species presumably developing as internal parasitoids of the bees and related aculeate Hymenoptera (stinging wasps) that the adult female flies attack in flight. The hundred or so species in the closely related family Stylogasteridae (sometimes included in the Conopidae) have long, harpoon-like eggs that they usually stab into orthopteroid prey. Many species of *Stylogaster* (the only genus in the family) hover over raiding masses of army ants (Neotropical) or driver ants (Afrotropical), waiting for the opportunity to dart an egg into roaches and other insects flushed out by the marauding ants.

This *Stylogaster* is hovering over a driver ant raid, waiting for the ants to flush out a host such as a cricket or cockroach.

Pyrgotidae (pyrgotid flies) with known biologies are nocturnal parasitoids of adult scarab beetles, which the female flies usually attack in flight. This family is considered to be entirely parasitic, although most of the 365 species in the family have never been reared and the larvae of the most "basal" subfamily (Teretrurinae) remain unknown. A related group of acalyptrate flies, sometimes treated as the family Tachiniscidae and sometimes as the subfamily Tachiniscinae within the Tephritidae, is also considered to be a parasitoid lineage, although only one of the 18 species in this rare group is associated with a host (it was reared once from an African moth pupa). Evidence for the parasitoid label often affixed to the small, extremely rare family Ctenostylidae is even more tenuous, as no species in this family has ever been reared. It seems to be a good guess that these pyrgotid-like nocturnal flies are parasitoids, but it is just a guess. Ctenostylids and pyrgotids are probably closely related to the Tephritidae (in the superfamily Tephritoidea), but small and supposedly parasitic families are also found in the superfamilies Sciomyzoidea and Ephydroidea. The Phaeomyiidae, a small group of five Old World species until recently treated as Sciomyzidae, are presumed to be millipede parasitoids because one species (the only one with known larvae) develops inside millipedes. Similarly, the small acalyptrate family Cryptochetidae is assumed to be entirely parasitic because known species are endoparasitoids of mealy bugs, even though one of the two genera in the family remains of unknown biology. Some dipterists consider the Cryptochetidae to be a specialized parasitic offshoot of the huge family Drosophilidae, a diverse group that otherwise lacks true endoparasitoids.

The family Sciomyzidae — the marsh flies or snail-killing flies — is made up almost entirely of species that develop on mollusks, usually as scavengers or predators but sometimes straddling the line between predation and parasitism (parasitoidism). Species that attack slugs, for example, start their larval lives as apparent endoparasitoids but kill and abandon their first hosts, moving on to complete their development as predators of further slugs. Similar dual strategies are reported for some other Sciomyzidae, but the family also includes several species that develop as true endoparasitoids inside single snails, ultimately pupariating inside the shell of the original host. Although the Sciomyzidae are generally aquatic, most of the parasitoid species in the family are members of the tribe Sciomyzini that attack only terrestrial snails.

The real explosion of endoparasitoids in the Diptera has taken place in one distinct lineage, the Calyptratae. Although this group includes important lineages of specialized parasites and predators, the most general larval habit across the Calyptratae is probably a mixture of saprophagy and facultative predation. Many Muscidae, for example, start out their larval life as microbial grazers and then move on to predation in later instars. Similar habits probably characterize the basic plan of other calyptrate families, including the Calliphoridae and Sarcophagidae, and must have predisposed the Calyptratae to develop endoparasitoid lineages because endoparasitoids are found in most calyptrate families (other than the families that have become specialized vertebrate parasites). Even the Anthomyiidae, a group otherwise largely characterized by saprophagy and phytophagy, includes a few parasitoid lineages such as the unusual grasshopper parasitoids in the genus *Acridomyia*. Parasitoid larvae are found in all three subfamilies of Sarcophagidae (flesh flies), and some of the largest genera in the family are parasitoids of other insects such as acridid grasshoppers and tenebrionid beetles. One flesh fly species (in a largely kleptoparasitic genus) is a significant endoparasitoid of adult Honey Bees; another is a parasitoid of cicadas, locating its hosts by sound, using an enlarged prosternum that functions like an ear. A similar prosternal "ear" is used by some Tachinidae to locate nocturnal grasshopper and cricket hosts.

The Tachinidae, with some 10,000 species, is the most significant lineage of endoparasitoids, and the family exhibits the whole range of host-seeking strategies, from planidial hunter-seeker larvae through to microtype eggs that get ingested by the hosts. Tachinids are mostly parasitoids of caterpillars, sawflies or beetles, but their hosts span more than a dozen arthropod orders. Some species attack a wide range of hosts and others show extreme host specificity; those that attack only a single pest species are often of value in biological control. Other calyptrate parasitoids include the Rhinophoridae, which develop as endoparasitoids of isopods, and several lineages of Calliphoridae. The best known calliphorid parasitoids are those in the genus *Pollenia*, which have active first-instar larvae that burrow into the soil to seek, penetrate and parasitize earthworms, but the hosts of most known parasitoid calliphorid groups are snails. At least three of the small subfamilies currently in the Calliphoridae include parasitoids that develop inside mollusk hosts before ultimately killing them. Larvae of one recently discovered Oriental calliphorid (in the subfamily Bengaliinae) develop in the head capsule of termites, and many other odd parasitoids undoubtedly remain to be discovered in this and other calyptrate groups.

Predaceous Adult Flies

Predaceous adults occur in relatively few fly lineages, and in that sense adult predation is much less general than predation by larval flies. Most predaceous lower Diptera belong to the Ceratopogonidae (biting midges), a group that includes bloodsucking adult parasites of vertebrates, bloodsucking adult parasites of invertebrates and predators of invertebrates. The difference between the latter strategies is small and really depends simply on whether the results of a bloodsucking attack are non-fatal or fatal, which in turn depends on the relative size of the host. Females of most genera of the subfamily Ceratopogoninae attack other flying insects similar in size to themselves, and are thus predators. Favorite targets are the mating swarms of

Efferia

Proctacanthus

Cyrtopogon

Ommatius

Stichopogon

Laphria

other lower Diptera, such as non-biting midges (Chironomidae), but some biting midges raid the male swarms of their own species for both a mate and a meal (generally the same lucky/unlucky individual). Some adult female Blephariceridae (net-winged midges) are predators as well, grabbing mayflies and other small insects with their modified hind tarsi and ripping in to their prey with serrated, knife-like mandibles.

Most Brachycera with predaceous adults are found in the Empidoidea and the lower Brachycera family Asilidae (robber flies). Members of both groups abound in almost all terrestrial habitats, with about 20,000 species adapted to feed on almost all kinds of other arthropods, ranging from the smallest midges and springtails to the largest beetles and dragonflies. Robber flies are all predators and all pierce prey with a long, syringe-like hypopharynx ("tongue"), injecting proteolytic and neurotoxic saliva prior to ingesting the prey contents. Empidoids, on the other hand, use a dagger-like labrum (upper lip) and associated epipharyngeal blades (on the lower surface of the labrum) to feed, usually on other invertebrates but sometimes on pollen and nectar. Adults of one major empidoid lineage (long-legged flies, family Dolichopodidae) prefer soft-bodied prey such as springtails, oligochaete worms or other flies, which they envelop between two sandpaper-like lobes (labella or labellar lobes) of the labium, seemingly slurping up the victim as they grip it between the lobes and puncture it with tooth-like epipharyngeal blades. Other empidoids variously specialize on such prey as emerging aquatic insects, bark insects or swarming flies, but perhaps the most unusual empidoid predators are female Empidinae, which cannot hunt for themselves and instead rely on prey brought to them by courting males.

Adults of higher Brachycera (Cyclorrhapha) are mostly consumers of liquid — or liquefiable — materials that they can mop up with their characteristic sponge-like labellum, but a few cyclorrhaphan lineages are either facultative or obligate predators. Among the Aschiza, the infinitely diverse family Phoridae includes a few predaceous adults. Some *Megaselia*, for example, have larvae that feed on figs, and adult females that prey on adult fig wasps by repeatedly piercing them and ingesting their body contents. Members of one other family of Aschiza, the small family Lonchopteridae (spear-winged flies), have also been seen eating small insects, but so-called predation in this group — as in many other Cyclorrhapha — is probably opportunistic and better described as scavenging. Several acalyptrate families, including Sciomyzidae, Somatiidae and Lauxaniidae, sometimes feed on dead insects, but confirmed predation by adult Acalyptratae is much more unusual; it is more or less restricted to a few species of Micropezidae (stilt-legged flies) in the subfamily Calobatinae and one large genus of Ephydridae (shore flies). Species of *Ochthera*, an unusual genus of shore fly equipped with massive raptorial front legs, are common predators of small insects along exposed shorelines or wet rocks along fresh waters over much of the world.

This *Ochthera* has grabbed a small chloropid with its raptorial forelegs. *Ochthera* is one of the few genera of acalyptrates with predaceous adults.

Calyptrate flies, including the familiar large blow flies, House Flies and flesh flies, are mostly saprophagous as adults, but a few lineages of Calyptratae are entirely predaceous. Adult flies in family Scathophagidae and the muscid subfamily Coenosiinae are common predators that puncture prey with a specially armed labellum.

Although few flies are true parasites of other invertebrates, some biting midges (Ceratopogonidae) have adults that suck blood from other insects. Most, like these females biting the wing veins of a green lacewing (Neuroptera), are in the genus *Forcipomyia*.

Coenosia adults are abundant predators in most parts of the world and are routinely seen with recently captured flies or other prey still impaled on their mouthparts. Adults of the mostly northern Holarctic family Scathophagidae are usually found near their larval habitats (usually plant stems, water or dung), where they can be locally abundant. One species, the Pilose Yellow Dung Fly (*Scathophaga stercoraria*) is often a conspicuously common predator in rural environments. Only a few predaceous calyptrate adults are found outside the Coenosiinae and Scathophagidae, but Old World tropical flies in the blow fly genus *Bengalia* depart from the usual adult calliphorid habits of lapping up liquids from carrion and dung, and instead actively hunt and kill other invertebrates or consume prey stolen from ants.

Parasitism and Kleptoparasitism

Relatively few flies are true parasites of other arthropods, that is, drawing sustenance from their hosts without killing them. The major exceptions are the biting midges commonly seen tapping the veins, joints or bodies of other insects. Otherwise, parasitism of other arthropods is almost unknown among adult flies and rare among larval flies. A few phorids are reported as developing in large beetle hosts, such as Passalidae, without killing them (and are thus parasites), and there have been a few rare reports of insects surviving after serving as hosts for larvae of Tachinidae. But these instances of dipteran endoparasites are exceptional, and reports of terrestrial dipteran larvae developing as ectoparasites remain open for testing. Larvae of the huge drosophilid genus *Cladochaeta*, for example, occur in the spittle masses made by nymphal spittlebugs in the genus *Clastoptera* (Cercopidae). Larvae are normally attached to the nymphs and some authors have assumed that they are parasites. In the absence of proof that the larvae are ingesting haemolymph, however, it seems equally plausible that the association between *Cladochaeta* maggots and the spittlebug nymphs is one of phoresy, not parasitism. Several aquatic midge larvae (Chironomidae) are ectoparasites of mayflies, stoneflies and other aquatic invertebrates.

Gall midges (Cecidomyiidae) are commonly seen hanging out in spider webs, often in large numbers. Most species that do this have no closer association with spiders, but a few are kleptoparasites that share the spider's prey.

Kleptoparasitism

Scarabs stowing their dung balls for future larval food, wasps stocking their burrows with paralyzed prey, bees provisioning nests with pollen, and predaceous invertebrates sloppily and slowly imbibing extra-orally digested prey all provide tempting opportunities for specialized food-stealing flies, or kleptoparasites. Kleptoparasitism is an unusual specialization, but it shows up a few times in all of the major groups of Diptera, including lower Diptera, lower Brachycera, Empidoidea, Aschiza, Acalyptratae and Calyptratae.

Most of the food thieves among the lower Diptera are gall midges and biting midges, and most of these steal only from spiders. Biting midges routinely imbibe haemolymph from other insects, so stealing a bit of haemolymph from another insect already immobilized and wrapped in silk by a much larger spider is not a big jump in habit. Still, the only biting midges I have seen doing this are attracted only to termites captured by one kind of Amazonian comb-footed spider that suspends its prey (a species of carton-nesting termite) in balls hung by silken threads. The dangling balls of strung-up termites are like feeding stations for the kleptoparasitic ceratopogonids. Gall midges (Cecidomyiidae) are much more common than biting midges in spider webs, but most just hang out in the webbing, often in aggregations and usually just ignoring the spider and its prey. One gall midge species, *Didactylomyia longimana*, is reported as consuming the liquefied prey of spiders — a doubly unusual phenomenon, since Cecidomyiidae adults don't usually feed at all. Some mosquitoes are also kleptoparasitic, stealing food from ants by hovering over their mouths and imbibing regurgitate, but otherwise kleptoparasitism in the lower Diptera is restricted to the Cecidomyiidae and Ceratopogonidae.

Kleptoparasitism is rare in the lower Brachycera and Empidoidea, but some species of *Microphor* (Dolichopodidae) hover over spider webs to steal prey. This seems only a slight departure from the adult predation that is the general rule in other Empidoidea. Kleptoparasitism also appears in a few lower Brachycera that feed on the prey of their hosts rather than the hosts themselves. Species in the bee fly (Bombyliidae) genus *Lepidophora*, for example, have larvae that feed as kleptoparasites in the nests of wasps that store paralyzed prey.

Flies in the acalyptrate family Milichiidae are often kleptoparasites, with either adult or larval stages that steal food from other invertebrates. Shown here are a *Phyllomyza* (LEFT) riding an ant back to its nest, presumably to lay eggs in or near the ant's foodstuffs, and several *Paramyia* (RIGHT) feeding on a stink bug in a spider's web.

Kleptoparasitism is relatively common in the Cyclorrhapha, starting with a variety of Phoridae (scuttle flies) that head up the interesting gang of thieves routinely found as kleptoparasites feeding in fungus gardens and stored food in ant and termite nests. Of these, among the most bizarre are the Old World tropical termite flies currently treated as the subfamily Termitoxeniinae. Termite flies look like normal fully winged Phoridae until they (mated females only) enter termite nests, where they shed their wings and often undergo such extreme abdominal expansion that they come to look more like termites than flies. These strange females led entomologists to treat termite flies as a separate family of flies before they were associated with their winged forms. Another group of termitophilous Phoridae (*Thaumatoxena*) has tiny, wingless females that are flat, drop-shaped and so un-fly-like they were first formally described as a new suborder of Hemiptera; they were long treated as their own family (Thaumatoxenidae) even once they were recognized as flies. Termite flies are tolerated by their hosts (usually fungus-gardening termites) probably in part because of appeasement substances they produce. Some kleptoparasitic ant-associated phorid flies are even less fly-like than termite flies; the adult females of one Oriental genus (*Vestigipoda*) are so similar to ant larvae I would need a microscope to tell them apart. Ants treat the *Vestigipoda* females as they would their own larvae.

Although most of the flies that steal food from social insects belong to the family Phoridae, a few acalyptrates also make their living as ant associates. Some adult Milichiidae, for example, steal food from foraging ants or ride the ants back to their nests to oviposit. In fact, the kleptoparasitic habits of a few species of milichiids have led some dipterists to suggest the name "freeloader flies" for this small family. Some develop as larvae in the fungus gardens of leaf-cutter ants, and adult milichiid flies often hitch rides on ants or their leaf-fragment loads as they stream back to their nests. Adults of a few ant-associated species solicit regurgitated food from ants, and others are kleptoparasites that steal food from feeding spiders, robber flies, predaceous bugs and other large insect predators. Similar habits are found in

Adults of the milichiid genus *Desmometopa* are often specialized kleptoparasites, swiftly attracted to Honey Bees captured by other arthropods such as robber flies or spiders, like the Cuban lynx spider shown here.

some Chloropidae, and small flies of one or the other of these two families are routinely found lapping up fluids leaking out of the victims of larger invertebrate predators. Milichiidae and Chloropidae that snatch snacks from the prey of larger predators are often quite specific, feeding only on captured Honey Bees (as in the milichiid genus *Desmometopa*) or only on stink bugs and other captured prey with strong chemical defenses (as in the chloropid genus *Olcella*).

Braulidae, small wingless flies related to the Drosophilidae, also snatch food from near their host's mouth, but in this case they are phoretic, remaining on the hosts (Honey Bees) as adults. Different sorts of foods are sought by flies in the acalyptrate family Sphaeroceridae, some of which are kleptoparasites with long legs modified to cling to adult dung-rolling scarabs. Adults in the kleptoparasitic sphaerocerid genus *Ceroptera* and in some kleptoparasitic species of *Norrbomia* mate and wait on their scarabaeid steeds until the opportunity arises to feed on and oviposit in the stored food (dung) when it is safely sequestered underground.

A few predaceous calyptrates (like *Bengalia*) are known to snatch food from other predators, but food thievery in the Calyptratae is otherwise mostly restricted to a couple of specialized genera of Anthomyiidae and one large lineage of the Sarcophagidae, the subfamily Miltogramminae, in which larvae usually develop as kleptoparasites on the paralyzed prey or stored pollen of hunting wasps and bees. Adult miltogrammines (and some kleptoparasitic Anthomyiidae) are sometimes known as "satellite flies" for their habit of flying at a fixed distance from bee or wasp adults, although some will ride on paralyzed prey as it is dragged to the host's burrow rather than following host adults in flight. Female flies usually deposit living larvae, either on the paralyzed prey before it is sequestered in the burrow, in the entrance to the burrow, or right on the pollen or prey stored deep within the nest. Some miltogrammine flies lay eggs on the prey for transport into the host's nest, an unusual behavior for the almost entirely larviparous (ovoviviparous) family Sarcophagidae. Miltogrammines are very abundant flies and can be a major cause of mortality among the aculeate Hymenoptera.

PART 2

CHAPTER 5

Origins & Distribution of the Diptera 95

CHAPTER 6

The Lower Diptera 107

CHAPTER 7

CHAPTER 8

5

Origins & Distribution of the Diptera

FLIES BELONG TO A LARGER GROUP OF INSECTS that includes scorpionflies (order Mecoptera) and fleas (order Siphonaptera), with which they share common ancestors that probably flew around the primeval Palaeozoic swamps some 300 million years ago. The first two-winged flies probably diverged from an aquatic scorpionfly-like ancestor not long after that, splitting off from a lineage that went on to become today's fleas and scorpionflies. But the fly lineage was more successful, eventually splitting into at least 50 times as many species as would ever evolve in the Mecoptera and Siphonaptera.

If species diversity and raw numbers of individuals are accepted as measures of success, then the Diptera are arguably the most successful of all animals. But how did this come about? When did the order diversify to the point where about one out of every half-dozen species is a fly? Fortunately, flies and their relatives are represented by a rich fossil record that, combined with an understanding of the phylogenetic relationships between the main fly lineages, allows us to address these questions by piecing together a 250-million-year-long history. Neal Evenhuis (2010) has contributed immensely to that task by cataloging some 3,125 species of flies known from amber or from compression fossils, including some from as early as the Triassic period and many from the Jurassic and beyond. Although flies were apparently rare in the early to mid-Triassic and represent only about 1 percent of insects known from that time, the order Diptera is thought to have originated early in the period, around 250 million years ago. The order then seems to have exploded in diversity very quickly, with the main lineages (infraorders) of lower Diptera already present in the fossil record by 200 million years ago.

Early Jurassic fossils show that the "higher Diptera" (suborder Brachycera) had also appeared by that time and were widespread before the planet's original landmasses broke apart and moved toward today's continental configuration. Most lower Diptera and the first (lower) Brachycera predate the breakup of the great southern continent, Gondwana, and show disjunct distributions today that reflect that great age.

Some of the fly families from the Jurassic and lower Cretaceous periods are now extinct. For example, a fossil lower brachyceran family, Cratomyiidae, was recently described from the lower Cretaceous in Brazil. Cratomyiidae were large flies with a very long proboscis and wing venation unlike that of any extant family; they were probably

OPPOSITE PAGE Members of the enormous crane fly family, Tipulidae, are typical lower Diptera, with the long antennae and delicate appearance that probably characterized the first flies.

With the exception of a single specimen of an extant Australian species, the family Valeseguyidae is known only from amber fossils, like this *Valeseguya disjuncta* from the Dominican Republic (photo courtesy of David Grimaldi, American Museum of Natural History).

related to the Xylomyidae and Stratiomyidae. Of the 159 families covered in the fossil fly catalog (Evenhuis 2010), 49 are now thought to be extinct. But, just as living coelacanths (a famous primitive fish) were found at a time when coelacanths were known only as fossils (1938), some fly families now presumed extinct may yet be found thriving in some forgotten corner of the globe. After all, the fly family Valeseguyidae would be known only from fossils but for a single specimen of an extant species from Australia. The routine discovery of new species, genera and even families of flies is a constant reminder of how little we really know about the planet's fly fauna. One such recently discovered fly — the dance fly (Hybotidae) genus *Alavesia* — had been known only from 100-million-year-old amber fossils from Spain and Burma prior to the discovery of a living species in Namibia in 2010.

The higher Brachycera (Cyclorrhapha) burst onto the scene much later, even though this well-defined lineage currently includes most of the families and the greatest ecological diversity in the order. The earliest Cyclorrhapha to appear in the fossil record are about 140 million years old, with most of the early records being Cretaceous fossils of basal lineages of higher Brachycera ("Aschiza"). The fossil record includes characteristic Cyclorrhapha larvae ("maggots" with a reduced head capsule) from middle Cretaceous amber, and uniquely cyclorrhaphan puparia from the same period, around 90 million years ago. The Schizophora, a distinct group that includes around a third of today's dipteran diversity, are unknown from the Cretaceous but had radiated rapidly by the Tertiary, exploding into numerous families of acalyptrate Diptera between 65 and 40 million years ago. The calyptrates, perhaps today's most familiar flies because of the synanthropic success of blow flies and House Flies, comprise the youngest major lineage of Diptera; they first appeared in the fossil record about 40 million years ago.

The numerous amber fossils now available, especially from the Baltic region and the Dominican Republic but also from older deposits such as those found in Canada and the Middle East, track the continued appearance and disappearance of many fly lineages over the last 130 million years or so. Amber, the fossilized resin of certain kinds of trees, often includes a great many small flies that were trapped millions of years earlier when it was fresh and sticky. Many of the genera known from Baltic amber (28 to 54 million years old) and Dominican amber (15 to 40 million years old) are the same as or obviously closely related to flies still on the wing today, although often with surprisingly different distributions.

Today's flies, like today's humans, are everywhere. Common, familiar and cosmopolitan species such as the ubiquitous House Fly (*Musca domestica*) give the impression that flies in general are widespread insects. However, relatively few fly species

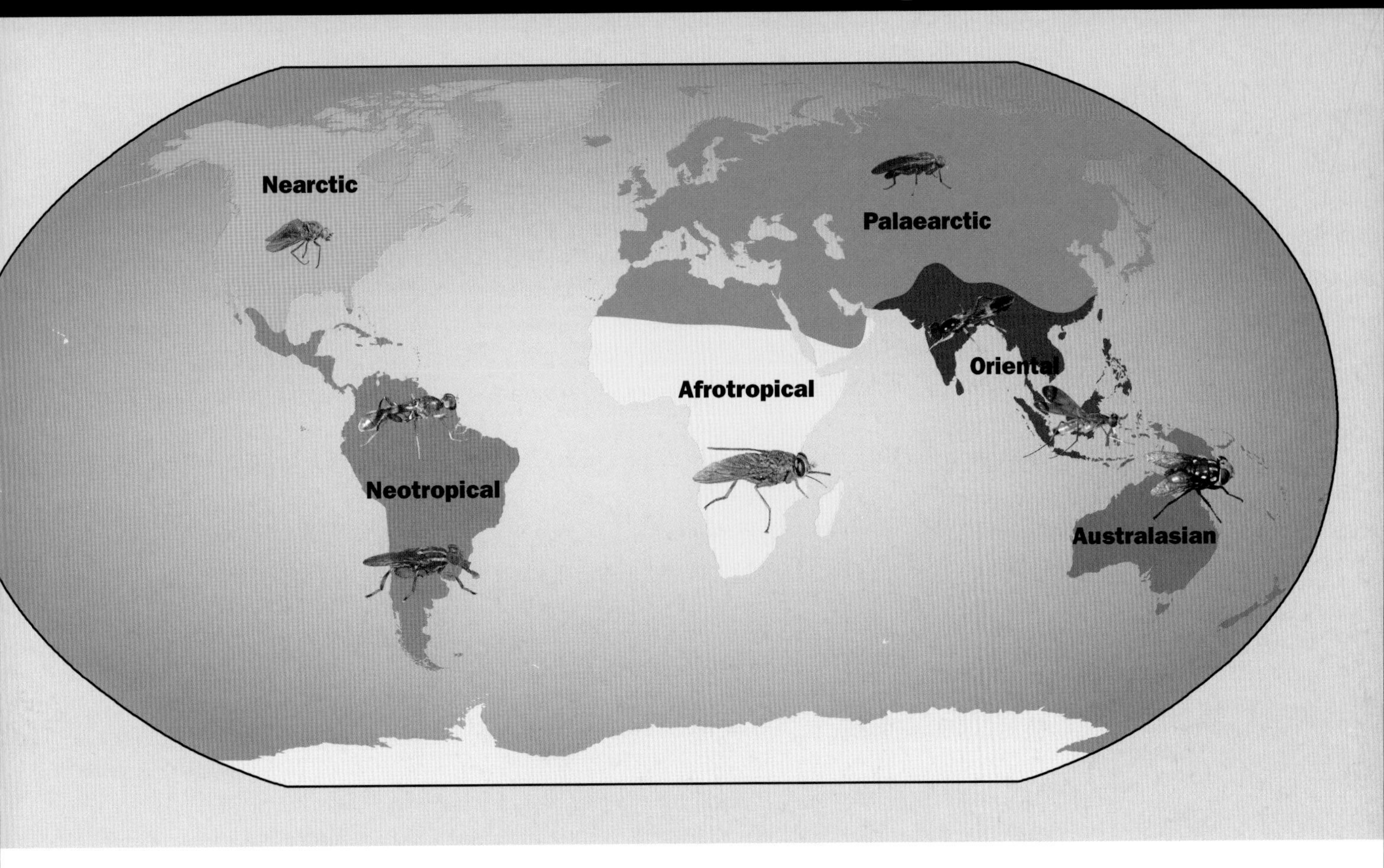

qualify as "cosmopolitan," and no fly species has attained a worldwide distribution without the accidental help of humans. At the family level it is a different story. Most of the common fly families occur in every continent except Antarctica, and even that inhospitable continent supports a few midges. Despite the many regional novelties emphasized throughout this book, students familiar with the main families of flies occurring in their home country will find that they can identify most flies encountered anywhere in the world, at least to the family level. Of course there are exceptions, especially in the isolated faunas of the south temperate countries and in the megadiverse tropics, but generally one must look very hard to find a fly that belongs to a narrowly distributed family.

Narrowly distributed families may be the exception, but narrowly distributed subfamilies are less unusual, and many, if not most, fly genera are endemic to one or a few biogeographic regions. Endemics, which are taxa naturally found only in particular areas, are always of interest because of the way they reflect the history and natural history of the planet. The main biogeographic regions of the world (see Figure 5.1) are

- the Nearctic region: North America, including temperate parts of northern Mexico;
- the Neotropical region: South and Central America and much of Mexico;
- the Palaearctic region: Europe and Asia;
- the Afrotropical region: Africa south of the Sahara Desert;
- the Australasian region: Australia, New Zealand and New Guinea; and
- the Oriental region: Southeast Asia, India and southern China.

These large regions are not clearly delineated by borders or walls but instead are divided by current or historical barriers such as oceans, mountains, deserts or climate. The names of the regions are routinely used as adjectives to refer to the distributions of plant and animal groups. For example, we can summarize the distribution of the fly family Pantophthalmidae in South America, Central America and Mexico by saying that the

Pantophthalmidae is a Neotropical family. Groups that occur in both the Neotropical and Nearctic regions are routinely referred to as having a "New World" distribution. The Palaearctic and Nearctic regions are very closely related, having many genera and even species in common, so these two regions are routinely lumped together as the Holarctic region. The fly family Scathophagidae is a characteristic example of a group with a mostly Holarctic distribution (most species are found in Europe, Asia or North America but a few species occur in other regions). Many other flies occur in the Palaearctic and contiguous Oriental and Afrotropical regions but not in the New World; such distributions are referred to as "Old World" distributions.

Oceanic Islands

The boundaries between regions are fuzzy, and some areas present particular obstacles to neat division of the world into half a dozen biogeographic regions. Oceanic islands, which have been colonized by dispersal from the major continental landmasses, often have remarkable faunas that are entirely distinct from those of the major regions.

The Hawaiian Islands, for example, are a chain of volcanic islands that began to emerge over a hot spot in the middle of the Pacific Ocean about 30 million years ago. Each island moved north on the Pacific plate as new islands appeared over the hot spot (the island of Hawaii is the youngest, at just over half a million years old). When they first poked their volcanic heads above the ocean, the islands were devoid of life, and they remained so until a series of rare, chance dispersal events brought colonizers from the nearest mainland, almost 4,000 kilometers away.

Evenhuis (2009) suggests that Hawaii's fantastic fly fauna evolved from at least 75 different instances of dispersal from the mainland, each followed by the establishment of a population on the islands. Those few lucky colonists have given rise to a modern fauna of more than 1,100 endemic fly species, over half of which are descended from three ancestral species of Dolichopodidae and Drosophilidae that underwent spectacular adaptive radiations into a wide range of niches on the islands. Other groups of Diptera with significant endemism on the Hawaiian Islands include a genus of Muscidae with more than 100 species and genera of Pipunculidae and Calliphoridae with 36 and 25 endemic species, respectively. The endemic genus of Calliphoridae, *Dyscritomyia*, contains several threatened or extinct species, illustrating the fragility of island faunas.

The Galápagos Islands, which are perhaps even better known for their endemic animals than are the Hawaiian Islands, are volcanic islands in the Pacific Ocean some 1,000 kilometers off the coast of Ecuador. Although younger than the Hawaiian Islands at 700,000 to four million years old, they are less isolated; their fauna of some 300 fly species is thought to have developed from about 200 successful colonizations (Sinclair, 2009). About a third of the species and five of the fly genera on the Galápagos are endemic, and none are currently known to have undergone spectacular adaptive radiations like those of the Hawaiian fruit flies and long-legged flies. Only a couple of fly groups (Dolichopodidae and Ulidiidae) appear to have more than a dozen endemic species on the Galápagos Islands. Most Galápagos flies have close relatives on the Ecuadorian mainland, but there are exceptions. For example, one endemic sphaerocerid (*Sclerocoelus galapagensis*) has its closest relative not in Ecuador, as you would expect, but on the remote Atlantic island of Tristan da Cunha.

The Juan Fernandez Islands, in the southern Pacific Ocean about 500 kilometers off the coast of Chile, are smaller and less familiar than the equatorial Pacific Galápagos Islands and the more northerly Hawaiian Islands, but they serve as my favorite example of a special oceanic island fly fauna. The best-known island in this small archipelago is Robinson Crusoe Island, so named because from 1703 to 1709 it was home to the shipwrecked Alexander Selkirk, the British seaman who became the inspiration for Daniel Defoe's *Robinson Crusoe*. Selkirk probably did not appreciate that the island supports about 250 species of flies, of which three-quarters are endemic. A few of the endemics are large and showy, such as the magnificent golden-faced flower fly *Sterphus aurifrons*, but most are inconspicuous flightless flies in two families (Sphaeroceridae and Ephydridae) and three genera, each of which includes at least one flightless species with narrow, spear-like wing rudiments. One of these genera, the widespread sphaerocerid genus *Phthitia*, has at least eight flightless species found nowhere in the world except the Juan Fernandez Islands.

Robinson Crusoe Island is itself only 4 million years old, and its smaller neighbor Santa Clara is about 5.8 million years old. So sometime in the few million years since the emergence of these volcanic islands, a *Phthitia* arrived from somewhere and diversified into the eight flightless species found there today. *Phthitia* is a worldwide genus with several species in mainland Chile, but the mainland Chilean species are not closely related to the island flies. Instead it seems that the closest known relative of the Robinson Crusoe *Phthitia* is another flightless island fly, *P. sanctaehelenae*, from the mid-Atlantic island of Saint Helena. Where did the common ancestor of these distant island endemics come from? Is it still waiting somewhere for an intrepid entomologist to collect and identify it, or has it been driven to extinction in some mainland home while its flightless island offspring have persisted through the millennia?

Most of the Sphaeroceridae on Robinson Crusoe Island are flightless, often with wings reduced to paddle-like or strap-like lobes that resemble the wings of several species of Hawaiian Dolichopodidae. The fact that flight has been repeatedly lost in different groups of flies on these remote oceanic islands is not all that remarkable. Wings, and the muscles to power them, are produced at the expense of energy that might otherwise be used for purposes such as egg production. Dispersal by flight is important in discontinuous or unstable environments, and escape by flight is no doubt strongly selected for in habitats with lots of predators (especially ants). In a stable, ant-free island forest surrounded by an infinite ocean, however, wings simply might not be a worthwhile investment.

Although special endemics and flightless species are regular features of oceanic island fly faunas, so too are unwelcome invasive species that often become disproportionately abundant once established. About a quarter of the fly species on the Galápagos Islands are introduced species, many of which pose threats to native biodiversity or to human and animal health. Oceanic fly faunas are also sensitive to invasive species from other orders: Magnacca and Price (2012) suggest that the decline of Hawaii's iconic "picture-winged" *Drosophila* — of which a dozen are formally listed as endangered or threatened — followed the introduction of an invasive yellowjacket wasp (*Vespula pensylvanica*) from the mainland.

Continental Islands

Continental islands — great fragments of formerly continuous landmasses — have very different histories and correspondingly different fly faunas than oceanic islands. Madagascar, for example, is usually treated as part of the Afrotropical region, but this huge Indian Ocean island has been separated from the continental landmasses (India and Africa) long enough to develop a remarkably unique flora and fauna. American dipterist Mike Irwin and colleagues recently (2003) provided a review of the Madagascar Diptera in which they estimate that 79 percent are endemic (found only there). Almost a fifth of the genera are endemic, and the genera that are shared with other regions show a mixture of affinities, often with close relatives in either the Oriental or the Afrotropical regions. In the robber fly family (Asilidae), for example, 82 percent of the genera occurring in Madagascar also occur in the Afrotropical region, and 29 percent are shared with the Oriental region. Another 29 percent of the genera are recorded from the Australasian region, 25 percent have been collected in both the Palaearctic and Nearctic regions, and 18 percent are found in the Neotropical region. This complex set of relationships is explained in part by Madagascar's ancient history as a continental landmass: like India, Africa, Australia and South America, it dates from the breakup of the southern hemisphere megacontinent Gondwana.

The history of Gondwana is so significant to a global view of dipteran distribution and diversity that it warrants a brief explanation here (readers interested in a detailed treatment of fly biogeography are referred to Cranston 2009).

The order Diptera had already begun to diversify by 250 million years ago, at a time when the terrestrial world comprised only a single landmass, called Pangaea. One of the major events in the history of the planet took place about 200 million years ago, when Pangaea split into the northern hemisphere's Laurasia and the southern hemisphere's Gondwana. When Gondwana itself started to sunder — about 170 million years ago — the fragments destined to become Antarctica, Madagascar, India and Australia became separated from Africa and South America, which in turn split apart as the South Atlantic Ocean opened between them, around 130 million years ago. New Zealand, Australia and South America remained connected by Antarctica until the late Cretaceous period, with New Zealand separating first. Australia started to separate about 80 million years ago and had finally spun off on its own by 40 or 50 million years ago. South America remained connected to Antarctica until much later, perhaps 40 million years ago. Madagascar and some smaller islands were linked to the Indian fragment until about 70 million years ago, and the Indian plate drifted north and "collided" with Asia about 45 million years ago (the Himalayas were created when the crusts north of this collision point buckled). These major events in the history of splits and joins in the southern half of the planet give some idea why so many groups of flies show interesting "Gondwanan" distributions. The best known Gondwanan disjunctions occur in lower Diptera (Nematocera) such as Chironomidae, Thaumaleidae, Blephariceridae and Perissommatidae, and in the oldest Brachycera lineages such as the Stratiomyidae, Xylophagidae, Pelecorhynchidae and Brachystomatidae.

Every taxon or lineage of flies (every species, genus, family and so forth) has a distribution, and each distribution tells a story about habitat and history. These stories are most interesting — and most useful — when enough is known about the relationships of the taxon to detect where its closest relatives occur. Disjunct, or separate, ranges of closely related taxa often reflect a history in which a common ancestor had its continuous range sundered by the appearance of a barrier such as an ocean, a mountain range or a desert. Older lineages of Diptera, for example, often include related species in Australia and southern South America, two main fragments of the formerly continuous southern continent Gondwana. These areas were connected by Antarctica until about 50 million years ago, before the ancestors of today's South American and Australian species in "Austral" groups such as the Perissommatidae became divided into separate South American and Australian lineages.

Although many of the common patterns in fly distribution can be explained by division of formerly continuous ranges (known as vicariance), ranges are always in flux, and species often cross barriers. This sort of "leakage" across a barrier such as an ocean or mountain range, followed by isolation and speciation on the other side of the barrier, can also result in important patterns of distribution. However, they tend to be less congruent from group to group than patterns of relationship and distribution following vicariance events (such as the drifting apart of continents) that divide whole faunas the same way. Entomologists often refer to these two general explanations for distributions as "dispersalist" and "vicariance" explanations. Gondwanan distributions, as discussed above, are usually explained by vicariance, while patterns of distribution on oceanic islands and recently glaciated landmasses (such as most of my home country, Canada) are obviously driven by dispersal. The diversity and distribution of flies of the world today reflect a 250-million-year history of vicariance, dispersal and speciation.

Biogeographic Regions and Diptera Distribution

Our knowledge of the fly faunas of different parts of the world varies tremendously from group to group and from region to region. The Holarctic faunas (Nearctic and Palaearctic regions) are by far the best known, and the tropical faunas (especially the Oriental and Neotropical regions) are by far the least known. Fly faunas of each major region were recently reviewed by various authors (in Pape et al. 2009), and the dipteran diversity of each region is detailed in the various regional Diptera manuals. Dipterological characteristics of the biogeographic regions are therefore only briefly summarized here.

Holarctic Region

The Holarctic region — the Nearctic region plus the Palaearctic region — takes up most of the northern hemisphere. Look at a globe (or Google Earth) from the top: the incomplete ring of land you see is the Holarctic region. Despite the big gap caused by the Atlantic Ocean (which has been there only 50 million years or so), it is easy to see that these two regions together largely circle the same latitudes. Add to that a history that has from time to time connected the eastern Palaearctic to the western Nearctic and the eastern Nearctic to the western Palaearctic, and you can see why the fly faunas of these huge regions have much in common. Both faunas originated largely in the ancient northern megacontinent Laurasia, which became divided into two separate continents about 100 million years ago. Those first northern continents did not correspond to the Nearctic and Palaearctic regions but instead were made up of Europe plus eastern North America (Euramerica) and Asia plus western North America (Asiamerica). The mid-continental seas that divided Euramerica from Asiamerica dried up tens of millions of years ago, coincident with the opening up of the Atlantic Ocean, but it is still useful to remember that today's Europe and Asia share a complex history with today's North America.

The **Palaearctic** region is more than twice as large as the Nearctic region, so, unsurprisingly, it has more flies. It also has more fly specialists, or dipterists, and a much longer history of research on flies, so the 45,000 or so species of flies known from the Palaearctic region is closer to a complete count than that for other regions. Almost all

OPPOSITE PAGE Many lineages of lower Brachycera and Empidoidea have distributions that reflect ancient connections between southern South America and the Australian region. The chiromyzine soldier flies **(TOP)** and ceratomerine dance flies **(BOTTOM)** are two such groups, and the South American species shown here are remarkably similar to relatives found in Australia.

RIGHT: Although one species occurs in the southern United States, the family Ropalomeridae is otherwise a Neotropical group. All 15 species of *Ropalomera*, the genus shown here, are restricted to the Neotropical region.

families and most genera found in the Palaearctic region are shared with the Nearctic region; some fly groups cross the desert barrier between the Palaearctic and Afrotropical regions; and several other groups spread across the mountains, plains and oceans that separate the Palaearctic from the Oriental region.

The **Nearctic** region has only half the land area of its Palaearctic counterpart, and slightly less than half the number of known fly species. Both the Nearctic and Palaearctic regions have a few families that are shared neither with each other nor with any other region, but these are all small groups of very restricted distribution. The Nearctic has the obscure families Apystomyiidae, known from only a few collections of adult flies in a small area of southern California, and Oreoleptidae, known only from larvae, pupae and a few adults reared from larvae and pupae collected in northwestern North America. The Palaearctic region can claim as its own the small aschizan family Opetiidae and the acalyptrate family Phaeomyiidae, which have fewer than ten species between them.

Each of these regions also has some elements shared only with neighbors to the south. For example, the essentially Oriental family Megamerinidae ranges north and west into the Palaearctic region but doesn't occur in the Nearctic region; the almost entirely Neotropical acalyptrate families Richardiidae and Ropalomeridae range north into the Nearctic region but don't occur in the Palaearctic. Some large and important fly families, including the Anthomyiidae and Scathophagidae, occur mostly in the Holarctic region and are diverse in both the Palaearctic and Nearctic. The *Manual of Nearctic Diptera* (McAlpine et al. 1981; 1987) includes keys to all the Nearctic genera, while the *Manual of Palaearctic Diptera* (Papp and Darvas 1997; 1998; 2000) includes keys to genera of most but not all of the Palaearctic Diptera families.

Neotropical Region

The Neotropical region has the most diverse and least-known fly fauna of any region. Although the current total of 31,000 known Neotropical fly species is far short of the 45,000 for the Palaearctic region, most Palaearctic species have already been named while the majority of Neotropical flies still remain to be discovered. It is not at all unusual to see revisions of Neotropical fly taxa in which 80 or 90 percent of the species are new to science, especially among relatively poorly known families of smaller flies such as Sphaeroceridae. The size and unique character of the Neotropical fly fauna is of course in part a function of great climatic and topographical diversity, but it is also the result of a long history of isolation bracketed by very early and relatively recent connections with other continents.

Before the breakup of the great southern hemisphere continent Gondwana, South America shared fly faunas with other Gondwanan continents, and those former connections are reflected today in close relationships between temperate South American flies and relatives in eastern Australia, Tasmania, New Zealand and South Africa. These relationships are best known in the older lineages of Diptera; similar though less well-documented relationships exist between tropical South American flies and their African relatives. Once South America became detached from other Gondwanan landmasses, it remained a huge, isolated island continent for tens of millions of years of tectonic turmoil and evolution. By the time South America was again connected to another

LEFT: The strictly Neotropical acalyptrate family Inbiomyiidae was first described only a few years ago, in 2006.

RIGHT: The huge flies in the Neotropical family Pantophthalmidae occur in tropical Mexico, Central America and South America.

continent — through the "new" Central America in the Pliocene (a mere three million years ago) — it supported the most unique flora and fauna the world will ever know. The impact of the Pliocene connection with North America and the resultant biotic interchange is well documented for Neotropical mammals, about half of which are descendants of northern immigrants; flies were probably similarly affected. The South American fly fauna is now an incredibly rich mixture, shaped by Gondwanan ancestry, influenced by northern connections and divided into a complex network of habitats, including variously interconnected or isolated mountains, grasslands, deserts, rainforests and temperate forests.

Several fly families or subfamilies are restricted or almost restricted to the Neotropical region. Inbiomyiidae, a family of tiny acalyptrates now known from various parts of Central and South America, was discovered and described only in 2006, but ten new species have been added to the genus *Inbiomyia* since then. Pantophthalmidae, which are among the largest and most spectacular of all flies, are restricted to the Neotropical region. Syringogastridae are elegantly ant-like flies found only in Mexico, Central America, Trinidad and South America; a 2009 revision of this family more than doubled the number of known extant species and added fossil species from Caribbean amber. The acalyptrate family Somatiidae, another group of attractive and distinctive flies, is known only from Central and South America, and one Chilean species in the lower Brachycera was recently described as its own, strictly South American family, Ocoidae (since renamed Evocoidae). There is also an extinct lower Brachycera family, Cratomyiidae, known from a single fossil species from Brazil. The recently published *Manual of Central American Diptera* (Brown et al. 2009; 2010), including generic keys (at least for Central American genera) and current synopses for all extant Neotropical fly families, has helped to stimulate a flurry of new work on the flies of Central and South America.

Afrotropical Region

The Afrotropical region, which includes Africa south of the Sahara as well as associated islands in the South Atlantic and western Indian Oceans, is home to about 20,000 known species of flies and at least as many undescribed species. The complex fly fauna of southern Africa is particularly rich in endemic genera or species and includes many Gondwanan elements with close relationships to other temperate Gondwanan faunas. The primitive dance fly genus *Homalocnemis* (sometimes treated as the subfamily Homalocneminae), for

RIGHT: *Adrama* is a small genus of fruit flies (Tephritidae) with an Oriental-Australasian distribution.

example, occurs in New Zealand, Chile and Namibia. Other examples of Gondwanan elements among Afrotropical flies include the snipe fly (Rhagionidae) genus *Atherimorpha* (South Africa, eastern Australia, southern South America) and the seepage midge (Thaumaleidae) genus *Afrothaumalea* (part of an otherwise Australian and Chilean clade). The vast equatorial rainforests seem to support fewer endemic flies of systematic significance, but some elements of this poorly known fauna have affinities to Neotropical rainforest flies, and other elements are shared with the Oriental region. The isolated mountain forests of eastern and southern Africa have a much higher level of endemism. As noted above, Madagascar is usually treated as part of the Afrotropical region, but many of the mostly endemic fly taxa of Madagascar have close relatives in the Oriental or Australasian region.

A few fly families are found only in Africa. Of these, the Glossinidae, or tsetse flies, are certainly the best known; the 23 extant species in this family of biting flies occur only in tropical Africa (and maybe southern Saudi Arabia). Extinct species, however, are known from Europe and North America. The recently described family Natalimyzidae is widespread in southern Africa and will probably turn out to be the largest family of flies endemic to Africa. The obscure acalyptrate family Marginidae is known from only a few specimens from Madagascar and Zimbabwe, and the mysterious hairy bat fly family Mormotomyiidae is known only from one cave in Kenya. The beetle flies (Celyphidae) occur only in the Afrotropical and Oriental regions, and the "nobody flies" (Neminidae) are known only from the Afrotropical and Australasian regions. A manual of Afrotropical Diptera, to include keys to all fly genera in the region, is in preparation (Kirk-Spriggs and Mostovski, http://afrotropicalmanual.net).

Australasian Region

The Australasian region, which includes Australia, New Zealand, New Guinea and a smattering of smaller islands, is best known for its highly endemic flora and unique vertebrate groups such

as kangaroos and koalas, but its fly fauna is no less unique. Endemism at the generic and specific level is very high, with many characteristic regional groups such as glowworms (*Arachnocampa*, eastern Australia and New Zealand) and antler flies (*Phytalmia*, New Guinea and north Queensland). The Australasian fauna is unusual even at the family level: two fly families (the nematoceran family Valeseguyidae and the aschizan family Ironomyiidae) are currently found only in Australia, and two families (the bat-associated calyptrate family Mystacinobiidae and the acalyptrate family Huttoninidae) are known only from New Zealand. Both of the Australia-only fly families are very old groups also known from Cretaceous amber deposits in the northern hemisphere. Many New Zealand and Australian fly lineages have their closest relatives on other parts of the former southern continent Gondwana; several groups, for example, are currently found only in eastern Australia, Tasmania and Chile. New Guinea remains inadequately explored as far as flies are concerned but seems to have a mixture of fly groups shared with the Oriental region and taxa in common with Australia.

The 19,000 or so species of flies currently known from the Australasian region probably

Most of the hundred or so species in the southern hemisphere horse fly genus *Scaptia* occur in the Australasian region, but this is also a common group in southern South America.

represent considerably less than half of the actual fauna. The Australasian flies are not covered by a single work comparable to the Diptera manuals available for some other regions, but Volume 2 of *The Insects of Australia* includes a key to families, and recent keys to Australian genera of many fly families can be found in the primary literature.

Oriental Region

The Oriental region is a diverse, largely tropical area between the Palaearctic region and Australasia; it includes India, southern China and most of Southeast Asia. This region is complicated both because of its current configuration as a jigsaw puzzle of isolated volcanic islands and continental fragments and because of its ancient history. India was once part of Gondwana but separated from that vast southern landmass very early (some 150 million years ago). It drifted north to collide with Asia around 100 million years later, possibly bringing with it some of the ancestors of today's characteristic Oriental fly lineages. About 90 percent of the almost 23,000 known Oriental fly species are found nowhere else, but most fly species of the region remain undescribed and much remains to be discovered about their diversity and relationships.

No widely accepted fly family is entirely unique to the Oriental region, but the small acalyptrate families Gobryidae and Nothybidae are mostly Oriental, with only a few records from the northern fringe of the Australasian region. Two other "characteristic" Oriental Diptera families — the acalyptrate families Celyphidae and Megamerinidae — also have species in the Afrotropical region (Celyphidae) or the Palaearctic region (Megamerinidae). Many parts of Southeast Asia remain virtually unexplored from a dipterological standpoint, and the literature is more scattered than is the case for other regions. Oosterbroek (1998) provided a key to the fly families of the region but there are no keys to the Oriental genera of most fly families.

To the south and east of the Oriental region a group of Indonesian islands forms a transitional zone, called Wallacea, between Borneo and Java (part of the Oriental region) and New Guinea (part of the Australasian region). Although the plants and animals of this transitional zone have a mixture of Australasian and Oriental affinities, many are endemic to Wallacea. The flies (other than mosquitoes and a few other families) of Wallacea are poorly known and would be rewarding to study.

6

The Lower Diptera

FLIES HAVE TRADITIONALLY BEEN DIVIDED INTO two suborders, the long-horned flies ("Nematocera") and the short-horned flies (Brachycera), but only the Brachycera form a natural group, characterized by their short antennae and several other novel adult and larval features. The Nematocera, in contrast, is not a real lineage but rather an artificial, or paraphyletic, group, with some members more closely related to the suborder Brachycera than to other so-called Nematocera. Paraphyletic groups are generally undesirable in classification systems, for the simple reason that they don't really exist — they are artificial assemblages rather than real branches of the tree of life. Despite that, and despite long-standing and almost universal acknowledgment of the artificial nature of the Nematocera, the name is still in widespread use. Why? One reason, of course, is that it is convenient and intuitive to refer to flies as belonging to one of two instantly recognizable suborders, even if one of them is artificial.

Another reason is that there is no consensus on an alternative subordinal classification that more accurately reflects phylogeny. That, in turn, stems as much from lack of agreement about what a "better" subordinal classification would comprise as from lack of agreement about the real relationships of the Nematocera. A recent brief proposal by Amorim and Yeates (2006) took seven of the widely used major subdivisions, or infraorders, of the Nematocera and argued for bumping them up to suborders, thus replacing the suborder Nematocera with the suborders Tipulomorpha, Psychodomorpha, Ptychopteromorpha, Culicomorpha, Blephariceromorpha, Bibionomorpha and Axymyiomorpha. This proposal has not gained much traction, probably in part because some of the proposed suborders contain only one or two families and are thus not very useful, and in part because the composition and status of the other "promoted" infraorders remain in dispute and unstable. So let's stay with the terms *Nematocera* or *lower Diptera* for all the non-brachyceran flies.

It is convenient, and not altogether wrong, to divide the lower Diptera into two subgroups: the terrestrial families and the aquatic or mostly aquatic families. The terrestrial families almost all fall into the single large infraorder Bibionomorpha, a lineage more closely related to the higher Diptera (Brachycera) than to other lower Diptera. The other lower Diptera, which could be called the "basal" lineages of the Diptera, comprise several aquatic or mostly aquatic groups,

OPPOSITE PAGE Although the 15,000 or so species in the crane fly family (Tipulidae) span a diversity of sizes and shapes, most are "typical" lower Diptera with long, multi-segmented antennae and a generally delicate appearance.

Figure 6.1 **The Main Subgroups of Lower Diptera**

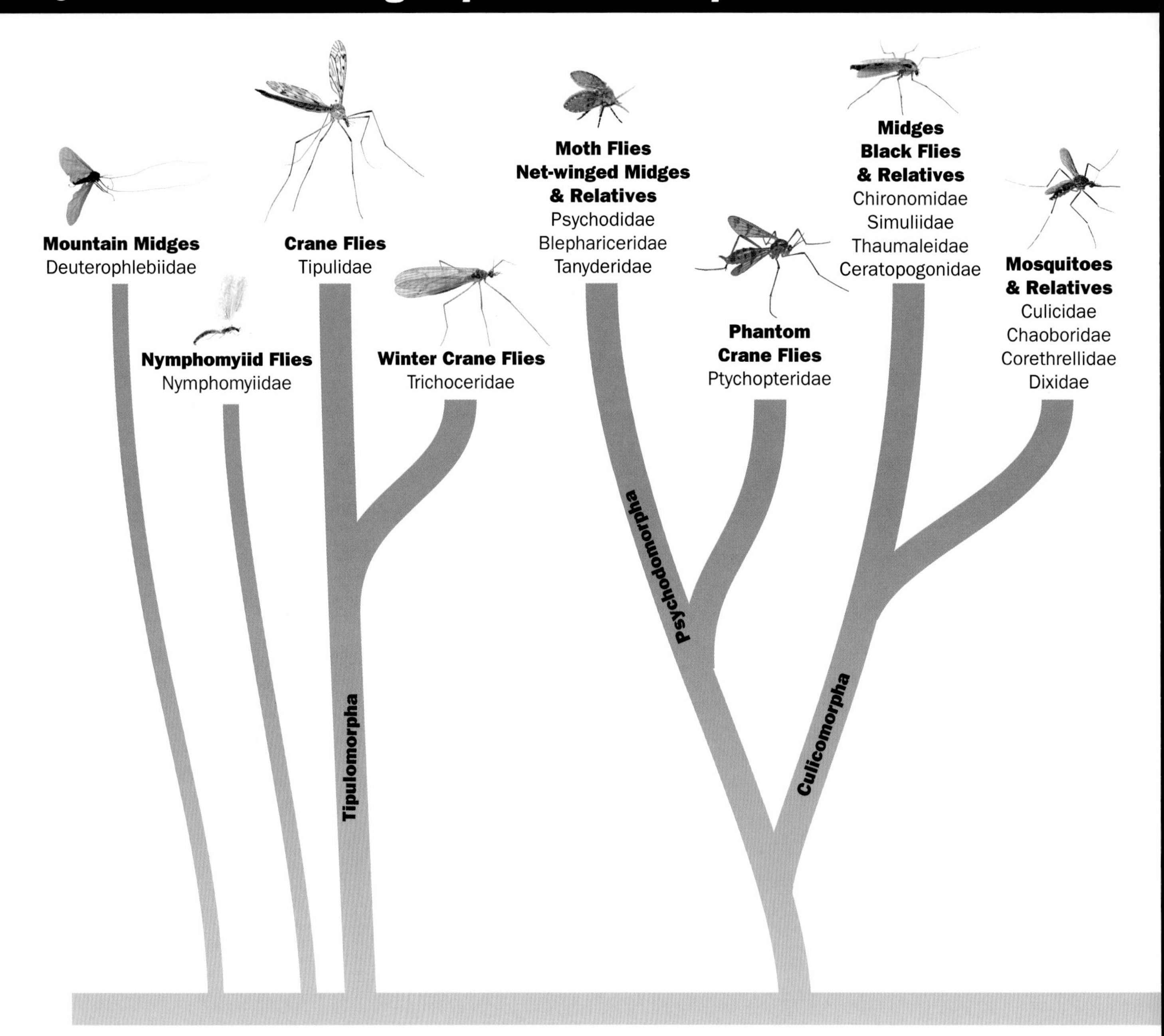

including the mosquitoes, midges and their relatives (infraorder Culicomorpha), the moth flies and their relatives (infraorder Psychodomorpha), the crane flies (infraorder Tipulomorpha) and a couple of unusual "primitive" aquatic families. This convenient division of almost all lower Diptera into aquatic families (several basal lineages) and terrestrial families (one lineage) is loosely based on the infraordinal classifications of Bertone et al. (2008) and Wiegmann et al. (2011), which are in turn very similar to those of Hennig (1973). Those interested in the debate over nematoceran higher classification should also consult the widely used classifications of Woodley et al. (2009) and Wood and Borkent (1989), who argue that the partially aquatic family Psychodidae belongs in a terrestrial lineage treated here as part of the Bibionomorpha.

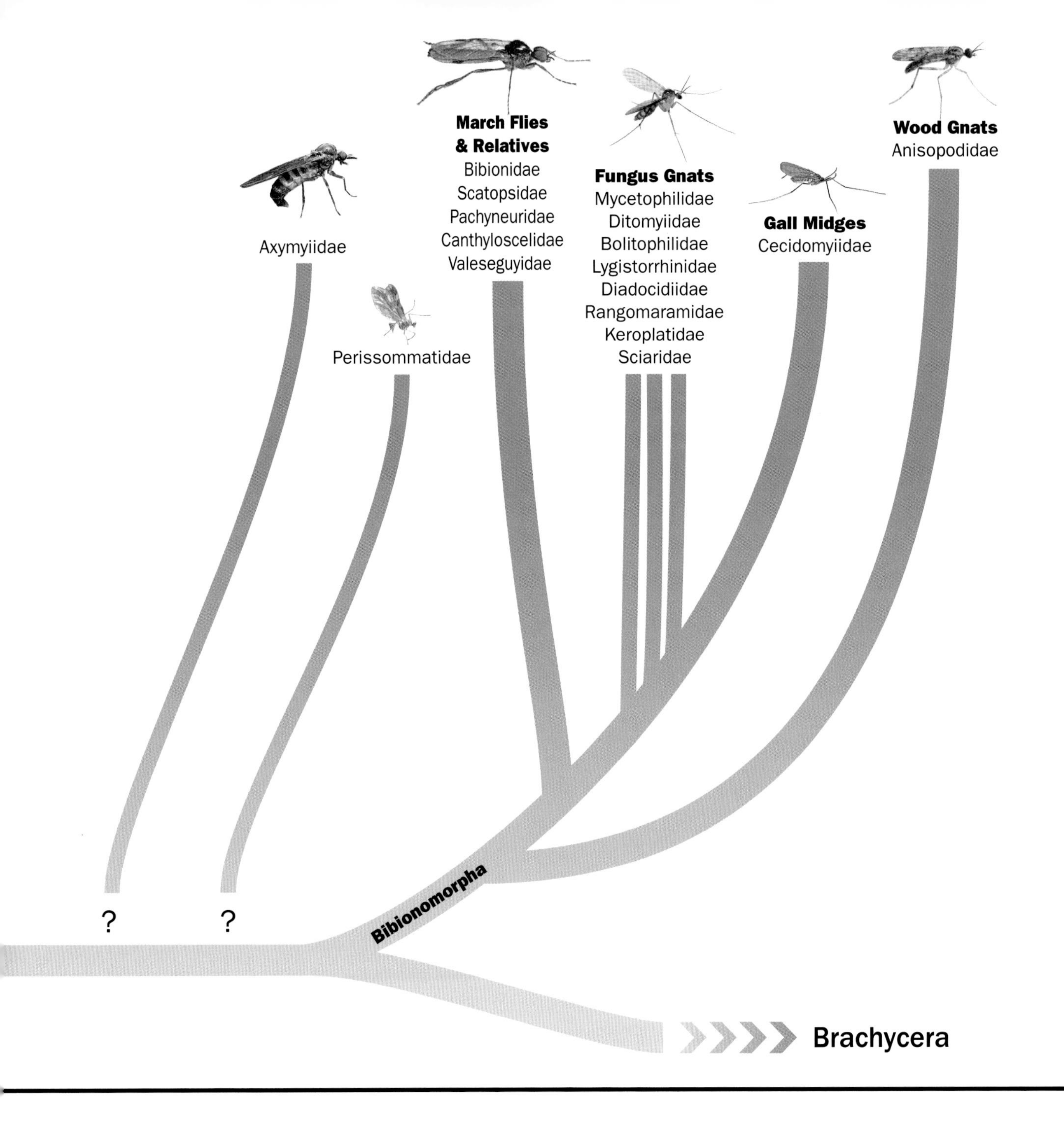

Aquatic Lower Diptera

Although only 10 percent or so of flies are aquatic, most aquatic insects are flies. The midge family Chironomidae alone often accounts for half or more of the animal species (and most of the individuals) sampled in stream and lake surveys; their larvae occupy a variety of aquatic environments ranging from ephemeral bits of moisture trapped at leaf bases to the deepest lakes and even the edge of the sea. Like almost all aquatic flies, midges leave the water as adults, and the ecological importance of their massive aerial swarms is tremendous, ending with the deposition of a significant biomass of dead flies in nearby environments. One paper recently reported the deposition of 135 kilograms of dead midges per hectare per day near a lake in Iceland, and dangerously dense midge swarms have been described as suffocating fishermen on east African lakes. These are exceptional examples, of course, but the abundance of flies in fresh waters is always significant, and that significance is largely due to the lower Diptera. Around 32 fly families are aquatic or mostly aquatic, with lineages scattered throughout the entire order and the main aquatic fly groups clustered as basal lineages of the lower Diptera.

The two families now widely treated as the most "primitive" or most "basal" Diptera — Deuterophlebiidae and Nymphomyiidae — develop only in cold running waters. Larvae of Deuterophlebiidae cling to rocks in torrential rivers with the aid of circlets of hooks (crochets) on lateral arm-like body extensions (abdominal prolegs); Nymphomyiidae clamber over moss-covered rocks with leg-like ventral body extensions. As far as is known, females of both families shed their wings after mating and head underwater to lay eggs and die. Both these rarely seen fly families have very restricted distributions: Deuterophlebiidae are found only in swift rivers in western North America and scattered localities in eastern and central Asia, while Nymphomyiidae are found in small, cold streams in eastern North America, as well as a few eastern Asian localities.

Aquatic lower Diptera other than the unique Nymphomyiidae and Deuterophlebiidae are usually divided into three infraorders: Tipulomorpha, Culicomorpha and Psychodomorpha. A couple of aquatic families, Ptychopteridae and Axymyiidae, don't fit neatly into this classification system. The Ptychopteridae are discussed here under Psychodomorpha and the Axymyiidae are treated at the end of the section as an unplaced "oddball" group.

Tipulomorpha

Crane Flies and Their Relatives

The infraorder Tipulomorpha is made up entirely, or almost entirely, of the family Tipulidae (crane flies), an enormous group that is sometimes split into multiple families. The small terrestrial family Trichoceridae is here treated as part of the Tiplulomorpha as well, although some evidence suggests that this little group of winter-active flies is not at all closely related to the Tipulidae, but is instead more closely related to other terrestrial families.

Tipulidae

Crane Flies

Every naturalist and field biologist is familiar with the delicate, long-legged flies known as crane flies or "daddy longlegs," an immense and superabundant group of lower Diptera found in a wide range of aquatic, semiaquatic and terrestrial habitats worldwide. Although crane flies range in size from tiny mosquito-like flies to gangly giants with legs that easily span a man's hand, their general appearance is that of large mosquitoes — a resemblance sometimes enhanced by the mosquito-like beak deployed by many nectar-feeding species. Seen from the top, the crane fly thorax has a distinctive V-shaped groove; crane flies further differ from superficially similar flies in lacking ocelli. Larvae also have a distinct appearance, with the head capsule excised at the back and retractable into the anterior end of the body. The posterior end of the body usually has a pair of conspicuous spiracles surrounded by lobes. Although most larvae with known biologies are saprophagous and eat microbe-rich organic matter (normally, decaying plant material) in wet environments, some crane flies are predaceous, fungivorous or phytophagous, and a few of the phytophagous species are considered major pests of turf. Some groups have become specialists in extreme environments such as caves, marine intertidal zones and deserts, but most occur in humid forests and wetlands. Most Tipulidae are unknown as larvae.

The tipulid lineage is one of the oldest and most diverse groups of Diptera, dating back to the Triassic period and now with more than 15,000 described species that almost all have a distinctive gangly-legged habitus rendering them easily recognizable as crane flies. Despite this, dipterists have disagreed for years about family-level nomenclature of the crane fly lineage. European authors have generally divided these long-legged flies into four families — Tipulidae,

Limoniidae, Cylindrotomidae and Pediciidae — as in the *Catalogue of the Crane Flies of the World* (Oosterbroek 2005), while most North American and British authors have treated the lineage as the single family Tipulidae, with two to four subfamilies. The latter system is followed here, in part because it renders the family easily recognizable by any student or naturalist, and in part because division of the group into four families was never defensible on phylogenetic grounds. The division of a broadly defined Tipulidae into four subfamilies, equivalent to the four families used in the world catalog, is also imperfect (Limoniidae is an artificial group whether it is recognized as the family Limoniidae or the subfamily Limoniinae). Still, it leads to an easily understandable classification that divides a monophyletic and easily recognized family into four familiar subfamilies, even though one of them is artificial and likely to be split into several subfamilies in future classifications.

Although the long beak of this *Geranomyia* (Limoniinae) imparts a mosquito-like appearance, there are no biting crane flies. These elongate mouthparts are used only for imbibing nectar.

Limoniinae The subfamily Limoniinae is a sort of grab bag into which dipterists have traditionally dumped an enormous mix of mostly small, slender crane flies generally characterized by elongate narrow wings and long, delicate legs. The more than 10,000 species in this group, however, span a wide range of morphologies and biologies. Adult limoniines usually occur near water, and many abound on flowers or low vegetation in wetlands or near water bodies; others occupy humid forests. Most larvae feed in the typical lower dipteran fashion of immersing themselves in decaying vegetation, where they consume a mixture of microorganisms and plant matter, but one significant lineage (*Limnophila* and its relatives) is made up of predators, and some species are fungivorous. Limoniine larvae occupy a wide range of habitats but are usually exceptionally abundant in thin, flowing films of water with lots of algae. Many limoniines, including most larvae of the mosquito-like genus *Geranomyia*, are aquatic and feed from the shelter of silken tubes. Some Neotropical and Pacific *Geranomyia*, however, are terrestrial; they maintain a high-humidity microenvironment by producing tubes or masses that resemble gelatinous raindrops on the upper surfaces of leaves. These are used as shelters from which the fly larvae feed on decomposing material (and, of course, associated microorganisms); pupation takes place in masses of jelly under leaves.

Larvae of the common aquatic genus *Antocha* also feed on detritus from the shelter of tubes, but silken tubes rather than jelly blobs, and on submerged rocks in flowing water rather than on leaves. Unlike *Geranomyia* and most other crane flies, *Antocha* larvae lack the posterior spiracles normally used to respire at the water surface, the surface of the jelly shelter or other points of contact with the open air. Instead, larvae of *Antocha* and a few related genera breathe directly through their skin. Similarly, the pupae of most crane flies breathe through open spiracles but *Antocha* draws oxygen directly from its running-water environment. In the swifter parts of many eastern North American rivers, almost every partially submerged rock is populated by larvae of *Antocha saxicola*; adult females are often seen crowding the waterline in order to insert their eggs into the algae covering the rocks.

One of the more unusual lineages traditionally treated in the Limoniinae is the Holarctic genus *Chionea* (snow flies), a group of about 40 species of wingless flies normally seen only during the colder months, slowly walking along (seeking mates?) on the relatively predator-free snow surface. Although these small crane flies lack wings, they do retain conspicuous halters; these prominent knobs easily distinguish *Chionea* species from the superficially similar *Boreus* (snow scorpionflies) active at the same time. Snow fly larvae are saprophagous, living among decaying leaves or feces in rodent burrows. Some authors now treat *Chionea*, along with the fully winged, normal-looking crane fly genus *Cladura*, in a separate subfamily and recognize further subfamilies for other distinctive groups now in the Limoniinae. Petersen et al. (2010) divided the group into several subfamilies based on a quantitative phylogenetic analysis using combined morphological characters and nuclear gene sequence data. It remains to be seen whether these suggested divisions become widely accepted.

Pediciinae The other subfamilies of Tipulidae are better defined than the Limoniinae and are generally distinctive groups. The "hairy-eyed crane flies" in the subfamily Pediciinae (or family Pediciidae) are the probable "basal lineage" of crane flies. This relatively small lineage of about 500 species includes several common, robust aquatic or semiaquatic species with larvae that prey on other aquatic invertebrates. Adults, such as the large and boldly patterned members of the Holarctic-Oriental genus *Pedicia*, are often conspicuous streamside insects. Members of one genus, *Ula*, are markedly different from other pediciines in habits and habitat; instead of being predaceous inhabitants of streams and rivers they are fungivorous inhabitants of terrestrial fungi. Pediciines can be distinguished from other tipulids by their hairy eyes, spurred tibiae and relatively complete venation.

Tipulinae The typical crane flies of the subfamily Tipulinae make up a distinctive lineage of about 4,300 species, including most of our relatively large crane fly species. Sometimes called long-palped crane flies because the last segment of the maxillary palpus is slender and relatively long or even whiplike, tipulines usually occur in damp places, where their larvae develop in wet decomposing plant material. Habitats frequented by typical crane flies, however, cover the extremes from wet to dry. A few notorious species in the dry group are phytophagous pests with larvae ("leatherjackets") that consume the roots of grasses and other plants. Leatherjackets, some of which have become secondarily widespread pests, belong to the worldwide genus *Tipula*. With 40 subgenera and around 2,000 named species, *Tipula* is one of the largest genera in the animal kingdom.

Cylindrotominae The subfamily Cylindrotominae is a small lineage of about 70 phytophagous species that develop on living terrestrial or aquatic plants, including mosses. Some *Cylindrotoma* species have caterpillar-like green larvae that are among the very few Diptera found feeding externally on foliage. Adults are relatively large, attractive crane flies with the abdomen characteristically constricted near the base, but these infrequently encountered insects are unfamiliar to most people. The subfamily Cylindrotominae is closely related to the subfamily Tipulinae.

Trichoceridae

Winter Crane Flies

The family Trichoceridae is a small group of about 160 species of delicate, uniformly dull brown flies that look like small Tipulidae with ocelli. Called winter crane flies because of their tipulid-like appearance and their habit of swarming during early or late winter months, when few other flies are on the wing, these common insects develop in compost and fungi. The relationships of the winter crane flies remain a matter of dispute among dipterists, even though a recent molecular study (Bertone et al. 2008) follows Hennig (1973) in placing these flies right where their appearance suggests: next to the Tipulidae. This is in marked contrast with the phylogenetic work of Wood and Borkent (1989), which treats the Trichoceridae as most closely related to some terrestrial families now treated as part of the Bibionomorpha.

Winter crane flies are terrestrial and also resemble Bibionomorpha in having ocelli as adults and in having both anterior and posterior spiracles as larvae, while crane flies lack larval anterior spiracles and adult ocelli.

The family Trichoceridae includes only five genera but is nonetheless sometimes divided into two subfamilies, Paracladurinae and Trichocerinae. The most commonly encountered species, including some widespread species found in both north and south temperate regions, are in the genus *Trichocera*. Some Australia–New Zealand Trichoceridae have their closest relatives in southern South America, reflecting the great age of the group and the ancient biogeographic connections between south temperate winter crane flies.

Culicomorpha

Mosquitoes, Black Flies, Midges and Their Relatives

The aquatic or mostly aquatic fly families other than Tipulidae and the oddball families Axymyiidae, Nymphomyiidae and Deuterophlebiidae seem to form one natural group made up of the large infraorders Culicomorpha and Psychodomorpha, along with the family Ptychopteridae.

The Culicomorpha is probably the least contentious and most clearly monophyletic infraorder of lower Diptera, as it is strongly characterized by adult and larval morphology and is also (unsurprisingly) recognized in all recent lower dipteran phylogenies based on molecular data. Because this infraorder includes most of the medically important blood-feeding and disease-transmitting families of lower Diptera, it is an especially well-known group. Most aspects of the Culicomorpha, including the phylogenetic relationships of its included families, have been thoroughly studied.

Virtually all students of culicomorph phylogeny agree that the four "mosquito-like" families Culicidae, Chaoboridae, Dixidae and Corethrellidae are closely related; they are treated together as the superfamily Culicoidea. The rest of the Culicomorpha — Thaumaleidae, Simuliidae, Chironomidae and Ceratopogonidae — are generally put into another superfamily, Chironomoidea, but dipterists disagree about the validity of this group and the details of relationships among its included families. Nonetheless, the following discussions of families in the infraorder Culicomorpha are grouped according to those two superfamilies.

Chironomoidea

Common Midges, Biting Midges, Black Flies and Seepage Midges

Chironomidae

Non-biting or Common Midges

Adults of non-biting or common midges in the family Chironomidae, which has about 8,000 named species and probably as many more unnamed, are among the most abundant aerial animals. Their larvae are dominant organisms in most aquatic environments in every part of the world, from the tropics almost to the poles, from the seacoast to mountaintops and from sizzling hot springs through to glacial freshets. Most aquatic Diptera species belong to the family Chironomidae, and since many midge species occur in dense populations, aquatic habitats often support more midges than any other group of animals. Although the delicate adults are relatively short-lived, the abundance of this group is reflected in the vast numbers of fuzzy-antennaed males that form conspicuous crepuscular swarms, and in the masses of midges often attracted to manmade light sources.

Midges are plentiful in both the Antarctic, where the wingless *Belgica antarctica* is an important endemic, and the Arctic, where enormous swarms of flying males are common features of the summer tundra. Sometimes even more spectacular swarms appear in warmer areas such as tropical Africa, where choking clouds of adult midges emerging from eutrophic lakes or enriched slow-flowing rivers can lead to serious allergy problems. Many midges, especially species of eutrophic waters, contain haemoglobins used as respiratory pigments in larvae but also present in adults — and in fragments of dead adults. The allergenic properties of these haemoglobins render some swarms a significant health hazard.

Midge swarms, almost invariably made up only of males of a single species, typically form over a visual marker such as an isolated rock or shrub, a tree branch or even a parked car. Different species prefer different kinds of markers and different heights, some swarming near the water surface and others over tall trees. Arriving females are immediately mated, forming pairs that drop out of the swarm to copulate on leaves or some other substrate. Males die shortly thereafter, and neither sex takes any food other than nectar, pollen or honeydew. Females of aquatic species (and most, though not all, midges are aquatic) usually deposit gelatinous egg masses or strings in or near the water; these normally expand or even "explode" when wet. The family Chironomidae is enormous and diverse, and almost any generalization about midge life histories will have exceptions. For example, several species are flightless and do not form swarms; many midge lineages are always or sometimes parthenogenetic (that is, they have no males); and some chironomids lay eggs individually rather than in masses. Other exceptional examples of midge life histories are mentioned below in the subfamily discussions.

Any sample of insects from almost any conceivable aquatic habitat will include some midge larvae, which are characteristically cylindrical and smooth-bodied, with a prominent head capsule, prolegs at both ends, and no open spiracles; their body colors vary, and some are bright red because of the presence of haemoglobin. Most midge larvae are aquatic consumers of algae or decomposing materials such as leaf litter or woody debris and their associated microbial communities in lakes or streams, but the family also includes several derived predaceous, parasitic and plant-feeding lineages. Aquatic species usually use salivary silk to make shelters, nets or tubes much like some caddisflies. Some, especially in the subfamilies Tanypodinae and Diamesinae, are free-living and some are commensal on other aquatic insects or parasitic on aquatic animals such as other insects, snails, fish and even freshwater sponges. The relatively few terrestrial chironomid species usually occur in moist, organically rich habitats such as rotting wood, leaf litter or fungi.

The pupal stage of Chironomidae is fairly brief (a few days at most) and usually takes place in a silk tube or shelter spun by the larva. Pupae of the subfamilies Tanypodinae and Podonominae, however, are free-living; they breathe at the surface using thoracic horns similar to those of mosquitoes and their relatives. Midges that pupate in silk tubes also have variously shaped thoracic horns, but rather than reaching the water surface they serve as tracheal gills that take up oxygen from water pumped through the tubes by the undulating pupa. The mature pupa swims to the surface (if it is not already there) and the adult quickly emerges into the air through a split in the back of the pupa. The pupal shells, or exuviae, are fairly persistent, and collection and identification of midge pupal exuviae can be an important component of aquatic biomonitoring programs.

The 8,000 or so known species and 450 or so genera in the family Chironomidae are divided among 11 subfamilies. However, almost all routinely encountered midges — and the great majority of species in the family — fall into the three subfamilies Chironominae, Orthocladiinae and Tanypodinae. Chironominae and Orthocladiinae are closely related and have somewhat reduced wing venation, in which the crossvein between the main central wing veins (the medial and cubital veins) is absent.

Chironominae Adult flies in the "typical" subfamily Chironominae have relatively long front legs with the first tarsomere (tarsal segment) longer than the foretibia. This subfamily includes the enormous and widely distributed genus *Chironomus*, familiar for extensive lakeside swarms of relatively large species such as the common Holarctic *Chironomus plumosus*, and well known for the blood-red pigment of common *Chironomus* larvae known as "bloodworms" and sometimes mass reared as pet food.

Another important chironomine is the genus *Polypedilum*, with around 450 described species found mostly in sediments, although *Polypedilum* species also inhabit water-holding plants (phytotelmata), mine in the leaves of aquatic plants,

Male midges often have characteristically plumose (feather-like) antennae, with a swollen, cup-like pedicel (second segment). This one is in the subfamily Chironominae, as you can tell by the very long first tarsomere on its front leg.

live in the pupal retreats of caddisflies and occur in a variety of small temporary waters. Of these, the most famous is *Polypedilum vanderplanki*, a sub-Saharan African midge that develops in tiny temporary pools in rock depressions, undergoes total desiccation when its habitat dries up, then rehydrates to resume development when it rains again. In its desiccated form it is able to withstand extraordinary extremes of temperature beyond anything likely to be encountered in nature; this is one of the best-known examples of cryptobiosis.

Another group of Chironominae associated with extreme conditions is the genus *Sergentia*, with several endemic species found up to one kilometer below the surface in the world's deepest freshwater body, Lake Baikal. A few other species in this large subfamily are among the few truly marine insects. Male adults of the Old World and Australasian genus *Pontomyia* usually emerge after sunset to spend their brief lives (a couple of hours) skating about on the surface of the sea in search of the wormlike wingless adult females. The tips of the male's second and third pairs of legs are its skates; the extremely long, curved first pair of legs barely skim the surface, like outriggers, while his bizarre paddle-like wings wave just above the water surface to provide propulsion. Once a *Pontomyia* male locates a wingless female, he picks her up — sometimes after extracting her from a pupa — mates while still in motion, and then drops her so she can lay eggs on coral or other objects sticking up out of the water.

Orthocladiinae The large subfamily Orthocladiinae, a heterogeneous and possibly artificial assemblage of midges — characterized by some dipterists as essentially the "residue" of the family after removal of other groupings — includes some of the most common and many of the most unusual midges. For example, the midges that seem to stray farthest from the aquatic heart of the family are dung-inhabiting species in the orthocladiine genus *Camptocladius*, and the best-known parasitic species in the family are in another orthocladiine genus, *Symbiocladius*. Only the very small first-stage (first-instar) *Symbiocladius* larvae are free-living; later instars are found only on the mayfly nymphs that they puncture to feed on haemolymph and tissue. *Symbiocladius* was long known only from Holarctic species (two in North America and one in Europe), but it was discovered in Chile in 1965 and then found in Australia in 1976. Perhaps the most unusual midges of all are the marine species; the subfamily Orthocladiinae includes the highly modified marine midges of the genus *Clunio*, found in the intertidal zone of rocky seashores throughout the world. *Clunio* males spend most of their brief adult lives frenetically searching for wingless females on rocks that get exposed only when the tide is exceptionally low. *Ichthyocladius* is another, aptly named group of specialized Orthocladiinae, with larvae that live on catfish.

Clunio, Symbiocladius, Ichthyocladius and *Camptocladius* are highly specialized groups, but the Orthocladiinae also includes some large genera with a variety of habits. *Cricotopus*, for example, is a large and widely distributed midge genus found in many different aquatic habitats, often in association with aquatic plants. One *Cricotopus* species is a pest that mines in rice plants, and some others are associated with benthic or floating algae. Some, in the aptly named subgenus *Nostocladius*, live in jellylike aquatic colonies of the colonial cyanobacterium ("blue-green alga") *Nostoc*. Each *Nostoc* colony is occupied by a single larva, which leaves its host only after pupation and eclosion to the adult stage. This relationship has been described as mutualism because the cyanobacteria supposedly benefit in some way from the presence of the midge larvae. The same might be said for a species of the widely distributed and diverse genus *Metriocnemus*, which is found in the trap fluids of pitcher plants (along with other flies such as the mosquito *Wyeomyia smithii*). Larvae of *M. knabi* feed on the decomposing carcasses of trapped insects, in turn contributing nutrient-rich feces to the plant's digestive fluids. Some other *Metriocnemus* species occupy accumulations of water — phytotelmata — in other kinds of fluid-holding plants.

Tanypodinae The subfamily Tanypodinae is the main lineage of predaceous Chironomidae, although some predators are also found in the subfamilies Chironominae and Orthocladiinae, and some Tanypodinae feed at least partly on detritus. Larval Tanypodinae are free-living insects, usually with well-developed mandibles anchored on a relatively long head and with relatively long antennae that can be retracted into the head. *Procladius* and *Tanypus* are among the more widespread and diverse genera of Tanypodinae; both develop in sediments in slow or still waters, including polluted waters with low oxygen levels. *Ablabesmyia* is another extremely widespread and speciose tanypodine genus. It is found in a wide variety of habitats but particularly abundant in and around large lakes, where the small, compact, distinctively striped, spotted and hairy-winged adults frequently fly to lights at night. *Ablabesmyia* larvae (and many other Tanypodinae) can develop on a diet of oligochaetes (aquatic worms), but some tanypodines apparently prefer to eat other midge larvae, benthic rotifers or detritus in phytotelmata. The subfamily Tanypodinae is most diverse at higher altitudes or latitudes: most of the world's 48 genera are found in the Holarctic region.

The subfamily **Diamesinae** is a relatively small group, but some species of this group can be quite abundant under cold conditions. In very early spring, for example, most of the midges seen out and about, on and over the snow, belong to this group, and the subfamily is also a dominant group of midges in the Arctic and along the edges of glaciers. Adults of the large, mostly Holarctic genus *Diamesa* frequently emerge in winter or early spring. Those of one *Diamesa* species are recorded as surviving air temperatures down to –20°C (–4°F) without freezing.

The subfamily **Prodiamesinae** includes only three, mostly Holarctic genera — *Prodiamesa, Odontomesa* and *Monodiamesa* — that have traditionally been treated (and still are by some) in the subfamily Diamesinae. One particularly common European species in this group, *Prodiamesa olivacea*, shows a high rate of deformities in contaminated water and is therefore used, along with *Chironomus* species, for biomonitoring.

The subfamily **Podonominae** is a small group (fewer than 200 species) associated with cool habitats such as wet or submerged mosses in small mountain streams or even freshets running right off glacial ice. This subfamily was known only from the northern hemisphere before Brundin documented a rich austral podonomine fauna in his groundbreaking 1966 monograph on the relationships of southern hemisphere midges. One genus, *Parochlus*, is particularly diverse in the southern hemisphere, with 50 species ranging from the Antarctic to the Andes, but only a single Holarctic species. Some southern African and Australian podonomines (*Archaeochlus* and *Austrochlus*) are unusual among "non-biting midges" in having adult mandibles similar to those of biting midges, although we don't yet know how they are used.

The subfamily **Chilenomyiinae** is an obscure and exceedingly rare south temperate subfamily. It includes only the odd Chilean species *Chilenomyia paradoxa*, which is known from a total of two adult specimens.

The subfamily **Buchonomyiinae** includes only one small genus, *Buchonomyia*, which has one species in each of the Palaearctic, Oriental and Neotropical regions. Larval habitat is unknown, but they presumably occupy large rivers.

The subfamily **Aphroteniinae** is a small and rarely encountered subfamily of three small southern hemisphere genera.

The subfamily **Telmatogetoninae** is a small seashore subfamily that includes two widely distributed genera: one, *Thalassomya*, entirely associated with salt water and one, *Telmatogeton*, with both marine and freshwater species. Several Hawaiian species of *Telmatogeton* develop in swift waters or splash zones near waterfalls. Like *Clunio* (subfamily Orthocladiinae), adults are most likely to be found as they scurry over wave-washed rocks; unlike *Clunio*, both sexes are winged. Larvae of most species make tubes in intertidal algae.

The subfamily **Usambaromyiinae** is a rare group recently described to hold a single, distinctively black-winged species from Chile.

Ceratopogonidae

Biting Midges

The biting midge family, Ceratopogonidae, is a diverse group of almost 6,000 named and at least as many unnamed species, best known for the biting habits of a relative few pest species known as "no-see-ums" or "punkies." Ceratopogonids are called "sand flies" in some areas, an unfortunate common name best applied to an entirely different group — biting Psychodidae in the subfamily Phlebotominae — and also applied to black flies (Simuliidae) in some countries. Female punkies have prominent biting mouthparts, and adults of both sexes look a bit like relatively small, squat common midges with short forelegs and overlapping wings.

Ceratopogonid larvae develop in a variety of habitats, ranging from lakes to treeholes and tank bromeliads. The most familiar larvae are smooth, slender nematode-like predators that abound in a variety of aquatic environments, while others are terrestrial or semiaquatic; some are covered with processes and have prolegs, unlike the common aquatic species. Adults are usually tiny and inconspicuous (mostly at the low end of 1–4 mm), but those that bite can be a huge nuisance. Even those tiny enough to be effectively invisible ("no-see-ums") bite painfully and can leave itchy welts. On the positive side, adults are important pollinators of some plants: without their appearance on Cacao flowers we probably would not have chocolate.

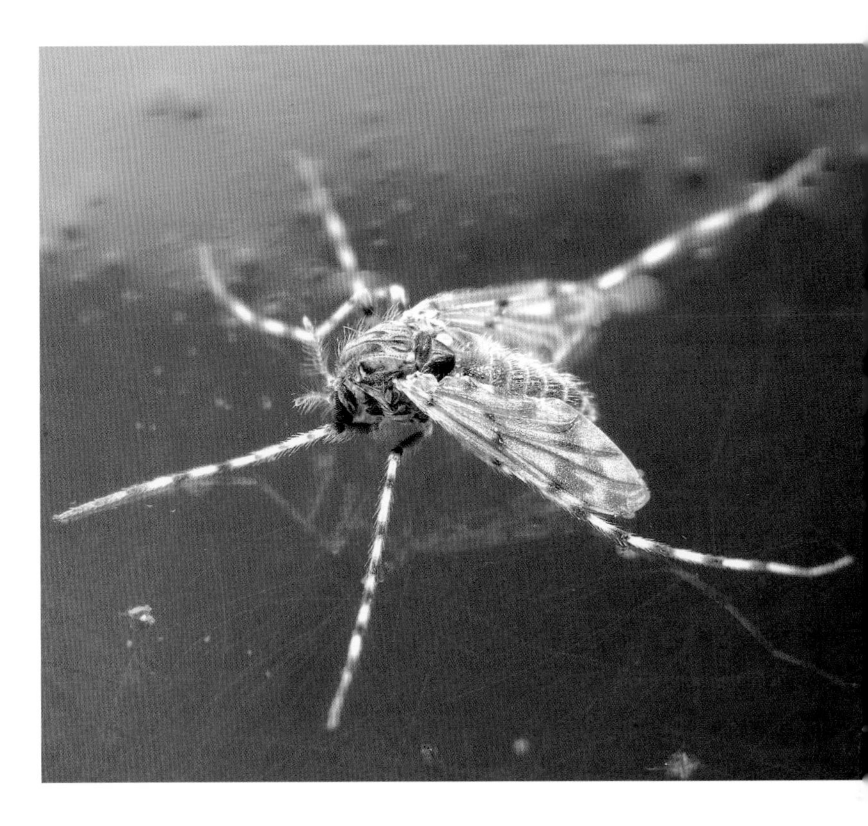

This tiny tanypodine midge (*Ablabesmyia*) has just emerged from a floating pupa in a large lake, where it developed as a predaceous larva feeding on other benthic invertebrates.

The extant Ceratopogonidae are divided into the subfamilies Leptoconopinae, Dasyheleinae, Forcipomyiinae and Ceratopogoninae.

Leptoconopinae Almost all species in the subfamily Leptoconopinae belong to one large, widely distributed genus, *Leptoconops* (with several subgenera). This ancient group also includes some entirely fossil genera with apt names such as *Fossileptoconops* and *Archiaustroconops*, as well as a mostly fossil genus, *Austroconops*, that includes several species from 120-million-year-old northern hemisphere amber as well as two living species in

Western Australia. *Leptoconops* females feed on the blood of various vertebrates and, according to Canadian ceratopogonid expert Art Borkent, sometimes bury themselves under a thin layer of sand after taking a blood meal. Larvae usually live in wet sand, especially along seashores, where they graze on bacteria, algae and fungi.

Dasyheleinae The subfamily Dasyheleinae includes only the widespread genus *Dasyhelea*. Species in this large genus are usually very small (less than a couple of millimeters) and have non-biting adult females with reduced mouthparts, used for feeding on nectar or perhaps honeydew. Larvae usually occur in small pockets of moisture or water, and some species are characteristic and abundant inhabitants of ephemeral rock pools, where their omnivorous feeding habits and resistance to desiccation allow them to thrive in pockets of water relatively free of predators and competitors. A few species are common in mangrove swamps and salt marshes.

Forcipomyiinae The subfamily Forcipomyiinae includes only two extant genera, *Forcipomyia* and *Atrichopogon*. Adult female forcipomyiines are frequently seen plugged into the wing veins or bodies of large insects. Those that specialize in sucking blood from blister beetles (Meloidae) are often abundant on — and apparently annoying to — common blister beetles such as *Epicauta* and *Meloe*; others are often seen on the wing veins of dragonflies, damselflies, lacewings, moths and butterflies. Other recorded hosts include katydids, stick insects, crane flies, spiders, stink bugs and some caterpillars. One recently discovered (and still unnamed) Ecuadorian *Forcipomyia* is a kleptoparasite, or food thief, that feeds only on termites freshly captured by a tiny spider that in turn seems to hunt only one species of termite. Another group of *Forcipomyia* species, the subgenus *Lasiohelea*, feeds on vertebrates, including lizards, frogs and, in the case of some Old World species, humans.

Larval Forcipomyiinae, which look quite different from other biting midge larvae because of their many body processes and their well-developed prolegs, occur in a variety of habitats ranging from aquatic to terrestrial. Most occur in organically rich fluids such as the gunk in tank bromeliads or the goo beneath the bark of recently dead trees, where they graze on microorganisms. Most have long hairs that secrete a sticky fluid, probably as a defense against ants.

Ceratopogoninae Most biting midges, including the notorious blood-feeding females in the genus *Culicoides*, are in the large subfamily Ceratopogoninae. Unlike biting relatives in the Leptoconopinae, which bite during the day, *Culicoides* species bite mostly in the evening or at night, and in doing so they become serious pests that transmit diseases of humans and livestock. A few members of the genus don't bite at all, and others have the unusual habit of obtaining second-hand blood by biting engorged mosquitoes, but for the most part the 500 or so species currently in the genus *Culicoides* bite vertebrates, including people, of course, but spanning a wide range of hosts from turtles to cattle. Other genera in the subfamily are mostly predators, often specializing in raiding the swarms of midges and other flies. In a few species with the remarkable habit of raiding male swarms of their own species, the raiding female mates with a lucky male before impaling him through the head, injecting him with proteolytic enzymes, and consuming his contents.

The subfamily Ceratopogoninae includes most aquatic Ceratopogonidae, including those with long, skinny, (pro)legless larvae routinely seen wriggling like nematodes or tiny snakes in a variety of aquatic and semiaquatic habitats. These smooth, seemingly unsegmented larvae have an elongate head capsule with the mouthparts directed forward, where they serve to impale invertebrate prey. A few larvae in the subfamily seem to go for the dipteran default strategy of microbial grazing, but most consume entire small organisms or chew their way into larger prey to eat them from the inside.

Simuliidae

Black Flies

Black flies are among the most readily recognized of insects in all life stages. The compact,

muscular-looking adults have a characteristic humpbacked appearance, short antennae and unusually heavy veins along the leading edge of the wing. Black fly larvae are unmistakable for their punching-bag shape: the fat end is equipped with hooks to cling to pads of labial-gland silk that the larvae usually place on solid substrates in flowing waters, while the narrower head end is normally equipped with foldable filters (labral fans) to strain food from the flow. Even the pupae, with their antler-like tracheal gills directed downstream, are distinctive. Pupae, all or partly ensheathed in a sac-like or slipper-like silken cocoon, adhere to substrates in moving water; emerging adults float to the surface in a bubble of air, like a silvery diving bell that explodes to release a winged adult as it pops above the water surface.

Although a few species are parthenogenetic, most form male swarms shortly after adult emergence. Females that enter the swarm are swiftly seized by males to form mating pairs that drop to the ground or to lower vegetation. Some species skip the swarms, instead mating on the ground immediately after emerging. Most (but not all) mated females require a blood meal before going on to lay eggs, which they usually do either by landing near the waterline to deposit strings or masses of eggs or by dropping eggs directly into the water while in flight. Males, of course, do not bite; they feed only on sugars.

Relatively few simuliid species are pests, but some of them can be abundant enough to kill animals by exsanguination, and black flies are often serious nuisance pests of humans. Biting females cut into the host's tissue with sawlike mandibles, inject saliva that prevents the blood from clotting and then feed on the oozing pool. I have often returned from early spring canoe trips with a continuous bloody welt around my waist or with my ears swollen from black fly bites. This, of course, is nothing to complain about compared with the issues that can follow black fly bites in some parts of the world: *Simulium* species can carry *Onchocerca volvulus*, the filarial nematode that causes human onchocerciasis, or river blindness. This is the only major human parasite carried by black flies, although many parasites of birds and other animals are carried by this group.

In my opinion, the best monographic treatment of any family of flies — and possibly the

Biting female black flies, like this *Simiulium tarsatum* from Costa Rica, cut through your skin, inject irritating anticoagulant saliva and feed on the resulting oozing pool.

best monographic treatment of any group of organisms — is a 937-page tome by Adler, Currie and Wood (2004). Although it deals specifically with the Simuliidae of North America, this book summarizes much of what is known of black fly biology and diversity.

The Simuliidae are divided into two subfamilies, but one of them, **Parasimuliinae**, is a small group of five non-biting species found only in the Pacific Northwest of North America, where their pale larvae develop in subterranean streams. All other black flies — around 2,000 species — are in the cosmopolitan subfamily Simuliinae, which is in turn divided into two tribes, the Simuliini and Prosimuliini.

Simuliinae Most Simuliinae, and in fact over 90 percent of all black fly species, are in the typical tribe ***Simuliini***. This group is in turn dominated by *Simulium*, the largest, most important and most widespread genus in the family. The *Simulium damnosum* species complex, made up of dozens of morphologically indistinguishable but cytologically distinct forms, includes several important vectors of onchocerciasis, and other *Simulium* species are among the worst nuisance pests throughout the Holarctic region and in South America. The biting black flies of the southern hemisphere belong in other genera, and most of the so-called "sand flies" of Australia and New Zealand are species of *Austrosimulium*. Some of the *Simulium* species that vector onchocerciasis in Africa cling to unusual substrates, including the bodies of other aquatic organisms such as crabs and mayfly nymphs. For example, in a 1971 investigation of onchocerciasis vectors, British dipterist Henry Disney discovered two new species of *Simulium* on African freshwater prawns, one living in the gill chambers and the other species attached to the bases of the prawn's head appendages. Other Simuliini that feed from unusual positions include the eastern North American genus *Ectemnia*, which has larvae that filter feed from the ends of long silken stalks. A few Australian species have similar habits.

The tribe ***Prosimuliini*** is a relatively small northern hemisphere group of five genera, dominated by the diverse genus *Prosimulium* but also including the small, specialized genera *Gymnopais* and *Twinnia*. Larvae of these genera have lost their labral fans and graze on algae rather than filter-feeding like most other black flies.

Thaumaleidae

Seepage or Trickle Midges

Adult seepage midges, which look a bit like small reddish brown non-biting black flies (Simuliidae) because of their stocky bodies and short antennae, are inconspicuous insects that normally stay close to the thin films of water in which their larvae develop. The larvae are only marginally more conspicuous than the adults, and then only to those who look closely at cool, shaded films of water flowing or seeping over steep faces of rock or clay. Seepage midge larvae, which have the general shape of common midge larvae (from which they differ in having open dorsal spiracles), are found in flowing waters no deeper than their bodies, where they can feed on diatoms on the substrate while breathing with spiracles open to the air. They often zip away with a sort of sidewinder motion when disturbed, which helps make them recognizable in the field. The family Thaumaleidae is a small group — about 175 species — probably most closely related to the Simuliidae. Seepage midges are best represented in the Holarctic region but a few species occur in the southern hemisphere, mostly in genera shared between Australia, South America and New Zealand. One unusual genus occurs in South Africa.

Culicoidea

Mosquitoes, Meniscus Midges, Frog Midges and Phantom Midges

Culicidae

Mosquitoes

When a mosquito lands on your arm, it is hard to resist the impulse to instantly flatten it, but if you take a minute to examine the delicate female attracted to your warmth, odor and carbon dioxide, there is much to admire. Her wings and body are decorated with scales much like butterfly scales, her eyes probably have a metallic sheen, and her long, sensitive mouthparts are a wonder to behold. Hidden inside the visible proboscis (a sheath-like lower lip) are sets of diabolical blades and needles that fit together into a slender tube (called a fascicle or syntrophium) used to puncture skin and to inject saliva and ingest blood. The blades are simply stretched-out knifelike mandibles and sawlike maxillae that slide back and forth like reciprocating saw blades to penetrate vertebrate tissues in search of a small blood vessel. The needles are the elongated hollow tongue (hypopharynx) that injects saliva to keep blood from clotting, and the straw-like labrum used to suck up the liberated blood.

The itch you feel after a mosquito has fed is a reaction to proteins in the mosquito's saliva; the more serious possible effects of a mosquito's bite derive from the microorganisms often injected with that saliva. As discussed in Chapter 3, disease organisms carried by mosquitoes incapacitate or kill millions of people per year. Even in the relatively disease-free north temperate countries mosquitoes are serious pests, as reflected in insect repellent sales that reach hundreds of millions of dollars per year in the United States alone. The mosquito hordes of subarctic areas are legend: I can personally attest to the speed with which exposed skin becomes packed with hungry snowmelt mosquitoes during the field season in Canada's Yukon Territory — although I've never tested the claim that mosquito bites can remove half of an unprotected person's blood in just over an hour!

All mosquitoes develop from aquatic larvae, found in a wide variety of still or very slowly moving waters where they are able to breathe through open posterior spiracles — normally at the water surface, although some species use aquatic vascular plants as living snorkels. Larvae, or "wrigglers," normally feed by using mouth brushes (projections of the labrum) to draw in or scrape up particulate food to be filtered by the mouthparts, but some feed on other mosquito larvae or similarly sized invertebrates. Pupae, or "tumblers," are comma-shaped, with a swollen thorax topped with a pair of trumpet-shaped breathing horns that are normally in contact with the water surface, although they can swim ("tumble") to concealment in bottom sediments if disturbed. Adult males are delicate non-biting insects that form aerial swarms in which they use conspicuously feathery antennae to detect the whine of arriving females. Mated females usually require a blood meal to support egg development, but some species can develop at least one batch of eggs without blood feeding. Eggs are deposited on the water, in the water or on substrates likely to be underwater in the future.

Because of the enormous importance of mosquitoes there is a profusion of books, monographs and websites on the group; it is quite possible that more is written on the Culicidae than all other Diptera put together. One of the best and most current sources of information on mosquitoes is the Mosquito Taxonomic Inventory website, http://mosquito-taxonomic-inventory.info/.

Anophelinae The family Culicidae is a large, worldwide group with more than 3,600 species divided between two subfamilies, the Anophelinae and Culicinae. The subfamily Anophelinae is a relatively uniform group, including only the small Australian genus *Bironella*, the small Neotropical genus *Chagasia* and the enormously important worldwide genus *Anopheles*. *Anopheles*, with some 460 species, is arguably the most dangerous of all animal genera because of the unique relationship between some *Anopheles* species and the parasites that cause malaria. *Anopheles* adults are usually attractively patterned flies, easily distinguished from other common mosquitoes by the long palpi that flank the proboscis.

Anopheles eggs are deposited singly on the water surface, where they float with the aid of a distinctive air-filled packet on each side. Larvae normally stay just under the water surface, suspended horizontally from the surface film as they use labral brushes to sweep streams of floating particles through their mouthparts. This feeding position requires an unusual 180-degree rotation of the head so that the ventral side of the head is facing up and the mouthparts are in position to filter out algae, bacteria and other organic matter from the surface film. *Anopheles* larvae lack the long siphon (the posterior breathing tube) seen in larval Culicinae; they hang out flush with the surface so the dorsal spiracles remain in contact with the air, rather than suspended by the siphon as in culicines.

Anopheles species vary widely in habits and habitats, normally developing in relatively clean water like lake margins and backwaters but sometimes preferring shaded or sunlit pools or different kinds of water bodies. Adult females normally feed in the evening or during the night; species differ in host preference, with some preferentially pestering humans and others associated with animals. Different *Anopheles* also differ in important behavioral details such as post-feeding behavior. Those that stay inside buildings and rest on walls, for example, are easily killed with residual insecticides, while those that disperse into the forest are harder to deal with. Only some species are good vectors for malaria, and some superficially similar species differ widely in their vector potential. The most devastating malaria carriers are in a complex of six or seven superficially similar African species together called the *Anopheles gambiae* complex. Although morphologically almost identical, species in this complex differ widely in habits and habitat, with some normally associated only with wild animals and others normally associated with humans. All species can be effectively kept off your sleeping body by a good bed net (preferably one treated with a contact insecticide), something to keep in mind if you are traveling or living in malarious areas.

Culicinae Most mosquitoes belong to the subfamily Culicinae and differ from Anophelinae in having short palpi as an adult and a long, snorkel-like posterior respiratory siphon as a larva. This large, diverse group is usually subdivided into tribes, of which the most important are Aedini (*Aedes* and relatives) and Culicini (*Culex* and relatives).

Aedini The large tribe Aedini (1,200 species) includes several of the most abundant, most invasive and most medically important mosquitoes. The infamous Yellow Fever Mosquito, *Aedes aegypti*, is now widespread in the warmer parts of the world; this originally Old World tropical species probably first arrived in the New World with the slave trade — and with the yellow fever virus. Despite periodic efforts at eradication, *Aedes aegypti* still persists along the eastern seaboard of the United States, and it has recently been joined by even more invasive congeners such as the Asian Tiger Mosquito, *Aedes albopictus*, a brightly pigmented fly that, like *Ae. aegypti*, is now a major vector of dengue. Both these species are container breeders, associated with tree holes in nature but happy to develop in old tires, flower vases in cemeteries and other rainwater-trapping objects. Another Asian *Aedes* on the march, *Aedes japonicus*, is known as the Japanese Rock Pool Mosquito because of its natural association with rock pools, but outside its home range it has adapted well to tires and other artificial containers. *Aedes* females lay single eggs, usually on substrates likely to become flooded in the future; eggs hatch in response to a combination of flooding and lowering oxygen tension that signals the presence of decaying material in which they can feed. The snowmelt mosquitoes that appear in enormous numbers in the northern spring develop from eggs laid the previous year in low-lying areas, where they are triggered to hatch by the organically rich spring meltwater.

Aedes is an enormous genus traditionally subdivided into several subgenera: the Yellow Fever Mosquito and the Asian Tiger Mosquito were placed in the subgenus *Stegomyia* and the Japanese Rock Pool Mosquito in the subgenus *Ochlerotatus*. This system worked well because the well-known name *Aedes* is a convenient handle for general communication about this widely recognized group, while subgeneric names provide more information about underlying relationships. Unfortunately, this easily understood system was destabilized about ten years ago when *Aedes* was split into two genera by elevating the subgenus *Ochlerotatus* to generic rank. Some scientists followed this proposal and started to refer to many familiar *Aedes* species as *Ochlerotatus* before the wisdom and necessity of this taxonomic change had been fully considered, while others kept using the name *Aedes* in the broad sense.

These male Rock Pool Mosquitoes (*Aedes atropalpus*) are just emerging from pupae floating on a small pool of water along a rocky lake shore.

Reinert (2000) provided solid phylogenetic evidence that *Aedes* in the old sense was not a natural group (it was paraphyletic with respect to other genera of Aedini), and Reinert and others have continued to suggest new generic concepts in the Aedini based on further phylogenetic analyses. The resulting "improved" classifications led first to a change from one genus (*Aedes*) to two genera (*Ochlerotatus* and *Aedes*) and ultimately to a system of 46 genera where before we had only one. These reclassifications reflect current ideas about mosquito phylogeny but at the same time have caused confusion and impeded communication about these important insects, illustrating the potential conflict between the goals of systematics (discovery and definition of monophyletic groups) and of practical classification.

The underlying phylogenetic work that led to these changes is fine, and it is certainly desirable to reflect phylogeny in classification as much as possible, but changing ideas about phylogeny should not automatically lead to disruptive changes to important and widely used names such as *Aedes aegypti* (which would become *Stegomyia*; see also the *Drosophila melanogaster* issue, Chapter 8). So I here use the name *Aedes* in the traditional sense, both for *Ochlerotatus* and for the other subgroups of *Aedes* that some scientists now treat as separate genera. It is useful to understand the phylogenetic relationships within this large group but it is easier for us to communicate about these important mosquitoes using the familiar name *Aedes*. The extra information provided by recent phylogenetic analyses can be reflected in subgeneric or species-group names.

Aedes is very closely related to the important New World genera *Haemagogus* — including the canopy mosquitoes that move yellow fever virus from monkey to monkey — and *Psorophora*. *Psorophora* includes some of the largest biting mosquitoes, most of which lay eggs in low areas where they hatch only when flooded. Larvae of some *Psorophora* are predators, consuming other mosquito larvae.

Like many mosquitoes, species of the New World genus *Wyeomyia* usually develop in small pockets of water, called phytotelmata, held in the leaves, stems or other parts of plants.

Culicini The tribe Culicini is dominated by the ubiquitous genus *Culex*, which has 768 species, including the vectors of various viruses that cause encephalitis, filarial worms that cause elephantiasis, and even protists that cause some kinds of malaria. The malarias carried by *Culex* and other culicine genera are common, with millions of cases in Europe and North America. But don't panic — the *Culex*-borne malarias are strictly bird diseases; *Anopheles* have a monopoly on the transmission of human malarias. Our most common *Culex* bite humans only as a last resort, normally preferring to bite birds. Some diseases, such as West Nile and St. Louis encephalitis, are normally carried from bird to bird by *Culex* bites but can also be moved from bird to human, either by errant *Culex* females or by other mosquitoes (known as bridge vectors) that indiscriminately molest both humans and birds.

Eggs of *Culex*, and the few other genera of Culicini, are deposited in rafts, unlike the single eggs of *Aedes* or the single floating eggs of *Anopheles*. Look for boatlike egg rafts of domestic *Culex* species in rain barrels, ditches or puddles; other species in the genus are found in a variety of still waters. Some *Culex* and most members of the related genus *Deinocerites*, for example, breed only in crab holes, while members of the *Culex* subgenus *Micraedes* develop in bromeliads. Larval Culicini, like *Aedes* and most other Culicinae, have a long posterior respiratory siphon and generally filter-feed on detritus, although species in the subgenus *Lutzia* consume other mosquito larvae; similar predaceous habits are found in some other mosquitoes such as *Psorophora* and *Toxorhynchites*. Adults of most Culicini have a more parallel-sided and blunter abdomen than the more common and generally larger and more distinctively patterned *Aedes* species. *Culex* adults can often be found throughout the year in temperate regions because some overwinter in houses and similar shelters.

Sabethini The tribe Sabethini is a large group (423 species) especially characteristic of the New World tropics but well represented in Australia and the Oriental region. Only a few species occur in the north temperate zone, but one of these, *Wyeomyia smithii*, is interesting because it ranges north into Canada, where it develops in the pools of water used as insect traps by insectivorous Northern Pitcher Plants (*Sarracenia purpurea gibbosa*); larvae overwinter in the frozen water. Another pitcher plant mosquito, *Wyeomyia haynei*, is found in the Southern Pitcher Plant (*Sarracenia purpurea venosa*) in the southeastern United States. Pitcher plant mosquitoes feed in the soup of half-digested prey inside the plant, apparently consuming enough protein to develop into adults that can lay eggs without bothering about a blood meal.

Other members of this New World genus do bite. Some of the 137 *Wyeomyia* species are annoying biters in Neotropical rainforests, where they develop in small water pockets in tank bromeliads, broken bamboo and similar habitats. Fortunately they are not known to transmit any human diseases. Some of the most striking Neotropical mosquitoes are in the related genus *Sabethes*, a group of 39 species that are usually resplendent with shimmering metallic scales and often further adorned by broad, feathery legs with paddles made up of outstanding scales. Like other Sabethini, these are forest mosquitoes that develop in pockets of water in plant material such as broken bamboo.

The aptly named "elephant mosquitoes" are magnificent non-biting flies once treated as their own subfamily but now as the Culicinae tribe **Toxorhynchitini**. All 80 or so species of elephant mosquito are in the cosmopolitan genus *Toxorhynchites*, a group of huge mosquitoes that develop as predators of larvae of other mosquitoes and other insects in pockets of water such as in tree holes or fallen coconuts. The probosces of adult *Toxorhynchites* are bent downward almost 90 degrees; they are obviously incapable of biting, although they are often seen nectar feeding.

The tribe **Mansoniini** includes only two closely related genera, *Mansonia* and *Coquillettidia*, which share the unusual habit of breathing through the tissues of aquatic vascular plants. The larval posterior respiratory siphon is modified into a spike used to penetrate the roots of aquatic plants, allowing the larva to stay submerged for its entire development period (similar habits are found in the small Old World tropical tribe **Ficalbiini**). The tribe Mansoniini is a widespread group that includes several vicious biters, usually encountered during the evening and on into the night.

The tribe **Uranotaeniini** includes only the genus *Uranotaenia*, a group of some 265 species found in every zoogeographic region but most common in the tropics. Larvae are unusual among the Culicinae for resting almost parallel to the water surface — somewhat like *Anopheles* but with a siphon — and are found in a variety of calm waters ranging from phytotelmata through to swamps and stream margins. These delicate, often beautifully marked mosquitoes are not known to carry any human diseases.

The tribe **Orthopodomyiini** includes only the widespread genus *Orthopodomyia*, with some 35 species found mostly in the tropics. A couple of species range north to Canada, where they breed only in tree holes and are not known to bite people.

The tribe **Culisetini** is a small tribe of only 37 species, all in one genus. *Culiseta* is found in all zoogeographic regions but is absent from South America and is most characteristic of cooler latitudes or altitudes. Larvae occur in a variety of habitats, including subterranean pools, but the most common species are found in marshes and ponds. Some species are local pests, and some North American species are of importance in the spread of encephalitis viruses.

The remaining tribes of Culicinae (**Hodgesiini, Ficalbiini** and **Aedeomyiini**) are relatively minor groups, each with only one or two mostly tropical genera made up of few and infrequently encountered species.

Dixidae

Meniscus Midges or Dixid Midges

Meniscus midge adults are delicate, long-legged midges that look like small (around 5–6 mm) crane flies but for the distinctive strongly arched forked vein (R_{2+3}) at the wing tip. They don't feed and don't fly far from the waters where their larvae develop, so they are not often encountered. The larvae are unmistakable for their U-shaped resting position, relatively long antennae, prolegs under the middle of the abdomen and large posterior spiracular plate. They lurk along the margins of moving or standing water, where they can often be flushed out onto the open water surface by splashing at the edge; once flushed, they swim along the surface by rapidly flexing the body. Dixid larvae feed in much the same fashion as *Anopheles* mosquitoes, by drawing surface film through the mouthparts, where organic material is filtered out.

This small family of fewer than 200 species is sometimes divided into three subfamilies, Paradixinae, Dixinae and Meringodixinae, but Borkent (2009) questions the validity of these divisions and they are not used here. The family occurs worldwide, except in Antarctica.

Corethrellidae

Frog Midges or Frog-Biting Midges

The frog midges are so called because almost all adult females in this family feed only on frogs, and apparently only on male frogs serenading potential mates. These small to minute flies home in on the singing male frogs, usually landing nearby before approaching the host on foot. Corethrellid larvae are similar to phantom midge larvae and occur in small pockets of water, sometimes underground but most often in tree holes or

Males of the large phantom midge genus *Chaoborus* often form huge swarms of millions of individuals. This one is from east Africa, where these abundant insects are known as lake flies.

at leaf axils, where they use their antennae to grab small arthropod prey in a rapid snap-trap motion.

Frog midges were until recently treated as a subfamily of Chaoboridae but are now recognized as a distinct family of over 100 named species — all in the genus *Corethrella* — the bulk of which were recently described in a magnificent monograph on the family by Canadian entomologist Art Borkent (2008). Many if not most of the specimens studied by Borkent were collected by luring them in with recordings of singing frogs. These elusive little flies are rarely encountered by other means.

The family is widespread but mostly tropical, including an unusual species in Australia (the only species without biting mouthparts) and another one in New Zealand (the "basal" species, possibly associated with some of the most primitive frogs in the world). The Corethrellidae, Culicidae and Chaoboridae are very closely related and were at one time treated together in the Culicidae.

Chaoboridae

Phantom Midges

The phantom midges form a small but widespread family of some 55 aquatic species in half a dozen genera. Adults are delicate, mosquito-like but non-biting midges most noteworthy for swarms of fuzzy-antennaed males often seen over lakes and ponds. Lights near such bodies of water are often mobbed by masses of midges, and emerging adults sometimes occur in immense swarms that stretch over miles and are made up of millions of individuals. Some east African phantom midges ("lake flies" in the genus *Chaoborus*, locally called *nkhungu* or *kungu*) emerge in such masses that swarming flies have been gathered and cooked into edible masses called "*kungu* cakes," described by Livingstone (1865) as tasting "not unlike caviare or salted locusts." *Chaoborus edulis* larvae are planktonic, occupying the water column of large lakes, but other larval Chaoboridae occupy all kinds of still waters, where — with the exception of a filter-feeding Australian species, *Australomochlonyx nitidus* — most use huge, prehensile antennae to grab small prey.

The Chaoboridae are sometimes divided into two subfamilies, although one of them, the subfamily **Eucorethrinae**, has only a single species, the North American *Eucorethra underwoodi*. An early spring inhabitant of temporary woodland pools, it preys on other insects, including mosquito larvae and adult insects unlucky enough to fall onto the water surface above a lurking *Eucorethra* larva. These voracious larvae seem to occur in sparse populations, in marked contrast with genera of the subfamily **Chaoborinae**, which often occur in very high densities.

Some chaoborine genera, such as *Mochlonyx* and *Cryophila*, inhabit small temporary bodies of water, but many *Chaoborus* species are found in permanent ponds and lakes, where they can build up enormous populations. *Chaoborus* is the largest, most widespread and best-known genus in the family; the transparent larvae of this genus are the archetypal "phantom midges." Unlike other members of the family, *Chaoborus* larvae have a closed respiratory system, that is, no open spiracles. They are able to float, ghost-like, up and down the water column with the aid of a couple of glistening silver kidney-shaped air sacs as they seek planktonic prey to grasp with huge, hairy, prehensile antennae. Some *Chaoborus* are restricted to relatively fish-free ponds, where they can be found hanging horizontally in the water column at any time of day; others are more typical of lakes and can be found near the surface only during the night, descending to the bottom of the lake about dawn.

Psychodomorpha

Net-winged Midges, Moth Flies and Their Relatives

Both the infraorder Psychodomorpha and its main included family, the moth fly family Psychodidae, have presented significant challenges to dipterists, and concepts of these taxa remain somewhat slippery. Differing opinions about the phylogenetic affinities of the partially aquatic family Psychodidae have led to radically different concepts of the infraorder Psychodomorpha. Borkent and Wood (1989), for example, include the entirely terrestrial families Trichoceridae, Perissommatidae, Anisopodidae, Scatopsidae and Canthyloscelidae along with the Psychodidae in the infraorder Psychodomorpha. Hennig (1973), on the other hand, recognized an almost entirely different infraorder Psychodomorpha by grouping the Psychodidae with the aquatic families Deuterophlebiidae, Nymphomyiidae, Ptychopteridae, Blephariceridae, Tanyderidae and Psychodidae. The recent analysis by Bertone et al. (2008) supported the inclusion of psychodids with the aquatic Diptera, grouping them with the Tanyderidae and Blephariceridae.

Here I follow the recent concept of the Psychodomorpha as made up only of the Tanyderidae, Psychodidae and Blephariceridae, although I suspect that the limits of this group will change as our understanding of nematoceran phylogeny improves.

Blephariceridae

Net-winged or Torrent Midges

Torrent midge larvae are among the most distinctive of all aquatic insects. Each of the six distinctly separated body sections of a blephariceridlarva is equipped with a ventral suction disk, allowing adhesion to rock surfaces buffeted by torrential water in rapids or cascades — the first body section is a composite of the head, thorax and first abdominal segment, and the remaining five sections are abdominal. The suction disks are so powerful that it takes some effort to pry these flat-bottomed larvae off their substrates; there is a palpable *pop* when each disk releases its hold. The oval pupae, which are held tightly to water-washed or -splashed rocks by ventrolateral adhesion disks rather than suction disks, hatch into delicate, crane-fly-like adults with long antennae and wings that display a signature network of cracks and folds, rather like shattered safety glass. Net-winged midge adults often hang like scorpionflies from vegetation or cling to wet rock faces near the rivers and streams in which their larvae live by scraping up diatoms and other microorganisms. Adult females of many lineages are predators, grabbing mayflies and other small insects with their modified hind tarsi.

The 322 species of Blephariceridae are divided into two subfamilies, Edwardsininae and Blepharicerinae, both of which are especially diverse in the southern hemisphere, although the family occurs throughout the world. The subfamily **Edwardsininae** is restricted to the southern hemisphere, including Madagascar, southern South America and Australian. Australia and Chile are home to particularly rich net-winged midge faunas, and there are close relationships between Australian and South American species in some groups, including *Edwardsina*. The subfamily **Blepharicerinae** is more widespread, occurring on all continents and many continental islands.

Until recently most dipterists considered torrent midges to be closely related to the mountain midges, Deuterophlebiidae — another

Most moth flies (Psychodidae) are small, squat flies that hold their hairy wings horizontally, rooflike, over the abdomen, giving them the general appearance of tiny moths.

group of rheophilic flies with larvae modified to stick to rocks — and both of these families were usually placed in an infraorder called Blephariceromorpha, sometimes along with the odd family Nymphomyiidae. Recent analyses, however, suggest that torrent midges are more closely related to the moth flies, while the Deuterophlebiidae are the basal lineage of the Diptera, and thus the sister group to all other flies.

Psychodidae

Moth Flies

Moth flies are familiar to most of us only from the dark, fuzzy-winged drain flies, *Clogmia albipunctata*, that appear on bathroom walls after emerging from the slimy stuff deep in your sink's drainpipe, although this cosmopolitan species is only one of almost 3,000 species in the family Psychodidae. Drain flies — and most other moth flies — are small, squat flies that usually hold their hairy wings horizontally, rooflike, over the abdomen, giving them the general appearance of tiny, long-antennaed moths. *Clogmia* and some *Psychoda* species are common in houses, but other moth flies are more likely to be spotted as they make short, erratic flights between shaded surfaces near springs, seeps, wetlands, mud and other microhabitats that combine water with microbe-rich organic matter. Since they are normally nocturnal, these small (usually 1–4 mm), fragile flies can also be found at lights during the night. Some *Psychoda* species can reach pest densities, especially in mushroom farms where they can be significant pests.

Perhaps the highest density of moth flies anywhere can be found in the trickling filters of sewage treatment plants, where *Psychoda* larvae often occur by the millions, feeding on films of microorganisms on the filters and preventing them from becoming clogged. The semiaquatic water-film microhabitat and microbe-grazing habits of *Clogmia* and *Psychoda* larvae are typical for the family, but other Psychodidae can be found in a variety of aquatic and semiaquatic habitats. For example, larvae of *Maruina* and a few other moth fly genera use ventral friction pads — superficially similar to the suction disks of torrent midges — to cling to rocks in waterfalls and torrential mountain streams. There they join other rheophiles such as torrent midges in the Blephariceridae, grazing on the surfaces of completely or intermittently submerged stones. Adult *Maruina* have relatively narrow, fringed wings that probably resist soaking as they lay eggs along the speeding, splashing waterline of rocks emerging from torrents. Little is known about the oviposition or mating behavior of other moth flies, but males of some blood-feeding species (sand flies) occupy and defend patches of skin on host animals as they wait for potential mates to come in search of a blood meal. Other species, such as *Thornburghiella albitarsis* in North American wetlands, form dense aggregations under broad leaves near larval habitats.

The relationships of the Psychodidae to other lower Diptera remain a matter of active debate; some place them among the aquatic lineages (as in Hennig 1973) and some put the Psychodidae with the terrestrial lineages (as in Wood and Borkent 1989). Bertone et al. (2008) argue for a close relationship between Tanyderidae (see page 130) and Psychodidae.

Psychodinae The Psychodidae are divided into six subfamilies, of which by far the largest is the Psychodinae — typical moth flies — with around 2,000 species, including several common and widespread species of *Clogmia* and *Psychoda*. This subfamily also includes many more specialized genera such as the rheophilic *Maruina* (western United States and the neotropics) and similarly torrent-adapted species of *Neotelmatoscopus* (Oriental) and *Neomaruina* (Africa). Most other psychodines develop in the usual range of organically rich semiaquatic habitats, but some are more specialized, such as the Neotropical larvae recorded from the leaf axils of bromeliads and others found in nests of *Azteca* ants.

Sycoracinae Adults of most moth flies have reduced mouthparts and cryptic habits, but female adults of two subfamilies, the Sycoracinae and Phlebotominae, are equipped to feed on vertebrate blood. The 35 or so species of the widely distributed subfamily Sycoracinae probably all attack reptiles and amphibians, although only a couple of species have been observed in action. One of these, *Sycorax silacea*, bites the European edible frog, *Pelophylax esculentus*, and can transmit a filarial parasite in the process. The only other species in the subfamily with known adult habits, the Ecuadorian *Sycorax wampukrum*, is associated with harlequin frogs (*Atelopus* spp.). Larvae of European sycoracines occur in wet moss along streams; the tropical genera are unknown as larvae.

Phlebotominae Members of the subfamily Phlebotominae, a well-known group with some 700 species worldwide, are known as "sand flies," a common name also applied to Simuliidae in some parts of the world and to Ceratopogonidae in others. Sand flies are of minor importance in North America and Europe, where they are normally associated with reptiles or small burrowing mammals, but many phlebotomine species in the Old and New World tropics bite humans and present important threats to human health. Although they can occasionally be abundant and conspicuous pests, these relatively slender, delicate flies often go unnoticed because they are small, skittish, quiet and crepuscular or nocturnal. Even unnoticed, however, their bites can have dire consequences.

A previous graduate student of mine once returned from a field trip to the Amazonian rainforest with a nasty-looking open lesion on his arm that refused to heal. He was astute enough to promptly visit a tropical medicine clinic, where a doctor confirmed that the lesion was a case of cutaneous leishmaniasis, caused by a protist injected into his arm with the bite of a phlebotomine sand fly; it was promptly treated with an antimony-containing compound. He was lucky, since untreated cutaneous leishmaniasis can progress to a leprosy-like disease that can cause serious tissue damage. The related visceral leishmaniasis is a much-feared and frequently fatal disease in both the New and Old World tropics. The sand flies responsible for leishmaniases (and several other diseases) are different in the Old and New World tropics: many species of *Lutzomyia* cause misery in the jungles and mountain valleys of the neotropics, and members of the large genus *Phlebotomus* bite vertebrates in southern Europe, Africa and Asia. I usually avoid insect repellents because of their dismaying effect on my camera equipment, but when I see sand flies I lather on the repellent before doing any nocturnal insect watching.

Larval sand flies, as the name suggests, can develop in drier areas than most other Psychodidae. They do, however, remain dependent on semiaquatic microhabitats such as deep soil moisture, condensation in cave cracks and crevasses, forest litter, animal burrows and artificial microhabitats in wells, stables and other moist, dark places rich in organic material. Several species of the Neotropical genus *Lutzomyia* breed in leaves on the forest floor, while the Old World genus *Phlebotomus* is often associated with drier conditions such as rodent burrows or farmyards.

Similarly terrestrial larval habits are found in the small subfamily **Bruchomyiinae**. It includes the genera *Nemopalpus* and *Bruchomyia*, which are sometimes considered the most "primitive" or basal moth flies and sometimes considered to be close relatives of the sand flies. Although adults in this small subfamily are superficially similar to Phlebotominae, they do not feed on blood. Some authors combine the Bruchomyiinae and Phlebotominae into a single group, sometimes treating them as a separate family (Phlebotomidae).

Even more "terrestrial" larvae are found in the subfamily **Trichomyiinae**. This small group has only a single widely distributed genus, *Trichomyia*, which is characterized by smooth, elongate larvae equipped with powerful mandibles for tunneling in rotten wood.

The subfamily **Horaiellinae** is a small group of aquatic Psychodidae comprising only the genus *Horaiella*, a rarely collected genus of four species associated with swift streams in northern India, Thailand, Nepal and China. At least some larvae of *Horaiella* are adapted for life in running water, reminiscent of *Maruina* and some other rheophilic Psychodinae.

Tanyderidae

Primitive Crane Flies

The primitive crane flies, as the name suggests, look just like large to medium-sized crane flies but for a few "primitive" characteristics, including wing venation in which the radial vein has the full complement of five branches (one or more branches are lost in most extant flies). This so-called primitive appearance is matched to a corresponding relict distribution, divided between temperate areas of the northern and southern hemispheres, where known larvae of this small family (about 40 species) live in flowing water. Tanyderids occur in southern Africa, Australia and New Zealand as well as both North and South America. They can be common in southern Chile and New Zealand, where larvae of some species are found in submerged wood or along gravelly stream bottoms, but they are relatively infrequently encountered elsewhere. A few species occur in the western United States; the only eastern North American species, *Protoplasa fitchii*, is an uncommon species that burrows in the gravelly bottoms of riffles and runs in small rivers. Adult tanyderids, often distinctive for their prominently patterned or spotted wings, are most likely to be seen hanging under bridges or from streamside vegetation. Like many other lower Diptera, they also show up at artificial lights.

Recent molecular analyses suggest a close relationship between Tanyderidae and Psychodidae (Bertone et al. 2008), but most dipterists (for example, Woodley et al. 2009; Wood and Borkent 1989) have followed Hennig (1973) in treating the Tanyderidae and Ptychopteridae as close relatives.

Ptychopteridae

Phantom or Fold-winged Crane Flies

The family Ptychopteridae is a small family of about 75 crane-fly-like species. Some of them earn the common name "phantom crane flies" because of their disruptive coloration, which gives them a ghost-like virtual invisibility as they float along with their black and white legs outstretched. In the common North American genus *Bittacomorpha* the first tarsomeres are inflated and filled with tracheae to aid the floating flight of these truly "phantom" crane flies. Not all ptychopterids are disruptively colored, however; many look like robust crane flies or large, slender fungus gnats. Larvae, which have long, telescoping tails, or "snorkels," for breathing at the water's surface, live as microbial grazers in seeps and mucky springs.

The phantom crane flies are divided into two subfamilies. **Ptychopterinae** are all treated as the large, widespread genus *Ptychoptera*, found almost everywhere except Australia and the neotropics. The subfamily **Bittacomorphinae** contains the small genera *Bittacomorpha*, with two Nearctic species, and *Bittacomorphella*, with four Nearctic and three eastern Asian species.

The family Ptychopteridae is widely considered one of the most "primitive" lineages of Diptera, but its exact relationships remain uncertain. Recent molecular studies suggest that it might be a basal lineage of the Psychodomorpha, but a few authors treat the Ptychopteridae as a separate infraorder, Ptychopteromorpha, sometimes along with the Tanyderidae.

Phantom crane flies in the genus *Bittacomorpha* are distinctive for their inflated tarsomeres and disruptively banded legs, held outstretched during their characteristic floating flight.

Aquatic Lower Diptera Families of Uncertain Infraordinal Placement

Deuterophlebiidae, Nymphomyiidae and Axymyiidae

The last three aquatic lower Diptera families, discussed below, are difficult to place in any of the major infraorders of lower Diptera. Two of these problematic families, Deuterophlebiidae and Nymphomyiidae, appear to be the basal lineages of all Diptera. The third, Axymyiidae, is simply an oddball group that does not fit neatly into the big infraorders. Some dipterists now treat each of these small, enigmatic lineages as its own infraorder.

Deuterophlebiidae

Mountain Midges

The 14 species in the essentially Holarctic family Deuterophlebiidae are known as mountain midges because of their usual association with rushing mountain rivers, where their remarkable flattened larvae use massive eversible lateral (splayed out to the sides), crochet-tipped abdominal prolegs to cling to water-buffeted rock surfaces. Pupae occur on the same slick stone substrates, usually in depressions on the rocks. Adults, which emerge very early in the morning to form short-lived swarms over rushing waters, are delicate little (less than 4 mm) midges with strikingly broad wings that usually display a silvery sheen. Males are spectacular for their enormous antennae, which extend up to five times the length of the body, while females have very short antennae. Both sexes have small heads with reduced mouthparts, in keeping with a brief adult life — less than a couple of hours for males — that does not involve feeding. Females go underwater to lay eggs, apparently shedding their wings along the way, but details of mating and oviposition are scarce for these infrequently observed flies.

Larvae of Deuterophlebiidae cling to rocks in torrential rivers by using circlets of hooks (crochets) on lateral arm-like body extensions (abdominal prolegs).

Deuterophlebia, the only genus in the family, occurs in Japan, Korea, central and eastern Asia, western North America and the Himalayas.

The phylogenetic placement of this unusual family remains contentious, although most recent authors have argued for a close relationship between Deuterophlebiidae, Blephariceridae and Nymphomyiidae, often treating them together as the infraorder Blepharicerimorpha (as in Woodley et al. 2009). This relationship seems intuitive because of the shared adaptations found in the similarly rheophilic Blephariceridae and Deuterophlebiidae — they even occur on the same rocks — and because of morphological similarities such as the long prolegs of Nymphomyiidae and Deuterophlebiidae. Recently, however, Bertone et al. (2008) made a strong case for interpreting these similarities as convergences, and for treating the Deuterophlebiidae as the sister group to all other Diptera.

Nymphomyiidae

Nymphomyiid Flies

Species in the small (seven species) Holarctic family Nymphomyiidae are associated with cold springs in which moss-covered rocks provide appropriate habitat for these rarely seen and truly bizarre little flies. The tiny (2 mm), elongate adults have clubbed antennae, odd eyes that wrap around the underside of the head, and thin, fringed, veinless wings that are shed soon after emergence. Deciduous wings are found elsewhere in the Diptera (in the Carnidae) and even more morphologically outlandish wingless adults occur in other groups, such as the termitophilous Phoridae, but the associated life history of nymphomyiids is the most incredible in the order. The winged adults form copulating pairs soon after emergence, which in some species — at least in the Japanese species *Nymphomyia alba* — is associated with huge crepuscular swarms. Pairs of mated adults then crawl into the water, where they end up as coupled wingless bodies surrounded by eggs on submerged mossy rocks. The eggs hatch into larvae that are almost as odd-looking as the adults, with eight long abdominal "prolegs" that impart an unmistakable habitus.

Although this small family has in the past been divided into three genera, Courtney (1994) thoroughly assessed the phylogeny and biology of the family and treated all seven species in the single genus *Nymphomyia*. Three species occur in northeastern Asia, one lives in the Himalayas, one is known only from Hong Kong and two are from eastern North America. Unsurprisingly, given the extremely specialized morphology of these little flies, the phylogenetic affinities of the Nymphomyiidae remain contentious. Popular points of view are that the family is related to the Culicomorpha, is close to the Blephariceridae or is the sister group to the rest of the Diptera — in other words, the most "primitive" fly lineage. Recent molecular phylogenies treat the Nymphomyiidae as the sister group to all flies other than Deuterophlebiidae.

Axymyiidae

Axymyiid Flies

The Axymyiidae is a small family of half a dozen north temperate species that develop only in partially submerged wet, rotting wood. Although the secretive, bibionid-like adults are rarely seen, the larvae can be located by looking for fairly firm, bark-free logs in continuous contact with the water or wet mud of small permanent woodland streams. Curls of wet sawdust on the wood surface indicate their presence, but you need to break apart the waterlogged wood to reveal the fat, big-headed and massive-jawed white larvae, each linked to the surface by a long, hard-tipped retractile respiratory siphon. Larvae are thought to feed on microorganisms in the wood for about two years before reversing direction and burrowing out to the wood surface to pupate and emerge in early spring. The relationships of the Axymyiidae remain a matter of debate; some dipterists place this enigmatic little family in its own infraorder, while others include it in the Bibionomorpha.

Terrestrial Lower Diptera

Bibionomorpha

March Flies, Fungus Gnats, Wood Gnats, Gall Midges and Their Relatives

The infraorder Bibionomorpha is a large, nebulously defined assemblage here treated simply as a group that includes almost all the terrestrial lower Diptera families, that is, all the entirely terrestrial lower Diptera families except Trichoceridae and possibly the Perissommatidae. There is good evidence that these terrestrial fly families and the higher Brachycera together make up a natural group — the Neodiptera of Michelsen (1996) — but there is some dispute over whether the Bibionomorpha is itself a "real" group or merely the basal lineages of the Neodiptera. Either way, I find it convenient to divide the lower Diptera into the basal aquatic or mostly aquatic lineages on one hand and the terrestrial Bibionomorpha on the other.

Bibionidae

March Flies and Lovebugs

Bibionidae, often called March flies (a name also applied to the very different family Tabanidae in Australia), are strikingly sexually dimorphic flies with big-eyed males and small-eyed females. These sluggish flies usually have relatively short antennae compared to most lower Diptera, and usually have swollen front legs armed with apical spurs or big spines. Adults are common on flowers and can be significant pollinators of fruit trees and some other plants, especially early and late in the season (thus the common name "March flies"). Although larvae of a few pest species feed on the roots of living plants, most bibionid larvae are saprophagous or phytosaprophagous, and some develop in huge numbers in decaying vegetable matter such as grassland thatch. Lovebugs (*Plecia nearctica*), for example, emerge in such numbers from cattle pastures in the southern United States that these small (7–8 mm) orange and black flies are considered significant pests during spring and late summer emergence periods. Even though adult Bibionidae are innocuous non-biting insects, the sheer number of fornicating flies fouling car windshields, pitting paint jobs and clogging up radiators renders Lovebugs a well-known Bible Belt nuisance. Marathon copulations in which the male serves as a mating plug, protecting his paternity, are reflected in other common names such as "honeymoon flies" and "double-headed bugs." Male *Plecia* are sometimes still hanging in there when the female is depositing her eggs on the soil surface. Most bibionids other than *Plecia* dig a chamber with their enlarged foretibiae, lay a cluster of eggs and then die within the chamber.

Bibionidae occur worldwide across a range of habitats from sea level to high *páramo*. The family is divided into either three or four subfamilies — depending on whether *Penthetria* is treated in its own subfamily — but five of the eight genera and most of the 750 or so bibionid species are in the subfamily Bibioninae. Most **Bibioninae**, including many common species, are in the genera *Bibio* and *Dilophus*; the former is a mostly temperate genus and the latter a widespread group characterized by an apical circlet of spines on the foretibia. The subfamily **Pleciinae** includes only the Lovebug genus *Plecia*, but this worldwide taxon is the largest genus of Bibionidae, with 252 mostly tropical species. The subfamily **Penthetriinae** includes only the genus *Penthetria*, a widespread genus of 32 species, most of which occur only in

Male Lovebugs (*Plecia nearctica*), like males of other Bibionidae, have relatively large heads made up mostly of eyes.

the Oriental region. The subfamily **Hesperininae**, comprising only the small Holarctic genus *Hesperinus*, is considered to be the "basal lineage" of the family; it looks quite different from other bibionids because of its long antennae and relatively delicate form. Known larvae of Hesperininae bore in rotting wood. Pleciinae, Penthetriinae and Hesperininae are treated as separate families (i.e., Pleciidae, Penthetriidae and Hesperinidae) by some authors.

Pachyneuridae

Pachyneurid Gnats

The small Holarctic family Pachyneuridae, which contains only five extant and two fossil species, is made up of elongate, long-legged flies that usually look a bit like crane flies. Larvae develop in rotting wood riddled with fungal mycelia, including tree stumps and fallen trees. The only North American species, *Cramptonomyia spenceri*, develops in alder twigs along the west coast from Oregon to British Columbia, emerging as adults in late winter to very early spring. Other members of the family are Palaearctic; they include the Siberian genus *Pergratospes*, the Japanese genus *Haruka*, and the more widespread European and Asian genus *Pachyneura*.

Pachyneurid gnats fall into two distinct lineages, leading some dipterists to divide this small group unnecessarily into two families — Pachyneuridae and Cramptonomyiidae — rather than treating the lineages as subfamilies. The subfamily **Pachyneurinae** has only one genus, *Pachyneura*, with one species in the Russian Far East and another that ranges across the Palaearctic region. The subfamily **Cramptonomyiinae** includes the other three genera mentioned above.

The Pachyneuridae are closely related to the Bibionidae, and these two families are probably sister groups.

Anisopodidae

Wood or Window Gnats

The Anisopodidae or wood gnats are familiar to anyone with a home composter full of the kind of wet fermenting material favored by the common wood gnat genus *Sylvicola*. These attractive little flies, which have pictured wings and the dimensions of a robust mosquito, lay gelatinous masses of eggs in the compost, where the saprophagous larvae develop. Most Anisopodidae are more likely to be found in sappy decaying wood or slime fluxes on wounded trees than in domestic compost. At least one species is recorded from living plants, some occur in sewage treatment plants, some members of the ubiquitous genus *Sylvicola* occur in dead bodies, and a few others have been reared from bromeliads and other plants that retain mixtures of fluid and decomposing organic material. *Sylvicola fenestralis*, the most common domestic wood gnat, is even implicated as a myiasis-causing species, as it occasionally shows up in the urogenital tract or lower intestines of humans. Adult *Sylvicola* often show up on the insides of windows, which is why they are sometimes called "window gnats."

This small family has a worldwide distribution but only about 150 species in seven genera, organized by different authors into two or three subfamilies that are recognized as separate families by some. The subfamily **Anisopodinae**, the "typical" wood gnats, contains the familiar genus *Sylvicola* and a variable number of other genera, of which two — *Olbiogaster* and *Lobogaster* — are sometimes carved off as the separate subfamily **Olbiogastrinae** (or even the separate family Olbiogastridae). The subfamily **Mycetobiinae** includes *Mycetobia* and a couple of related genera associated with sap flows on wounded trees.

The relationships of the Anisopodidae remain in question but the family is sometimes considered to be the sister group (the most closely related lineage) to the higher Diptera or Brachycera. I here follow Hennig (1973) and Bertone et al. (2008) in treating the Anisopodidae as part of the Bibionomorpha, and the whole infraorder Bibionomorpha as the sister group to the Brachycera (Figure 6.1, page 108).

Scatopsidae

Black Scavenger Flies

Black scavenger flies are small (usually 1–3 mm), compact flies with relatively short antennae, somewhat like black flies (Simuliidae) but without the humped back and the biting mouthparts. Although this is a significant group, with about 325 named (and at least as many unnamed) species, scatopsids are poorly known. They are familiar to most entomologists only from a couple of ubiquitous saprophagous species, *Scatopse notata* and *Coboldia fuscipes*, that inhabit various kinds of decaying material. The cosmopolitan species *Scatopse notata* in particular can be common in home composters and other kinds of vegetable waste, but most scatopsid species seem to be associated with smaller pockets of wet decay such as tree rot holes or phytotelmata. Larval *Rhexoza*, for example, often occur by the thousands in and under the wet, fermenting bark of cut poplars or other recently dead deciduous trees, and larval *Akorhexoza* are associated with decaying cacti from Central America to the southwestern United States. Scatopsids can also be abundant flower visitors, and *Swammerdamella* adults are sometimes seen in the flowers of wetland plants such as Bog Bean (*Menyanthes trifoliata*). At least

The most frequently encountered Anisopodidae belong to the common wood gnat genus *Sylvicola*. Members of this large genus occur worldwide.

one genus of scatopsid, *Colobostema*, seems to be associated with ants.

The Scatopsidae are divided into four subfamilies, with 30 of the 33 genera occurring in the large "typical" subfamily **Scatopsinae**, which includes the familiar cosmopolitan *Scatopse* species as well as the widely distributed *Swammerdamella* and *Rhexoza* mentioned above. Some scatopsines have interesting intercontinental distributions. *Ferneiella*, for example, is Palaearctic and Australian, *Reichertella* is Holarctic and Australian, and *Holoplagia* occurs in regions other than the Nearctic and Afrotropical. The small subfamily **Apistinae** is an entirely Holarctic subfamily of only two genera, *Apistes* and *Arthria*. The biology of this distinctive group remains unknown beyond the observation that females use their characteristic spinelike foretibiae to burrow in the sand, sometimes dragging along an attached male (see photograph 5 on page 176). The subfamily **Psectrosciarinae** also includes only a couple of genera, the widespread *Psectrosciara* and *Anapausis*, a mostly Palaearctic genus with a south temperate subgroup, a subgroup in Central America, an Afrotropical and Oriental subgroup and a Holarctic subgroup. The subfamily **Ectaetiinae** comprises only the Holarctic and Neotropical genus *Ectaetia*. Known *Ectaetia* larvae develop in decaying wood, sometimes under the bark of twigs or small branches of deciduous trees.

The family Scatopsidae is closely related to the small families Canthyloscelidae and Valeseguyidae.

Canthyloscelidae

Canthyloscelid Gnats

Flies in the small family Canthyloscelidae, which has only 16 species, look like members of the closely related Scatopsidae except for the greater number of antennomeres (10 to 14) and palpal segments (four). These rarely seen flies have a scattered or "relict" distribution, with species in the Holarctic region, southern South America and New Zealand. Most Canthyloscelidae are north temperate in distribution, but *Canthyloscelis* is a southern hemisphere genus with a disjunct distribution, including different subgenera in New Zealand and southern South America. Larvae develop in wet decaying wood and have a close association with ancient forests.

The species currently included in the Canthyloscelidae are sometimes treated as two families (or subfamilies), Synneuridae (or **Synneurinae**, including *Synneuron* and *Exiliscelis*) and Canthyloscelidae (or **Canthyloscelinae**, including *Hyperoscelis* and *Canthyloscelis*). Amorim (2000), however, showed that this classification did not correspond with phylogeny: *Synneuron* and *Exiliscelis* are not each other's closest relatives. He also provided support for the idea that the Canthyloscelidae and Scatopsidae are closely related and together form a single lineage.

Valeseguyidae

Valeseguyid Gnats

Valeseguyidae is one of the smallest fly families, especially rare and unusual in that all but one of the known specimens of the family are fossils of extinct species. The only extant species in the family, *Valeseguya rieki*, is known only from a single male specimen collected in southern Australia, originally described by Colless (1990) as an unusual member of the Anisopodidae subfamily Mycetobiinae. A fossil species from Caribbean amber was later described in the same genus (Grimaldi 1991). In 2006 a related fossil species from Burmese amber was discovered and described as another new species in a new genus, *Cretoseguya*, by Amorim and Grimaldi; they erected the new family Valeseguyidae to hold their new genus (with one species) along with *Valeseguya* (with two species).

Amorim and Grimaldi (2006) also made the case for a close relationship between Valeseguyidae, Canthyloscelidae and Scatopsidae, demonstrating that the latter two families form the sister group to the former. Together these three families are now treated as the superfamily Scatopsoidea, although they could as easily be treated as subfamilies of a monophyletic family Scatopsidae.

Perissommatidae

Perissommatid Gnats

The five species in the small family Perissommatidae are rarely collected because adults appear only in the winter, when most insects (and most insect collectors!) are in diapause, and because they appear in only a few high forests in a few widely separated localities in Australia (four species) and Chile (one species). Larvae occur in fungi and moist leaf litter; the adults are small (less than 2 mm), long-winged flies with distinctively divided eyes (that is, they have four eyes). They are apparently active only in the colder months, when swarms of at least one Australian species have been observed in shafts of light above the foliage.

Although these odd little flies are usually treated as part of the Bibionomorpha, there is no consensus regarding their exact relationships. Wiegmann et al. (2011) treat perissommatid gnats as the sister group to the Bibionomorpha plus the Brachycera — in other words, as the sister group to the Neodiptera. Unsurprisingly, some dipterists now treat the perissommatids as their own infraorder, Perissommatomorpha.

Cecidomyiidae

Gall Midges

The gall midge family is an enormous group of tiny flies with about 6,000 named species, and with so many more still awaiting discovery that it qualifies as the least-known and perhaps even the largest of all fly families. It is no wonder this immense group remains as much as 90 percent unknown, because most adult gall midges are tiny, short-lived, delicate flies that require special care to collect, preserve and study. Furthermore, even a well-preserved adult gall midge specimen can be impossible to identify without additional information such as the host and (in the case of phytophagous species) the type of damage caused or type of gall induced. And not all gall midges induce galls. Despite their common name, Cecidomyiidae show a range of habits and habitats commensurate with the size of the group. The ancestors of the family were almost certainly saprophagous, and so we see that the basal lineages (the subfamilies Lestremiinae and Porricondylinae) are generally saprophagous (or fungivorous), developing in fungi or wet pockets of decay such as compost, the sappy bark of dead trees, and rotting wood. One South American species, *Termitomastus leptoproctus*, even occupies the nests of termites.

The plant-feeding habit is characteristic of the largest subfamily, Cecidomyiinae, and the habit of inducing host plants to create a fly-specific structure — the gall — in which the larva lives and feeds is characteristic of a subset of the Cecidomyiinae. The subfamily Cecidomyiinae also includes a few parasitoids and many predators, some of which are sold as biological control agents. The bright orange predaceous larvae of *Aphidoletes*, for example, are familiar sights in aphid colonies both in greenhouses and in the wild. Other predaceous gall midge larvae feed on mites, aphids, coccoids and other arthropods, while larvae of several species are internal parasitoids of aphids and psyllids. Other larvae in the family, especially the *Miastor* species found under bark, are well known for reproducing without developing to the adult stage (paedogenesis). In contrast to the adults, which often exist for only a few hours, cecidomyiid larvae are often long-lived, with lifespans of one to three years. They are also generally distinctive, with the head capsule reduced and a sclerotized "sternal spatula" under the prothorax.

The family Cecidomyiidae is divided into four or five subfamilies; most are large, widespread groups, but one is the obscure and rarely encountered subfamily **Catotrichinae**, comprising only eight extant species in Australia and the Holarctic region. The vast majority of cecidomyiid species belong in the mostly phytophagous subfamily **Cecidomyiinae**. This diverse subfamily includes mycophagous, predaceous and parasitic subgroups, but most cecidomyiines feed on specific plant hosts and, with the exception of secondarily widespread pest species such as the Hessian Fly (*Mayetiola destructor*), most have restricted distributions that are limited by the distributions of their host plants. Saprophagous gall midges in the other subfamilies tend to have wider ranges, and genera in the Lestremiinae and Porricondylinae are often widespread.

Larvae of the most primitive, or basal, group in the Cecidomyiidae, the subfamily **Lestremiinae**, are mycophagous or saprophagous, developing in decaying vegetation, rotting wood, plant wounds

Most adult gall midges (Cecidomyiidae) are tiny, short-lived, delicate flies that require special care to collect, preserve and study; as a result, the great majority of species remain undescribed or unidentifiable.

or fungi. The subfamily Lestremiinae was recently (Jaschhof and Jaschhof 2009) divided into two subfamilies, with the subfamily **Micromyinae** being created for most (47) of the genera previously placed in the Lestremiinae. Members of this group are mycophagous and are found in mushrooms and rotting logs. Larvae in the subfamily **Porricondylinae** have similar habits, usually occurring in soil and decaying wood, where they feed on fungal mycelia and bacterial films. One of the best-known members of this subfamily is the previously mentioned genus *Miastor*.

Although the derived subfamily Cecidomyiinae has exploded into the most important lineage of phytophagous Diptera, the ground plan feeding habit of the family Cecidomyiidae is saprophagy or mycophagy, as in the closely related fungus gnat lineages. Most dipterists agree that the gall midges belong phylogenetically with the fungus gnat group of families ("Mycetophiliformia" or Mycetophilidae *sensu lato*; see below). They treat the gall midges plus the fungus gnats as the superfamily Sciaroidea, but there is no consensus regarding which group of fungus gnats is most closely related to the gall midges. One possibility, supported by a phylogeny published by Amorim and Rindal (2007), is that the family Cecidomyiidae is the sister group (the most closely related lineage) to the fungus gnat group, Mycetophilidae *sensu lato*. Wiegmann et al. (2011) treat the Cecidomyiidae as a lineage arising in the middle of the fungus gnat group, as a close relative of the Sciaridae and Lygistorrhinidae. Although gall midges as a whole are poorly known and difficult to identify, the plant-feeding species of North America are treated in a thorough and accessible book by Raymond Gagne (1989).

Fungus Gnats

The fungus gnats, traditionally treated as the two families Sciaridae (dark-winged fungus gnats) and Mycetophilidae (fungus gnats), comprise a diverse and abundant group of some 8,000 species of generally small, somewhat humpbacked flies most often associated with fungal tissue, spores or hyphae. Dipterists have long suspected that the Mycetophilidae, as defined in the old and broad sense (*sensu lato*), was not a natural group and that some "mycetophilids" were more closely related to the Sciaridae than to other Mycetophilidae. Fungus gnat experts have opted to solve this problem by splitting the former Mycetophilidae (Mycetophilidae *sensu lato*) into a more narrowly defined Mycetophilidae (Mycetophilidae *sensu stricto*) and several small, superficially similar families, while at the same time continuing to refer to all fungus gnats as the Mycetophilidae *sensu lato*. The Mycetophilidae *sensu lato* are divided below into multiple families according to current standard practice, with the largest of these families retaining the name Mycetophilidae.

Fungus gnat taxonomy is a very active field because several outstanding dipterists currently specialize in the group, and because collections of these common insects have grown exponentially with the growing use of mass trapping techniques such as pan traps, flight intercept traps and Malaise traps. For example, when Kjaerandsen and Jordal (2007) ran one Malaise trap and one window trap in a Norwegian forest for a year, they were able to extract a total of 23,000 specimens and 315 species of fungus gnats from just those two traps.

Mycetophilidae

The family Mycetophilidae *sensu stricto*, a world-wide group with more than 4,000 named species and at least double that number awaiting formal description, is the most common and diverse group of fungus gnats. Mycetophilids are dominant insects in moist forests and also abound in wetlands, caves and other habitats, where their larvae usually feed on mycelia, spores or fruiting bodies of fungi. Many species are specific to particular genera or families of fungi, especially fleshy "mushrooms," although some species develop in bracket fungi and a few are associated with Ascomycetes and Myxomycetes. Larvae in several groups construct sticky silken webs under bark or fungi, and in some cases these larvae have moved from mycophagy to predation, feeding on other insects rather than fungal spores. A very few species in the group are reported as developing on mosses and liverworts, and larvae of cave-dwelling *Speolepta* species graze on mineral precipitates and associated microorganisms on the cave walls and ceilings. Perhaps the most unexpected habit in the group is found among Australian *Planarivora* larvae, which are parasitoids of land planarians.

Pupae of mycetophilids are sometimes enclosed in dense cocoons, as in many mycetophilines, and are sometimes loosely supported in simpler silken structures, as in the *Speolepta* that pupate dangling from the ends of silken threads attached to cave ceilings. Adult mycetophilids are easily found in damp, dark places ranging from rock overhangs to mammal burrows, and they are occasionally encountered in aggregations of thousands of flies under stream banks or in similar humid refuges. Some can be found in numbers overwintering under bark. Several species are common flower visitors and may be significant pollinators; *Gnoriste*, for example, is a genus of strikingly long-beaked fungus gnats that regularly occur on flowers.

The Mycetophilidae are divided into different subfamilies by different dipterists. Here I follow the relatively simple system used by Vockeroth (2009) and divide the family Mycetophilidae into just two subfamilies, Sciophilinae and Mycetophilinae. The "typical" subfamily **Mycetophilinae** is best represented in the Holarctic region and is dominated by the widely distributed genus *Mycetophila* (around 700 species, in all zoogeographic regions). The more heterogeneous subfamily **Sciophilinae** also includes some large, widespread genera, such as *Sciophila* (around 150 species, in every region but the Australasian), *Mycomya* (around 400 species, in all zoogeographic regions) and *Leia* (around 150 species, in every region), as well as several genera found in the Australasian region and South America and a few groups found only in the Holarctic region. Some dipterists divide the Sciophilinae into three or more subfamilies (including Gnoristinae, Sciophilinae and Leiinae), but a phylogenetic analysis by Rindal et al. (2009) indicated that these are not natural groups. A further group, made up of the small cosmopolitan genus *Manota*

Fungus gnats, such as this *Leia*, have a characteristic habitus with long coxae and long leg bristles. Flies in the fungus gnat family Mycetophilidae are among the most common flies in forested areas, especially in temperate regions.

and three small Oriental genera, is sometimes treated as a subfamily of Mycetophilidae, sometimes as a family (Manotidae) and sometimes as part of the Sciophilinae.

Ditomyiidae

The family Ditomyiidae is a small (about 100 species) lineage of fungus gnats best represented in the Australasian region and South America; three genera are restricted to South America and two are split between the Australasian region and South America. About a quarter of the world's Ditomyiidae are endemic to New Zealand. The family is not known from Africa, only three species occur in the Oriental region, and only two genera — *Ditomyia* and *Symmerus* — occur in Europe and North America. Most known ditomyiid larvae (two species of *Ditomyia*) are associated with bracket fungi (*Polyporus*). Some *Symmerus* species have been reared from elm logs but this may not be a typical habit for the family, since recent analyses suggest that *Symmerus* is misplaced in the fungus gnat group and is more closely related to the Pachyneuridae and Bibionidae (Bertone et al. 2008).

Bolitophilidae

The family Bolitophilidae is a small group of slender, long-legged fungus gnats found almost entirely in the Holarctic region, where their mycophagous larvae are usually associated with specific fungus hosts. Like most fungus gnats, bolitophilids are associated with cool, moist habitats such as stream margins in forests. Adults are most likely to be found in spring and fall. The only extant genus in the family, *Bolitophila*, is mostly Palaearctic, although about 20 of the 64 species in the genus are Nearctic, and a couple of species recently discovered in the mountains of Taiwan extend the known range of *Bolitophila* to the Oriental region. The Bolitophilidae are closely related to the Diadocidiidae and Keroplatidae.

Lygistorrhinidae

The family Lygistorrhinidae is a small group of about 35 extant (and several extinct) species of long-beaked fungus gnats with distinctively reduced wing venation. Most species are restricted to tropical or subtropical areas, especially the Old World tropics, and are rarely encountered by naturalists or general collectors. Most of the species in the family have been discovered and described in the past 25 years, and many of the recently described species were taken in Malaise traps. While the growing use of these extraordinarily efficient insect traps has certainly contributed to fungus gnat taxonomy, trapped specimens tell us little about biology. Larvae and larval habitats of the Lygistorrhinidae remain unknown.

Most recent works on fungus gnat phylogeny have treated the Lygistorrhinidae and the Mycetophilidae *sensu stricto* as very closely related, but the recent phylogeny of Wiegmann et al. (2011) posits the Cecidomyiidae and Sciaridae as the closest relatives to the Lygistorrhinidae.

Diadocidiidae

The family Diadocidiidae is a small group of yellowish or brown fungus gnats currently comprising 24 species, with 9 in the Palaearctic region and the balance scattered across the Nearctic, Oriental, Australasian and Neotropical regions. Known larvae of these rarely encountered flies are associated with decaying wood, where they develop on or in encrusting fungi, feeding from the shelter of a slimy silk tube on the fungus and ultimately spinning a silken cocoon. The family includes only a single genus, *Diadocidia*, although some dipterists include the enigmatic fungus gnat genus *Freemanomyia* (one Chilean species) in the Diadocidiidae. *Heterotricha* has also been treated with the Diadocidiidae in the past but is now in the Rangomaramidae (Amorim and Rindal 2007). *Diadocidia* (and thus the Diadocidiidae) is closely related to the Keroplatidae; the two groups are sometimes treated as sister families and sometimes considered to be part of a larger group that includes the Diadocidiidae, Keroplatidae, Ditomyiidae and Bolitophilidae.

Rangomaramidae

The family Rangomaramidae was recently described (by Jaschhof and Didham 2002) to hold the unusual New Zealand genus *Rangomarama*. It was later expanded to include *Heterotricha* and related genera from Nepal, Africa, Madagascar and South America, as well as *Ohakunea* and related genera from South America and the Australasian region (Amorim and Rindal 2007). Jaschof (2009) treats *Ohakunea* and related genera as "unassigned to any family-level taxon currently recognized in the Sciaroidea," and it seems likely that this group too will eventually be elevated to family rank. Amorim and Rindal (2009) treat Rangomaramidae (including the *Ohakunea* group) as the sister group to the Mycetophilidae and Lygistorrhinidae, but Jaschof (2009) suggests that the *Ohakunea* group is perhaps closer to the Sciaridae. Nothing is known of the biology of these obscure little flies, and none of the genera of Rangomaramidae, including the *Ohakunea* group, are known as larvae. Like most fungus gnats, adults are found in cool, moist areas rich in rotting wood.

Keroplatidae

The family Keroplatidae is a large lineage that includes around 800 species of fungus gnats. European researchers have long considered the Keroplatidae a separate family, but until recently most North American entomologists treated the group as a subfamily of Mycetophilidae. Although some species are mycophagous, many keroplatids have predaceous larvae that kill their prey by using oxalic acid, produced in their salivary glands and deposited as droplets on a web. Larvae of many species feed on fungal spores trapped in webs or sticky silk tubes under bracket fungi and other fungi, but keroplatids don't normally develop inside mushrooms as do many mycetophilids. Keroplatid tubes or webs can also be found under bark, in worm burrows, in moss, around grass roots and in caves.

Keroplatidae are divided into three subfamilies: Arachnocampinae, Keroplatinae and Macrocerinae.

Arachnocampinae Species of the subfamily Arachnocampinae occur only in Australia

Some Keroplatinae, like this Canadian *Keroplatus militaris*, develop as larvae in mucous tubes on fungi or under bark, while larvae of other species make webs to trap fungal spores or small animals.

and New Zealand, where their bioluminescent larvae are known as "glowworms." Glowworms are predaceous larvae that trap prey — usually small flies attracted to the bioluminescence — in sticky threads dangled like fishing lines beneath them and periodically reeled in to consume the captured items. Best known for the famous cave-inhabiting populations of New Zealand Glowworm, *Arachnocampa luminosa*, this subfamily includes nine species (five of which were newly described in 2010) and is common in caves and in shady moss banks in eastern Australia and New Zealand. The genus *Arachnocampa* is very distinct from other Keroplatidae, and it might be closely related to the genus *Diadocidia* (currently in its own family, Diadocidiidae).

Keroplatinae The subfamily Keroplatinae, which is characterized by distinctively compressed (flattened) antennae, includes most of the genera and species in the Keroplatidae. The subfamily Keroplatinae as a whole is cosmopolitan, as are some genera (*Keroplatus* and *Heteropterna*), but most of the genera have more limited distributions. This subfamily includes both mycophagous and predaceous species; known larvae of one lineage — *Orfelia* and related genera — are predators, and some, including the southeast North American *Orfelia fultoni* (known as "dismalites"), are bioluminescent. Unlike larvae of *Arachnocampa*, which use a single blue-green bioluminescent organ to attract prey to a sticky dangling thread, larvae of *Orfelia* use paired sets of front and back brilliant blue bioluminescent organs to lure insects into a sticky web. Larvae of related species in the tribe Orfeliini prey preferentially on ants, and some live in the domatia that plants produce specifically to house colonies of ants. Fungus gnats have been reported living in ant-inhabited domatia in Africa, Central America and Sri Lanka, where these myrmecophilous keroplatine larvae use silk to capture live ants.

Macrocerinae The subfamily Macrocerinae, characterized by distinctively elongate antennae (especially in the males), includes eight genera, but about 185 of the 215 or so species in the subfamily are in the cosmopolitan genus *Macrocera*. *Macrocera* species are variously reported in the literature as mycophagous (Rindal et al. 2008), predaceous (Matile 1989) or facultatively predaceous (Madwar 1935); it seems likely that they feed both on fungal spores and on small invertebrates captured in their sticky silk. Madwar describes *Macrocera* larvae — found living in webs of saliva under the bark of damp logs near London — as resembling small earthworms, each gliding "with greatest ease and even with rapidity over the even surface of fungoid growth covering the bark. In doing so, it is assisted by the fine threads which it emits so as to bridge over these inequalities. It moves forwards and backwards with equal facility. Sometimes it turns by reversing its head and gliding along its side." The larvae of cave-inhabiting *Macrocera* are known to North American cavers as "monorail worms" because of the way they glide smoothly up and down their silken tracks.

Dark-winged fungus gnats (Sciaridae) often form frenetic mating aggregations, like this one in Costa Rica.

Sciaridae

Dark-winged or Black Fungus Gnats

Flies in the very common family Sciaridae are generally small, dark midges characterized by eyes that meet on top of the head to form a narrow bridge between the ocelli and the antennae. Most of the 2,400 or so named species develop in soil, but sciarids also occur in a variety of other habitats, including plant tissue and rotting wood. The most familiar dark-winged fungus gnats are undoubtedly the cosmopolitan *Bradysia* species that emerge from potting soil as greenhouse and household pests. *Lycoriella* species are also particularly common; two species are even established at Antarctic research stations, and one species, *Lycoriella mali*, is a major pest in mushroom houses. Larvae of other sciarids can be found in aggregations under bark, and some form migrating masses that move along the forest floor in spectacular conveyer-belt-like columns. The so-called "armyworm" species that show this odd behavior — *Sciara militaris* and some similar *Sciara* species — reportedly die without pupating following their group processions. To my knowledge nobody has provided an adequate explanation for the phenomenon, but analogous mass maggot movements in the higher flies are usually just shifts to optimal sites for pupation.

Larvae of another sciarid species (*Cratyna perniciosa*) pupate in groups of three to five (always including both males and females) in common cocoons; the pupae yield immobile adults that mate and oviposit in the cocoon (Steffan 1975). Some other sciarids are parthenogenic and bypass pupae and adults altogether.

Adult sciarids are generally weak-flying insects found near the larval substrates, but a few visit flowers to feed on nectar. Several sciarid species have short-winged or entirely wingless females.

Sciaridae are poorly known, especially for regions other than the Palaearctic, and thousands of species still await description. The sister group (the most closely related lineage) to the Sciaridae is either all the rest of the fungus gnats, as argued by Amorim and Rindal (2007), or a subgroup of the Mycetophilidae *sensu lato*, as suggested by most recent analyses.

A Photographic Guide to the Lower Diptera

FAMILY TIPULIDAE (TRUE CRANE FLIES) Subfamily Cylindrotominae ❶ Although the crane fly subfamily Cylindrotominae is a relatively uncommon group, ***Liogma nodicornis*** can be abundant in eastern North American wetlands, where their larvae occur among mosses. The other half-dozen species in this genus are Palaearctic. **Subfamily Limoniinae** ❷ ❸ Larvae of the common aquatic genus ***Antocha*** feed on detritus from the shelter of tubes. This female is ovipositing on a marl-covered rock in an eastern Canadian stream. ❹ ❺ Some authors now treat *Chionea* (along with some fully winged normal-looking crane flies) as a separate subfamily, Chioneinae. These ***Chionea valga*** (a female and a male) are walking on the snow in eastern Canada, but many other wingless species of this genus are found throughout the Holarctic region. ❻ This ***Conosia angustissima*** is hanging out among marsh grasses in Tanzania. Other members of the same genus occur throughout the Old World tropics and the Australasian region. ❼ This ***Dactylolabis montana*** is laying eggs in an alga-covered seep on a Canadian cliff. The small larvae surrounding her are probably of the same species. Congeners occur throughout the Holarctic region. ❽ ***Epiphragma fasciapenne*** is a widespread and common species in North America, where it is typical of moist woodlands in early June. Larvae of this large, widely distributed genus occur in rotting wood. ❾ Larvae of the aptly named green crane fly ***Erioptera chlorophylla*** develop in damp soil in eastern North America. This enormous, widespread genus is divided into several subgenera; this species is in the subgenus *Erioptera*. Some *Erioptera* pupae breathe by tapping aquatic plants with sharp processes surrounding their thoracic spiracles.

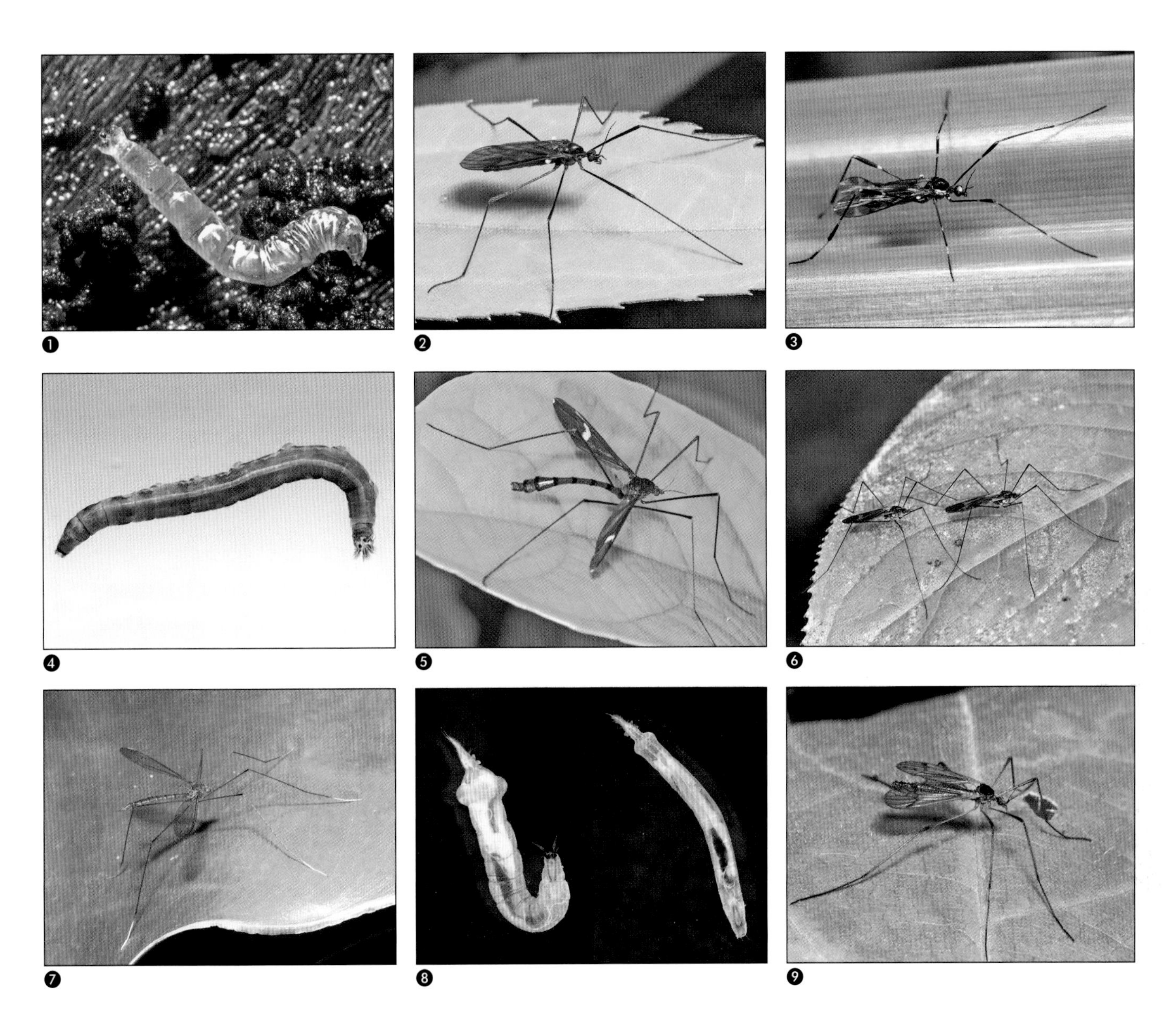

FAMILY TIPULIDAE (continued) ❶ ❷ ***Gnophomyia tristissima***, here seen in both larva and adult forms, is one of the most common subcortical (under-bark) crane flies in eastern North America. Other members of this large, widespread genus have a variety of habits. None of the 91 Neotropical *Gnophomyia* species are known as larvae, but an Indonesian species has been reared from decaying flower bracts and a New Zealand species lives in organic material in epiphytes. ❸ ***Gymnastes*** is a mostly Old World tropical genus. This one is from Tanzania. ❹ The head capsule of this aquatic crane fly larva (***Helius*** spp.) is partly retracted into the body, but the horizontally opposed mandibles are clearly visible. Four long lobes, each fringed with water-repellent hairs, surround the posterior spiracular disk. *Helius* species occupy diverse aquatic and semiaquatic habitats, and some are known only from caves. ❺ ❻ ❼ ***Hexatoma*** is a large, widespread genus. Shown here are species from Vietnam (black and white), Ecuador (two dark flies) and Costa Rica (orange). ❽ Known ***Hexatoma*** larvae are predators in the sandy or gravelly bottoms of streams or stream margins. Part of the abdomen can be expanded into a spherical anchor to enable locomotion through the substrate. ❾ Larvae of the large, widespread genus ***Limnophila***, like this ***L. (Prionolabis) rufibasis*** from Canada, are predators in wet substrates along the margins of wetlands and streams.

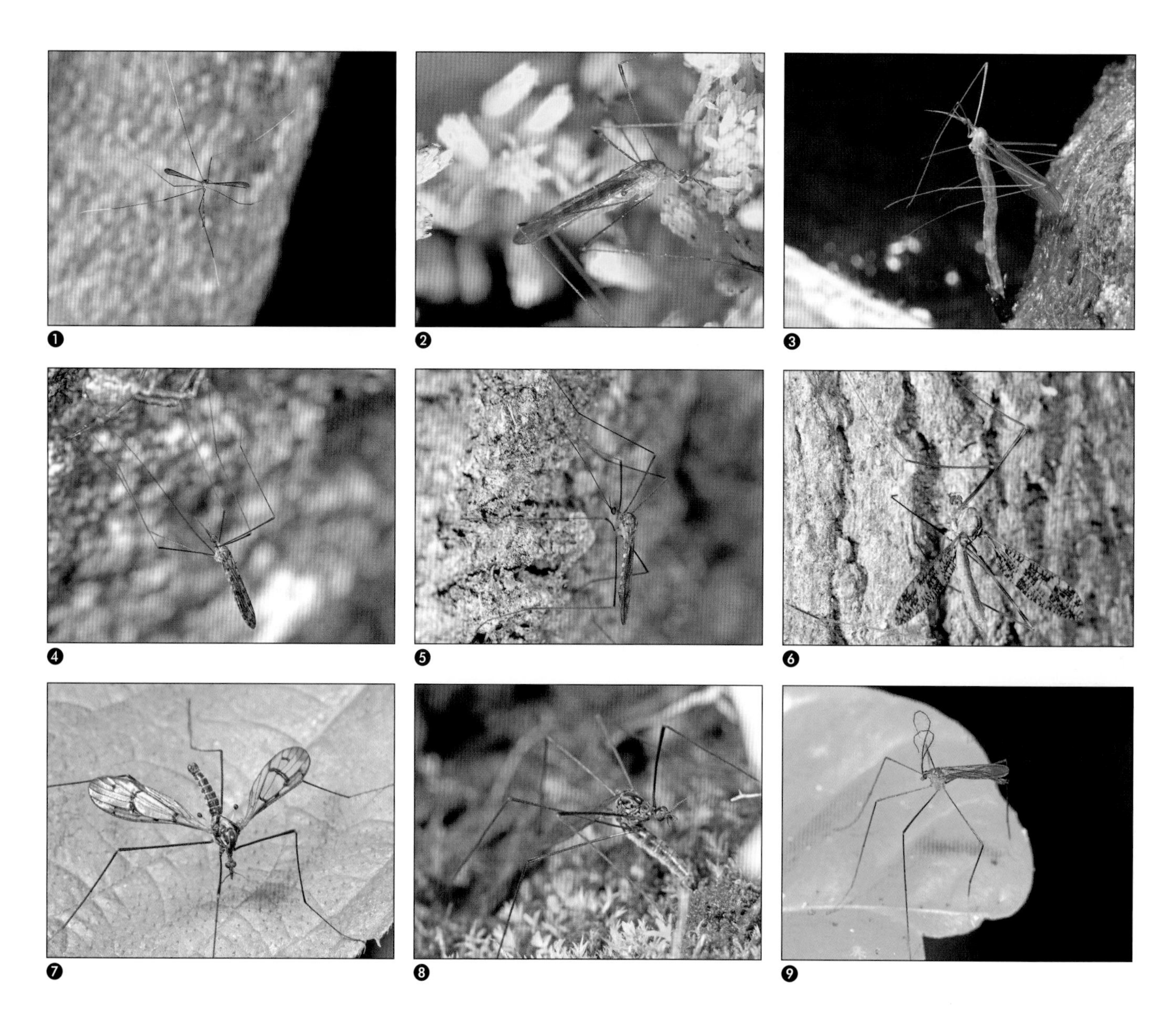

FAMILY TIPULIDAE (continued) ❶ Like many Nematocera, some Limoniinae hang out in spider webs. ***Limonia*** is an enormous, widespread genus divided into several subgenera; this is a subgenus ***Dicranomyia*** species from Africa. *Dicranomyia* and other subgenera of *Limonia* are sometimes treated as separate genera, and *Dicronomyia* is in turn sometimes divided into multiple subgenera. ❷ ❸ Most larvae of the mosquito-like *Limonia* subgenus ***Geranomyia*** (sometimes treated as a seperate genus) are aquatic and feed from the shelter of silken tubes. One of these North American flies is spattered with pollen; the other is emerging from its pupa along a rock in the middle of a small river. ❹ ❺ The mosquito-like mouthparts of the small crane flies in the widespread subgenus ***Geranomyia*** are used for taking nectar from flowers, but these east African species are on a tree trunk and a cliff. Known larvae in this group feed on algae and sometimes live in gelatinous tubes on wet surfaces. ❻ This ***Limonia (Neolimnobia)*** is on a tree trunk in Ecuador. ❼ ❽ These ***Limonia (Peripheroptera)***, a male and an ovipositing female, were seen at around 3,000 meters elevation in Ecuador. *Peripheroptera* is sometimes treated as a subgenus of *Dicranomyia*. ❾ ***Polymera***, like this Bolivian species, often stand out for their unusually long antennae.

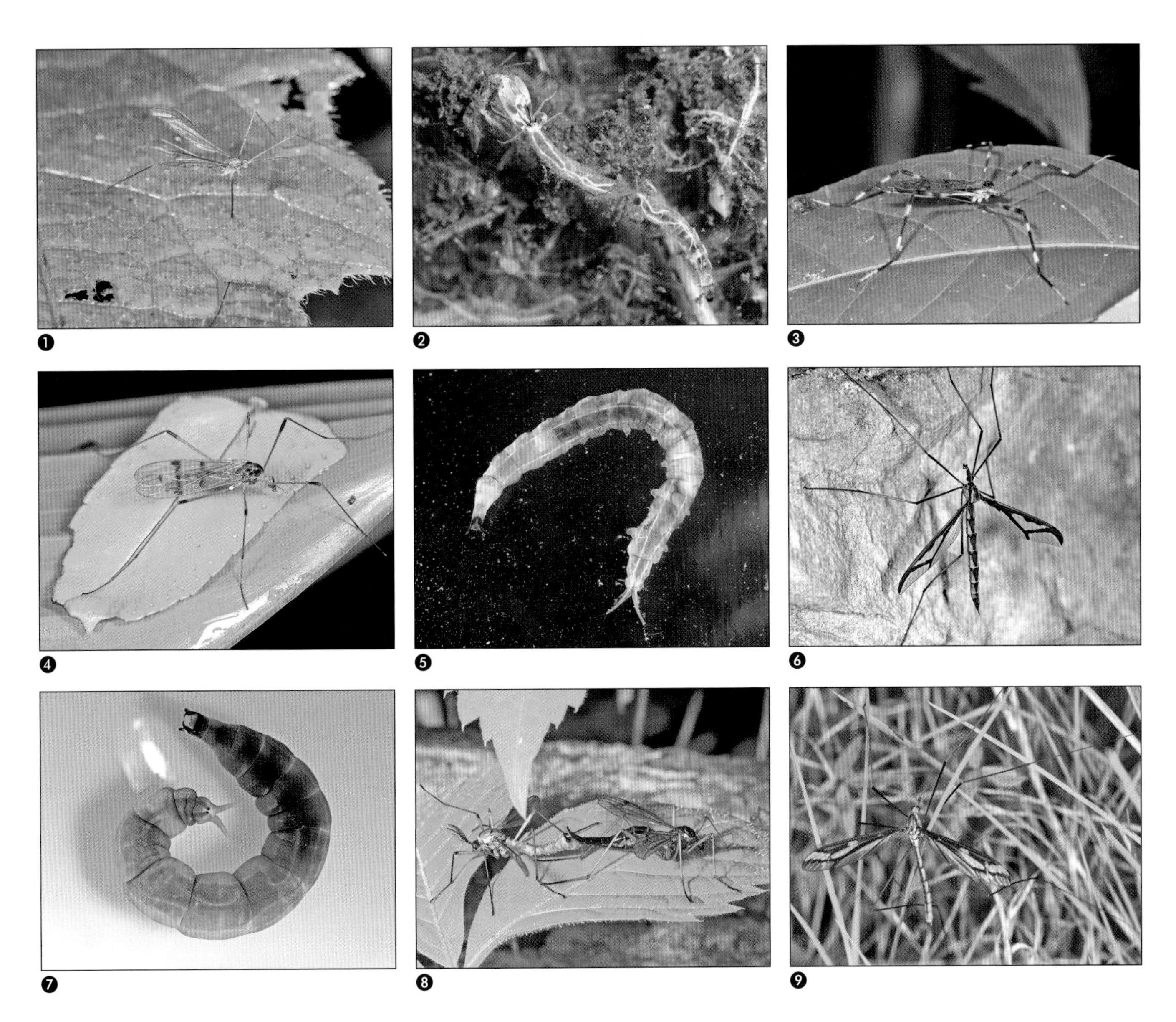

FAMILY TIPULIDAE (continued) ❶ This Costa Rican ***Polymera*** male has strikingly curved antennae. Known larvae of this mostly New World genus are predators of other invertebrates in mud. ❷ Larvae of ***Pseudolimnophila***, which develop on detritus in swamps and ponds, have long-haired ventral lobes on the posterior spiracular disk. This is a North American species, but the genus is mostly Old World. ❸ Larvae of the New World genus ***Sigmatomera***, like this one from Ecuador, are found in tree holes, where they prey on mosquito larvae and other organisms. ❹ This distinctively polished Costa Rican crane fly belongs to the large genus ***Teucholabis***. Most species are unknown as larvae, but one Nearctic species has been reared from under bark. **Subfamily Pediciinae** ❺ The predaceous larvae of ***Dicranota*** are common stream insects, characteristic for their prominent prolegs and the long ventral lobes on the posterior spiracular disk. *Dicranota* is a large, widespread genus. ❻ ❼ The large and boldly patterned adults of the Holarctic-Oriental genus ***Pedicia*** are often conspicuous streamside insects; larvae are predators in streams. This is the Nearctic ***P. albivittata***. **Subfamily Tipulinae** ❽ The black female of this eastern North American pair of ***Ctenophora dorsalis*** has a long, blade-like ovipositor (seen here projecting over the orange male's genital capsule), which is used to insert eggs into fallen trees, often through holes left by wood-boring beetles. This is the only North American species in the subgenus ***Tanyptera***. ❾ ***Clytocosmus*** is a small genus (half a dozen species) of crane flies found only in Australia.

FAMILY TIPULIDAE (continued) ❶ ❷ Members of the large, widespread genus ***Dolichopeza*** often hang out in crowds in shaded areas such as overhanging banks. Those shown here are in a northern Canadian outbuilding. ❸ ❹ Like most genera of this very old family, the genus ***Megistera*** has a broad distribution. Shown here is an adult ***M. filipes*** awkwardly resting on a leaf in Africa, and a larval ***M. longipennis*** pulled out of a hollow stem from which it was feeding on floating algae in a southeastern U.S. pond. ❺ ❻ ***Nephrotoma*** is a huge worldwide genus, with more than 500 species. Shown here are an ovipositing female of an African species and a copulating pair of a Canadian species. ❼ This aquatic ***Tipula*** larva has its posterior spiracles surrounded by six lobes fringed with water-repellent hairs, here suspending the larva from the water surface. ❽ Crane fly larvae, like this aquatic ***Tipula***, can withdraw the head capsule into the body like a turtle pulling its head into its shell; this one has its head completely withdrawn. The long, pale lobes below the posterior spiracular disk are called anal gills, although they serve for osmoregulation rather than respiration. ❾ With 40 subgenera and around 2,500 named species, *Tipula* is one of the largest genera in the animal kingdom. These ***Tipula (Yamatotipula) caloptera*** are massing on leaves trailing in the larval habitat, a small, clean stream in eastern Canada.

FAMILY TIPULIDAE (continued) ❶ This elegantly long legged ***Tipula (Indotipula)*** is resting under a leaf in the Philippines. Larvae of this genus develop in mud or semiaquatic habitats. ❷ ❸ ***Tipula (Tipula) paludosa*** (shown here mating) and ***T. oleracea*** (a single female shown here) are two similar and closely related European *Tipula* species recently introduced from Europe to North America, where they are now serious pests that feed on the roots of grasses. ❹ This attractive North American species, ***T. nobilis***, is placed in the subgenus ***Nobilotipula***. ❺ The *Tipula* subgenus ***Eumicrotipula*** is a characteristic Neotropical group, with 260 or so species found from Mexico south to Chile. The *Eumicrotipula* shown here are clustering over a small stream in central Chile, but larval habits remain unknown for the entire subgenus. ❻ These brightly colored Vietnamese crane flies belong to the Oriental *Tipula* subgenus ***Formotipula***. ❼ This crane fly female laying eggs in moss along one of the North American Great Lakes is in the *Tipula* subgenus ***Savtshenkia***. Crane flies are often seen in this position as they insert their eggs using a long, pointed abdominal tip. **FAMILY TRICHOCERIDAE (WINTER CRANE FLIES)** ❽ ❾ Winter crane fly adults look much like delicate Tipulidae but differ in having ocelli. They occur in both north and south temperate regions, where they typically fly on cold days in late fall and early spring. Larvae usually develop in decomposing fungi or plant material, and have been reared from a range of substrates, including penguin droppings. These ***Trichocera*** larvae were among a dozen in a decomposing bracket fungus in eastern Canada.

FAMILY CHIRONOMIDAE (MIDGES) Subfamily Diamesinae ❶ ❷ ***Diamesa***, like this female and this fuzzy-antennaed male walking on snow during a Canadian winter, is the most common winter-active midge genus in north temperate countries. **Subfamily Tanypodinae ❸** Larvae of midges in the subfamily Tanypodinae have relatively long antennae that can be retracted into sockets in the head capsule. **❹** Like other Tanypodinae, this ***Ablabesmyia*** pupa is free-living and has earlike respiratory horns on its thorax. The distinctively pigmented wings of the developing adult are visible through the wing sheaths. **❺** ***Ablabesmyia*** is a common genus of distinctively colored midges found in a wide variety of aquatic habitats throughout the world. This one is floating on a rock pool on the Canadian shore of Lake Huron. A Neotropical *Ablabesmyia* was currently described as living in freshwater sponges. **❻** This male ***Coelotanypus scapularis*** was one of many swarming along the Ontario shores of the Great Lakes, where it probably developed as a predator of oligochaetes and other invertebrates in the lake's deep, muddy bottom. **Subfamily Orthocladiinae ❼** The distinctively black-and-white-winged species of ***Chasmatonotus*** don't form swarms, and males lack the plumose antennae of most other midges. This is a mostly Nearctic genus with one Japanese species; larvae are unknown. **❽** ***Clunio*** is a genus of marine midges with winged males that spend their very brief lives between tides in a frenetic search for wingless females. **❾** This close-up of an exposed mussel bed in southern Chile shows two ***Clunio*** winged males, both trying to mate with a wingless female (barely visible under the in-focus male).

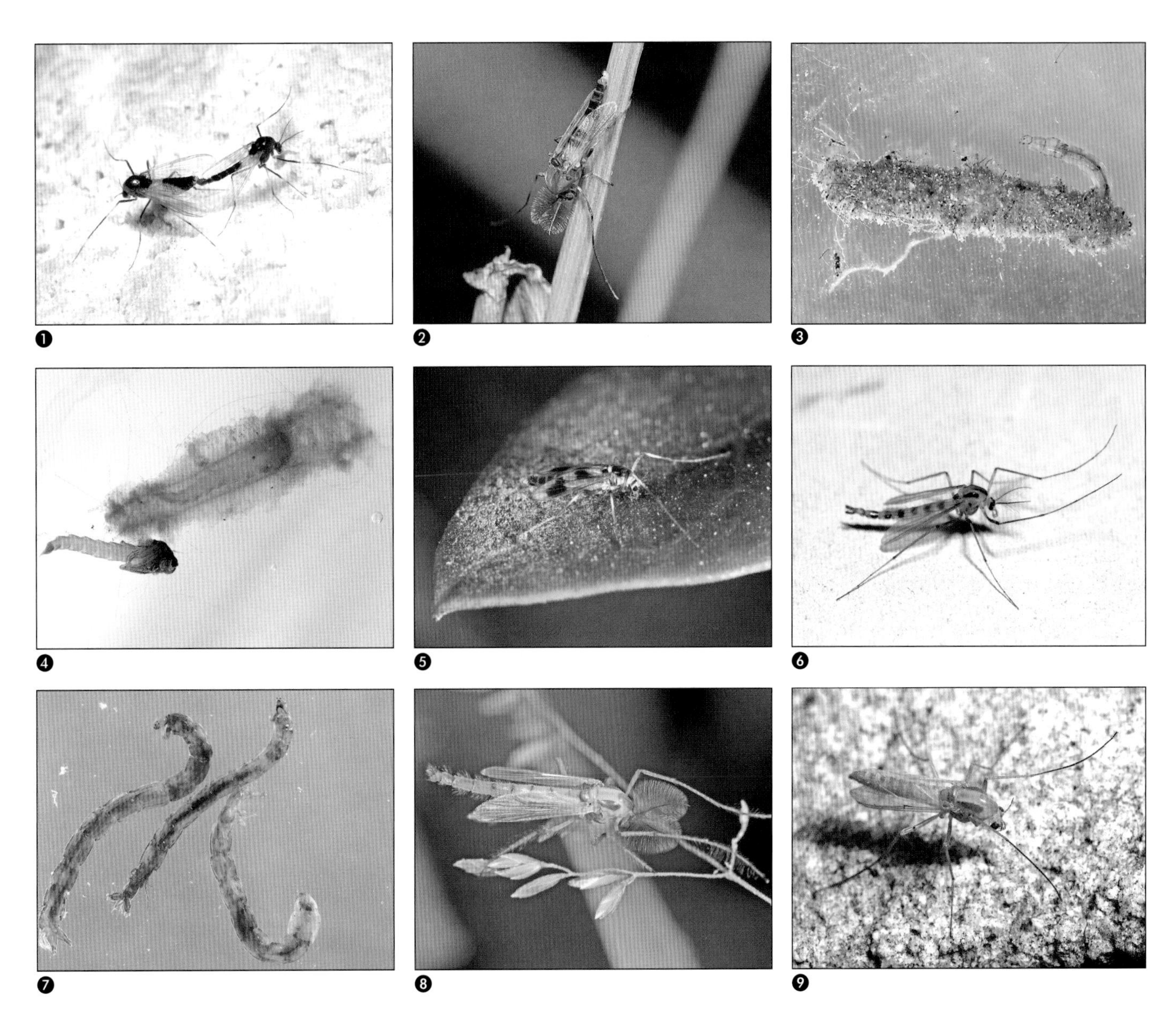

FAMILY CHIRONOMIDAE (continued) ❶ This pair of small ***Cricotopus*** shows typical sexual dimorphism, with the male's antennae relatively large and plumose. *Cricotopus* is one of the largest and most widely distributed midge genera. ❷ *Cricotopus* species, like this male ***C. trifasciatus***, are very common midges found in a wide variety of aquatic habitats. ❸ ❹ Most Chironomidae larvae make tubes or shelters from salivary secretions (silk), often with the addition of sand or sediment from the larval environment. ❺ ***Stictocladius***, like this Chilean female, is one of the few midge genera known only from the southern hemisphere. ❻ This eastern North American ***Xylotopus par*** is one of only two species in the genus; the other occurs in Burma. Larvae mine in submerged rotting wood. **Subfamily Chironominae** ❼ Larvae of ***Chironomus*** and related Chironominae are known as bloodworms because of the red haemoglobins they use for oxygen uptake. ❽ This chironomine male shows the long first tarsomere typical of the subfamily, the feathery antennae typical of male midges, and the longitudinally grooved postscutellum that is a good diagnostic character for most Chironomidae. ❾ This chironomine female shows the relatively short, simple antennae typical of female midges.

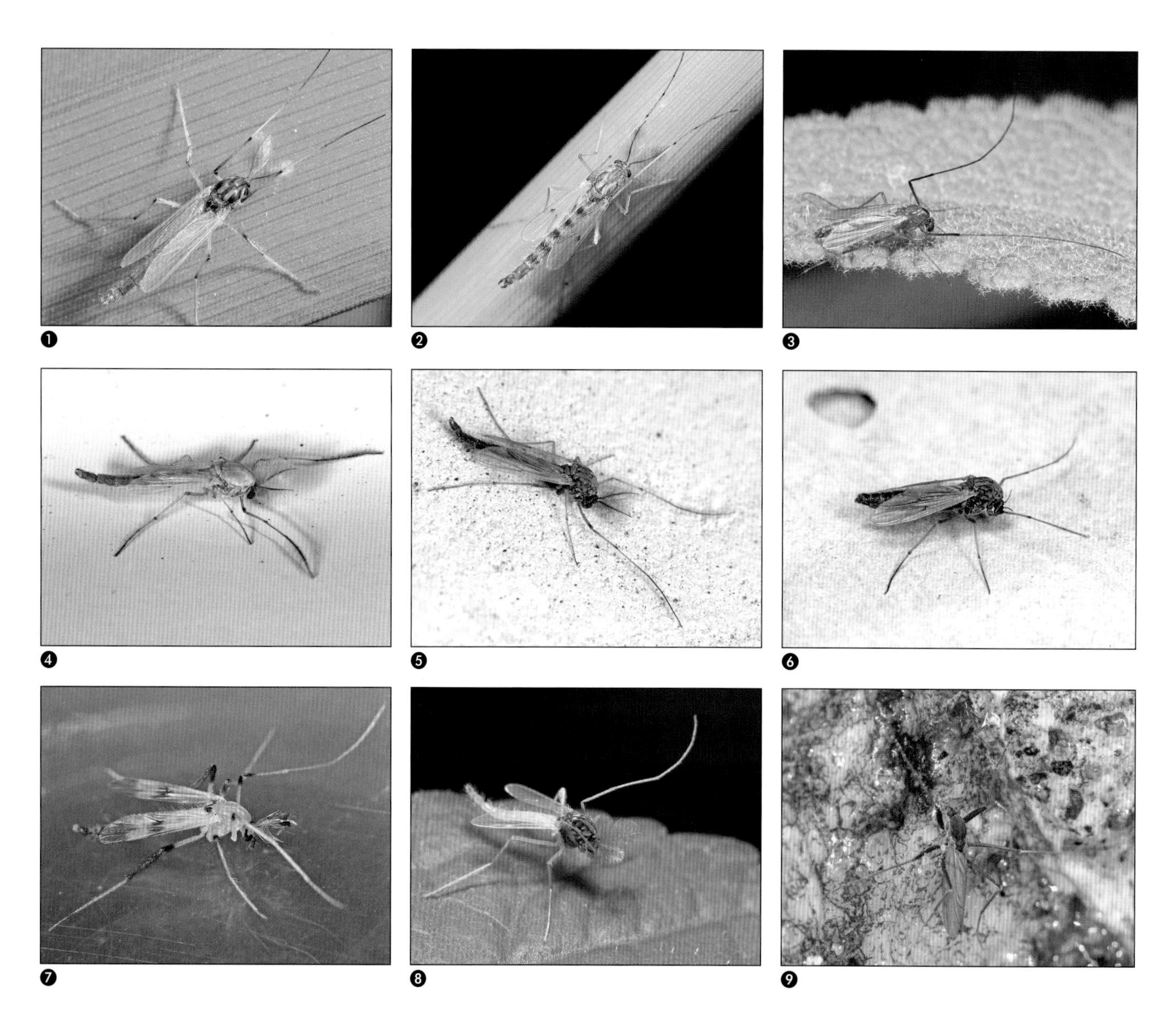

FAMILY CHIRONOMIDAE (continued) ❶ ❷ Males of the large genus ***Chironomus*** often form huge swarms; these were among thousands swarming and resting on the foliage near an eastern Canadian lake. Members of this cosmopolitan (except Antarctica) genus occur in a wide variety of standing waters, including polluted waters. ❸ This Tanzanian ***Dicrotendipes nigrolineatus*** has a very long first tarsomere, typical of the Chironominae, and simple antennae, typical of most female midges. Members of this widely distributed genus develop in standing waters. ❹ This male Tanzanian ***Kiefferulus brevibucca*** probably developed in the sediments in a nearby ditch, a typical habitat for members of this widespread genus. ❺ ❻ This pair of ***Paratendipes*** illustrates the typical sexual dimorphism of the family, with the male's antennae feathery and the female's antennae simple. Both these flies are resting under a concrete bridge over a Canadian stream. ❼ This colorful male ***Polypedilum*** is one of around 450 described species in this worldwide genus. ❽ Larvae of the worldwide genus ***Stictochironomus*** mine in submerged wood and leaves. **Subfamily Telmatogetoninae** ❾ Although some Pacific species of ***Telmatogeton*** occupy freshwater habitats, Holarctic species such as this western North American ***T. alaskensis*** are entirely intertidal and can withstand regular immersion in salt water.

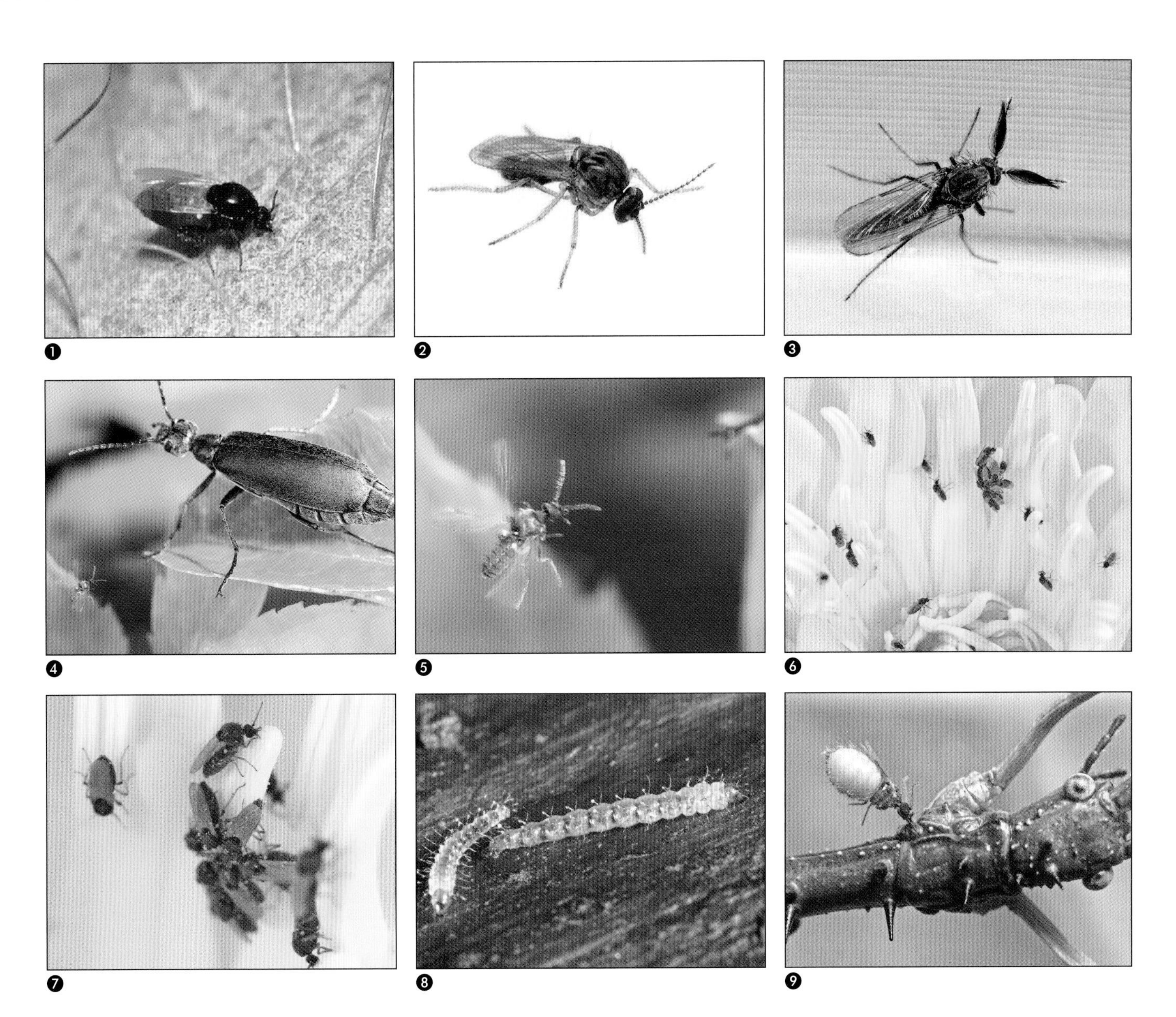

FAMILY CERATOPOGONIDAE (BITING MIDGES) Subfamily Leptoconopinae ❶ This tiny ***Leptoconops*** is biting my knee on a Western Australian beach. Larvae of this widely distributed genus develop in wet sand. **Subfamily Dasyheleinae** ❷ This tiny female ***Dasyhelea*** was one of several reared from a wet tree hole in a southern Ontario hardwood forest. ❸ This male ***Dasyhelea*** was one of many attracted to the lights at a Costa Rican mountain lodge. **Subfamily Forcipomyiinae** ❹ ❺ This common blister beetle (*Epicauta* spp.) is being molested by a female fly in the specialized ***Atrichopogon*** subgenus ***Meloehelea***. Blister beetles are often mobbed by these little biting midges. ❻ ❼ Ceratopogonidae are important pollinators of several plants and many, like these Canadian ***Atrichopogon***, can be found in large numbers as they "prey" on pollen grains in water-lily flowers. ❽ Larvae in the subfamily Forcipomyiinae, like these under the bark of a fallen pine tree, look quite different from other biting midge larvae because of their many body processes or tubercles and their well-developed prolegs. The long hairs secrete droplets of sticky fluid, probably as a defense against ants. ❾ Neotropical stick insects (Phasmatodea) are often parasitized by biting midges in the ***Forcipomyia*** subgenus ***Microhelea***, locally known as "stick ticks." This one is stuck into the foreleg base of an Ecuadorian stick insect. Stick ticks are of course not ticks, nor do they feed only on stick insects; some also bite katydids.

FAMILY CERATOPOGONIDAE (continued) ❶ ❷ ❸ ❹ These female biting midges engorging themselves on the body and antenna of Ecuadorian stick insects (*Pseudophasma* spp.) are "stick ticks" (***Forcipomyia*** subgenus ***Microhelea***). ❺ Biting midges in the genus ***Forcipomyia*** are commonly seen imbibing blood (haemolymph) from the wing veins of dragonflies or damselflies, as seen here. ❻ ❼ ***Forcipomyia*** species commonly bite the wing veins of green lacewings (Neuroptera). ❽ The wing veins of this southwestern American Pygmy Blue (*Brephidium exile*) are yielding enough blood to feed a couple of adult female ***Forcipomyia***. ❾ This South American coreid bug is being bugged by a small ***Forcipomyia*** biting through its scutellum. Male and female adults of many *Forcipomyia* species are common flower visitors, and some are significant pollinators of economically important plants. Larvae of this large genus occur in a variety of terrestrial, semiaquatic and aquatic habitats.

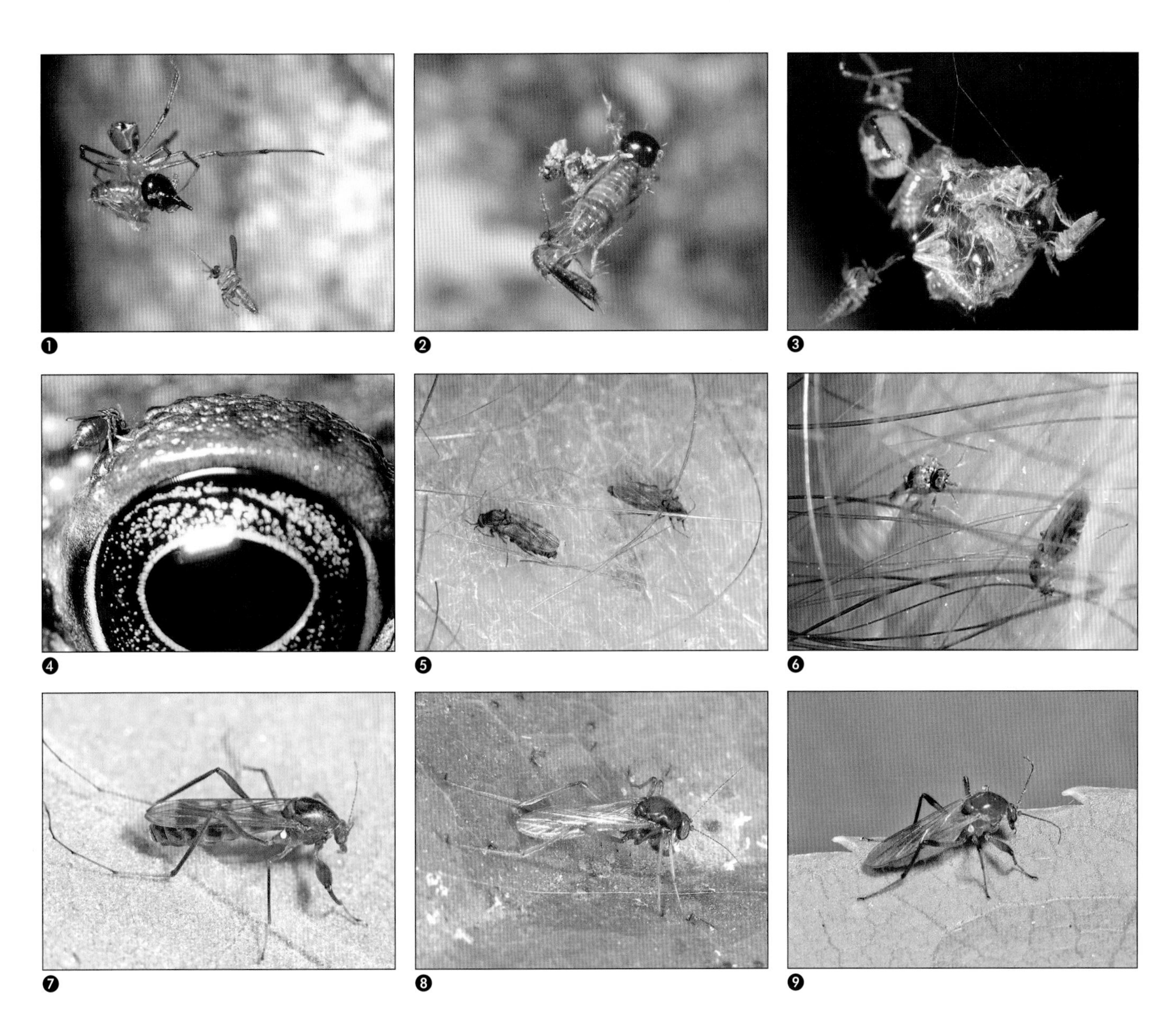

FAMILY CERATOPOGONIDAE (continued) ❶ ❷ ❸ This undescribed South American ***Forcipomyia*** is a kleptoparasite (food thief) associated with a spider (*Episinus* spp., Theridiidae) that raids damaged termite (*Nasutitermes ephratae*) nests and suspends its prey from silk threads dangling from the nest. These images show one fly biting the hind end of a suspended termite, and a couple of midges feeding on the spider's accumulated pendant of termites. ❹ ***Forcipomyia fairfaxensis***, seen here sucking blood from a Green Frog, is known to transfer viruses from frog to frog. Only one group of *Forcipomyia* species (the subgenus *Lasiohelea*) attacks vertebrates; other species bite invertebrates only. **Subfamily Ceratopogoninae** ❺ ❻ ***Culicoides*** is the most important genus of biting midge, and most midge species that bite people belong to this group of "no-see-ums." They are small enough to be almost invisible, but their bite can cause a burning itch. ❼ The North American ***Heteromyia pratti*** is an unusually distinctive biting midge, and relatively large at about 3 mm. *Heteromyia* females are predaceous on other small insects, especially swarming flies. ❽ This Tanzanian biting midge is a species of ***Johannsenomyia***, a genus most diverse in the Old World tropics but also with a couple of Nearctic and Neotropical species. ❾ This eastern North American fly is a ***Palpomyia***, a common and widespread genus of predaceous biting midges.

FAMILY CERATOPOGONIDAE (continued) ❶ The smooth, nematode-like aquatic larvae of ***Palpomyia*** have an elongate head capsule with mouthparts directed forward, where they serve to impale invertebrate prey. ❷ This strikingly elongate Ecuadorian ceratopogonine belongs to the Neotropical genus ***Paryphoconus***. ❸ This ***Probezzia*** has just emerged from a floating pupa in a water sample taken from a Canadian trout stream. Larvae of *Probezzia* occur in similar habitats over much of the world. **FAMILY SIMULIIDAE (BLACK FLIES) Subfamily Simuliinae: Tribe Simuliini** ❹ Black flies usually lay eggs while in flight over the water, but several species, such as this Canadian ***Simulium vittatum***, lay strings of eggs while landed on objects along the waterline. ❺ ❻ Black fly larvae, such as these ***Simulium venustum*** buffeted by the flow in a Canadian stream, use foldable fan-like strainers (labral fans) to filter particles out of flowing waters. Each larva is hanging on by using a circlet of anal hooks stuck into a pad of salivary silk it has previously "spat" onto the leaf surface. The close-up photo shows a black fly larval head with the labral fans unfolded. ❼ Black fly pupae are found underwater, adhering to the same objects as larvae. Shown here is a pupal exuvium (empty pupal shell). ❽ ❾ Most black flies form aerial swarms, but these ***Cnephia dakotensis*** were among many mating along the edge of a small Canadian stream. These common North American black flies normally mate in streamside aggregations shortly after emergence.

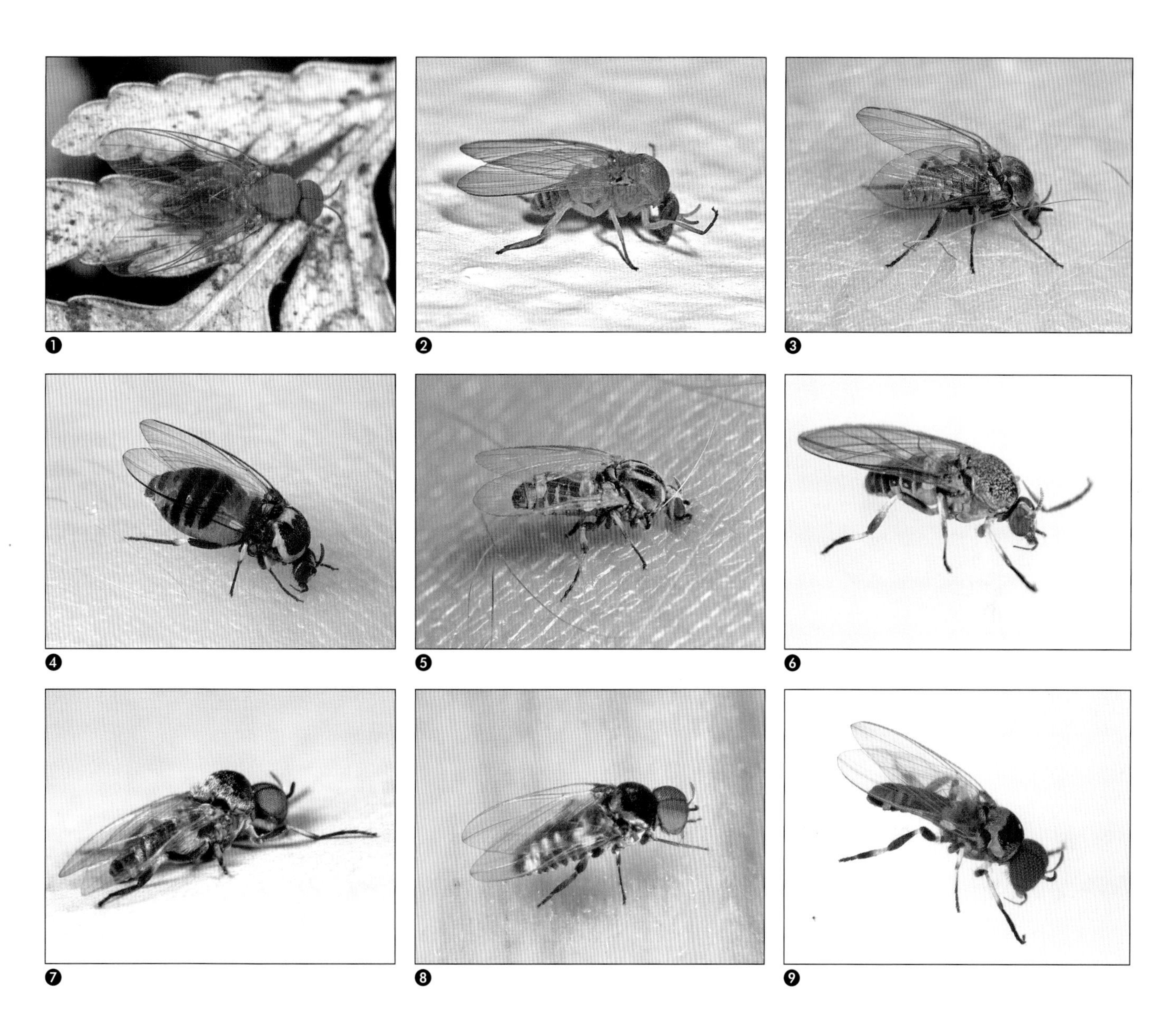

FAMILY SIMULIIDAE (continued) ❶ ❷ ***Gigantodax*** is a large Neotropical genus with almost 70 species traditionally treated as part of the tribe Prosimuliini but now considered to be more closely related to *Simulium*. This is a Chilean species that probably feeds on birds. ❸ The Chilean species ***Paraustrosimulium anthracinum*** was until recently treated as the only South American member of the Australia–New Zealand genus *Austrosimulium* (the genus that includes all New Zealand black flies). It is now treated as a separate genus of one species but its close relationship to *Austrosimulium* still reflects the close zoogeographic relationship between southern South America and parts of Australia and New Zealand. ❹ ❺ ❻ ***Simulium*** is a worldwide genus with hundreds of species and dozens of subgenera. The female *Simulium* species shown here are the aptly named ***S. bicoloratum*** in Ecuador (subgenus ***Ectemnaspis***), a biting ***S. nemorale*** in Chile (subgenus ***Pternaspatha***) and a ***S. tarsatum*** from Costa Rica (subgenus ***Hemicnetha***). ❼ ❽ ❾ Black fly males are non-biting insects with large eyes and often with striking vestiture or other adornment. The hirsute male is ***Simulium donovani*** from the American southwest, the white-spotted male is ***S. metallicum*** from Costa Rica and the silver-banded male is an unidentified Vietnamese *Simulium*.

FAMILY SIMULIIDAE (continued) Tribe Prosimuliini ❶ ❷ The voracious females of ***Prosimulium mixtum*** can make an early spring fishing trip in eastern North America a distressingly desanguinating experience. Of course, this male (with the characteristically enlarged eyes of male black flies) cannot bite. ❸ This female *Prosimulium* (probably ***P. fulvum***) from western North America is one of several species that don't fit the family's common name (black fly) very well. **Subfamily Parasimuliinae** ❹ ***Parasimulium*** is a small genus of five non-biting species found only in the Pacific Northwest of North America, where their pale larvae develop in subterranean streams. Shown here is ***P. crosskeyi***, which develops in underground flows in the Columbia River Gorge of Oregon. This female was captured on vegetation near the outflow and photographed in a tent. **FAMILY THAUMALEIDAE (SEEPAGE OR TRICKLE MIDGES)** ❺ This ***Androprosopa*** adult is standing over a larva of the same species on a cliff along the California coast. ❻ Ideal thaumaleid habitat: collecting seepage midge larvae on a spring seep on Ontario's Niagara Escarpment in Canada. ❼ Seepage midge larvae have open anterior and posterior spiracles on short processes, allowing them to breathe even as they feed on diatoms on the substrate beneath thin films of water. ❽ This ***Androprosopa*** female is laying eggs in a mossy seep. **FAMILY CULICIDAE (MOSQUITOES) Subfamily Anophelinae** ❾ ***Anopheles*** larvae lie horizontally along the water surface with their posterior spiracles open to the air. In this position the head has to be rotated 180 degrees (so the ventral side is up) in order to feed, an operation shown by these images of an entire floating larva (head unrotated), the head and thorax of a larva rotating its head, and the anterior part of the body of the same larva with the head rotated into feeding position.

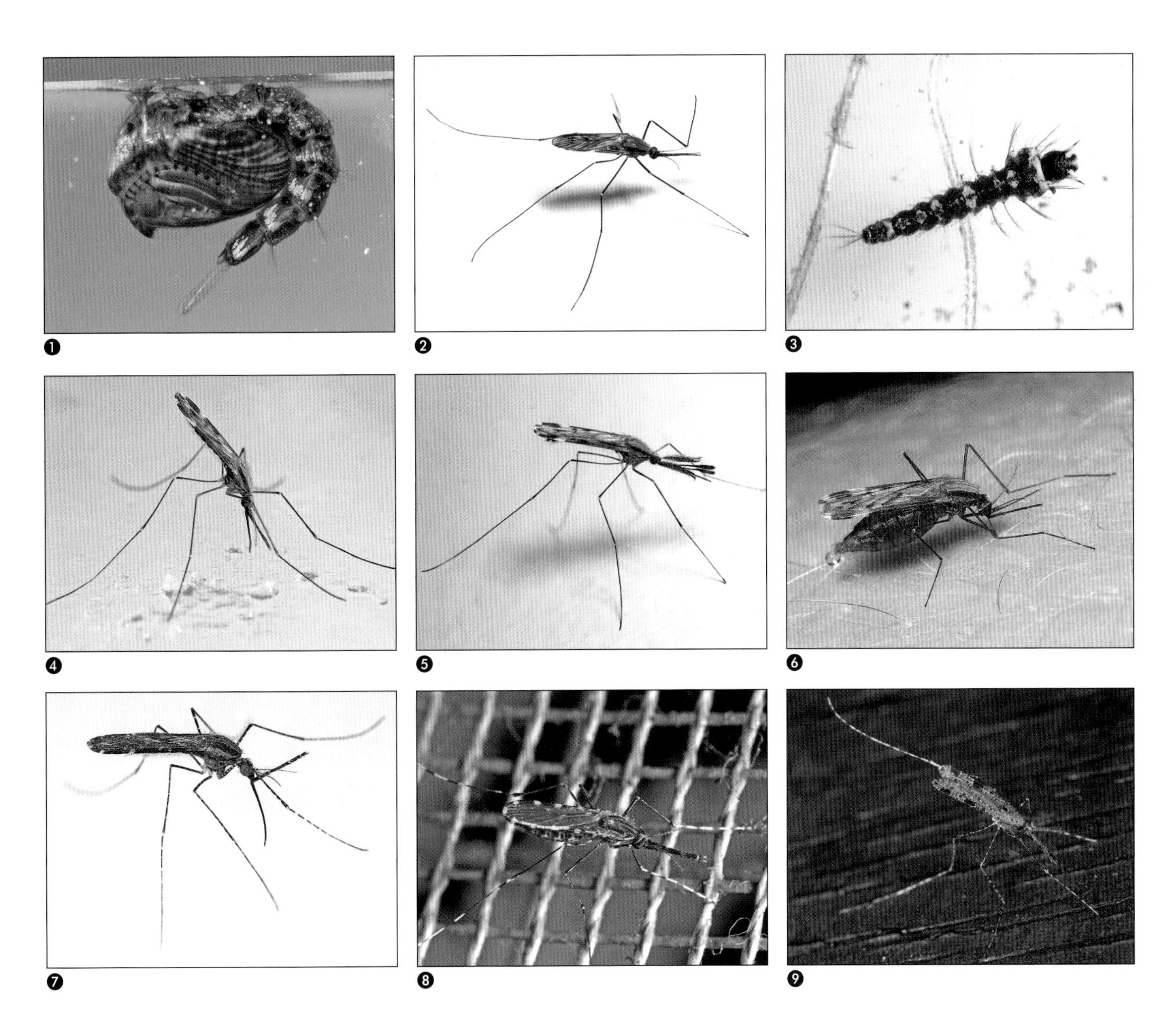

FAMILY CULICIDAE (continued) ❶ This ***Anopheles*** pupa has its thoracic respiratory horns open to the water surface and its developing adult eyes and appendages visible through the pupal skin. ❷ ❸ ***Anopheles quadrimaculatus*** is a common Nearctic species ranging from Canada to Mexico. This female was overwintering in a basement, and the larva was in the surface film of a small pond, filter-feeding with its head rotated 180 degrees. ❹ ❺ The female of this pair of ***Anopheles punctipennis***, a common species throughout North America, is taking a typical tail-up resting position. ❻ *Anopheles* often rest right after biting, pumping out copious amounts of blood-flecked urine as they concentrate the blood meal. This is ***An. punctipennis***. ❼ This African ***Anopheles*** looks threatening with her palps swung up and her proboscis exposed. The long palps are characteristic of the genus. *Anopheles* is a huge genus of about 450 named species worldwide. ❽ This ***Anopheles neivai*** female, on the inside of a window screen in a Central American cabin, is full of my blood. Fortunately, *An. neivai* is not known as a health hazard in Central America, but it is a potentially significant vector of human malaria in South America, where it has been associated with both *Plasmodium vivax* and *P. falciparum*. ❾ This attractive Ecuadorian *Anopheles*, probably ***An. malefactor***, was among many attracted to the lights at an Amazonian research station. This species is not known to transmit malaria. *Anopheles* includes over a hundred Neotropical species, but only a few species (especially *Anopheles darlingi*) are important vectors of malaria.

FAMILY CULICIDAE (continued) Subfamily Culicinae: Tribe Aedini ❶ ***Aedes (Ochlerotatus) stimulans*** is an extremely abundant, and annoying, spring mosquito in eastern North America. This male and female were among thousands resting on low vegetation near a maple swamp. ❷ ❸ ❹ Three unwelcome immigrants to the New World: ***Aedes (Ochlerotatus) japonicus*** (**Japanese Rock Pool Mosquito**, here half full of blood), ***Aedes (Stegomyia) aegypti*** (**Yellow Fever Mosquito**, here emerging from a pupa at the water surface) and ***Aedes (Stegomyia) albopictus*** (**Asian Tiger Mosquito**, here in a discarded paint can). ❺ A close-up of the water in this discarded paint can beside an eastern U.S. roadside shows larvae and newly emerged adults of the **Asian Tiger Mosquito** (***Aedes (Stegomyia) albopictus***). This originally Southeast Asian species is now found in Europe, Africa, North and South America, and the Caribbean, where it is a serious invasive pest with potential to transmit West Nile virus, Yellow fever virus, St. Louis encephalitis, dengue fever and Chikungunya fever. ❻ ❼ ❽ ❾ These small rock pools along an eastern Canadian lakeshore are home to healthy populations of the **Rock Pool Mosquito**, ***Aedes (Finlaya) atropalpus***, seen here as larvae feeding at the bottom of a pool, as larvae and a pupa resting at the surface, and as newly emerging males on the surface of the pool. This native Nearctic species also breeds in artificial containers such as tires, throughout much of eastern North America.

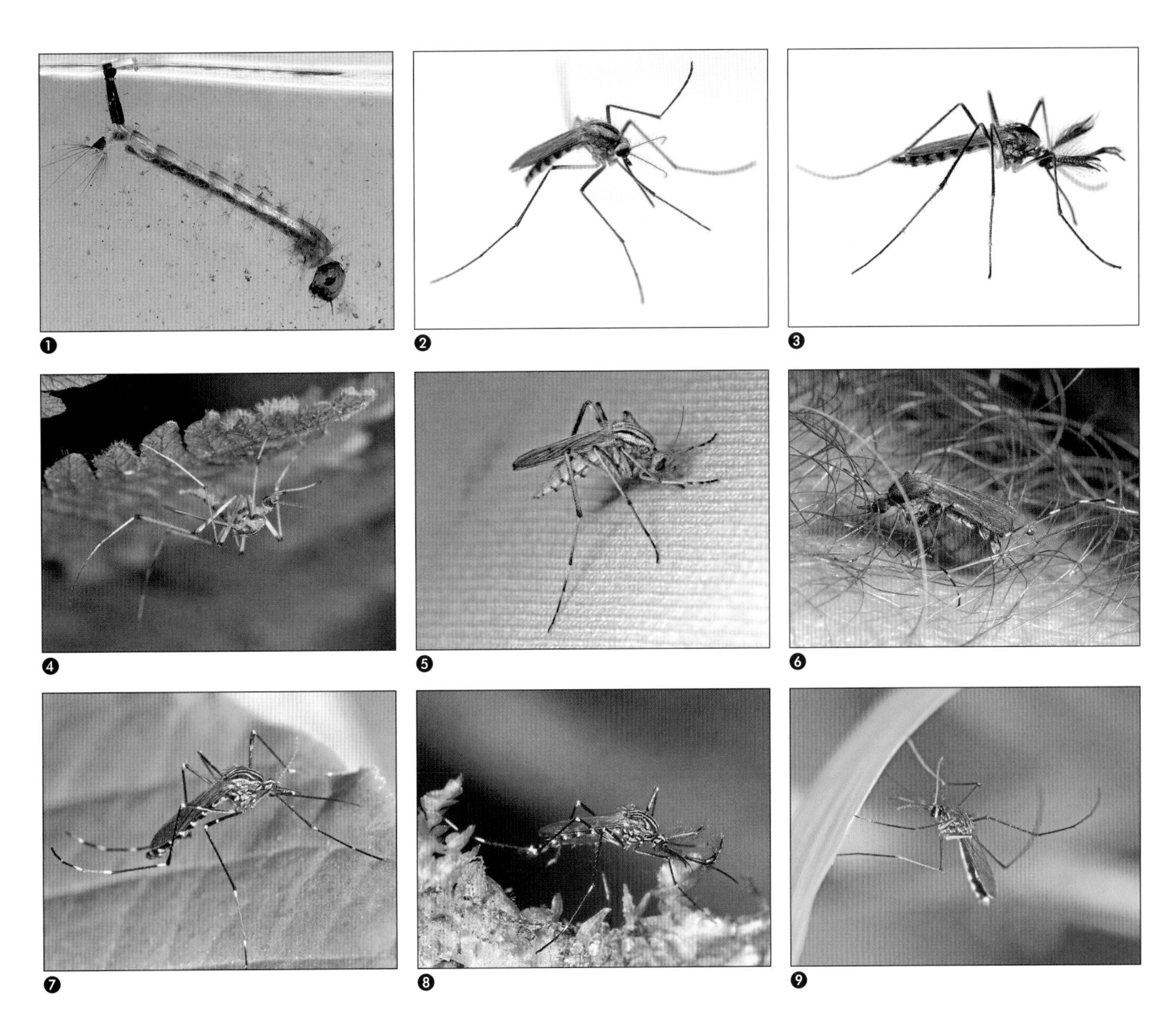

FAMILY CULICIDAE (continued) ❶ ❷ ❸ The **Eastern Tree Hole Mosquito**, ***Aedes (Ochlerotatus) triseriatus*** (seen here as a larva, a female and a male), is a common tree-hole mosquito throughout eastern North America. Larvae can also develop in artificial containers such as abandoned tires; eggs hatch in spring when the container (or tree hole) is flooded. This forest mosquito feeds on both birds and people and can carry encephalitis viruses between species. ❹ Most biting mosquitoes in lowland Amazon rainforest belong to relatively few species of *Psorophora* and *Aedes*, like this pretty ***Aedes (Ochlerotatus) fulvus*** in typical resting position under a fern leaf. ❺ Australia has a large and diverse mosquito fauna, with more than 600 species in around 70 genera or subgenera. As elsewhere in the world, the largest genus is *Aedes*. This is ***Aedes (Ochlerotatus) vittiger***. ❻ The **Eastern Saltmarsh Mosquito**, ***Aedes (Ochlerotatus) sollicitans***, is a vicious biter along the eastern seaboard of the United States and Canada. This species sometimes occurs in saline habitats, such as ditches contaminated by road salt, far from the coast. ❼ ❽ Larvae of many mosquitoes, including species in the *Aedes* subgenus ***Howardina***, develop in tank plants (bromeliads) in Neotropical cloud forests. The male and female *Howardina* shown here are from the high Andes of Ecuador. ❾ This female ***Armigeres manalangi*** from the Philippines probably developed in stagnant water trapped in a tree hole, flower bract or similar forest cavity. This Australasian-Oriental genus is divided into two subgenera; shown here is a species in the larger subgenus (*Armigeres*), which has 41 species.

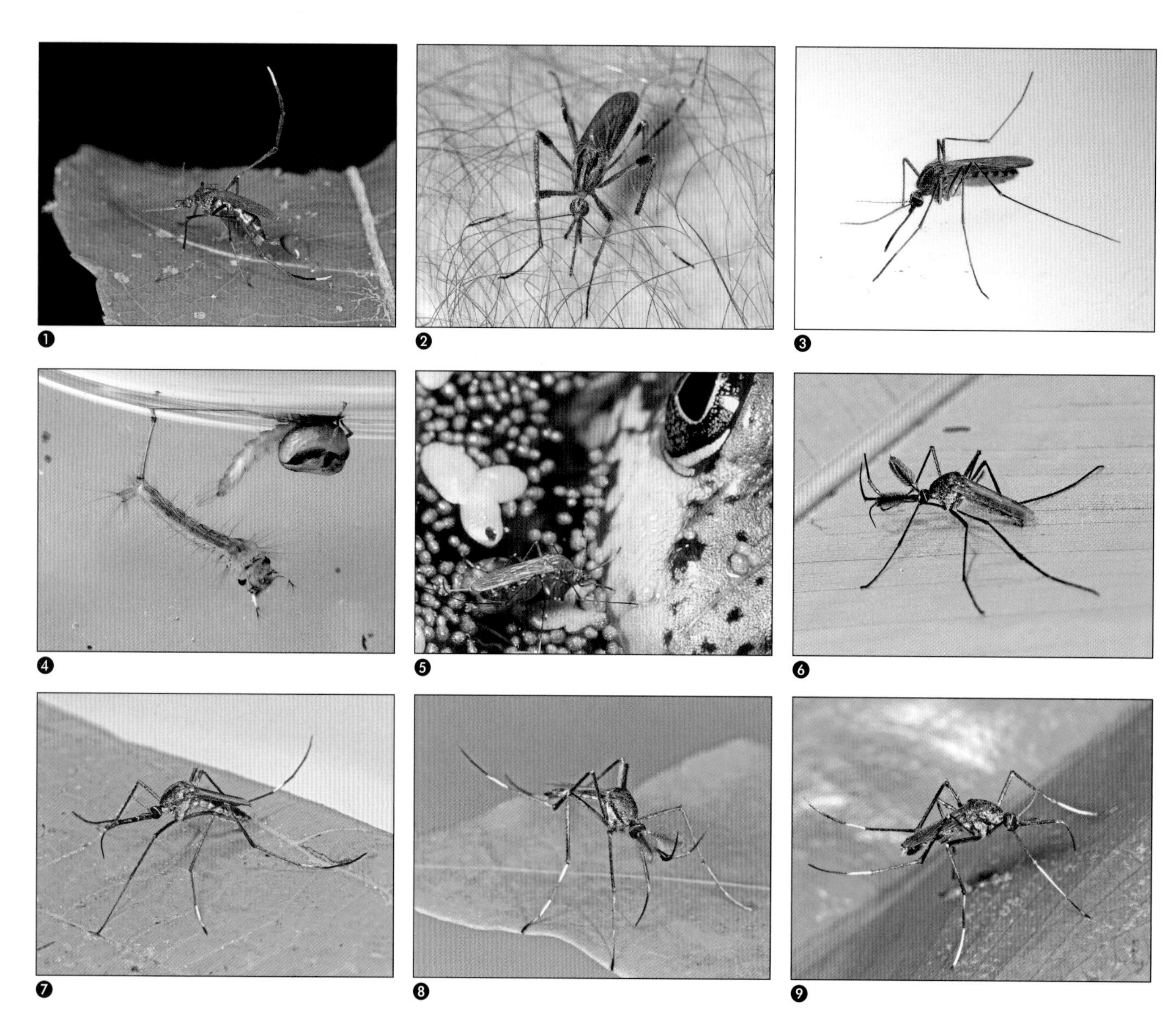

FAMILY CULICIDAE (continued) ❶ The genus ***Psorophora*** is a New World genus of 48 species, mostly associated with floodwater pools of Central and South America. This is a South American species, photographed in Bolivia. *Psorophora* species are closely related to *Aedes* (*Ochlerotatus*) species. ❷ ***Psorophora ciliata***, like other members of the subgenus *Psorophora*, are predaceous at least as later-stage larvae. The large and robust adults ("gallinippers") are vicious biters found throughout the neotropics and northward to southeastern Canada. **Tribe Culicini** ❸ **Northern House Mosquitoes** (***Culex pipiens***) spend the winter as adults, and this one was found midwinter in a Canadian house. This species prefers to bite birds but can move viruses that cause encephalitis from birds to man. Similar *Culex* species are pests worldwide. ❹ ❺ *Culex* larvae and pupae are found in a variety of still waters, including in tree holes and artificial containers, but these ***Culex territans*** immatures were found in the same pond as this frog being bitten by a *C. territans* adult. This widespread Holarctic species prefers frog blood but occasionally bites other vertebrates, including people. **Tribe Toxorhynchitini** ❻ ❼ Adult ***Toxorhynchites***, or elephant mosquitoes, have a trunk-like proboscis bent down at almost 90 degrees and are obviously incapable of biting, although they are often seen nectar-feeding. The genus (the only one in the tribe) is widespread, but more diverse in tropical areas. This feathery-antennaed male and bent-beaked female are ***T. (Lynchiella) haemorrhoidalis*** from the rainforests of Bolivia. ❽ ❾ Elephant mosquitoes, genus ***Toxorhynchites***, are non-biting mosquitoes that develop as predators in tree holes and similar habitats worldwide. Shown here are a male and female of ***T. splendens*** from Vietnam.

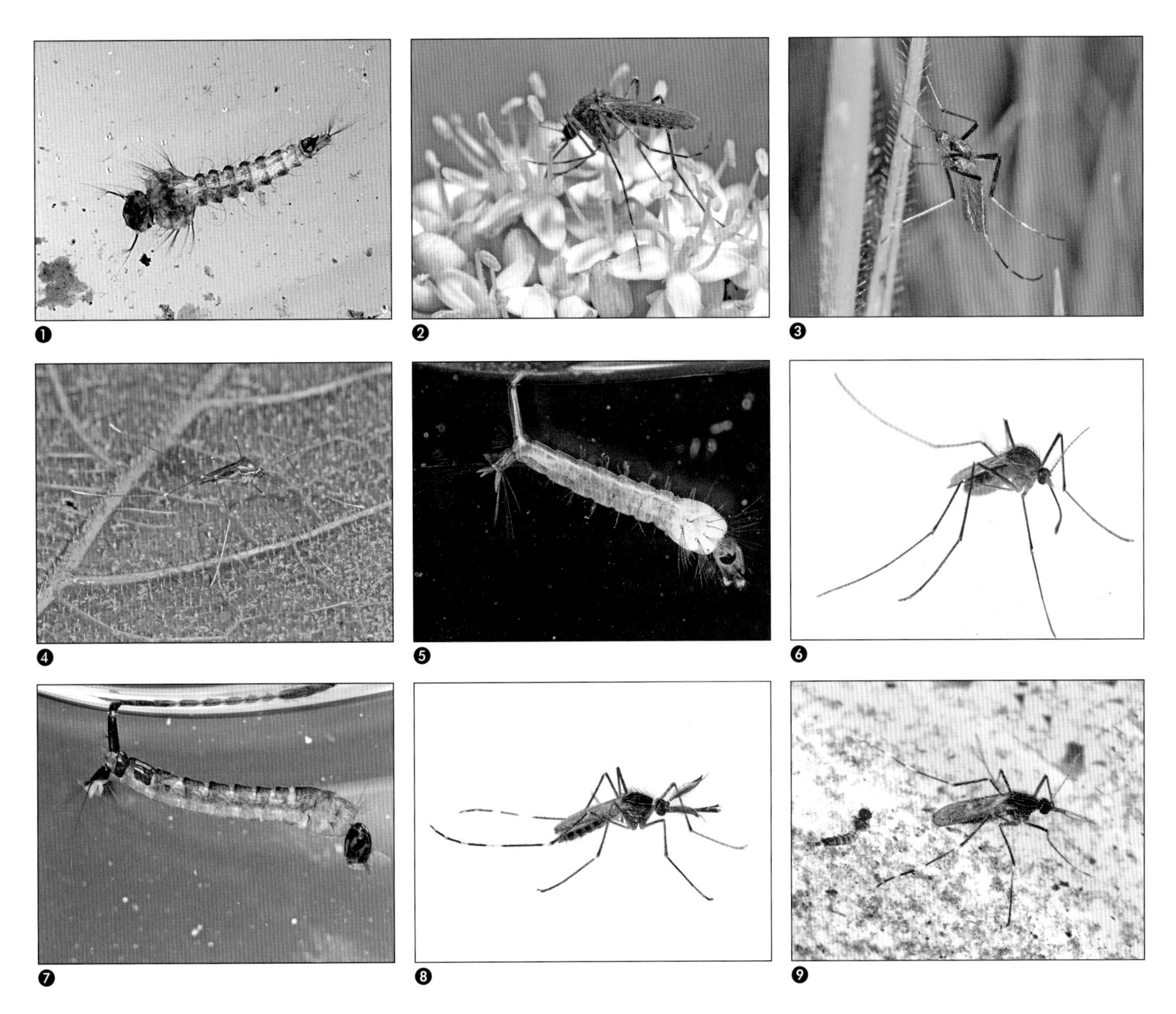

FAMILY CULICIDAE (continued) Tribe Mansoniini ❶ ❷ *Coquillettidia perturbans* is a widespread North American species that can be a vicious nocturnal biter near large marshes, where the larvae live in mud around the roots of aquatic plants. The larval breathing tube (posterior respiratory siphon) is modified into a spike used to impale the plant tissue to breathe. In older literature this species is in the genus *Mansonia*. ❸ This large mosquito photographed in an African marsh is a ***Coquillettidia***, probably ***C. flavocincta***. Species of the widespread genus *Coquillettidia*, like those in the related genus *Mansonia*, respire by puncturing the roots of emergent aquatic plants with their spike-like posterior respiratory siphon. **Tribe Uranotaeniini** ❹ The genus ***Uranotaenia*** occurs in every zoogeographic region, but most, like this delicate female from the mountains of Costa Rica, occur in tropical countries, where their larvae develop in ground pools, swamps and lakes. Adult females usually bite frogs. ❺ ❻ ***Uranotaenia sapphirina*** is the only species of this mostly tropical genus to range north to Canada. **Tribe Orthopodomyiini** ❼ ❽ ❾ The tribe Orthopodomyiini includes only the widespread genus ***Orthopodomyia***, with some 35 species, found mostly in the tropics. This larva and male of ***O. signifera*** are from tree holes in southern Canada; the female standing on the water surface near a small larva is from a tree hole in the southeastern United States.

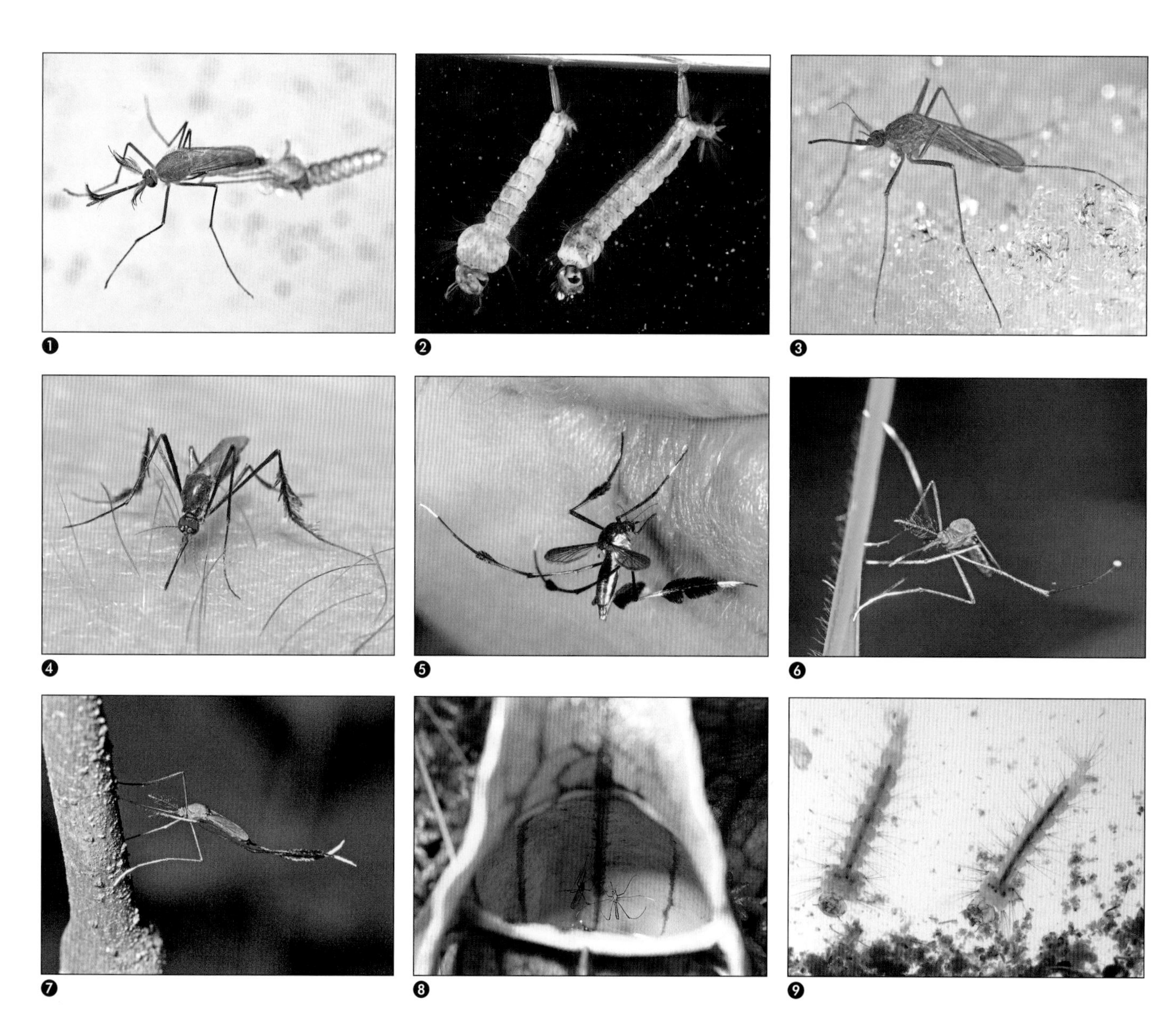

FAMILY CULICIDAE (continued) Tribe Culisetini ❶ ❷ ❸ ***Culiseta*** is a small but widespread genus often associated with cooler habitats. This species, ***C. inornata***, occurs throughout North America and ranges north into the Arctic. Seen here are a female newly emerged on the water surface, larvae from a marsh pool, and a male walking on the snow in early spring. **Tribe Sabethini ❹** This Bolivian ***Sabethes*** probably developed as a larva deep in a tree hole. Adult *Sabethes* sometimes engage in elaborate courtship displays using their fringed hind legs. Both sexes are similarly ornamented. ❺ This face-sucking Ecuadorian ***Sabethes (Sabethes) tarsopus*** is one of 37 species in this exclusively Neotropical genus. According to Ralph Harbach, the expert on this genus, they usually bite people near the nose. ❻ ❼ The closely related sabethiine genera *Sabethes* and *Wyeomyia* are common day-active mosquitoes in Neotropical forests. These ***Wyeomyia*** were photographed in Ecuadorian rainforest. ❽ ❾ Although *Wyeomyia* is a mostly Neotropical genus, ***Wyeomyia smithi*** ranges into northern North America as a specialized inhabitant of pitcher plants, like this *Sarracenia purpurea* in a Canadian bog. The mosquitoes in this water-holding pitcher are laying eggs in the water, where their larvae will develop.

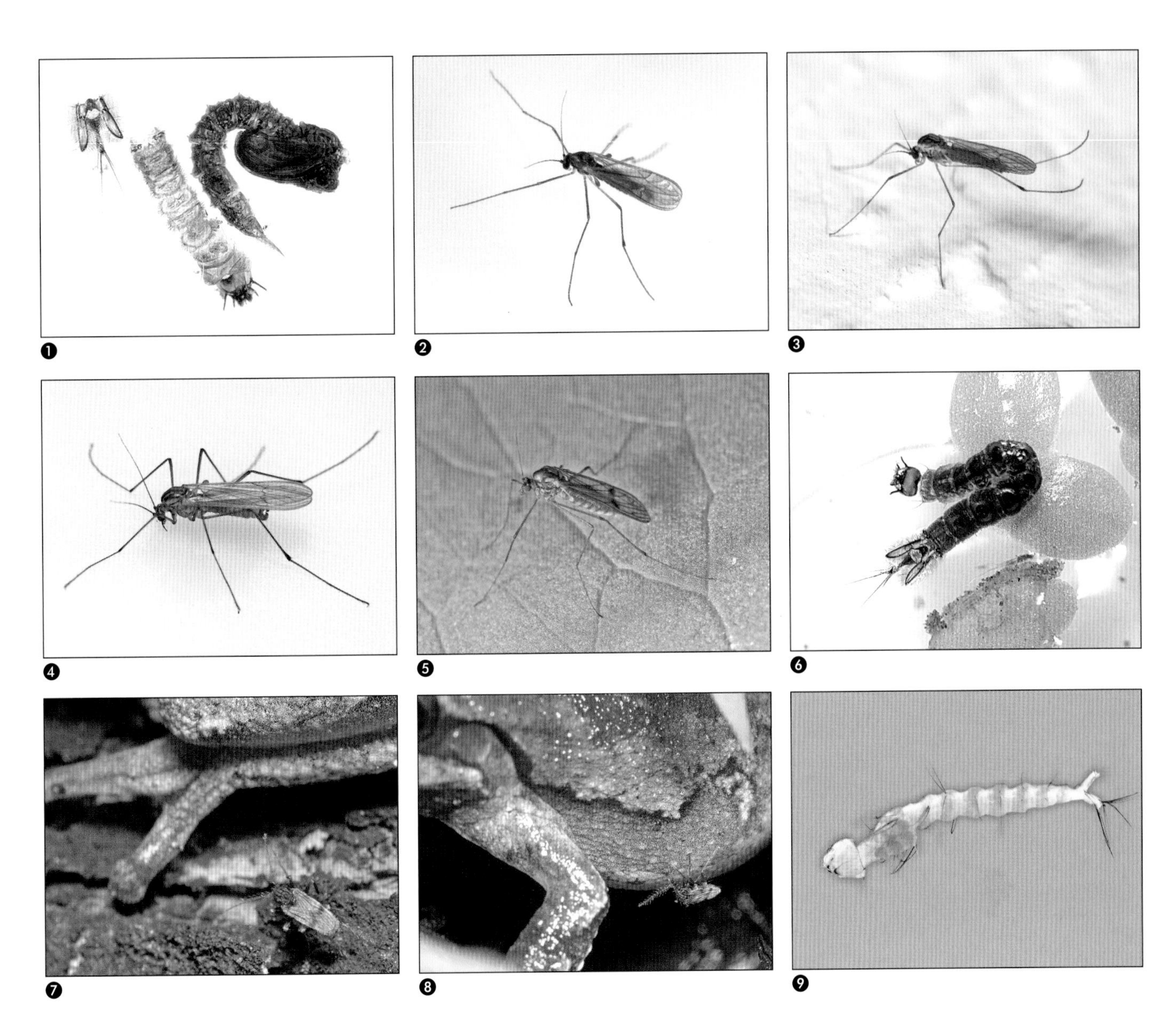

FAMILY DIXIDAE (MENISCUS MIDGES) ❶ ❷ The pupa from which this ***Dixa repanda*** emerged is shown here beside fragments of the larval exuviae. The larva was collected at the margins of an Ontario stream. ❸ This ***Dixa brevis*** was attracted to a light in the American southwest. ❹ This ***Dixella cornuta*** was reared from larvae living in a spring pool in a Canadian wetland. ❺ This ***Dixella nova***, from a Canadian bog, shows the arched and forked wing vein (R_{2+3}) typical of the family. ❻ This ***Nothodixa*** larva from Chile is taking a characteristic U-shaped position on the surface of a small stream. *Nothodixa* is a southern hemisphere genus similar to the northern genus *Dixa*; it also occurs in Australia and New Zealand. **FAMILY CORETHRELLIDAE (FROG MIDGES)** ❼ ❽ ***Corethrella*** females bite only singing frogs, usually landing near the frog before making a final approach to the host on foot, like this ***C. wirthi*** approaching and then biting a Pinewoods Tree Frog in a South Carolina swamp. ❾ ***Corethrella*** larvae occur in small pockets of water such as tree holes or leaf axils, where they use their antennae to grab small arthropod prey — including young mosquito larvae — with a rapid snap-trap motion (this is a preserved specimen).

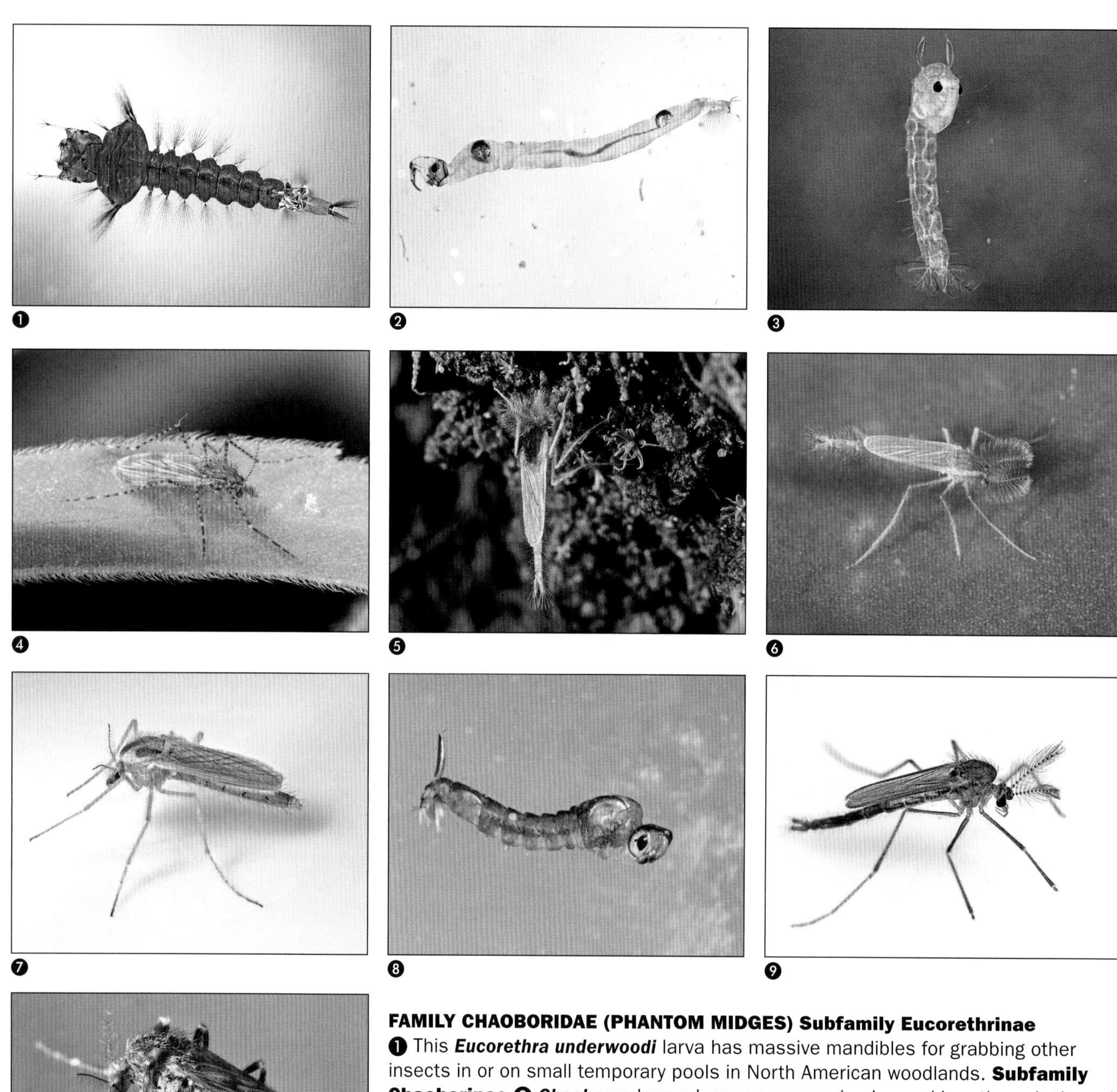

FAMILY CHAOBORIDAE (PHANTOM MIDGES) Subfamily Eucorethrinae ❶ This ***Eucorethra underwoodi*** larva has massive mandibles for grabbing other insects in or on small temporary pools in North American woodlands. **Subfamily Chaoborinae** ❷ ***Chaoborus*** larvae have no open spiracles and breathe only through the skin. The two silvery bubbles are used to control descent and ascent in the water column, and the enormous antennae are used to grasp prey. ❸ Pupae of ***Chaoborus*** have "horns" on the thorax used to breathe at the surface, much like mosquito pupae, but the abdomen is held straight, unlike the curved abdomen of a mosquito pupa. The things at the tip of the tail are swimming paddles. ❹ This is a ***Chaoborus ceratopogones***, one of the "lake flies" that occur in large numbers in African lakes. Lake flies (this and related African *Chaoborus* species) often make up most of the individual organisms in a lake. ❺ This is a male Tanzanian "lake fly" (***Chaoborus*** spp.). Male lake flies often form huge swarms of millions of individuals. ❻ ❼ These ***Chaoborus*** adults were among thousands emerging from a northern Canadian lake. The male is the one with the plumose antennae. ❽ ❾ ❿ ***Mochlonyx*** is an inhabitant of temporary woodland (vernal) pools. Larvae have grasping antennae like *Chaoborus* but a posterior breathing tube like most mosquitoes. Adult females are relatively compact, with simple antennae compared to the male.

FAMILY BLEPHARICERIDAE (NET-WINGED OR TORRENT MIDGES) Subfamily Blepharicerinae ❶ This recently emerged ***Agathon comstocki*** is resting on a rock in a torrential western American river. *Agathon* is a Holarctic genus, with most species in western North America but others in Russia, Japan and Korea. ❷ Each of the distinct body sections of this ***Blepharicera*** larva from California is equipped with a ventral suction disk, holding it tightly to the rock surface in a rushing river. ❸ ❹ This ***Blepharicera*** male and these two females were among hundreds hanging from vegetation near a large eastern U.S. river. The Holarctic genus *Blepharicera* is the largest genus in the family, with more than 50 species. ❺ This ***Neocurupira campbelli*** is taking a typical pose on a rock just above the waterline in a swift New Zealand stream. ❻ ❼ ❽ This larva of an undescribed genus of torrent midge is stuck to a rock, head upstream, among a bunch of black fly larvae (heads downstream) in a Chilean stream. The pupae are from a nearby river, and the adult is on a flower close to the larval habitat. ❾ The leading authority on torrent midges, Gregory Courtney of Iowa State University, here demonstrates ideal blephariceridhabitat in the mountains of Costa Rica.

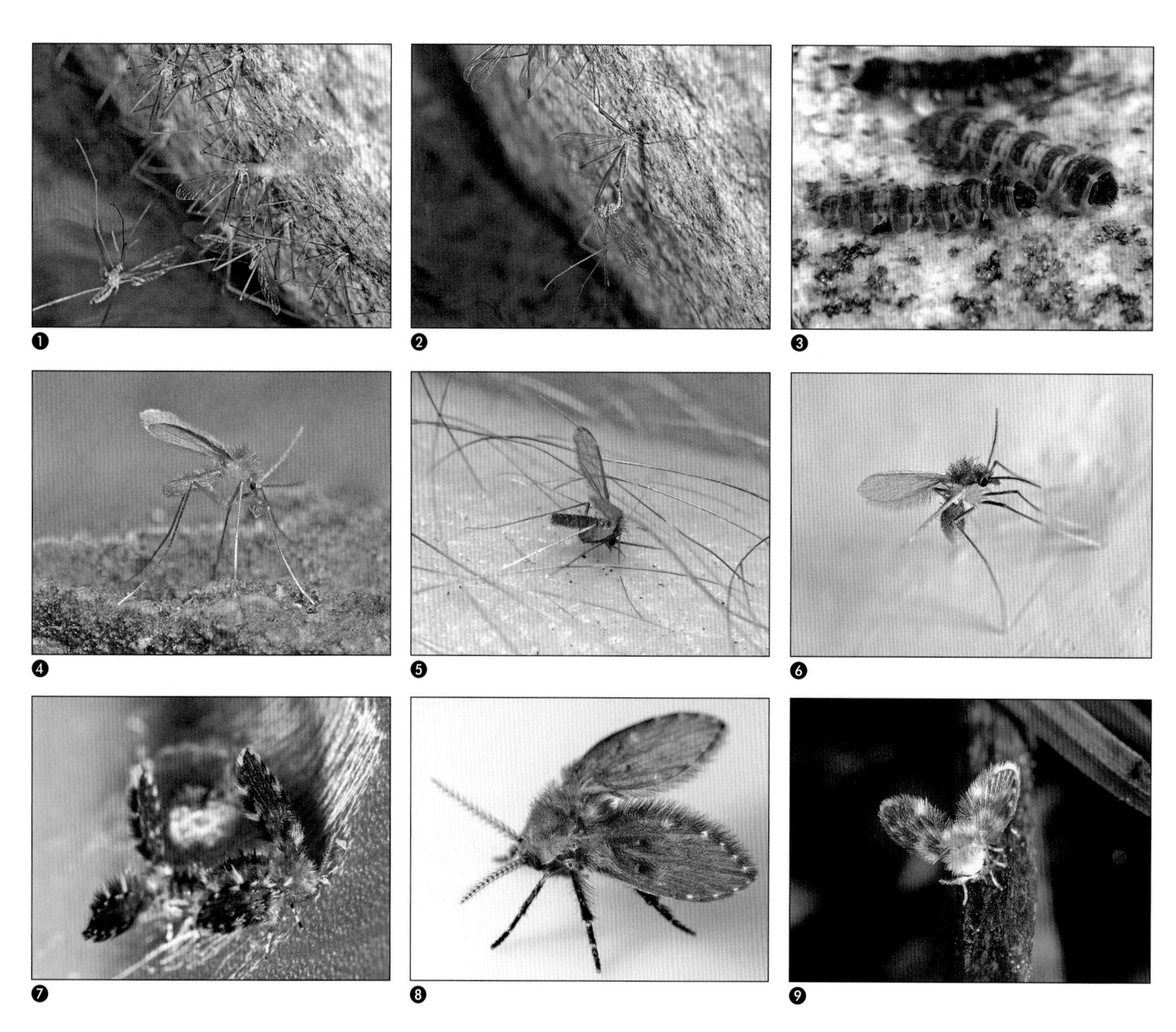

FAMILY BLEPHARICERIDAE (continued) Subfamily Edwardsininae ❶ ❷ ❸ These ***Edwardsina*** are aggregating and mating on a rock in a torrential stream in Chile. The two larvae were adhering to a submerged rock in the same rapids. *Edwardsina* has a distribution divided between southern South America and southeastern Australia. **FAMILY PSYCHODIDAE (MOTH FLIES) Subfamily Phlebotominae ❹ ❺ ❻** ***Lutzomyia*** is a large genus of tiny "sand flies," including the carriers of leishmaniasis and other diseases in the New World. One of these was among many on the wall of a cave in Cuba, one is biting my arm in Bolivia and one is on the whitewashed wall of a research station in the Ecuadorian Amazon. **Subfamily Psychodinae ❼** ***Alepia*** is a Neotropical genus of 11 species, including one known from larvae living in the twig nest of an *Azteca* ant. **❽** ***Clogmia albipunctata*** (the **Large Drain Fly**) is a ubiquitous domestic species that breeds in scummy bathroom drains and other biofilms around the world. **❾** This ***Clytocerus americanus*** is standing on some debris over a small eastern North American seep, where it probably developed.

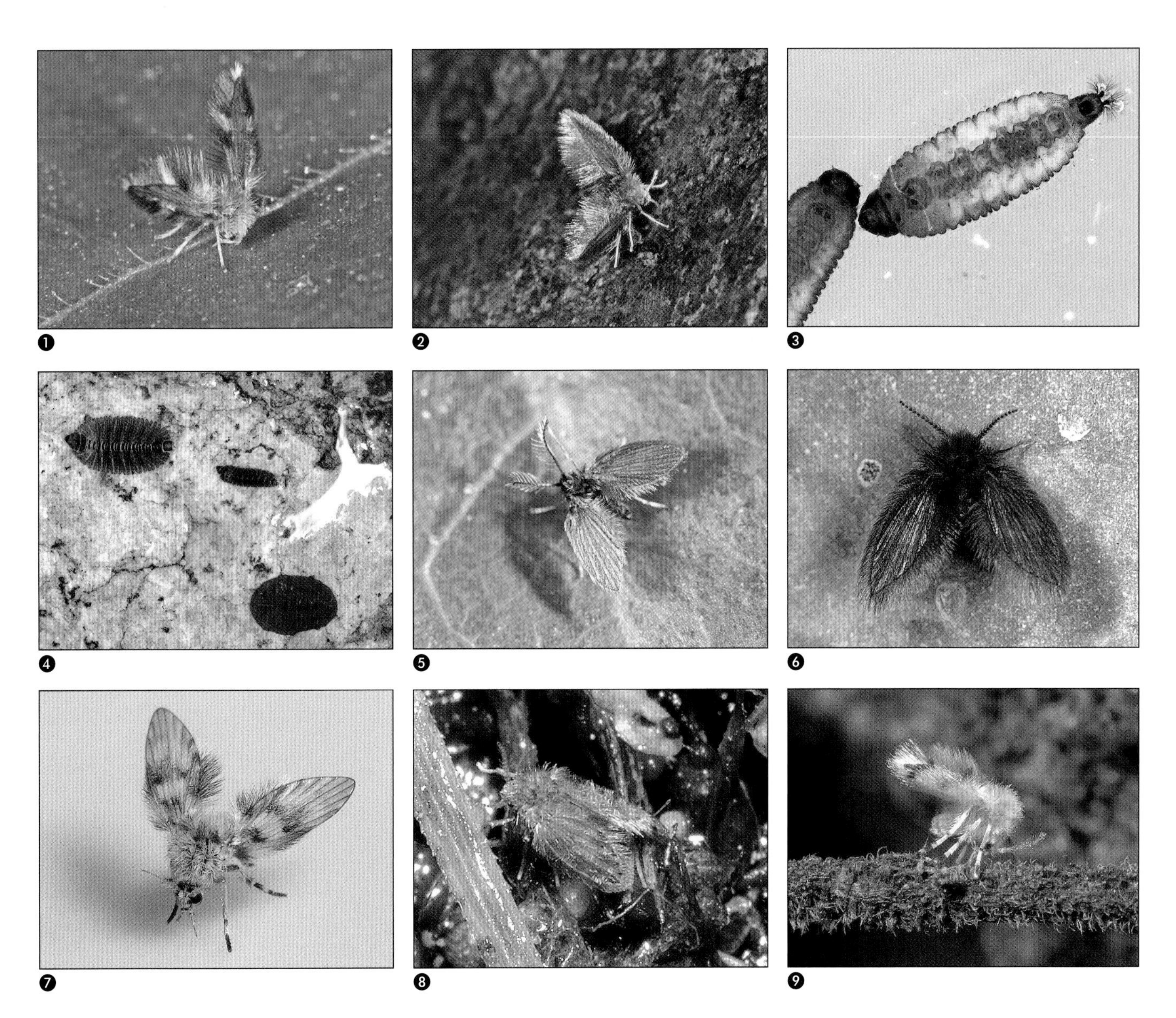

FAMILY PSYCHODIDAE (continued) ❶ ❷ These ***Maruina*** adults from Costa Rica have relatively narrow, fringed wings that probably resist soaking, a useful property for this female laying eggs along the splashing waterline of a river rock. ❸ ❹ Larvae of ***Maruina*** use ventral friction pads, superficially similar to the suction disks of torrent midges, to cling to rocks in torrential mountain streams. The larva seen here in ventral view is from California; the two different species seen clinging to a submerged rock are from Costa Rica. ❺ ***Nemoneura*** is a small southern South American genus of four named species; this one is from Chile. ❻ Moth flies, like this Canadian ***Neoarisemus niger***, are often found resting under bridges or overhangs over small, mossy streams. ❼ ***Pericoma*** is one of the more common aquatic psychodid genera, and adults are often found along the margins of streams and debris along lakes. This ***P. signata*** is one of 24 Nearctic species in this very large, widespread genus. ❽ ***Psychoda alternata*** (the **Small Drain Fly**) is a common insect of household drains, although it is now less common in houses than the Large Drain Fly. This and other *Psychoda* species are also found in other semiaquatic habitats, such as stream margins and trickling filters in sewage plants. ❾ This delicate little moth fly from Costa Rica is not identified, but it shows the typical habitus of the Psychodinae.

FAMILY PSYCHODIDAE (continued) ❶ This little silver-spotted moth fly is a ***Setomima*** from Tanzania. Most members of this small genus are Afrotropical or Oriental, but two species occur in the Caribbean. ❷ ***Thornburghiella albitarsis*** adults often cluster in large numbers under the broad leaves of plants in and around springs. These are under the leaf of a Marsh Marigold in eastern Canada. **Subfamily Trichomyiinae** ❸ The subfamily Trichomyiinae includes only the single, widely distributed genus ***Trichomyia***, known species of which develop in rotten wood. Shown here is ***T. kenricki*** from Chile. **FAMILY TANYDERIDAE (PRIMITIVE CRANE FLIES)** ❹ ❺ This Chilean primitive crane fly (Tanyderidae) was found over a cold stream, presumably the larval habitat. ❻ Larval Tanyderidae, such as this western Canadian ***Protanyderus margarita***, develop in the sandy or gravelly margins of large streams. **FAMILY PTYCHOPTERIDAE (PHANTOM OR FOLD-WINGED CRANE FLIES) Subfamily Ptychopterinae** ❼ ❽ ***Ptychoptera*** is a large, widespread genus that develops in organically rich wetlands, where the rat-tailed larvae embed themselves in sediment while using their tail-like siphons to breathe at the surface. **Subfamily Bittacomorphinae** ❾ ❿ ***Bittacomorpha*** includes only two species, both restricted to North America, where their rat-tailed larvae can be very common in shallow flowing water. Adults float through the air with the aid of air sacs in their inflated outstretched legs. The larvae shown here were cleaned up and photographed in an aquarium; they would normally be covered with muck.

FAMILY DEUTEROPHLEBIIDAE (MOUNTAIN MIDGES) ❶ ***Deuterophlebia*** larvae use lateral, crochet-tipped abdominal prolegs to cling to water-buffeted rock surfaces in rushing waters of Japan, Korea, central and eastern Asia, western North America and the Himalayas. ❷ Mountain midge (***Deuterophlebia***) adult males never land; they spend their brief lives swarming over the swirling rapids in which they develop. Females can crawl beneath the water and oviposit on current-exposed rocks. This male, with its spectacular antennae, was trapped in a spider web. ❸ The spectacularly long antennae characteristic of mountain midge males can be seen looped in front of this flying ***Deuterophlebia***. This individual was netted from a swarm, then photographed upon release. **FAMILY NYMPHOMYIIDAE (NYMPHOMYIID FLIES)** ❹ The family Nymphomyiidae was known only from this one Japanese species (***Nymphomyia alba***) until the 1960s, but Nymphomyiidae are now known from seven extant species (two eastern Nearctic and five eastern Asian/eastern Palaearctic) and one fossil species (from Baltic amber). This photograph of mating ***N. alba*** was taken by Seiji Sato. ❺ Larvae of Nymphomyiidae are found among moss in small, cool streams, like the mountain stream in North Carolina where this tiny (less than 2 mm) specimen was washed off a moss-covered rock. Photo by Keith Bayless. **FAMILY AXYMYIIDAE (AXYMYIID FLIES)** ❻ This ***Axymyia furcata*** female is ovipositing in a log half-submerged in a Canadian stream. ❼ This ***Axymyia furcata*** larva was exposed by breaking up a half-submerged log in a Canadian stream. The head is down and the hard-tipped breathing tube is up. ❽ ❾ Axymyiid larvae are found in firm dead wood partially submerged in flowing water, like the log shown here in eastern Canada. Their presence in this log is advertised by the typical curls of wood-filled frass always found on the surface of axymyiid-inhabited logs.

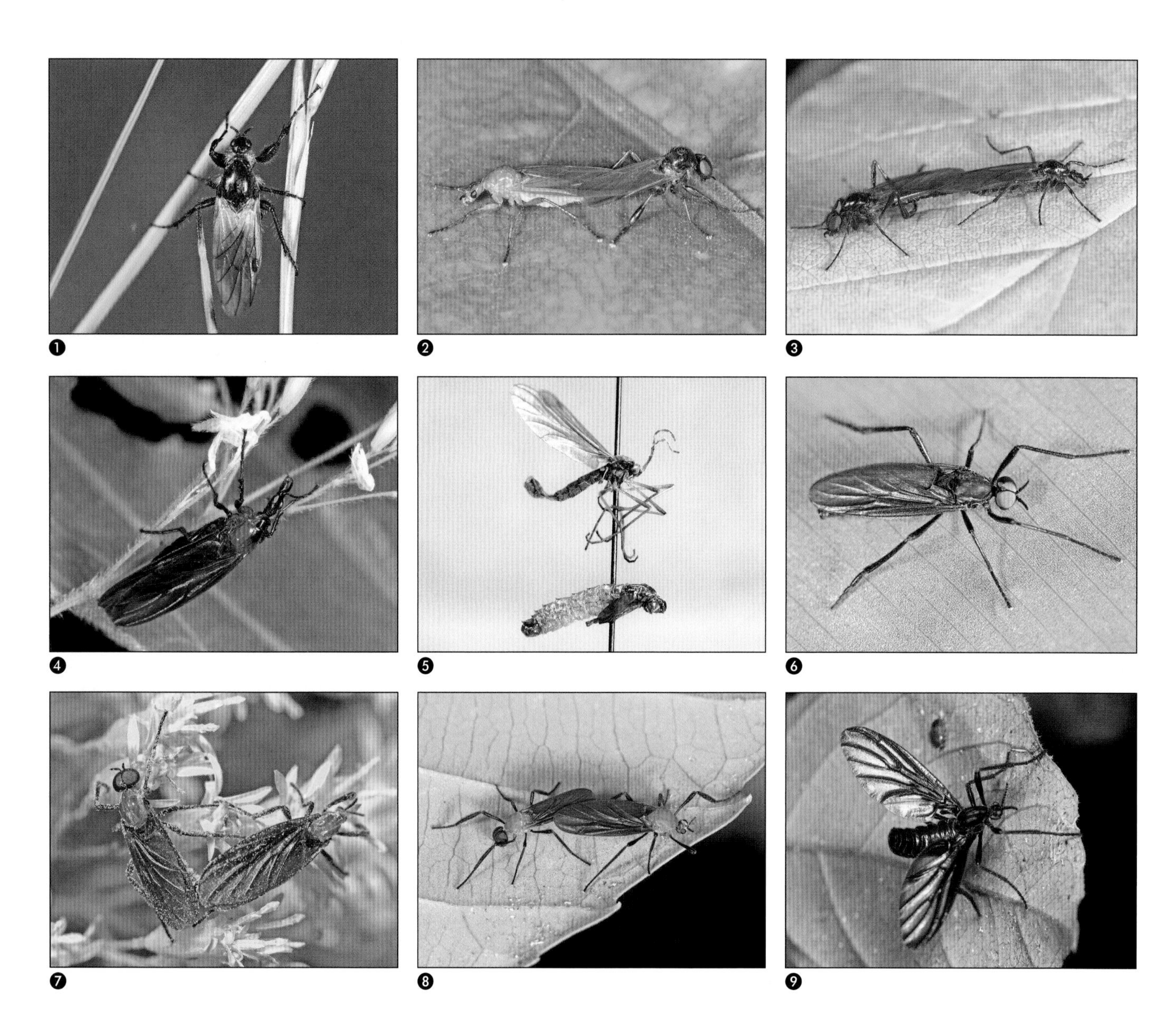

FAMILY BIBIONIDAE (MARCH OR ST. MARK'S FLIES) Subfamily Bibioninae ❶ ***Bibio*** is a large worldwide genus of almost 200 species, mostly described from the Holarctic region. This female was among dozens clinging to grass in an eastern North American meadow. ❷ Bibionidae, like this pair of ***Bibio*** in Vietnam, are characterized by long copulations during which the big-eyed males protect their paternal investments by staying "plugged in." ❸ Like many flies that form male swarms, male bibionids have large heads with massive eyes. These ***Bibio*** are mating on late fall foliage in eastern Canada. ❹ Members of the large (200 species), widespread genus ***Dilophus*** have a distinctive circlet of spines at the apex of the fore tibia. **Subfamily Hesperinae** ❺ This specimen of ***Hesperinus brevifrons*** is pinned over the exuviae (cast pupal skin) from which it emerged, with the free wing cases and legs clearly visible. The pupa was found in a rotting log in Canada's Yukon Territory. **Subfamily Penthetriinae** ❻ This black male from Costa Rica is a typical member of the mostly Old World tropical genus ***Penthetria***. **Subfamily Pleciinae** ❼ This pollen-covered pair of ***Plecia nearctica*** (**Lovebugs**) will stay *in copula* for a long time, probably much of their adult lives. These harmless flies can be conspicuously abundant in the southern United States, where they breed in cattle dung. ❽ ***Plecia*** is a large, widespread but mostly tropical group that includes more than 250 species worldwide. Shown here is a pair from the Philippines. ❾ *Plecia* such as this Amazonian species (probably ***P. plagiata***) differ from other Bibionidae in having the third long wing vein (Rs) forked, with the upper branch of the fork short and oblique. This species looks a bit like a dark-winged fungus gnat (Sciaridae) but differs in having two closed wing cells.

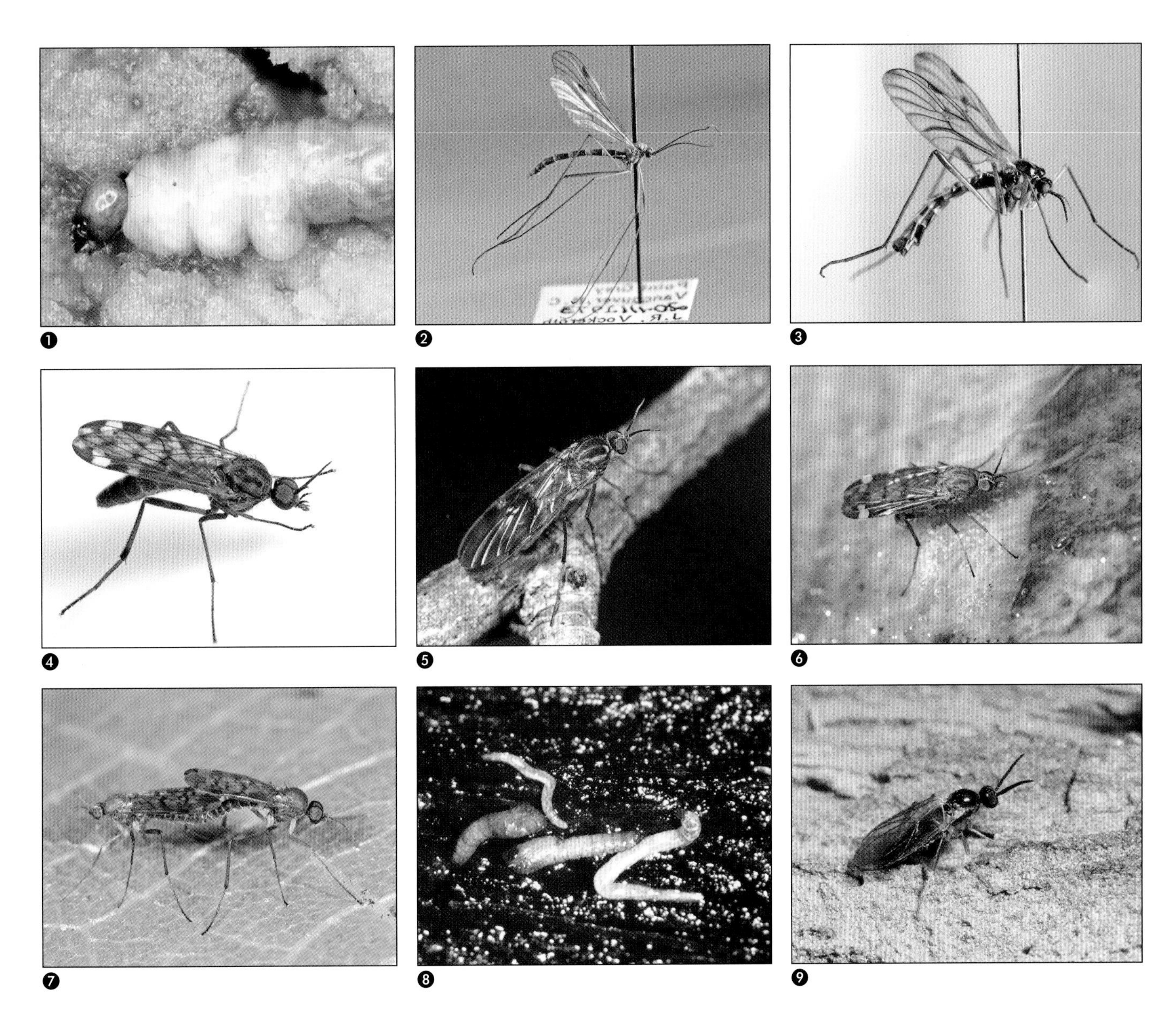

FAMILY PACHYNEURIDAE (PACHYNEURID GNATS) Subfamily Cramptonomyiinae ❶ ❷ ***Cramptonomyia spenceri***, the only species in the genus and the only Nearctic pachyneurid, develops in red alder twigs on the Pacific coast of North America (larval close-up courtesy of Greg Courtney). Adult ***Cramptonomyia*** look much like some crane flies (Tipulidae), from which they are easily distinguished by their prominent ocelli. **Subfamily Pachyneurinae** ❸ ***Pachyneura fasciata*** ranges across the northern Palaearctic region, from Sweden to Japan. **FAMILY ANISOPODIDAE (WOOD OR WINDOW GNATS) Subfamily Anisopodinae** ❹ ❺ ❻ ❼ ***Sylvicola*** is a cosmopolitan genus often found on tree sap, compost and similar materials. Shown here are species from Australia, Chile (on a twig), Canada (on tree sap) and Central America (*in copula*). Larvae of this genus occur in a wide variety of decaying organic matter, ranging from tank bromeliads to compost heaps. **Subfamily Mycetobiinae** ❽ ❾ These ***Mycetobia divergens*** developed in the bleeding sap of a wounded poplar tree in my Canadian backyard. *Mycetobia* is a worldwide genus, but this is the only Nearctic species.

FAMILY SCATOPSIDAE (BLACK SCAVENGER FLIES) Subfamily Scatopsinae ❶ Small scatopsids in the genus ***Rhexoza*** often occur at high densities under the bark of recently felled trees. The several larvae in this sweet sap under the bark of a Canadian poplar tree are ***R. similis***. The other insect in this image is a small springtail, Collembola. ❷ ***Scatopse notata*** is a widespread compost species, and this was one of thousands emerging from garden compost in eastern Canada. ❸ These larval, pupal and adult ***Coboldia fuscipes*** were among hundreds in a rotting mushroom in Greece. This cosmopolitan species is sometimes a pest of cultivated mushrooms. ❹ These tiny ***Swammerdamella*** were among many in the flowers of Bog Bean (Buckbean) plants (*Menyanthes trifoliata*) in a Canadian fen. Each flower had dozens of flies, although only two are visible in this photo. **Subfamily Apistinae** ❺ Females in the rarely encountered genus ***Apistes*** use their characteristically produced spinelike foretibiae to burrow in the sand, in this case dragging an attached male along. **Subfamily Ectaetiinae** ❻ ***Ectaetia***, the only genus of Ectaetiinae, is a Holarctic and Neotropical genus associated with dead wood. This is the North American ***E. clavipes***. **Subfamily Psectrosciarinae** ❼ Most members of the genus ***Anapausis***, like this ***A. soluta***, are European, but the genus occurs worldwide. **FAMILY CANTHYLOSCELIDAE (CANTHYLOSCELID GNATS)** ❽ Members of the Holarctic genus ***Synneuron*** are small flies superficially similar to Scatopsidae, from which they differ in having a multi-segmented palpus. This ***S. annulipes*** was photographed in the Czech Republic by Jan Sevcik. ❾ ***Hyperoscelis*** is a small Palaearctic genus of three rarely seen species; this ***H. eximia*** was photographed in Russia by Dmitry Gavryushin. The superficially similar genus ***Canthyloscelis*** has a south temperate distribution, with different subgenera in New Zealand and southern South America.

FAMILY VALESEGUYIDAE (VALESEGUYID GNATS) ❶ Valeseguyidae are among the rarest of flies; all but one of the known specimens of the family are fossils of extinct species. This is a Dominican amber fossil of ***Valeseguya disjuncta*** (courtesy David Grimaldi, American Museum of Natural History). **FAMILY PERISSOMMATIDAE (PERISSOMMATID GNATS)** ❷ Perissommatid gnats, like this ***Perissomma bellissimum*** from Australia and this clear-winged ***P. congrua*** from Chile, are minute midges currently known from a few sites in Australia and Chile although the family is also known from Northern Hemisphere fossils. ❸ Perissommatid gnats, like this ***Perissomma fusca*** from Australia, have the eyes divided so that they appear to have four. Adults of these rarely encountered flies swarm during the colder months. Larvae have been found in mushrooms and moist leaf litter. **FAMILY CECIDOMYIIDAE (GALL MIDGES)** ❹ ❺ Gall midges are usually minute insects with reduced wing venation. This female is using her telescoping abdomen to oviposit in a log; the male is walking on a mossy rock. ❻ ❼ ❽ ❾ Gall midges commonly hang out in spider webs, presumably for safety from predators. This habit is worldwide: the flies shown here are in Tanzania, Arizona (two images) and Costa Rica (with the web-making spider). Several distant lineages of Cecidomyiidae have similar habits.

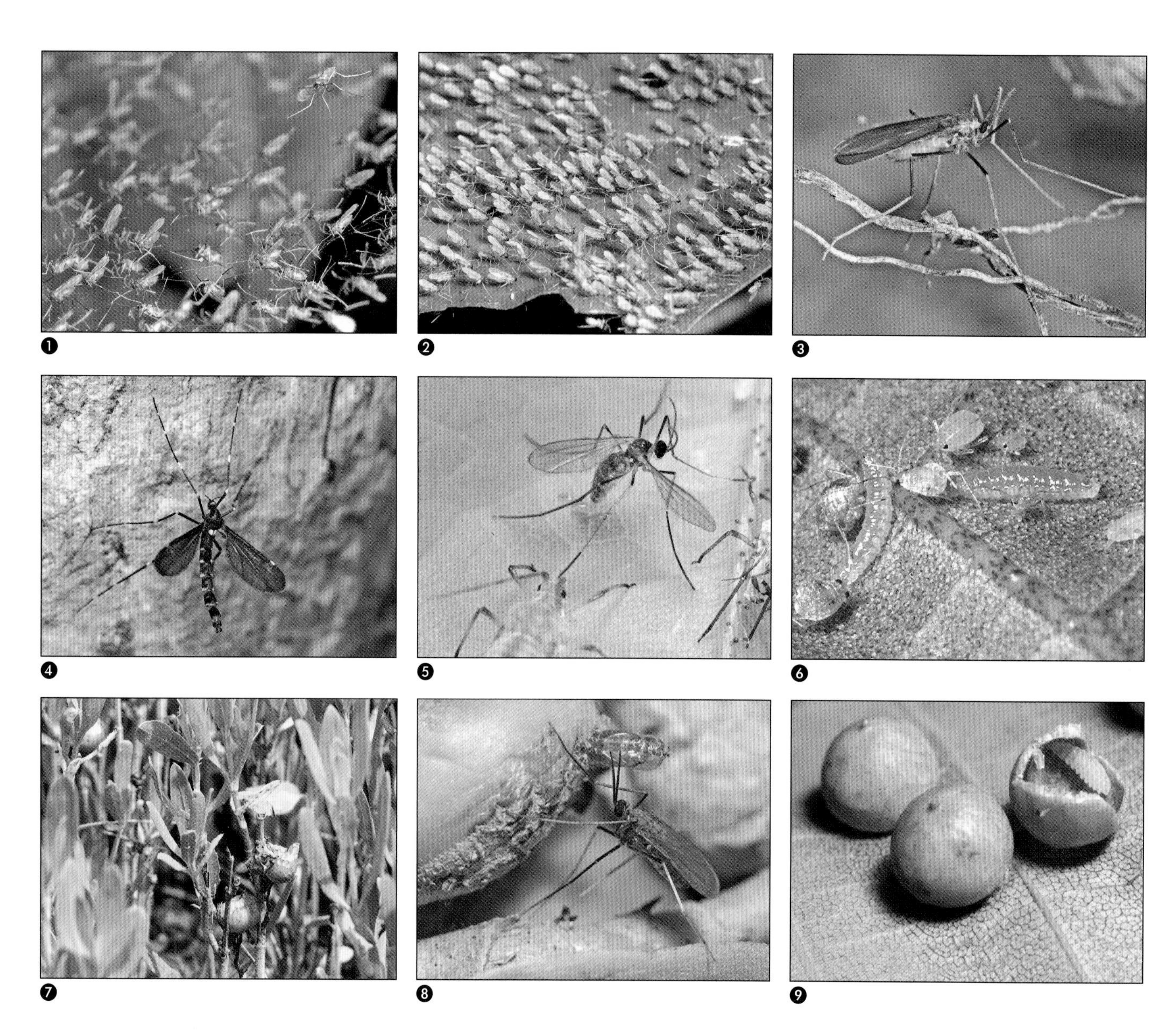

FAMILY CECIDOMYIIDAE (continued) ❶ ❷ Gall midges often aggregate in dense masses that arise as a cloud when disturbed. This mass of African cecids is sharing a spider web, while these Costa Rican flies are almost entirely covering a broad leaf. **Subfamily Cecidomyiinae** ❸ Most Cecidomyiinae — indeed, most gall midges — are delicate insects with very few adult features that can be used for identification. Identification of gall midges usually requires information about larval biology, such as host plant and gall type. ❹ Members of the tribe **Alycaulini** are unusually large (mosquito-sized) and relatively colorful, with banded legs and marked scale patterns on the abdomen. This is an unidentified Central American species. ❺ ❻ Predaceous gall midges in the genus ***Aphidoletes*** are used for the biological control of aphid pests. Shown here are an adult female, laying eggs, and some characteristically orange larvae feeding on young aphids. ❼ ❽ Species in the large, widespread genus ***Asphondylia*** are responsible for a variety of galls, like these conspicuous swellings on the salt-marsh plant called Sea Oxeye (*Borrichia frutescens*), induced by ***A. borrichiae***. *Asphondylia* larvae feed on fungi in the gall (the fungi are introduced to the plant with the egg). Pupae poke out through the gall before emerging to adult; the exposed part of the pupal shell is visible behind this adult. ❾ All 14 named species (and at least as many unnamed species) of the hickory gall midges, ***Caromyia***, make galls on the leaves of hickory trees (*Carya* spp.) in the Nearctic region. One of these galls has been cut away to expose the fly larva.

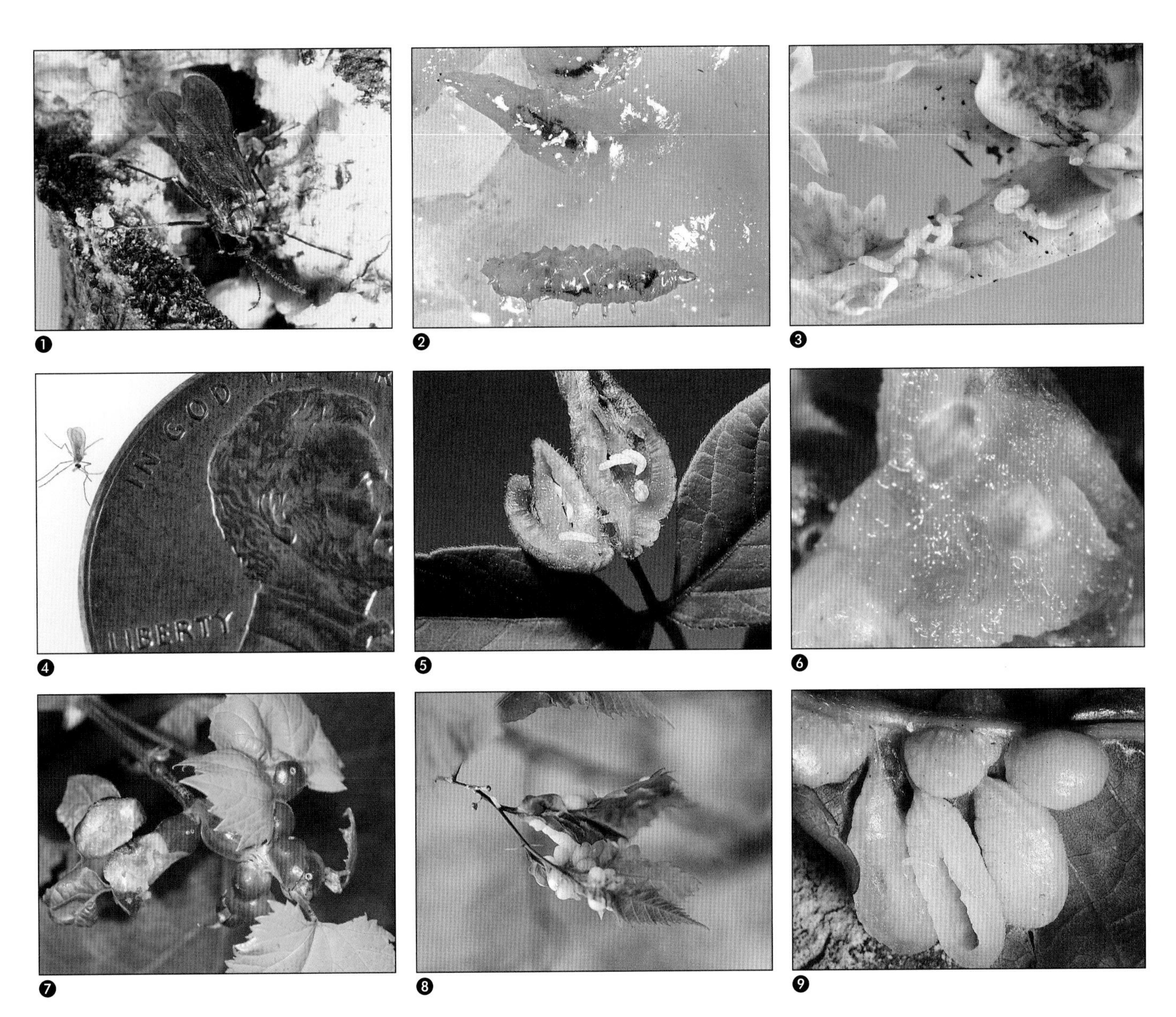

FAMILY CECIDOMYIIDAE (continued) ❶ ❷ Larvae of the genus ***Cecidomyia*** develop in masses of resin on pine trees (like the species shown here) or other conifers. ❸ ❹ ***Contarinia nasturtii***, the **Swede Midge**, is a pest of cabbage, broccoli and other crucifers. Shown here are larvae in a broccoli shoot and an adult beside an American penny in eastern North America, where this originally European pest is a relatively recent arrival. Larvae feed gregariously in a slurry of tissue dissolved by salivary secretions, later leaping away to pupate in a cocoon in the soil. ❺ ***Contarinia*** is a very large, heterogeneous genus found on a wide range of hosts. This Nearctic species, ***C. negundinis***, forms galls on Box Elder (*Acer negundo*). ❻ The orange larva that induced these succulent-looking red grape galls is probably ***Janetiella brevicauda***, although other species have been reared from this type of gall. ❼ ***Janetiella brevicauda*** is one of several closely related Nearctic species that make distinctive galls on grapes (*Vitis* spp.). ❽ ❾ Members of the Holarctic genus ***Macrodiplosis*** form fold-like galls on oak and chestnut leaves. Shown here are galls and larvae of ***M. castaneae*** on chestnut (*Castanea*).

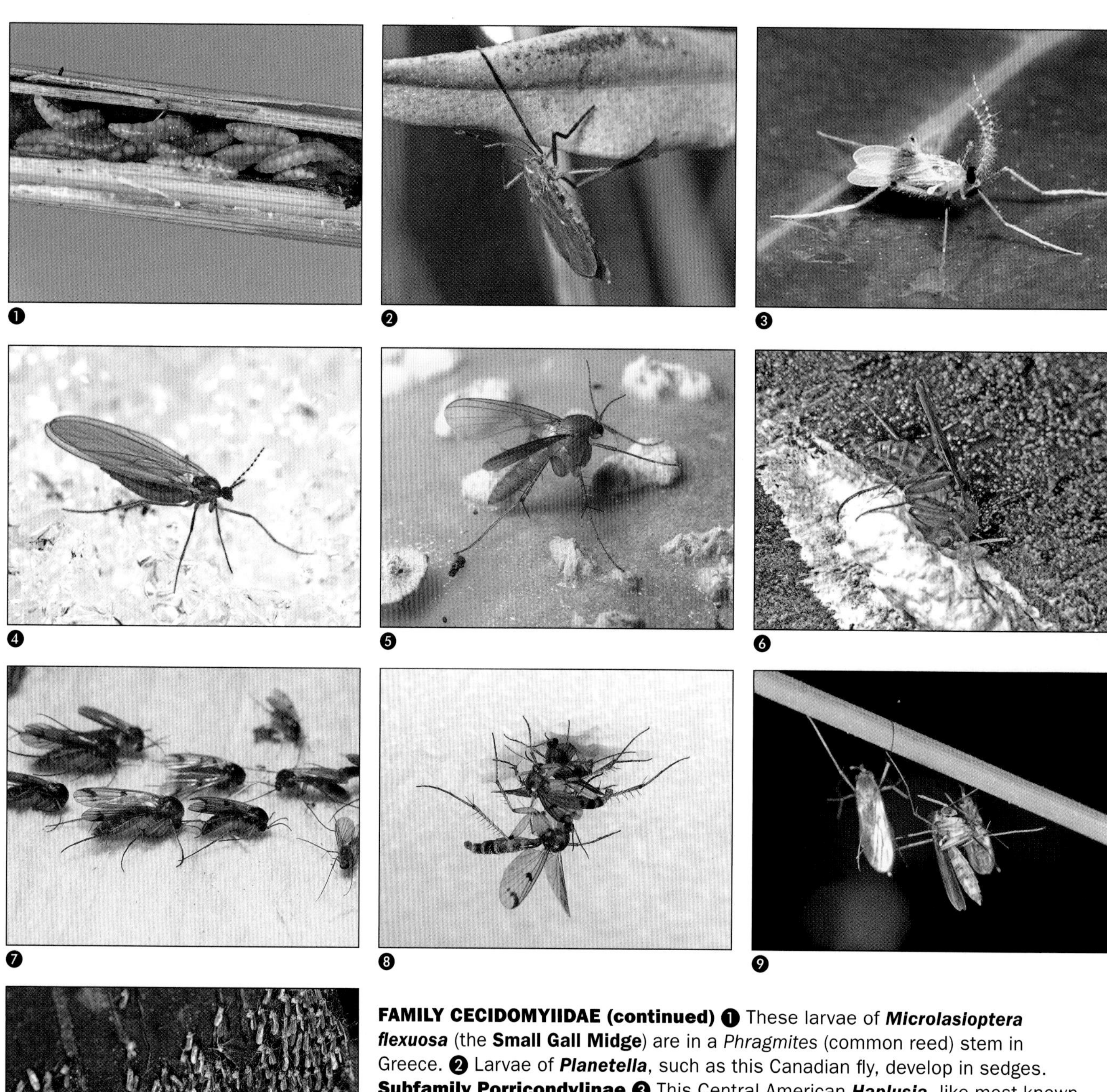

FAMILY CECIDOMYIIDAE (continued) ❶ These larvae of ***Microlasioptera flexuosa*** (the **Small Gall Midge**) are in a *Phragmites* (common reed) stem in Greece. ❷ Larvae of ***Planetella***, such as this Canadian fly, develop in sedges. **Subfamily Porricondylinae** ❸ This Central American ***Haplusia***, like most known Porricondylinae, probably developed as a fungus-feeding larva in leaf litter or other decaying plant material. The only known Central American species in the genus is *H. cincta*. **Subfamilies Lestremiinae and Micromyinae** ❹ Lestremiines and micromyines (until recently in the single subfamily Lestremiinae) are tiny, but generally distinguishable from other gall midges by their ocelli and their tarsi with five segments (tarsomeres), of which the first is very long. Other cecidomyiids lack ocelli and either have four tarsomeres or a very short first tarsomere. This female ***Eucatocha betsyae*** (subfamily Micromyinae) was walking on a patch of early winter snow in the western United States. **FAMILY MYCETOPHILIDAE (FUNGUS GNATS) Subfamily Mycetophilinae** ❺ ❻ ❼ ❽ ***Mycetophila*** is the largest and most common fungus gnat genus, with around 700 species scattered over all zoogeographic regions. Shown here are a Canadian species on the orange cap of an *Amanita muscaria* mushroom, a Costa Rican species feeding at a bird dropping, a group of overwintering adults exposed by peeling bark from a Canadian tree, and two males trying to mate with a single female under a Canadian shelf fungus. ❾ ❿ The distinctive subfamily Mycetophilinae is usually divided into two tribes, the **Mycetophilini** for *Mycetophila* and its relatives, and the **Exechiini** for *Exechia* and its relatives. These are ***Exechia***, photographed in a massive aggregation in an Australian rock crevasse. Fungus gnats often form huge aggregations in crevasses or other shaded pockets.

FAMILY MYCETOPHILIDAE (continued) Subfamily Sciophilinae ❶ ***Aneemia*** is a mostly Holarctic genus, but some species range into the neotropics and Old World tropics. Larvae are associated with either decaying wood or wood-decay fungi. This is a southwestern U.S. species. ❷ ***Boletina*** is a large, mostly Holarctic genus of about 100 species, including some that are active in cold seasons. This one is walking on the snow in Canada. ❸ Most species of ***Clastobasis***, like this pair from Tanzania, occur in the Old World tropics, but the genus was recently recorded from the Neotropical region as well. ❹ This distinctively bicolored ***Docosia*** is hunkering down on a tree trunk in the eastern United States, but other species in this genus occur in the Palaearctic, Neotropical and Oriental regions. *Docosia* is sometimes included with *Leia* in the Leiinae. ❺ This ***Gnoriste megarrhina*** is using its long proboscis to probe for nectar in the floral tube of a Piggyback Plant (*Tolmiea menziesii*) in California. *Gnoriste* and related genera are sometimes treated as a subfamily, Gnoristinae. ❻ Princess ***Leia*** and some putatively related genera are sometimes treated as a separate subfamily, Leiinae. *Leia* is a very large, widespread genus often found resting among the shaded lower branches of deciduous trees. ❼ ***Leptomorphus***, a widespread genus of about 30 species, occurs on every continent except Australia (and Antarctica). Larvae spin webs under encrusting fungi or bracket fungi, eating the spores that fall into the webs. Shown here is the eastern North American ***Leptomorphus bifasciatus***. ❽ ***Mycomya*** is a very large (around 400 species) group found in every zoogeographic region. Some classifications put *Mycomya* in a separate subfamily, Mycomyinae, which may be more closely related to the Mycetophilinae than to other "Sciophilinae." ❾ This ***Sciophila*** is resting under an encrusting fungus on a fallen log in Canada; larvae were found nearby in a loose matrix of sticky silk strands. *Sciophila* is a large, widespread genus with more than a hundred species; like most fungus gnats, it is most diverse in the Holarctic region.

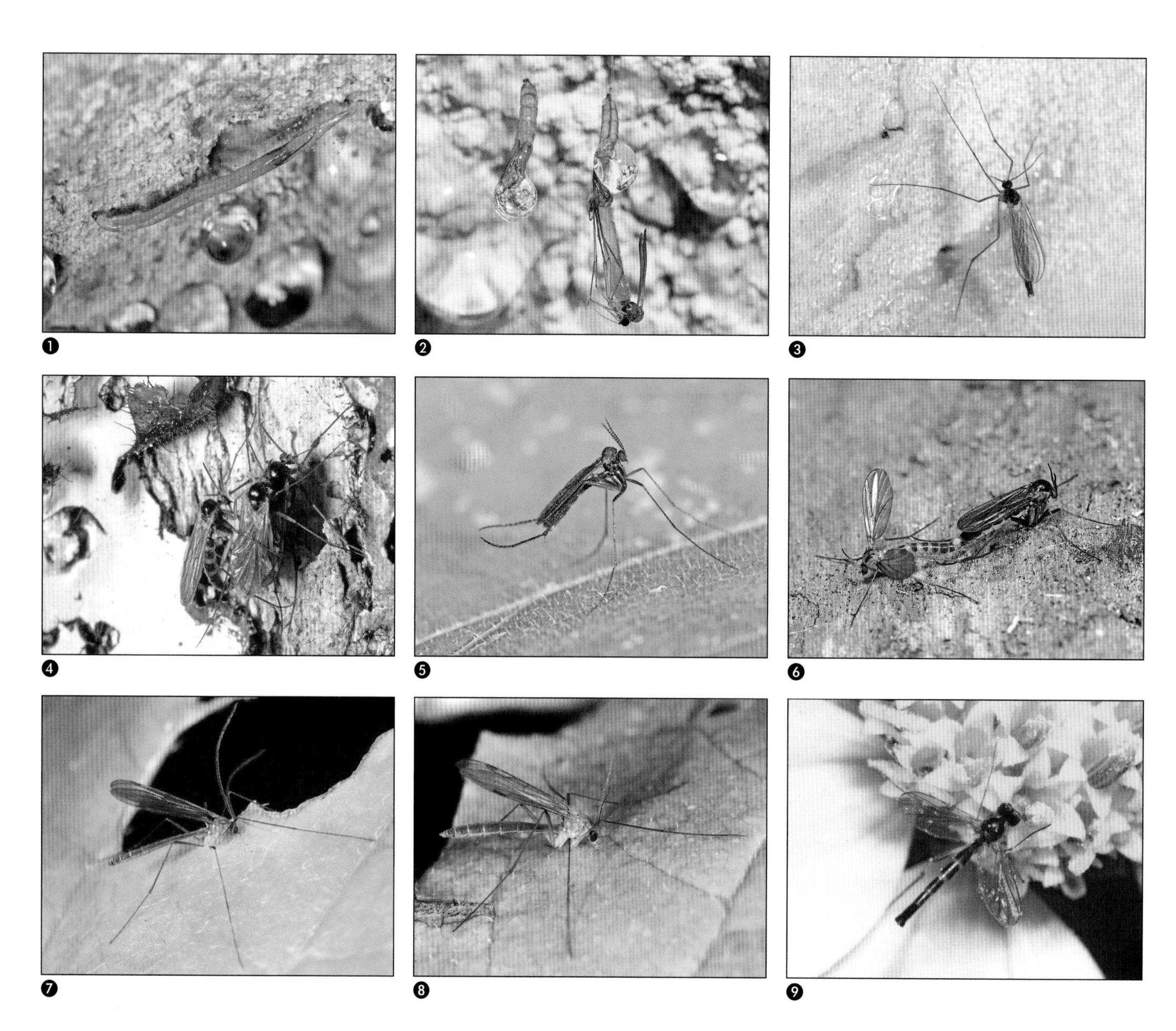

FAMILY MYCETOPHILIDAE (continued) ❶ ❷ ❸ The recently described ***Speolepta vockerothi*** lives in Canadian caves, where its larvae graze along the wet cave ceiling while suspended in a slimy silk harness. Pupae are hung from the ceiling by a silken thread. **FAMILY DITOMYIIDAE (FUNGUS GNATS)** ❹ ***Ditomyia*** is a small Holarctic genus that has been reared from bracket fungi (*Polyporus*). These two males are competing for a single female on a tree trunk in the eastern United States. ❺ The Neotropical genus ***Rhipidita*** is formally known only from a couple of Brazilian species, but this Costa Rican example is one of many undescribed species found elsewhere in South and Central America. ❻ Larvae of the Holarctic genus ***Symmerus*** have been reared from "dead but sound" hardwood logs. This pair was on a woodpile in my Canadian backyard. **FAMILY BOLITOPHILIDAE (FUNGUS GNATS)** ❼ ❽ ***Bolitophila***, the only genus in the family Bolitophilidae, is a mostly Holarctic group of about 60 slender, long-legged fungus gnats, usually associated with specific fungus hosts. The flies shown here were attracted to honey spray along a snowy December stream margin in western Canada. **FAMILY LYGISTORRHINIDAE (FUNGUS GNATS)** ❾ Lygistorrhinidae is a small family of about 25 species found scattered throughout the world but mostly restricted to tropical and subtropical regions. This is ***Lygistorrhina sanctaecatharinae***, the only Nearctic species in the family.

FAMILY DIADOCIDIIDAE (FUNGUS GNATS) ❶ All ten species in the small family Diadocidiidae are in the genus ***Diadocidia***, which occurs in the Holarctic, Neotropical and Australasian regions; known larvae live in mucous tubes under rotting logs. This ***D. ferruginosa*** was photographed in Europe by Czech dipterist Jan Sevcik. **FAMILY KEROPLATIDAE (FUNGUS GNATS) Subfamily Arachnocampinae** ❷ ***Arachnocampa luminosa***, the **New Zealand Glowworm**, fishes for small flies and similar prey by dangling silken threads studded with sticky globules. **Subfamily Keroplatinae** ❸ The long-beaked keroplatids in the genus ***Asindulum***, like this Canadian ***A. montanum***, are most often seen at flowers. Larvae spin webs under fungus-covered logs, consuming small invertebrates and spores caught in the web. ❹ Keroplatine larvae sometimes live in webbing on fungus-infested logs, where they feed on small invertebrates or fungal spores. This eastern North American larva is probably ***Cerotelion johannseni***, the only North American member of this mostly Palaearctic genus. ❺ ❻ ❼ Some larval Keroplatinae, like this ***Keroplatus militaris***, live in mucous tubes on fungi or under bark, while others make webs to trap fungal spores or small animals. Shown here are a larva, a pupa in a silk cocoon and an adult hanging from a shelf fungus. Some *Keroplatus* species, such as the Palaearctic *K. tipuloides* found in groups in glutinous webbing under polypore fungi, are bioluminescent. ❽ ❾ These silken strands hanging from a Costa Rican cavern look much like those produced by the New Zealand Glowworm, but the larvae of this species (possibly ***Neoditomyia aeropiscator***; seen here in a horizontal slime tube along the cavern ceiling) are not bioluminescent.

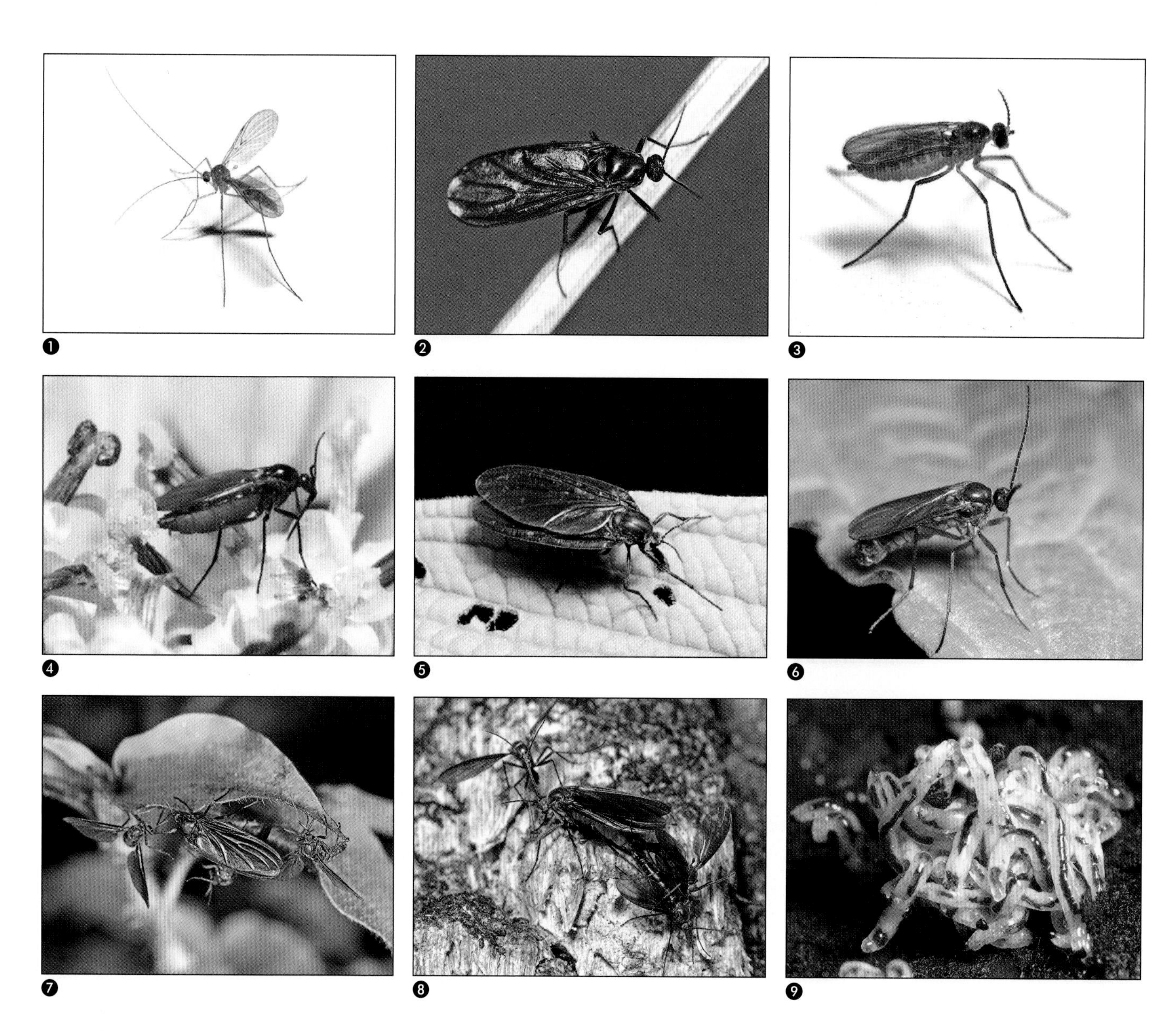

FAMILY KEROPLATIDAE (continued) Subfamily Macrocerinae ❶ The long antennae referred to by the name ***Macrocera*** make the 200 or so species in this cosmopolitan group fairly distinctive. Cave-dwelling *Macrocera* larvae are known as "monorail worms" because of the way they smoothly glide up and down silken tracks on cave walls and ceilings. **FAMILY SCIARIDAE (DARK-WINGED OR BLACK FUNGUS GNATS)** ❷ Sciaridae are usually tiny, delicate, dark-winged flies with the eyes forming a continuous "bridge" on top of the head. ❸ Although almost all sciarids are dark flies with dark wings, a few, such as this Tanzanian fly, are entirely pale, and the females of several species have pale or reddish abdomens. ❹ The mostly Nearctic genus ***Eugnoriste*** has a long beak used for feeding in flowers. This is ***E. occidentalis***. ❺ As the species name suggests, this Central American ***Hybosciara gigantea*** is a large fly with conspicuously expanded wings. This small genus has a couple of species in the neotropics and one in Taiwan. ❻ The "typical" genus ***Sciara*** is found worldwide; this Canadian specimen is probably ***S. humeralis***. ❼ ❽ Sciarid females are often mobbed by several males in mating aggregations, such as this swarm of black sciarids in Costa Rica and the several males trying to mate with an black and orange female (possibly ***Odontosciara nigra***) in the United States. ❾ Several sciarid species slide along as rolling masses of larvae, sometimes forming conspicuous undulating balls slowly moving along the forest floor. This ball of larvae was found under the bark of a rotten log.

7

The Lower Brachycera and Empidoidea

LIKE THE "NEMATOCERA," OR LOWER DIPTERA, the lower Brachycera form an easily recognized but arguably artificial group, and just as the lower Diptera are basal to the rest of the Diptera, the families traditionally treated as lower Brachycera are basal to the rest of the Brachycera. The classification of those basal lineages into useful groups has proven difficult, so much so that the tumultuous taxonomic history of the lower Brachycera has left us with a range of alternative higher classifications. The enormously influential classification systems followed in *Handbuch der Zoologie* (Hennig 1973) and slightly modified for the first volume of the *Manual of Nearctic Diptera* (McAlpine et al. 1981) divided the lower Brachycera into two apparently distinct infraorders, the Tabanomorpha or Homeodactyla (having "feet" with three broad, flaplike pads, one in the middle and one under each tarsal claw) and the Asilomorpha (usually with two tarsal pads, one under each tarsal claw). This system was convenient and intuitive for most people interested in classifying and identifying flies, but, alas, it has fallen victim to the rigors of the rigid cladistic approach to classification that Hennig himself pioneered. As a result, multiple conflicting classifications have sprung up around alternative interpretations of the phylogenetic relationships between the lower brachyceran families.

The least controversial change to the traditional system was eviction from the Asilomorpha of the dance flies and their relatives, the Empidoidea, following recognition that they are more closely related to higher Brachycera, or Cyclorrhapha (such as blow flies and flower flies), than to other families traditionally treated as lower Brachycera. Empidoidea and Cyclorrhapha share many morphological and developmental similarities (only three larval instars, for example) and dipterists have long been in general agreement that the Empidoidea and Cyclorrhapha together make up one of the main branches of the dipteran tree; this branch is called the Eremoneura. Since it includes almost half of all Diptera, the Eremoneura is really more of a main trunk than a branch.

The flies that remain in the lower Brachycera after removal of the Empidoidea can still be divided into Tabanomorpha (three tarsal pads) and Asilomorpha (two tarsal pads), even though some recent papers have suggested significantly different classifications. These alternative classifications have been championed mostly because of disagreement about the relationships and composition of the Asilomorpha. Different publications posit

OPPOSITE PAGE The family Asilidae is an entirely predaceous lineage of lower Brachycera. The two pulvilli (tarsal pads) of each "foot," typical of Asilomorpha and Eremoneura, are large and distinct on this fly.

Figure 7.1 **The Main Subgroups of the Lower Brachycera**

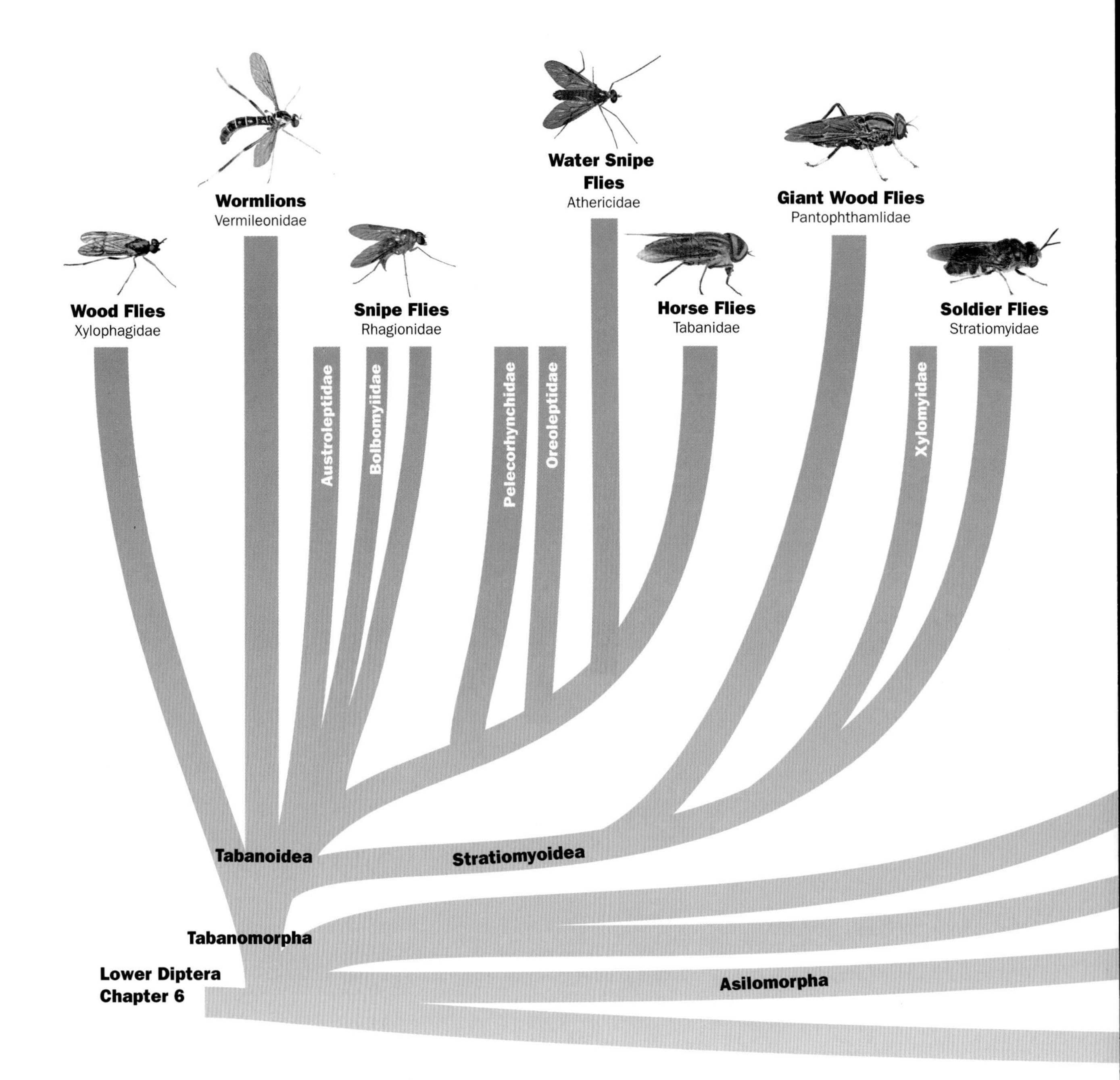

almost every possible permutation and combination of relationships between the asilomorph lineages and other Brachycera, but recent proposals by Woodley et al. (2009) and Wiegmann et al. (2011) illustrate two popular interpretations and the resultant conflicting classifications. Woodley et al. (2009) combine the Nemestrinidae, Acroceridae and the families treated here as Asilomorpha with the Eremoneura to form the infraorder Muscomorpha (a name previously used as a synonym of Cyclorrhapha). This concept of Muscomorpha, first proposed by Woodley in 1989, is widely accepted, but a more recent paper (Wiegmann et al. 2011) argues the alternative view that the Asilomorpha comprise a single lineage more closely related to the Tabanomorpha than to the Eremoneura. In this paper Wiegmann and 26 influential coauthors indicate a single

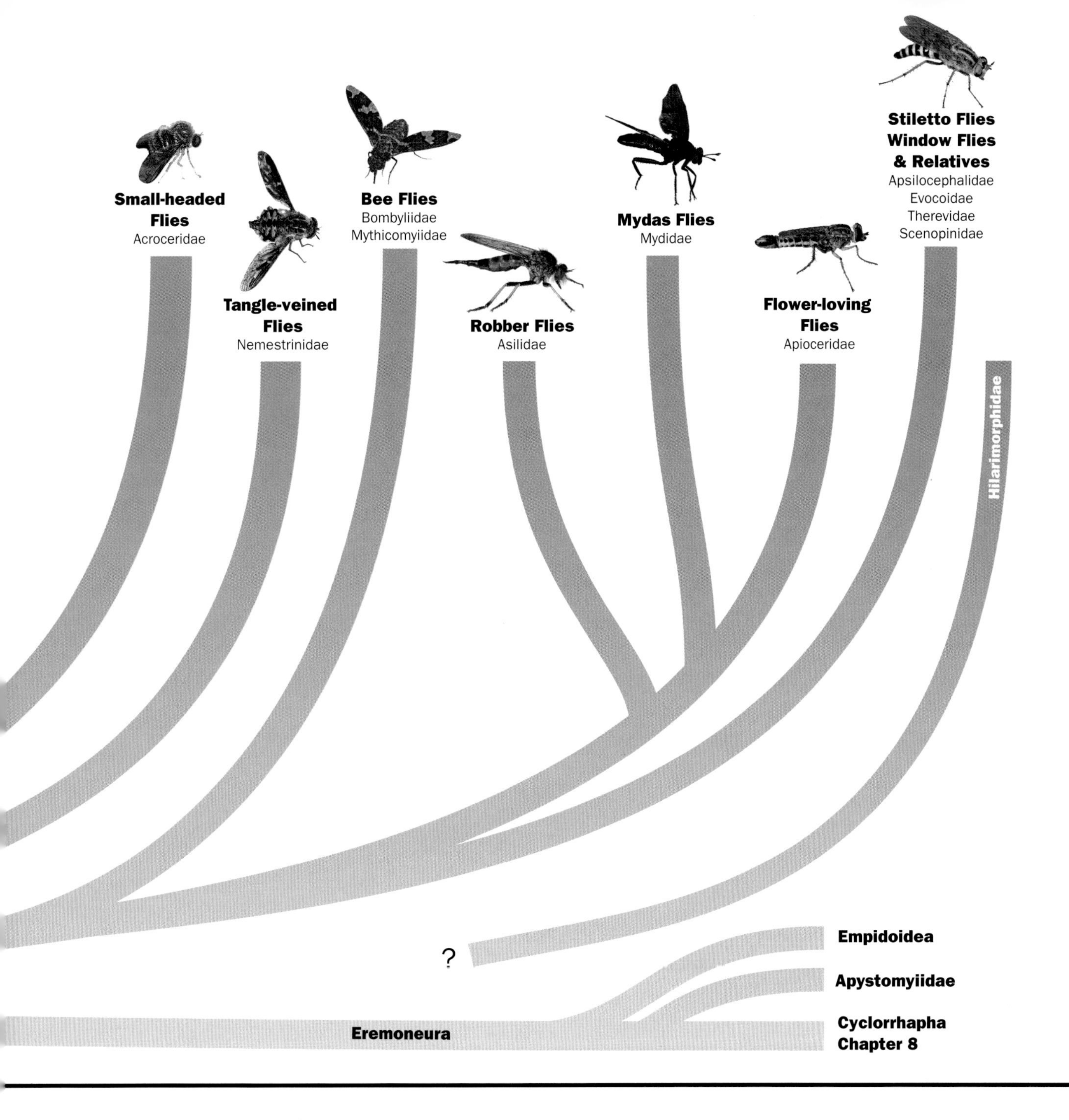

lineage for all non-Eremoneura Brachycera, and they call it "Orthorrhapha" (an old term previously used for all non-Cyclorrhapha Brachycera, including the Empidoidea). These alternative classifications reflect fundamental disagreements over whether or not the lower Brachycera is a monophyletic or "real" group. Here I skirt the entire issue by simply dividing the suborder Brachycera into the Eremoneura on one hand and the lower Brachycera on the other, much the way we divide the entire order Diptera into the Brachycera on one hand and the lower Diptera on the other. This leaves open the question of whether the lower Brachycera form a natural (monophyletic) group or not.

The lower Brachycera are here divided into the Asilomorpha and the Tabanomorpha; some authors further divide the Tabanomorpha into two or three infraorders: Stratiomyomorpha, Tabanomorpha and sometimes Xylophagomorpha. This proliferation of infraorders complicates the higher classification of flies and seems unnecessary in view of morphological and molecular data supporting a broadly defined Tabanomorpha (Wiegmann et al. 2003; Yeates 2002; Yeates and Wiegmann 2005), although Wiegmann and colleagues (2011) now suggest that the soldier flies and their relatives (Stratiomyoidea) might be more closely related to Asilomorpha than to horse flies and their relatives (Tabanoidea). Despite this dispute over the possible paraphyly of the group, the infraorder Tabanomorpha is used here in the broad sense of McAlpine et al. (1981), for all lower Brachycera other than the Asilomorpha. Tabanomorphs all have a pulvilliform empodium, which means that they have the usual pads (pulvilli) under their tarsal claws plus another pad between the claws. Most Brachycera other than the Tabanomorpha have a bristle-like empodium and thus have only two tarsal pads. The lower Brachycera with only two tarsal pads are here treated as Asilomorpha, although, as discussed below, this too might be an artificial group.

To reiterate then, the Brachycera are here organized into the Tabanomorpha, Asilomorpha and Eremoneura, with the Eremoneura further divided into the Empidoidea and the Cyclorrhapha. One odd species, *Apystomyia elinguis*, does not fit into any of these groups and so is shoehorned in between the Empidoidea and the Cyclorrhapha as its own one-species family, Apystomyiidae. The term "lower Brachycera" has previously been used to refer to all of these groups except the Cyclorrhapha, but it is here used to include only the Tabanomorpha and Asilomorpha — in other words, all Brachycera other then Eremoneura.

Tabanomorpha

Horse Flies, Soldier Flies, Snipe Flies and Other "Homeodactyla"

Members of the infraorder Tabanomorpha (the Homeodactyla of some authors) are relatively large-bodied lower Brachycera with feet that end in three broad pads — the pulvilli and a broad empodium — in contrast to the two tarsal pads of other Brachycera. They also usually have conspicuous tibial spurs and antennae with more than four flagellomeres. There remains some disagreement about the relationships between the very different groups included in the Tabanomorpha, but the main tabanomorph lineages are here treated as the superfamilies Tabanoidea (Tabanidae, Athericidae, Rhagionidae, Oreoleptidae, Pelecorhynchidae, Xylophagidae and Vermileonidae) and Stratiomyoidea (Pantophthalmidae, Xylomyidae and Stratiomyidae). These lineages are sometimes treated as separate infraorders and, as discussed further below, the families Vermileonidae and Xylophagidae are often excluded from the Tabanoidea. The problematic parasitic families Nemestrinidae and Acroceridae, traditionally treated as a separate superfamily Nemestrinoidea and often treated as part of the Asilomorpha, are included here with the Tabanomorpha, although the origins and phylogenetic affinities of these two unusual families remain intriguingly uncertain.

Tabanoidea

Tabanidae

Horse Flies, Deer Flies, Clegs and Their Relatives

The family Tabanidae is a common and familiar worldwide group of some 4,400 named species of generally stout, fast-flying insects ranging in size from just a few millimeters up to about 3 centimeters. Females of most species bite, slicing skin with their stabbing maxillae and scissors-like mandibles before imbibing the blood so liberated from ruptured capillaries. A few species feed only on pollen and nectar, and several of the flower-visiting species have spectacularly long mouthparts. Some Tabanidae use their long mouthparts to deeply (and painfully!) penetrate vertebrate skin in search of blood, while others are strictly nectar feeders. Male tabanids, easily distinguished from the females by their large, contiguous eyes and non-biting mouthparts, are sometimes common on flowers, but males of many species are encountered rarely or only at hilltop lek sites. Others have the intriguing habit of hovering motionless, often at or near eye level, as they await conspecific females.

Scaptia lata, a large biting fly in the Tabanidae subfamily Pangoniinae, occurs in Argentina and Chile, where it is known as Tabano Negro or Coliguacho. Note the three tarsal pads characteristic of the Tabanomorpha.

Once mated, female Tabanidae usually deposit masses of eggs on objects overhanging potential larval habitats, into which the larvae drop upon hatching. Look for the distinctive glutinous heaps of hundreds of tabanid eggs on foliage overhanging water or wetlands, especially on emergent vegetation in ponds. Larvae are generally subterranean predators that use fang-like mandibles to impale prey, immobilizing their victims with a paralytic poison produced in special glands and delivered through poison canals in the mandibles. Similar poison canals are found in related Athericidae and Pelecorhynchidae (and presumably Oreoleptidae). Most tabanid larvae live in aquatic substrates or wet ground, where they consume other invertebrates, but some of the larger horse fly larvae are known to ambush small vertebrates (such as toads) from underfoot, dragging them underground like a scene from a horror movie. Some species are entirely terrestrial and some are associated with specialized aquatic or semiaquatic habitats such as tree holes and bromeliads. Mature larvae usually pupate with the front end of the pupa near to, or sticking out of, the surface of the substrate (although some pupate in mud cylinders); muddy pond margins are often briefly but conspicuously studded with the pupal exuviae of recently emerged tabanids.

A few horse flies are crepuscular or nocturnal, but most tabanid adults are diurnal feeders that seek their hosts primarily by sight and are therefore not effectively deterred by insect repellents. Nor are they always stopped by layers of clothing that would shut out lesser biting flies: I have returned from the field many times with my T-shirt stuck to my shoulders by caked blood from deer fly bites, and it's always a shock when a long-beaked species nails me right through a heavy canvas shirt. Fortunately, Tabanidae are not in the same league as lower Diptera when it comes to the transmission of disease; relatively few human health problems are associated with horse flies and deer flies. As anyone who has tried to swat one of these swift flies can attest, however, tabanids are nervous feeders that tend to visit hosts repeatedly rather than engorging at a single sitting like a mosquito. This gives them a dangerous potential to move disease organisms such as anthrax and tularemia quickly from host to host, even though the pathogens do not develop in the fly. Remarkably, only a few disease organisms have a specific relationship with tabanids, in contrast with the way several pathogens depend on mosquito vectors. One of the few such relationships is that between deer flies and the filarial worm that causes loa loa in west and central Africa. Even though Tabanidae on the whole are not significant vectors of human disease, their numbers can be high enough to make them annoying nuisances, interfering with outdoor activities and limiting tourism in areas with large tabanid populations.

The family Tabanidae is divided into four subfamilies, the Tabaninae (horse flies and clegs), Pangoniinae (long-tongued horse flies), Chrysopsinae (deer flies and their relatives) and Scepsidinae.

Tabaninae The typical species in the subfamily Tabaninae are the familiar horse flies in the genera *Hybomitra* and *Tabanus*, relatively large and robust flies that usually lack conspicuous wing markings. Horse flies, as the name suggests, often preferentially bite large animals, and they can be serious pests of cattle, horses and other livestock, although many are also annoying and painful pests of people. The enormous genus *Tabanus*, with more than a thousand species, includes a diversity of large, conspicuous flies in every zoogeographic region. Living individuals (but not dried specimens) of most Tabanidae have brightly colored or patterned eyes. Some *Tabanus* are called "greenheads" because of their brilliant green eyes; *Tabanus nigrovittatus*, for example, is an exceptionally vicious North American species known as the Salt Marsh Greenhead.

This subfamily also includes the conspicuously pigmented species of the genus *Haematopota*, known as clegs. *Tabanus* and *Haematopota* both include significant livestock pests that can cause enough annoyance and blood loss to stress out the animals, leading to weight loss and lower milk yields. Many other Tabaninae are relatively rare. *Atylotus*, for example, includes a number of infrequently collected, poorly known species associated with North American sphagnum bogs.

Pangoniinae The worldwide (but mostly tropical and southern hemisphere) subfamily Pangoniinae includes the long-tongued horse flies, many of which have spectacular beaks used to probe for nectar in deep-throated flowers. Some of them coevolved with particular plants or guilds of plants that are now dependent on these flies for pollination. Not all of the long-beaked Pangoniinae are able to bite, but those that do can be most impressive: a Central or South American *Scione* driving its syringe-sized beak through your shirt and into your arm is a remarkable sight (once!). Many of the most important biting flies of the southern hemisphere belong to this subfamily. For example, species of the large genus *Scaptia* are abundant biting flies in Australia (where Tabanidae are known as March flies) as well as in South America. One of Chile's most spectacular flies is the bumble bee–sized, brilliant black and orange *Scaptia lata*, a beautiful but vicious biting pangoniine.

The few North American pangoniines are generally uncommon non-biting species, often with unusual habits. *Apatolestes*, for example, is a southwestern American genus that includes at least one species, *A. actites*, that oviposits in crustacean burrows in beach habitats. *Stonemyia* species range north to Canada, where they can be common on flowers, but all species of the genus remain unknown as larvae. *Goniops* includes a single eastern North American species, *G. chrysocoma*, in which the female has the remarkable habit of laying eggs in a mass under a leaf, then spending the rest of her life straddling the eggs, guarding them from parasitoids such as *Telonomus goniopsis* (a parasitic wasp known only from *Goniops* eggs).

Unguarded egg masses of other tabanid species are usually frequented by different *Telonomus* (Scelionidae) species and other hymenopteran egg parasitoids in the family Trichogrammatidae. These tiny wasps seem to be the major natural enemies of horse and deer flies.

Chrysopsinae The subfamily Chrysopsinae is a group of generally boldly pigmented Tabanidae dominated by the huge and widespread genus *Chrysops*, whose hundreds of species are widely known as deer flies, a name also sometimes used for the closely related genus *Silvius*. Deer flies are relatively small, attractive flies with patterned wings, colorful bodies and iridescent eyes strikingly patterned in reds and greens. Their larvae often occur at higher densities than horse flies, possibly because they are less cannibalistic or more omnivorous than their larger relatives in the Tabaninae, but feeding habits of chrysopsine larvae are poorly known. They probably prey on other invertebrates, as do the better-known larvae of Tabaninae.

Scepsidinae The subfamily Scepsidinae is a problematic group of three or four odd species found on beaches in southern Africa and Brazil. These fuzzy gray seashore flies have greatly reduced mouthparts and behave more like Therevidae than Tabanidae, perching out in the open on the sand and making short flights when disturbed. Work by Shelah Morita and colleagues suggests that at least some of the species now treated as Scepsidinae are related to Chrysopsinae.

Rhagionidae and Bolbomyiidae

Snipe Flies

The family Rhagionidae has historically been a sort of taxonomic trashcan for the Tabanoidea, serving as a receptacle for a wide variety of lower brachyceran genera of uncertain relationships. The Vermileonidae, Athericidae and Pelecorhynchidae — in fact, all tabanoid families except for Tabanidae itself — have been part of the Rhagionidae at some time or another. Some groups were included in the Rhagionidae more by default than on the basis of a clearly documented close relationship to *Rhagio* and other "true" Rhagionidae.

Adult Rhagionidae, called "snipe flies" for reasons unfathomable to me, are often colorful, conspicuous flies, most familiar for the *Chrysopilus* and *Rhagio* species that periodically abound on foliage or tree trunks. Both of these large genera have been reported as predators, an unsubstantiated assertion uncritically repeated in many field guides and textbooks. Although the larvae of both genera are predators of soft-bodied soil invertebrates, the adults do not appear to feed. Males of some North American *Rhagio* take a "hunting" position — usually head down on tree trunks — but they are hunting for females, which are usually seen on or near the ground, rather than prey items. Even more conspicuous are the large, brilliantly pigmented eastern North American species *Chrysopilus ornatus* (Ornate Snipe Fly) and *C. thoracicus* (Golden-backed Snipe Fly), which take provocatively visible positions on lush low vegetation as they too seek mates, not prey.

Chrysopilus and *Rhagio* (and thus most Rhagionidae) rarely visit flowers, but a few snipe flies are specialized nectar feeders. For example, the long-tongued species in the small genus *Arthroteles* are part of a guild of long-tongued pollinators associated with long-tubed South African flowers. Other Rhagionidae have mouthparts modified for piercing skin, and two genera, the Holarctic genus *Symphoromyia* and the Australian genus *Spaniopsis*, are blood feeders. Neither of these biting snipe fly genera are confirmed disease vectors or major pests, although some blood-feeding species of *Symphoromyia* (so-called Rocky Mountain Black Flies) sometimes reach nuisance densities in parts of western North America. Some other *Symphoromyia*, including species in eastern North America and the Palaearctic region, apparently don't blood feed. Larval *Symphoromyia* have been reported both as moss eaters and as predators; larval *Ptiolina* also reportedly have unusual (for Tabanomorpha) vegetarian habits, apparently feeding on mosses and liverworts. Most other rhagionid larvae, except for the predaceous larvae of the common genera *Rhagio* and *Chrysopilus*, remain unknown.

Species in the Holarctic genus *Symphoromyia* (Rhagionidae) have blood-feeding females, but this big-eyed fly is a newly emerged male.

As might be expected of a family with such a checkered taxonomic history, the Rhagionidae has been variously defined and variously splintered into subfamilies, some of which are defined differently by different authors. For example, Nagatomi (1982) divided the group into four subfamilies: Glutopinae (now usually treated as part of the Pelecorhynchidae), Rhagioninae, Spaniinae and Austroleptinae. Stuckenberg (2001), on the other hand, treated the latter three groups as the separate families Rhagionidae, Spaniidae and Austroleptidae and included the Pelecorhynchidae in the Rhagionidae. Kerr (2010) provides the most recent rhagionid reshuffle, and his classification is more or less followed here, even though it differs from Stuckenberg (2001) and recently published taxon lists in recognizing the Pelecorhynchidae as a separate family and treating the Spaniidae as part of the Rhagionidae.

The subfamily **Rhagioninae**, the least specialized subgroup of Rhagionidae, has traditionally been home to the great majority of the 700 or so species in the family, including 165 species in the mostly northern hemisphere genus *Rhagio* and more than 300 species in the exceptionally widespread genus *Chrysopilus*. However, Kerr (2010) suggests that *Chrysopilus* is more closely related to the subfamily **Spaniinae** than to other Rhagioninae and should therefore be put in its own subfamily, **Chrysopilinae**. In Nagatomi's concept of the Spaniinae, the largest genus (with 22 species) is the Holarctic genus *Ptiolina* and the second largest (with seven species) is the Australian genus *Spaniopsis*; in Kerr's concept the Spaniinae include the Holarctic blood-feeding genus *Symphoromyia*.

Whether all the above genera and their less speciose relatives are treated as one subfamily (Rhagioninae), two subfamilies (Rhagioninae and Spaniinae) or three subfamilies (Rhagioninae, Chrysopilinae and Spaniinae) does not change the fact that they together form one natural group that we might consider the "typical" Rhagionidae. Almost all the snipe flies encountered by insect collectors or photographers will belong to this large and relatively uncontroversial lineage. Some dipterists treat the Spaniinae as a separate family, but that seems unjustifiable in view of the close relationship between Rhagioninae and Spaniinae and the fuzzy boundaries between these groups. One of the largest genera of Rhagioninae, the archaic-looking genus *Atherimorpha*, with 50 species, shows a classic Gondwanan distribution spanning Australia, southern South America and South Africa. *Atherimorpha* is closely related to *Arthroteles*, a genus endemic to South Africa.

Most of the debate about the limits of the Rhagionidae centers on a few atypical genera that fit into neither the Rhagioninae nor the other subfamilies of "typical" Rhagionidae. One such "atypical" genus is *Arthroceras*, which comprises a few poorly known species in North America, Japan, Siberia and China. The placement of this odd group, still unknown from larvae, is tentative, but it is most often treated as a rhagionid and as the only genus in the subfamily **Arthroceratinae**.

The atypical genera *Bolbomyia* and *Austroleptis* are probably the most controversial "rhagionid" genera, and it is likely that neither genus really belongs in the Rhagionidae. *Bolbomyia* is a rarely collected Holarctic genus sometimes abundant in woodland clearings or on early spring flowers but still unknown as larvae. This small, empid-like genus was first described in the Xylophagidae (in 1850, based on an amber fossil) and was later bounced back and forth between the Xylophagidae and the Rhagionidae while dipterists argued about its placement. On the basis of male genitalic structure, Sinclair et al. (1994) showed that *Bolbomyia* was more closely related to Tabanidae than to Rhagionidae, but Stuckenberg (2001) still treated this genus as the subfamily Bolbomyiinae in his concept of Rhagionidae. Kerr (2010) suggests a separate family, **Bolbomyiidae**, for *Bolbomyia*; this well-substantiated suggestion is now followed by many dipterists.

The other controversial genus currently treated by some authors as a snipe fly is *Austroleptis*, an enigmatic south temperate group of eight species split between Australia and Chile. Although Nagatomi (1982) treats this genus (in its own subfamily) in the Rhagionidae, others (for example, Stuckenberg 2002) have treated it as a separate family, and some have suggested that it is more closely related to the Xylophagidae than to the Rhagionidae (Sinclair et al. 1994). Kerr (2010) provides evidence that *Austroleptis* is closely related to the Rhagionidae but nonetheless treats it as the separate family **Austroleptidae**. Neither the family Austroleptidae nor the family Bolbomyiidae is discussed as a separate section below, but they are indicated as separate families on Fig. 7.1. *Austroleptis* adults have been noted anecdotally as emerging from dead wood, but the larvae remain unknown. Larval characteristics are significant in the classification of lower Brachycera families, so the eventual discovery of larvae of *Austroleptis* (and *Bolbomyia* and *Arthrocerus*) will undoubtedly lead to reconsideration of the placement of these "problem" genera.

Athericidae

Water Snipe Flies

The family Athericidae is a group of about 120 species of aquatic flies treated as part of the superficially similar Rhagionidae until Stuckenberg (1973) documented the many morphological attributes they share with horse flies and deer flies. Further evidence for the close relationship between tabanids and athericids has accumulated since, and now the only concession to the old idea that *Atherix* and its relatives were rhagionids (snipe flies) is the common name "water snipe flies."

Known larvae of water snipe flies are distinctive aquatic predators, with prominent prolegs and other body processes, and adults of the most common north temperate genus, *Atherix*, are short-lived non-biting flies. Adults of some genera, however, are biting flies that feed on blood from both cold-blooded and warm-blooded animals. Stuckenberg (2000) records two species of *Atrichops* as attacking frogs in Japan and three species of *Suragina* as biting birds in Africa, humans and horses in Mexico and cattle in Japan. He also cites records of *Dasyomma* biting cattle in Chile, and I can add from personal experience that *Dasyomma* females sometimes persistently bite entomologists doing fieldwork in southern Chile. I've also given blood to species in the widespread genus *Suragina* — the largest

Females of the eastern North American *Atherix lantha* deposit eggs in communal masses on objects over the flowing water in which their predaceous larvae develop.

genus of Athericidae — which has about 40 Old World, mostly tropical species plus a smattering of Australian and Neotropical species of questionable relationship to their Old World congeners. *Dasyomma*, with 32 species, also has a split distribution, with centers of diversity in southern South America and Australia. The family Athericidae is sometimes divided into two subfamilies, with *Dasyomma* all by itself in the subfamily **Dasyommatinae** and other genera in the subfamily **Athericinae** (Stuckenberg 2000).

The major north temperate genus of water snipe fly is *Atherix*, a non-biting genus with 22 species, including the common eastern North American *Atherix lantha*, well known for its habit of laying eggs in communal masses under bridges or other structures that overhang flowing water. Masses of *Atherix* eggs mixed with dead female flies can be conspicuous, routinely reaching fist size or larger. *Atherix* is a mostly Holarctic group but also includes a couple of Neotropical and one Oriental species. The only other athericid genus with more than three species is *Atrichops*, mentioned as the frog-biter above. *Atrichops* has nine named species, mostly Palaearctic but with one extant species in each of the Oriental and Afrotropical regions, and one extinct species from the Nearctic region. The remaining few extant athericid genera are smaller, less frequently encountered and mostly restricted to small areas. Several athericids are known only from fossils.

Vermileonidae

Wormlions

William Morton Wheeler's 1930 book *Demons of the Dust,* one of the great classics of insect natural history, devotes some 377 pages to the peculiar pit-making predators known as antlions and wormlions. These distantly related groups of insects have converged on the construction of conical pits used to trap other small invertebrates, and most naturalists are familiar with the pits made by larvae of the widespread and common antlions (order Neuroptera) found in sand or loose soil surfaces worldwide. Wheeler, however, devotes most of his book to a detailed account of the infrequently encountered and oddly distributed wormlions (flies in the family Vermileonidae) — the true "demons of the dust" — that make their pits in dry, dusty soil or fine sand. He observed captive vermileonid females laying eggs one by one by "thrusting the extensible tip of the abdomen down into the sand." Each oviposition site becomes a larval pit. Prey are encouraged to take a one-way trip down the sloping sides of the pit: the waiting wormlion flicks its body, flinging dirt and collapsing the pit's wall to bring its victim within striking range. The strike itself is swift as the wormlion wraps its body around the victim, immobilizing it. The corpses of past prey are later flicked out of the burrows. Every wormlion pit I have seen has been protected from the rain by an overhanging rock or other shelter; larval Vermileonidae sometimes occur in high densities in such appropriately sheltered pockets of loose soil.

The wormlion family, Vermileonidae, comprises some 60 generally rare species known from scattered areas of the Holarctic, Afrotropical, Oriental and Neotropical regions. New World wormlions are represented only by one undescribed Central American species, a couple of uncommon species in Mexico, a few isolated records from Cuba and Jamaica and a couple of elusive species from the western United States. The New World species belong to the mostly Holarctic genus *Vermileo*; other wormlion genera are mostly distributed in the Oriental tropics or southern Africa, although some species of the mostly African genus *Lampromyia* also occur in southern Europe. Vermileonids were discovered for the first time in China in 1979 and in Madagascar in 2002. The Madagascar species, put in its own genus *Isalomyia*, was described from a single reared fly (Stuckenberg 2002). Oddly, wormlions are absent from Chile and Australia, countries that are otherwise home to diverse and interesting tabanomorph faunas and seemingly rich in appropriate habitats. Adult vermileonids are rarely seen in nature; most species are known only from reared material. Rearing them requires patience, as they can take more than a year to develop from seemingly mature larvae to adults.

The family Vermileonidae is one of the "problem" lineages among the lower Brachycera, and dipterists have long debated its relationships. There seems to be some consensus, following Sinclair et al. (1994) and others, that the Vermileonidae is an early diverging lineage of Tabanomorpha, but there are still dissenting views. We can at least be certain that this remarkable group of flies, with its many unique behavioral and morphological attributes, is itself a monophyletic and distinctive group.

Pelecorhynchidae

Pelecorhynchid Flies

Species in the genus *Pelecorhynchus* are among the largest and most attractive of all tabanomorphs, and their aesthetic appeal is magnified by an intriguingly limited range comprising small parts of eastern Australia, Tasmania and Chile. These rarely seen non-biting flies have the general appearance of large horse flies dressed in formal velvet, a resemblance reflected in their inclusion in the family Tabanidae prior to creation of the family Pelecorhynchidae by Mackerras and Fuller in 1942. Mackerras and Fuller not only named the group but thoroughly reviewed it, combining the Chilean and Australian species for the first time into the single genus *Pelecorhynchus*. Their comprehensive and fascinating monograph describes the adult morphology from stem to stern and covers larvae and pupae to a level of detail available for very few groups of flies even today.

Larvae of *Pelecorhynchus* are predators in wet soils, while the adults are nectar feeders most often found on mountain flowers, where they frequently bury themselves in the flowers to the point where they are not easily disturbed. Males of some species, in both Chile and Australia, hover like the males of some horse flies and some tangle-veined flies, remaining motionless in one spot and returning to the same spot after being disturbed. Others have different hovering habits, constantly moving rather than holding a single position. The genus *Pelecorhynchus* includes about 30 Australian and half a dozen Chilean species, but all are so uncommon that to see one alive — especially one of the gaudier species, such as the boldly striped Chilean *Pelecorhynchus biguttatus* or the rich red velvet–topped Australian *P. fulvus* — would be an exciting event. This is one of the very few fly families I have never seen on the wing, and *Pelecorhynchus* sits at the very top of my list of insect taxa I would most like to see alive.

Two North American genera are now usually included in the same family group as *Pelecorhynchus*, but the placement of *Glutops* and *Pseudoerinna* in the Pelecorhynchidae is based almost entirely on larval morphology. Teskey (1970) moved *Glutops* from the Xylophagidae to the Pelecorhynchidae on the basis of his careful comparison of *Glutops* and *Pelecorhynchus* larvae, but since then some authors have argued for alternative placements, including a separate family, Glutopidae, for *Glutops*; a subfamily, Glutopinae, in the Rhagionidae for *Glutops* and the similar *Pseudoerinna*; or a subfamily, Pelecorhynchinae, in the Rhagionidae for *Pelecorhynchus*, *Glutops* and *Pseudoerinna*. *Glutops* and *Pseudoerinna* species are smaller and less brilliantly colored than *Pelecorhynchus*, and north temperate in distribution in contrast to the south temperate *Pelecorhynchus*. The 11 species of *Glutops* occur mostly in western North America, where their larvae can be abundant in gravelly mountain streams, but the genus occurs in Japan and Russia as well. *Pseudoerinna* has a similar distribution pattern but only two species.

Oreoleptidae

Oreoleptid or "Mountain Leptid" Flies

The Diptera, and perhaps especially the lower Brachycera, are riddled with unsolved mysteries, not only among the untapped riches of the tropics but also in even the most accessible temperate habitats. The remarkable discovery of a new family of tabanomorph flies, the Oreoleptidae, in a well-studied part of Canada was recently announced in a paper titled "Discovered in Our Backyard: A New Genus and Species of a New Family from the Rocky Mountains of North America" (Zloty et al. 2005). The naming of a new taxon that represents a previously unnamed family is always exciting news, no less so in this case because the taxon had been known for many years before it was given a species name and its own family. The aquatic larvae of *Oreoleptis* had been noted repeatedly in collections from Montana, Alberta and the Yukon and had long been considered to belong to an undescribed genus of Athericidae (Webb 1994). Even now, adults of *Oreoleptis* (and thus the family Oreoleptidae) have yet to be collected in the field; the description of *Oreoleptis torrenticola* — the only species in the family — is based on adults reared from larvae collected in torrential streams and rivers and from pupae collected from riverbanks.

Although the larvae are very much like Athericidae, adult *Oreoleptis* have wing venation and other morphological characters that exclude them from both the Athericidae and the closely related Tabanidae. Zloty and co-authors (2005) concluded that *Oreoleptis torrenticola* must therefore be an early diverging lineage in the group that includes Tabanidae and Athericidae, so they created the new single-species family Oreoleptidae. The family name is from the Greek word for "mountain" combined with the Greek for "slender," which was previously used as the root of the family name Leptidae, an old name once in wide use for the snipe flies and water snipe flies. The common name used here — "mountain leptid flies" — thus reflects both the known habitat and the nomenclatural history of this smallest of families. Not all dipterists agree that a separate family is needed for this single species; some suggest that it is "probably nothing more than an athericid that has lost its synapomorphy" (Thompson 2009).

Xylophagidae

Wood Flies

The family Xylophagidae, with some 136 species in 10 genera, is sometimes treated in its own superfamily or even its own infraorder. This is in part because of uncertainty about the relationships between Xylophagidae and other Tabanoidea, and in part because of the age and heterogeneous nature of the family Xylophagidae itself. Some of the genera now included in the Xylophagidae have in the past been treated as separate families (Rachiceridae, Exeretonevridae, Coenomyiidae, Heterostomidae) and some are clearly very old lineages.

Coenomyiinae The subfamily Coenomyiinae includes the small genera *Arthropeas, Coenomyia* and *Dialysis*, which occur over more or less the same Holarctic and Oriental range as the much more common *Xylophagus*. These convex, robust flies look so different from the relatively slender Xylophaginae they are often treated in the separate family Coenomyiidae.

The only xylophagid genus in Australia is *Exeretonevra*, a group of four sluggish species found in eastern Australia and Tasmania. The classification of *Exeretonevra* has been a matter of debate for more than a hundred years, and the genus has been placed in several different lower brachyceran families by different authors. *Exeretonevra* species can be common in eastern Australia but immature stages remained unknown until an Australian student, Chris Palmer, reared larvae of this enigmatic genus from eggs laid by captured gravid females. Information garnered from those newly available larvae allowed Palmer and Yeates (2000) to confirm that *Exeretonevra* is the only Australian representative of the Xylophagidae, and to suggest that it belongs in the subfamily Coenomyiinae.

Both the Australian *Exeretonevra* and the Holarctic members of the Coenomyiinae are sluggish flies with a similar appearance in the field;

in my experience they usually look a bit stunned as they perch with their tarsi held at odd angles or projecting over the edges of their perches, with their bulky bodies close to the substrate rather than raised as in most other flies.

Xylophaginae The subfamily Xylophaginae is dominated by the large Holarctic genus *Xylophagus*, which includes some of the most common subcortical (under-bark) Diptera in north temperate forests. The distinctive *Xylophagus* larvae, which are mostly white with a long, shining brown head, are easily found under the bark of a variety of trees, where they prey on other subcortical larvae. Adults are elongate, mostly black, rather sluggish flies common on logs and tree trunks in spring. Other Xylophaginae belong to the strange-looking genus *Rachicerus*, distinguished from all other Brachycera by its apparently multi-segmented antenna with a pectinate (comb-like) flagellum. *Rachicerus* is a diverse genus with species in most parts of the world, including a handful of species in New Guinea and nine in the neotropics, but none in Australia or Africa.

One odd xylophagid species remains difficult to classify in a subfamily and is even a problem to include confidently in the family Xylophagidae. The Chilean *Heterostomus curvipalpus*, the only species in its genus, is a mysterious black fly with orange legs, wings and antennae. Nagatomi (1992) suggested that *Heterostomus* is closely related to the odd Australian genus *Exeretonevra*, and treated both genera as separate families (Heterostomidae and Exeretonevridae). Most dipterists now treat *Heterostomus* tentatively as a member of the Xylophagidae, pending the discovery of larvae or further species.

Stratiomyoidea

Soldier Flies and Their Relatives

The Pantophthalmidae, Xylomyidae and Stratiomyidae have long been recognized as close relatives, and recent morphological and molecular studies consistently support this grouping. Known larvae of this superfamily are saprophagous or phytophagous, unlike the predaceous larvae that characterize other lower brachyceran lineages.

Members of the genus *Rachicerus* (Xylophagidae) differ from all other Brachycera in having an apparently multi-segmented antenna with a pectinate flagellum.

Stratiomyidae

Soldier Flies

The Stratiomyidae, or soldier flies, form a huge, worldwide group of some 2,800 species ranging in size from just a couple of millimeters to around 3 centimeters in length and exhibiting an impressive range of colors and shapes. Despite all that, the family is fairly cohesive and usually distinctive for wing venation with a small discal cell, although some soldier fly adults lack wings altogether. Many soldier flies, especially the large and colorful species of the subfamily Stratiomyinae, are common flower visitors; adults of some groups aggregate near larval food sources, where they occasionally form large swarms.

Larvae, as discussed below under the subfamily headings, occur in a wide range of terrestrial and aquatic habitats but are especially common in decomposing plant material and under the bark of recently fallen trees. Like larvae of the closely related Xylomyidae, soldier fly larvae inflate the skin of the last larval stage to form a shelter in which to pupate, analogous but not homologous to the puparium in which Cyclorrhapha pupate. Larvae of both Stratiomyidae and Xylomyidae have tough, sandpaper-like skin studded ("shagreened," like sharkskin) with wart-like calcium deposits.

The taxonomy of the Stratiomyidae seems to be in better shape than that of most other lower Brachycera, thanks to good regional treatments (such as Rozkozny 1982, 1983) and an apparently stable higher classification. The following discussion is based on the classification in the *World Catalog of Stratiomyidae* (Woodley 2001) and also draws heavily on Woodley 2009.

Stratiomyinae The subfamily Stratiomyinae includes the "typical" soldier flies, familiar to most people as big, bright flower-visiting species such as those in the large genera *Stratiomys* and *Odontomyia*, many of which are mimics of bees and wasps. The aquatic larvae of stratiomyines are common filter feeders in a variety of lotic and (mostly) lentic waters, where their elongate larvae taper to a posterior breathing tube ending in a coronet of water-repellent hairs. This is the largest subfamily of soldier flies. Although most are aquatic, some Stratiomyinae are specialists in narrow or restricted habitats, including hot springs, madicolous (seepage) habitats and brackish water.

Raphiocerinae The almost entirely Neotropical subfamily Raphiocerinae is a small group of fewer than 50 species, many of which are distinctively elongate, brightly colored flies with patterned wings. Recent molecular phylogenetic work suggests that the Raphiocerinae are closely related to some Stratiomyinae.

Pachygastrinae Species in the subfamily Pachygastrinae are usually associated with fallen trees. Rearing of larvae found under bark can often yield otherwise rarely encountered species of these attractive little flies. The group is divided into a large number of genera — more than any other soldier fly subfamily — of varying size, color and shape. Most are small, angular species.

Hermetiinae The subfamily Hermetiinae is best known for the Black Soldier Fly, *Hermetia illucens*, a common species in warmer parts of the world. It breeds in a wide variety of wet decomposing materials ranging from pit toilets and decaying cadavers through to the intestinal tracts of living humans. Black Soldier Flies are now routinely used to process waste materials such as manure, with the larvae in some cases being harvested and sold for fish food. The specific name *illucens* refers to the clear windows near the base of the Black Soldier Fly's abdomen, which give the abdomen a petiolate appearance like that of a wasp; similar windows are found in several species of the large genus *Hermetia*. Most other Hermetiinae are relatively large, wasp-like species with much more restricted breeding habits. Some, such as the southwestern American *Hermetia comstocki*, are mimics of brightly colored sympatric wasps.

Sarginae The large subfamily Sarginae includes some of the most common and conspicuous soldier flies. *Sargus*, for example, comprises some 90 species of typically elongate metallic green flies commonly seen hovering or swarming above decaying fruits and compost, in which the larvae develop. *Sargus* is most diverse in the neotropics, but this widespread genus includes some common Holarctic and Old World tropical species. The huge worldwide genus *Ptecticus* also includes a few abundant synanthropic compost species such as the yellowish *P. trivittatus*, as well as a diversity of tropical forest species. These elongate flies are often seen staking out small but prominent male territories or aggregating on fallen fruit and damaged trees.

Chrysochlorininae The subfamily Chrysochlorininae is a small group of large flies superficially similar to some Hermetiinae, but with the last antennal segment drawn out into a thin, arista-like stylus. This subfamily is one of many small lower brachyceran lineages with species in South America and related lineages in Australia. Larvae of one species in this subfamily, *Chrysochlorina haterius*, have been collected from bromeliads.

Beridinae The subfamily Beridinae is a large and widely distributed group of almost 40 genera, the most common of which are small metallic green or blue flies with several long spines or tubercles

Species in the huge genus *Ptecticus*, in the subfamily Sarginae, are among the most common Stratiomyidae in much of the world, in habitats ranging from temperate compost heaps to tropical forests. This one is from Vietnam.

on the scutellum. Adults often feed at flowers; males of several species, such as the common Nearctic *Actina viridis*, form large mating swarms, often in shafts of sunlight in forests. Larvae are terrestrial, found in moist decaying plant matter or moss. Frequently encountered genera include the metallic green to blue species in the genus *Beris* (mostly Holarctic but ranging to Southeast Asia) and the similar Holarctic genus *Actina*. The more widespread genus *Allognosta* includes about 40 species found throughout the Holarctic region and the Old World tropics. Most are relatively dark and squat; some species have large translucent patches on the abdominal tergites.

Chiromyzinae The subfamily Chiromyzinae is a soldier fly lineage native to Australia and South America, especially cool temperate areas such as Chile and Tasmania, where the small, big-eyed males and much larger females of these slow-moving, dull-colored flies are conspicuous elements of the fauna. Chiromyzines are distinctive for an elongated abdomen, a spineless scutellum and atrophied mouthparts; females of some Australian *Boreoides* species are entirely wingless. The biology of this group is as unusual as its appearance and distribution: some species of the subfamily are among the very few herbivorous lower Brachycera and are the only lower Brachycera considered to be agricultural pests. Most known larvae of the family feed on grasses; some *Inopus* species are pests of sugarcane in Australia and have been accidentally introduced to California and New Zealand, where they attack pasture grasses and corn. Another species, *Tana paulseni*, is reported as a pest of wheat, rye and other grasses in Chile.

Members of the subfamily **Nemotelinae**, like those of Stratiomyinae, are common on flowers, but they are much less likely to be noticed because of their relatively small size and dull colors. Flies in the large genus *Nemotelus* often occur on the same flowers as Stratiomyinae and, like Stratiomyinae, Nemotelinae have aquatic larvae.

The subfamily **Parhadrestiinae** was first described by Woodley (1986), who showed that it is the most basal lineage of Stratiomyidae and

Most species of Xylomyidae belong to the large, almost cosmopolitan genus *Solva*, and are commonly found on or around fallen trees.

thus the sister group to the rest of the family. This very rare group of two extant Chilean species (and an extinct species in Canadian amber) is characterized by fusion of the antennal flagellomeres into one or two segments.

The subfamily **Antissinae** is a small group generally split between the Australasian and Neotropical regions, although one Holarctic genus, *Exodontha*, is "provisionally" treated as part of this subfamily (Woodley 2001).

The remaining subfamily, **Clitellariinae**, seems to be the inevitable taxonomic trashcan, serving as a repository for a few groups that don't fit into any of the well-defined subfamilies. This group includes *Cyphomyia*, one of the most charismatic fly genera common in the neotropics. These large, conspicuously colored flies — frequently distinctive for a yellow head or yellow head and shoulders contrasting with a blue-black body — often abound around decaying matter such as damaged cacti or recently cut or fallen trees, in which their larvae develop. *Cyphomyia* occurs in southern North America and the Old World tropics as well, but most of the almost 100 species are Neotropical.

Xylomyidae

Xylomyid or Wood Soldier Flies

The family Xylomyidae is a small group closely related to the Stratiomyidae. Most of the 138 described species of this family are in the large, almost cosmopolitan genus *Solva*, a remarkably uniform and distinctive group of relatively small (5–6 mm), compact, almost bare flies with slightly swollen and armed hind femora. They often have yellow legs or other yellow markings contrasting with an otherwise blackish body. *Solva* species can be found on tree trunks and cut wood almost anywhere in the world except New Zealand; I have seen remarkably similar *Solva* species, often laying eggs under the bark of fallen trees, in Africa, Australia, South America and North America.

The genus *Arthropeina* is a small, poorly known Neotropical group of mostly yellow flies with a *Solva*-like habitus. *Xylomya* species, on the other hand, are larger, slender, ichneumonid wasplike or sawfly-like flies, often very brightly marked. Eight *Xylomya* species occur in North America and several more species occur in the Palaearctic, Neotropical and Oriental regions.

Known larvae live (as the prefix *xylo* suggests) in wood, sometimes in roots but usually under bark, where they are similar to Stratiomyidae in having a shagreened appearance, due to deposits of calcium carbonate in the integument. Like Stratiomyidae, but unlike any other lower Brachycera, xylomyid flies pupate within the skin of the last larval instar. This and many other morphological attributes suggest that the Xylomyidae and Stratiomyidae are closely related.

Pantophthalmidae

Pantophthalmid or Giant Wood Flies

The small family Pantophthalmidae comprises only two genera and 20 species, all Neotropical. Pantophthalmids look like outrageously huge soldier flies, or perhaps enormous horse flies, with body lengths running to almost 5 centimeters. Despite their intimidating appearance, these infrequently encountered giants have small mouthparts, like the closely related Stratiomyidae and Xylophagidae, and they don't share the blood-feeding habit of the superficially similar Tabanidae. The larvae too have more in common with saprophagous soldier fly larvae than with the predaceous larvae of horse flies. Pantophthalmid larvae are muscular wood-borers equipped with a solid head capsule and massive mandibles used to bore into living trees. Although giant wood flies are considered forest pests in Brazil, they normally attack only unhealthy trees, feeding on fermenting sap in their burrows rather than on the wood itself.

Adults are relatively uncommon; most of those I have seen were attracted to lights or resting inconspicuously on trunks of trees. Zumbado (2006) reports that Costa Rican species prefer latex-loaded trees such as figs or those with a mucilaginous sap such as kapok — he describes tree trunks peppered with hundreds of holes drilled by pantophthalmid larvae. The larvae pupate in their burrows, wriggling partway out before emergence and sometimes leaving empty pupal exuviae hanging out of the holes. Zumbado also says that trees full of larval Pantophthalmidae are noisy: the larvae produce a sound that can be heard meters away.

Nemestrinidae and Acroceridae: "Superfamily Nemestrinoidea"

Tangle-veined Flies and Small-headed Flies

The tangle-veined flies (Nemestrinidae) and small-headed flies (Acroceridae) are distinctive flies with unusual parasitoid larvae and adults that resemble the other groups treated here as Tabanomorpha in having three tarsal pads (that is, with a pulvilliform empodium). Both the Nemestrinidae and Acroceridae lay huge numbers of eggs away from their hosts. Eggs hatch to active first-instar larvae, or planidia, that hunt out host arachnids or insects in which they develop as internal (or, very rarely, external) parasitoids. Remarkably, given the similar biologies of these two families, most recent phylogenetic studies have suggested that the Nemestrinidae and Acroceridae are not really closely related. Beyond that, there is little agreement about their phylogenetic affinity. Some recent phylogenies (Wiegmann et al. 2011) treat them as widely separated lineages related to the groups treated here as Tabanoidea and Stratiomyoidea, although both families have been treated by many authors (following Hennig 1973) as related to the Bombyliidae (which are often also parasitoids with active first-instar larvae). Some dipterists (Yeates 2002, Woodley et al. 2009) treat the Nemestrinidae and Acroceridae as basal lineages next to a clade, or lineage, that includes the Asilomorpha and the higher Diptera (the Muscomorpha of Woodley 1989). Despite — or perhaps because of — such uncertainty about the phylogenetic affinities of Nemestrinidae and Acroceridae, many dipterists continue to follow Hennig (1973) in grouping these two families together in a probably artificial superfamily called Nemestrinoidea.

Nemestrinidae

Tangle-veined Flies

Nemestrinidae, or tangle-veined flies — so called because of the way the veins near the tip of the

Tangle-veined flies in the genus *Hirmoneura* develop as parasitoids of larvae and pupae of subterranean beetles.

wing are curved forward and apparently tangled — look superficially like similarly stout, fuzzy bee flies (Bombyliidae), from which they differ most obviously in having three tarsal pads. These generally uncommon flies are extremely fast fliers, most likely to be observed as they dip, hummingbird-style, into long-tubed flowers or as they hover, seemingly motionless, in midair mating displays, often to the accompaniment of a loud, high-pitched buzz. Some of the long-beaked species are important pollinators of long-tubed flowers in Chile and South Africa. *Moegistorhynchus longirostris*, part of the remarkable guild of specialized pollinators of long-tubed flowers endemic to South Africa's Western Cape, has a proboscis that can exceed 80 millimeters in length, in striking contrast to the many species in the family that have vestigial mouthparts.

Other tangle-veined flies are diverse in South Africa as well; although the family occurs in every biogeographic region, it is concentrated in southern South America, Australia and South Africa. All known species are parasitoids. Female flies usually have a tapered telescoping ovipositor used to lay large numbers (thousands) of eggs that hatch into active planidial larvae able to seek out and penetrate their hosts, usually grasshoppers and katydids (Orthoptera), although members of the subfamily Hirmoneurinae parasitize the larvae and pupae of subterranean beetles. Tangle-veined flies that attack Orthoptera are important natural biological control agents for locusts and other pest grasshoppers.

The distribution of tangle-veined flies is unusual, with some genera occurring only in widely separated parts of the planet. Such disjunctions reflect the great age of this family, which was already diverse more than 150 million years ago, as indicated by Jurassic fossils of half a dozen extinct species in two subfamilies.

Nemestrininae The 275 or so extant species of Nemestrinidae are divided into five subfamilies, of which the "typical" subfamily Nemestrininae is the largest and most widespread. This group includes the long-tongued African *Moegistorhynchus* as well as the large austral genus *Trichophthalma*, which has 45 species in Australia and another 21 in southern South America. Although tangle-veined flies are generally uncommon, some Chilean and Australian species of *Trichophthalma* can be locally abundant: Paramanov (1952) describes Australian species as occurring "in enormous numbers," with "hundreds of males hovering in the shade of eucalyp trees" producing a "constant humming sound." Almost all Palaearctic tangle-veined flies belong to the large nemestrinine genus *Nemestrina*, and most African species in the family belong to the nemestrinine genus *Prosoeca*.

Tangle-veined flies in the subfamily **Hirmoneurinae** deviate from the usual nemestrinid habit of attacking grasshoppers and seem to parasitize only larvae and pupae of scarab beetles. This small group, with about 50 species scattered over most of the planet except Africa and Australia, was divided into nine genera by Bernardi (1976), but the subfamily is still widely treated as comprising the single genus *Hirmoneura*.

The subfamily **Atriadopsinae** is a mostly Old World group but also occurs in Australia (eight species) and the Neotropical region (one species). A Japanese species in this group has been reared from katydids (Tettigoniidae).

The subfamily **Trichopsideinae** is also a mostly Old World group, but it includes an Australian species and five New World species. The latter occur in grasslands from Costa Rica to western North America, with two species, *Trichopsidea clausa* and *Neorhynchocephalus sackenii*, ranging north to Canada. Members of this subfamily parasitize grasshoppers (Acrididae). The subfamily **Cyclopsideinae**, a small group named by Bernardi (1973) for the unusual genus *Cyclopsidea*, is known only from a couple of specimens of one rare Australian species. *Cyclopsidea* is thought to be the most "primitive" of extant nemestrinid lineages, and is probably the sister group to other living tangle-veined flies.

Acroceridae

Small-headed Flies

Acrocerids, called "small-headed flies" for obvious reasons, constitute one of the most distinctive groups in the entire order. Although there is a surprising range of body shapes and sizes (3–20 mm) in the family, the comically small head is a consistent family signature, converged upon only by a few bee flies and dance flies — which, of course, differ obviously from acrocerids in having only two tarsal pads. Like Nemestrinidae, all known Acroceridae lay enormous numbers of eggs that hatch into active first-instar larvae, or planidia, able to hunt down appropriate invertebrate hosts. However, unlike Nemestrinidae, which parasitize beetles, katydids and grasshoppers, small-headed flies attack only arachnids.

Acrocerids apparently have enough going on in their tiny heads to dump batches of eggs near appropriate spider hosts, sometimes even in their webs. Newly hatched larvae lucky enough to penetrate spider hosts — often through leg joints — sometimes move to the book lung area, where they enter a period of arrested development (diapause) that may last years while the spider feeds and grows. When the small-headed fly larva molts to its final stage, it swiftly consumes the contents of the doomed host, often first inducing the spider to obligingly spin a protective web that will serve as a silken shelter for the fly pupa. Some acrocerid species are solitary parasitoids that have their hosts to themselves; others are gregarious and share their spiders with several other larvae. The Acroceridae is a very old group, known from Jurassic fossils of species similar in form to extant flies.

The 500 or so species of Acroceridae occur in every part of the world, but most are known from only a few specimens or only from specimens reared from captive spider hosts. They are rare flies in most regions, although North American species are sometimes locally abundant on prominent rocks or other objects such as discarded antlers of deer and caribou, and species of one genus (*Eulonchus*) can be common on flowers in the western United States. Small-headed flies are not at all rare in southern South America, where you can find about a quarter of the world's named species, including many conspicuous long-tongued pollinators of long-tubed flowers. Fat, metallic *Lasia* species are, for example, regularly seen hovering over bright *Alstroemeria* lilies in central Chile, and the smaller, grotesquely humpbacked *Megalybus* species are common pollinators of other flowers in the same places.

Acrocerinae The Acroceridae are divided into three subfamilies, Acrocerinae, Panopinae and Philopotinae. The subfamily Acrocerinae currently includes the large, cosmopolitan (or almost cosmopolitan) genera *Ogcodes* and *Acrocera*, which are small, compact, relatively inconspicuous flies that don't visit flowers but often perch on prominent objects on the ground. Most acrocerids in North American insect collections belong to these genera, which are parasitoids of araneomorph spiders (a derived spider lineage that includes orb-weavers). Members of the small Nearctic and Caribbean genus *Turbopsebius* are specialized parasitoids of funnel-weaving spiders (Agelenidae), and species of the worldwide genus *Pterodontia* have been reared from the commonly encountered orb-weaver, wolf and jumping spider families (Araneidae, Lycosidae and Salticidae). *Ogcodes*, a worldwide genus of 110 species and the largest and most widespread group of acrocerids, attacks a wide range of araneomorph spiders. *Acrocera* comprises fewer species but attacks a similar range of hosts in every continent except Australia (and Madagascar). A recent molecular study (Winterton et al. 2007) suggests that, unlike the other subfamilies, Acrocerinae is an artificial group, and that *Acrocera* and *Sphaerops* form a basal lineage that is the sister group to the rest of the family Acroceridae. *Sphaerops* is a small Chilean genus with the unusual habit of developing outside its spider host, unlike the endoparasitic habits of all other Acroceridae. According to Winterton and colleagues, other genera now included in the Acrocerinae — including the large genus *Ogcodes* — appear to be more closely related to the Panopinae.

Panopinae The mostly large, robust flies in the distinctive subfamily Panopinae develop as internal parasitoids of mygalomorph spiders (the group that includes tarantulas and trapdoor spiders). The brightly metallic and long-tongued species of *Lasia* (Neotropical) and *Eulonchus* (Nearctic) are typical Panopinae. These relatively conspicuous flies can be common visitors to some long-tubed flowers such as the *Alstroemeria* lilies of central Chile. Other members of the subfamily, such as the Chilean genus *Arrhynchus*, have reduced mouthparts and are not known as flower visitors. According to Winterton et al. (2007), the most primitive panopines are in the endemic New Zealand genus *Apsona*.

Small-headed flies in the subfamily Philopotinae often have a grotesquely hunchbacked appearance, like this Chilean *Megalybus*. The head is facing down on this individual, and the part facing up and left, looking remarkably like a spider cephalothorax, is the back of the mesothorax.

Philopotinae Members of the bizarre subfamily Philopotinae are unmistakable flies, more elongate than other acrocerids and with an improbably extreme hunchbacked body shape. There are relatively few host records for the Philopotinae, but known hosts are all araneomorph spiders. According to Winterton et al. (2007), the relationships between philopotine genera reflect continental plate movements during the Cretaceous period; the basal lineages are in New Zealand (the endemic genus *Helle*) and New Caledonia (the genus *Schlingeriella*). The distinctive New World genera of this family, including the southern South American *Megalybus* and its northern Neotropical counterparts *Philopota* and *Quasi*, have their closest relatives in the Afrotropical region, from which they must have been separated since the middle Cretaceous, about 115 million years ago.

Asilomorpha

Robber Flies, Bee Flies and Their Relatives

The infraorder Asilomorpha includes three main lineages or groups of families: the bee fly lineage, including only the Bombyliidae and Mythicomyiidae; the robber fly group of families, including the Asilidae, Apioceridae and Mydidae; and the stiletto fly group of families, comprising the Therevidae, Scenopinidae, Evocoidae and Apsilocephalidae. The latter two groups are here treated together as the superfamily Asiloidea, leaving the bee fly lineage in its own superfamily, Bombylioidea. Although these three groups are each well-defined and recognizable clusters of similar families, the relationships between them are controversial; it is possible that the Asilomorpha as a whole is not a natural group. The discussions below are therefore clustered according to the three main "asilomorph" lineages, two in the Asiloidea and one comprising the Bombylioidea.

Asiloidea Part One

The Robber Fly Lineage

Asilidae

Robber Flies

The robber fly family, Asilidae, is not only one of the largest (more than 7,000 species) and most familiar groups of Diptera, it is also one of the most popular, vying with the flower flies and bee flies as the fly family most sought out by amateurs and photographers. These exclusively predaceous flies range from delicate insects measuring only 2 or 3 millimeters through to 6-centimeter-long monsters able to tackle dragonflies and bees. All are easily recognized by a long body, a forward-facing beak and a signature tuft of hairs, or mystax, above the beak. Many species scan for prey from a ground-level vantage point, but others routinely perch conspicuously on leaf tips, logs or other sun-splashed objects, where they use their large, well-separated eyes to spot appropriately sized insects or spiders. Most home in on flying adult insects, darting out to impale them with a stout beak.

Prey are penetrated by a needlelike hypopharynx ("tongue"), usually through the neck or some other weak spot such as the eye or between the elytra, and immobilized and liquefied by an injection of saliva that contains nerve poisons and proteolytic enzymes. Some of the more robust robber flies will inflict a painful bite if you confine them in your hand, but these well-armed flies rarely attack vertebrates other than the odd unlucky hummingbird. Asilid larvae are also predators, digesting prey — frequently soil-dwelling scarab beetle larvae — extra-orally before ingesting the liquefied tissues.

Most female robber flies work their entire abdomen down into soil, sand or wood in order to deposit eggs, but a few lay eggs on plants from which newly hatched larvae will drop to the ground below; some of our most common woodland species develop in decaying wood or under bark. One species common in backyards and gardens lays its eggs one at a time while flying, and several species coat their sticky eggs with a protective shell of sand grains before oviposition, as do some bee flies.

Asilidae are most diverse in open areas but several are forest dwellers. Some of the most common and conspicuous forest species, in the large genus *Laphria*, are almost indistinguishable from bumble bees at first glance. Many flies are remarkable bee and wasp mimics, presumably using their intimidating appearance to deter sting-shy predators; robber flies are often marvelous mimics of spider wasps, bumble bees, leafcutter bees and others. Bumble bee–like *Laphria* are a common sight along woodland paths, where they sit on exposed leaves or twig tips and scan discrete search areas for potential prey to impale with a bladelike beak (robber flies other than *Laphria* and a few related genera have a cylindrical beak). After darting out in pursuit of a flying insect, often a beetle, they usually return to the same perch.

Some robber flies engage in complicated courtships during which males hover and display over perching females, often adding flash to their dash with ornaments such as long fringes on the hind legs or flashy silver genitalia. Males of many species take a more direct approach, simply chasing down females in flight.

Robber flies are often seen with prey impaled on their beaks — sometimes large or well-defended prey such as bees, stink bugs or rove beetles. Such asilid feasts are routinely attended by smaller flies that take advantage of the lengthy time these large predators spend liquefying and imbibing the body contents of their prey. These small flies, called kleptoparasites, are attracted to a mixture of robber fly saliva, prey blood and prey defensive chemicals; they often appear quickly when the larger robber flies impale appropriate prey.

The higher classification of the Asilidae has been notoriously unstable, with different authors using conflicting systems with anywhere from four to 14 subfamilies, often differently defined by different authors and often representing artificial (non-monophyletic) groups. The subfamilies used here follow Torsten Dikow's (2009a, 2009b) classification based on his exhaustive study of robber fly morphology and molecules (see also Dikow's website, http://www.asilidaedata.tdvia.de/classification_dikow2009.html). The discussion below also draws on Eric Fisher's (2009) excellent recent overview of the biology and classification of Neotropical Asilidae, although Fisher follows a slightly different subfamily classification than Dikow.

Trigonomiminae Flies in the small subfamily Trigonomiminae — "goggle-eyed robber flies" — are generally small flies that commonly perch at the ends of twigs or sedge stems while they use their disproportionately large, bulging eyes to seek prey such as midges and springtails. The many distinctive little species of *Holcocephala* are common goggle-eyed robber flies throughout the New World. Other genera treated as Trigonomiminae occur almost worldwide, although the subfamily is absent from Australia and some data suggest that the Old World tropical genus *Damalis* doesn't really belong in the subfamily. The subfamily Trigonomiminae is currently considered to be the sister group to the rest of the Asilidae.

Laphriinae With the exception of the basal goggle-eyed robber fly subfamily Trigonomiminae, the Asilidae can be divided into two lineages: one that includes only the large and familiar subfamily Laphriinae and another made up of all other robber flies. The best-known members of Laphriinae are the spectacular bumble bee mimics in the widespread genus *Laphria* and related genera. Other members of the subfamily include the common shoreline robber flies in the genus *Laphystia* (previously treated as a separate subfamily) and a variety of taxa associated with fallen logs, often in light gaps in forested areas. The small species of *Atomosia*, for example, are among the most common robber flies on fallen logs in the New World and are among the several laphriine genera that develop as predators in beetle burrows in wood.

Most laphriines, including the familiar *Laphria* species, are woodland insects that develop in dead wood, in contrast with other robber flies, which generally frequent open areas ranging from deserts to grasslands and which develop from soil-dwelling larvae. Some authors divide the large genus *Laphria* into multiple genera, but, broadly defined, this genus includes a huge diversity of charismatic flies from southern Australia to northern North America.

Asilinae The "typical" robber fly subfamily Asilinae — which includes the Apocleinae of earlier classifications — is a huge worldwide group comprising most of the conspicuous, large-bodied robber flies outside the Laphriinae. *Promachus* and *Proctacanthus* species are large insects often seen hunting from open ground such as dunes or clearings in north temperate regions; Neotropical species in the large cosmopolitan genus *Promachus* are more often seen on or among the foliage of trees. Larvae feed on scarab larvae in the soil, while adults of *Promachus* and related genera prey heavily on bees and wasps. The most commonly collected north temperate Asilidae — *Asilus*, *Machimus* and *Neoitamus* — belong to this subfamily, along with the giant Australian robber flies in the genus *Blepharotes* and rarely collected, spectacular genera such as the southwestern and Central American genus *Wyliea*. One of the two species of *Wyliea* is a remarkable mimic of the large spider wasps known as tarantula hawks, and the other is an equally striking mimic of Central American scarab-hunting wasps. Most asilines lay eggs in the soil with the aid of structures on

Robber flies in the large genus *Laphria* usually ambush flying insects like this captured calyptrate, often returning to the same perch from which they spotted their prey.

toothed plates (acanthophorites) at the tip of the female abdomen, but some, such as *Mallophora*, *Promachus* and related genera, lay eggs in masses on branch tips and vegetation.

Ommatiinae The subfamily Ommatiinae is closely related to the Asilinae and comprises a relatively small group of genera dominated by the huge, widespread genus *Ommatius*. Distinctive for their apparently pectinate antennae (because of the bristles on the anterior surface of the antennal stylus), these small to medium-sized flies are among the more common robber flies in tropical and subtropical regions. *Ommatius* is found in warmer areas throughout the world, while other genera of the subfamily are mostly Old World. They are most likely to be seen perching on the tips of twigs and similar prominent plant parts.

Leptogastrinae The delicate, crane fly–like robber flies in the subfamily Leptogastrinae — the "slender robber flies" — are so distinct from other Asilidae that they have in the past been treated as a separate family (Leptogastridae), although recent phylogenetic studies have placed this group either as the sister group to other Asilidae or as a lineage deep within the Asilidae, close to the Asilinae and Ommatiinae. Slender robber flies have a slow, almost helicopter-like flight that contrasts with the jetlike flight of other robber flies. They usually snatch their prey off leaf surfaces or tree trunks — in contrast with other robber flies, which usually grab flying prey — and then perch at the tip of a leaf or twig to consume the lower dipteran, springtail or other small captured insect.

Stichopogoninae The subfamily Stichopogoninae is a small group of generally inconspicuous robber flies characteristic of open sandy or rocky surfaces. The best-known genera in the subfamily are the widespread genus *Stichopogon* and the Holarctic genus *Lasiopogon*. *Stichopogon* species, especially the abundant and exceptionally widespread New World species *S. trifasciatus*, are common on shorelines, where they perch on the ground. *Lasiopogon* species have similar habits but are less often collected, probably because most have relatively northern ranges and short flight periods very early in the season (*Lasiopogon* is the most northerly-ranging robber fly genus).

Dasypogoninae According to Dikow (2009) the rest of the robber fly subfamilies group together into a single lineage, but they are nonetheless still split into several small subfamilies. The largest and most widely distributed of these, the subfamily Dasypogoninae, is found worldwide but is especially diverse in the Neotropical and Australasian regions. Most members of this group have a curved spur at the tip of the leg that is apparently used to clamp wasp prey around their narrow waists. The most familiar New World flies in this group are the numerous (78 species) gangly-legged members of *Diogmites*, which are often seen hanging from vegetation by a single front leg while consuming a bee or wasp (or occasionally a dragonfly or fly). Similar behavior is found in some related dasypogonines.

Dasypogon, the type genus of the subfamily, is absent from the New World but widely distributed from Australia to the Palaearctic region. Geller-Grim (1998) described the biology of *Dasypogon diadema*, one of the most common robber flies in central Europe, demonstrating that it lays eggs in clutches in the soil protected by a sand coating he called a "sand cocoon." He also showed that, as is typical for the subfamily, more than 70 percent of the prey items were wasps or bees.

Brachyrhopalinae The subfamily Brachyrhopalinae now includes species that were treated as Dasypogoninae and Stenopogoninae in earlier classifications (and some current ones). The group as a whole is not particularly distinctive, but some of its genera are relatively well known for both their abundance and their courtship behavior. *Cyrtopogon*, for example, is a large, mainly Holarctic genus common in coniferous forests, where they are major predators of other flies. *Cyrtopogon* males perform courtship displays on logs or other solid substrates, while males of the related genus *Heteropogon* approach twig-perching females while on the wing, hovering in front of the female to perform elaborate courtship displays. Other common twig perchers in this subfamily include the tiny *Holopogon* species often seen at the ends of twigs, to which they will return repeatedly after disturbance. *Cyrtopogon* and *Heteropogon* are mostly Holarctic genera, but this subfamily occurs worldwide (the type genus, *Brachyrhopala*, is Australian).

Dioctriinae The small subfamily Dioctriinae includes *Dioctria hyalipennis*, a very common Holarctic species until recently known as *D. baumhaueri*. Arguably the only asilid introduced from Europe to North America, these nondescript slender robber flies can easily be found on shrubby vegetation in almost any city park in eastern North America, and are often seen feeding on small wasps impaled through their anal or oral orifices. Females drop their eggs while on the wing, in contrast with the careful oviposition habits of most other asilids. This large, mostly Palaearctic genus includes a few other Nearctic species and a single species that ranges into the neotropics.

The subfamily **Willistonininae** is a recently described subfamily that includes a couple of Nearctic genera, a couple of Afrotropical genera and one genus that occurs in both Africa and the Palaearctic region. Although this group is small and poorly known, some members can be common in open sandy areas.

The remaining robber fly subfamilies are made up of genera previously treated as part of the subfamily **Stenopogoninae**, which was widely recognized as an artificial assemblage before Dikow (2009) suggested dividing it up into the subfamilies Stenopogoninae, Bathypogoninae, Phellinae, Tillobromatinae and Willistonininae. This reclassification leaves only a dozen genera in the Stenopogoninae, although *Stenopogon*, with 160 mostly northern hemisphere species — typically found perching on low shrubs or on the ground in dry areas such as the American southwest and Mexico — remains one of the larger and more frequently encountered genera in this assemblage. Other Stenopogoninae are found mostly in drier areas of the Holarctic and Afrotropical

Cyrtopogon is a large, mainly Holarctic genus common in coniferous forests. This female has worked her abdomen into the soil to lay eggs.

regions. The small subfamilies **Bathypogoninae** and **Phellinae** show interesting Australia/southern South America distributions, while the **Tillobromatinae** are restricted to southern Africa and South America.

Apioceridae

Apiocerid or Flower-loving Flies

Despite the name "flower-loving flies," members of the small and generally uncommon family Apioceridae are rarely seen on flowers. They are much more likely to be seen perching on the ground, much like superficially similar gray and white robber flies (Asilidae) and Therevidae. For this reason, and because the family's best-known member — the Delhi Sands Flower-loving Fly, *Rhaphiomidas terminatus abdominalis* — was recently transferred from the Apioceridae to the Mydidae, I prefer to call members of this family simply "apiocerids." Australia seems to have the richest apiocerid fauna (with about 70 species), but members of this small family are also found in arid areas of southern South America, South Africa, western North America and Mexico, where they apparently develop as predators in soil. Unlike the robber flies, adult apiocerids are not predators; if they feed at all, it is only on nectar or honeydew.

Apiocerids are very closely related to mydas flies (Mydidae), from which they differ mostly in having shorter antennae and a two-segmented rather than one-segmented maxillary palpus. Both Mydidae and Apioceridae are old groups, known from fossils formed more than 110 million years ago. All 140 or so species in the family Apioceridae are currently in one genus, *Apiocera*.

Mydidae

Mydas Flies

Where I live, in eastern Canada, the only species of mydas fly (*Mydas clavatus*) is our largest and most spectacular fly, with a body length of around 3 centimeters, strikingly long antennae, black wings and a black body marked with orange. Elsewhere in the world, several of the 461 species of mydas flies reach similar or even more heroic

The southwestern American genus *Opomydas* is one of two Nearctic genera in the small mydas fly subfamily Ectyphinae.

proportions: the South American *Gauromydas heros* attains a length of 6 centimeters — arguably the largest fly in the world. *Gauromydas heros* females are occasionally seen laying eggs in the nests of South American leafcutter ants, where the larvae of these big flies presumably feed on ant brood. The rainforest range of *Gauromydas heros* is atypical of mydas flies, most of which inhabit the kinds of hot, dry habitats preferred by the closely related apiocerid flies. Like apiocerids they are non-biting; some feed at flowers, especially males and especially the species with long beaks, while some species have reduced mouthparts and do not feed at all. Also like apiocerids, mydas flies usually lay eggs in dry, sandy soil, where the larvae develop as predators, although some hunt beetle larvae in rotting wood.

Mydidae occur in every region of the world; the total of less than 500 species is divided into 11 subfamilies plus the inevitable handful of unplaced species (Dikow 2010; http://www.mydidae.tdvia.de). The largest subfamily, **Syllegomydinae**, is almost entirely African, and members of four other subfamilies also occur there, making Africa — especially southern Africa — the centre of mydid diversity. The "typical" subfamily **Mydinae** is the second largest group, with 84 New World species, including many huge species such as *Gauromydas heros*. The next largest group, the subfamily **Leptomydinae**, has a more scattered distribution, with species in North America, Central America, South America, Europe, Asia and Africa.

The remaining subfamilies are small, with between two and 30 species each, often with interesting disjunct distributions. The subfamilies **Diochlistinae** and **Apiophorinae**, for example, are both split between southern South America and the Australasian region, and the subfamily **Ectyphinae** has two genera in southern Africa and two in the southwestern United States. The subfamily **Megascelinae** shows a classic Gondwanan distribution: one genus in Australia, one in Chile and one in southern Africa. The subfamily

Rhopaliinae occurs in both the neotropics and the Old World but not in the Nearctic region. Other subfamilies have more restricted distributions. The subfamily **Anomalomydinae** comprises only two Australian species, and the subfamily **Cacatuopyginae** includes only six Oriental species.

The subfamily **Rhaphiomidinae** occurs only in the western United States and nearby Mexico, and contains only the 22 species in the genus *Rhaphiomidas*. This genus was treated as Apioceridae until Sinclair et al. (1994) demonstrated that its closest relatives were in the family Mydidae. Yeates and Irwin (1996) moved it into the Mydidae, where it is now the only genus in the subfamily Rhaphiomidinae. These family-level shifts caused a bit of general confusion, since *Rhaphiomidas* is the group that includes the Delhi Sands Flower-loving Fly, *R. terminatus abdominalis*, a species very much in the public eye because of its contentious listing as an endangered species (of Apioceridae) by the United States Fish and Wildlife Service in 1993.

The Delhi Sands Flower-loving Fly is a subspecies apparently restricted to an isolated area of inland sand dunes in densely populated parts of California's San Bernardino and Riverside counties. This area, known as the Delhi Sand Dunes, once covered about 140 square kilometers, but development has left species endemic to this special habitat barely hanging on in the few hundred remaining acres, which are fragmented into several small patches. The Delhi Sands Flower-loving Fly is a large, spectacularly long-beaked fly seen on or over flowers during its short summer flight season. It clearly warrants the protection it is now receiving, despite protests from those who don't like the idea of a lowly fly standing in the way of development.

Asiloidea Part Two

The Stiletto Fly Lineage

Therevidae

Stiletto Flies

The family Therevidae comprises about a thousand described species, distributed throughout the world but most diverse in Australia and most common in arid regions or on sandy seacoasts. The long, wiry larvae are normally found in dry soils, where they actively hunt and kill invertebrate prey. Adults are not predators, despite the suggestive common name "stiletto flies" and their superficial similarity to the much more common flies in the well-known predaceous family Asilidae (robber flies). Adult stiletto flies are occasionally seen on flowers, where they presumably take nectar, but they are neither common flower visitors nor important pollinators. Although the ecological role of their predaceous larvae is probably significant, not much more is known about the biology of Therevidae. Adults are encountered infrequently, although desert species often occur along the margins of intermittent streams or other small patches of water, and males of a few species form mating swarms or leks. Some appear to be mimics of spider wasps or other stinging Hymenoptera.

The Therevidae are very closely related to the Scenopinidae (window flies), a relationship reflected in some remarkable similarities, including long, slender, snakelike larvae with body segments secondarily divided to give an apparent 17 body segments. These two families are in turn closely related to the tiny families Evocoidae and Apsilocephalidae, with which they form a distinctive subgroup of the Asiloidea. Four therevid subfamilies are recognized: the widespread subfamilies Phycinae, Xestomyzinae and Therevinae, and the southern hemisphere subfamily Agapophytinae.

Therevinae Most of the diverse, frequently encountered genera of stiletto flies — such as *Thereva*, *Cyclotelus*, *Acrosathe*, *Spiriverpa* and *Ozodiceromyia* — belong in the large, cosmopolitan subfamily Therevinae. Of these, the Holarctic genus *Thereva* is common in forests, while the other genera are more characteristic of sand dunes and similar habitats. *Ozodiceromyia* is the most diverse genus in the New World, with many undescribed species in a range of habitats. Australia has a distinct and highly diverse fauna of Therevinae that includes three genera, dominated by the genus *Anabarhynchus*, the largest therevid genus in

Stiletto flies, such as this Nearctic species of *Acrosathe*, look a bit like robber flies (Asilidae) but lack the distinctive mystax and beak of that entirely predaceous group. Only the larvae of Therevidae are predators.

the world, with more than 200 species. Australian therevid genera are otherwise either included in the endemic Australian subfamily **Agapophytinae** or left unplaced as to subfamily.

The subfamily **Phycinae** is a small group with species in western North America as well as in the Old World and New World tropics. The small subfamily **Xestomyzinae** occurs in southern Africa, Madagascar and the New World, with the only New World genus, *Henicomyia*, ranging from the southwestern United States to Brazil. There remain several genera that therevid taxonomists currently treat as unplaced, so higher classification of the Therevidae (along with that of the closely related Scenopinidae, Evocoidae and Apsilocephalidae) is likely to continue to change.

Scenopinidae

Window Flies

Scenopinids, or "window flies," are small, generally dark flies with unmarked wings and distinctively short antennae that apparently end in a fingerlike single segment, or flagellomere. The family Scenopinidae is best known for one widespread synanthropic species, *Scenopinus fenestralis*, called the Window Fly because it is so frequently found on the inside of house windows. Larvae of *Scenopinus fenestralis* and the related *S. glabrifrons* are predators and presumably develop on dermestid larvae and other insects inside buildings, where newly emerged adults are likely to be attracted to the nearest window. Although the common name "window flies" is applied to the whole family, very few of the 350 or so species in this worldwide group have been collected at windows. All scenopinid larvae are probably predators, and most records are from nests of rodents, termites or birds. Some scenopinid species are associated with wood-boring insects, but most are found in semiarid habitats and remain unknown as larvae.

The Scenopinidae are sometimes divided into subfamilies: either three (Caenotinae, Scenopininae and Proratinae), following Yeates (1992), or two (Scenopininae and **Proratinae**), following the opinion of Nagatomi et al. (1994) that the Caenotinae belong in the Proratinae. Most members of the family belong to the **Scenopininae** in any case, as the other subfamily (or subfamilies) includes only a few unusual groups that were previously treated as Bombyliidae. About half the known species in the family are in the genus *Scenopinus*, but this is a poorly known group and many species remain to be described.

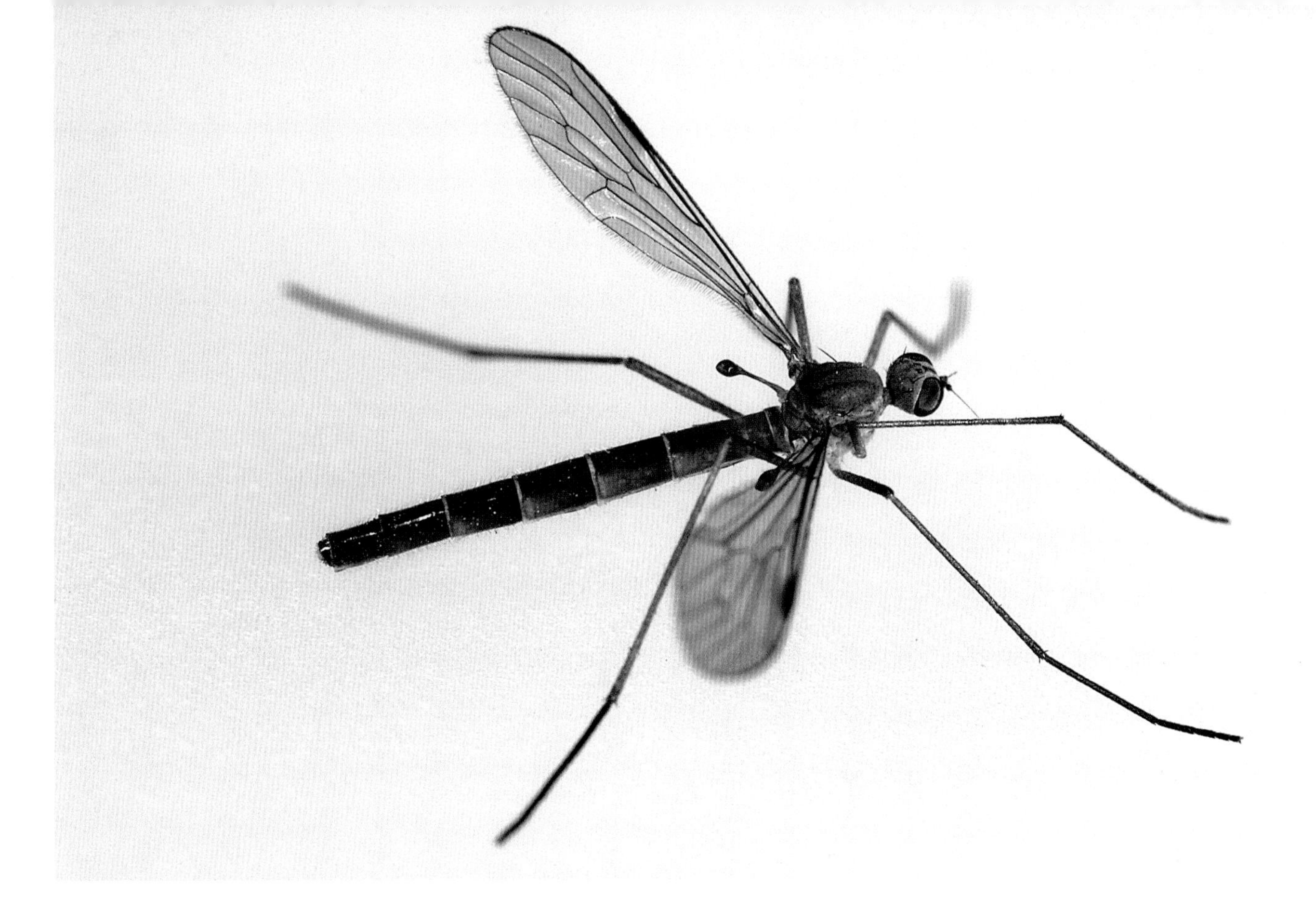

Evocoa chilensis, the only species in the family Evocoidae, is known only from a small area around Santiago, Chile.

The Scenopinidae are closely related to the Therevidae and might even be a derived lineage within the Therevidae (Woodley 2009).

Apsilocephalidae

The family Apsilocephalidae is another one of those odd little lineages of interest to dipterists and biogeographers but unlikely to come to the attention of most biologists. Originally in the Therevidae, this small group was given its own family by Nagatomi et al. (1991). Some subsequent work has indicated that Apsilocephalidae is either the sister group to the Therevidae or (more likely) basal to a lineage that includes both Therevidae and Scenopinidae. The Apsilocephalidae include two uncommon Tasmanian and a recently described New Zealand species, one rarely collected southwestern North American species, and a couple of species known only from fossils in the Holarctic and Oriental regions. The Nearctic species *Apsilocephala longistyla* has been collected around burrow entrances, and a Tasmanian species, *Clestenthia aberrans*, has been collected in an abandoned farm building (which, I might add, I have visited twice without seeing one of these rare flies). Otherwise nothing is known about the biology of these "almost" therevids.

Evocoidae

The family Evocoidae was created in 2003 by Yeates et al. as the family Ocoidae, to hold a new genus and species of odd slender, therevid-like fly collected in Malaise traps — the tentlike traps often used to survey flies — in the vicinity of Santiago, Chile. They named the genus *Ocoa* after the nearest town to the type locality, but when it turned out the name was already being used for another group of animals, Yeates et al. (2006) changed the name of the genus to *Evocoa*, and thus changed the name of the family to Evocoidae. (The *Ev* part of the new name honors a sponsor of their research program, the prominent dipterist Evert Schlinger.) I later visited the type locality for *Evocoa chilensis* — thus the type locality for the family — along a dry creekbed in Chile's Campaña National Park, near Ocoa, in hopes of seeing this remarkable fly alive. The photo included above, showing a fly found on the mesh of my Malaise trap during that visit, is to my knowledge the only living evocoid anyone has seen, so it follows that we know nothing about the biology of this species.

Evocoa chilensis is treated as a separate family, rather than as a member of the Therevidae, because phylogenetic analysis indicates that it is a basal lineage like the Apsilocephalidae, and possibly closely related to the Apsilocephalidae. Recognition of a separate family for this species is in keeping with the current practice among lower Brachycera specialists of recognizing new families for species or small groups of species that do not fit well into established families.

Bombylioidea

Bee Flies and Micro Bee Flies

Typical bee flies are conspicuous for their robust, fuzzy (or scaly) bodies and their outstretched and frequently patterned wings, but this huge worldwide family spans a tremendous variety of shapes and sizes, ranging from tiny species not much more than a millimeter in length up to giants with a wingspan measured in centimeters. A few earn the common family name "bee flies" by closely mimicking bumble bees, other bees or stinging wasps, but most are beelike only in being relatively robust, fuzzy diurnal flower visitors. Almost all adult females are obligate pollen feeders, and adults of both sexes are often seen hovering over flowers in search of nectar.

Larval Bombyliidae are generally hidden away as they consume a variety of underground insects, such as larval bees, larval beetles, caterpillars, ground-living fly larvae, antlion larvae and grasshopper eggs. Most of the world's 5,000 or so bee fly species are ectoparasitoids — their larvae attach to a single host and consume it from the outside — but a few are endoparasitoids, developing inside a single host, and some are predators (consuming many hosts) or kleptoparasites (consuming the host's food). Most bee flies attack hosts found in burrows or nests on or in the ground, especially solitary wasps and bees. The larvae of tiger beetles, antlion larvae, tsetse fly puparia and grasshopper egg pods are other subterranean targets of bee fly larvae. Hosts are usually attacked as relatively immobile stages (mature larvae or pupae), and most bee flies spend a period as an active host-seeking first-instar larva (as in Acroceridae and Nemestrinidae) before transforming into a sessile parasitoid. This was described with unparalleled flourish by the great naturalist Jean-Henri Fabre over a hundred years ago. He wrote of newly hatched bee fly larvae penetrating the masonry-like nest of a solitary bee before molting to immobile ectoparasitoids of bee larva therein. He remarked on the parasitoid larva's mouth as a minute "bowl shaped opening" applied to the bee larva in a "perfidious kiss" through which the bee fly "inhaled" the host's body fluids without visible perforation, reducing it to a flaccid sac before molting to a pupa armed to drill out the nest and molt to an adult fly.

Bee flies occur on every continent except Antarctica, but the diversity of the group is highest in the Afrotropical and Palaearctic regions, with the largest number of species (1,411) known from the Afrotropical region. As you might predict, given the general biology of the family, most species of Bombyliidae are associated with relatively open, dry areas. Many of the larger genera, such as *Exoprosopa*, *Anthrax* and *Villa*, include hundreds of species each and are common in most zoogeographic regions. But the bee flies also show extreme regional endemism and include hundreds of small, localized genera divided among some 16 subfamilies, of which six are strictly Old World.

Most of the currently recognized bee fly subfamilies were used by Frank Hull in his monumental 1973 book *Bee Flies of the World* and later largely supported by an important study of bee fly classification by David Yeates (1994).

Yeates's classification is used here as a framework for a brief overview of bee fly subfamilies, with a few modifications following the *World Catalog of Bombyliidae* (Evenhuis and Greathead 1999). Most notably, as explained below, the Usiinae in Yeates's classification are split here into the Usiinae and Phthiriinae, as in Hull's classification, and the subfamily Ecliminae is recognized here as a separate group rather than as part of the Bombyliinae. A recent molecular analysis (Trautwein et al. 2011) indicates that seven out of the 15 subfamilies in Yeates's classification are non-monophyletic, so the definitions and limits of bee fly subfamilies are likely to change as these new data are reconciled with morphology.

Micro bee flies, like this *Mythicomyia*, are tiny (1–3mm) bee flies that feed on pollen as adults and prey on subterranean insects as larvae.

Mythicomyiidae

Micro Bee Flies

The micro bee flies form an ancient group widely recognized as the basal bee fly lineage, or the sister group to all other bee flies. Micro bee flies are treated by some authors as the separate family Mythicomyiidae and by others as the subfamily Mythicomyiinae. Both of these treatments are consistent with the widely accepted idea that micro bee flies are closely related to but distinct from other bee flies. Those in favor of separate family status emphasize the great age of the group, as indicated by middle Jurassic fossils, and those who prefer subfamily status point out that including micro bee flies in the Bombyliidae results in a perfectly good monophyletic group.

Either way, the micro bee flies form a worldwide group of about 350 tiny (hence "micro") bee flies ranging from well under a millimeter up to about 3 millimeters in length. These inconspicuous, relatively bare, often humpbacked flies are not often collected, but they can be common on flowers. Larvae of the few species with known life cycles prey on subterranean hosts, including grasshopper egg pods and bee larvae. One species develops in ant nests, presumably preying on ant brood.

Micro bee flies remain poorly known, as illustrated by the recent (2002) description of the new New World genus *Pieza* by Neal Evenhuis. Seven of the 11 species in the genus were new, including the amusingly named *Pieza deresistans, Pieza pi, Pieza rhea*, and *Pieza kake*.

Bombyliidae: Part One

The Lower Bee Flies

The subfamily **Oligodraninae** is an obscure basal bee fly lineage that includes only the small Palaearctic genus *Oligodranes*, a group of small flies that resemble those in the more widespread and common genus *Apolysis*. Hosts of *Oligodranes* remain unknown and the only host information for Apolysis is a record of an African species parasitizing a pollen wasp (Vespidae) pupa. *Apolysis*, with almost 100 species, is currently treated as part of the subfamily **Usiinae**, but its similarity to *Oligodranes* is such that several species currently in *Apolysis* were until recently included in *Oligodranes*. In fact, bee fly expert Neal Evenhuis named two southwestern North American species *Oligodranes humbug* and *Oligodranes zzyzxensis* in 1985 and then later (1999) moved them to the genus *Apolysis*. In the meantime (1990) he had published a thorough study of the Usiinae and created the subfamily Oligodraninae to hold a few species of *Oligodranes* that he recognized as basal to the rest of the family. Such changes are necessary to reflect changing ideas about phylogenetic relationships. Work by Evenhuis also suggests that the Phthiriinae, treated as part of the Usiinae by Yeates (1994), should be treated as a separate subfamily, as in the world bee fly catalog by Evenhuis and Greathead (1999).

Phthiriinae The subfamily Phthiriinae is a noteworthy group because these small, attractive, long-tongued flies are familiar to naturalists the world over as the genus *Phthiria*, even though the 60-plus extant species in *Phthiria* (as currently defined) are either Old World or Chilean, and the genus no longer includes species in North America or Australia. The apparent disappearance of the genus *Phthiria* from these regions is real in the sense that "true" *Phthiria* are known from Oligocene fossils in North America but unknown from extant North American species, and artificial in the sense that the extant Nearctic and Australian species previously placed in *Phthiria* have simply been moved to more narrowly defined genera.

Thus the North American species once known by the charming name *Phthiria relativitae* (try saying this out loud) is now in the New World genus *Poecilognathus* and has the far less interesting name *Poecilognathus relativitae*. There is an irony here, since the species was discovered and named *Phthiria relativitae* by Neal Evenhuis in 1985, and its shift to *Poecilognathus* resulted from a later paper published by the same author. This species name is properly recorded in the technical literature as "*Poecilognathus relativitae* (Evenhuis)," with the parentheses around the author's name indicating that the species is now in a different genus than when it was first described. Evenhuis, by the way, is famous for combining meticulous taxonomic work with a bit of tongue-in-cheek nomenclature (as you may have gathered from names such as *Phthiria relativitae* and *Oligodranes zzyzxensis*). My favorite Evenhuis name is *Iyaiyai*, which he applied to an obscure genus of fossil phantom midge from China, and which I say over and over again as I smack my forehead, trying to make sense of the higher classification of Bombyliidae! Only a few *Phthiria* have been reared, in each case from small moth hosts.

Toxophorinae The subfamily Toxophorinae, the "slender bee flies," is a cohesive group made up almost entirely of species placed in three large, globally distributed, distinctive genera: *Systropus*, with well over a hundred strikingly slender, long-legged, sphecid wasp–mimicking species; *Geron*, a commonly encountered group of some 130 species of small flies that look a bit like hump-backed mosquitoes, but with a long proboscis used for nectaring, not blood feeding; and *Toxophora*, a smaller group of about 50 somewhat stouter humpbacked species. *Systropus* and *Geron* are not only unlike other bee flies in appearance, they also have atypical life cycles for a family made up mostly of species that develop as ectoparasitoids of subterranean hosts. Larvae of *Systropus* are endoparasitoids of larvae and pupae of Limacodidae (slug caterpillars), while known *Geron* larvae are parasitoids of larvae and pupae of other moths, including Noctuidae, Psychidae, Pyralidae and Tortricidae (bee fly host data are reviewed by Yeates and Greathead [1997]). *Toxophora*, as far as is known, have more typical bee fly habits and develop as ectoparasitoids in the nests of solitary bees and wasps.

Heterotropinae The subfamily Heterotropinae currently includes only one odd Old World

Systropus is a distinctive, widely distributed bee fly genus of about 170 species; known larvae are endoparasitoids of larvae and pupae of Limacodidae (slug caterpillars).

genus, *Heterotropus*, although a number of other genera have been treated as heterotropines in the past. Little is known about the biology of *Heterotropus* beyond records of larvae "sifted from soil" and presumed to be free-living predators. The 44 species now included in this group are bright yellow and black flies that occur only in the Palaearctic and Afrotropical regions. Morphological and molecular data strongly suggests that the heterotropines, like the micro bee flies (Mythicomyiidae), form an "early diverging lineage" relative to other bee flies (Sinclair et al. 1994, Trautwein et al. 2011). Both groups seem to have predaceous (as opposed to parasitic) larvae, as we would expect of basal bee fly lineages.

Bombyliidae: Part Two

The Higher Bee Flies or Sand Chamber Flies

Higher bee flies share a set of remarkable adaptations that allow females to gather sand into an abdominal chamber. Here the sticky, thin-walled eggs are coated with sand prior to being released into the hostile environment in which they will hatch as host-seeking first-instar larvae. Female flies with sand chambers have hairy, uniquely modified abdominal sclerites (tergite and sternite 8) that serve to scoop sand up into the chamber. The lineage of bee flies characterized by this remarkable sand-scooping and egg-coating equipment is referred to as the "sand chamber subfamilies" or "higher bee flies" and, presumably because this is an enormous lineage that includes thousands of species, it is divided into several subfamilies.

Bombyliinae Even with division of the higher bee flies into several subfamilies, most of the "typical" robust, fuzzy species remain in the subfamily Bombyliinae, or "typical bee flies." This group seems to be a default subfamily for typical bee flies that don't fit elsewhere, and it is thus an artificial (non-monophyletic) group of superficially similar higher bee flies (Trautwein 2009). The genus *Bombylius*, with more than 350 species found

Female higher bee flies are often seen touching down to gather sand into an abdominal chamber where the eggs are given a protective sand coating. Larvae of this genus (*Anastoechus*) are predators of grasshopper egg pods.

almost everywhere except South America and Australia, contains the most familiar of the true bee flies, including the type species of the family, *Bombylius major*. This common Holarctic species is one of the most conspicuous early spring insects in North America, Europe and Asia. Adults are commonly seen visiting spring flowers, inserting a strikingly long beak while hovering, humming-bird-style, over the blooms. Like other known *Bombylius* species, hovering *B. major* females flick their sand-coated eggs into the burrows of solitary ground-nesting bees. Larvae enter host cells and later, after a period of inactivity, become ectopar-asitoids of immobile, prepupal bee larvae.

Several other Bombyliinae also attack solitary bees, and the western North American *Heterostylum robustum* has been considered a pest species because of its impact on beneficial bees needed for the pollination of forage crops. Other bombyliines, especially the large genus *Systoechus* and the similar *Anastoechus*, are beneficial insects that develop on the egg masses of pest locusts and other short-horned grasshoppers. Although bombyliines generally develop on concealed hosts and are most often described as ectoparasitoids, the line between parasitism, predation and klepto-parasitism is a bit fuzzy. Some consume more than one host and are thus predators, while others feed either on host provisions (as kleptoparasites) or consume only a single host larva (as parasitoids).

Bombyliinae are "typical" higher bee flies, but not all members of the subfamily have sand chambers. Two such genera, the western North American *Lordotus* (29 species) and *Geminaria* (two species) were moved from the Bombyliinae to a new subfamily, Lordotinae in Yeates's 1994 bee fly classification. *Lordotus* are fuzzy bee flies sometimes seen hovering over the entrances of solitary wasp nests as they deposit eggs, after the fashion of other bee flies in the subfamily Bombyliinae. They were placed in a new subfamily by Yeates because they have highly modified ovipositors and because they lack the sand chamber characteristic of the Bombyliinae. Evenhuis and Greathead (1999), however, point out that the sand chamber has been secondarily lost in other lineages of typical bee flies, and return *Lordotus* and *Geminaria* to the Bombyliinae.

The small Holarctic and Neotropical subfamily **Crocidinae** is superficially similar to the Bombyliinae — both groups differ from other higher bee flies in that the back of the head is flat rather than concave — and was previously treated as a tribe of Bombyliinae. Nothing is known about the biology of this uncommon subfamily, and two of the nine genera are known from only one specimen each.

The small (50 species) worldwide subfamily **Ecliminae**, treated as part of the Bombyliinae by Yeates and Lambkin (2004), includes the elongate scaly flies in the genus *Lepidophora*. *Lepidophora* larvae develop in the nests of solitary wasps, where they consume the paralyzed prey rather than the developing wasp larvae; they are thus kleptoparasites.

Anthracinae Species in the remaining higher bee fly subfamilies, of which by far the most important is the subfamily Anthracinae, differ from Bombyliinae in that the back of the head is distinctively concave as opposed to flat, but they are otherwise typically robust higher bee flies. Anthracinae is the largest subfamily of Bombyliidae, including almost half of all bee fly species and many of the largest and most widespread genera, such as *Exoprosopa*, *Anthrax* and *Villa*. *Exoprosopa*, with 400 or so species, is found in every region of the world and includes a conspicuous variety of often strikingly patterned bee flies. The relatively few *Exoprosopa* with any host data have been reared from cocoons of solitary ground-nesting wasps (Sphecidae, Pompilidae and Tiphiidae).

For some reason the equally widespread but somewhat smaller (240 species) genus *Anthrax* is much more frequently reared; it is known to develop as an ectoparasitoid on larvae and pupae of a variety of wasps and bees as well as a few other kinds of subterranean larvae. One *Anthrax* species, *A. georgicus*, is a well-known parasitoid of tiger beetle larvae — which of course occur in burrows in the same localities as ground-nesting solitary bees and wasps — and another species of *Anthrax* has been reared from an antlion (Neuroptera). Attacking antlions in their pits is a habit more characteristic of a related genus, *Dipalta*, larvae of which attack antlion pupae (antlions pupate in a cocoon in their underground pit). *Villa*, with a worldwide distribution and species diversity similar to *Anthrax*, develops inside its hosts — caterpillars, or occasionally fly and beetle larvae — and is thus endoparasitic, in contrast with its ectoparasitic relatives. The large, showy bee flies in the genus *Poeciloanthrax* are also endoparasitoids, in the caterpillars of noctuoid moths. Other habits known in this subfamily include predation of grasshopper egg pods, parasitism of tsetse fly puparia and hyperparasitism of the pupae of parasitoids from other orders.

Lomatiinae Probably the most distinctive of the remaining higher bee fly families is the subfamily Lomatiinae, the "long-winged bee flies." Members of this cohesive group are particularly diverse in Australia, where members of the large genera *Comptosia* and *Aleucosia* are among the most conspicuous and frequently encountered bee flies; closely related genera occur in South America and Africa. The biology of the Lomatiinae is poorly known, but the few species that have been reared are associated with ants, primitive moths and tenebrionid beetles. The remaining higher bee fly subfamilies are relatively minor and rarely encountered groups.

Empidoidea and the Enigmatic Families Apystomyiidae and Hilarimorphidae

Flies in the superfamily Empidoidea, affectionately known as "doids" by those who study them, are common flies the world over and are often the most abundant small (generally 2–10 mm) insect predators on leaves, tree trunks, shorelines and moist or wet surfaces. Most prey preferentially on small, soft-bodied invertebrates, but some are regular flower visitors and a few genera feed by puncturing pollen grains rather than preying on other invertebrates. Larvae, with a few exceptions discussed below, are predators too, occurring in wet soils, under bark, in algae and in some aquatic and semiaquatic substrates.

The almost 12,000 empidoid species form a single lineage widely accepted as the sister group to the Cyclorrhapha (the Empidoidea and Cyclorrhapha together make up a lineage termed the Eremoneura) and are traditionally divided into the Empididae (dance flies) and Dolichopodidae (long-legged flies). This traditional classification has long been recognized as artificial because parts of the "old" Empididae were more closely related to Dolichopodidae than to other Empididae. Chvála (1983), in his monograph on Scandinavian dance flies, addressed this issue by treating some of the subfamilies of Empididae (Hybotinae, Atelestinae, Empidinae and Microphorinae) as separate families. Although Chvála justified his new classification in part on a now-refuted hypothesis that the Cyclorrhapha arose from within the Empidoidea as sister group to *Atelestes* and related genera, his work ignited debate and stimulated several subsequent papers by dipterists from all over the world, most of whom endorsed Chvála's system of five empidoid families.

The most exhaustive treatment of empidoid phylogeny and classification to date is a monograph by Sinclair and Cumming (2006), who used morphological characteristics to provide a firm foundation for division of the Empidoidea into the families Atelestidae, Hybotidae, Empididae, Brachystomatidae and Dolichopodidae, leaving a few genera unassigned as to family (see below). In this landmark paper, Microphoridae were formally subsumed into the Dolichopodidae, reflecting a long-recognized relationship, and the monophyly of the Empidoidea as a whole was confirmed. A more recent review of the Empidoidea, using only molecular characters, supported most of Sinclair and Cumming's groups but argued for inclusion of the Brachystomatidae as part of the family Empididae rather than as a separate family (Moulton and Wiegmann 2007). The discussion of empidoid families here follows the classification of Sinclair and Cumming (2006).

Dolichopodidae

Long-legged Flies

"Typical" members of the family Dolichopodidae — slender, long-legged, relatively bare and brilliantly metallic green flies — seem to thrive in every garden or park and can be easily found on low, moist foliage anywhere in the world. A closer look will reveal some of the 7,000 or so world dolichopodid species on a wide variety of surfaces, including rocks, tree trunks, logs, seacoasts and mudflats. Although the most familiar species are shining green, with conspicuously large male genitalia slung under the body, the family includes several groups of dull-colored or yellowish species, and a few lineages lack the characteristic well-hung appearance that usually distinguishes male long-legged flies from other empidoids. Males of many Dolichopodidae are adorned with conspicuous flag-like expanded feet, silver patches, expanded antennae, decorated wings or other garish ornamentation used to signal potential mates.

All adult dolichopodids are predators that pursue soft-bodied prey such as springtails, oligochaete worms or flies, which they envelop between two sandpaper-like lobes (labella) of the labium (lower lip), seeming to slurp up the victims as they are gripped by the lobes and punctured by the labrum (upper lip) with its tooth-like epipharyngeal blades on the lower surface. This is in marked contrast to most other empidoids, which impale their prey with dagger-like mouthparts, which in turn contrasts with the way robber flies stick their victims with just a syringe-like hypopharynx ("tongue"). A few long-legged flies feed on nectar and pollen, but this is a more unusual habit in the Dolichopodidae than in other empidoids. One genus of Dolichopodidae, *Melanderia*, is sometimes described as having jaws, but the mandible-like structures of this western North American genus are simply strikingly modified labella.

Larval dolichopodids, like most larval lower Brachycera and Empidoidea, are almost all predators in soil, under bark and in similar habitats. The one known exception to this rule is the genus *Thrypticus* (subfamily Medeterinae), which includes at least some species that develop as stem miners in aquatic or semiaquatic plants (all records are from monocots). Last-instar larvae of dolichopodids spin a loose silken cocoon mixed with debris. Pupae are enclosed in the cocoon but have a pair of large respiratory horns that project from the top of the thorax through the silken enclosure.

The subfamily classification of the Dolichopodidae has had a tortuous history, leaving a bewildering legacy of alternative subfamily divisions. The higher classification of the family remains complicated and is likely to change as more and more species of this enormous family are described and put in phylogenetic context, but some of the 17 widely used subfamilies are robust and predictive on a worldwide basis. Others, such as the Sympycninae, remain unstable and are treated differently by different taxonomists. The classification used here follows Pollet and Brooks (2008).

Microphorinae and Parathalassiinae The subfamilies Microphorinae and Parathalassiinae are now treated as basal lineages of the

Dolichopodidae, but historically these small, empidid-like, rarely encountered flies have been treated either as separate families or as part of the Empididae. *Microphor*, the more common of the two genera of **Microphorinae**, includes one Tasmanian and several Holarctic species. Some species, including the European *Microphor crassipes* and the eastern North American *M. obscurus*, have been reported as kleptoparasites that steal prey from spider webs; *Microphor* can also be common on some flowers such as false Solomon's seal (*Maianthemum* spp.). The other microphorine genus, *Schistostoma*, is a group of just over 20 species of small, dull grayish flies found on sandy shorelines, mostly in Europe and Asia but with a few species in North America and southern Africa.

Dolichopodidae in the subfamily **Parathalassiinae** are found in similar habitats, usually either seacoasts or wet surfaces near bodies of fresh water. European, Australian and North American species of the widespread parathalassiine genus *Microphorella* occur along freshwater margins, while the Southeast Asian species are recorded from sandy seashores. The other five genera in this subfamily are also small and rarely collected flies associated with seashores the world over.

All dolichopodids other than Microphorinae and Parathalassiinae — which are sometimes treated together as the single subfamily Microphorinae — form a distinctive monophyletic group corresponding to the traditional concept of the Dolichopodidae.

Dolichopodidae in the genus *Melanderia* are sometimes described as having jaws, but the mandible-like structures referred to are simply strikingly modified labellae (probably derived from the labial palpi).

Dolichopodids, such as this South American *Condylostylus* masticating a scuttle fly, are common predators of small, soft-bodied arthropods, including other flies.

Dolichopodinae The subfamily Dolichopodinae, the largest subfamily of long-legged flies, includes the "typical" long-legged flies such as those in the huge genera *Hercostomus* and *Dolichopus*. Some of the more than 500 *Dolichopus* species are very common, and members of this genus are usually conspicuous anywhere near water in north temperate countries. *Dolichopus* males are often conspicuously adorned with swollen bits and pieces such as silver and expanded tarsomeres, used as signaling devices in elaborate courtship displays. Courtship displays in the Dolichopodidae, which usually include vigorous wing- and leg-waving and sometimes spectacular backflips or other gymnastic feats, can be easily observed against the backdrop of the broad leaves, water, mud or other substrates on which they display. This contrasts with the more general ("primitive") mating behavior of other empidoids, many of which are "dance" flies with courting rituals played out in aerial swarms.

Although *Dolichopus* is absent from the neotropics, related dolichopodines such as *Pelastoneurus* and *Tachytrechus* are diverse in Central and South America. Both genera also range north to Canada, where the distinctively spot-winged *Tachytrechus vorax* is the characteristic dolichopodid of large expanses of wet sand around the Great Lakes. Other *Tachytrechus* species can be found on wet sand in Africa and elsewhere. Marine shorelines in Africa and Sri Lanka are home to another attractive sand-loving dolichopodine genus, the green-eyed, yellow-bodied *Argyrochlamys*, some species of which are able to stay active on dry, sun-seared Indian Ocean beaches.

Sciapodinae The subfamily Sciapodinae, with around 1,500 species, is the second largest subfamily of Dolichopodidae and includes many of the commonly encountered attractive flies characteristically seen standing on forest foliage, often with attractively patterned wings held outstretched at an angle from the body. Bickel (2009) suggests that the tropics are home to an enormous number of undescribed sciapodines, pointing to his 2002 revision of *The Sciapodinae of New Caledonia,* in which 53 of the 55 included species

were new to science. The Sciapodinae include the brilliantly metallic green dolichopodids that hunt minute prey on foliage in gardens and city parks, far from the humid habitats on which most dolichopodids depend. One such familiar genus, *Condylostylus*, includes more than 260 mostly New World species and is found from northern Canadian city parks all the way to Bolivian jungles. The more than 200 named species in the similar genus *Chrysosoma* are more common in the Old World tropics and Australia (Bickel 1994).

Sympycninae The subfamily Sympycninae is one of the largest subfamilies of Dolichopodidae, with almost 1,000 species, but the limits of the subfamily and its included genera are ill-defined. The type genus, *Sympycnus*, is a large group found in all biogeographic regions, as are some of the other large genera currently included in this subfamily. The genus *Campsicnemus* has a history that parallels that of the more famous Hawaiian *Drosophila*, as both genera have undergone spectacular adaptive radiations on the Hawaiian archipelago. The 163 named Hawaiian *Campsicnemus* are thought to represent only about half of the actual fauna, and many species remain known only from single collections. Hawaiian *Campsicnemus* species occupy an extraordinary range of habitats and include water skaters, species found in leaf litter, species associated with ferns and species of the high canopy (Evenhuis 2009). Mainland *Campsicnemus* species, generally north temperate in distribution and entirely absent from the neotropics, occupy a much narrower range of mostly shoreline habitats.

One of the most remarkable genera currently included in the subfamily Sympycninae was described by Runyon and Hurley (2003), who described the new genus and new species *Erebomyia exalloptera* on the basis of southwestern American specimens with very differently shaped left and right wings. Such extreme directional asymmetry is otherwise unknown among insects (or any other winged organisms), and since the asymmetry in *E. exalloptera* occurs only in males and is highly consistent from specimen to specimen, Runyon and Hurley speculated that it must play an important role in courtship. They later reviewed the genus *Erebomyia* and added another three new species to the genus, all with symmetrical but modified wings (Hurley and Runyon 2009). Although there are very few specimens of *Erebomyia* in major collections, the genus is fairly common on vertical to overhanging rock surfaces in dark cavities along creeks in Arizona canyons. Hurley and Runyon observed wing-fanning as part of the male courtship behavior of two species (*E. exalloptera* and a species with symmetrical wings), and speculated that wing-fanning in the dark habitats of *Erebomyia* generates distinctive sounds according to wing shape and ornaments. Sadly, their 2009 review of the genus did not appear in print until six months after the untimely death of the senior author.

Diaphorinae The subfamily Diaphorinae is a large and important group, with around 800 described species and several diverse and commonly encountered genera. The mostly Holarctic genus *Argyra* is among the most easily recognized — and most attractive — genera in the family. Species of *Argyra* are usually partially covered by silvery pruinosity that flashes conspicuously in the right light. The much smaller metallic green flies in the diverse diaphorine genus *Chrysotus* can also appear as silvery flashes in the right light, but in this case the reflective surfaces are on the male palpi. The large diaphorine genera *Asyndetus* and *Diaphorus* are among the most common dolichopodids wordwide; *Diaphorus*, with 266 or so species, is especially common in both New World and Old World tropical forests. At least one Pacific *Diaphorus* species is parthenogenetic (no males required) and some *Asyndetus* species develop in crab holes, but otherwise life history and habitat information is available for relatively few diaphorines (and, indeed, for relatively few Dolichopodidae).

Medeterinae The subfamily Medeterinae is one of the more easily located groups of long-legged flies. One needs only look at the nearest tree trunk to find adults of these small, compact, relatively dull-colored Dolichopodidae standing around with their characteristic head-up "woodpecker" stance as they scan the surface for minute prey such as springtails (Collembola), mites or gall midges (Cecidomyiidae). Females of the common cosmopolitan genus *Medetera* can often be seen slipping their soft, telescoping ovipositors into small openings through the bark — usually holes left by egg-laying beetles — where their larvae are well-known predators of bark beetle larvae. Most of the 300 described species of *Medetera* are north temperate flies, but the Medeterinae includes diverse lineages throughout the world.

Systenus, known species of which develop in rot holes and sap runs on hardwood trees, is also widespread, with Neotropical, Holarctic, Australasian and Oriental species. My favorite medeterine is the tiny (just over 1 mm) *Papallacta stenoptera*, recently described by Bickel (2006) entirely on the basis of specimens I collected in 1979 in a cold, high Ecuadorian *páramo*. This little fly has broad flags at the end of its antennae and its wings are reduced to short, very narrow straps. *Papallacta stenoptera* is the only dolichopodid in the entire New World with such reduced wings, although similar wing reduction is known from several Pacific island dolichopodids. Interestingly, *Ceratomerus apterus*, the only known flightless member of the empidoid family Brachystomatidae, was also described from specimens I discovered at that same high Andean site.

Wing reduction in *Papallacta* is unusual, but perhaps less unusual than the habitat shift that partially defines the medeterine genus *Thrypticus*. All known larvae in this large (86 species), cosmopolitan genus are stem miners that bore into living plant tissue, and all species in the genus use a uniquely derived sclerotized blade for piercing and ovipositing within plant tissue (Bickel and Hernandez 2004). Although similar structures and habits occur convergently in some lineages of Cyclorrhapha (for example, Tephritoidea), no other lineage of the Empidoidea or lower Brachycera has similar structures or habits; this appearance of an apparently phytophagous lineage in the middle of a huge predaceous family is quite remarkable. *Thrypticus* species are most diverse in the neotropics, where they have been reared from several groups of aquatic and semiaquatic monocotyledonous plants, including the invasive water hyacinth (*Eichhornia crassipes*) that has become a pest in several parts of the world. Several *Thrypticus* species feed on water hyacinth, so they have potential as biological control agents for use against this invasive weed.

Hydrophorinae The subfamily Hydrophorinae is familiar to most naturalists because of the conspicuous water-skating habits of the genus *Hydrophorus*, a widespread and common fly on water surfaces almost everywhere but the neotropics. These silvery long-legged flies skate across the water surface in search of prey and mates, and they can often be seen *in copula* on the surface of a small puddle, tide pool or creek backwater. Members of the water-skating genus *Hydrophorus* occur on a wide range of calm water surfaces, both fresh and salt, but seashores are the preferred habitat for many other hydrophorine genera. The genus *Thambemyia*, for example, includes about 30 species native to Hawaii, China and Japan, where they can be found feeding on midges among the barnacles encrusting wave-splashed shoreline rocks. One of the Japanese species, *T. borealis*, has recently established itself at a number of seaports in the United States as well as on the Pacific coast of South America, where these brightly patterned black and white flies are now abundant near some ports.

Another distinctive hydrophorine genus of the rocky intertidal zone is the western North American "mandibulate" genus *Melanderia*. The genus was described as new by J.M. Aldrich in 1922, back to back with a paper on mandible substitutes in the Dolichopodidae written by the morphologist R.E. Snodgrass. The great morphologist introduced his paper by saying that "a first view of the face of *Melanderia mandibulata* gives one a decided shock, followed by a desire to discover by what morphological trick the fly so cleverly imitates the features of a mandibulate insect." He then pointed out that although the sharp, polished "mandibles" could be worked in and out in true mandibular fashion, they were but lobes of the labium.

Related coastal Hydrophorinae include the distinctive eastern Nearctic genus *Hypocharassus*, which is a characteristic dolichopodid found on the sort of intertidal muck that can quickly swallow the rubber boots of an intrepid fly photographer. *Hypocharassus*, one of the more distinctive dolichopodids because of its bottle-opener-shaped antennae, can also be incredibly abundant on mats of seaweed. However, not all hydrophorine genera are associated with marine habitats; some, such as the large and attractive *Liancalus*, occur on inland seeps and others are found in a variety of other freshwater habitats.

Neurigoninae Species in the subfamily Neurigoninae, like those of the Medeterinae, are most often found on tree trunks, but the most common tree-trunk genus in this group, *Neurigona*, could hardly look more different from common Medeterinae. Instead of being small, squat and dull like *Medetera*, *Neurigona* species are elegant, delicate long-legged flies, often with brilliantly green eyes. About 75 percent of the species in this subfamily are in the cosmopolitan genus *Neurigona*, with the remaining species occurring in mostly small Neotropical groups.

Hydrophorinae occur in a variety of marine and freshwater environments, often skating on the water surface like this *Hydrophorus*. The bare spots on the side of the abdomen are abdominal plaques, where muscles were attached to the pharate adult still inside the pupal shell, allowing it to move the pupal shell prior to emergence. Abdominal plaques are found in lower Brachycera and Empidoidea but not higher Brachycera.

The entirely New World subfamily **Enliniinae** includes only two genera and around 100 described species. These minute flies are very poorly known, and hundreds of Neotropical species undoubtedly await discovery. Look (very closely) for adult enliniines hovering along wet rocks near seeps or small waterfalls, where they can be common and sometimes occur in communities of several species. With an average body length of less than 1 millimeter, it is likely that this group — termed "microdolichopodids" by Harold Robinson (1975) — will remain poorly known for many years to come. Robinson (1969) documented 53 species of the microdolichopodid genus *Enlinia* from Mexico alone, indicating that an enormous diversity of this genus awaits collection elsewhere in the neotropics.

The small subfamily **Rhaphiinae** is dominated by the distinctive genus *Rhaphium*, a mostly Holarctic genus with almost 200 species, of which only a few occur in the tropics. *Rhaphium* species are sexually dimorphic: males are unmistakable for their broad, bladelike antennae. The other half-dozen or so genera in this subfamily are small and far less common than *Rhaphium*.

The subfamily **Achalcinae** is a small group of about half a dozen genera and fewer than 70 described Holarctic, Neotropical and Australasian species. The Holarctic-Neotropical genus *Achalcus* (23 species) occurs in marshes, while the mostly Neotropical and Australasian *Australachalcus* (24 species) breeds in tree holes. According to Pollet and Brooks (2008) there are diversity "hot spots" for this subfamily in New Zealand and Australia, and the group includes a number of as yet undescribed genera.

The subfamily **Plagioneurinae** exists as a taxonomic home for only a single species, the New World species *Plagioneurus univittatus*. This attractive, superficially *Dolichopus*-like species ranges from South America to Canada.

The remaining subfamilies are small groups and not routinely encountered. The subfamily

Stolidosomatinae is a small, mostly Neotropical subfamily of only 29 species, mostly restricted to rainforests. The subfamily **Babindellinae** includes only two species, both Australian. **Peloropeodinae** is a small, somewhat obscure subfamily that includes several genera sometimes included in the subfamily Sympycninae. The small subfamily **Xanthochlorinae** includes only the single genus *Xanthochlorus*, a small Holarctic group that, according to Pollet and Brooks (2008), is associated more with dry deciduous forest than the usual damp haunts of most dolichopodids. Yang et al. (2006) add two more subfamilies, **Antyxinae** and **Kowmunginae**, to hold a few Australian and Oriental species supposedly related to the Babindellinae and Enliniinae.

Empididae

Empidid Dance Flies

The family Empididae currently comprises some 3,000 species of relatively long-legged flies varying widely in size (1–10 mm) and appearance, although the three main subfamilies, Empidinae, Hemerodromiinae and Clinocerinae, are well defined, each having a characteristic appearance, behavior and natural history. A fourth and poorly defined subfamily, Oreogetoninae, is currently included in the Empididae as a matter of convenience, even though its included genera are probably basal empidoid lineages and not phylogenetically part of the Empididae. Older classifications put all empidoids other than the Dolichopodidae in the family Empididae, but that broad concept of the family is no longer used.

A few genera of Empididae remain of uncertain placement within the family and are not currently classified in a subfamily. One such group, referred to by Sinclair and Cumming (2006) as the "*Ragas* group," includes the unusual fire-adapted genus *Hormopeza*. This genus has been recorded from South Africa, Australia and Brazil but is best known from Europe and North America, where these otherwise rare flies sometimes occur in large swarms of males and females around smoky campfires. I've seen them land and mate on sizzling-hot rocks within inches of glowing coals around campfires in northern Ontario, where they feed voraciously on other flies apparently attracted to the smoke. *Hormopeza* and the "true smoke flies" in the Platypezidae (*Microsania*) often occur together; both have similar antennal pits, presumably to help home in on hot habitats.

Both males and females of the fire-adapted empidoid genus *Hormopeza* are attracted to smoky campfires, where they mate and prey on other flies, including smoke flies (*Microsania*).

Empidinae The subfamily Empidinae includes many of the largest and most familiar dance flies, well known for their conspicuous swarms and ritual transfer of nuptial gifts from males to their prospective mates. Some of these relatively broad-winged, often long-beaked flies (especially *Empis* species) are frequent flower visitors, but many can be more reliably located in their characteristic habitats or hunting grounds. Species in the large genus *Hilara* and some closely related genera, for example, can easily be found in swarms over ponds, lakes, rivers and streams. Males hold their legs outstretched as they swiftly skim along the water surface in search of prey, often in groups of hundreds or thousands. Catching these frenetic flies is a challenge because they fly so close to the water, but a captured male can be recognized as a member of this group (the tribe Hilarini or "balloon flies") by the huge, swollen first tarsal segment on his foreleg. This swollen fore basitarsus contains a silk gland made up of several glandular cells that produce silk for use as "gift wrap" by the courting male. Males usually use the silk, released through ejector-like hollow hairs under the basitarsus, to wrap a small captured insect into a balloon-like package before it is presented to the female as a "nuptial gift," ensuring his acceptance by the female and influencing the duration of the copulation.

Females throughout the Empidinae seem unable to hunt for themselves; if they require protein for egg development they can obtain it only from nuptial gifts of prey captured by males. Nuptial gifts are not always packaged in silk like those of the balloon flies in Hilarini, and when the females are given silk packages they do not always include prey. Some balloon fly males consume the prey themselves, leaving only a fragment attached to the silk balloon to present to the female; others skip the prey altogether and present prospective mates with nothing but a silk balloon, a fluffy seed or some other inedible object. One undescribed Australian *Hilarempis* (a close relative of *Hilara*), for example, skims diatoms off the water surface and binds the algal mass in silk to make a presumably inedible vegetarian gift with which to regale the female. Spinning silk from the front feet is a trick almost unique to *Hilara* and some closely related species in the tribe Hilarini, although convergently similar protarsal silk glands occur in another insect order, the Embiidina (webspinners).

The subfamily Empidinae is divided into two tribes, the **Hilarini** and the **Empidini**. Although the common name "balloon flies" is often applied to the whole subfamily, species in the tribe Empidini lack tarsal silk glands, and only a few of them — in the genus *Empis*, subgenus *Enoplempis* — produce balloons, presumably from saliva. Most Hilarini belong to the genus *Hilara*, while the overwhelming majority of species in the Empidini are in the genera *Rhamphomyia* and *Empis*. The number of undescribed species in these genera can only be guessed at, but it is huge. Bickel (2009) has suggested that the 300 or so named species of *Hilara* balloon flies represent less than 10 percent of the actual world fauna.

Empidini Both *Empis* and *Rhamphomyia* include hundreds of undescribed species, and of the 800 or so described species only a handful are well known. Most species with known behaviors form large swarms of dancing males bearing nuptial gifts of captured insects. Females typically arrive at the swarm and choose a mate, forming a pair that leave the swarm to hang out — literally, using the male's long front legs — on nearby vegetation. Mating takes place while the females are busy with the nuptial gift. Although females are normally the choosy sex throughout the animal kingdom, when males bring a valuable resource to the table the sex roles can get reversed, as has happened several times in the genus *Rhamphomyia*. Some species, such as the common northeastern North American *R. longicauda*, form swarms of females trying to look as large and fecund as possible in order to attract a choosy, gift-bearing male. Each swarming female strives to stand out by displaying strikingly feathered legs while at the same time inflating her abdomen to look jam-packed with mature eggs. Males, in contrast, are simple and unadorned.

Although not as well known as the big north temperate genus *Rhamphomyia* or the more widespread genus *Empis*, some related genera in the tribe Empidini occur in the tropics. Swarms of male *Porphyrochroa*, their metallic blue abdomens flashing in the odd shaft of sunlight, are commonly encountered in Neotropical cloud forests, where males and females of the same genus are also found on flowers. Males and females of the closely related sinuate-bodied *Macrostomus* are sometimes seen feeding at extrafloral nectaries, and males of both genera are probably predaceous, although the biology of these Neotropical lineages is virtually unknown. Other genera of Empidini are small and south temperate, occurring in southern South America, South Africa, New Zealand and Australia.

Hilarini Although the enormous balloon fly genus *Hilara* occurs worldwide, related genera are mostly restricted to south temperate countries. Some, like the long-legged Neotropical *Aplomera* species that sweep up prey with their stout hind legs, range through much of South America, but most other genera of Hilarini occur in Australia, South

Swarms of male *Porphyrochroa*, their metallic blue abdomens flashing in the odd shaft of sunlight, are common in Neotropical cloud forests.

Africa, New Zealand or southern South America. Several empidid genera and species groups within the larger genera have intriguing split distributions, with close relatives in different south temperate countries that were once part of the ancient continent of Gondwana.

Hemerodromiinae The subfamily Hemerodromiinae, in contrast to the Empidinae, is a distinctive group of elongate flies with enlarged mantid-like raptorial forelegs widely separated from the midlegs. Hemerodromiinae are usually associated with water, and adults — especially those of *Hemerodromia* — often occur on wet rocks or the lower surfaces of bridges crossing clean streams. Members of this subfamily don't form aerial swarms, but mating and oviposition sometimes take place in dense aggregations on wave-splashed rocks. The Hemerodromiinae are divided into two tribes, Chelipodini and Hemerodromiini. Some ***Chelipodini*** have been reported as having terrestrial larvae and a few species are associated with rocky marine shores. Known ***Hemerodromiini*** are aquatic and can be important predators of black flies and other lotic larvae.

Clinocerinae The subfamily Clinocerinae is a distinctive, well-defined and easily recognized group of medium-sized empidids normally associated with shorelines or emergent rocks, where the adults are often major predators of emerging aquatic insects. Known larvae are aquatic. The group is worldwide and includes some large, widespread genera; several of the more "primitive" genera occur in the southern hemisphere. The phylogeny and taxonomy of this group are exceptionally well known, thanks to a series of monographs by the Canadian empidologist Bradley Sinclair.

Oreogetoninae (or Homalocnemiidae, Oreogetonidae and unplaced Empidoidea) The subfamily Oreogetoninae is one of those "grab-bag" groups currently used as somewhere to stick a few awkward lineages that don't fit well into any empidoid family. With all the recent shuffling and reclassification of the 12,000 or so known empidoid species, it is not surprising that a couple of dozen species remain to be satisfactorily placed. It is even less surprising that the problematic species can be loosely described as basal empidoid lineages, since it always seems to be the "basal lineages" that complicate classifications. Sinclair and Cumming (2006) treat these disputed species in three species groups that they leave unplaced as to family and subfamily. Moulton and Wiegmann (2007) retain all three species groups in the subfamily Oreogetoninae of the family Empididae, which is where they were before Sinclair and Cumming's 2006 study. Despite evidence that such a broadly defined Oreogetoninae is neither a natural (monophyletic) group nor part of a monophyletic Empididae, its disparate subgroups are mentioned here under the Empididae as a matter of convenience, and because it seems premature to toss them out until the evidence is unequivocal

The aquatic genus *Oreogeton* is one of those lineages of Empidoidea that does not fit well in any of the familiar families, and may warrant recognition as a separate family (Oreogetonidae). It is currently treated as part of the Empididae.

and there are other appropriately defined families into which to toss them.

There have been proposals for new taxa to hold some of these problematic flies. For example, the great empidid specialist James Edward Collin assigned the genus *Homalocnemis* to its own subfamily, Homalocnemiinae, as early as 1928. This oddly distributed south temperate (New Zealand, Chile, Namibia) genus is predaceous but, like the Atelestidae, lacks the epipharyngeal blades most empidoids use to puncture their prey. This suggests that *Homalocnemis* represents one of the oldest extant lineages of the Empidoidea, and it was thus excluded from the Empididae by Sinclair and Cumming (2006). Some authors are now treating *Homalocnemis* as a separate family, the **Homalocnemiidae**.

The genus *Iteaphila* and the closely related *Anthepiscopus* also lack epipharyngeal blades. These genera form another possibly "basal" lineage of entirely flower-feeding empidoids, some of which are very common early spring flower visitors in the Holarctic region. A third "problem" lineage currently treated as part of the Oreogetoninae comprises the genus *Oreogeton* and some close relatives. They form a Holarctic group considered by Sinclair and Cumming (2006) to be the oldest empidoid lineage with epipharyngeal blades. The immature stages of *Oreogeton*, a small genus of about a dozen species, are aquatic. Some dipterists now recognize a separate family, **Oreogetonidae**, for *Oreogeton*.

Hybotidae

Hybotid Dance Flies

The family Hybotidae is referred to as the hybotid dance flies because this group was long grouped with all other non-dolichopodid Empidoidea in the family Empididae, an artificial group almost universally known as dance flies. The "old"

Empididae is now broken up into the Hybotidae, Empididae, Brachystomatidae and Atelestidae, with 99 percent of the species split between the two huge families Hybotidae and Empididae. The name "dance fly" was never descriptive for this whole group, since it refers to the dancing aerial swarms characteristic of only some members of the subfamily Empidinae in the Empididae. A few hybotids form small hunting swarms, but mating in Hybotidae normally occurs on the ground or other surfaces, as is also true for Dolichopodidae and three out of four subfamilies of Empididae. Still, using the name "hybotid dance flies" reflects the way many entomologists and naturalists continue to think of all non-dolichopodid empidoids as "dance flies." Hybotidae is a diverse family of almost 2,000 species grouped together on the basis of important but relatively inconspicuous characters, such as the male genitalia and a unique gland near the base of the foretibia. The four subfamilies are each fairly uniform in both appearance and habitats.

Tachydromiinae The generally small, compact flies in the subfamily Tachydromiinae are usually fast-running insects reluctant to leave the tree trunks, beach sand, rocks or other substrates on which they run down minute prey. In fact, if you put a bottle right over one of the common tree-trunk *Tachydromia* species, it will dash around on the surface rather than flying up into the bottle, as would almost any other fly. This subfamily is divided into three tribes, of which only the Tachydromiini and Drapetini are routinely encountered (***Symballophthalmini*** includes only one genus, *Symballophthalmus*). The eight genera of ***Tachydromiini*** include the very common genus *Platypalpus*, which can be found hunting on the leaves of almost any wild or domestic plant, as well as the distinctive *Tachypeza* and *Tachydromia* species so reluctant to leave their solid wood or rock hunting surfaces.

Drapetini The 20 or so genera in the tribe Drapetini are relatively stout flies associated with various habitats, including leaf litter and beaches. *Chersodromia* is a marine shoreline genus, often with reduced wings, that hunts flies in wrack (seaweed) and related seashore habitats. The similar *Stilpon* and *Drapetis* are voracious hunters of leaf-litter arthropods; they are often attracted to dung, carrion and other materials that attract the Sphaeroceridae and similar small flies these genera prefer to eat. I have even seen females of *Drapetis* riding on leaves carried by leafcutter ants in Costa Rica, a vantage point from which they can hunt incoming ant-decapitating flies (Phoridae) and other prey.

Hybotinae The subfamily Hybotinae includes several genera of flies with holoptic females — that is, their eyes meet on top of the head — of which the most common are somewhat humpbacked flies with a stout, forward-pointing beak. The large genus *Hybos*, distinctive for its fat, hairy hind legs, is a common group almost everywhere in the world. The genus *Syneches*, one of the prettiest hybotines because of its pigmented wings and the female's big red eyes, is another diverse cosmopolitan group in this subfamily. Some hybotines are well known for their swarming habits although, unlike empidines, they mate on the ground. Skinny-legged humpbacked species in the common genus *Bicellaria* are the most common swarming hybotines in the Holarctic region.

The subfamily **Ocydromiinae** is smaller and less commonly encountered than the similar Hybotinae. Members of the Holarctic genus *Ocydromia* are shining black (or black and yellow) species with the unusual habit of larvipositing — rather than laying eggs — on dung, where the larvae prey on other invertebrates. Females of *Leptopeza*, a more widespread genus of Ocydromiinae, are slender flies with a long, almost stinger-like, telescoping ovipositor.

The subfamily **Oedaleinae**, often treated as a tribe in the subfamily Ocydromiinae, is a small group of four genera with mouthparts directed obliquely forward. This group includes the most abundant of all flower-visiting empidoids in the Holarctic region, the appropriately named genus *Anthalia*. *Anthalia* species are tiny flies that often occur by the hundreds on single plants, where they feed on pollen.

The subfamily **Trichininae** is a small group of two genera — the small Holarctic genus *Trichinomyia* and the more widespread genus *Trichina* — of uncertain placement. Sinclair and Cumming (2006) recognize the Trichininae for these small and rather nondescript flies "mostly for convenience," because they share some primitive similarities. They treat one other genus, the southern African *Stuckenbergomyia*, as a hybotid dance fly unplaced to subfamily, showing that in the Hybotidae, as in many families, classifications remain works in progress as we discover more taxa and clarify their relationships.

Brachystomatidae

Brachystomatid Flies

The family Brachystomatidae is a heterogeneous empidoid family of about 175 species, grouped together mostly on the basis of sexually specific abdominal characters, especially of the female abdomen (Sinclair and Cumming 2006). The Brachystomatidae are divided into three distinct subgroups, the subfamilies Ceratomerinae, Brachystomatinae and Trichopezinae. A molecular phylogenetic study by Moulton and Wiegmann (2007) suggested that all three of these subfamilies should be treated as part of the family Empididae rather than as part of a separate Brachystomatidae, but morphological data do not support a monophyletic Empididae that includes these groups. Most empidoid specialists continue to treat the Brachystomatidae as a separate family as defined by Sinclair and Cumming (2006).

Ceratomerinae The subfamily Ceratomerinae is an austral group of three genera, mostly found in rainforests of Australia, New Zealand and southern South America but with a few isolated records from the northern Andean countries. Within this limited range, these attractive little flies with characteristically elongate antennal bases are usually found on wet surfaces near flowing water, although one New Zealand species group occurs on flowers and some species are associated with alpine wetlands or other habitats. Ceratomerines are assumed to be predaceous, but except for one report of a *Ceratomerus* scavenging on squished flies, this has yet to be confirmed.

The family Brachystomatidae is divided into three subfamilies. This Peruvian *Heterophlebus* belongs to the small subfamily Trichopezinae.

One of the most unusual species in the family Brachystomatidae is the wingless species *Ceratomerus apterus*, the only known flightless member of the family. This little fly is known only from one high (above 4,000 m), cold *páramo* in Ecuador, where I first found it in 1979 and where I collected many more specimens in 1999 and 2002. *Ceratomerus apterus* remains known only from those collections and was not formally named until Bradley Sinclair revised the South American *Ceratomerus* in 2010, describing nine new species, including *C. apterus*. Another remarkable flightless empidoid, the dolichopodid *Papallacta stenoptera*, was described from the same collections. Our knowledge of many important fly lineages often depends on specimens from one or a few collection events — a tantalizing indication of the new dipterological discoveries that will flow from future fieldwork, and a reminder of how important it is to have the specimens from that fieldwork properly housed in major insect collections where they can be made available to the international community of taxonomists.

The subfamily **Trichopezinae** is a widespread lineage of around a dozen genera. It includes some well-known north temperate species as well as a few tropical and south temperate groups. Some Trichopezinae have been observed feeding on insect larvae, and others have been seen using their forelegs to hold small midge prey.

The subfamily **Brachystomatinae** is a small mostly north temperate group with only a few species outside the Holarctic region, in the Oriental, Afrotropical and southern Neotropical regions. Little is known about the biology of this group of small predators.

Atelestidae

Atelestid Flies

The family Atelestidae includes a few small (2–3 mm), nondescript, dull-coloured, uncommon empidoids of note primarily because of their position as a basal lineage within the Empidoidea. Atelestids were originally misplaced in the Platypezidae, and then moved to the Empididae once it was realized that they were more closely related to Empidoidea than to the Platypezidae. This small group of about ten extant species in four genera — plus several extinct genera and species — is now widely accepted as a distinct family, and as the sister group to the rest of the Empidoidea, based on both molecular and morphological evidence.

Sinclair and Cumming (2006) divided the Atelestidae into two subfamilies, the Nemedininae and the Atelestinae. Extant species in the subfamily **Nemedininae** are known only from the Palaearctic region, while the four genera of the subfamily **Atelestinae** currently occur in both north and south temperate regions. The best-known genus, *Atelestus*, is Palaearctic, while *Acarteroptera* is Chilean and *Meghyperus* occurs in both the Palaearctic region and the western United States. The most primitive and most remarkable genus of Atelestinae is *Alavesia*, recently discovered alive and well in Namibia although previously known only from Cretaceous amber fossils (Sinclair and Kirk-Spriggs 2010). Unsurprisingly, little is known about the biology of these little flies, and their larvae remain unknown. Adult atelestids lack the epipharyngeal blades that most empidoids use to puncture prey; they are probably obligate pollen feeders.

Hilarimorphidae and Apystomyiidae

Hilarimorphid Flies and Apystomyiid Flies

The Hilarimorphidae (hilarimorphid flies) and Apystomyiidae (apystomyiid flies) are small, enigmatic lineages that have only recently attracted attention from entomologists. Prior to a study of the genus by Webb (1974), *Hilarimorpha*, the only extant genus in the family **Hilarimorphidae**, included seven described species, based on fewer than two dozen specimens. Hilarimorphids are not conspicuous flies: Webb uses the appropriate adjective "nondescript" to refer to these small, empid-like insects. Although rarely noticed, they can sometimes be netted in numbers from willow foliage along narrow streams in spring. Webb obtained enough new specimens to describe 22 new species and to redefine the family in his 1974 paper, and many more specimens have since accumulated in major insect collections.

Despite that, *Hilarimorpha* remains a phylogenetic mystery. Even in the latest and largest molecular analyses it is treated variously as the sister group to the Asilomorpha, the sister group to the Acroceridae, the sister group to the Eremoneura, the sister group to the Asilomorpha plus the Eremoneura, a close relative of the Bombyliidae, or simply unplaced. Most species of the family Hilarimorphidae are North American, but some occur in the Palaearctic and Southeast Asia. *Hilarimorpha* remains the only extant genus in the family, although Grimaldi

Atelestidae, like this Chilean *Acarteroptera*, probably feed only in pollen.

and Cumming (1999) described an extinct genus from Cretaceous amber, and some authors have included the problematic species *Apystomyia elinguis* in the Hilarimorphidae.

The odd little southwestern American *Apystomyia elinguis* is an unremarkable-looking fly with the dimensions of a microbombyliid (Mythicomyiidae) but bearing a close similarity to *Hilarimorpha*. The name *Apystomyia* is from the Greek words *a* ("without"), *pystos* ("learned") and *myia* ("fly") — loosely translated as "a fly of which nothing is known." That was an accurate comment when the great American dipterist Melander coined the name in 1950, and *Apystomyia* has continued to pose a puzzle. Essentially nothing is known about the biology of either *Apystomyia* or *Hilarimorpha*, and larvae remain unknown for both genera (and thus both families).

Nagatomi and Lui (1994) moved *Apystomyia* from the Hilarimorphidae into its own family, **Apystomyiidae**, indicating that it is neither a hilarimorphid nor closely related to the Bombyliidae. The latest word on the position of *Apystomyia* comes from a recent molecular phylogenetic study (Trautwein 2009; Trautwein et al. 2010) that provides evidence that it is the sister group to the rest of the Cyclorrhapha. That would make it a sort of higher dipteran missing link — heady stuff for a tiny, obscure fly known from only a handful of specimens collected in southern California! Morphological evidence, by the way, places *Apystomyia* as the sister group to the entire Eremoneura.

A Photographic Guide to the Lower Brachycera and Empidoidea

FAMILY TABANIDAE (HORSE FLIES, DEER FLIES, CLEGS) Subfamily Chrysopsinae ❶ Male Tabanidae, like this Nearctic deer fly (***Chrysops***), are non-biting flies with large eyes that meet along the top. Most are rarely seen, although females of the genus Chrysops are often abundant in open woodland, especially in north temperate regions. ❷ ❸ Female *Chrysops*, like this pale ***Chrysops celatus*** and this darker ***C. niger***, often bite humans, despite their common name "deer flies." ❹ This pollen-spattered North American deer fly is ***Chrysops cincticornis***. Like many flies, tabanids utilize natural plant sugars such as honeydew or nectar to fuel their flight. ❺ This deer fly (***Chrysops excitans***) is depositing an egg mass on horsetail (*Equisetum*) in a northern Canadian marsh. The eggs will soon turn black, and the hatching larvae will later drop into the water. ❻ The huge genus *Chrysops* occurs throughout the world; this is an African member of the ***Chrysops laniger*** group. **Subfamily Pangoniinae** ❼ ❽ Pangoniinae are particularly diverse in southern South America. Shown here are a relatively short-beaked ***Caenopangonia brevirostris*** (on a leaf) and ***Chaetopalpus annulicornis***, from Chile. ❾ ***Esenbeckia*** is a genus of about 20 mostly Neotropical Pangoniinae. This is one of the three species in the genus that ranges north to Arizona.

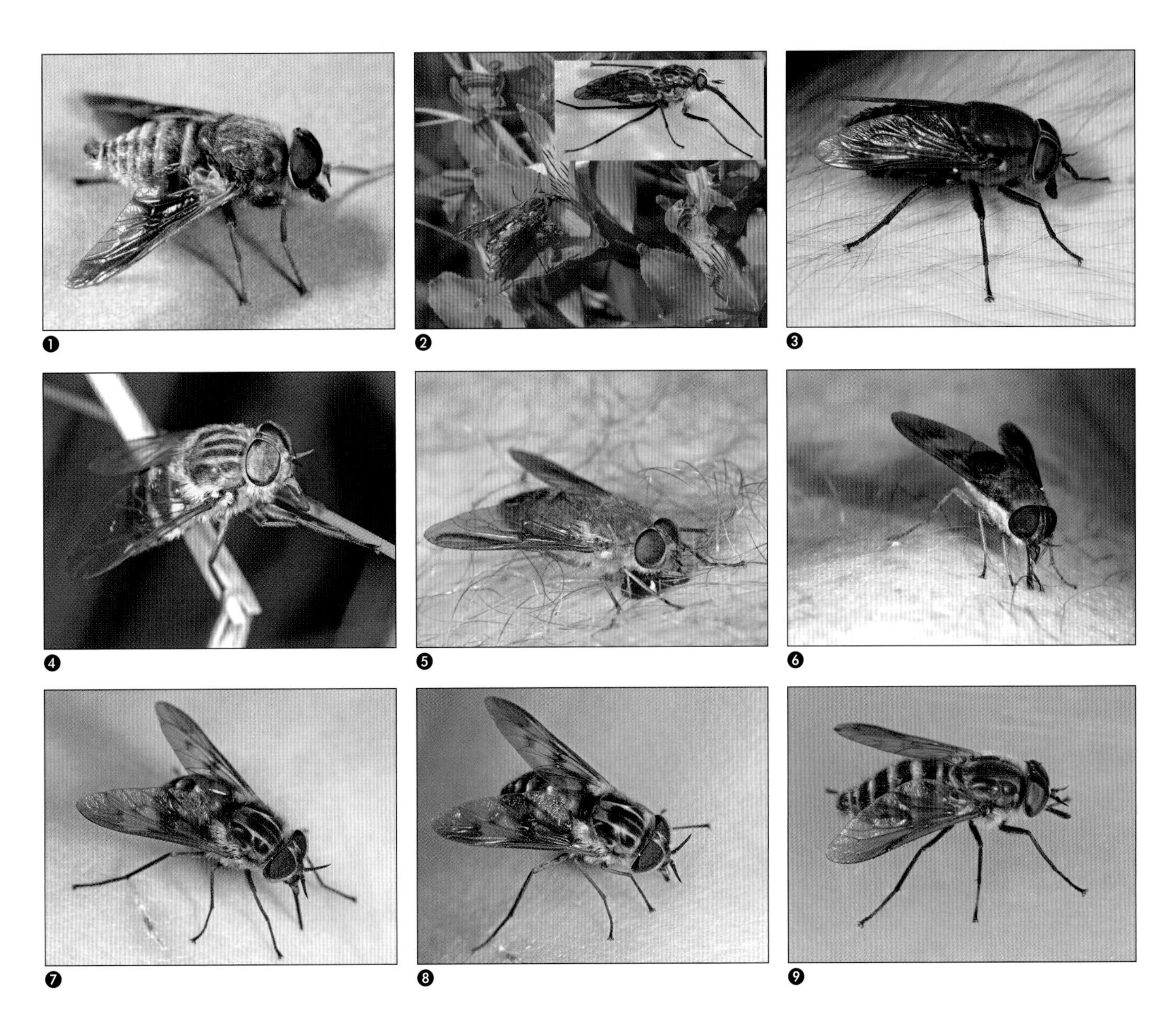

FAMILY TABANIDAE (continued) ❶ This is a male of the rarely encountered North American genus ***Goniops***, which includes only the single species ***G. chrysocoma***. Females of this species guard their eggs, staying with the egg mass until death. ❷ The long-beaked tabanid ***Mycteromyia conica*** is a common visitor to *Alstroemeria* flowers in central Chile. ❸ ❹ The spectacular red and black ***Scaptia (Scaptia) lata*** is a biting fly that can be common in Chile and parts of Argentina. As one might expect for this robust fly of almost 2 cm in length, its bite packs quite a punch. The *Scaptia* subgenus *Scaptia* also occurs in Australia, and this green-eyed fly perched on a twig is ***S. (S.) patula*** from New South Wales. ❺ ❻ The large genus ***Scaptia*** includes some of the most common biting tabanids in both southern South America and Australia. Shown here are two Australian species, ***S. (Pseudoscione) viridiventris*** (feeding through a pool of blood) and ***S. (Pseudoscione) quadrimaculata*** (just starting to penetrate my leg). ❼ ❽ ***Scione*** is a common Neotropical genus of 42 species, some of which are common biting flies able to penetrate deeply into their hosts by using impressively long mouthparts, as demonstrated in these two images taken a few seconds apart. ❾ *Stonemyia* is a Holarctic genus sometimes common on flowers but still unknown as larvae. This is a ***Stonemyia tranquila*** from northern Canada.

FAMILY TABANIDAE (continued) Subfamily Scepsidinae ❶ ❷ These ***Adersia oestroides*** behaved more like Therevidae than Tabanidae, making short, swift flights on a sand beach (in Dar es Salaam, Tanzania). *Adersia* is a small genus of half a dozen species found in coastal eastern and southern Africa. **Subfamily Tabaninae ❸** The long antennae and elongate body of this Amazonian ***Acanthocera*** give it an unusually wasp-like habitus for a tabanid. ❹ The Neotropical genus ***Agelanius*** is considered to be part of the most primitive group within the tribe Diachlorini. ❺ Although the genus ***Atylotus*** is widespread in the Holarctic region and ranges into the Afrotropical and Oriental regions, it is relatively uncommon throughout its range and most species are known from relatively few, mostly northern, localities. This is a male of one of the 14 Nearctic species, which tend to stay near the wetland soils in which the larvae develop. ❻ *Catachlorops* is a large Neotropical genus divided into two subgenera, the South American subgenus *Catachlorops* and the more widespread subgenus *Psalidia*. This is ***Catachlorops (Psalidia) umbratus*** from Costa Rica. ❼ ***Dasybasis*** is a diverse genus in both Australia and southern South America; this is the Chilean ***D. pruinivitta***. ❽ ***Diachlorus*** is a mostly Neotropical genus of 27 species including one (the Yellow Fly, *D. ferrugatus*) that occurs as a pest in the southeastern United States. This is a South American species, ***D. bicinctus***. ❾ The large genus ***Dichelacera*** (70 species) occurs throughout the neotropics, usually only in the forest canopy; larvae remain unknown. This is ***D. regina*** from Costa Rica.

FAMILY TABANIDAE (continued) ❶ The strikingly speckle-winged and bright-eyed species of ***Haematopota***, known as clegs or stouts, are abundant seasonal pests in some areas. This species, ***H. americana***, occurs across North America but is uncommon in the eastern half of the continent. ❷ ❸ ***Haematopota*** is a large and widespread genus occurring in the Holarctic, Oriental and Afrotropical regions. These are male ***H. nobilis*** from Tanzania and an unidentified female from Vietnam. ❹ ❺ The large genus ***Hybomitra*** is the most diverse group of horse flies in the northern parts of the Holarctic region. These two biting females (***H. criddlei*** and ***H. illota***) were photographed in eastern Canada, where *Hybomitra* is the most common genus of horse fly. ❻ This Ecuadorian ***Stibasoma*** was photographed high in the primary rainforest canopy, where it possibly developed as a predator in the aquatic or semiaquatic habitat inside tank-type bromeliads on tree branches. Most of the 18 species in this Neotropical genus, including this one (probably ***S. festivum***), are apparent orchid bee mimics in the canopy-associated subgenus *Stibasoma*. ❼ The massive eyes of this ***Tabanus superjumentarius*** from eastern North America mark it as a harmless male. Male *Tabanus* are infrequently seen, although they sometimes visit flowers or honeydew-spattered leaves. ❽ ❾ ***Tabanus atratus***, the **Black Horse Fly**, is a common, widespread Nearctic horse fly that often attracts attention because of its large size and striking appearance. Shown here are a male and a brightly colored larva.

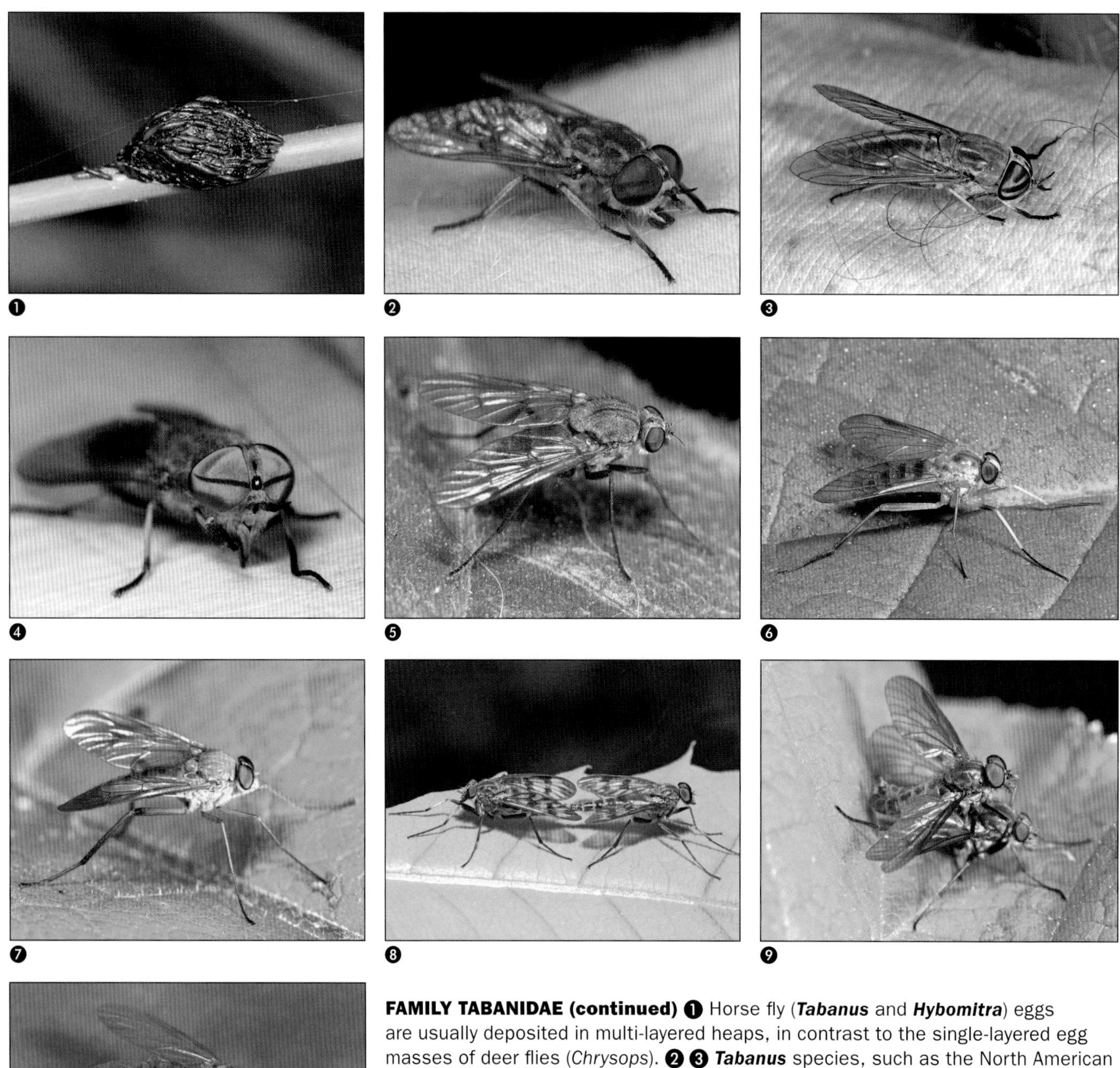

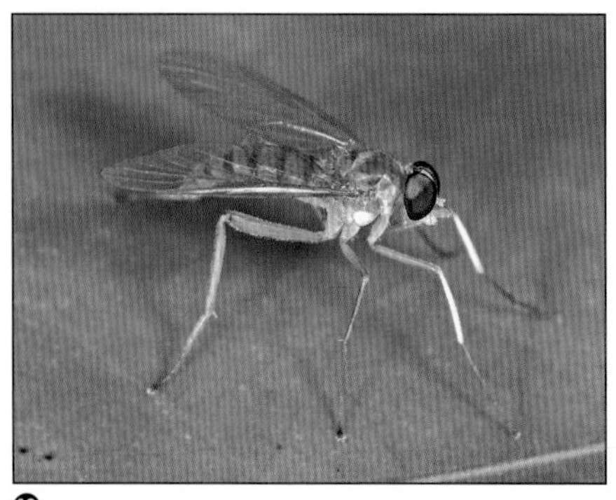

FAMILY TABANIDAE (continued) ❶ Horse fly (***Tabanus*** and ***Hybomitra***) eggs are usually deposited in multi-layered heaps, in contrast to the single-layered egg masses of deer flies (*Chrysops*). ❷ ❸ ***Tabanus*** species, such as the North American ***T. marginalis*** (brown eyes) and ***T. similis*** shown here, are the most common biting horse flies throughout most of the world. ❹ The striking eye color of this ***Tabanus quinquevittatus*** (and other Tabanidae) is created by stacked layers in the facet lenses that act as interference reflectance filters. It fades to black in dead specimens when the eyes dry out and the layers collapse. **FAMILY RHAGIONIDAE (SNIPE FLIES) Subfamily Rhagioninae** ❺ One of the largest genera of Rhagioninae, the archaic-looking genus ***Atherimorpha***, with 50 species, shows a classic Gondwanan distribution, with species in Australia, Patagonia and South Africa. ❻ ❼ The mostly northern hemisphere genus *Rhagio* includes around 165 described species. Shown here are a female ***Rhagio hirtus*** and a male ***R. tringarius***, both common in eastern North America, although the latter species (introduced from Europe) seems to be replacing the native *R. hirtus*. ❽ Males of the common North American ***Rhagio mystaceus*** are often seen perched on tree trunks; females are more commonly seen on leaf litter or foliage. ❾ This is a pair of mating ***Rhagio*** from western North America. Note the male's large, contiguous eyes. ❿ ***Rhagio*** is a large, widely distributed and relatively uniform genus. This one is from northern Vietnam.

FAMILY RHAGIONIDAE (continued) Subfamily Spaniinae ❶ This ***Symphoromyia*** male has just emerged from a mossy stream bank in eastern North America. Larvae of this Holarctic genus usually occur in mossy substrates; female adults feed on blood. ❷ Although ***Symphoromyia*** occurs across North America, these biting flies only reach pest status in the west, where they are known as "Rocky Mountain black flies." **Subfamily Chrysopilinae** ❸ ❹ ***Chrysopilus*** is a common worldwide genus of about 300 named species. Shown here are a male from Tanzania and a female from Chile. ❺ This is a pair of mating ***Chrysopilus proximus*** from Canada (the male is the one with large eyes). ❻ ❼ ***Chrysopilus*** includes several very common species, some of which occur in large numbers on low foliage in early spring. Shown here are two of the more distinctive North American species, a female of ***C. ornatus*** and a brighter male of ***C. thoracicus***. **Subfamily Arthrocerinae** ❽ ***Arthroceras*** is a small and rarely encountered Holarctic genus. Nothing is known of its biology. This specimen of the northeastern North American ***A. leptis*** is in the Canadian National Collection of Insects. **Subfamily Bolbomyiinae** ❾ ***Bolbomyia*** is an enigmatic genus currently treated as a subfamily of Rhagionidae, although recent work suggests that the genus should be treated in its own family, Bolbomyiidae. Look for members of this small Holarctic genus in deciduous woods in early spring.

FAMILY ATHERICIDAE (WATER SNIPE FLIES) Subfamily Athericinae ❶ ❷ ❸ ❹ Females of the mostly Holarctic genus ***Atherix*** lay their eggs in communal masses over running water, staying with the egg masses until they die. Hatching larvae drop into the water. This is the common eastern North American ***A. lantha*** (a male on a plant, a live female on a streamside rock, a mass of eggs and dead females on a bridge, and a predaceous larva). ❺ ❻ ❼ ***Suragina*** is the largest and most widespread genus of Athericidae, with about 40 species mostly found in warmer parts of the Old World. Shown here are two from east Africa and one from Vietnam. ❽ ❾ The small genus ***Atrichops*** is currently found only in the Old World, but fossil species are known from the United States. Most species are African, although a few occur in the Palaearctic region and two species are known from Thailand. Shown here is an undescribed species from Vietnam.

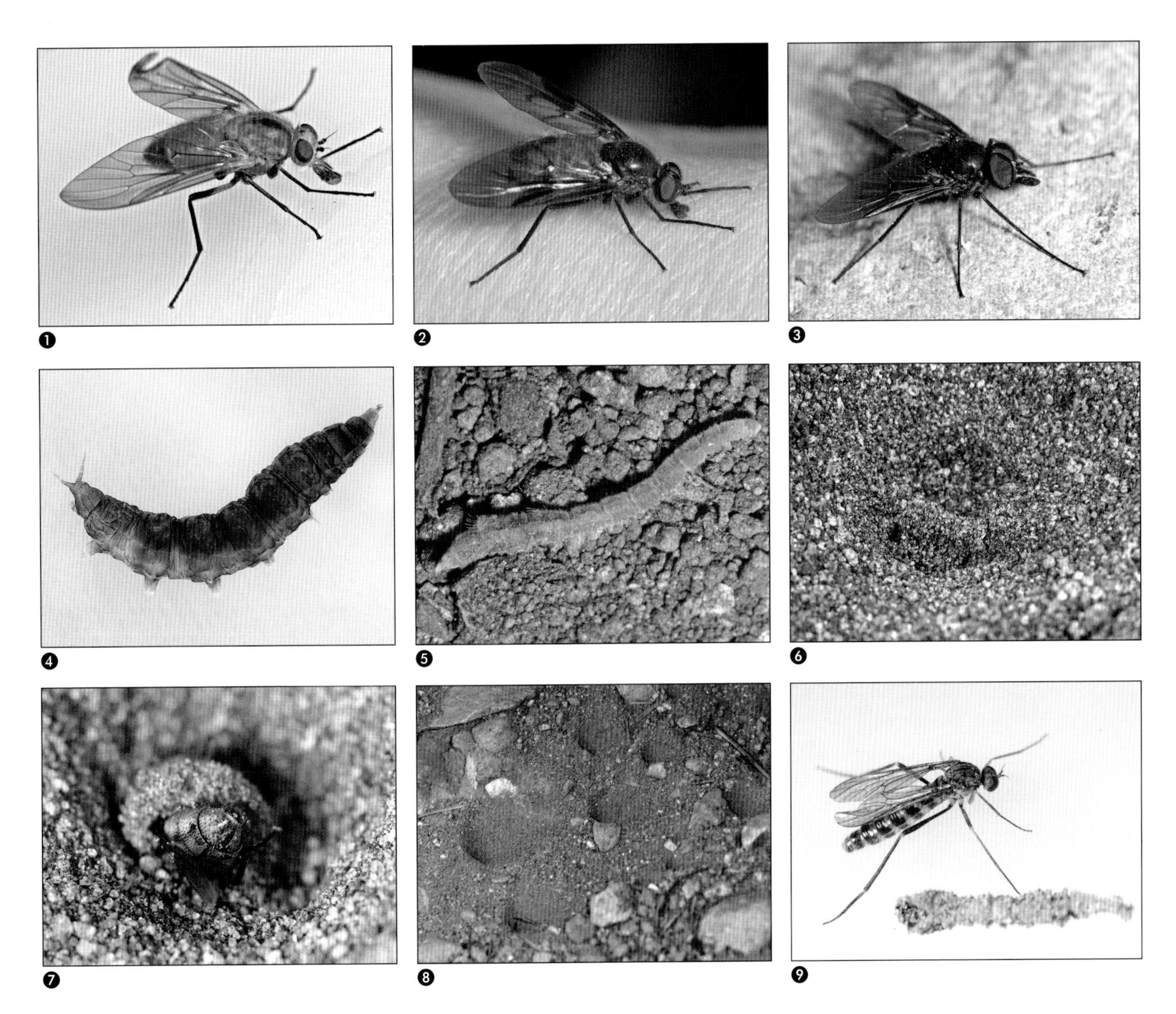

FAMILY ATHERICIDAE (continued) Subfamily Dasyommatinae ❶ ❷ ❸ ❹ ***Dasyomma***, the only member of the subfamily Dasyommatinae, occurs in southern South America and Australia. These two females, this male and this aquatic larva are all from Chile. **FAMILY VERMILEONIDAE (WORMLIONS)** ❺ This wormlion (an undescribed species of ***Vermileo***) larva was pulled out of its pit under the overhang of a cliff in Costa Rica for a photograph. ❻ Look closely at this conical sand pit to see the telltale convex silhouette of a wormlion (***Vermileo*** larvae) waiting in ambush. It will disappear into the dust at the least disturbance (including your shadow), but it will strike like a snake if a small insect stumbles into the pit. ❼ This muscid fly was dropped into a ***Vermileo*** pit, where it was swiftly encircled and struck by the waiting wormlion. ❽ ❾ These wormlion (***Vermileo opacus***) pits were among hundreds under the shelter of a bridge abutment in Oregon, and the adult was reared from a larva in one of these pits.

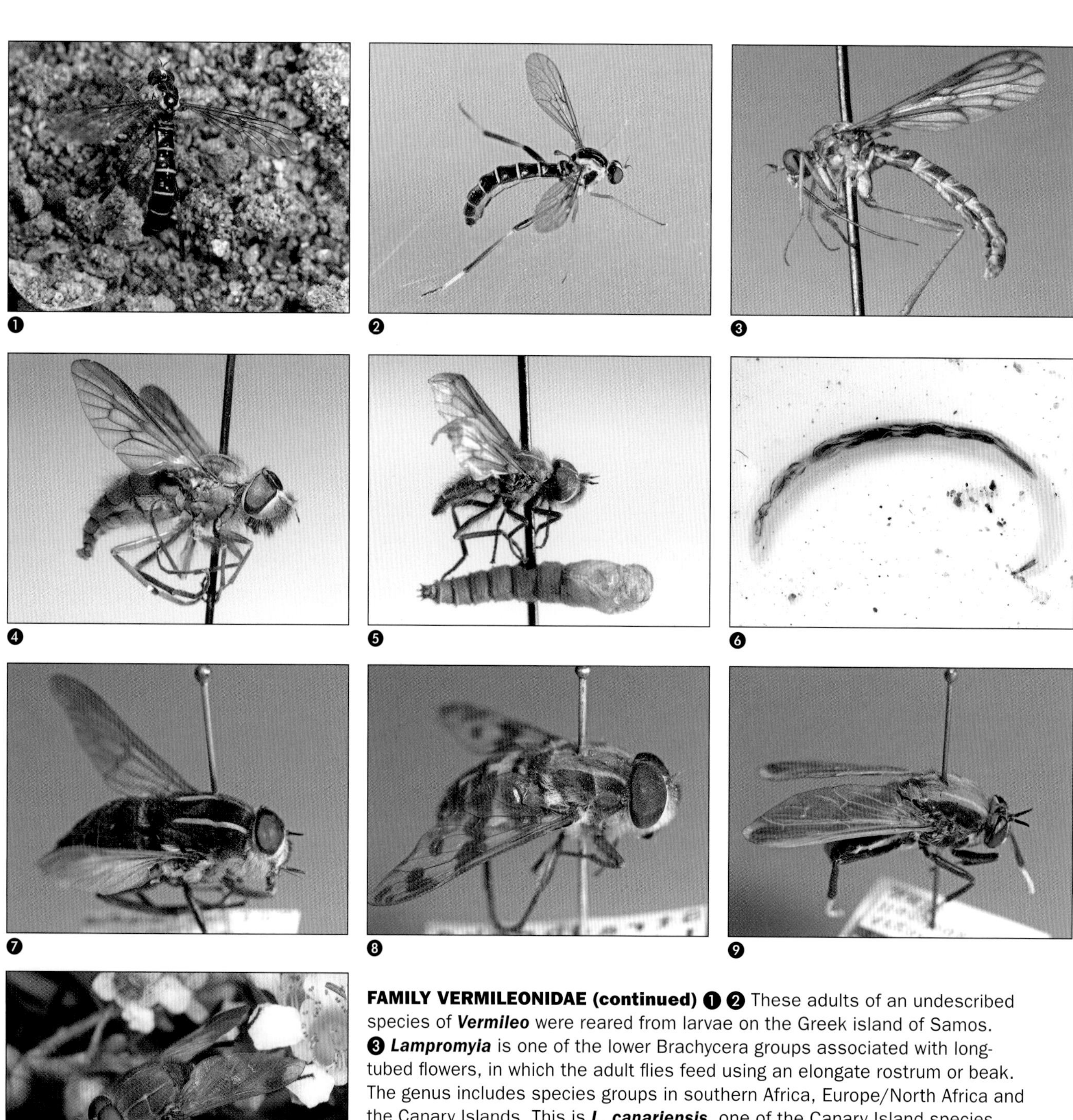

FAMILY VERMILEONIDAE (continued) ❶ ❷ These adults of an undescribed species of ***Vermileo*** were reared from larvae on the Greek island of Samos. ❸ ***Lampromyia*** is one of the lower Brachycera groups associated with long-tubed flowers, in which the adult flies feed using an elongate rostrum or beak. The genus includes species groups in southern Africa, Europe/North Africa and the Canary Islands. This is ***L. canariensis***, one of the Canary Island species. **FAMILY PELECORHYNCHIDAE (PELECORHYNCHID FLIES)** ❹ ***Glutops***, the only pelecorhynchid genus to range north to Canada, is a rarely collected genus. This eastern North American ***G. singularis*** was photographed in the Canadian National Collection of Insects in Ottawa. ❺ This western North American ***Glutops rossi*** is pinned over the shell (exuvia) of the pupa from which it emerged. ❻ Larvae of the Holarctic genus ***Glutops*** (this is probably *G. rossi*) can be common in small, clear mountain streams. ❼ ❽ ❾ The spectacular flies in the genus ***Pelecorhynchus*** are divided between southern South America and Australia, where they sometimes abound on *Leptospermum* flowers. These specimens in the U.S. National Museum in Washington are ***P. biguttatus*** (two stripes, Chile), ***P. personatus*** (spotted wings, Queensland) and ***P. fulvus*** (orange wings, New South Wales). ❿ This ***Pelecorhynchus*** (probably *P. fusconiger*) was photographed in Australia by Paul Zaborowski.

FAMILY AUSTROLEPTIDAE (AUSTROLEPTID FLIES) ❶ ❷ These pinned specimens (a male and female of ***Austroleptis rhyphoides*** in the Australian Museum) belong to an enigmatic genus sometimes treated as a rhagionid and sometimes as a separate family. **FAMILY OREOLEPTIDAE (MOUNTAIN LEPTID FLIES)** ❸ This is one of the type specimens used to describe the family Oreoleptidae, and the species ***Oreoleptis torrenticola***, in 2005. Like all known specimens of *Oreoleptis*, this fly was reared from aquatic larvae; nobody has reported seeing adult Oreoleptidae in the wild. **FAMILY XYLOPHAGIDAE (WOOD FLIES) Subfamily Xylophaginae** ❹ ***Exeretonevra***, with four species, is known only from Australia and is the only Australian xylophagid genus. ❺ ***Heterostomus***, like the Australian *Exeretonevra*, is an unusual fly of uncertain relationships. This is the only known species of the genus, ***H. curvipalpus*** from Chile. ❻ ❼ ***Rachicherus*** is unusual among the Brachycera for its multi-segmented antennae. Shown here are an unidentified (probably undescribed) black species from Costa Rica and a ***R. obscuripennis*** from the eastern United States. Most members of this large genus are Oriental. ❽ This ***Xylophagus (Xylophagus) lugens***, standing beside its pupal exuvium, developed as a predator under the bark of a dead maple tree in eastern North America. ❾ ***Xylophagus (Archimyia) reflectens*** is commonly seen on fallen trees in eastern North America.

FAMILY XYLOPHAGIDAE (continued) ❶ ❷ Xylophagidae, like this western North American ***Xylophagus (Xylophagus) gracilis***, develop as predators under the bark of dead trees. Larvae have an elongate conical, strongly sclerotized head capsule and posterior spiracles surrounded by a plate with hook-like processes. **Subfamily Coenomyiinae** ❸ ***Coenomyia*** is a Holarctic-Oriental genus of uncommon, robust Xylophagidae. Most species are Old World, but this is the Holarctic ***C. ferruginea***. ❹ ***Dialysis*** is a Holarctic-Oriental genus and, with around 20 species, the largest genus in the subfamily. This is one of about half a dozen western North American *Dialysis* species. **FAMILY STRATIOMYIDAE (SOLDIER FLIES) Subfamily Clitellariinae** ❺ ***Campeprosopa*** is a small Oriental genus of three named species; this strikingly long-spined fly is ***C. longispina*** from Vietnam. ❻ All eight species in the genus ***Chordonota***, like this Peruvian species, are Neotropical. ❼ ❽ ***Cyphomyia*** is a common and charismatic genus found routinely on cut or fallen trees and other kinds of decaying plant material in the neotropics. Females of many, like this one from Ecuador, have metallic blue bodies and contrasting orange or yellow heads, possibly to warn predators of their unpalatability. Males have the head taken up largely by eyes, but some (like this Costa Rican species) attain a similar appearance with a bicolored thorax. ❾ Most (about 75 species) ***Cyphomyia*** species are Neotropical, but the genus occurs worldwide. This ***C. erecta*** is in an Arizona streambed, but the same species has been reared from rotting cactus debris in a Mexican packrat nest.

FAMILY STRATIOMYIDAE (continued) ❶ ***Dieuryneura***, with only one species (***D. stigma***), is known from the southwestern United States and northern Mexico. Larvae have been found in decomposing desert spoon (*Dasylirion wheeleri*).
❷ ***Diaphorostylus*** is a small Neotropical group of four species; this one is from Ecuador. ❸ ***Euryneura elegans***, seen here in Costa Rica, is the most widespread and frequently encountered member of the New World genus *Euryneura*, which ranges from the southwestern United States to Argentina. Larvae of the genus have been reared from rotting roots. **Subfamily Antissinae** ❹ Almost all members of the small subfamily Antissinae, like this Australian ***Anacanthella angustifrons*** photographed by Shaun Winterton, occur in southern South America or Australia. ❺ ***Cyanauges*** is a small genus from southern South America. **Subfamily Nemotelinae** ❻ ❼ ***Nemotelus*** is a huge genus of tiny flies that breed in aquatic or semiaquatic habitats and frequently appear on flowers as adults. Most species are Old World (Palaearctic and Afrotropical) but this is a North American species. **Subfamily Sarginae** ❽ The small genus ***Himantigera*** is entirely Neotropical; this one is from Costa Rica. ❾ ***Merosargus*** is a diverse genus common throughout the Neotropical region. This wasp-like species is from Costa Rica.

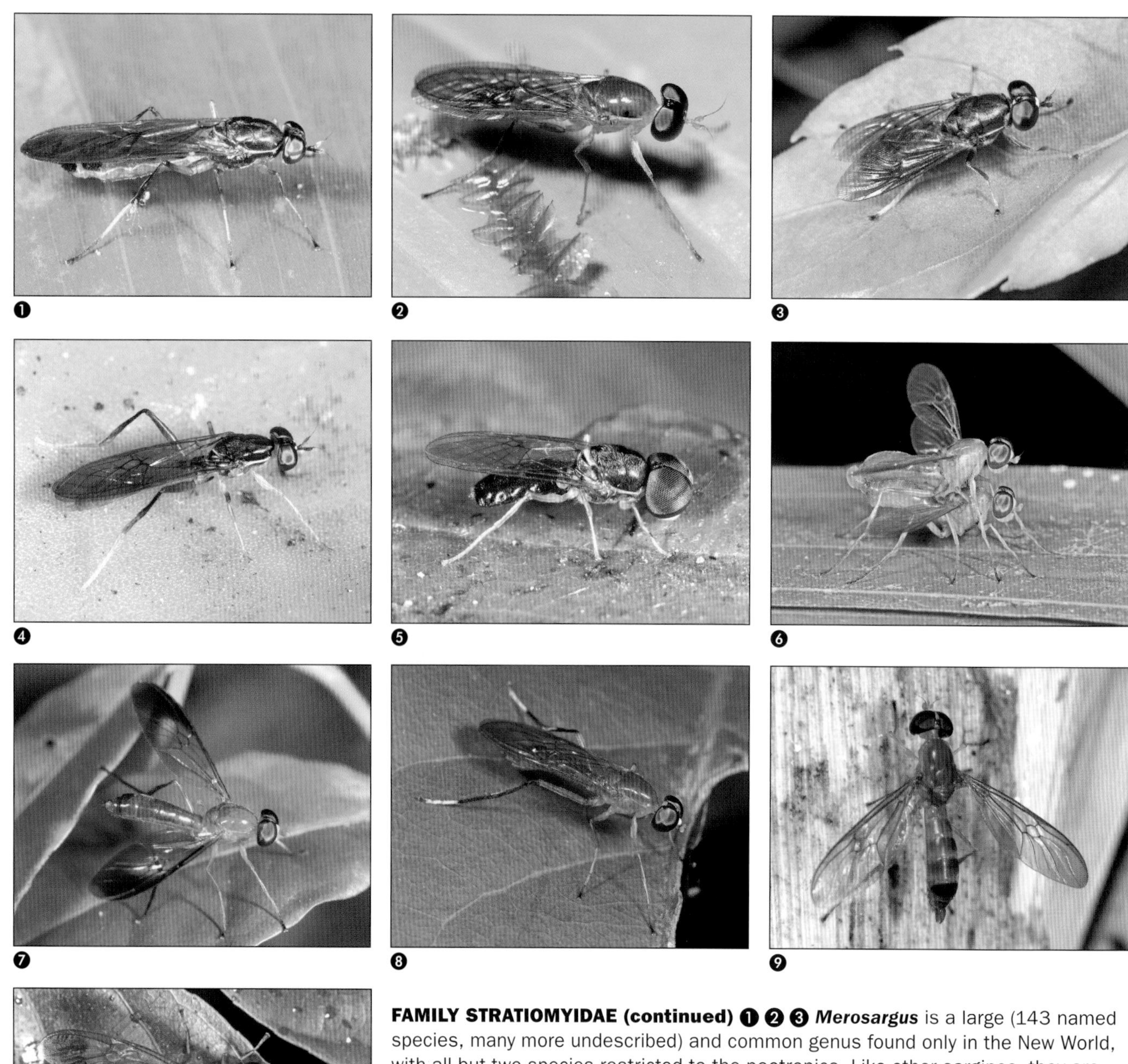

❶ ❷ ❸ ❹ ❺ ❻ ❼ ❽ ❾ ❿

FAMILY STRATIOMYIDAE (continued) ❶ ❷ ❸ ***Merosargus*** is a large (143 named species, many more undescribed) and common genus found only in the New World, with all but two species restricted to the neotropics. Like other sargines, they are associated with decaying vegetation, and some develop in *Heliconia* bracts. Shown here are a dark Ecuadorian species (note the mite on the hind leg) and two species from Costa Rica. ❹ Adults of this ***Merosargus*** species appeared in large numbers as soon as a large, bromeliad-covered tree crashed to the ground in the Amazonian rainforest. Larvae probably develop in the fallen bromeliads. ❺ ***Microchrysa*** is a large, widespread genus often common around compost, dung and other abundant sources of decaying plant material. Shown here is a male from Tanzania. ❻ ❼ ❽ ***Ptecticus*** is a large, widespread genus that abounds in both the New World and Old World tropics. Larvae are most often found in decaying fruit. ❾ ***Ptecticus trivittatus*** is a common Nearctic species frequently found in backyard compost and similar habitats in central and eastern North America. ❿ ***Ptecticus*** is an enormous and common genus, often conspicuous in behavior and appearance. This large Oriental species (seen here in Vietnam) is ***P. aurifer***, an apparent mimic of similarly colored spider wasps.

FAMILY STRATIOMYIDAE (continued) ❶ ***Sargus*** is a cosmopolitan genus with more than 100 species, many of which are common on compost and other decaying plant material. Shown here is the widespread Nearctic species ***S. decorus***. **Subfamily Chiromyzinae** ❷ ❸ Members of the distinctive yet "primitive"-looking subfamily Chiromyzinae, like these Chilean ***Chiromyza***, have atrophied mouthparts and don't feed as adults. ❹ The sluggish adults of ***Inopus*** are common flies in Australia, where the larvae of some species are pests of sugarcane. One species (the Sugarcane Soldier Fly, *I. rubriceps*) was accidentally introduced to California and New Zealand, where it has become a pest of pasture grasses. ❺ ***Boreoides tasmaniensis*** is a wingless chiromyzine from Tasmania. **Subfamily Beridinae** ❻ ***Actina*** is a mostly Holarctic genus with a few species that reach the Oriental region. Not much is known about the biology of these slender, metallic green flies, but the males of this Nearctic species (***A. viridis***) form conspicuous swarms. ❼ The male of this pair of ***Allognosta*** has large, contiguous eyes. *Allognosta* is a mostly Oriental-Holarctic genus that ranges to Africa and New Guinea, but this species (***A. fuscitarsus***) is Nearctic. ❽ ***Chorisops*** is a small Old World genus; this one is from Greece but the genus is more diverse in the Oriental region. Known larvae are terrestrial or semiaquatic. **Subfamily Parhadrestiinae** ❾ The subfamily Parhadrestiinae is a small subfamily named in 1986 to hold two genera of obscure, tiny and rare soldier flies. This is the type specimen of one of the two species in the type genus, ***Parhadrestia***. Too small to pin, the fly is glued to the end of a paper point that shares a pin with a red holotype label and a plastic capsule holding the dissected genitalia in a bit of glycerin.

FAMILY STRATIOMYIDAE (continued) ❶ ***Hadrestia*** is a small group of three southern South American species; this one is from Chile. ❷ Like similar beridines elsewhere, males of the Neotropical genus ***Oplachantha*** sometimes swarm in light gaps. There are 20 species in the genus; this one is from Ecuador. **Subfamily Chrysochlorininae** ❸ ***Chrysochlorina*** is a New World genus ranging from Florida south to Brazil. This Bolivian species is a convincing mimic of a sympatric vespid wasp. ❹ This vespid wasp (*Angiopolybia*) occurred with the fly in the previous frame. **Subfamily Hermetiinae** ❺ ❻ Several wasp-mimicking species of the large (more than 75 species) genus ***Hermetia***, such as ***H. comstocki*** and ***H. concinna***, breed in decaying plants, such as agave and sotol, in the southwestern United States. Most other *Hermetia* species are Neotropical, but species occur throughout much of the world. ❼ The **Black Soldier Fly, *Hermetia illucens***, is a common species in warmer parts of the world, where it breeds in a wide variety of wet decomposing materials, including feces. ❽ This Neotropical ***Hermetia flavipes*** has a window-like transparency near the base of its abdomen, much like the more widespread Black Soldier Fly (*H. illucens*). The transparency creates the illusion of a wasp waist in flight, presumably protecting the fly from sting-shy predators. **Subfamily Pachygastrinae** ❾ ***Aulana*** is a small Oriental-Australian genus of half a dozen species; this is one from the Philippines.

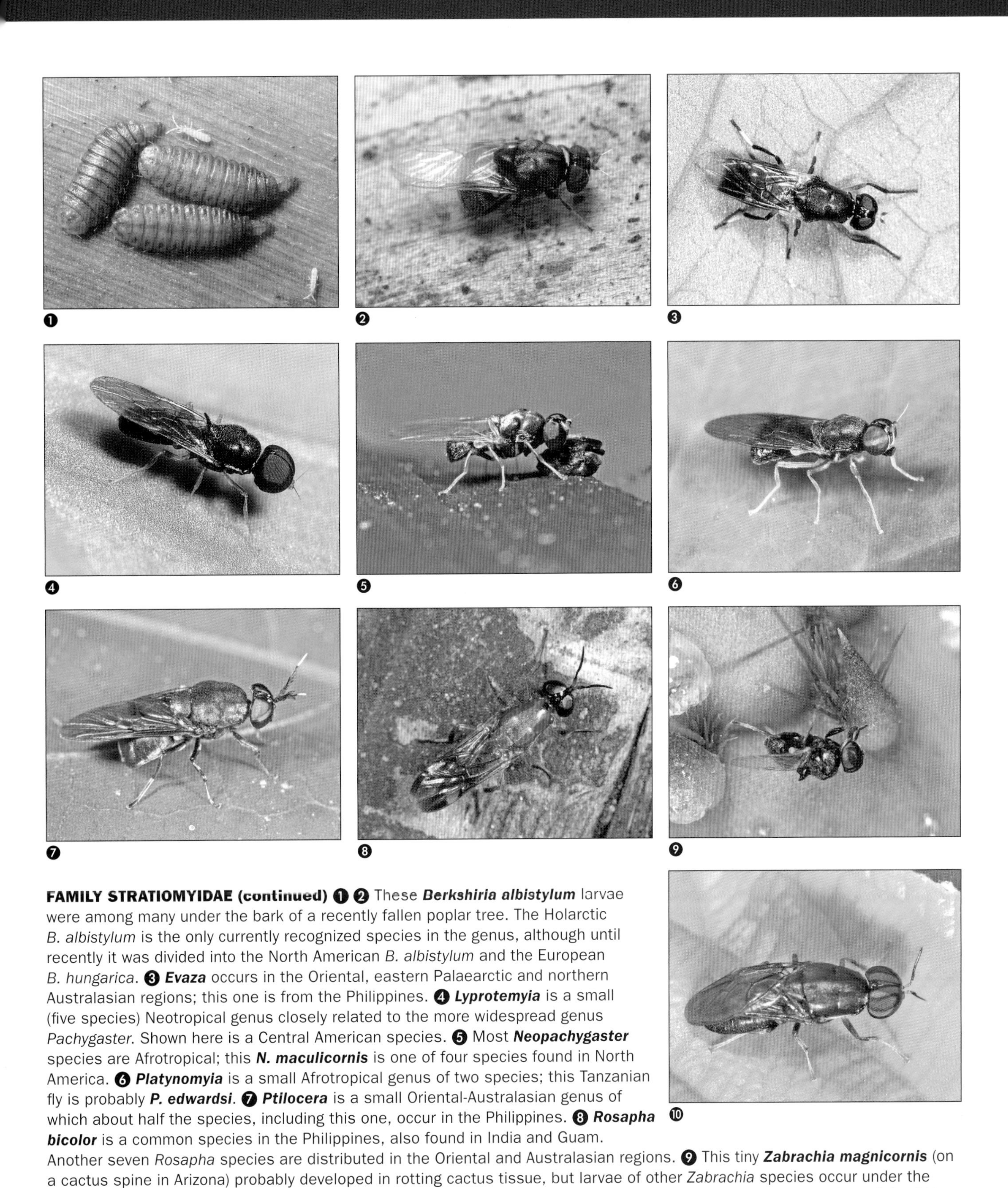

FAMILY STRATIOMYIDAE (continued) ❶ ❷ These ***Berkshiria albistylum*** larvae were among many under the bark of a recently fallen poplar tree. The Holarctic *B. albistylum* is the only currently recognized species in the genus, although until recently it was divided into the North American *B. albistylum* and the European *B. hungarica*. ❸ ***Evaza*** occurs in the Oriental, eastern Palaearctic and northern Australasian regions; this one is from the Philippines. ❹ ***Lyprotemyia*** is a small (five species) Neotropical genus closely related to the more widespread genus *Pachygaster*. Shown here is a Central American species. ❺ Most ***Neopachygaster*** species are Afrotropical; this ***N. maculicornis*** is one of four species found in North America. ❻ ***Platynomyia*** is a small Afrotropical genus of two species; this Tanzanian fly is probably ***P. edwardsi***. ❼ ***Ptilocera*** is a small Oriental-Australasian genus of which about half the species, including this one, occur in the Philippines. ❽ ***Rosapha bicolor*** is a common species in the Philippines, also found in India and Guam. Another seven *Rosapha* species are distributed in the Oriental and Australasian regions. ❾ This tiny ***Zabrachia magnicornis*** (on a cactus spine in Arizona) probably developed in rotting cactus tissue, but larvae of other *Zabrachia* species occur under the bark of coniferous trees in Europe and North America. ❿ ***Tinda*** is a small genus of soldier flies found in Australia and the Old World tropics; this one is from Vietnam.

FAMILY STRATIOMYIDAE (continued) Subfamily Raphiocerinae ❶ ***Heptozus*** includes only two described species, both Neotropical. This one was standing on a bromeliad leaf in Costa Rica, and perhaps developed in the decomposing organic material at the base of the leaf. **Subfamily Stratiomyinae** ❷ The attractive little soldier flies in the genus ***Caloparyphus*** are often associated with shallow water such as seeps and mossy springs, where their flattened, tough-skinned larvae are sometimes seen sprawling on the submerged leaves of aquatic plants. This Nearctic genus and the related New World genus *Euparyphus* belong to the tribe Oxycerini, a group traditionally treated as part of the Clitellariinae. Shown here is ***C. decemmaculatus***. ❸ ❹ ***Euparyphus*** larvae are found in thin films of water or madicolous habitats throughout North and South America. This larva is in a clay-bank seep in Canada, and the adult is on the bank of a small stream in New Mexico. ❺ Species in the New World genus ***Hedriodiscus*** are common flower visitors as adults and aquatic as larvae. This is ***H. varipes***, a widespread North American species. ❻ ***Hoplitimyia*** is a small New World genus of ten species that mimic sympatric stinging Hymenoptera (vespid wasps). ❼ ***Myxosargus*** is a New World genus of a dozen species, possibly more closely related to species currently in the Raphiocerinae than to other Stratiomyinae. ❽ ❾ Stratiomyine larvae are typically aquatic, breathing through posterior spiracles surrounded by a fringe of water-repellent hairs, clearly visible on these ***Odontomyia***.

FAMILY STRATIOMYIDAE (continued) ❶ ❷ ***Odontomyia*** is a very large (more than 200 species) worldwide genus. The colorful adults are common on flowers, especially in temperate regions. Shown here are the North American ***O. cincta*** (on goldenrod) and an unidentified New Zealand species. ❸ ***Promeranisa*** is a poorly known group of four Neotropical species. ❹ ***Psellidotus*** is a *Hedriodiscus*-like genus ranging from Canada to southern South America; this is ***P. elegans*** from Chile. ❺ ***Rhaphiocerina*** is a Japanese genus with only one species, ***R. hakiensis***, here seen in a garden not far from Tokyo International Airport. ❻ ❼ ***Stratiomys*** is the most familiar large soldier fly genus in north temperate countries, where these bee-sized flies are regular flower visitors. Their larvae are aquatic, with elongate posterior segments. This is the North American ***S. meigenii***. **FAMILY XYLOMYIDAE (WOOD SOLDIER FLIES)** ❽ ❾ ❿ Members of the large worldwide genus ***Solva*** are uniform in appearance, as shown by these African, North American and Philippine examples. Larvae develop under the bark of dead or damaged trees.

FAMILY XYLOMYIDAE (continued) ❶ ***Xylomya*** species are wasp-like flies found in much of the northern hemisphere. This ***X. terminalis*** is one of eight North American species in the genus. **FAMILY PANTOPHTHALMIDAE (GIANT WOOD FLIES)** ❷ This ***Pantophthalmus chuni*** is on a tree branch in the Yasuni Biosphere Reserve, Ecuador. All but one of the 20 species in the small, entirely Neotropical, family Pantophthalmidae are in the genus *Pantophthalmus*. Larval *Pantophthalmus* bore in the trunks of living trees. ❸ This ***Pantophthalmus planiventris*** was found on a large tree in primary rainforest along the border between Peru and Bolivia. **FAMILY NEMESTRINIDAE (TANGLE-VEINED FLIES) Subfamily Hirmoneurinae** ❹ This Chilean ***Hirmoneura***, like all adequately known species of this widespread subfamily, probably developed as an internal parasitoid of an immature scarab beetle. *Hirmoneura* is sometimes treated as the only genus in this subfamily, but some authors divide the 50 or so species in this group up into multiple narrowly defined genera. **Subfamily Nemestrininae** ❺ ❻ Some tangle-veined flies, like these Chilean ***Trichopthalma lundbecki***, have very long mouthparts and are important pollinators of long-tubed flowers. ❼ ***Trichopthalma*** species, like this female from Australia, develop as internal parasitoids of grasshoppers. ❽ ❾ ***Trichopthalma*** includes 45 species in Australia and another 21 in southern South America. Shown here are two Chilean species.

FAMILY NEMESTRINIDAE (continued) Subfamily Trichopsidoinae ❶ Larvae of the small and uncommon genus ***Trichopsidea*** parasitize grasshoppers. This is the only Nearctic species (***T. clausa***), but other species occur in the Old World and Australia. ❷ ***Nycterimorpha*** is a small Australian-Oriental genus containing half a dozen described species. This Australian species was photographed by Paul Zaborowski. **FAMILY ACROCERIDAE (SMALL-HEADED FLIES) Subfamily Acrocerinae** ❸ Species in the large, cosmopolitan genus ***Ogcodes*** are the most commonly seen small-headed flies. These odd-looking spider parasitoids are sometimes seen on prominent objects such as discarded deer antlers or isolated rocks. Eggs are often deposited on dead twigs, where newly hatched larvae seek out jumping spider hosts. ❹ ***Pterodontia***, like this Costa Rican species, are widespread but rarely encountered. Species with known habits lay eggs on tree trunks, where hatching larvae hunt for orb-weaver, wolf or jumping spider hosts. ❺ Species of ***Turbopsebius***, like this ***T. sulphuripes*** from Canada, have the antennae on top of the head instead of the bottom as in similar *Ogcodes* species. Females oviposit in the funnel webs of agelinid spiders, and larvae make their way down the silken strands in search of hosts. **Subfamily Panopinae** ❻ ***Panops*** is endemic to Australasia (earlier records of *Panops* from Brazil were of a *Lasia* species). This image is by Paul Zaborowski. ❼ ***Eulonchus*** is a North American genus that oviposits in flight, leaving the first-instar larvae to seek spider hosts (possibly trapdoor spiders). Shown here is ***E. sapphirinus*** from California. ❽ ❾ The long-tongued species of ***Lasia*** range from Central America to Chile, where they are part of the pollinator guild associated with *Alstroemeria* lilies (shown here). Like other panopine acrocerids, they parasitize mygalomorph spiders and are occasionally reported as pests by tarantula enthusiasts.

FAMILY ACROCERIDAE (continued) ❶ ❷ The aptly named ***Lasia rufa*** is one of the more common, and most attractive, of Chilean small-headed flies. Like other members of the Panopinae, it is a parasitoid of mygalomorph spiders. ❸ ***Ocnaea*** species parasitize mygalomorph spiders ("tarantulas" and their relatives) from California south to Chile. This ***O. vittata*** was photographed on a Chilean fencepost, but adults of these short-tongued flies are much less frequently encountered than their long-tongued flower-visiting relatives. ❹ This larva was found suspended in silk beside the dead body of a tarantula (*Eupalaestrus campestratus*) from Paraguay. About 20 days later it developed into an adult small-headed fly, probably an ***Ocnaea***. (Photo and rearing by Tom Patterson.) **Subfamily Philopotinae** ❺ ❻ ❼ The strikingly humpbacked species of ***Megalybus*** are restricted to Chile, where they parasitize spiders in the family Amaurobiidae. ❽ ***Philopota*** ranges from Mexico to Argentina, but this undescribed species is from Costa Rica where it is probably a parasitoid of an araneomorph spider. ❾ These different-looking Central American ***Philopota*** were captured *in copula*. Note the huge calypters characteristic of this bizarre family of spider parasitoids.

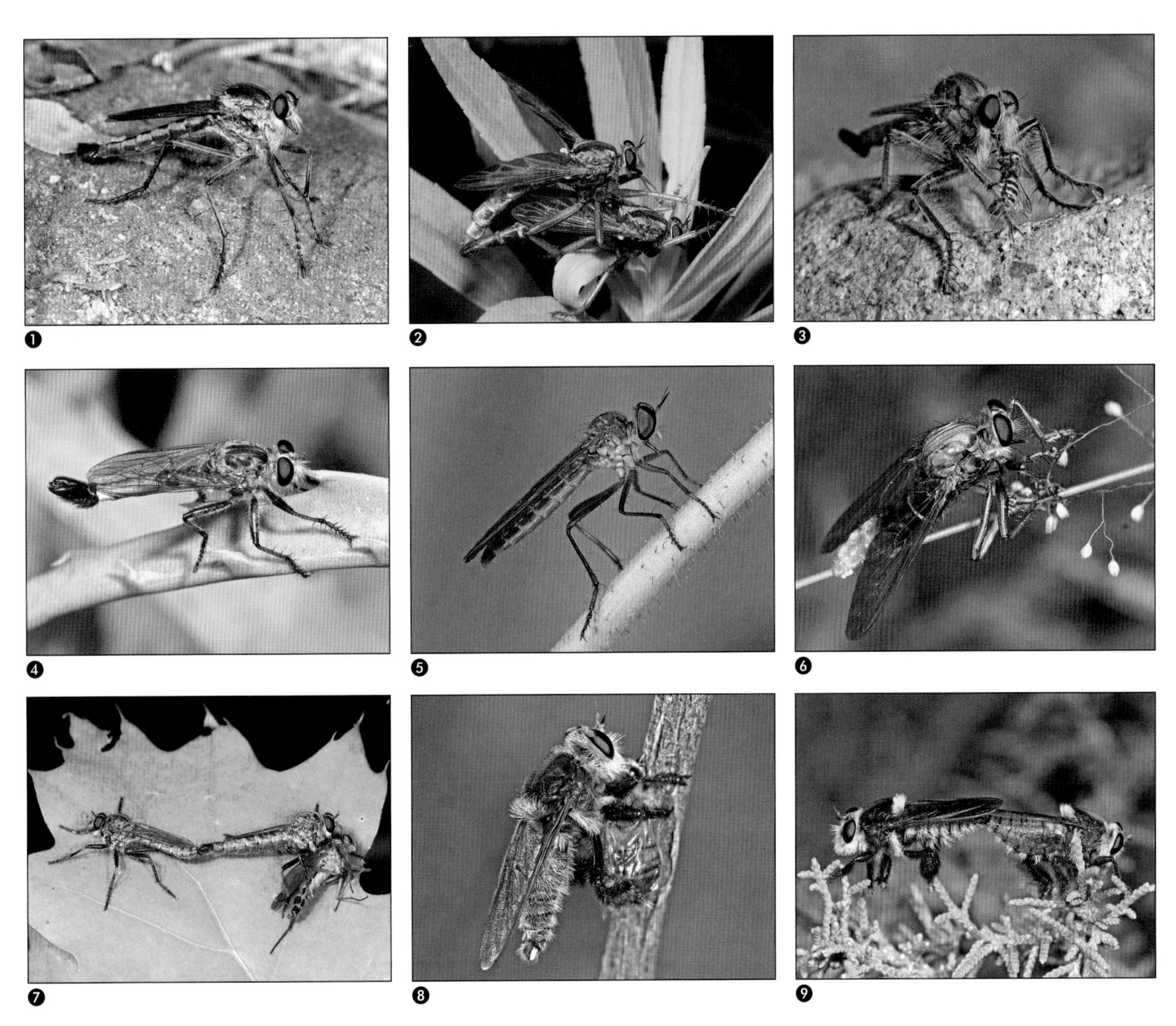

FAMILY ASILIDAE (ROBBER FLIES) Subfamily Asilinae (including Apocleinae) ❶ The large flies in the African genus ***Alcimus*** are among the more conspicuous (dipteran) predators of the African savannah. ❷ The generic name ***Asilus*** has in the past been used for a wide range of medium-sized robber flies, including many now placed in other genera such as *Machimus*. As now defined, the genus is mostly Old World, but it still includes this eastern North American species (***A. sericeus***). ❸ ❹ ***Efferia*** is the largest New World robber fly genus, ranging from Canada to southern South America and with some 230 named species. Shown here are a southwestern American species (in the ***E. carinata*** group) perched on the ground, as is typical for Nearctic species, and a Cuban species perched on vegetation, as is typical of Neotropical species. ❺ This delicate little ***Glaphyropyga*** is perched on a stem deep in the gloom of a high-altitude Ecuadorian cloud forest. This small genus is entirely Neotropical. ❻ Robber flies have a variety of oviposition habits, sometimes laying eggs in the soil and sometimes laying them in or on vegetation. This large Vietnamese ***Hoplopheromerus*** has just deposited eggs on an improbably slender stem. *Hoplopheromerus* is a small Old World (mostly Oriental and Afrotropical) genus; this species is probably undescribed. ❼ ***Machimus*** is a large genus that includes many of the more common medium-sized robber flies throughout the Holarctic region, as well as a few species found at higher altitudes in the tropics. The female of this pair is eating a *Rhagio* (Rhagionidae). ❽ ❾ These southwestern American ***Mallophora*** show the typical robust, bee-like appearance of this large New World genus, species of which are often seen darting out from bare branches in pursuit of bees and similar prey.

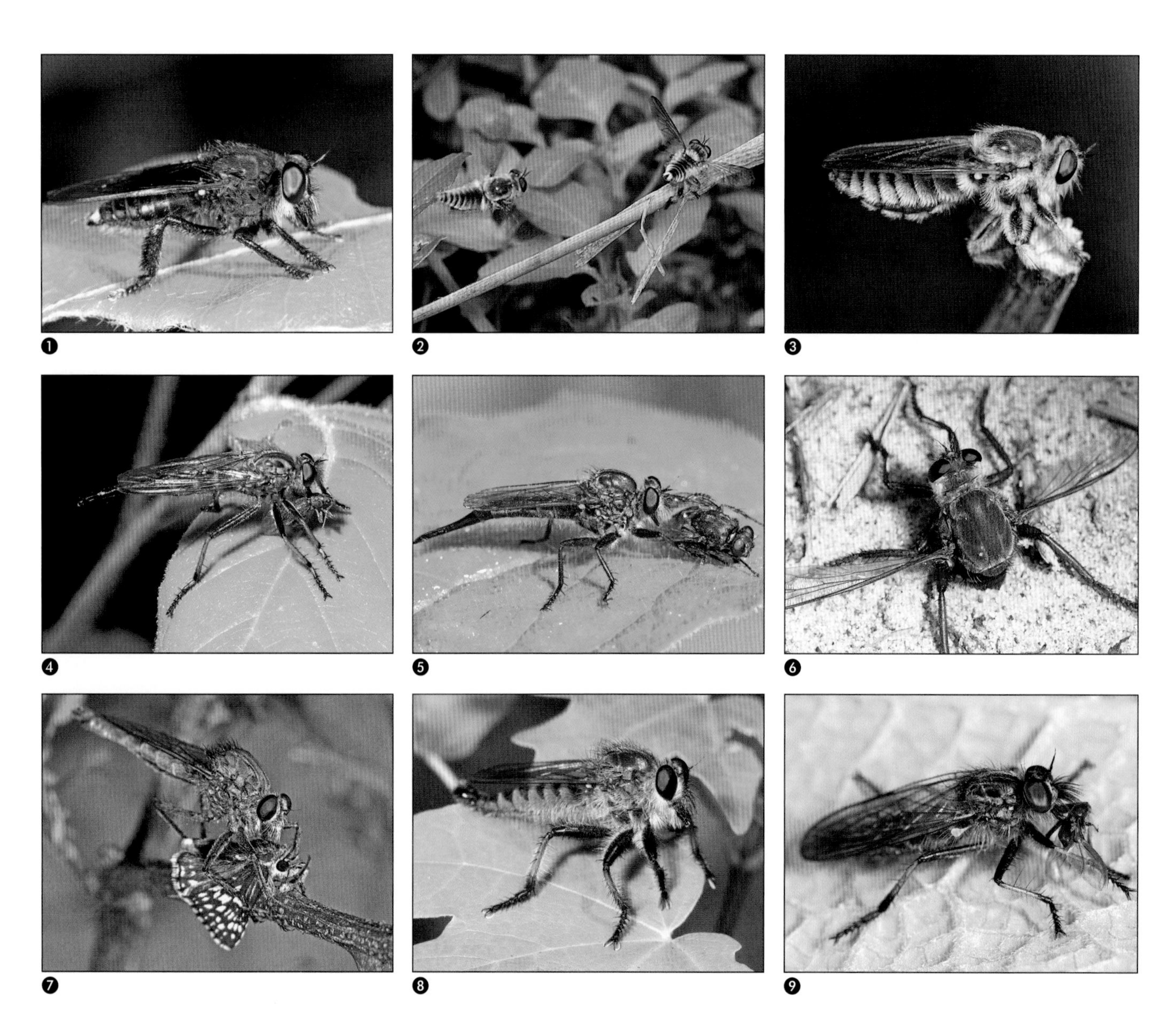

FAMILY ASILIDAE (continued) ❶ Most ***Mallophora*** are Neotropical, and this black species (possibly ***M. atra***) is from Amazonian Ecuador. ❷ ❸ ***Megaphorus*** is a Nearctic genus of small (around 11 mm), robust flies that usually hunt small bees and other Hymenoptera from perches on low vegetation. Members of this genus have the remarkable habit of laying eggs in foamy masses on grass stems, from which the hatching larvae drop to the ground. ❹ ❺ ***Neoitamus*** is a large genus of medium-sized, somewhat nondescript robber flies found in most parts of the world but absent from the neotropics. Females have an elongate, compressed abdominal tip used for inserting eggs into flower heads or leaf sheaths, from which the hatching larvae drop to the ground. ❻ ❼ This ***Proctacanthus hinei*** is laying eggs with her abdomen deep in an eastern Canadian sand dune. This southwestern American fly is eating a skipper butterfly. ❽ ***Promachus*** is a large genus, widely distributed but most diverse in the Holarctic region. ❾ This Chilean ***Stizolestes*** has impaled a small bark beetle between its elytra, having chased it down in flight before returning to a perch. All eight species of *Stizolestes* are from Chile.

FAMILY ASILIDAE (continued) ❶ ❷ These ***Threnia*** species, from a high-altitude forest in Ecuador, are feeding on a disproportionately small dolichopodid fly and a robust sarcophagid. *Threnia* is an entirely Neotropical genus. **Subfamily Brachyrhopalinae** ❸ Many robber flies oviposit by working the entire abdomen into the soil, like this female ***Cyrtopogon falto*** from Canada. ❹ This male ***Cyrtopogon falto*** is eating a hybotid dance fly (*Platypalpus*). *Cyrtopogon* is a large, widespread genus; this is a common Nearctic species. ❺ Most of the 75 or so Nearctic ***Cyrtopogon*** species are found in western North America; this one (***C. marginalis***) is one of the few eastern species. ❻ ***Heteropogon*** is a mostly Holarctic genus of around 50 species. Most of the Nearctic species are western, but this one (***H. macerinus***) occurs in the eastern United States. ❼ ***Holopogon*** species are small flies that perch in a characteristic way at the tips of twigs. Males are often seen hovering over perched females, performing a courtship display. This genus of about 70 species is mostly Holarctic but includes about a dozen Neotropical species. **Subfamily Dasypogoninae** ❽ ❾ ***Diogmites*** is a large (78 species) and commonly encountered New World genus ranging north to Canada but most diverse in the Neotropics. *Diogmites* species are well known for "hanging out" with prey, dangling from a foreleg while dining. This is a Bolivian species.

FAMILY ASILIDAE (continued) ❶ ***Neoderomyia fulvipes*** is the only species in this Chilean genus. ❷ The bare, strikingly humpbacked species of the small Neotropical genus ***Pseudorus***, like this Ecuadorian fly, develop in the burrows of wood-boring beetles in dead trees. ❸ ***Saropogon*** is a large, widespread genus found in most parts of the world but conspicuously absent from some (such as eastern North America and Central America); it is most diverse in the southern Palaearctic and northern Afrotropical regions. This is ***S. gayi*** from Chile. **Subfamily Dioctriinae** ❹ ❺ ❻ ***Dioctria hyalipennis*** is a very common Holarctic species and was possibly introduced from Europe to eastern North America. This species is most often seen eating small Hymenoptera, but these individuals have captured a big-headed fly, a small scraptiid beetle and a minute parasitic wasp. *Dioctria* is a large, mostly Holarctic genus commonly seen on low vegetation, including weeds and shrubs in disturbed areas. **Subfamily Laphriinae (including Laphystiinae)** ❼ As the name suggests, species of the large New World genus ***Atomosia*** are among the smallest robber flies. This western North American ***A. mucida*** has captured a small muscid on some low foliage, but other species often hunt on bare logs or tree trunks. ❽ Species in the Neotropical genus ***Atractia*** are small robber flies commonly seen on fallen trees, where they hunt bark beetles such as platypodids and scolytids. ❾ This ***Cerotainiops omus***, perched on the scree of a desert mountain in Arizona, is one of half a dozen species in this southwestern Nearctic genus.

FAMILY ASILIDAE (continued) ❶ ***Dissmeryngodes*** is a South American genus known mostly from Brazil, although this one is from Amazonian Ecuador. ❷ ❸ The small species of the Neotropical genus ***Eumecosoma*** typically scan for prey from the tips of low vegetation. One of the Ecuadorian species shown here has captured a small crane fly. ❹ This Tanzanian ***Goneccalypsis*** is feeding on a gall midge (Cecidomyiidae) barely larger than its beak. *Goneccalypsis* is a small genus of four African species. ❺ This ***Hexameritia micans*** was one of many feeding on a concentration of jumping plant lice (Psyllidae) on a small shrub. Both species in this small genus are from Chile. ❻ This Bolivian ***Lampria*** has captured a small braconid wasp. Some 20 species of *Lampria* occur throughout the New World; this species (***L. clavipes***) ranges through much of Central and South America. ❼ This Bolivian ***Lampria dives*** is exhibiting typical behavior for the genus, perching on the end of a sun-splashed leaf in a lowland rainforest. ❽ ❾ The robust and colorful species of ***Laphria*** are common and conspicuous flies throughout the Holarctic region. Shown here are two similar Nearctic species, ***L. sericea***, feeding on an ichneumonid wasp, and ***L. index***, using its flattened, blade-like beak to impale a click beetle between its elytra.

FAMILY ASILIDAE (continued) ❶ This eastern North American ***Laphria flavicollis*** is feeding on an anthomyiid fly. ❷ ***Laphria saffrana***, one of the most spectacular Nearctic robber flies, is a characteristic species of the Atlantic coastal plain, where it mimics the Southern Yellowjacket (*Vespula squamosa*). ❸ Although ***Laphria*** are best known as the "bee-like" robber flies because of the many species that mimic bumble bees, some species are relatively dull in color. Shown here is one such species, the Nearctic ***L. sicula***, eating a flower fly. ❹ Members of the mostly Holarctic genus ***Laphystia*** are common on sand dunes and beaches; this ***L. flavipes*** is feeding on a stiletto fly (Therevidae). ❺ ***Pagidolaphria*** is a group of large and beautiful robber flies previously treated as part of the huge genus *Laphria*. This Vietnamese fly is probably ***P. reinwardtii***, a species described from Java and Sumatra in 1828 and rarely seen since. ❻ The upturned needle-like proboscis of this Central American ***Pilica formidolosa*** is typical for the genus. *Pilica* is entirely Neotropical, with about a dozen species. ❼ ***Strombocodia*** includes only two species, both South American and both tiny (about 5 mm) relative to other asilids. This species (***S. elegans***) is known from lowland rainforests of Bolivia and Peru. ❽ This Tanzanian ***Trichardis*** was hunting from the ground in a savannah game park. Most other members of this Old World genus are also Afrotropical. ❾ This enormous robber fly, ***Wyliea mydas***, is an uncommon southwestern American species that looks and acts like a tarantula hawk (*Pepsis* spp., family Pompilidae), which occurs in the same places. The species name refers to a similarity between these robber flies and other *Pepsis* mimics in the family Mydidae. The only other *Wyliea*, *W. chrysauges*, is a scoliid wasp mimic found in Mexico and Guatemala.

FAMILY ASILIDAE (continued) Subfamily Leptogastrinae ❶ ❷ Slender robber flies are delicate, inconspicuous flies such as this ***Mesoleptogaster*** from the Philippines (on a twig tip) and this ***Leptogaster*** female from Tanzania. ❸ This slender robber fly, ***Tipulogaster glabrata***, occurs throughout eastern North America, where it can be found picking small prey off low vegetation. The only other species in this small genus is Neotropical. **Subfamily Ommatiinae ❹ ❺** This perching male and ant-eating female from Vietnam belong to the genus ***Emphysomera***, which until recently was treated as a synonym of *Ommatius*. ❻ ❼ Distinctive for their apparently pectinate antennae, ***Ommatius*** species are among the more common robber flies in warmer areas throughout the world. Shown here are species from Bolivia (eating a rove beetle) and Costa Rica (eating a richardiid fly) ❽ This ***Pygommatius*** female from the Philippines is eating a blackfly (Simuliidae). The Old World tropical genus *Pygommatius* was only described in 2003. ❾ Although ***Ommatius*** is mostly a genus of the tropical and subtropical regions of the world, this huge genus does occur in temperate regions. Shown here is the rare southwestern American species ***O. bromleyi*** eating a tiny lower dipteran.

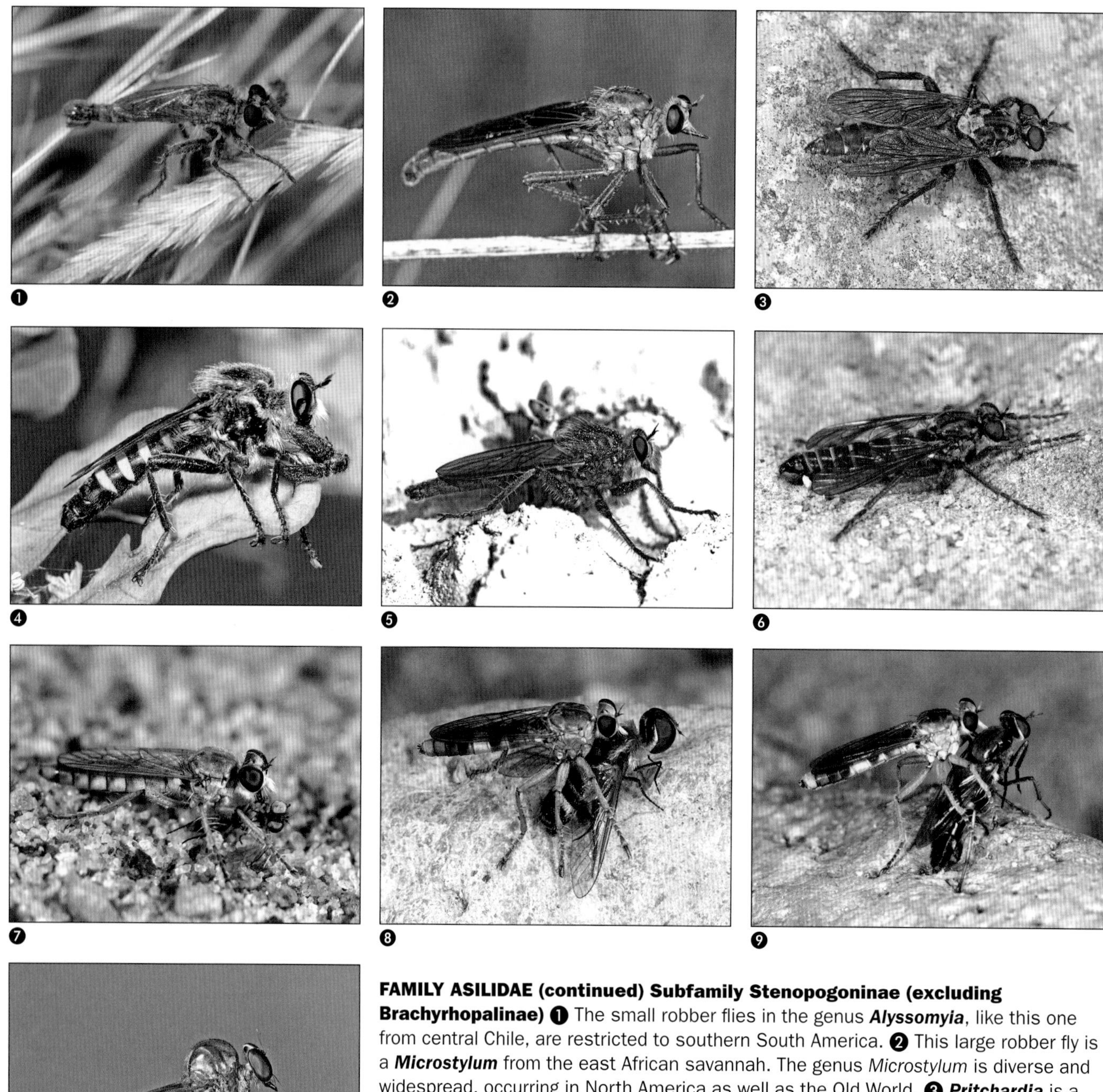

FAMILY ASILIDAE (continued) Subfamily Stenopogoninae (excluding Brachyrhopalinae) ❶ The small robber flies in the genus ***Alyssomyia***, like this one from central Chile, are restricted to southern South America. ❷ This large robber fly is a ***Microstylum*** from the east African savannah. The genus *Microstylum* is diverse and widespread, occurring in North America as well as the Old World. ❸ ***Pritchardia*** is a small southern South American genus of five species. This is ***P. hirtipes***, from Chile. ❹ The mostly African genus ***Scylaticus*** is widespread in the Old World and southern South America but is apparently absent from North America and Central America. This is ***S. lugens*** (originally described as a *Dasypogon*), one of five Chilean species in the genus. ❺ This ***Stenopogon*** perched on the sun-blasted sand of a Nevada dune is one of more than 160 species in this large, mostly Holarctic genus. **Subfamily Stichopogoninae** ❻ ***Lasiopogon***, the most northerly robber fly genus, includes around 70 species in the Holarctic region, where they are most often found near shorelines in early spring. This recently discovered species, ***Lasiopogon marshalli***, abounds on exposed rock flats of the New River in the U.S. state of Virginia. ❼ This ***Stichopogon argenteus*** has captured a small fly (Dolichopodidae) on a Canadian lakeshore. ❽ ❾ ***Stichopogon*** species occur worldwide but the genus is most diverse in the Old World; it includes relatively few New World species. This is ***S. trifasciata***, the most common *Stichopogon* from Canada south to Mexico. The southwestern American flies shown here have captured a bee fly (Bombyliidae) and a stiletto fly (Therevidae). **Subfamily Trigonomiminae** ❿ This strikingly humpbacked robber fly from Vietnam belongs to the large Old World tropical genus ***Damalis***.

FAMILY ASILIDAE (continued) ❶ ***Haplopogon*** is a small genus with about half a dozen southwestern American species and one Russian species. This teneral specimen has just emerged from the desert soil of Tucson, Arizona, and is expanding its wings on a cactus spine. ❷ The small "goggle-eyed" robber flies in the genus ***Holcocephala*** are often abundant hunters of small prey such as springtails and midges in wet meadows. All 40 or so described and probably as many undescribed *Holcocephala* species occur in the New World. This Bolivian fly has captured a very small rove beetle. **Subfamily Willistonininae** ❸ ***Ablautus coquilletti*** is a small western Nearctic species difficult to spot on its white sand habitat. Other members of this small genus also occur in western North America. **FAMILY APIOCERIDAE** ❹ Most members of the small, uniform family Apioceridae occur in Australia, but others occur in arid areas of southern South America, South Africa, western North America and Mexico, where they apparently develop as predators in soil. This is ***A. painteri***, from the southwestern United States. **FAMILY MYDIDAE (MYDAS FLIES) Subfamily Apiophorinae** ❺ The subfamily Apiophorinae occurs in southern South America and Australia. Shown here is a member of the Australian genus ***Miltinus (M. maculipennis)*** photographed in Western Australia by Jean Hort. **Subfamily Anomalomydinae** ❻ The subfamily Anomalomydinae includes only two species, both Australian. Shown here is ***Anomalomydas mackerrasi***, photographed in Western Australia by Jean Hort. **Subfamily Rhopaliinae** ❼ Shown here is a ***Perissocerus arabicus*** from the United Arab Emirates, photographed by Drew Gardner. **Subfamily Mydinae** ❽ ***Mydas xanthopterus*** is a striking mimic of the large and well-armed tarantula hawks (*Pepsis* spp., Pompilidae) that abound in the same southwestern American range. Other members of this genus occur from Canada south to Brazil (although the genus is sometimes defined in a restricted sense to include only species from Mexico, the United States and Canada). ❾ ***Mydas ventralis*** is oddly named, since the distinctive orange patch is on the dorsal side of the abdomen. This is one of the several large and attractive *Mydas* species of the American southwest.

FAMILY MYDIDAE (continued) Subfamily Leptomydinae ❶ The 20 species of the New World genus ***Nemomydas***, such as this ***N. venosus*** from New Mexico, are characteristic of open, dry, sandy areas. **Subfamily Diochlistinae** ❷ ***Mitrodetus*** is a small southern South American genus of eight species. Shown here is a pair of ***M. dentitarsis*** from Chile. **Subfamily Ectyphinae** ❸ ***Opomydas*** is a small genus restricted to the American southwest and Mexico. This is ***O. townsendi*** from New Mexico. **Subfamily Megascelinae** ❹ The subfamily Megascelinae has one genus in Australia, one in Chile and one in southern Africa. This is ***Megascelus melanoproctus***, one of five species in the Chilean genus. **Subfamily Rhaphiomidinae** ❺ ***Rhaphiomidas*** includes the infamous endangered species *R. terminatus* (Delhi Sands Flower-loving Fly) as well as another 22 species found in the southwestern United States and Mexico. This is a specimen of ***R. acton*** from California. **FAMILY THEREVIDAE (STILETTO FLIES)** ❻ Therevidae and the closely related Scenopinidae have long, slender, snake-like predaceous larvae with the segments secondarily divided so that they seem to have a total of 17 body segments. This unidentified larva is from Tanzania. **Subfamily Agapophytinae and Taenogera genus-group** ❼ The genus ***Agapophytus***, like the rest of the subfamily Agapophytinae, occurs only in Australia. This is ***A. quatiens*** from Tasmania. ❽ ***Ectinorhynchus***, a New Zealand/Australia genus of about 15 species, is one of a number of therevid genera not placed to subfamily; instead it is currently treated as the *Taenogera* genus-group. This one is from Tasmania. **Subfamily Phycinae** ❾ The subfamily Phycinae is a small group with some species in western North America and others in the Old and New World tropics. Shown here is a member of the New World (mostly southwestern Nearctic) genus ***Pherocera***, photographed by Shaun Winterton.

FAMILY THEREVIDAE (continued) Subfamily Therevinae ❶ The Australian therevid fauna is dominated by the genus *Anabarhynchus*, the largest therevid genus in the world. This is ***Anabarhynchus kampmeierae***, named after Diptera informatics specialist Gail Kampmeier. ❷ Stiletto flies, such as this Nearctic species (***A. vialis***) in the Holarctic genus ***Acrosathe***, look a bit like robber flies (Asilidae), but adult therevids lack the biting mouthparts of their predaceous relatives. Larvae are subterranean predators. ❸ ***Breviperna*** is a small genus of two species, one in the American southwest and one in Mexico. This ***B. placida*** was one of many stiletto flies drinking at an intermittent stream during a dry month in the American southwest. ❹ ***Cyclotelus rufiventris*** is a widespread eastern Nearctic species, common in open sandy areas, where the larvae presumably prey on subterranean invertebrates. Most of the species in this New World genus are described from South America. ❺ ❻ *Ozodiceromya*, like many therevids, often concentrate around limited water sources in their otherwise arid habitats. This male and female ***Ozodiceromya metallica*** were among half a dozen species of Therevidae found on the same stretch of intermittent stream in Arizona. ❼ This undescribed species in the diverse New World genus ***Ozodiceromya*** was one of two new species of the genus found (with *O. metallica*) along an intermittent stream in Arizona (both are under study by therevid specialist Stephen Gaimari). ❽ ***Pachyrrhiza*** is a small southern South American genus with only two species. This one (probably ***P. pictipennis***) is drinking along a small stream in arid central Chile. ❾ The western American ***Pallicephala pachyceras*** is one of several similar North American stiletto fly species in which the males are relatively fuzzy and white.

FAMILY THEREVIDAE (continued) ❶ ***Pallicephala pachyceras*** is a western American stiletto fly found in all the Pacific coast states. ❷ ***Penniverpa*** is a mostly Neotropical genus, but this southeastern American species (***P. festina***) is perched on a South Carolina leaf. ❸ ***Peralia*** is a small Chilean genus with only three described species. This one was visiting the edge of a stream in the dry Mediterranean habitat near Santiago, Chile. ❹ ***Spiriverpa*** is a small Holarctic genus with about half a dozen species, including this ***S. albiceps*** from northern Ontario. ❺ ❻ Unlike most therevids, which are blisteringly fast fliers that usually perch on the ground, species in the widespread genus ***Thereva*** are usually relatively sluggish and perch on foliage. This female (with separated eyes) is from Greece; the male is from California. **FAMILY SCENOPINIDAE (WINDOW FLIES)** ❼ Most Scenopinidae are rarely encountered, but ***Scenopinus fenestralis*** is a cosmopolitan species often found on the insides of windows. Larvae are probably predators of dermestid beetles. ❽ ***Pseudatrichia*** is a New World genus of some 39 species mostly found in western North America, where their larvae develop as predators in the tunnels of wood-boring beetles. Shown here is a specimen of ***P. punctulata*** from Brazil. ❾ Although most scenopinid genera are restricted to a single zoogeographic region, species of the small genus ***Metatrichia*** occur worldwide. This one was photographed in western Canada by Werner Eigelsreiter.

FAMILY EVOCOIDAE ❶ This ***Evocoa chilensis*** was found at the type locality near Ocoa, Chile. This species, the only member of the family, is otherwise known only from dead specimens taken in Malaise traps. **FAMILY APSILOCEPHALIDAE** ❷ This ***Apsilocephala*** was captured near a burrow in southern California and photographed by Shaun Winterton. **FAMILY BOMBYLIIDAE (BEE FLIES) Subfamily Anthracinae** ❸ ***Anthrax albofasciatus***, like many species in this huge worldwide genus, parasitize hunting wasps that build mud nests. Females are often seen hovering around wasp nest sites, popping down periodically to place an egg in the nest. ❹ ❺ Adult females of ***Anthrax georgicus*** are commonly seen hovering over the entrances to tiger beetle larval burrows and ovipositing near the host beetle larvae. ❻ Members of the large, mostly Holarctic genus ***Aphoebantus*** lay uncoated eggs in sandy soils, where their larvae develop as predators of grasshopper eggs. This one is from Arizona. ❼ This eastern North American ***Chrysanthrax lepidotoides*** is ovipositing in bare soil, where her active first-instar larvae will locate and parasitize a larval beetle (probably a scarab in the genus *Phyllophaga*). *Chrysanthrax* is a New World genus. ❽ ***Dipalta banksi*** is a parasitoid of antlion larvae in eastern North America. The only other species in the genus is a widespread Nearctic and northern Neotropical species. ❾ ***Exoprosopa*** species are large, conspicuous flies often seen on flowers. Larvae are hyperparasitoids, attacking other insects that parasitize subterranean larvae. This is a huge worldwide group of around 350 species.

FAMILY BOMBYLIIDAE (continued) ❶ ***Hemipenthes sinuosa***, like other members of this mostly Holarctic genus, are hyperparasitoids that develop on the larvae of other parasitoids inside their hosts. ❷ ❸ This ***Hyperalonia chilensis*** is laying eggs in the nest of a solitary wasp, in which her larvae will develop as ectoparasitoids on the wasp larvae. All five species of *Hyperalonia* are South American. ❹ ***Lepidanthrax*** is a western American genus that ranges south to Panama. Most (38) of the 50 species in this genus were described in a single paper by Jack Hall in 1976. ❺ Known larvae of the large cosmopolitan genus ***Ligyra*** are parasitoids or hyperparasitoids of ground-nesting wasps such as Scoliidae and Sphecidae. Shown here is ***L. orientalis*** from the Philippines. ❻ ***Paranthrax*** is a small genus found only in Chile. ❼ ***Poecilanthrax tegminipennis***, like other species of this entirely New World genus, are parasitoids of soil-dwelling caterpillars (cutworms, army worms) in the Noctuoidea. ❽ Larvae of the New World genus ***Rhynchanthrax*** are ectoparasitoids of white grubs (larval scarab beetles). ❾ Known larvae of the large, widespread genus ***Thyridanthrax*** are predators of locust eggs. Shown here is a southwestern American species.

FAMILY BOMBYLIIDAE (continued) ❶ ❷ ***Villa*** is a large, difficult genus including many parasitoids of moth larvae. Shown here are ***V. alternata*** in a Canadian gravel pit and a southwestern American species resting on some dry grass. *Villa* is one of the largest and most widespread bee fly genera (266 species). ❸ ***Xenox tigrinus*** is an eastern North American species often seen hovering over wood riddled by the nests of host carpenter bees (*Xylocopa* spp.). The other four species in the genus occur in the southwestern Nearctic and northern Neotropical regions. **Subfamily Bombyliinae** ❹ Species in the mostly Holarctic genus ***Anastoechus*** larvae are predators of grasshopper egg pods. This is an eastern American species. ❺ Known species in the large, mostly Holarctic genus *Bombylius* are parasitoids in the nests of solitary bees. Shown here is the North American species ***Bombylius mexicanus*** nectaring on pinflowers. ❻ ❼ Bombyliidae, including the large genus ***Bombylius***, are particularly diverse in the American southwest. Shown here are a mating pair of an unidentified species and a hovering individual of the American ***B. comanche*** species group, both from Arizona. ❽ The mostly African genus ***Bombylella*** is also found in the Palaearctic and Oriental regions. ❾ The aptly named ***Bombylella delicata*** is seen here gathering sand grains in her sand chamber; these will coat eggs that she will oviposit while hovering. ❿ ***Euchariomyia*** is a small Oriental genus of only three species, including the relatively widespread and common ***E. dives***.

FAMILY BOMBYLIIDAE (continued) ❶ ***Heterostylum*** species are parasitoids of bees, and some members of this New World genus are considered pest species because of their impact on beneficial bees needed for the pollination of forage crops. ❷ ❸ Males of ***Lordotus*** sometimes aggregate at lek sites, where they defend positions likely to be visited by incoming females. The same lek sites are often used by sequential generations. Shown here are a nectaring female and a perching male. ❹ ❺ Members of the Nearctic genus ***Lordotus*** lack the sand chamber of similar higher bee flies. They use a telescoping abdomen to oviposit into soil near the ground nests of spheciform wasps (Sphecidae and Crabronidae), where their larvae burrow down in search of a host. ❻ This Australian ***Meomyia*** is scooping sand into her sand chamber, an abdominal cavity in which her eggs will be coated with sand before being released into the environment. ❼ ❽ *Parasystoechus* and *Systoechus* larvae are predators of grasshopper egg pods. Shown here are ***Parasystoechus flavescens*** from Chile (purple flower) and ***Systoechus vulgaris*** from Canada. *Systoechus* is a large, widespread but mostly Palaearctic and Afrotropical genus; *Parasystoechus* is a small group of four southern South American species. ❾ ***Triploechus*** is a small New World group of about ten species distributed mostly in southern South America and the southwestern United States; this is ***T. heteronevrus*** from Chile.

FAMILY BOMBYLIIDAE (continued) Subfamily Ecliminae ❶ The elongate scaly flies in the genus ***Lepidophora***, like this eastern Canadian ***L. lutea***, develop as kleptoparasites in the nests of twig-nesting crabronid wasps. ❷ ❸ The slender species of ***Thevenetimyia*** are characteristic of wooded areas, where they probably develop as parasitoids of wood-boring beetle larvae. The genus is almost cosmopolitan, but the flies shown here are a ***T. funesta*** from the eastern United States (yellow flower) and an unidentified species from California (white flower). **Subfamily Lomatiinae** ❹ ***Comptosia*** is a diverse genus with around 140 species currently restricted to Australia, New Guinea and Indonesia; but also known from Eocene fossils in Europe and North America. These are common and conspicuous flies, but their hosts remain unknown except for one rearing from a hepialid moth larva in Tasmania. ❺ ❻ The small genus ***Ogcodocera*** includes one Neotropical species, one southwestern American species (***O. analis***, shown here on a twig) and one northeastern American species (***O. leucoprocta,*** shown here on a blade of grass). Hosts remain unknown. ❼ Members of the large Old World genus ***Petrorossia*** are often seen hovering around the entrances of nest holes of solitary bees. This Philippines species is gathering sand into her sand chamber. ❽ ❾ This delicate little ***Plesiocera*** was almost invisible as it landed on the hot sand in a dry part of Tanzania. *Plesiocera* is a small, mostly African group that ranges to southern Europe.

FAMILY BOMBYLIIDAE (continued) Subfamily Phthiriinae ❶ ***Neacreotrichus*** is a small, mostly Nearctic genus (previously treated as a subgenus of *Phthiria*) that ranges south to Central America. These two are using their long beaks to nectar in a long-tubed Arizona flower. ❷ Known larvae of the large, widespread genus ***Phthiria*** are predators of grasshopper egg pods. Shown here is a species from Greece. ❸ ❹ These Chilean ***Phthiria*** — ***P. homochroma***, sharing a yellow flower with a small tachinid, and ***P. austrandia***, on a yellow and white flower — are using their legs to pick pollen off their beaks. ❺ ❻ ***Poecilognathus sulphurea*** ranges across North America and south to Mexico. ❼ Most members of the entirely New World genus ***Poecilognathus*** (previously treated as a subgenus of *Phthiria*) occur in the southwestern United States. This is an unnamed species from Arizona. **Subfamily Toxophorinae** ❽ Most of the distinctively slender species in the genus ***Dolichomyia*** occur in Chile, although Venezuela and the southwestern United States are each home to one species and there is a similar genus (*Zaclava*) in Australia. Shown here is a mating pair from Chile. ❾ The distinctively humpbacked species of the large, cosmopolitan genus ***Geron*** are common flower visitors almost everywhere, even on oceanic islands. This one is from the southwestern United States.

FAMILY BOMBYLIIDAE (continued) ❶ Known larvae of the large, widespread genus ***Geron*** are parasitoids of small caterpillars, including larval snout moths (Pyralidae) and leafrollers (Tortricidae). Shown here is a species from Australia. ❷ ❸ The slender, wasp-like species of ***Systropus*** are parasitoids of slug caterpillars (Limacodidae). *Systropus* is almost cosmopolitan but is absent from Europe. These flies are from Vietnam, but *Systropus* species are remarkably similar the world over. ❹ Species in the cosmopolitan genus ***Toxophora***, such as these eastern Canadian ***T. amphitea***, are found in wasp nests (Vespidae, usually Eumeninae), where they are either parasitoids of wasp larvae or hyperparasitoids attacking other parasitoids of the wasps. **Subfamily Usiinae** ❺ ***Apolysis*** is a large Holarctic-Afrotropical genus of small, inconspicuous bee flies such as this minute male from the American southwest. Larvae remain unknown. **Subfamily Cythereinae** ❻ The distinctively wooly flies in the odd little genus ***Pantarbes*** are found only in the Nearctic region, where they are sometimes common in the desert areas of California, New Mexico and Arizona. **FAMILY MYTHICOMYIIDAE (MICRO BEE FLIES; sometimes treated as subfamily Mythicomyiinae of the family Bombyliidae)** ❼ ❽ Micro bee flies form a worldwide group of about 350 species of tiny flies ranging from well under a millimeter long up to about 3 mm. Shown here are two female ***Mythicomyia*** from Arizona. ❾ Micro bee flies are relatively bare flies sometimes common on flowers. Male ***Mythicomyia***, like this one from the American southwest, are holoptic (with big eyes that meet on top of the head) and strongly humpbacked.

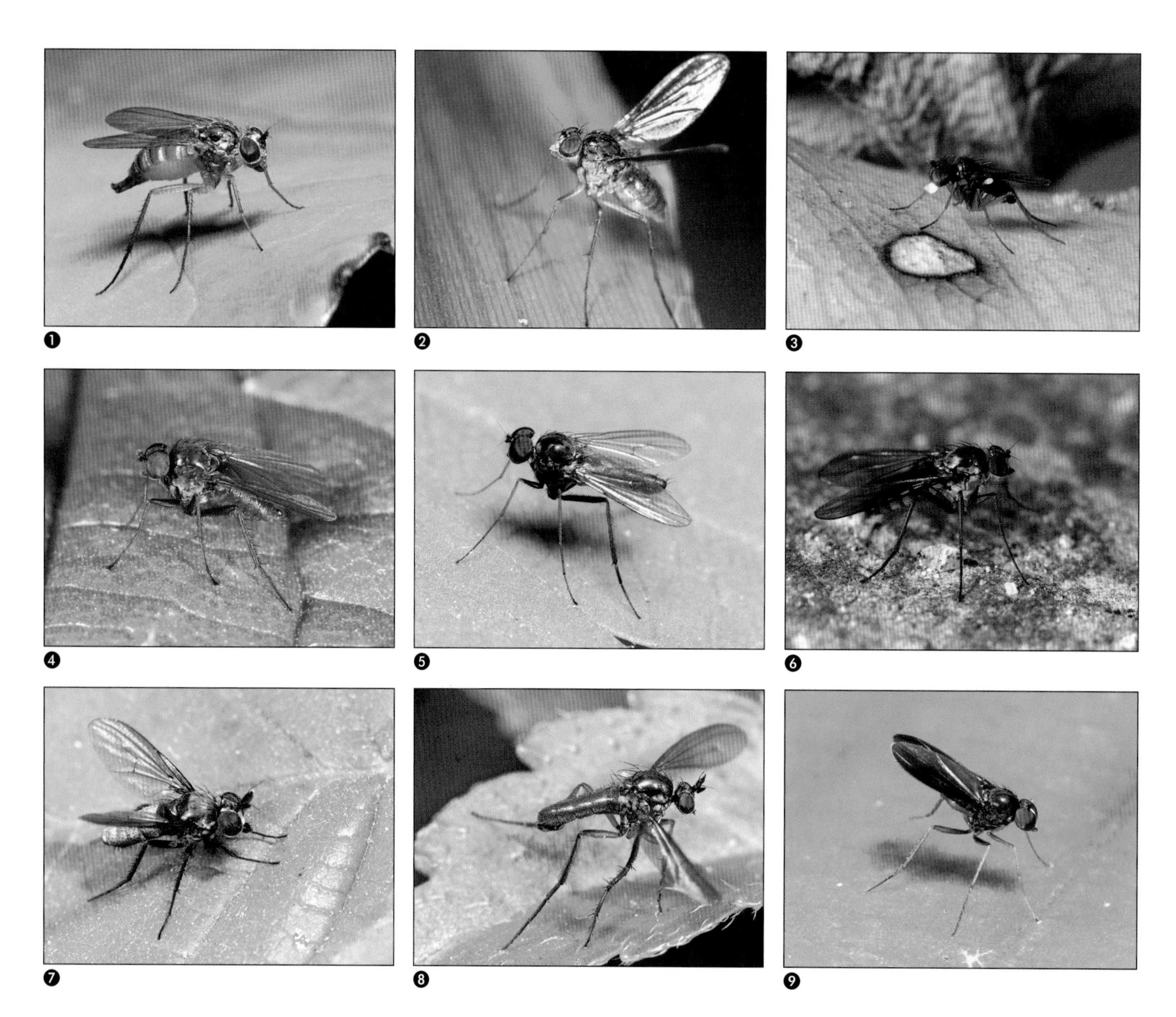

FAMILY DOLICHOPODIDAE (LONG-LEGGED FLIES) Subfamily Diaphorinae ❶ ***Argyra*** is a genus of about 100 species mostly found in the Holarctic region (such as this one from Canada), although a few occur in the Neotropical, Oriental and Afrotropical regions. The silvery abdomen alluded to in the generic name is visible only in the right light. ❷ ***Asyndetus*** is a common group with about 100 species worldwide; this one is from Tanzania. Some species develop in crab holes, probably preying on other insects in the same habitat. ❸ ***Chrysotus*** is a common and widespread genus of some 285 species, but these tiny and generally nondescript species escape the attention of most naturalists. Males of some species, such as this one seen in a Canadian butterfly house, flash disproportionately large silvery reflective palpi. ❹ ❺ ***Diaphorus*** is a large (266 species) worldwide genus of relatively dull-colored dolichopodids with a superficial resemblance to snipe flies (Rhagionidae). This dark-eyed fly is from eastern Australia; the red-eyed individual is from Canada. ❻ ***Lyroneurus*** is a Neotropical (and southern Nearctic) genus until recently treated as part of the large genus *Chrysotus*. This one is from an Ecuadorian cloud forest. ❼ ***Pseudargyra*** is a small genus of four Neotropical species with a silvery abdomen like that of *Argyra*. This one is on a water-splashed leaf beside a small waterfall in southern Chile. ❽ This ***Somillus*** male has elaborately sculptured legs that undoubtedly play a role in courtship. This is an odd little genus of two species found only in Chile. ❾ ***Symbolia*** is an entirely Neotropical genus of 16 species; this one is from Ecuador.

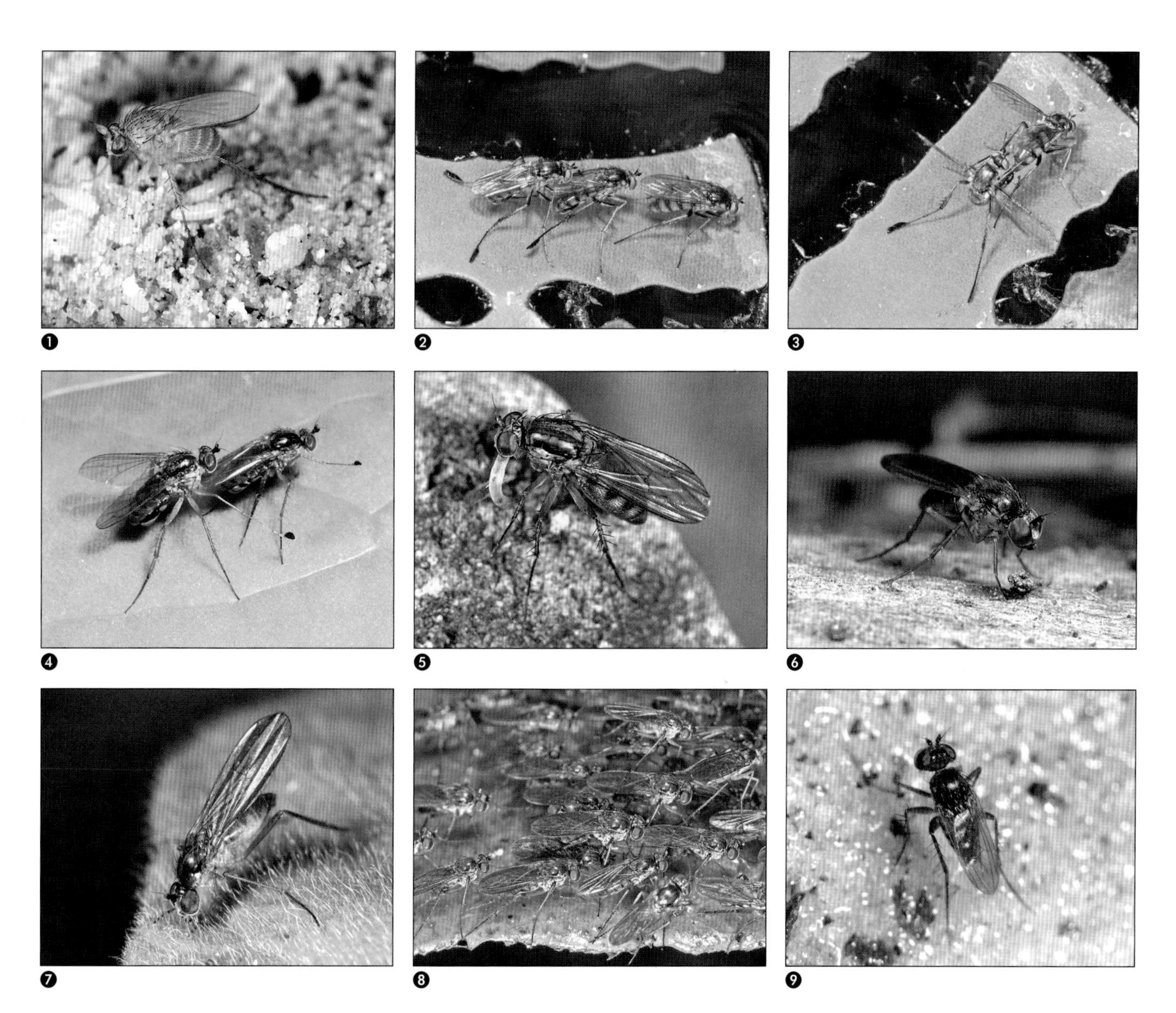

FAMILY DOLICHOPODIDAE (continued) Subfamily Dolichopodinae ❶ *Argyrochlamys* is a small genus of four species found on beaches in the Afrotropical region, such as this one in the Tanzanian city of Dar es Salaam. ❷ ❸ Male Dolichopodidae often have ornaments used in courtship displays; *Dolichopus* species frequently have feather- or flag-like ornaments on the front, middle or hind legs. These male ***Dolichopus remipes*** are lining up behind an unadorned female on a lily pad in a Canadian lake; the successful male flew over the female and showed her his hind tarsi before unslinging his impressively large genitalia. ❹ ***Dolichopus*** is by far the most diverse genus of Dolichopodidae, with more than 450 species in the Holarctic region and another 100 or so scattered around the rest of the world. The genus is especially speciose in the Nearctic but absent from the Neotropics. ❺ ***Dolichopus*** species often seem to slurp up oligochaete worms the way a child might ingest a strand of spaghetti, enveloping prey between their labellae (the lobes of the labium, or lower lip) while puncturing tissues with epipharyngeal blades. ❻ ***Gymnopternus*** species are distributed much like those of the much larger genus *Dolichopus*, with most (75) species in the Nearctic and fewer in the Palaearctic region. ❼ ***Hercostomus*** is an important Old World genus, with most of its 470 species in the Old World tropics. ❽ These female flies were among hundreds of a ***Hercostomus*** species found aggregating on a few leaves above a small stream in the Eastern Arc Mountains of Tanzania. When I took this photo in 2009, the species was known only from a single specimen (the holotype). ❾ ***Pelastoneurus*** is a large (more than 100 species) genus of relatively small dolichopodids found mostly in the warmer parts of the world, although this one was photographed on an Atlantic beach in the eastern United States.

FAMILY DOLICHOPODIDAE (continued) ❶ ***Lichtwardtia*** is a mostly Afrotropical group with a couple of Oriental and Australasian species. This one was attracted to lights on an African wall. ❷ Many dolichopodids have elaborate courtship rituals. This male Chilean ***Tachytrechus*** followed a female around a mudflat where she was foraging, continually posturing and buzzing his wings. Members of this large (more than 150 species) and widespread genus are often common on wet beaches. ❸ The male of this pair of ***Tachytrechus*** (probably ***T. sanus***) on the Pacific coast of North America has spectacularly ornamented antennae, with a bicolored flag at the tip of the arista. ❹ This African ***Tachytrechus*** is eating a small arthropod (perhaps a grasshopper nymph pulled from its egg) by crushing it between the labellae and puncturing it with the sharp blades of the epipharynx. **Subfamily Hydrophorinae** ❺ ❻ Calm puddles or pools of water are usually frequented by water-skating flies in the genus ***Hydrophorus***. A few *Hydrophorus* species occur south of the equator, but most of the 118 species in the genus, such as this one on a Canadian lake and this pair on a Florida pond, are found in the northern hemisphere. *Hydrophorus* are often seen in pairs because some species exhibit "mate guarding," in which males continue to hold the females after mating. ❼ ***Hypocharassus*** adults, distinctive for their "can-opener" antennae, are often abundant on intertidal mud, where their larvae develop. Two species occur on eastern North American seashores and two more are found in China and Taiwan. This one is from the Gulf Coast of Florida. ❽ ❾ ***Liancalus*** species are among the largest dolichopodids, often exceeding 11 mm in length. Known larvae occur in algal mats in seeps and under waterfalls, and adults are usually found on nearby vertical rock surfaces. Most of the 19 *Liancalus* species are Old World, but four species occur in the Nearctic region, including the eastern ***L. genualis*** (shown here in copula) and the western ***L. limbatus***.

FAMILY DOLICHOPODIDAE (continued) ❶ ***Melanderia mandibulata*** is a remarkable fly sometimes described as having jaws, although the mandible-like structures are simply strikingly modified labellae. All four *Melanderia* species occur on rocky marine shores of western North America. ❷ ***Paraphrosylus***, like the related genera *Melanderia* and *Thambemyia*, occurs only on rocky seacoasts. All six species of *Paraphrosylus* occur on the west coast of North America. ❸ ***Thambemyia*** is a mostly Palaearctic-Oriental genus of coastal flies that lay eggs and hunt prey (mostly other fly larvae) among barnacles in the intertidal zone. Shown here is ***T. borealis*** on the coast of Peru, where the species was probably accidentally introduced from its original Japanese range. **Subfamily Medeterinae** ❹ This ***Medetera*** has her ovipositor slid into a beetle hole on a Canadian tree. *Medetera* larvae are significant predators of bark beetles (Curculionidae, Scolytinae). ❺ The small, compact flies in the cosmopolitan genus ***Medetera*** are ubiquitous on tree trunks, where they usually take a characteristic "heads up" stance, such as the one shown by this southwestern American fly scanning the surface of a wet rock. **Subfamily Microphorinae** ❻ ***Microphor*** adults are sometimes kleptoparasites that steal prey out of spider webs, but this eastern Canadian species can be reliably found feeding on the flowers of False Solomon's Seal (*Maianthemum racemosum*). *Microphor* species occur from Tasmania to northern Europe. ❼ This western North American ***Schistostoma*** is on a willow leaf along a sandy riverbank, a typical habitat for this small, mostly Palaearctic genus. **Subfamily Neurigoninae** ❽ ❾ ***Neurigona*** species are delicate dolichopodids most often seen on tree trunks, where the adults feed on small, soft-bodied arthropods. Some (not all) species lay eggs in beetle borings, where the larvae can prey on wood-boring insects. This genus of about 150 species occurs worldwide; the female and male illustrated here are from eastern Canada.

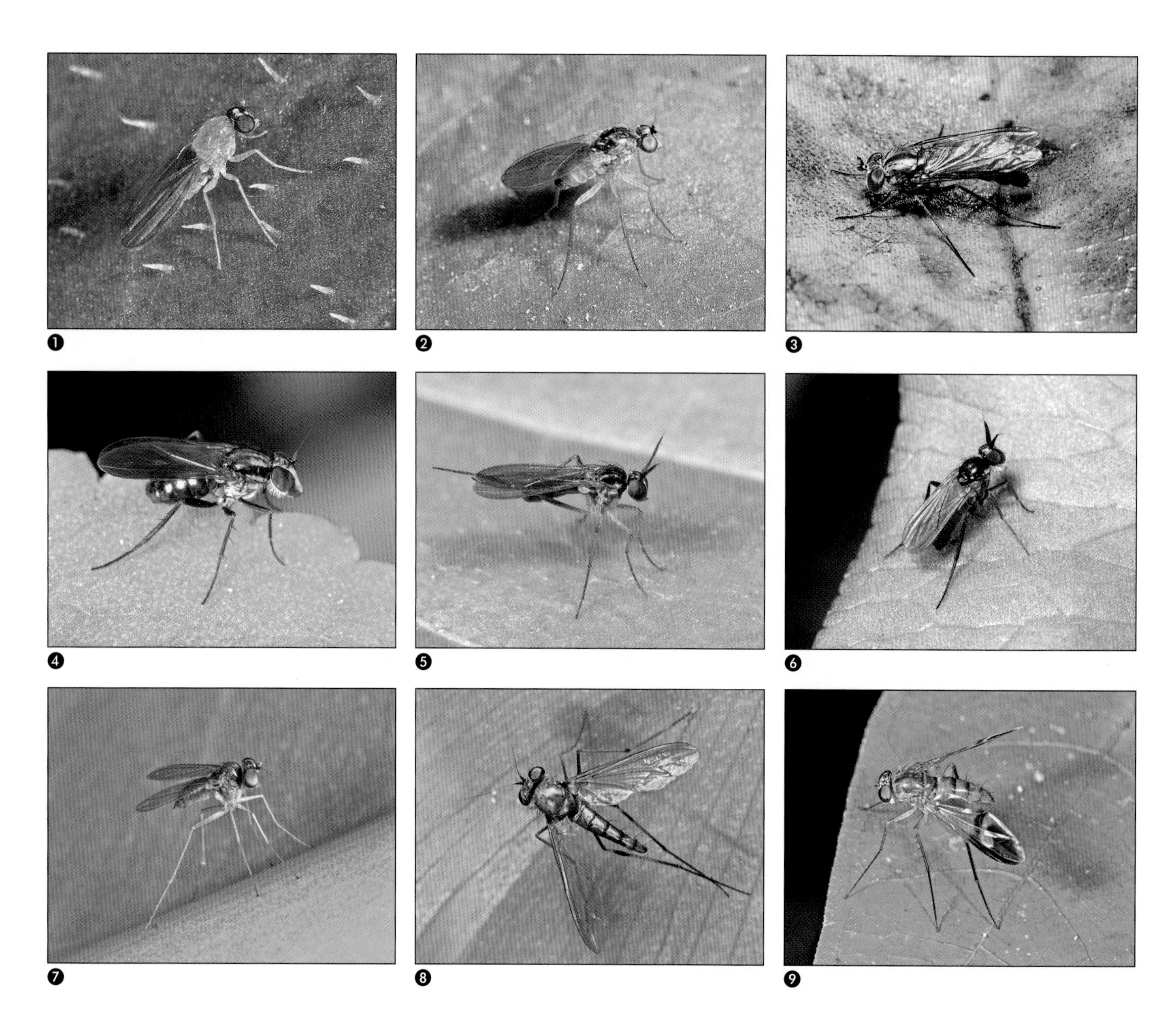

FAMILY DOLICHOPODIDAE (continued) Subfamily Peloropeodinae ❶ ❷ The subfamily Peloropeodinae is a small, somewhat obscure subfamily that includes genera sometimes treated as part of the subfamily Sympycninae. These two tiny flies (an orange fly from Costa Rica and a green and orange Chilean species) are in the genus ***Chrysotimus***, a widely distributed genus of 67 species. **Subfamily Plagioneurinae ❸** The Plagioneurinae include only a single genus and only the single species ***Plagioneurus univittatus***, which ranges through much of the New World from Uruguay north to Canada. **Subfamily Rhaphiinae ❹** ***Rhaphium*** is a mostly Holarctic genus with almost 200 species, of which only a few occur in the tropics. This female is from eastern Canada. **❺ ❻** ***Rhaphium*** species are sexually dimorphic, and males are unmistakable for their broad, blade-like antennae. This slender male (grooming) is from Africa; the other one is from eastern North America. **Subfamily Sciapodinae ❼** This tiny Vietnamese dolichopodid belongs to the very large and widespread genus ***Amblypsilopus***. Most of the 275 named species in this genus are Old World tropical or Pacific, but the genus occurs in all zoogeographic regions. **❽** ***Chrysosoma*** does not occur in the New World, but it abounds almost everywhere else. This African species is one of more than 200 species in the genus. *Chrysosoma* and its similar, mostly New World counterpart *Condylostylus* both have antennae that seem to have an arista, like higher Diptera, but the arista-like apical stylus is not homologous with the arista. **❾** *Condylostylus* includes more than 260 mostly New World species and ranges from northern Canadian city parks all the way to the Bolivian jungles, where this ***Condylostylus caesar*** was photographed.

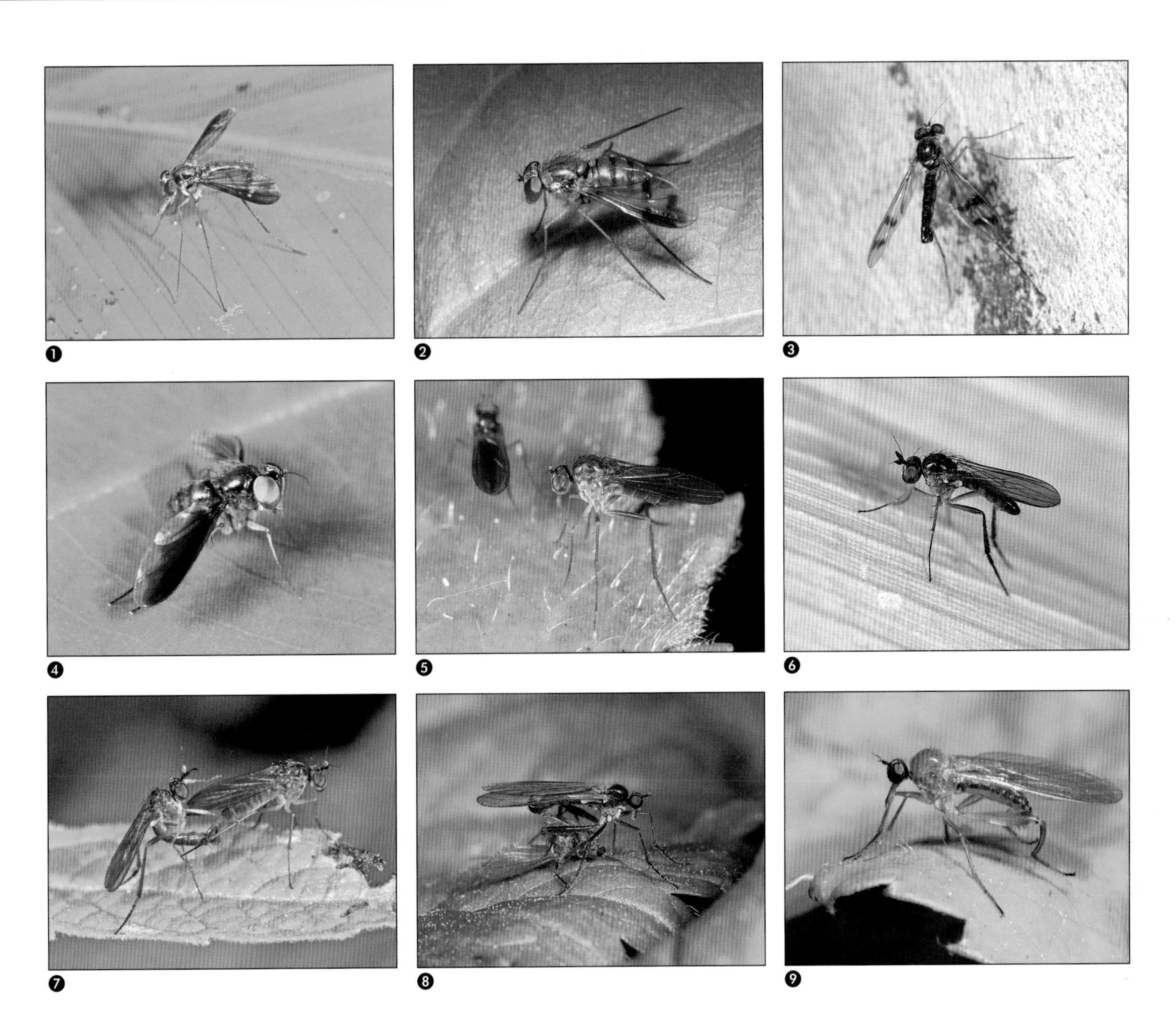

FAMILY DOLICHOPODIDAE (continued) ❶ ❷ ***Condylostylus*** is the biggest genus of Sciapodines, and one of the most common flies on foliage in parks, gardens and forests throughout the New World (although the genus occurs in the Oriental and Afrotropical regions too). Shown here are species from Ecuador (eating a small arthropod) and Canada. ❸ ***Heteropsilopus*** is an Australasian-Oriental genus often found on tree trunks. ❹ This colorful Vietnamese dolichopodid belongs to the large Australasian-Oriental genus ***Krakatauia***, originally described partly on the basis of specimens from the famous Indonesian volcanic island of Krakatoa. **Subfamily Sympycninae** ❺ ❻ ***Sympycnus*** is a large genus found in every biogeographic region. One of these flies is standing on a leaf over a stream in central Chile; the other is on a blade of dry grass in the American southwest. ❼ ***Calyxochaetus*** is a New World genus of 28 species, most of which, such as this eastern Canadian species, occur in the Nearctic region. **FAMILY EMPIDIDAE (EMPIDID DANCE FLIES) Subfamily Empidinae: Tribe Hilarini** ❽ ❾ Males in the Neotropical genus ***Aplomera*** use their stout hind legs to sweep up other flies and similar prey, but females (like most or all Empidinae) are unable to hunt for themselves. These two species are from southern Chile.

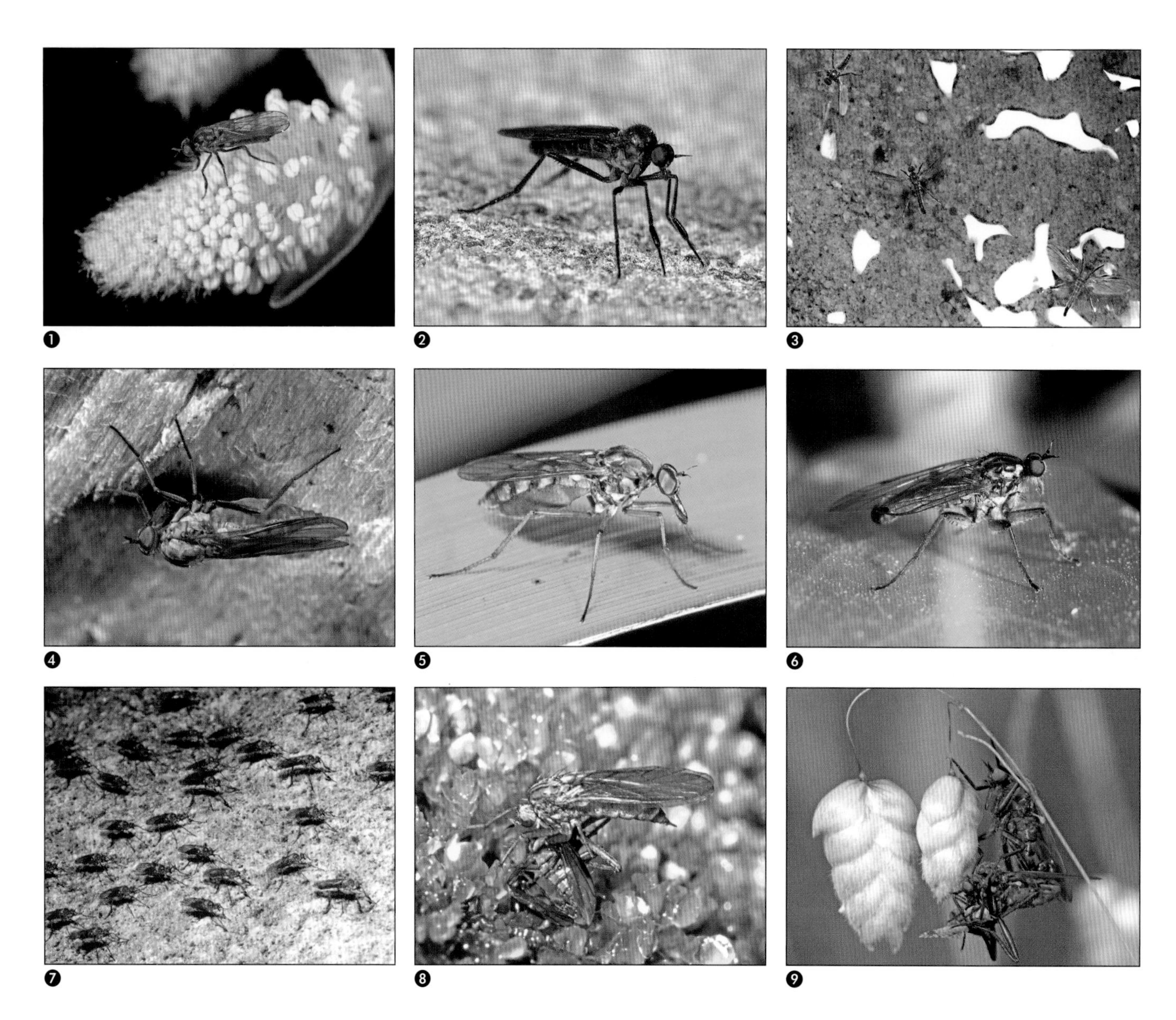

FAMILY EMPIDIDAE (continued) ❶ ❷ Males of ***Hilara*** and related genera have remarkable silk-producing structures in the swollen first tarsomere of their front legs. This male on a willow flower is from Canada; the one on a streamside rock is from Chile. ❸ Males of ***Hilara*** hunt small prey, usually flying a swift pattern millimeters over the surface of the water. Tarsal silk is used to wrap the prey as a nuptial gift, and sometimes the silk alone constitutes the nuptial gift. ❹ Females of ***Hilara*** do not hunt for themselves, and they can often be found in large numbers underneath overhangs near water surfaces patrolled by the more conspicuous hunting males. This was one of dozens under a fallen log over a Canadian stream. ❺ The Australian empid fauna is enormously diverse and still incompletely known. This female fly from eastern Australia, for example, is an undescribed species in an undescribed genus (due to be named soon, though, by Australian empidoid specialist Daniel Bickel). ❻ ❼ ***Hilarempis*** and some related Hilarini are usually found on rocks or vegetation in or near flowing water. This group is particularly diverse in the south temperate region, especially New Zealand, but this male *Hilarempis* is from south Chile. The mixed group of males and females is on a rock in a New Zealand stream. **Tribe Empidini** ❽ Although most empidine females are unable to hunt for themselves, this ***Empidadelpha*** female from New Zealand is obviously an exception. *Empidadelpha* is a small group with an austral distribution (three species in New Zealand and one in Chile). ❾ Most empidine females, including all ***Empis*** species, are unable to hunt for themselves and rely on nuptial gifts from courting males. The female of this California couple is engaged with her nuptial gift (a tephritid fly) while mating with a male hanging from the stem above.

FAMILY EMPIDIDAE (continued) ❶ *Empis* is an enormous worldwide genus of around 500 species, with some species common on flowers almost everywhere. ❷ ❸ These two species, the orange-collared ***Empis fulvicollis*** and the long-legged ***Empis liberalis***, are from Chile. ❹ ❺ Empididae are generally rare in the lowland tropics, but ***Macrostomus*** is a characteristic empid of Neotropical rainforests. One of these images shows a Bolivian species feeding at an extrafloral nectary on an *Inga* leaf; the other is from the Ecuadorian Amazon. ❻ ❼ ***Porphyrochroa*** is a Neotropical genus most characteristic of higher-elevation cloud forests, where swarms of these little metallic flies can often be seen glinting in shafts of sunlight in the forest. These are from Ecuador. ❽ ❾ ***Rhamphomyia longicauda*** is a common Nearctic species that forms massive crepuscular swarms of elaborately adorned females. Swarming females such as this one inflate their abdomens to look as fecund as possible, and to increase the chances of being chosen by prey-bearing males arriving at the swarm. This male has captured a small caddisfly.

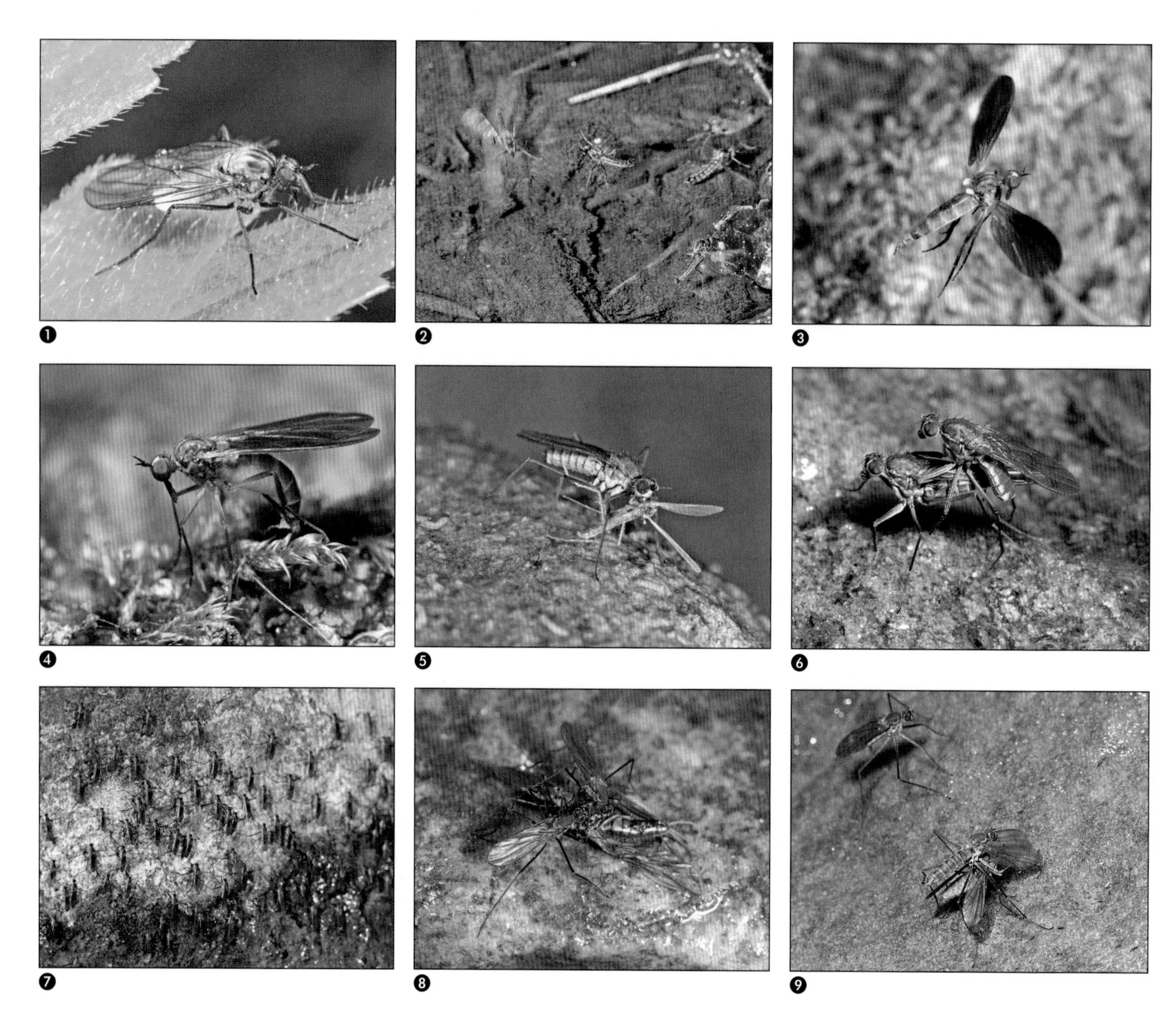

FAMILY EMPIDIDAE (continued) ❶ ❷ Males of some *Rhamphomyia* (in the subgenus ***Megacyttarus***) form swarms close to the surface of flowing water, much like males in the tribe Hilarini. Shown here are a female and some swarming males of a *Megacyttarus* from western North America. ❸ ❹ ***Rhamphomyia*** is an enormous, mostly Holarctic group closely related to *Empis*. This female is hovering over, and then ovipositing in, a moss- and lichen-covered rotting log in a Canadian wetland. **Subfamily Clinocerinae** ❺ Clinocerine adults, such as this ***Trichoclinocera*** eating a chironomid midge, are common predators around running waters throughout the Holarctic region, occupying more or less the same niche held by the brachystomatid subfamily Ceratomerinae in Australia and New Zealand. Larvae are predators too, eating other aquatic insect larvae. ❻ Clinocerines don't bother with the elaborate courtships found in the Empidinae. The male of this pair of ***Clinocera binotata*** performed swiftly, without conspicuous male displays or other precopulatory behavior. ❼ Clinocerine adults often cluster on rocks near the waterline. This vertical rock in the middle of an eastern North American river was massed by thousands of ***Trichoclinocera***. ❽ ***Trichoclinocera*** species differ from *Clinocera* in having short bristles along one of the wing veins (vein R). These adults seem to have crossed the line between predation and scavenging as they gang up on a dead mayfly subimago. ❾ This male ***Trichoclinocera*** seems to be exhibiting ritualized mate-guarding behavior on a wet rock in an eastern Canadian stream, straddling the female and flexing his wings over her head, although mate guarding has not been documented in this group.

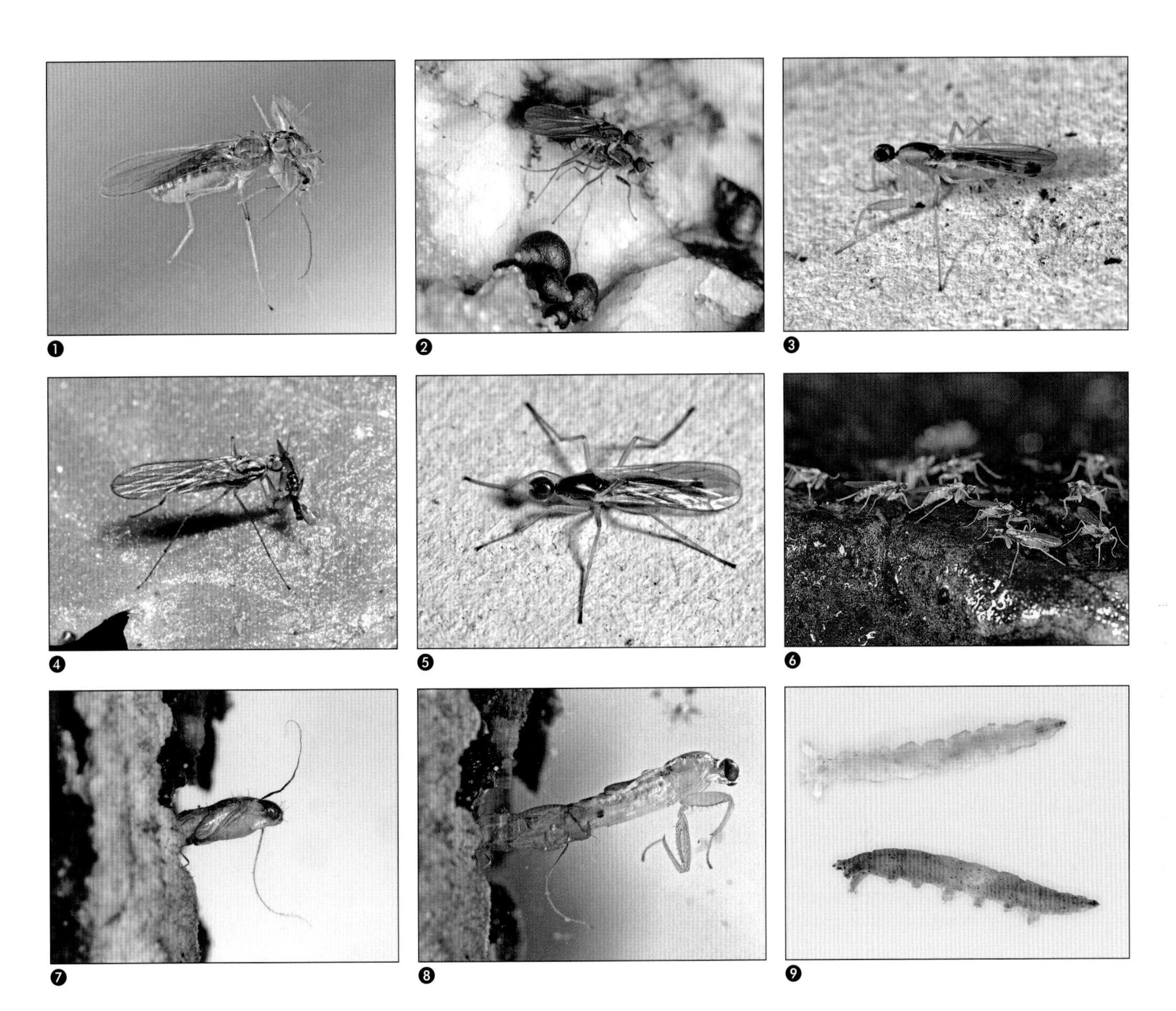

FAMILY EMPIDIDAE (continued) Subfamily Hemerodromiinae ❶ ***Chelipoda*** species, such as this Costa Rican fly eating a chironomid midge, have a long, thin, arista-like style at the end of the antennae, unlike similar genera. Like the similar genera *Chelifera* and *Hemerodromia*, adult flies grab prey with their front legs and larvae are aquatic predators, mostly in streams. ❷ Although most ***Chelipoda*** species are associated with fresh water, this Peruvian species is intertidal. ❸ ***Chelifera*** species are generally associated with small headwater streams, where their predaceous larvae live concealed in debris. This adult male is resting under a bridge over a small Canadian stream. ❹ ***Cladodromia*** has an interesting disjunct distribution, with about half a dozen species, including this one, in New Zealand and the rest of the genus in southern South America. ❺ This ***Hemerodromia*** male is resting under a concrete bridge over a small Canadian stream. ❻ Although most ***Hemerodromia*** species are associated with streams and rivers, a few species develop in lakes. These females were among hundreds laying eggs on wave-splashed rocks along the Canadian shores of Lake Huron. ❼ ❽ ***Hemerodromia*** pupae can often be found projecting from the submerged woody debris in which the associated larvae developed. This adult is emerging under water before floating to the surface. ❾ Larvae of ***Hemerodromia*** are common aquatic predators, distinctive for their seven abdominal prolegs and divided posterior lobes.

FAMILY EMPIDIDAE (continued) ❶ ❷ ***Neoplasta***, like the other hemerodromine genera illustrated here, is a widespread group associated with clean streams. One of the species shown here was preying on other insects attracted to night lights on a white concrete wall in Costa Rica; the other is on foliage over a Canadian stream. **Subfamily Oreogetoninae and unplaced empidoid genera** ❸ ❹ ***Hormopeza*** is the "smoke dance fly" genus, normally seen only around smoky fires. The four males seen here encircling one female were among many courting on a sizzling-hot rock bordering a northern Canadian campfire, and this female from the same spot is eating what appears to be a *Microsania* (smoke fly, Platypezidae). *Hormopeza* is currently treated as "Empididae of uncertain placement," as it cannot be placed in any of the established subfamilies; larvae remain unknown. ❺ ***Philetus***, with two rarely collected western North American species is currently treated as of "uncertain placement," without obvious affinity to the established subfamilies. ❻ ❼ The Holarctic genus ***Iteaphila*** and the more widespread ***Anthepiscopus*** do not fit well in any empidoid family, and taxonomists currently treat them together as a group of "uncertain placement." They are often common on flowers (like the similarly small hybotids in the genus *Anthalia*); this *Iteaphila* is on a strawberry blossom in the southeastern United States, while this *Anthepiscopus* is underneath a flower on the west coast of the United States. Both of these genera lack the epipharyngeal blades that most other empidoids use for feeding on other insects. ❽ ❾ The Holarctic genus ***Oreogeton*** does not fit well in any empidoid family, and most empidoid specialists currently treat it as of "uncertain placement." Some dipterists now recognize a separate family, **Oreogetonidae**, for *Oreogeton*. Larvae are predators in streams, and both of these flies were photographed near streams in northern California.

FAMILY HYBOTIDAE (HYBOTID DANCE FLIES) Subfamily Oedaleinae ❶ ❷ ***Anthalia***, like these on a yellow buttercup flower and on white wood sorrel flowers on the American west coast, are small pollen-eating flies. **Subfamily Ocydromiinae ❸** This minute ***Chvalaea*** repeatedly returned to the same twig-tip perch in a high-altitude Ecuadorian forest, much like the hunting strategy of some Asilidae. *Chvalaea* was first described for a single European species, but it is now recognized as a widespread group found in the Australasian, Neotropical, Oriental and Palaearctic regions. This is probably an undescribed species, as the genus is unknown from Ecuador. **❹ ❺** ***Leptopeza*** is a small Holarctic genus of delicate dance flies that includes a couple of northern species occurring in both Canada and Europe. This male, on a leaf in California, is ***L. disparilis***, and the female on the white background is ***L. borealis***. **❻** ***Ocydromia*** species, like this ***O. glabricula*** from eastern Canada, deposit living larvae in dung. As with most other hybotids, both adults and larvae are predators. **Subfamily Hybotinae ❼** ***Bicellaria*** species, with their slender legs, humped back and small beak, are common predators of minute insects such as gall midges. This is an eastern North American species, but the genus is widespread. **❽** ***Hoplocyrtoma*** is a small genus of four species: two in Japan, one in western North America and this one, ***H. femorata***, from eastern North America. **❾** ***Hybos*** is a large worldwide genus characterized by a stout, forward-facing beak and somewhat swollen and hairy hind legs. This is a pair of the North American species ***H. reversus***.

FAMILY HYBOTIDAE (continued) ❶ ❷ ❸ ***Syneches*** species are strongly humpbacked flies commonly attracted to lights. Two of these, a Canadian fly with strongly marked wings and a Costa Rican species with yellow legs, were feeding on other small flies around lights at night; the third was on a leaf in a Philippines garden. **Subfamily Tachydromiinae** ❹ This ***Crossopalpus*** was hunting on the open sand of a beach in the city of Dar es Salaam, Tanzania, a habitat more typical of the related genus *Chersodromia*. *Crossopalpus* is a worldwide genus that was until recently treated as a subgenus of *Drapetis*, a similar genus of stout little tachydromiines. ❺ ❻ ***Chersodromia*** species are tiny, frenetic flies that run incessantly over the sand and debris on North American, European and African seacoasts. Even the winged ones, such as this eastern North American ***Ch. inusitata*** eating a canacid, are reluctant to fly. Some, such as this short-winged western American species, are unable to fly. ❼ ❽ ❾ Species of the worldwide genus ***Drapetis*** are robust but tiny stout-beaked flies often found hunting other flies on forest floors. However, one of the two Costa Rican species shown here rides on leaf fragments carried by leaf-cutting ants (*Atta*). I've speculated that it benefits the ants by eating ant-decapitating flies (Phoridae), but this remains an undescribed species and larvae remain unknown.

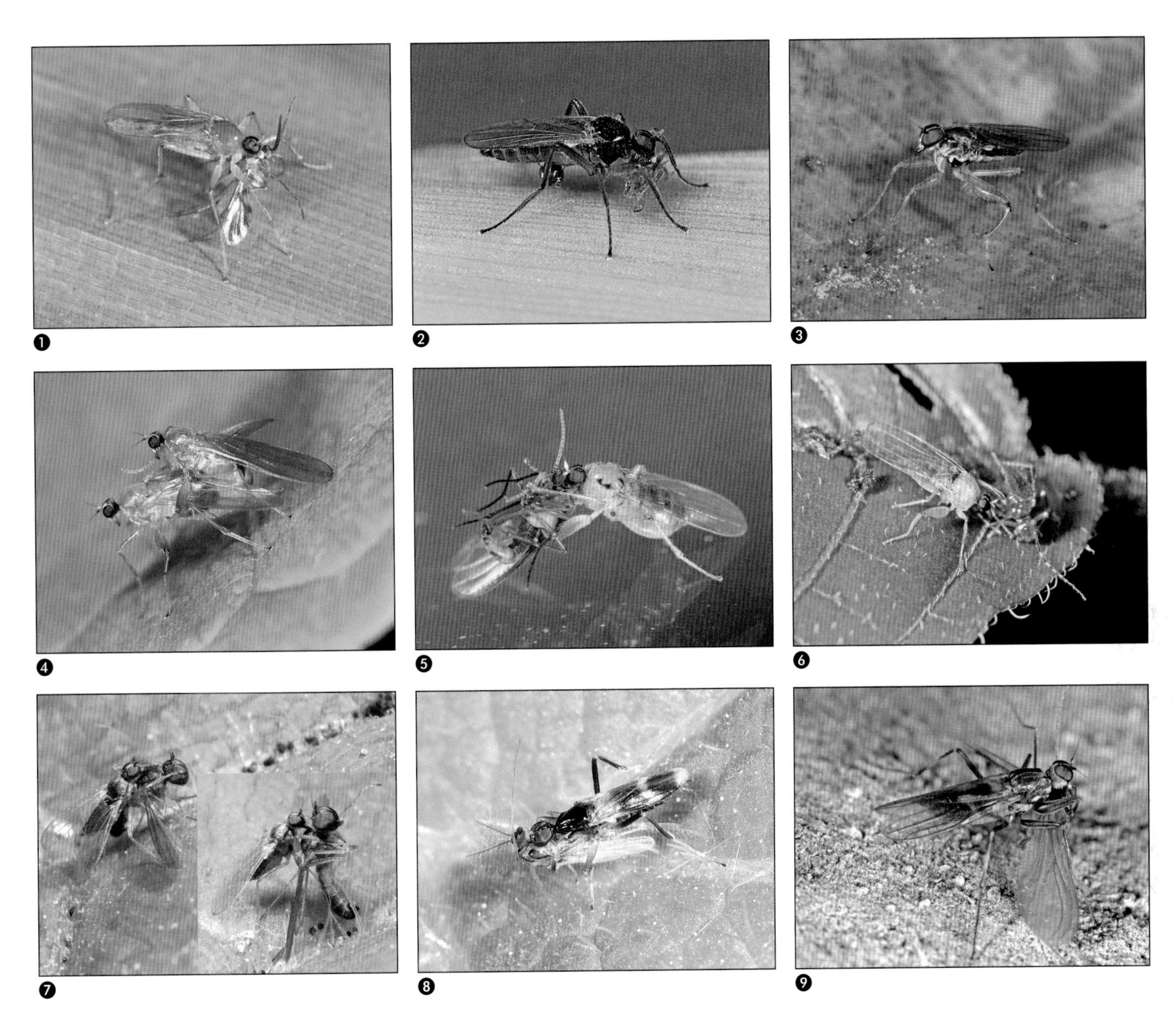

FAMILY HYBOTIDAE (continued) ❶ ❷ ***Elaphropeza*** is a circumtropical genus of about 160 described species, often seen on leaves and in grass; a few species also occur in the north temperate countries. This pale African species has captured a small nematoceran fly; this black Bolivian species seems to have impaled a mite. ❸ ❹ ***Platypalpus*** is an enormous (hundreds of species) and common genus, with several species that abound on low foliage in parks and gardens. Although most named species are north temperate, the genus occurs worldwide. The single individual here is from the Philippines and the mating pair is from Canada. ❺ ❻ ***Platypalpus*** species, such as these from Costa Rica, grab their prey with swollen middle legs. ❼ ***Platypalpus*** species are often abundant garden predators. This one is approaching a small long-legged fly (Dolichopodidae) from behind and then impaling it through the soft tissue behind the head. ❽ This North American ***Tachydromia*** has impaled a small midge (Chironomidae). The tiny flies in this large worldwide genus are common swift-running predators on tree trunks, foliage and rocks; they are very reluctant to fly. ❾ Like the superficially similar *Tachydromia*, species in the widespread genus ***Tachypeza*** are fast-running predators that are reluctant to leave the tree trunks or rocks on which they hunt. This western North American species has impaled a nematoceran fly.

FAMILY BRACHYSTOMATIDAE (BRACHYSTOMATID FLIES) Subfamily Brachystomatinae ❶ This eastern Nearctic ***Brachystoma (Blepharoprocta) serrulatum*** has captured a small nematoceran fly. *Brachystoma* is the largest and most widely distributed of the three genera in this small, mostly north temperate subfamily. **Subfamily Trichopezinae ❷ *Heterophlebus*** is a small genus of southern South American flies, one of only a dozen or so genera in the heterogeneous subfamily Trichopezinae. Nothing is known about the biology of *Heterophlebus*, but this one is flying over a cold Chilean stream. **❸ *Apalocnemis*** is a Neotropical group of about 30 species, mostly from South America, although this male is in a high cloud forest in Costa Rica. **Subfamily Ceratomerinae ❹ ❺** Members of the small southern hemisphere subfamily Ceratomerinae have a distinctively elongate antennal base and long, narrow wings. ***Ceratomerus*** has a classic "trans-Antarctic" distribution, with most species occurring along cool streams in wet forests of Australia, New Zealand and southern South America. The other two genera in the subfamily are similarly distributed. These two are seen on a stream margin in Australia and a flower in New Zealand. **❻ *Ceratomerus apterus***, the only known member of the Brachystomatidae with reduced wings, is known only from one high, cold Ecuadorian *páramo*. This is one of the type specimens, on which the description and definition of this species are based. **FAMILY ATELESTIDAE (ATELESTID FLIES) ❼** This ***Acarteroptera licina***, one of two species in this exclusively Chilean genus, is using its characteristically long beak to feed at a flower just north of Santiago, Chile. **FAMILY APYSTOMYIIDAE ❽ *Apystomyia elinguis***, from southern California, is the only member of the Apystomyiidae and is possibly the sister group to the rest of the Cyclorrhapha. **FAMILY HILARIMORPHIDAE ❾** Species in the obscure Holarctic-Oriental family Hilarimorphidae can apparently sometimes be netted in numbers from willow foliage along narrow streams in spring, but the group remains rarely collected and (to my knowledge) never photographed in the wild. This is ***Hilarimorpha mandana*** from the western United States.

8

The Higher Brachycera or Cyclorrhapha

THE GREAT MAJORITY OF FLY FAMILIES, MOST FLY genera and almost half of all named fly species belong to the Cyclorrhapha, the biggest subgroup of the suborder Brachycera. This enormous group of flies is characterized by the habit of pupating inside the shelter of a seedlike structure, called a puparium, derived from the inflated and hardened skin of a mature (third-instar) larva. Adult cyclorrhaphan flies ultimately escape from the puparium by popping off an end cap, which is separated from the rest of the puparium by a circular line of weakness, or seam — the source of this group's name: *cyclor* means "circular" and *rhapha* means "seam." Most Cyclorrhapha, including House Flies, fruit flies, blow flies and tens of thousands of other kinds of flies, facilitate their emergence from the puparium by opening a fissure at the front of the head and pumping out a relatively large, spiny, airbag-like structure — the ptilinum — to pop off the cap.

All flies that use a ptilinum to push out of the puparium belong to the **Schizophora**, a group descriptively named for the conspicuous fissure or "schism" that arches over the antennae like a massive scar, where the deflated ptilinum was withdrawn into the head of the adult fly. A few basal lineages of Cyclorrhapha that predate the origin of the ptilinum — and thus lack the ptilinal fissure — are usually treated together as the "**Aschiza**," even though they probably don't form a natural (monophyletic) group. A very few Aschiza (some African Phoridae) have a ptilinum-like structure that is not homologous with the ptilinum of the Schizophora, and, conversely, some Schizophora (genus *Sepedon*, family Sciomyzidae) have lost the ptilinum, instead popping open the puparium by using the proboscis, the lower parts of the face and the antennae. These are exceptional cases, however; we can generally view the ptilinal fissure as a reliable confirmation of membership in the Schizophora.

The Schizophora (that is, all Cyclorrhapha except for those few basal lineages that lack a ptilinal fissure) form a natural group that can be conveniently divided into two different-looking subgroups, the calyptrates and the acalyptrates. **Calyptrates** are generally big, bristly flies characterized by a large flap of membrane, or calypter, under the base of the wing and a distinctive "button" — a small, round swelling called the greater ampulla — on the side of the thorax. House Flies and blow flies are typical calyptrates. The **acalyptrates**, on the other hand, make up a phenomenally diverse and probably artificial

OPPOSITE PAGE These Andean flies belong to widespread genera (*Allograpta* above and *Toxomerus* below) in the largest family of Aschiza, the Syrphidae or flower flies.

Figure 8.1 The Main Subgroups of Higher Brachycera, or Eremoneur

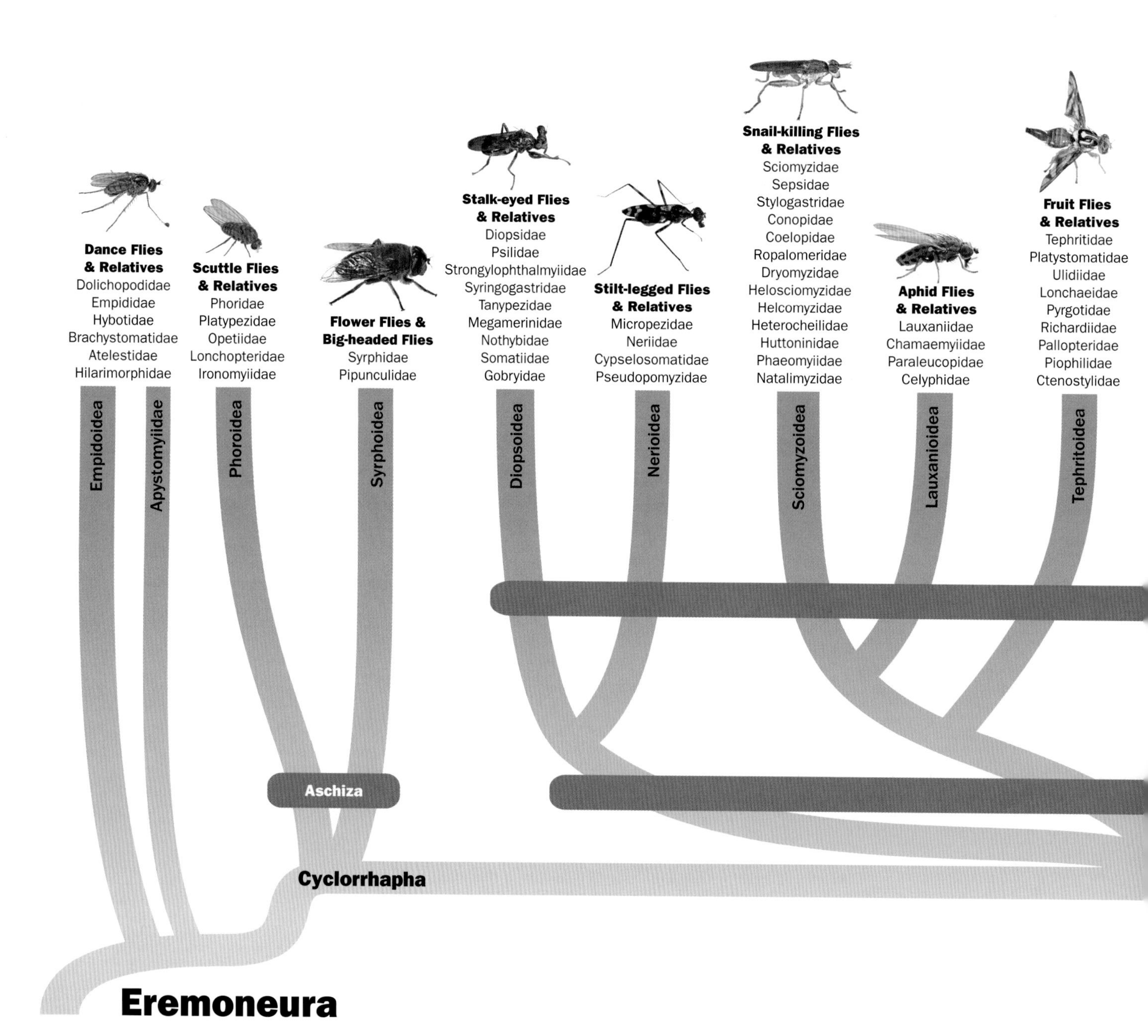

(non-monophyletic) group of generally smaller flies with small, inconspicuous calypters and usually without a greater ampulla.

The Cyclorrhapha is divided into the widely used and easily recognized groups Aschiza, Calyptratae and Acalyptratae, even though of these only the Calyptratae represent a well-defined monophyletic group. These three groups are used below to organize our overview of the Cyclorrhapha into three parts.

The Aschiza

The families of Aschiza form a heterogeneous cluster of families that can be conveniently partitioned into two groups: the flower fly group of families (superfamily Syrphoidea) and the scuttle fly group of families (superfamily Phoroidea). However, the higher classification of Aschiza is highly contentious, and each of these groups is

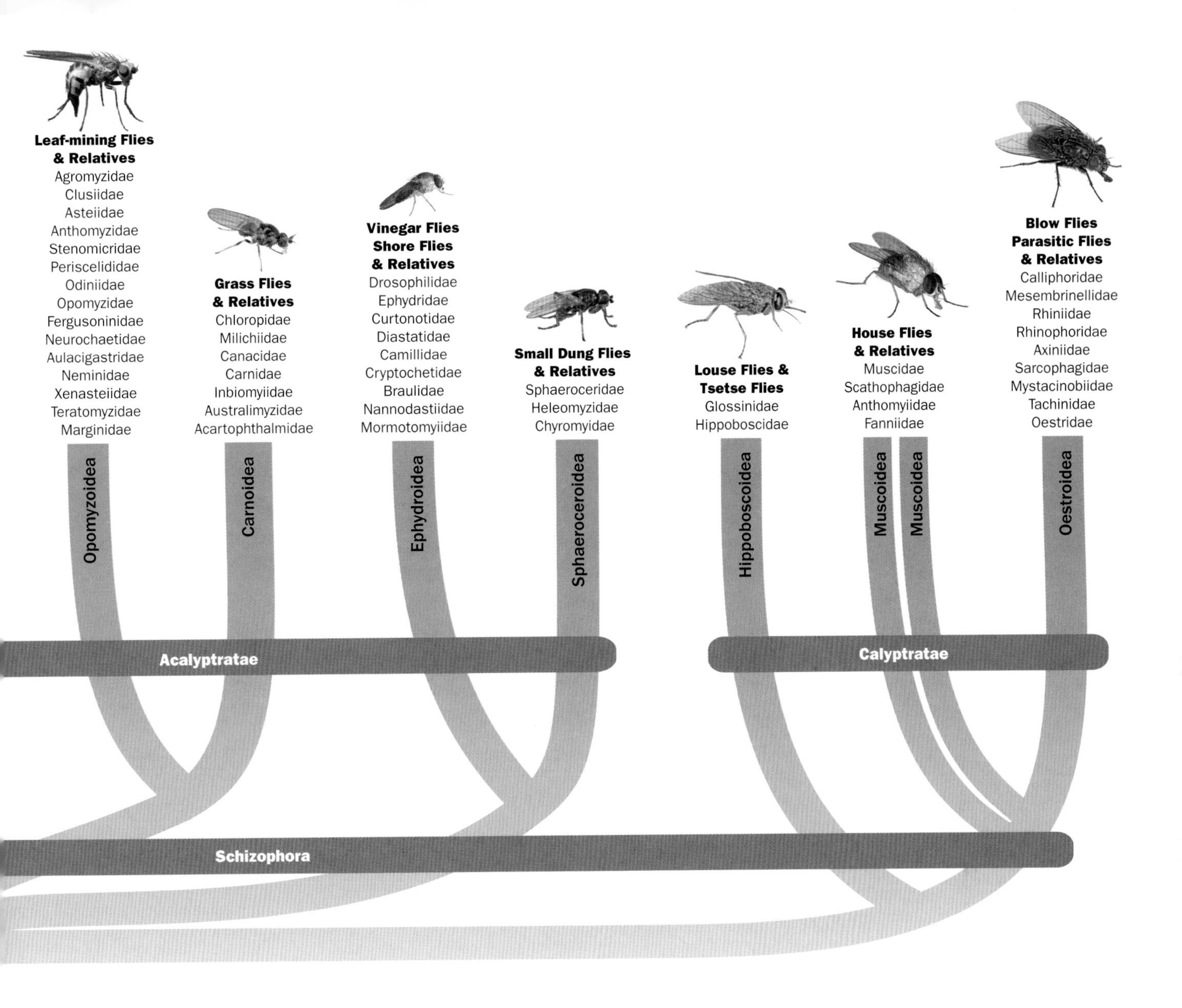

considered artificial by some authors and natural by others. Many dipterists treat the Platypezidae and Opetiidae as basal lineages outside the above superfamilies, while others treat them as part of the superfamily Phoroidea (sometimes referred to as the superfamily Platypezoidea). The superfamily Syrphoidea has long been considered a natural group made up of the two similar families Syrphidae and Pipunculidae, but some recent research suggests that the Pipunculidae, or "big-headed flies," are more closely related to the Schizophora than to Syrphidae.

Although the Aschiza as a whole is difficult both to characterize and to organize into a meaningful higher classification, most of the included families are highly distinctive and easily recognized. They are grouped here into the two superfamilies Phoroidea (Platypezidae, Opetiidae, Lonchopteridae, Phoridae and Ironomyiidae) and Syrphoidea (Pipunculidae and Syrphidae).

Phoroidea

Flat-footed Flies, Enigmatic Opetiids, Spear-winged Flies, Ironic Flies and Scuttle Flies

Platypezidae

Flat-footed Flies

The Platypezidae are called "flat-footed flies" because females of most species and males of many have flattened and sometimes spectacularly ornamented hind tarsi. These small (usually 2–5 mm, rarely up to 10 mm) flies develop in living fungi but are most often seen as frenetic adults scooting around leaf surfaces with a characteristic jerking motion. Males, which form aerial swarms, have conspicuously enlarged and contiguous eyes, unlike the small and separated eyes of females. Large aerial male swarms of one subfamily, Microsaniinae (smoke flies), are routinely seen in plumes of smoke from dying campfires, while other Platypezidae form relatively small male swarms unassociated with smoke. Most of the 250 or so species in this small family are in the relatively common and diverse subfamilies Platypezinae and Callomyiinae; the other two subfamilies (smoke flies and stinkhorn flies) together make up a smaller lineage.

The Old World genus *Opetia*, formerly treated as part of the Platypezidae, is now widely treated as the separate family Opetiidae. The Platypezidae and the Opetiidae are considered by many to be the most "primitive" Cyclorrhapha, but there are conflicting hypotheses about the relationships of these so-called basal cyclorrhaphan lineages. Recent studies place them close to the Phoridae and Ironomyiidae.

Melanderomyiinae and Microsaniinae The subfamily **Melanderomyiinae** as currently defined includes only one species, the North American *Melanderomyia kahli* (Stinkhorn Fly). Stinkhorn Flies are rarely collected but can be common on their only known hosts, stinkhorn mushrooms in the family Phallaceae. The closely related subfamily **Microsaniinae** is much more widespread, being found throughout the world, but paradoxically much less understood. *Microsania*, the only genus in the subfamily, includes about 25 species known as "smoke flies." Although it is not hard to attract a swarm of these little flies to a smoky fire of wood or plant debris, little else is known about the group: they have never been reared or found as larvae. It seems a reasonable guess that *Microsania* find their mates in smoky swarms and then lay eggs in the fungi that appear after forest fires, since all Platypezidae with known larvae develop in fungi. *Microsania* adults are never seen running around the surfaces of leaves like other Platypezidae; in fact they are almost never seen at all, except when swarming in smoke. Swarms of *Microsania* are often accompanied by an associated dance fly, *Hormopeza* (Empididae), that seems to be a specialized predator of smoke flies. Like *Microsania*, the smoke dance flies are rarely seen except when they appear in plumes of smoke.

Platypezinae and Callomyiinae The subfamilies Platypezinae and Callomyiinae occur throughout most of the world, although neither subfamily occurs in New Zealand (where the only platypezid is a *Microsania* species). Members of both subfamilies are fungivorous flies normally found in woodlands, where they frequent sun-dappled foliage on which they presumably scavenge for honeydew. Males form small swarms as an essential precursor to mating, which usually takes place on nearby foliage once a female has come into the swarm and coupled with a male. Females use a telescoping ovipositor to insert eggs between the gills or in the pores of living fungi, and each species usually develops only in a particular species or genus of fungus, although some species of fungus are hosts to several platypezid species.

According to Chandler's (2001) definitive treatment of the family, members of the subfamily **Callomyiinae** are probably confined to polypore and encrusting fungi. The nominate genus of the subfamily, the Holarctic genus *Callomyia*, feeds on the moist surfaces of fungus-encrusted substrates (usually rotting logs), but species of the more widespread calomyiine genus *Agathomyia* and the two named species of *Bertamyia* develop inside polypore fungi. *Bertamyia notata* is a stunning silver and black species found throughout the New World from Argentina north to Canada; the only

Male Platypezidae, like this *Lindneromyia*, have conspicuously enlarged and contiguous eyes, unlike the small and separated eyes of the females.

other species currently in the genus is known only from South Africa. Some members of the subfamily **Platypezinae** also feed in or on polypores, but most occur in soft fungi, especially gill fungi. The most widespread platypezine is the worldwide genus *Lindneromyia*, species of which develop in a wide variety of soft fungi in both tropical and temperate forests.

Opetiidae

Opetiid Flies

The family Opetiidae is one of the smallest (one genus, five species) and most narrowly distributed (Palaearctic only) fly families, but it is of special interest because some dipterists have suggested that it is the sister group to all the rest of the Cyclorrhapha. These rather nondescript small (about 3 mm), dark flies resemble primitive Platypezidae in the genera *Microsania* and *Melanderomyia*. However, antennal structure — only a single basal aristomere as opposed to two in the otherwise similar Platypezidae — simple legs, unusual genitalia and other features argue for treatment of this group outside the Platypezidae. One species, *Opetia nigra*, is fairly common in Europe but larvae remain undescribed, and larval habitat remains unknown for this and other species in the family. *Opetia nigra* has been reared from rotting birch, but the larvae were not noted prior to emergence and the exact microhabitat remains a matter of speculation. *Opetia* adults have been collected at lights, in pitfall traps and — unlike similar Platypezidae — at flowers.

Lonchopteridae

Spear-winged Flies

Lonchopteridae, called "spear-winged flies" because of their somewhat tapered and pointed wings, form a small, mostly north temperate lineage of Aschiza. All 60 or so world species of spear-winged flies are superficially similar; they usually occur in moist meadows or wet grasses, although a couple of European species are aquatic or semi-aquatic in or among springs, seeps or wet rocks on

Lonchoptera bifurcata is the common spear-winged fly of North America and much of the world, where it thrives in disturbed habitats such as ditches. This facultatively parthenogenetic species normally occurs as female-only populations.

both freshwater and marine shorelines. Larvae are little-known but seem to have the default maggot habit of consuming decaying organic matter and associated microorganisms. Adults are occasional consumers of pollen and nectar, but they are probably more often opportunistic scavengers. I've seen lonchopterids eating other insects, but it seems unlikely that they hunt active live prey.

There is no doubt that the Lonchopteridae form a natural group; in fact all members of the family share so many attributes they are usually treated as a single genus, *Lonchoptera*. The relationships of the spear-winged flies to other Cyclorrhapha, on the other hand, are fuzzy to the point where different authors have given this small family almost every possible phylogenetic position within the Aschiza. Lonchopteridae have been traditionally treated as the basal lineage of Cyclorrhapha (the sister group to the rest of the Cyclorrhapha), but most dipterists now put the spear-winged flies in the superfamily Phoroidea, as the sister group to the Phoridae plus Ironomyiidae. Some recently published phylogenies based on molecular characters place the Lonchopteridae as sister group to either the Opetiidae or to a larger group comprising the Opetiidae, Phoridae, Ironomyiidae and Platypezidae. Yet another recent molecular study concludes that the spear-winged flies are the sister group to the Phoridae.

The family Lonchopteridae is absent from the Neotropical and Australasian regions except for the widely distributed *Lonchoptera bifurcata*. Probably European in origin, *L. bifurcata* is the common spear-winged fly of North America and much of the world, including Australia and New Zealand, where it thrives in disturbed habitats such as ditches and vegetable gardens (larvae have been found in wet material between the leaves of cultivated cabbages). Such synanthropic habits undoubtedly contributed to this species' widespread distribution, but another factor was probably its propensity to reproduce by parthenogenesis (without males). Males of *L. bifurcata* appear only very rarely, in contrast to other species of spear-winged flies, which routinely occur as both males and females.

The other five species of Lonchopteridae found in North America, including three species only recently discovered and described (Klymko and Marshall 2008), are generally uncommon flies associated with undisturbed native habitats, especially in northern or montane areas. Two of the North American species appear to be naturally Holarctic, occurring in northeast Asia as well as the far northwest of North America. Many of the named species in the family occur in the Palaearctic region, with about 15 known from China and a few more recently described from Africa. Numerous Oriental and African species undoubtedly remain to be described.

Ironomyiidae

Ironic Flies

The name *Ironomyia*, first applied to this group of flies by Arthur White in a 1916 paper on the Brachycera of Tasmania, combines the Greek word for fly (*myia*) with the Latin word *ironia*, which refers to dissimulation: "to say one thing while meaning the opposite." It is hard to discern what is ironic about these superficially platypezid-like flies, but their distribution could be described as enigmatic. The three extant species, all in the genus *Ironomyia*, are entirely Australian, found only in eastern Australia and Tasmania (D.K. McAlpine 2008), but the 15 described fossil species (in several genera) are known only from the northern hemisphere (J.F. McAlpine 1967). The biology of the family is poorly known, though adults are not uncommon in eucalyptus forest and larvae have been reared from eucalyptus leaf litter. Adults can be locally abundant on low foliage and tree trunks.

Phoridae (and Sciadoceridae)

Scuttle Flies

The family Phoridae is like a biodiversity iceberg. The part of the iceberg above the taxonomic ocean is made up of about 4,000 named species found almost everywhere and doing almost everything possible for an insect to do: parasitism, kleptoparasitism and parasitoidism, saprophagy, fungivory, predation, phytophagy. The hidden part of the iceberg is the estimated 30,000 or so unnamed and unstudied species that represent an almost unimaginable diversity of habits and habitats. Morphological diversity is of course high in such a large group, made up as it is of wingless flies, winged flies, phoretic flies and flies characteristic of an immense spectrum of different habitats, but most phorids are nonetheless easily recognized as such by a typical humpbacked habitus and a characteristic jerky or scuttling motion — hence the name "scuttle flies." Wing loss is common in the group, but in phorids with wings almost all the heavy veins are squeezed into the basal front half of the wing. The small head is usually covered with large, backward-pointing bristles that give the appearance of a slicked-back hairstyle, a punky impression often enhanced by the presence of large, spiny palpi sticking out from the mouthparts like armed clubs.

Another widely used common name for the Phoridae is "coffin flies," which really refers to just one widespread species, *Conicera tibialis*. Like it or not, coffin flies are apt to find you in your final resting spot, where they will lay eggs on your nutritious cadaver, initiating a subterranean population able to spend several generations in your coffin. Some other phorids breed in meat, damaged fruit and other things that you might ingest along with the fly eggs or larvae, so it is also possible to share your body with phorids even before your demise. All life stages of phorids have been recovered from the fresh feces of individuals unlucky enough to ingest maggoty fruit, but this is unusual, as most phorids are unlikely to routinely survive ingestion. A few occasionally infest people through wounds or the eyes, but most attack other invertebrates, not vertebrates.

Although the Phoridae are aptly characterized as a "family of flies whose diversity of larval lifestyles is apparently without rival among insect families" (Disney 1994), most phorids seem to be specialized predators or parasitoids. Among the most important and interesting parasitoids in the family are the large groups known as ant-decapitating flies and bee-killing flies. Ant-attacking phorids in the genera *Apocephalus* and *Pseudacteon* are often major enemies of ants, which react to the presence of these little flies with what appears to be a panic response. Given the range of possible outcomes of attention from parasitic phorids, ants have every reason to "fear" these little flies.

One possible outcome, as its common name suggests, is that the fly's visit will result in the ant losing its head. When you see a carpenter ant head pop off for no apparent reason, the odds are that a phorid fly female previously inserted a single egg into the hapless insect, where it hatched into a tiny maggot that fed on its host's internal contents. After consuming the contents of the ant's head, the maggot produced enzymes that dissolved the connective tissue between the ant's head and thorax, effectively decapitating it. The severed ant head provides a tough shelter in which the phorid can pupate before emerging as an adult fly about three weeks later. Look for adult female ant-decapitating flies hovering around ant trails or nests, waiting for an opportunity to dart in and nail another hapless victim with a fatal egg.

Several aspects of ant behavior provide conspicuous evidence of the importance of ant-parasitizing phorids. Foraging leafcutter ants, for example, are protected from the ever-present parasitic phorids by specialized smaller worker ants that "ride shotgun" on top of the leaf pieces conspicuously carried by these ubiquitous Neotropical insects. Many other ants simply stop foraging when parasitic phorids appear on the scene. This is the case among fire ants, including the aggressive imported fire ant *Solenopsis invicta*. Imported fire ants originated in South America, but these aggressive stinging insects have now spread into the southern United States, where they outcompete native ants and represent a serious pest problem. Several species in the phorid genus *Pseudacteon* attack fire ants in Brazil, but no native phorids attack these foreign ants in North America. Entomologists are hoping that importation of ant-decapitating flies from Brazil to the United States will disrupt foraging by imported fire ants, thus giving native ants — which are not attacked by these particular ant-decapitating flies — a competitive upper hand. Imported ant-decapitating flies were released for the first time in Florida in 1997, and five species of *Pseudacteon* have since become firmly established as major natural enemies of fire ants in the southeastern United States.

The diversity of ant-associated phorids is especially high in the tropics, where many species are attracted to injured or dying ants. Females of one South American species use a strikingly long proboscis to penetrate an injured ant's anus and feed inside its digestive tract before doing an about-face to insert an egg into the ant's abdomen (Brown and Feener 1993). Larvae of these long-beaked phorids develop as parasitoids inside host ants, one larva per ant.

Dozens of the more than 200 ant-associated scuttle fly species are scavengers or predators in the nests or middens of army ants, where they are the most abundant myrmecophilous insects, often occurring by the thousands in single colonies. Some, such as the wingless females of *Aenigmatopoeus* species associated with Old World army ants, can be seen running among columns of emigrating or (sometimes) foraging ants, while others are known only from the ants' middens or, more rarely, from their bivouacs. Myrmecophilous scuttle flies are often highly specialized, with larvae or occasionally adults that are fed or groomed by their hosts. The legless, wingless adult females of the aptly named genus *Vestigipoda* are astonishingly convincing mimics of ant larvae and are fed and tended by the workers of their army ant hosts.

Ants are not the only Hymenoptera attacked: many bees (plus a few wasps) also serve as hosts for parasitic phorids. Bumble bees and Honey Bees are attacked by a few phorid species, but most bee-parasitizing phorids are specialized parasitoids of stingless bees (family Apidae, tribe Meliponini). For example, the brightly colored, relatively large (3–6 mm) phorids in the genus *Melaloncha* (bee-killing flies) are major parasitoids of bees throughout the neotropics. Brown (2005) describes some as first curling the ovipositor forward under the body so that the tip extends below the head, and then advancing to inject an egg into the doomed bee's head or thorax. Although mostly associated with native Neotropical bees,

Melaloncha species also attack Honey Bees in Central and South America. Some *Apocephalus* species attack bees in much the same way that others attack ants, and *A. borealis* is a newly recognized pest of Honey Bees in the southwestern United States. Until recently, this species had been known only as a parasitoid of bumble bees.

Some of the most specialized groups of scuttle flies are associated with termites: almost 200 species of phorid either parasitize termites or live as specialized scavengers or guests in termite nests. Among the oddest termite-parasitizing phorids are the Southeast Asian species aptly called "con flies" because of their remarkable termite-duping behavior. The females of con flies, which are atypical members of the widespread genus *Diplonevra*, deceive worker termites into following them away from the colony. The flies, attracted to breached termite nests, land near the termite workers. An arriving con fly then prods a termite and somehow induces it to follow her out of the nest, far from the safety of the colony. The fly then somehow immobilizes the termite, lays an egg in its abdomen, covers it with soil particles, and proceeds to guard her comatose parasitized victim. Two kinds of con flies were described as new species by Disney (1986); the above account is adapted from his magnificent 1994 overview of the Phoridae, in which he also deals at length with other phorid–termite associations.

Associations between Phoridae and termites were further explored by Dupont and Pape (2008), who recognize four general types of termite–scuttle fly relationships: exploitative, generalist, protective and guest types. The exploitative category includes parasitoids such as con flies and a wide variety of scavengers referred to by Disney (1986) as the "ghoul-guild," for their common habit of opportunistically feeding on termites rendered vulnerable by damage to their nest. Both males and females of this type of phorid have fully developed wings, but females of other termite-associated phorids (like many other phorids) either lack wings entirely or shed their wings prior to entering the termite colony. Wingless females often rely on the winged males to carry them around *in copula* (this is known as intraspecific phoresy), resulting

Diplonevra is a large, widely distributed scuttle fly genus routinely reared from dead arthropods, This is *D. bifasciata*, a widespread Oriental species. Phorids range in size from 0.4mm to over 6mm; this is one of the larger species.

in a bewildering level of sexual dimorphism in this family. Some of the wingless females belong to Dupont and Pape's generalist group, able to develop in different kinds of termite colonies or in an even wider variety of habitats. Wingless *Puliciphora* females, for example, are common even in areas that lack termites, occurring in a wide variety of decomposing material such as compost, ant nests and leaf litter. Generalists don't have a close relationship with their termite hosts and usually avoid contact with them.

The most interesting and extreme of the termite-associated phorids are the true termitophiles whose special strategies for survival in termite colonies represent the protective and guest types. The former are most obviously implemented in species with tanklike ("limuloid") wingless females able to hunker down and protect themselves from their termite hosts. The most specialized of these wingless females look so unlike flies they were first formally described as a new suborder of Hemiptera; even once they had been recognized as flies they were long treated as their own family, Thaumatoxenidae. We now recognize them simply as very specialized wingless, well-armored female phorids. Females of guest-type ant-associated phorids, in contrast, are often conspicuously soft and vulnerable, presumably protected by their physical and chemical similarity to their hosts. They often have expanded termite-like bodies and specialized bristles with porous bases that ooze exudates attractive to host termites. Females of one of the main lineages of truly termitophilous scuttle flies, the subfamily Termitoxeniinae, shed their wings before entering the termite nests and undergo further transformation in form once integrated into the host colony: newly arrived females often have a small abdomen (termed "stenogastric") but soon develop an enormously expanded abdomen ("physogastric").

Sexual dimorphism and extreme female body forms are not restricted to termite-associated scuttle flies. For example, the tiny phorids in the genus *Wandolleckia* are phoretic on giant African snails, where the wingless females have been repeatedly recorded as apparently feeding on mantle secretions. According to Disney (1994), only the females are known from snails, although the fully winged males have been reared from snail dung. However, I have seen fully winged males, wingless females of male-like dimensions, and females with enormously distended abdomens all on the slimy surfaces of *Achatina* snails in Tanzania.

Higher classification of this enormous worldwide family remains somewhat unsettled, with the huge majority of species currently placed in the two large subfamilies Phorinae and Metopininae. The exact composition of these subfamilies, and the subfamily placement of other species, is a matter of dispute. There is now general agreement that the small south temperate group previously treated as the family Sciadoceridae belongs in the Phoridae (where it is now treated as the subfamily Sciadocerinae), but otherwise the subfamily classifications followed by leading experts in the field have been quite different. For example, Disney (1994) recognized the five subfamilies Phorinae, Aenigmatiinae, Thaumatoxeninae, Termitoxeniinae and Metopininae in contrast with Brown's (1992) classification of the family into Hypocerinae, Phorinae, Aenigmatiinae, Conicerinae and Metopininae. Brian Brown is just about to publish a new phorid higher classification using the four subfamilies Sciadocerinae, Termitoxeniinae, Metopininae and Phorinae. This system, attractive for its simplicity as much as for the fact that it is supported by a large set of morphological and molecular characters, is followed here.

Metopininae The Metopininae is the largest subfamily of Phoridae, if only because it includes the enormous worldwide genus *Megaselia*, which in turn includes about half the species in the family. *Megaselia*, certainly one of the most taxonomically intractable of all animal genera, remains made up mostly of undescribed species. The known species are often very common and include the ubiquitous *M. scalaris*, one of the most widespread and polysaprophagous of all insects. Disney (1994) cites records of this species from the usual range of decaying plant and animal material as well as some truly unusual substrates, including beets pickled in vinegar, a snake preserved in alcohol, boot polish and a pot of blue paint. There is even a report of *M. scalaris* breeding in food supplies at a research station in Antarctica.

Other *Megaselia* species with known larvae include specialized parasitoids or predators of snails and slug eggs, spider eggs, millipedes, mealy bugs, beetle pupae and fly larvae and pupae. *Megaselia* includes a wide variety of economically important species, including beneficial predators and parasitoids as well as species (*M. halterata* and others) that are pests in commercial mushroom houses. Yet another member of the subfamily Metopininae, *Metopina ciceri*, develops in the root nodules of chickpeas. This remarkable species has dimorphic females that have either fully developed wings and eyes or reduced wings and atrophied eyes. Another *Metopina*, *M. pachycondylae*, has a kleptoparasitic larva that wraps itself around the neck of an ant larva, where it can intercept and steal food meant for its host. Other ant-associated members of this subfamily include the ant-decapitating flies and the anus-probing ant parasitoids discussed above. The relatively large,

The genus *Megaselia*, with thousands of species, is one of the largest, most biologically diverse and taxonomically difficult genera in the entire animal kingdom.

common slug-associated phorids in the genus *Gymnophora* also belong in this group.

Phorinae The subfamily Phorinae includes an enormous variety of scuttle flies previously placed in the subfamilies Hypocerinae, Aenigmatiinae and Conicerinae. The Coffin Fly, *Conicera tibialis*, is in this group, as is the common mammal-burrow genus *Anevrina* and the common velvety black species in the typical genus *Phora*.

Termitoxeniinae Adult females of these bizarre flies are extraordinary in that they undergo remarkable changes as adults, not just shedding their wings to facilitate invasion of their host's nests (nests of Old World fungus-gardening termites) but also showing growth in various body parts (head, abdomen, hind legs) as they integrate into the termite colony. This kind of post-metamorphic growth is extraordinary for an adult holometabolous insect like a fly. Mature females often closely resemble their host termites, and apparently produce exudates that attract or appease their hosts. This small (about 60 species) subfamily is so unusual there is still no general agreement about where it fits into the higher classification of the Phoridae.

Sciadocerinae The subfamily Sciadocerinae includes only two extant species, *Sciadocera rufomaculata* and *Archiphora patagonica*, although there are additional species in three extinct genera known only from Baltic amber. These relatively large, striking phorids have traditionally been treated as a separate family, Sciadoceridae, but they are now treated as a probable basal lineage of the Phoridae. *Sciadocera rufomaculata* is a common carrion fly in Australia and New Zealand, and *Archiphora patagonica* occurs in southern Chile.

Syrphoidea

Flower Flies and Big-headed Flies

Syrphidae

Flower or Hover Flies

The Syrphidae are the butterflies of the fly world, attracting an enthusiastic following among amateur and professional entomologists because of their bright colors, easily observed behavior, morphological diversity and apparent ubiquity. The 6,000 or so named species of syrphids range from just a few millimeters in length to well over a couple of centimeters, and from nondescript little black flies through to stunningly beautiful mimics

The Drone Fly, *Erisalis tenax*, is one of the most widespread and familiar flower flies. Larvae, called rat-tailed maggots, often abound in pockets of dirty water such as ditches and puddles.

of bees and wasps. Syrphidae have for years been the most widely collected of all Diptera; they are now also the most popular group of flies among insect photographers. Although much of this entomological attention is directed at the more conspicuous adult flies, even the larvae of Syrphidae are relatively well studied. The range of larval habits in the family seems to correspond reasonably well with its division into the three wordwide subfamilies Eristalinae, Syrphinae and Microdontinae.

Eristalinae Almost half the species in the family Syrphidae, including most members of the large subfamily Eristalinae, have larvae that fit the general maggot model of munching microbes from within a mass of organic matter, such as wet rotting wood, decomposing vegetation, mucky bits of water or tree holes that combine both rotting wood and water. Like most maggots, eristaline larvae breathe through a pair of posterior spiracular lobes, but, as in other Syrphidae, the lobes are fused into a single tube. In at least one aquatic *Chrysogaster* species the tube is modified to puncture the roots of vascular aquatic plants to obtain air; the more normal arrangement is for the fused lobes to form a variously elongated snorkel that serves to access open air above the larvae's aquatic or semiaquatic habitats. In several species, such as the common Drone Fly or Rat-tailed Maggot, *Eristalis tenax*, the tube is long and retractile, indeed resembling a rat's tail.

Drone Flies often abound in microbe-rich microhabitats such as contaminated ditches or even puddles in piles of manure. Some other eristalines, including the distinctively long-snouted *Rhingia*, carry this trend to the extreme by inhabiting dry droppings of large animals. European species of *Rhingia* have been reported as laying eggs on foliage over dung, into which the hatching larvae drop. The only North American *Rhingia* species, *R. nasica*, presumably has similar habits but, remarkably, the habits of this common species have never been confirmed. Habits of the many tropical *Rhingia* species also remain unknown.

Most members of the subfamily Eristalinae have big-mouthed larvae that filter-feed in fermenting tree sap, tree holes or other wet microhabitats rich in decaying plant material. *Eristalis*, the type genus of Eristalinae, is a diverse genus with a range of stout-bodied species, including several striking mimics of bumble bees and Honey Bees. One of the latter, the aforementioned Rat-tailed Maggot or Drone Fly, *E. tenax*, is probably the most widespread and familiar flower fly. Its affinity for the pockets of foul fluids or ooze so often associated with animal agriculture have long made this species a familiar associate of mankind, a familiarity that in ancient times led to instructions for producing bees from the bloated bodies of deceased livestock. These bee recipes presumably stemmed from the double misconception that Drone Flies were bees and that they were spontaneously generated by fetid fluids in putrefying carcasses. The Greek philosopher Aristotle put to rest the first misconception when he coined the name *diptera* for the two-winged insects some 2,400 years ago, also pointing out that "no two-winged insect has a sting in the rear." A belief in the spontaneous generation of living things, however, persisted for another 1,600 years or so after Aristotle's time.

Even though Drone Flies look more like harmless male bees, or drones, than well-defended stinging females, some experimental evidence shows that their beelike appearance does indeed deter predators. Many other syrphids are such perfect wasp and bee mimics in appearance, sound and behavior that there is little doubt they benefit from the mimicry, even though experimental evidence is often lacking. Drone Flies have spread to the ends of the earth from their European origin. They are now common from New Zealand to Canada, where they seem particularly abundant just before mated females start to settle into sheltered spots for the winter (most other flower flies overwinter as larvae). The robust adult flies are easy to find, especially at yellow flowers, around which they often exhibit stereotypical mating behaviors. Males sometimes defend limited mating territories and court flower-visiting females by "singing" to them with a specific hum as they hover above prospective mates.

Other *Eristalis* species have slightly different strategies, often darting out at high speed to intercept potential mates rather than courting them at flowers. Another common, widespread garden-variety eristaline noted for its swift pursuit of mates is the tiny compost flower fly *Syritta pipiens*. Males demonstrate an astonishing aerial agility as they hover, dart and even fly backwards in search of females on flowers. They also seem to show engaging pluck by darting after any insect of almost any size that appears in their search area, tracking a fairly consistent 10 centimeters behind it until the target lands on a flower. The *Syritta* male then dives on his target, quickly forcing a mating if it turns out to be a female of his own species. Females of other syrphid species usually choose mates based on swarming, displays or other factors, but mating in *Syritta pipiens* seems to be entirely male-driven (the term "rape" even appears in literature discussing mating behavior in this species). Eggs are deposited in decomposing material such as compost bins.

Saprophagous larval habits such as those of *Eristalis* and *Syritta* are typical for the Eristalinae, but some species in the subfamily invade living plant tissues such as flower bulbs, usually turning them into oozing masses of decomposing material in which they feed. One eristaline genus, the large genus *Cheilosia*, includes species with larvae that feed on living fungi as well as those that live in the stems and leaves of living plants. The plant-feeding species include a leaf miner, stem miners, root miners and species that feed on sap under the bark of living conifers. Some plant-feeding *Cheilosia* are even considered pests, joining the eristaline bulb feeders in the genera *Merodon* and *Eumerus* as generally unwelcome members of a family dominated by beneficial species. Other *Cheilosia* are considered beneficial for their habit of feeding on noxious weeds; one such species was even deliberately transplanted from Europe to Canada in an attempt at biological control of pest thistles.

A few eristaline species also consume animal material. Some Southeast Asian species (of *Nepenthosyrphus*) eat insects trapped in pitcher plants, and a number of North American and European species consume dead insects and other debris in wasp nests. The only eristalines to have moved on to actively hunting and killing living animal prey, however, are found in the nests of social Hymenoptera; they include some of the *Volucella* species found in North American and European wasp and bee nests. Some *Volucella*, such as the big, beelike *V. bombylans*, develop as scavengers on debris, such as dead wasps and moribund brood, in yellowjacket wasp nests. Yellowjacket colonies in temperate regions are annual and are abandoned in the fall, leaving dead larvae and other debris for scavengers such as *Volucella bombylans* larvae, which remain in the nest. Other *Volucella* species — at least *V. pellucens* — have apparently crossed the line from feeding on dead brood to feeding on living brood in wasp nests, and have thus become predators. Predaceous lineages seem to have arisen independently at least three times in the Syrphidae: at least once in *Volucella*, one or more times within the Microdontinae and once in the ground plan of the subfamily Syrphinae (including the tribe Pipizini, formerly placed in the Eristalinae but now treated as a basal lineage of Syrphinae).

Syrphinae Although it makes up somewhat less than half the family Syrphidae, the subfamily Syrphinae includes the majority of the common and familiar flower fly species such as the conspicuously bright black and yellow, light-bodied "garden-variety" syrphids. Syrphine larvae are almost all predators — usually of aphids or related plant-sucking bugs — although some species feed on pollen as well as more mobile protein, and a couple of Neotropical lineages in the large genus *Allograpta* appear to have entirely abandoned aphid-eating in favor of plant-eating. Similarly, larvae of a couple of species in the large (around 150 species) genus *Toxomerus* feed on grass pollen rather than animal prey, although the most widespread and abundant species in this genus are well-known aphid predators.

The blind, land leech–like larvae of common garden syrphines are best observed at night as they grope actively among clusters of aphids on exposed stems and leaves, sequentially entangling soft-bodied victims in sticky salivary secretions, impaling them with mouthhooks and usually lifting them off their host plants to imbibe the body contents. Unlike most maggots, which live deep in their foodstuffs and have a correspondingly sallow, Morlock-like complexion, many syrphine larvae are exposed among their foliage-feeding prey; they are often attractively camouflaged by greens or yellows, created by blood pigments or fat deposits that show through the translucent skin. Other syrphines seek soft prey in concealed places, feeding on root aphids underground, on gall-forming aphids within the galls, or even in the subterranean nests of ants tending honeydew-producing aphids. A very few feed on insects other than soft-bodied Hemiptera, venturing into consumption of thrips, caterpillars and even the larvae of other insects, including ants, beetles and flies such as mosquitoes and crane flies.

Perhaps the most unusual of the predaceous Syrphinae are the aquatic *Ocyptamus* species recently discovered in the water pockets of Neotropical tank plants (Bromeliaceae) and described as feeding on a variety of other insect larvae found in the same specialized aquatic habitat (Rotheray et al. 2000). Another *Ocyptamus*, recently discovered by Onanchi Ureña and Paul Hanson (2010) right on the campus of the University of Costa Rica, is like other syrphines in that it feeds among aggregations of soft-bodied bugs — whiteflies, in this case. But instead of feeding on the whiteflies, as you might expect, this *Ocyptamus* species ambushes and eats adult flies attracted to the honeydew produced by the whiteflies.

One of the best-known species of Syrphinae is *Syrphus ribesii*, a very common and widespread species that follows the smell of honeydew to leaves or stems on which aphids cluster. Eggs are laid among the aphids, where the fly larvae later feed voraciously, casting about with their blind, tapered heads to grasp soft-bodied aphids using mouthhooks that work against the syrphine maggot's apically stiffened lower lip, or labium. Hapless victims are held aloft until partially consumed and then are cast aside. The syrphid larva feeds in this wasteful manner on dozens of doomed aphids before pupariating (among its prey) to transform into a medium-sized, broad-bodied black and yellow flower fly. These are abundant flies, often seen (or heard) when males

aggregate in search of females along woodland trails or in similar spots. Perching males make a distinctive buzz as they warm up by vibrating their thoracic muscles; swarming males sing a slightly different tune as they dance in shafts of sunlight that passes right through their translucent, apparently empty abdomens.

Syrphus ribesii larvae feed on a wide range of aphid species that feed in dense colonies, including many pest aphids. Many of the rarer syphines can develop only if particular host aphids are available, but larvae of the most common species are less fussy, and at least a few — some *Platycheirus* species, for example — are apparently able to get by feeding on rotting plant material if aphids are not available. Some *Platycheirus* adults, along with species in the similar, closely related genus *Melanostoma*, also have somewhat unusual feeding habits, preferentially consuming the pollen of normally wind-pollinated plants such as grasses. *Platycheirus* is one of the most common and diverse flower fly genera in temperate regions, with about 70 species occurring in Canada and about a dozen in New Zealand (but none in Australia). About half the species in the genus occur in the Far North and a handful occur in cooler habitats in the neotropics. Unsurprisingly for such a northern group, many species are Holarctic, occurring in both North America and Europe or Asia.

Microdontinae Members of the enigmatic subfamily Microdontinae, at least those for which larvae have been observed, live in the nests of ants, where their dome-shaped, mollusk-like larvae creep slowly around the colony like little tanks. The appearance of microdontine larvae is so slug-like that these odd maggots have been formally described as mollusks on four separate occasions! Second- and third-instar larvae of the few adequately documented species feed on ant larvae or eggs (or in some cases only immobile

Syrphinae are major pollinators as adults and important predators as larvae. The halters are normally pale but in some species, like this *Melanostoma mellinum*, they turn bright blue-green in females ready to lay eggs.

prepupal larvae and pupae); there are also reports of a Brazilian species feeding on aphids or related bugs living in ant nests. The mouthparts of known microdontine larvae seem to be uniquely modified for predation, but it seems likely that the predaceous species are descended from scavenging ancestors. Some species are thought to be scavengers at least as first-instar larvae, but this remains to be adequately documented.

Microdontine syrphids don't normally occur on flowers and are usually rare compared to other Syrphidae. The sluggish adults are most often found near the host nests, although I have seen groups of male microdontines hovering high overhead in the neotropics, and Reemer (2012) reports both hovering and apparent lekking behavior in other members of the subfamily. Although many new genera are about to be described (Reemer 2012), over half of the 565 or so Microdontinae species currently belong to the worldwide genus *Microdon*, and almost half are Neotropical in distribution. The rest occur mostly in other tropical regions, with relatively few species in temperate regions, as is also true for the ants on which microdontines depend. The subfamily Microdontinae is considered to be the sister group to the rest of the family, and thus the "basal lineage" of the Syrphidae (Ståhls et al. 2003).

Pipunculidae

Big-headed Flies

"Big-headed flies," which look like small syrphids with outlandishly enlarged heads, have long been considered to be close relatives of the Syrphidae, with which they share a remarkable aerial agility. Pipunculids can even hover within the confines of a folded-over insect net, a feat probably beyond most so-called hover flies and a bit of behavior that correlates with the hunting habits of this entirely parasitic family. Although life histories have been worked out for only a tiny fraction of the 1,400 described species of Pipunculidae, big-headed flies can often be seen hovering low among dense vegetation, where they use their enormous eyes to search for hosts or mates. Females of almost all species with known life histories actively hunt out leafhoppers and related Hemiptera (Auchenorrhyncha) to parasitize, injecting eggs into their doomed hopper hosts with a scimitar-like ovipositor.

The only known exception is the genus *Nephrocerus*, a group of relatively large, syrphid-like pipunculids recently found to parasitize adult crane flies (Tipulidae) rather than the hopper hosts utilized by other members of the family (Koenig and Young 2007). Female *Nephrocerus* attack relatively large adult crane flies — almost invariably females as well — injecting an egg into each host. Larvae develop quickly as internal parasitoids, killing their crane fly hosts before leaving to enter a prolonged diapause in the soil. Adults do not appear until almost a year later, in the following spring. Pipunculidae occur throughout the world but *Nephrocerus* is almost entirely a north temperate group, with only a few records from the northern part of the Neotropical region. Remarkably, given the striking appearance, unusual biology and phylogenetic significance of the genus, *Nephrocerus* was poorly known until very recently. Four of the six North American species, for example, were first discovered and named in 2005.

The family Pipunculidae is divided into three subfamilies, with most species either in the subfamily **Chalarinae** or the much larger subfamily **Pipunculinae**. The unusual genus *Nephrocerus* is treated, usually along with the poorly known southern South American genus *Protonephrocerus*, as the separate subfamily **Nephrocerinae**. The Nephrocerinae have some similarities to Pipunculinae — both groups have lost their ocellar bristles, for example — and they are usually considered to be closely related, despite continued debate about their exact relationships and the plausibility of a host shift from leafhoppers to crane flies.

The Acalyptrates

The acalyptrate flies, or Acalyptratae ("without calypters"), form an important, widely recognized assemblage of mostly small flies, including all Schizophora that lack the defining characters of the Calyptratae. Some acalyptrate flies appear to be more closely related to calyptrate flies than to other acalyptrates, which would render the Acalyptratae an artificial assemblage: a grade rather than a clade, like the old suborder Nematocera. Despite the trend away from using formal names for such artificial groups, the term "acalyptrate" remains a convenient and widely used term to refer to the Cyclorrhapha that are neither Aschiza nor Calyptratae. Unsurprisingly, given the probable artificial nature of the group, the Acalyptratae lack obvious unifying features and the group includes an enormous variety of flies.

About half of all families of Diptera are acalyptrates. Only four of these, Drosophilidae, Agromyzidae, Chloropidae and Tephritidae, rank among the 20 largest families of flies, but the total number of described acalyptrate species still exceeds 30,000. Most acalyptrate families are relatively small; only half a dozen approach or exceed a couple of thousand described species. These figures are skewed by the enormous number of undescribed species, especially in relatively nondescript groups such as the Sphaeroceridae, in which thousands of species still await description.

Although the acalyptrate families are here arranged by superfamily, the superfamilies are a bit fuzzy around the edges: only a few are clearly defined monophyletic groups. They are used here because most are useful assemblages of superficially similar families, despite disputes about their phylogenetic reality, and because the best of them are indeed predictive groups of families linked by inherited biological, morphological and molecular characteristics. Several families of acalyptrates are difficult to place in a superfamily and have been treated differently by different authors; they are placed in Fig. 8.1 according to either traditional or recently published placements. Alternative interpretations are discussed in the text. Families are ordered by size, from largest to smallest, within the superfamily treatments below. Most of the unplaced families are small and obscure, and therefore appear at the ends of the superfamily treatments.

OPPOSITE PAGE This wasp-like microdontine flower fly (*Mixogaster*) was attracted to honeydew on a leaf in Ecuador. The big-headed fly (*Eudorylas*) in the lower photo has landed on a dead leaf to drink from an intermittent desert stream in the American southwest.

Snail-killing flies in the widespread genus *Sepedon*, like this *S. aenescens* from Vietnam, are distinctive for their prominent antennae and unique among the Schizophora for their reduced ptilinal suture.

Sciomyzoidea

Snail-killing Flies and Their Relatives

Sciomyzidae

Marsh or Snail-killing Flies

Among the most characteristic and charismatic of wetland insects the world over, sciomyzids are generally yellowish or orangish flies commonly seen perching prominently on pondside vegetation, usually with an alert head-down stance. With the notable exception of an African species that feeds on oligochaetes, all Sciomyzidae with known life cycles are predators, parasitoids or scavengers that consume mollusks or their eggs. The family includes several common species that attack terrestrial snails or slugs, but most sciomyzids occur in moist to aquatic environments, where they develop as parasitoids or predators of aquatic or semiaquatic snails or sphaeriid clams such as pea mussels (fingernail clams). In part because of the potential use of these flies for biological control of pest snails — especially snails that serve as intermediate hosts for the *Fasciola* and *Schistosoma* flukes that cause fascioliasis and schistosomiasis — the biology of Sciomyzidae has been relatively well studied, and life cycles are known for a remarkable 40 percent or so of the described species. A few species have been moved around the world for biological control, and some, such as *Sepedomerus macropus* and *Sepedon aenescens* in Hawaii, are now established outside their native ranges. The biology of the Sciomyzidae is thoroughly reviewed in a recent book by Knutson and Vala (2011).

The family Sciomyzidae is currently divided into only two subfamilies, the small subfamily Salticellinae and the large, cosmopolitan subfamily Sciomyzinae. The families Huttoninidae, Phaeomyiidae and Helosciomyzidae are sometimes treated as subfamilies of Sciomyzidae, but the relationships between these groups and the groups here treated as Sciomyzidae remain inadequately resolved. They are treated as separate families below.

Salticellinae The subfamily Salticellinae includes only one extant genus, *Salticella*, with one species in southern Europe and North Africa and one in southern Africa. The European species *S. fasciata* can apparently develop as a facultative scavenger although it normally oviposits directly on living snails, often relatively large coastal snails. The southern African member of the subfamily, *Salticella stuckenbergi*, is the world's largest sciomyzid at about 15 millimeters.

Sciomyzinae The diverse (more than 600 species) subfamily Sciomyzinae occurs throughout the world, although most species are associated with the relatively cool climates of north and south temperate countries. Genera in this subfamily vary widely in size, appearance and habits and have traditionally been classified into two tribes, the ***Sciomyzini*** and the ***Tetanocerini***. One aquatic African sciomyzine species, *Sepedonella nana*, has been recorded as developing on oligochaetes (Vala et al. 2000), but all other species in the subfamily apparently consume only mollusks. Many of the larger and more familiar genera, such as *Sepedon*, *Dictya* and *Tetanocera*, are made up entirely or mostly of species with aquatic larvae that attack relatively large aquatic snails, either as predators that kill multiple hosts or as parasitoids that develop in a single host. A few species are specialists on fingernail clams (pea mussels), and at least one genus, *Anticheta*, develops only on snail eggs. The genus *Tetanocera* is particularly diverse and includes species that develop in terrestrial slugs; at least one species in the genus can immobilize its slug hosts by using a salivary neurotoxin (Rozkosny 1998).

Sepsidae

Sepsids or Ant-like Scavenger Flies

Sepsidae are called "ant-like scavenger flies" because most are shiny black saprophagous flies with a prominent spherical head and a constricted abdomen that imparts a decidedly ant-like appearance. These small (2–7 mm) acalyptrates are often superabundant around fresh cow or horse dung but also occur on other decomposing materials, such as carrion, compost, sewage sludge and decomposing kelp (wrack) on north temperate seashores. The Sepsidae are thought to be closely related to the Coelopidae and Ropalomeridae, and recent work (Su et al. 2008) suggests that Ropalomeridae and Sepsidae are sister groups.

Orygmatinae The sepsids found in seaweed belong to the wrack-restricted Holarctic species *Orygma luctuosum*, which is traditionally treated as a separate subfamily, Orygmatinae — considered to be the basal, or most primitive, lineage of the family. *Orygma luctuosum* adults are robust, flattened Atlantic seashore flies that look and act less like typical sepsids than true seaweed flies in the family Coelopidae. Larvae are saprophagous maggots that occur in dense masses in moist, well-decayed pockets of wrack.

A recent phylogenetic analysis of the Sepsidae (Su et al. 2008) showed that molecular data support treatment of the Old World genus *Ortalischema* in the subfamily Orygmatinae rather than as a basal lineage of the Sepsinae, as previously suggested on the basis of morphological studies (Meier 1995). Both species of *Ortalischema* are relatively inconspicuous, short-winged flies that breed in horse dung; neither adults nor larvae show any obvious morphological similarities to *Orygma*.

Sepsinae The remaining 375 or so sepsid species are usually placed in the more derived subfamily Sepsinae, a heterogeneous and possibly artificial group generally divided into three or four tribes: ***Saltellini***, ***Sepsini***, ***Toxopodini*** and ***Palaeosepsini***. Most sepsines are shiny black insects that strut and dance around on dung and surrounding foliage, alternately flicking or waving their wings. They can be strikingly sexually dimorphic: males are often conspicuously adorned with sex-specific structures such as abdominal brushes and modified forelegs. Some of the modifications serve as tactile or visual stimulation for the other sex and others provide species-specific mechanisms by which the males hang on to their mate's wings. It is probably because of the impressive array of morphological and behavioral traits shown by these common, easily reared flies that there is now a rich literature on the morphology, evolution and sexual behavior of Sepsinae. For example, Puniamoorthy and colleagues (2009) maintained 27 sepsid species in laboratory cultures, where they used video recordings to demonstrate that each species had its own mating profile and that behavioral characters evolved faster than sex-specific morphological traits.

Not all sepsines are generalists that lend themselves to laboratory culture. *Themira lucida*, for example, is a scavenger in yellowjacket nests, and some other *Themira* species specialize on waterfowl droppings. Nor do they all conform to the ant-like shape typical of the group. *Saltella sphondylii*, for example, is a Holarctic cow dung–breeding species that has a relatively broad abdomen, rendering it a bit atypical for the subfamily. *Saltella* males guard their mates after copulation, while many other sepsid males guard females before mating; some species indulge in violent precopulatory struggles as females apparently try to mitigate the impact of copulation on their longevity. *Saltella sphondylii* is atypical in that copulations decrease male longevity rather than female longevity (Martin and Hosken, 2004).

Other sepsids engage in even more conspicuous activities, in some cases drawing attention by forming astonishingly dense aggregations of frenetic adults. One European species, *Sepsis*

Most Stylogastridae hover over the swarm fronts of army ant raids, watching for opportunities to ram their harpoon-like eggs into potential hosts flushed out by the ants.

fulgens, is famous for forming late-season masses of tens of thousands of flies that mill about over a few square meters of bramble bushes before disappearing at the onset of winter. According to Pont (1987), *S. fulgens* swarms seem to hang together for no apparent purpose — without mating, although both sexes are present, and without feeding or laying eggs. He suggests that the swarms are involved with hibernation, serving to mark sites at which males and females can meet the following spring after spending the winter as adults. Both sexes have large scent-producing glands with which they mark the swarm site, often imparting a strong odor that has been compared to lemon thyme. Similar sweet aromas are released when various sepsid species are crushed; this has led one entomologist to suggest that the common name for the family should be "scented flies."

Conopidae and Stylogastridae

Thick-headed Flies

Conopidae and Stylogastridae, sometimes called "thick-headed flies" because of their relatively chunky heads, are conspicuous, well-known insects whose morphologies and life histories are markedly different from all other flies. In fact they are so markedly distinct from other acalyptrates that dipterists have generally been divided between the view that Conopidae and Stylogastridae form a "basal" group of Schizophora (perhaps the sister group to all other Schizophora), and an opposing view that they are closely related to the Tephritoidea or some other acalyptrate superfamily. I have treated thick-headed flies with the superfamily Sciomyzoidea here following the recent phylogeny of Wiegmann et al. (2011), although other recent papers have suggested other superfamily affinities and there is a good case to be made for recognizing a separate superfamily, Conopoidea, for these unusual flies.

The family **Stylogastridae** is a widespread but remarkably uniform group comprising about a hundred species of slender flies characterized by a very long beak normally held folded under the head. *Stylogaster*, the only genus in the group, is mostly tropical (Afrotropical and Neotropical) but ranges north through the Nearctic region all the way to Canada. Most tropical species characteristically hover over the swarm fronts of army ant raids, watching for opportunities to ram their harpoon-like eggs into potential hosts flushed out by the ants. Known hosts of this group are orthopteroids such as crickets and cockroaches, but Afrotropical species are also known to occasionally (accidentally?) inject eggs into other hosts, such as relatively large calyptrate flies (mostly Muscidae). Despite the striking appearance and (literally) striking habits of *Stylogaster*, some authors treat the genus as a subgroup of Conopidae. The Stylogastridae and Conopidae are probably sister groups (each other's closest relatives), but each of these very distinct groups is a well-defined natural group on its own. It is therefore a matter of opinion whether they should be treated together (as the Conopidae) or separately (as the Stylogastridae and Conopidae). I prefer to treat them separately.

The 800 or so described species of the family **Conopidae** are scattered over all biogeographic regions, where females of species with known biologies are diurnal dive-bombers that attack bees and related aculeate Hymenoptera (stinging wasps) in flight. These often brightly colored wasplike or beelike parasitoid flies are frequently seen around flowers, where they await opportunities to insert a single egg into the abdomen of an appropriate adult aculeate. Larval conopids develop inside the bee or wasp, ultimately killing the host before pupating in its abdomen. Conopids in general are not particularly host specific: some species are recorded from a wide range of host aculeate Hymenoptera, sometimes including both wasps and bees.

In contrast to the "harpoon" host-hunting strategy of *Stylogaster*, female conopids normally pinion and pry open potential hosts using a robust abdomen with abdominal sternites 5 and 6 modified into a can-opener-like structure called a theca. The theca allows female conopids to use a projection behind sternite 5 to grip the host bee or wasp against sternite 6 while inserting an egg into the upper side of its abdomen. Unlike *Stylogaster* species, which are remarkably uniform in appearance throughout the world, conopids vary widely in size and shape and can closely resemble their host bees and wasps.

In contrast to stylogastrids, which are only occasionally seen feeding on nectar with their extremely elongate mouthparts, conopids are common at flowers and are routinely seen nectaring at blossoms visited by superficially similar wasps and bees.

Species in the conopid subfamilies **Conopinae** and **Zodioninae** have long mouthparts that stick out straight in front of the head, while those in the subfamilies **Myopinae** and **Dalmanniinae** usually have a long beak jointed in the middle so it can be folded back under the head while at rest. The latter three subfamilies always differ from the large subfamily Conopinae in having an antenna with a dorsal arista rather than the unusual (for acalyptrate flies) terminal stylus that characterizes the antennae of flies in Conopinae. About 600 of the 800 or so species in the family belong in the Conopinae.

In addition to the widespread subfamilies discussed above, the subfamily **Notoconopinae** was recently named for the single Australian species *Notoconops alexanderi*. This species, known only from two specimens collected in New South Wales, was newly described and placed in its own new genus and subfamily by Margaret Schneider in 2010.

Coelopidae

Seaweed Flies

All 30 or so species of the family Coelopidae are seaweed specialists that abound in the piles of brown algae (wrack) that build up in the supralittoral zones of marine coastlines. Coelopids are found on most temperate shorelines but are strangely absent from the wrack-rich coasts of southern South America, where they seem to be replaced by superficially similar flies in the related family Helcomyzidae. Both Coelopidae and Helcomyzidae are medium-sized flattened, dark-bodied flies with an ability to move swiftly among mucilaginous masses of decaying algae and a propensity to take flight instantly when exposed.

Although they breed only in shoreline seaweed, coelopids sometimes build up in enormous numbers, often aggregating by the thousands on sheltered rock surfaces near their larval habitats and occasionally blowing inland to appear in nuisance numbers in maritime towns. The breeding habits of seaweed flies lend themselves to laboratory rearing indoors in jars of seaweed, and one of many easily cultured species, the Holarctic *Coelopa frigida*, has been widely used as a laboratory animal for studies of genetics and ecology. Among other things, this has led to the discovery that regulation of the extraordinary range of body size in this species involves a complex chromosomal inversion system. Body size is in part associated with quantity and quality of larval food but is also driven by sexual selection: bigger males are better able to force matings with females following a characteristic pre-mating struggle. Males are thus larger and much more variable in size than females, so much so that they have sometimes been erroneously named as different species or even different genera.

The Coelopidae are superficially similar to some Sepsidae. The sepsid genus considered to be closest to the groundplan of that family, the genus *Orygma*, is remarkably coelopid-like in both habitat and habitus.

The Coelopidae are divided into two subfamilies, the "primitive" subfamily **Lopinae**, which contains only the Australian species *Lopa convexa*, and the relatively diverse subfamily **Coelopinae**, comprising the rest of the species in the family. The Coelopinae are in turn divided into about a dozen genera, with more than a third of the species in the genus *Coelopa*. Of the remaining genera, one, *Coelopina*, is found on the west coast of North America; one, *Malacomyia*, is shared between Europe and North Africa; and the others are restricted to Australia and/or New Zealand. David McAlpine (1991b) reviewed the Australian genera, naming most of them with pithy (and pronounceable) generic names such as *Rhis* and *This* (his office door once had a picture of one of these seaweed flies with the notation "look at *This*").

Ropalomeridae

Ropalomerid Flies

The family Ropalomeridae is an almost entirely Neotropical group, with only one of the 27 included species, *Rhytidops floridensis*, ranging north to Florida. Ropalomerids are handsome, relatively large (often over 10 mm) flies, most distinctive for a bulldog-like stance that is enhanced by stout, swollen legs and a jowly head (a tall head with big cheeks). Newly fallen trees sometimes attract good numbers of these relatively

Adult Ropalomeridae are most often seen perching on trunks or logs near the tree wounds or decaying wood where the larvae develop, but this one was lured to a plastic card coated with sugar.

uncommon flies; adults are often seen perching on trunks or logs near the tree wounds or decaying wood where the larvae develop. This small family is divided into eight genera; most species are in the genera *Ropalomera*, with 16 species, and *Willistoniella*, with four species. Adult morphology and molecular data suggest that ropalomerids are probably most closely related to the larger and much more widely distributed family Sepsidae (Feng-Yi Su et al. 2008), although larval characters suggest that Sepsidae and Coelopidae are more closely related (Meier 1996).

Dryomyzidae

Dryomyzid Flies

The family Dryomyzidae is a mostly Holarctic group of about 25 species generally associated with decomposing material, although *Oedoparena glauca*, a dryomyzid fly common on the Pacific coast of North America, is a predator of barnacles. Each larva, hatched from an egg deposited on a closed barnacle at low tide, enters the opening barnacle as the tide comes in, eats the contents of the now closed and submerged barnacle, and then moves to another barnacle at a later low tide. A second, more recently described species of *Oedoparena* occurs in the same area and presumably has the same habits as *O. glauca*; a third species in the genus occurs in Japan. Other Dryomyzidae, mostly in the Holarctic genus *Dryomyza*, develop in a variety of rotting material such as bird droppings, dead bodies, fungi and dung. *Dryomyza anilis* (often treated as *Neuroctena anilis*) is an exceptionally common insect in much of eastern North America as well as in Europe. Four other genera are sometimes recognized in the family: *Steyskalomyza*, with one species from Japan, the more widespread *Paradryomyza* and *Pseudoneuroctena*, both described in 1987 for species previously in *Dryomyza*, and the small Holarctic genus *Dryope*.

Dryomyzids are generally medium-sized orange flies with a superficial similarity to common Heleomyzidae and Scathophagidae. Although there is no consensus about the phylogenetic affinities of this family, dryomyzids are probably closely related to the Sciomyzidae and Helosciomyzidae. The Helcomyzidae, previously treated as part of the Dryomyzidae, are now considered to be closer to the Coelopidae.

Helosciomyzidae

Helosciomyzids or Comb-winged Flies

The Helosciomyzidae, a small group comprising only 27 or 28 species of medium-sized, sciomyzid-like flies, is the only acalyptrate family with a mostly south temperate distribution split between southern South America and the Australasian region. This sort of distribution is common in ancient lineages of lower Brachycera and Nematocera that were widespread on Gondwana before it broke up into today's southern land masses, but it is unusual in the supposedly younger Acalyptratae. Only two species of Helosciomyzidae are known from South America, one known only from a single Brazilian specimen and the other known from a couple of Chilean localities; both belong to a distinctive South American genus, *Sciogriphoneura*. Other helosciomyzid genera have restricted ranges in Australia and New Zealand and its associated islands, the Snares and Auckland islands. One species has been raised in captivity on a diet of dead insects; others have been reared from fungi and carrion, but it is not known if they were eating other fly larvae or the decomposing material itself. Larvae of one New Zealand species, *Helosciomyza subalpina*, have been reported as predators of ant larvae.

Helosciomyzids were once treated as part of the superficially similar Sciomyzidae, from which they differ most obviously in having rows of spines on the costa, or leading edge of the wing, similar to many Heleomyzidae. Griffiths (1972) was the first to treat the group as a separate family, a decision supported in a complete revision of the group by Barnes (1981). The relationships between Helosciomyzidae and other Sciomyzoidea remain a matter of debate, but most evidence points to a close relationship with the Coelopidae, Huttoninidae and Helcomyzidae. One species of Helosciomyzidae, the Australian *Cobergius vittatus*, is associated with seacoasts like the related Coelopidae and Helcomyzidae, but most are associated with inland habitats.

Helcomyzidae, like this Chilean *Paractora*, occur on piles of seaweed (wrack) on ocean shores in widely separated parts of the world. Helcomyzids share the wrack with similar Coelopidae in Europe, southern New Zealand and western North America, but in southern South America they replace Coelopidae as the dominant large wrack flies.

Heterocheilidae and Helcomyzidae

Heterocheilids and Helcomyzids

The small families Heterocheilidae and Helcomyzidae are strictly coastal groups, each with interesting intercontinental disjunctions. The two species of **Heterocheilidae** are known only from seaweed-strewn European seacoasts and piles of stranded kelp on the Pacific coast of North America; the family **Helcomyzidae** also occurs on the seacoasts of widely separated regions, but with a north–south as well as an east–west disjunction. In addition to three species of *Helcomyza* split between Europe and western North America — similar to the distribution of the two species of *Heterocheila* — the Helcomyzidae include two south temperate genera. The one species of *Maorimyia* is a huge, hairy fly found only in New Zealand, and the half-dozen or so species of *Paractora* occur mostly in southern South America, where they seem to replace Coelopidae as the dominant large wrack flies. Others are found on South Georgia and the Falkland Islands.

Both the Heterocheilidae and Helcomyzidae have been bounced around from family to family by various dipterists, at times treated together in the single family Helcomyzidae and at times treated as part of the Dryomyzidae. D.K. McAlpine (1991) argues that the Helcomyzidae are closely related to the Coelopidae, but he treats the Heterocheilidae as a group of uncertain relationships.

Huttoninidae

Huttoninids

The recently recognized family Huttoninidae is one of two families of flies entirely restricted to New Zealand (the other is the bat fly family Mystacinobiidae). This small group of only nine species — eight *Huttonina* and one *Prosochaeta* — was until recently treated as a subfamily of Sciomyzidae because of a superficial resemblance between these groups. The Huttoninidae remain unknown as larvae; however, there is no evidence that they share the mollusk-eating habits that characterize the Sciomyzidae.

Phaeomyiidae

Millipede-killing Flies

The family Phaeomyiidae, which contains only five species and two genera, was treated as part of the Sciomyzidae until Griffiths (1972) gave it family status because of fundamental differences between the male genitalia in this group and homologous structures in the Sciomyzidae. There remains some disagreement about the status of Phaeomyiidae: some still treat it as a subfamily (Phaeomyiinae) of Sciomyzidae and others argue that it in fact belongs in the sciomyzid subfamily Sciomyzinae. The biology of one species, *Pelidnoptera nigripennis*, is known, and it is a parasitoid of millipedes, not mollusks, as would be expected of a sciomyzid. Eggs are deposited on the millipede's body and newly hatched larvae burrow in through the membrane between the host's sclerotized segments.

This small group is one of only two fly families endemic to the Palaearctic region (the other is the small family Opetiidae). *Pelidnoptera nigripennis* now also occurs in Australia, where it was deliberately introduced for biological control of an accidentally introduced pest millipede of European origin. The Phaeomyiidae include four dark-colored species of *Pelidnoptera* and one bright yellow species in the genus *Akebono*, recently described from Japan (Sueyoshi, Knutson and Ghorpade 2009).

Natalimyzidae

Natalimyzid Flies

The family Natalimyzidae is a recently described group of small, slender yellowish to brown flies known only from Africa. When the family was formally described by David Barraclough and David McAlpine in 2006, it became the second family of flies to be described solely on the basis of previously undescribed African species, following the description of the Marginidae in 1991. Unlike the rare and obscure family Marginidae, however, Natalimyzidae are well represented in collections and widespread in Africa, with some 20 species found in South Africa, Zimbabwe, Kenya and Nigeria. These rather nondescript little yellowish flies were first collected in 1911 and have been informally recognized as a distinct family since the 1950s. They were first formally brought to the attention of the scientific community by Miller (1984) and explicitly recognized as a new family in a published key by Barraclough (1995) more than a decade before it was formally named.

It is a credit to all involved that the formal naming and description of the family was delayed until enough information and enough specimens were available to support the thorough justification of the family by Barraclough and McAlpine (2006). Although their analysis left no doubt that natalimyzids are distinct from other sciomyzoids, they concluded that "no definite sister group relationship with any of the families in the wider Sciomyzoidea" could be suggested, leaving it as *incertae sedis* — "of uncertain placement" or, more loosely interpreted, "an unsolved mystery." Larvae remain undescribed but are probably microbial grazers, since natalimyzids have been reared from decaying grass (Miller 1984).

Sphaeroceroidea

Sphaerocerids, Heleomyzids and Their Relatives: From the Ubiquitous to the Obscure

Sphaeroceridae

Sphaerocerid Flies or Small Dung Flies

Few, if any, insects are as deserving of the term "ubiquitous" as sphaerocerids, since these small, generally dull-colored little acalyptrates are among the most common insects in all reasonably humid terrestrial environments. You can demonstrate this for yourself by putting a bowl or pan of soapy water in your backyard (or in a wetland, forest or field anywhere in the world) to see for yourself; many of the insects that soon drown in your "pan trap" will be sphaerocerids, easily recognized as such by the distinctively shortened first tarsomeres on their hind legs. Pan trapping is an effective way to collect many kinds of flies, and single traps embedded in humid habitats routinely take dozens of sphaerocerid species, especially if the pan or surrounding material smells of decay.

Larval sphaerocerids are microbial grazers in a variety of humid, bacteria-rich environments such as dung, carrion and various types of decomposing plants and fungi. Many species appear to be polysaprophagous and can be found in large numbers in a diversity of decomposing materials, ranging from compost heaps to tree falls, but others are highly specialized. One Neotropical species develops only in the microbe-rich goo around the excretory glands of land crabs; other species are associated with army ants, bromeliads, caves, mudflats, seaweed, fungi, mammal burrows and dung-rolling scarab beetles. The 1,339 known species of Sphaeroceridae were catalogued in 2001 but the taxonomy of this group is changing rapidly; another 34 genera and more than 300 species have been described since then. Thousands of sphaerocerid species remain undescribed.

The Sphaeroceridae of North America and Europe are well known, but these important and abundant flies remain effectively unidentifiable for other regions. The first-ever key to the genera of Neotropical Sphaeroceridae (Marshall and Buck 2010) appeared in the recent *Manual of Central American Diptera*, but several genera in that key are recognized simply as "undescribed genus A," "undescribed genus B" and so forth, with formal names and descriptions due to follow in detailed papers dealing with all known specimens of the new genera. Papp (2008) took a very different approach to dealing with the poorly known sphaerocerid fauna of the Old World tropics; he recently published a single paper formally naming a staggering 25 new genera of Limosininae — mostly Afrotropical and Oriental — from the Old World tropics. Many of the Old World tropical genera are recognizable only on the basis of the male genitalia, many are based on only one or a few specimens, and nothing is known of their relationships to one another or to genera in other parts of the world. The situation is little better in the Australasian region, although Richards (1973) described half a dozen new genera or subgenera for the region and provided keys to the Australian species known at that time. The endemic Australian sphaerocerid fauna is made up entirely of Limosininae, probably most closely related to Oriental species, and remains mostly unidentifiable and largely undescribed.

Limosininae Sphaeroceridae are divided into five subfamilies, but the overwhelming majority of the species are currently placed in the huge, diverse subfamily Limosininae. Members of this subfamily are extraordinarily abundant flies in all parts of the world. Limosinines in the wrack-restricted genus *Thoracochaeta* abound on seaweed-strewn seashores, *Coproica* species usually outnumber all other flies in manure and most animal droppings, and dozens of other limosinine genera dominate decomposing vegetation and other moist, microbe-rich environments. Some specialized groups of Limosininae can be found clinging to millipedes and scarab beetles, waiting to oviposit in the dung deposited or gathered by their hosts. Others occur in bromeliads, the leaf sheaths of bamboo and other plants, carrion of various sizes, caves, mammal runs, mushrooms and ant nests. The subfamily occurs in most terrestrial environments the world over, from mangrove swamps to mountaintops and from farmyards to fern forests on isolated islands. Peatlands (bogs and fens) are home to many Limosininae, including some found in no other habitats.

Dozens of limosinine lineages have independently become short-winged or completely wingless; diverse clades of flightless Limosininae can be found on mountaintops, in ancient forests and on isolated islands throughout the world. One of the most exceptional clusters of ground-bound limosinines is found on Robinson Crusoe Island, an isolated part of the Juan Fernandez Archipelago, about 500 kilometers off the coast of Chile. Although this island is small enough to walk across in a morning, it is home to a rich fauna of about a dozen endemic flightless short-winged species in two normally fully winged limosinine genera, *Phthitia* and *Leptocera*. The much larger islands of New Zealand are also home to several endemic *Phthitia*, but they are all fully winged; the variety of wingless sphaerocerid flies in New Zealand is made up mostly of undescribed species in the genera *Biroina* and *Howickia*. High mountains throughout the world are populated by a remarkable diversity of flightless Limosininae, largely undescribed but comprising a significant proportion of all flightless flies. Although Limosininae are generally small (1–4 mm), one wingless species is the giant of the family. *Anatalanta*, an isolated lineage of two species known only from the subantarctic islands, grows large in the guano of Crozet, Heard and Kerguelen islands, where its larvae can exceed 10 millimeters in length. The slightly smaller but equally unusual wingless genus *Siphlopteryx* is found on the same islands. Many of the wingless sphaerocerids that abound in the mountains of Africa and South America are minute flies, almost springtail-like both in size and movement.

Although only a few species in the subfamily Limosininae are known as larvae, some of the known species are highly habitat specific. *Acuminiseta pallidicornis* is phoretic on West African millipedes, on which the adult flies ride until the opportunity arises to oviposit in the millipede's frass. Both Old World and New World *Ceroptera* species are similarly phoretic on dung-rolling scarab beetles; the fly larvae develop as kleptoparasites in the buried dung balls, often along with other kleptoparasitic sphaerocerids in the subfamily Copromyzinae (genus *Norrbomia*). *Pterogramma cardisomi* is another limosinine specializing on uniquely situated waste products — larvae of this species are known only from the excretory glands of land crabs in Costa Rica. Some of the most specialized species of Limosininae belong to genera otherwise found in a wide variety of habitats. One species of the large genus *Coproica*, for example, is restricted to Gopher Tortoise (*Gopherus polyphemus*) burrows, and another is known only from rotting cacti. Other *Coproica* are cosmopolitan species that abound on a less exotic decomposing material in agricultural environments; *C. hirtula*, for example, almost invariably occurs by the millions in poultry houses. These minute flies rarely stray far from the manure and thus do not become pests like the larger Muscidae found in the same environment.

Copromyzinae The subfamily Copromyzinae is made up mostly of relatively large species with a generally more "primitive" appearance than the more diverse Limosininae. This group includes some common, secondarily widespread species found on large animal feces. *Copromyza equina*, for example, is probably the most common large sphaerocerid on farmland throughout the world; other widespread synanthropic species occur in the genera *Norrbomia* and *Lotophila*. *Copromyza* and *Norrbomia* both include many localized species throughout the Holarctic region as well as in Africa and the Oriental region. Many specialized copromyzine lineages occur in the Holarctic and (especially) the Afrotropical regions, but the few copromyzine species in Australia and South America are all introduced synanthropic species.

Archiborborinae Most of the South and Central American species previously treated as Copromyzinae belong to a distinctive entirely Neotropical lineage now treated as the subfamily Archiborborinae. Although most members of the subfamily are relatively large, heavily sclerotized and fully winged flies, it also includes numerous flightless species in the extreme south of South America, high in the Andes and on windswept islands including the Falklands and South Georgia. One remarkable group of Archiborborinae, the genus *Frutillaria*, is restricted to the Valdivian rainforest of Chile and adjacent Argentina. *Frutillaria* species are minute flightless flies with wings reduced to strap-like or pin-like processes. All 15 species have restricted ranges within the remaining fragments of the originally extensive Valdivian forest; some are known only from a single locality. Other Archiborborinae are common insects of the cool, wet forests of south temperate South America and higher altitudes throughout the neotropics.

Tucminae The subfamily Tucminae is noteworthy mostly because this small group of two species in the genus *Tucma* is considered to be the sister group to the rest of the family. Both species, which apparently occur only in the mountains of northern Argentina, have several "primitive" characters that have been lost in other Sphaeroceridae.

Sphaerocerinae The distinctive subfamily Sphaerocerinae is a small group made up of relatively large, squat, flattened flies, often armored or ornamented with spines, spurs, teeth and even windows (translucent patches on the abdomen). Although best known for the stout-legged *Sphaerocera curvipes*, common on horse manure in most temperate countries, this subfamily includes distinctively different groups in the New and Old Worlds. Only two members of *Sphaerocera* are endemic to the New World — one in Central America and the Caribbean, one in the northwestern Nearctic region — and just nine of the 32 species in the closely related genus *Ischiolepta* occur in the New World, again including a secondarily widespread species, some northwest North American species and a single species in Central America. Otherwise, the Sphaerocerinae are more or less divided into a diversity of Old World genera plus a single distinctive New World lineage made up of three very closely related genera, *Parasphaerocera*, *Mesosphaerocera* and *Neosphaerocera*. Species in these genera, which often have conspicuous window-like dorsal abdominal patches, range throughout the New World, but most are Neotropical.

Homalomitrinae The species of the subfamily Homalomitrinae are extremely rare and are thought to be associated with New World army ants, although no definite relationship has been established. These bizarre, pale flies have fully developed but weakly veined wings that look unlikely to function in flight, yet the majority of known specimens of the half-dozen species in this small subfamily were collected in Malaise traps — used to intercept and retain flying insects — in Central and South America.

Heleomyzidae

Wombat Flies, Cave Flies and Others

Heleomyzidae comprise a bit of a grab-bag family made up of a number of distinct lineages of dubious relationship. It is, in fact, such a loose assemblage that some authors divide it into five or six families, and one leading expert on the group has suggested it should be combined with almost all the rest of the Sphaeroceroidea — including the enormous family Sphaeroceridae — into a single family called Heteromyzidae (D.K. McAlpine 2007). Little would be gained by such an aggregation of families except to create a huge family that would be even less recognizable than the current Heleomyzidae, while at the same time reducing the huge and distinctive family Sphaeroceridae to a subfamily. It is more likely that the Heleomyzidae will ultimately be made into a natural group by reclassification of some of the subfamilies listed below, once their relationships to one another and to other families are better understood. While there are many exceptions, heleomyzids are generally flies of cold or dark places, most likely to be encountered in the spring or late fall and at high or low latitudes. A number of lineages are associated with caves, mammal burrows or fungi, but some common genera are associated with dung, carrion or birds' nests, and a few occur in other specialized microhabitats; an Australian species, for example, is associated with the tunnels of wood-boring insects. As usual for acalyptrates, most heleomyzid species are unknown as larvae.

Most but by no means all of the 700 or so species of Heleomyzidae have the leading edge of the wing, the costa, armed with conspicuous spines that dwarf the normal costal bristles; many are medium-sized flies of a yellowish orange color. Otherwise the family is tremendously diverse with no obvious unifying features, and dipterists have disputed the classification of the Heleomyzidae for decades. Even the relatively distinctive groups here treated as subfamilies are the subjects of sometimes rancorous disagreement: recent works treat the same groups as tribes, subfamilies or even separate families. Despite all that, there is general agreement that the following "subfamilies" represent real groups at some level, and that they are probably part of the broader group treated here as the superfamily Sphaeroceroidea. That still leaves lots of room for new work on the taxonomy of this difficult group. The subfamilies below correspond closely with the subfamily concepts of McAlpine (2007), although the Heleomyzinae and Heteromyzinae are more narrowly defined and Suilliinae is treated as separate from the Heleomyzinae.

Like most Heleomyzidae, this *Anorostoma alternans* has a row of short, spine-like bristles along the costa (the leading edge of the wing).

Heleomyzinae Heleomyzines are "typical" heleomyzids, and this subfamily includes several genera of mostly medium-sized yellow to brown flies. Carrion flies in the genus *Neoleria* are among the more common Heleomyzinae, and other genera are abundant cave insects. *Scoliocentra defessa*, for example, is a common cave fly in eastern North America. Heleomyzinae are mostly north temperate, with a few species accidentally introduced to Australia and New Zealand. Some authors define the Heleomyzinae more broadly, including the Heteromyzinae and Suilliinae.

Suilliinae The subfamily Suilliinae is a group of common, generally yellowish brown, medium-sized flies associated with macrofungi. Although they form one of the most distinctive subgroups of Heleomyzidae because of their unusual oblique orbital plates, Suilliinae are sometimes treated as a subgroup of the Heleomyzinae. Suilliinae occur in both temperate and tropical regions but the subfamily is most diverse in the cooler parts of the Holarctic region. Although normally associated with fungi, some Suilliinae are phytophagous, and one — the Garlic Fly, *Suillia univittata* — is a pest of garlic and onions in Europe. No other Heleomyzidae are of known economic importance, although truffle-feeding *Suillia* species are reportedly tracked by people seeking those coveted underground fungi.

Tapeigastrinae The subfamily Tapeigastrinae can be thought of as the Down Under counterpart to the Suilliinae, because these handsome orangish flies are as frequent on mushrooms in Australia as are the superficially similar Suilliinae on mushrooms in Europe and North America. All 12 species are Australian endemics that develop in the fruiting bodies of fungi.

Rhinotorinae Rhinotorinae, frequently treated as the family Rhinotoridae, is a small and distinctive group of relatively squat, stout-legged flies with mottled or patterned wings and sometimes with the top of the head distinctly sunken. One species occurs in the southern United States, but otherwise these unusual flies are restricted to the Neotropical and Australasian regions. Although most of the species occur in South America, the only rhinotorines for which larvae are known are members of the Australian genus *Cairnsimyia*, described from beetle burrows in fig trees.

Trixoscelidinae The subfamily Trixoscelidinae, often treated as the separate family Trixoscelididae, is a small group of flies generally characterized by forwardly shifted ocellar bristles, and often by patterned or pigmented wings. The best known and largest genus in the subfamily is the widely distributed genus *Trixoscelis*, a group of small (usually 2–3 mm) flies found in open, sandy, short-grassed areas, including disturbed areas such as cemeteries. Immature stages remain unknown for the subfamily, but a couple of British species have been reared from bird nests. The subfamily Trixoscelidinae has been traditionally interpreted as a mostly Holarctic and African group, with no Australasian or Oriental species and few Neotropical species, but the limits of the group are disputed; McAlpine (2007) includes a few endemic New Zealand and Australian genera (*Fenwickia, Waterhouseia, Pentachaeta*) in his concept of the Trixoscelidinae.

Cnemospathidinae The subfamily Cnemospathidinae, sometimes treated as the family Cnemospathididae, is a heterogeneous assemblage of generally obscure south temperate groups, including some peculiar Australasian flies and some poorly known groups found in southern South America. This subfamily includes the wombat flies, genus *Borboroides*, now a well-known group thanks to the meticulous studies of David McAlpine. All 23 species of *Borboroides* occur in Australia, where several species are associated with marsupial dung and some species seem to be dependent on wombat dung. One species of wombat fly described by McAlpine (2007) has the lyrical name *Borboroides musica*, referring to the elaborate stridulatory organ on its front legs. J.R. Malloch first described the genus *Borboroides* in the Sphaeroceridae in 1925, and there remains a good possibility that at least some members of the Cnemospathidinae are more closely related to Sphaeroceridae than to other Heleomyzidae. Some of the southern South American genera currently placed in the Cnemospathidinae are especially similar to "primitive" sphaerocerids, further suggesting that the Heleomyzidae is an artificial group. One widespread species of Cnemospathidinae, *Prosopantrum flavifrons*, now occurs in the Australasian and Palaearctic regions as well as South Africa and southern South America, although it is probably native to South America. Most related species are South American, but several very similar *Prosopantrum* species were recently described from Africa.

The subfamily **Borboropsinae** is a small north temperate group treated by some as a family, Borboropsidae. It includes only the small Holarctic genus *Borboropsis* along with the more recently described *Nidomyia*, the single species of which was recently discovered in the nests of European birds. Borboropsinae are rarely encountered small (less than 3 mm), sphaerocerid-like flies that lack the costal spines characteristic of many heleomyzid lineages.

The subfamily **Chiropteromyzinae** is another very small Holarctic group, again treated by some authors as a separate family, Chiropteromyzidae. Like the Borboropsinae, these are small (2 mm or less), rarely encountered flies associated with either bird nests or bat guano. The three species of *Neossos* are associated with different kinds of bird nests in North America.

The subfamily **Diaciinae** includes two South American genera, *Diacia* and *Dichromya*, and one Western Australian genus, *Amphidysis*. Although these are distinctive flies with black bodies and orange heads, they are rarely encountered and little known. Some evidence suggests that at least *Dichromya* is viviparous, but larvae and larval habitats remain unknown.

The subfamily **Heteromyzinae** is a mostly Holarctic group of three genera: the small (one European species) seashore genus *Tephrochlaena* and the common genera *Tephrochlamys* and *Heteromyza*. Both of the latter genera occur around human habitations, and the dung-breeding *Tephrochlamys rufiventris* is now secondarily widespread and established in farmyards from Europe to New Zealand. *Heteromyza atricornis*,

Heleomyzidae, like this Costa Rican *Suillia inens*, often have the leading edge of the wing armed with stout bristles. Members of this subfamilly, Suilliinae, have characteristic orbital plates that angle away from the eye rather than running parallel along the inner eye margin.

like several other heleomyzids, is a characteristic cave fly. Some authors (for example, Griffiths 1972) have elevated this group to family level; D.K. McAlpine (2007) promoted an exceptionally broad concept of the group as a family, Heteromyzidae, that includes all Heleomyzidae and all Sphaeroceridae.

Even after allocating most heleomyzid genera to the many subfamilies above, a couple of genera remain unplaced. The most significant of these is *Cinderella*, a group of tiny flies transferred to the Heleomyzidae by Hennig (1969). These poorly known flies, with several species in southern South America, Central America and the southern United States, were put in their own tribe by D.K. McAlpine (1985), but they don't fit in any of the subfamilies above and they fit only uneasily in the family Heleomyzidae. The same is true for the unusual southwestern American genus *Paraneossos*, originally in the Trixoscelididae but put by McAlpine (1985) in the mainly neotropical tribe Gephyromyzini. We can expect the classification of this family to change significantly as the heleomyzid faunas of key regions (especially southern South America) become better understood.

Chyromyidae

Golden-eyed Flies

Chyromyidae are small, mostly pale-bodied flies that, at least when alive, have contrastingly bright, iridescent eyes — reflected in their common name "golden-eyed flies." The golden shimmer of iridescence that marks the eyes of living chyromyids, however, disappears in pinned specimens because the color is created by light interference between narrowly separated cuticular layers that collapse as they dry.

Golden-eyed flies are generally treated as part of the superfamily Sphaeroceroidea, but the relationships of this small group remain uncertain and little is known about their biology. Larvae are probably bacterial grazers like most acalyptrates; some species have been reared from debris in the nests of birds and mammals, bat caves, tree hollows and similar pockets, mostly in otherwise dry environments such as dunes and grasslands. Chyromyidae are rarely collected, and this group was one of the least-studied fly families until a recent series of papers by Martin Ebejer, one of which (2009) described 27 new African species and five new genera. This is a remarkable novelty rate in a family that formerly included only four genera and a total of slightly more than 100 species, most of which were discovered only in the past few years. Ebejer divides the family Chyromyidae into two subfamilies, **Chyromyinae** and **Aphaniosominae**, both of which occur in most parts of the world.

Mormotomyiidae

Hobgoblin Flies

The only species in the enigmatic family Mormotomyiidae is one of the rarest flies in the world and arguably represents the insect family with the smallest known distribution. *Mormotomyia hirsuta* (loosely translated as "hairy terrible fly" or "hairy hobgoblin fly") is known only from one cleft in one small rocky outcrop on top of one small hill near the town of Ukazi, about 230 kilometers east of Nairobi on Kenya's Highway A3. The adults, which are relatively large (House Fly–sized) flies with minute eyes and bizarre little strap-like wings, look like long-legged and extremely hairy spiders. We wouldn't know about these remarkable flies at all — and we would thus probably be entirely unaware of one of the rarest and most phylogenetically contentious families of flies — if a Mr. H.B. Sharpe had not been present and curious when heavy rain flushed a mass of bat dung, fly eggs, larvae, puparia and hairy adult flies from that obscure Kenyan rock cleft in the mid-1930s. Sharpe collected all stages and noted that the hairy adult flies seemed to "float like feathers" when dislodged from above. He sent the specimens to a specialist, E.E. Austen, who coined the name *Mormotomyia hirsuta* and suggested that it was an acalyptrate related to the family Sphaeroceridae.

The next time someone made a collection of Hairy Hobgoblin Flies was in 1948, when V.G.L. van Someren collected a long series. He sent them to another entomologist, F.I. van Emden, who suggested that these strange flies might not be acalyptrates at all but, rather, "transitional between calyptrates and acalyptrates" and probably related to the Scathophagidae, many of which are similarly hairy yellowish flies. This idea that these flies are transitional was more or less echoed and given greater precision by Hennig (1971), who suggested that Mormotomyiidae might be the sister group to the Calyptratae. These apparently disparate opinions about the phylogenetic affinities of mormotomyiids are not as contradictory as they seem at first, since the acalyptrates are probably not a monophyletic group, and some acalyptrate families are probably more closely related to calyptrates than to other acalyptrates.

Van Emden's 1950 paper on that second collection of Hairy Hobgoblin Flies recorded their larvae as developing in bat dung and speculated that the adult flies live by scavenging secretions from bats' bodies. That was about where things stood in the summer of 2010, when South African dipterist Ashley Kirk-Spriggs invited me to join a collecting trip to Burundi by way of Nairobi, where he was to meet up with Kenyan entomologist Robert Copeland, who would lead the trip. Nairobi is only a few hours' drive from Ukazi, so I suggested that we tack on a side trip in search of the missing Mormotomyiidae. That suggestion was enthusiastically received — it turned out that Copeland knew the place and was keen to go. Unfortunately I had to pull out of the trip at the last minute, so I missed the excitement of visiting the almost mythical Ukazi Hill and being there to see firsthand that the Hairy Hobgoblin Flies were still alive, well and abundant. Kirk-Spriggs and Copeland collected lots of larvae, pupae and adults in and around bat dung at the cave entrance, where they described these long-legged, wingless flies as running around "like little sun scorpions." The new specimens they collected have allowed for careful reconsideration of the phylogenetic affinities of these rare flies. The current consensus is that they are indeed acalyprates, perhaps more closely related to the Ephydridae than to the Sphaeroceridae.

Tephritoidea

True Fruit Flies and Their Relatives

Although a true ovipositor derived from abdominal appendages has been lost in the Diptera, several fly lineages expedite egg delivery by using an ovipositor-like stiffened or telescoping abdominal tip. The superfamily Tephritoidea is a group of families in which the female's seventh segment has become stiffened (through the combination or fusion of the seventh tergites and sternites) to form a rigid tube, or oviscape, used to deliver eggs into fruit, wood, stems or other substrates. The part of the abdomen that does the actual piercing and egg-guiding, called the aculeus, is derived from the next (eighth) segment and is normally not visible because it is retracted or telescoped into the oviscape. Most tephritoids, other than the exceptional parasitic Pyrgotidae and Tachiniscinae, insert their eggs into living or dead plant material.

The huge majority of species in the Tephritoidea belong to a single, well-characterized lineage made up of the four closely related families Tephritidae, Platystomatidae, Ulidiidae and Pyrgotidae, which are mostly attractively colored flies with boldly patterned wings. A few smaller tephritoid families are of somewhat contentious relationship both to one another and to the "main" lineage of Tephritoidea. These families — Lonchaeidae, Piophilidae, Pallopteridae and Richardiidae — are generally treated as basal lineages of the superfamily Tephritoidea and are thus sometimes referred to as the "lower Tephritoidea" (Korneyev 2000). The nocturnal, exceptionally rare flies in the enigmatic family Ctenostylidae have also been included with the Tephritoidea in the past, but this placement is now in dispute (D.K. McAlpine 1990; Korneyev 2000). I have included the mysterious Ctenostylidae in this section because they are traditionally treated with the tephritoids and do not fit easily into any other superfamily.

Tephritidae

True Fruit Flies

The Tephritidae — called "true fruit flies" to avoid confusion with the "fruit flies" or vinegar flies in the family Drosophilidae — form an enormous worldwide family with more than 4,600 named species, most of which develop on a limited range of plant hosts. Many tephritids occur only on a single host species, genus or family, although some of the serious pests in the family are highly polyphagous and attack hundreds of different plants. The family Tephritidae is of immense importance mostly because of a few major agricultural pests, but also because some weed-eating species have proven useful as biological control agents. Although almost all tephritids develop in plants — usually in living tissues ranging from (of course) fruit through to stems, roots, shoots and flowers — the family also includes some extraordinary parasitoids in the rarely encountered subfamily Tachiniscinae, a few predaceous species, a couple of species found in termite nests and several species (including the famous "antler flies") that develop in microbially rich damaged plant tissue such as decomposing wood. Development in rotting material is probably the ancestral habit of the family, and microbial grazing remains a widespread strategy among tephritid larvae. Even familiar fruit-feeding species such as the Apple Maggot and Medfly have larvae equipped with pharyngeal filters used to concentrate and ingest microorganisms in the pockets of decaying fruit created by their feeding activities.

Tephritinae Species in the nominate subfamily of Tephritidae, Tephritinae, are almost all flower feeders, gall formers or stem miners associated with composite flowers (plants of the family Asteraceae). Only a few groups of Tephritinae, most notably the common genus *Eutreta*, have secondarily shifted to other host plants. Perhaps the high host specificity in this subfamily explains its exceptional diversity, with around 2,000 species worldwide. Most tephritine larvae occur in flower heads, but some develop in the stems, roots or (rarely) leaves of a narrow range of host plants and pupate in the same plant tissue, in marked contrast to the fruit-infesting fruit flies in the subfamilies Dacinae and Trypetinae, which leave their decomposing hosts to pupate in the ground. Many tephritines induce their hosts to form galls, such as the familiar ball galls induced on goldenrod (*Solidago* spp.) by the North American

Larvae of the Thistle Gall Fly (*Urophora cardui*) develop inside characteristic swellings on thistle stems. This is one of several species of Tephritidae deliberately introduced from Europe to North America for biological control of invasive weeds.

Goldenrod Gall Fly, *Eurosta solidaginis*, and the conspicuous galls of *Rachiptera limbata* that abound on *Baccharis* shrubs in Chile. Like other species in the small Neotropical genera *Rachiptera* and *Strobelia*, *Rachiptera limbata* secretes a liquid that hardens into a globular protective structure outside its gall. These protective structures are the bumpy, brilliant white blobs that are so conspicuous on the *Baccharis* shrubs that carpet the hillsides of central Chile.

Most of the Tephritidae of proven or potential value as biological control agents belong to the subfamily Tephritinae, and many tephritine species have been moved around the world to assist with control of introduced weeds. The European *Urophora affinis*, *U. cardui*, *U. quadrifasciata*, *U. sirunaseva*, *U. solstitialis* and *U. stylata*, for example, have been introduced to North America for biocontrol of noxious weeds such as thistles and knapweeds, and *Eutreta xanthochaeta* (Lantana Gall Fly) has proven to be a cost-effective way to control *Lantana camara* in Hawaii. Tephritine fruit flies are occasionally pests of economically valuable Asteraceae such as sunflowers and safflowers, but in the balance this is a beneficial group. A few gall-inhabiting tephritines have shifted from feeding on plant tissue to feeding on other insects in galls. One such species, the European *Euphranta toxoneura*, develops as a predator in the willow leaf galls induced by *Pontania* sawflies.

Dacinae Most tephritine fruit flies now found outside their native ranges are beneficial species, but the accidental movement of species in the subfamilies Dacinae and Trypetinae from country to country is another story. These subfamilies include some of the most serious pests of fruits. The subfamily Dacinae is a mostly Old World and Pacific group, although some species in the notorious dacine genera *Bactrocera*, *Dacus* and *Ceratitis* threaten fruit production and distribution worldwide. Members of these major fruit-feeding genera

The huge (over 500 species) Old World tropical genus *Bactrocera* includes many important and invasive fruit pests. Several are in the *B. dorsalis* complex, which includes the Oriental Fruit Fly, the Asian Fruit Fly and dozens of similar species such as this *B. occipitalis* from the Philippines.

threaten economic loss not only because of their direct impact on fruit but also because many countries are justifiably concerned about accidental import of exotic fruit fly pests. As a result, the presence of these flies can close export markets upon which fruit producers depend. Countries such as New Zealand, which currently has no fruit fly pests, and Chile, which has few, must be especially vigilant about preventing accidental introductions through infested fruit. This explains why you can expect a punitive fine if you get caught casually "smuggling" an innocent fresh fruit snack when you fly into one of those countries. Those deflated cherries or battered oranges you must discard on arrival could otherwise cost your destination country millions — even hundreds of millions — of dollars.

Illustrations of the impact of invasive fruit flies abound, ranging from the well-known story of the Medfly, *Ceratitis capitata* (see also Chapter 2), through to newer occurrences such as that of the recently discovered Asian Fruit Fly, *Bactrocera invadens*, an Asian species that showed up in Africa in 2003 to become a serious new pest of mangoes and other fruit. The Old World tropical genus *Bactrocera* is the most damaging genus in the entire family (perhaps in the whole order), and it now has many secondarily widespread species that feed on an enormous variety of fruits. The 520 species of *Bactrocera* include around 40 serious pests, mostly Oriental or Australasian in origin. Some of the more important are the Oriental Fruit Fly and other members of the *B. dorsalis* species complex; the Melon Fly, *B. cucurbitae*; the Olive Fruit Fly, *B. oleae*; the Queensland Fruit Fly, *B. tryoni*; and the Peach Fruit Fly, *B. zonata*. When invasive *Bactrocera* species such as the Melon Fly and Oriental Fruit Fly appear outside their established ranges, for example in the southern islands of Japan, regional authorities wisely spare no expense to eradicate them — often through release of massive numbers of sterilized males.

The genus *Dacus* is a mostly Afrotropical group

very similar to and closely related to the genus *Bactrocera* (some species currently in *Bactrocera* used to be treated as *Dacus*). However, the 250 or so species of *Dacus* are more host specific and largely restricted to fruits and flowers of Cucurbitaceae. Despite this narrow host range, *Dacus* does include some pests, such as the Pumpkin Fly, *D. bivittatus*. The genus *Ceratitis*, another originally Old World tropical group in the Dacinae, includes the widespread polyphagous Medfly, *C. capitata*, along with another 77 species, of which about ten are pests. Although the Medfly is a major pest or potential pest throughout the warmer parts of the world, most *Ceratitis* are still restricted to Africa, where some members of this important genus have yet to be formally described and named.

Trypetinae The Dacinae and Trypetinae are both large, economically important subfamilies of more than 1,000 species each but, in contrast to the mostly Afrotropical and Oriental subfamily Dacinae, more than half of the species in the subfamily Trypetinae occur either in the Neotropical or Holarctic regions. The Holarctic and Neotropical genus *Rhagoletis* includes 69 species, some of which are the most important fruit fly pests of temperate fruit crops, including apples (attacked by the Apple Maggot, *R. pomonella*), cherries (attacked by the European and Eastern Cherry Fruit Flies, *R. cerasi* and *R. cingulata*) and blueberries (attacked by the Blueberry Maggot, *R. mendax*). These host-specific pests produce one generation per year, closely tracking the seasonality of their host plants. Polyphagous fruit-infesting species, in contrast, usually breed continuously.

Anastrepha, the largest genus of Trypetinae and the most economically important genus of fruit fly in the neotropics, includes about 200 species, of which almost 50 have been recorded from cultivated plants. Of these, only 15 — including the Mexican Fruit Fly, *A. ludens*; the West Indian Fruit Fly, *A. obliqua*; and the South American Fruit Fly, *A. fraterculus* complex — are currently considered significant pests. A closely related Neotropical genus, *Toxotrypana*, also includes a few pest species, of which the most familiar is the Papaya Fruit Fly, *Toxotrypana curvicauda*, a Neotropical species accidentally introduced to Florida about 100 years ago. The spectacular oviscape of this species looks like an enormous sting; combined with their bright yellow and black color, this makes female Papaya Fruit Flies resemble unusually well-armed vespid wasps.

Tachiniscinae Flies in the subfamily Tachiniscinae, sometimes treated either as part of the Pyrgotidae or as the separate family Tachiniscidae, are appropriately named for the way at least the typical genera resemble big, bristly tachinid flies (family Tachinidae), not only in general appearance but apparently also in biology. The only species of tachiniscine about which we have any life history information is an African species that was reared from a moth pupa and presumed to be a parasitoid (all Tachinidae are parasitoids, mostly of Lepidoptera). Despite this remarkable convergence in body form and biology, the structure of Tachiniscinae, especially the female abdomen, leaves little doubt that they belong in the Tephritoidea; they are now widely considered a basal lineage of the family Tephritidae, along with the subfamilies Blepharoneurinae and Phytalmiinae. The 18 known species are scattered over the Neotropical, Palaearctic, Afrotropical and Oriental regions, but they are rarely encountered and poorly known. Some but not all Tachiniscinae are robust, tachinid-like flies, and one African species is a vespid wasp mimic. Some data suggest that tachiniscines are more closely related to pyrgotids than to tephritids, and should therefore be treated as a separate family.

Blepharoneurinae The small subfamily Blepharoneurinae currently includes only 34 species in five genera, but experts estimate that the Neotropical genus *Blepharoneura* alone has at least 200 species. This estimate, based on an extrapolation from the number of known species associated with well-studied hosts to the number of expected species on little-studied hosts, reflects the remarkable specificity of known *Blepharoneura* species. Like all known members of the subfamily, *Blepharoneura* develops only on host plants in the family Cucurbitaceae; unsurprisingly, many species are associated with single species of host plants. More remarkably, several host plant species support a number of closely related *Blepharoneura* species, with different species being found on different plant structures. Species of one plant genus, *Gurania*, sometimes host different but closely related species of *Blepharoneura* in their male and female flowers; the Ecuadorian species *Gurania spinulosa* is host to six *Blepharoneura* species, of which five develop in the calyces of flowers of only one sex and one develops in both male and female flowers (Condon et al. 2008). *Blepharoneura* species are usually attractive flies with broad, brightly patterned wings used in elaborate inter- and intrasexual displays, sometimes at leks (aggregation sites not associated with a food resource) or lek-like aggregations under leaves. Adults also have the habit of using a spiny labellum to rasp away at plant tissue. Some other Tephritidae, like many Agromyzidae, feed at juices liberated by oviposition punctures, but using mouthparts to rupture plant tissue is unusual for adult flies.

Phytalmiinae Many tephritids exhibit intersexual displays by which males flash females with body parts such as boldly patterned wings, and many have elaborate intrasexual displays by which males aggressively challenge other males. The most spectacular of such displays are undoubtedly those of the "antler flies," so called because males of some species in this group defend their territories from other males with outrageously large antler-like processes arising from their cheeks. Like most genera in the subfamily Phytalmiinae, antler flies are saprophagous, developing in recently damaged or dead tissues of host plants.

Phytalmia species, the best known and most spectacular of the antler flies, develop in fallen rainforest trees in New Guinea and northern Australia, where the males stake out appropriate oviposition sites and defend them against competing males. Females arriving at the site must mate with the territory-holding male before ovipositing (this sort of resource-defense mating system is also found in other groups of flies such as the Neriidae). Cheek processes are found in various other Tephritidae and other acalyptrates, including Platystomatidae, Ulidiidae, Richardiidae and Clusiidae, but some species of *Phytalmia* outdo all other acalyptrates with their astonishingly large and elaborate antlers. Antler development varies within a species, and interactions between well-endowed males and smaller males usually lead to displacement of the smaller male without physical contact. When evenly matched males duke it out for a bit of territory, the antlers are used in pushing contests.

Platystomatidae

Signal Flies

The Platystomatidae earn the name "signal flies" for their sometimes vigorous use of prominently pattered wings in elaborate courtship behavior, similar to the behavior of some Tephritidae and Ulidiidae. As is true for a few members of related tephritoid families and many members of the distantly related family Micropezidae, the courtship behavior of signal flies can also often involve "kissing," in which mates lock lips for extended exchanges of oral fluids. Males of some platystomatid species transfer oral fluids to the female's thorax and then imbibe it themselves; others ingest anal fluids produced by the female. Several have enlarged or modified heads used to defend their territories from other males. Males of some *Achias* species in Australia and New Guinea, for example, have tremendously widened heads with the eyes sitting at the ends of long stalks, superficially like those of stalk-eyed flies (Diopsidae). Other species have spectacular cheek processes reminiscent of the antler flies (Tephritidae).

Most of the approximately 1,200 named signal fly species are found in the Australasian, Oriental and Afrotropical regions. Most of the few dozen New World species belong to either the enormous worldwide genus *Rivellia* or the New World genus *Senopterina*. Species of *Rivellia* — at least, those with known biologies — develop inside the root nodules of legumes, feeding on a mixture of plant tissue and nitrogen-fixing bacteria. The impact of *Rivellia* species on economically important legumes such as soybeans (attacked by *R. quadrifasciata*, the Soybean Nodule Fly, in the United States), peanuts and pasture legumes can be

Bromophila (Platystomatidae), one of Africa's most recognizable insect genera, has the startling habit of ejecting bright yellow fluid from its mouth. This is one of the very few acalyptrate flies without ocelli. The only named species in this genus is *B. caffra*.

significant. Most other Platystomatidae are probably saprophagous, grazing bacteria from a range of rotting plant material such as wood, bulbs, fruit, stems, flowers, dung and fungus. A few unusual species are predators of other insects or their eggs, and some species develop in carrion.

Even though signal flies are usually conspicuous and attractive flies, many species remain undescribed. The Australasian platystomatid fauna is exceptionally diverse; estimates given in D.K. McAlpine's 2001 review of the Australasian genera of signal flies suggest that the 500 or so described Australasian species represent only about half of the actual fauna. Most of the 34 Australasian genera are found nowhere else in the world.

The family Platystomatidae is divided into four subfamilies, the Platystomatinae, Plastotephritinae, Scholastinae and Trapherinae.

Platystomatinae The subfamily Platystomatinae, the largest of the platystomatid groups, includes the cosmopolitan genus *Rivellia* as well as a wide range of genera endemic to different zoogeographic regions. Males and females of many platystomatines look different from each other, but males of some species in the genus *Achias* — which has more than 100 Australasian species — take sexual dimorphism to an extreme with their outlandishly broadened heads, used in male–male territorial battles. *Achias rothschildi*, from New Guinea, probably has the widest head (up to 55 mm) of any insect (D.K. McAlpine 2001). Other species of *Achias*, and many other platystomatines, including some Neotropical *Senopterina*, are wasp mimics. Several platystomatines are so garish there is little doubt they are using conspicuous colors to warn potential predators of their unpalatability. *Bromophila*, for example, is one of Africa's most recognizable insect genera because of its blue-black wings, huge black body, contrasting red head and startling habit of ejecting bright yellow fluid from its mouth. Larvae apparently develop on the roots of trees loaded with toxic compounds that ultimately protect the conspicuously sluggish adult flies from predation. *Bromophila caffra* is currently the only described species in the genus.

Scholastinae The subfamily Scholastinae includes the nominate genus *Scholastes*, which has about 15 Pacific and Oriental species. Several *Scholastes* species develop in coconuts, and the adults can be common on the fallen fruit. Five of the six species of the scholastine genus *Lenophila* are considerably more specialized and localized, being normally found only on Australian grass trees (*Xanthorrhoea* spp.), where they probably develop among decaying material at the base of the leaves. Adults of the genus exhibit elaborate intersexual interactions on grass tree leaves; I have seen males of the Western Australian *Lenophila nila* stake out positions on leaves, where they court arriving females with wing waving and prolonged "kisses." As in other platystomatids, the transfer of oral fluids starts before mating and continues during copulation, despite the considerable head twisting this demands of the female.

The mostly Afrotropical subfamily **Plastotephritinae** includes several species in which males have bizarrely shaped heads, often with an almost comical appearance derived from jowl-like hyperinflation of the lower cheeks. None of the 90 known species of plastotephritines are known as larvae, and mating behavior remains unknown for the whole subfamily. Other groups of flies with similarly modified heads usually have males that fight each other to retain territory, either as part of resource-defense behavior among species that develop in relatively rare and restricted microhabitats (as in phytalmiine tephritids) or as part of lekking behavior in species that develop in more common but scattered microhabitats (as in some Clusiidae and Richardiidae). There is a clear challenge here to seek out and observe these pretty flies in their Oriental (2 species), Australasian (10 species) or African (78 species) habitats.

Even less is known about the remaining platystomatid subfamily, the **Trapherinae**. This small group is restricted mostly to the Afrotropical (five genera) and Oriental (seven genera) regions, although one genus, *Phylax*, was recently described from Fiji (D.K. McAlpine 2001).

Widened or oddly distorted heads occur in several acalyptrate families, including Ulidiidae. This Peruvian species has prominent cheek processes; some other ulidiids sport spectacularly broadened heads, but only in the males.

Ulidiidae

Picture-winged Flies

The "picture-winged flies" were until recently known as the family Otitidae, but most dipterists now refer to this group using an older family-group name, Ulidiidae, and further divide this well-known cluster of conspicuous flies into the subfamilies Otitinae and Ulidiinae. Most of the 700 or so species of picture-winged flies are innocuous insects that breed in decomposing plant material, but a few are pests. For example, the Cornsilk Fly, *Euxesta stigmatias*, a tropical species widespread from Florida and the Caribbean through to southern South America, sometimes crosses the line between saprophagy and phytophagy when it infests ears of sweet corn. It has become a serious pest in the southern United States, where infestations of these pretty little metallic green flies can render entire fields unsuitable for harvest. Half a dozen other species of the large (more than 100 species), mostly Neotropical genus *Euxesta* have been reported as damaging crops, and the family

Ulidiidae also includes a few further pest species such as *Tritoxa flexa*, the Black Onion Fly, and *Tetanops myopaeformis*, the Sugarbeet Root Maggot. Most picture-winged flies, however, are more noteworthy for their dancing, wing waving and other conspicuous courtship behavior.

The male of the now widespread species *Physiphora alceae* (previously known as *P. demandata*), for example, carries out a complex courtship sequence that involves drumming on the female with his forelegs, vibrating his body and waving his wings; the female sometimes responds by placing her proboscis on the male's back and pulling him backwards in a spiral dance (Alcock and Pyle 1979). Another spectacular display is provided by the North American *Callopistromyia annulipes*, a common fly on recently cut deciduous logs, where these stunning little flies can be seen strutting around with their wings twisted vertically and held out from the body. Larvae of this and other wood-associated Ulidiidae probably feed on frass in beetle borings in the wood.

Otitinae The subfamily Otitinae is a mostly New World and Palaearctic group that includes several wood-associated genera such as *Callopistromyia*, *Pterocalla* and *Pseudotephritis*, as well as some important plant-associated groups. The type genus, *Otites*, is usually associated with reeds or sedges, *Tritoxa* species are associated with onions and *Tetanops* includes species that develop in the roots of sugar beets. Others, such as *Delphinia* and *Seioptera*, prefer decomposing organic material, including dung.

Ulidiinae The subfamily Ulidiinae is a mostly tropical group that encompasses some of the most unusual biologies and morphologies in the family. Males of the widespread Neotropical *Plagiocephalus latifrons*, for example, have spectacularly widened heads that can exceed twice the body length (females are "normal"). Some Ulidiinae are associated with particular plants, as is the case with *Stictomyia longicornis*, a cactus-loving species with the unusual habit of bending its wings over its body to give it a superficially beetle-like appearance (this kind of presumed beetle mimicry also occurs in the Sphaeroceridae, Rhiniidae, Drosophilidae and Lauxaniidae). The mostly African ulidiine genus *Physiphora* includes several saprophagous species found in manure and similar substrates; one of the widespread synanthropic species in this genus is common on carrion, as is the Oriental *Pseudeuxesta prima*. Phytophagous species in this subfamily include the eastern Nearctic *Eumetopiella rufipes*, which lays eggs on *Echinochloa* grass stems; the newly hatched larvae systematically sever the peduncle so that the inflorescence on which they need to feed stays soft and succulent, enclosed in its sheath. The huge genus *Euxesta* is also well known for plant-infesting members such as the Cornsilk Fly, *E. stigmatias*, although this diverse group includes many saprophagous species found in dung and decaying vegetation. Larvae of other ulidiines, such as the abundant wetland species of *Chaetopsis*, are sometimes secondary invaders in monocot stems already damaged by other insect larvae, and are sometimes phytophagous in young plants.

Lonchaeidae

Lance Flies

Lonchaeidae, sometimes called "lance flies" because of the lance-like female oviscape, which conceals a long ovipositor, are small, clear-winged flies common throughout most of the world but most diverse in north temperate regions. In contrast to the impressive variety of shapes, patterns and sizes in most other tephritoid families, most Lonchaeidae are uniformly stout, usually metallic green to bluish black flies of around 4 or 5 millimeters in length. The combination of a distinctive habitus and signature black (or dark brown) halters makes the lance flies among the easiest families of acalyptrates to recognize. Some are very common; *Lonchaea* species are among the most regular visitors to dead or dying wood, where females are routinely seen slipping their long abdominal tips under bark or into beetle burrows to lay eggs.

Although most lance flies are saprophagous, a few of the 500 or so species so far described in the family are pests of seed cones, fruits or vegetables. For example, the North American Fir Cone Maggot, *Earomyia aquilonia*, feeds on the seeds of fir and tamarack, sometimes totally destroying the seeds, and several species of Neotropical *Dasiops* and *Neosilba* attack living plants, including the developing flower buds and fruits of passionfruit. One of the many plant-associated *Dasiops* species, *D. alveofrons*, was recently recognized as a pest of apricots in California, and a species of *Neosilba*, *N. perezi*, is a shoot miner in cassava. The Australian lonchaeid fauna is dominated by shining *Lamprolonchaea* species, of which the most common is a pest species called the Metallic Green Tomato Fly, *Lamprolonchaea brouniana*. Old World lance flies include *Silba adipata,* the Black Fig Fly, on figs in the Mediterranean region and South Africa, and *Silba gibbosa*, on citrus flower buds and flowers in Southeast Asia. Several other members of the large (about 90 species), mostly Old World tropical genus *Silba* appear only after fruit flies (Tephritidae) have invaded and damaged host fruit. Similar habits are found in the large, mostly Neotropical genus *Neosilba*.

Lance flies like this *Lonchaea* are regular visitors to dead or dying wood, where larvae of many species develop under the bark; some members of this large, widespread genus breed in decaying fruit.

Perhaps because of their uniform appearance, lance flies have not been as thoroughly investi gated as other tephritoids, except for a few well-studied regional faunas. Tropical Lonchaeidae remain very poorly known, as indicated by Frank McAlpine and George Steyskal's assertion (1982) that the 15 known species of *Neosilba* represented a fraction of the real fauna and that at least 60 species awaited description (they are still waiting). Larvae are unknown for most species in the family and little is known about behavior, although adults of at least some species form swarms. Swarming is uncommon among acalyptrate flies, being more or less restricted to the families Lonchaeidae and Milichiidae, both of which swarm in shafts of sunlight in forest openings. Adults of some British species have been recorded as possible kleptoparasites, feeding on prey captured by garden spiders (Dobson 1992).

Although there are only nine genera in this superficially homogeneous family, it is divided into two subfamilies, Dasiopinae and Lonchaeinae. The subfamily **Dasiopinae** contains the single genus *Dasiops*, a mostly Holarctic genus of about 40 species, of which several feed on the flowers and fruits of passionfruit (*Passiflora*) plants and several feed on cacti. The subfamily **Lonchaeinae** is dominated by the common genus *Lonchaea*, a worldwide but mostly Holarctic genus of about 140 species, including the most commonly encountered wood-associated lance fly species. Some *Lonchaea* larvae are thought to be predators of wood-boring beetle larvae, but there are many records of this genus from fruits and vegetables. It seems likely that fungi and frass are major dietary items for many wood-inhabiting species.

Pyrgotidae

Scarab-killing Flies

The attractive, moderately large (usually over 10 mm) flies in the family Pyrgotidae are relatively rarely seen, at least in part because the daily activity of these unusual parasitoids usually corresponds with that of their presumed hosts, adult scarab beetles. Since their hosts tend to be crepuscular or nocturnal, most collection records of Pyrgotidae are based on specimens attracted to lights. Even at that, the generally low number of pyrgotid specimens in collections hardly reflects the diversity of this family, which currently includes about 365 described species in around 55 genera. The apparent rarity of pyrgotid flies might also simply reflect the relative rarity of most host-specific parasitoid species and the high proportion of pyrgotid species restricted to relatively poorly documented tropical faunas. Pyrgotidae occur throughout the world, but only nine species occur in the Nearctic region and only 22 species occur in the Palaearctic.

The Pyrgotidae are characterized as parasitic flies that attack adult scarab beetles by injecting eggs between the beetle's elytra when its abdomen is exposed during flight, but this is based on only relatively few species, almost all from the small Holarctic fauna. The most commonly collected North American genera, *Sphecomyiella* and *Pyrgota*, for example, parasitize June beetles (the very common genus *Phyllophaga*) and the northeastern North American *Pyrgotella chagnoni* can be seen attacking chafers (*Dichelonyx*) around dusk. Curran, in his classic 1934 book on North American Diptera, wrote: "I suspect that *Pyrgotella chagnoni* Johnson is parasitic on species of *Dichelonyx* but my suspicion is based merely upon the fact that I have observed this species commonly in open woods where the adult beetles were very common." Curran's guess was correct; the association of *P. chagnoni* with *Dichelonyx* was confirmed by Canadian dipterist Kevin Barber, who showed me these very beautiful flies darting after chafers on basswood trees during warm June evenings in southern Ontario.

Flies in the family Pyrgotidae are parasitoids of adult beetles in the scarab family such as the chafers (*Dichelonyx*) chased by this eastern North American species (*Pyrgotella chagnoni*).

A few other species have been reared from scarab beetle adults in Africa and Southeast Asia, but the great majority of pyrgotids, including all the Neotropical species and all members of the subfamily Teretrurinae, have never been reared. Despite that, similarities between the needle-like abdominal tip in known parasitic species and the abdominal tips of species of unknown biology suggest that all Pyrgotidae attack adult beetles. Also, adult females of a few Australian pyrgotid species have been observed in apparent pursuit of adult scarab beetles; one was reported as ovipositing through a beetle's anus. However, no Australian species has been reared from its host. Those few pyrgotids with known life histories consume the adult beetle from the inside and pupate in the empty shell of the dead host.

The family Pyrgotidae has been variously subdivided by different taxonomists; the current classification, proposed by D.K. McAlpine (1990) and refined by Korneyev (2004), recognizes two subfamilies plus two or three small genera of "uncertain placement." The great majority (about 350 species) of scarab-killing flies belong in the subfamily **Pyrgotinae**, which is found throughout much of the world but is most diverse in the Old World tropics. This group is absent from New Zealand but includes about 70 species in Australia. The subfamily **Teretrurinae** is usually defined to include only the poorly known southern South American genera *Teretrura* and *Pyrgotosoma*, but Mello and Lamas (2010) argue that this subfamily also includes some endemic Australian genera. Very few higher Diptera groups show this sort of split south temperate distribution. The higher classification of Pyrgotidae is far from resolved, and some unusual "basal" genera, such as the one southern South American species of *Descoleia*, don't fit in either subfamily.

The rare and enigmatic family Ctenostylidae was treated as a subfamily of Pyrgotidae (Lochmostyliinae) until recently, but these rarely collected nocturnal flies are now considered to be only distant relatives of the Pyrgotidae.

Richardiidae

Richardiid Flies

Although the family Richardiidae is one of the smaller groups of tephritoids, at only 175 described species, these attractive flies are both conspicuous and conspicuously diverse throughout the neotropics. Dung and other decomposing material in South or Central America almost invariably attracts a variety of brightly colored richardiids, usually with banded wings and occasionally with outlandishly broadened heads, bizarrely expanded cheeks or disproportionately expanded hind legs. Richardiids rule in the neotropics but the family becomes rare as one heads north into the Nearctic region, where we find fewer than ten species in the United States and only a single species, *Sepsisoma flavescens*, in Canada. The family does not occur at all outside the New World.

Only a few species in this family have been reared, but most richardiids are probably saprophagous, grazing microorganisms from moist microhabitats rich in decaying plant material. *Beebeomyia* species, for example, are well-known inhabitants of the wet, sticky interface between bracts of *Heliconia* flowers. Larval *Beebeomyia* feed in flower parts, and I've seen adult females slipping their long, tapered abdominal tip between the layers of flower buds of *Heliconia* and other water-holding flowers. A few other richardiid species have pushed the boundary between saprophagy and phytophagy, attacking living plant tissue; at least one species, the Pineapple Worm, *Melanoloma viatrix*, is a plant pest. This species emerged as a pest of pineapple fruits in northern South America only in the past few decades, but it is now reported to be causing significant yield losses in Colombia and Venezuela.

Most richardiids are conspicuous flies that use their boldly patterned wings and other ornaments in a variety of interesting inter- and intrasexual interactions. Although few detailed observations are available, the spectacularly broad-headed *Richardia telescopica* males undoubtedly use their modified heads in male–male agonistic interactions, as do the lesser-known *Richardia* species with elaborate genal (cheek) processes. Males in the genus *Setellia* perform ritualized displays on leaf surfaces that serve as lek sites (mating arenas away from any larval or adult food source). Some richardiids, such as those in the genus *Sepsisoma*, are amazing ant mimics, presumably avoiding predation by looking like unpalatable or well-defended ants. *Sepsisoma* males, like most richardiids, have conspicuous male–male agonistic interactions that involve ritualized displays of wings and other body parts.

The family Richardiidae is divided into two subfamilies, with most species in the typical subfamily **Richardiinae**. Most Richardiinae have their hind femora armed with ventral spines, at least near the apex. The small subfamily **Epiplateinae** includes only three genera, two of which, *Omomyia* and *Automola*, are among the few richardiid genera to occur in North America. *Automola rufa* breeds in rotting cacti and *Omomyia* (two species) have been found on dung, wet dead wood and various other moist organic material; both genera occur in the American southwest. *Omomyia* is a problem genus, originally described in the family Coelopidae and since bounced around between the families Thyreophoridae, Pallopteridae and Richardiidae.

Pallopteridae

Flutter Flies

The "flutter flies," so called because of the way some males in this family hold their wings out

Richardiidae, usually recognizable by the sharp spines visible under the hind femur, are common throughout the Neotropical region.

from the body and rapidly vibrate them, comprise a small family of tephritoid flies mostly restricted to temperate climates. The majority of the 70 or so species in the family occur in the Holarctic region, with about half the described species restricted to the Palaearctic region. The family is also well represented in some parts of the southern hemisphere, with nine endemic species in New Zealand and many more in the extreme south of South America (including several species on the Falkland Islands). Pallopterids are apparently absent from Australia and New Guinea.

Flutter fly adults are often conspicuous, diurnally active flies, and some, such as the North American *Toxonevra superba*, can be common and conspicuous as they display on foliage late in the season (and often only late in the day). Larvae occur in plant stems, shoots and flower heads or under bark, where they have been reported as either phytophagous or at least facultatively carnivorous. Some species are certainly phytophagous, including a Japanese species, *Temnosira czurhini*, recently reported as a leaf miner and stem borer in *Huperzia*, a primitive plant in the club moss order (Lycopodiales); a number of European species have been reported as damaging grass shoots. Although many other species are associated with particular plants such as thistles, umbelliferous plants or some kinds of trees, it is not always clear whether they are feeding on plant tissue. At least some species are confirmed predators of beetle larvae under bark, and one western North American species consumes phytophagous gall midges in the cones of Douglas fir. The biology of most Pallopteridae species remains unknown; none of the southern hemisphere species has been reared.

The Pallopteridae are divided into two subfamilies, the Pallopterinae and the Eurygnathomyiinae. The subfamily **Eurygnathomyiinae** includes only the European *Eurygnathomyia bicolor*, a rarely collected species with a distinctively spinose costa. This odd species was treated as a separate family, Eurygnathomyiidae, by Griffiths (1972), who also argued that all other Pallopteridae belonged in a broadly defined family Tephritidae. The biology of *E. bicolor* is unknown. The subfamily **Pallopterinae** is a more heterogeneous group and is the usual home for all other species in the Pallopteridae, even though some, including the only extant species of *Morgea* (from Israel) and the only species of *Aenigmatomyia* (from Chile) are markedly different from other species in the family.

Papp (2011) excluded *Morgea* and *Aenigmatomyia* from the Pallopteridae (without putting them in any other family) and in the same paper described a new family, Circumphallidae, to hold a newly described species based on only one specimen recently collected in Vietnam. This species, *Circumphallus significans*, will key to Pallopteridae, and in the absence of evidence that it is more closely related to some other family or group of families, I prefer to treat it as part of the Pallopteridae. Perhaps, once we better understand the relationships between the genera currently in the Pallopteridae and other groups of Tephritoidea, it will prove necessary to redefine the Pallopteridae, and maybe then we will have to move some species to other families. But for now I prefer not to recognize another small acalyptrate family for one obscure specimen of indeterminate relationships.

Piophilidae

Skipper Flies

Members of the small family Piophilidae are called "skipper flies" because of the well-known leaping behavior of their larvae, a remarkable mode of maggot motion long recognized in the common stored-product pest *Piophila casei*, the Cheese Skipper. Larval leaping, which serves to propel the maggots away from decaying substrates to nearby pupation sites, also occurs in some other flies, although it is best known in the Piophilidae. Observers since the 17th century have remarked on the ability of a *Piophila casei* larva to form a loop by grabbing its tail with its head and then straightening out like a spring to make prodigious leaps from the surfaces of stored food products such as cheese, smoked fish or cured ham, on which these larvae develop.

Cheese Skippers are now cosmopolitan and common insects on a variety of substrates ranging from roadkill to rotten fish, but their association with cheese remains interesting and normally unwelcome. There is, however, an unusual specialty cheese produced only with the help of Cheese Skippers. The Sardinian sheep's-milk cheese *casu marzu* ("rotten cheese") is the result of fermenting pecorino cheese with the help of deliberately introduced Cheese Skipper maggots. A drippingly soft cheese writhing with maggots must surely be an acquired taste, and my appetite is not whetted by reports of diners shielding their *casa marzu* sandwiches with their hands to prevent skipping maggots from hitting them in the face. Cheese Skippers are among the flies that most frequently cause enteric myiasis (see Chapter 3), so if you do consume cheese primed with piophilids, watch out for enteric issues induced by living maggots wriggling around in your intestines and ultimately appearing — alive or dead — in your feces. Even if you don't consume Cheese Skippers when they are alive, they may well consume you when you are not, as *Piophila casei* is a common species on human cadavers.

Most of the 70 or so species in the globally distributed family Piophilidae are associated with high-protein dead food such as carrion, and mostly with the older, drier stages of decomposition. An exception is one unusual species, *Neottiophilum praeustum* (European Nest Skipper), larvae of which suck the blood of nestling birds, much like some Muscidae and some Calliphoridae. *Neottiophilum* and the related Holarctic genus *Actenoptera* (biology unknown) are usually treated as the subfamily **Neottiophilinae**, following J.F. McAlpine's (1989) classification of the Piophilidae, although some authors still treat this group at the family level. The other 66 or so species in the family are treated as the subfamily Piophilinae and occur throughout the world, although native species are restricted mostly to north and south temperate areas. The bulk of the species in the family occur in the Holarctic region. J.F. McAlpine (1989) argued that the family Pallopteridae is most closely related to the Piophilidae, and that these two groups, together with the Richardiidae, comprise a monophyletic lineage within the Tephritoidea.

Piophilinae The subfamily Piophilinae includes the "bone skippers" in the related genera *Centrophlebomyia* and *Thyreophora*, some of which are rarely encountered Old World species associated with large carcasses such as dead horses, but only after they have become largely dry skin and bones. One species of bone skipper, *C. anthropophaga*, had not been recognized since its description in 1830 when Michelsen (1983) reported it from the carcass of a horse in Kashmir. Another, the strikingly orange-headed *T. cynophila*, had not been seen since 1850 and was considered extinct until two papers published in 2010 independently recorded it as extant in Spain (Carles-Tolrá et al., Martín-Vega et al.). These spectacular blue-black and orange flies are active only in the winter and develop only in the marrow of the crushed or broken bones of dead large animals, which is probably why entomologists overlooked them for the past century and a half — the remains of large animals with crushed bones are rarely seen lying around the forest these days, and not too many entomologists look for flies in the colder months. *Thyreophora* and its relatives have been treated in the past as a separate subfamily or even a separate family (Thyreophoridae).

The North American antler flies recently given the name *Protopiophila litigata* are also in the subfamily Piophilinae. These remarkable little flies were the subject of Russell Bonduriansky's thesis work at the University of Guelph, Ontario; he discovered them breeding in the upper surfaces of discarded moose antlers and named them *litigata* to reflect their penchant for spectacular turf battles on the antler surfaces (1995). Bonduriansky described the males as readily attacking flies vastly larger than themselves, or "even the tip of a biologist's pen," as they defend mating territories or search the antlers for arriving females. The successful male remains on the female's back while she lays eggs in the antler, fending off rivals with his wings. Larvae develop in the porous bone matrix inside the antler, later pupariating at the surface.

One of my favorite piophilines is *Prochyliza xanthostoma*, a Nearctic species that appears in abundance in early spring when melting snow exposes the carcasses of winter-killed animals. Males of *P. xanthostoma* have long antennae and a greatly elongated head that is used in ritual male–male battles for territory, during which these striking insects stand on their hind legs and engage in vigorous head-butting contests. Another piophiline of note (if only because its image adorns a University of Guelph entomology T-shirt!) is the Holarctic genus *Amphipogon*, named for its "two beards": tufts of hair that stick out from the underside at the opposite ends of the male's body. Nobody has recorded the behavior of *Amphipogon*, and the role of its spectacular adornment remains to be discovered.

Ctenostylidae

Enigmatic Ctenostylid Flies

Ctenostylids are among the rarest and strangest of all flies. These nocturnal flies lack ocelli and mouthparts, the females have uniquely fanlike aristae and oddly transparent facial "windows," and both sexes have a proportionally huge head and broadly banded wings. This group was until recently treated as a subgroup of the Pyrgotidae, and it has been generally assumed that Ctenostylidae, like Pyrgotidae, are parasitoids. The truth, however, is that we know almost nothing about the life histories of these bizarre flies. A couple of female specimens of South American species were found to have first-instar larvae in their abdomens, so the family is presumably larviparous, but there are no host data for any species in the family. One possible clue to the biology of the family is a large, convex prosternum, somewhat reminiscent of the grossly inflated prosternum used by tachinids in the genus *Ormia* to locate singing Orthoptera hosts.

The few ctenostylid specimens in the world's insect collections were taken almost exclusively from light traps in the Oriental, Afrotropical and Neotropical regions, although a species was recently reported from Korea, which is part of the Palaearctic region. The family includes only 13 species, scattered among seven genera and known mostly from only one or two specimens per species. All the ctenostylid specimens ever collected would probably fit into a single insect-pinning box.

Although they are strikingly similar to Pyrgotidae and were previously treated as part of that family, D.K. McAlpine (1990) reassessed the evidence for this placement. He concluded that Ctenostylidae were probably not closely related to Pyrgotidae, because they retain structures — different parts of the male and female abdomens — lost in a common ancestor of Pyrgotidae and related tephritoids. If the Ctenostylidae and Pyrgotidae were closely related, then these lost structures would have had to reappear in the Ctenostylidae, which seems unlikely. D.K. McAlpine concluded that the evidence for including the Ctenostylidae in the Tephritoidea was "at best ambiguous" but suggested continued treatment of the family as part of the Tephritoidea anyway, since evidence for a relationship with any other superfamily is lacking. Ctenostylidae remain an enigma — one of those tantalizingly rare groups that beg further investigation and promise surprising new discoveries about fly relationships and biology.

Nerioidea (Micropezoidea)

Stilt-legged Flies and Their Relatives

Micropezidae

Stilt-legged Flies

The family Micropezidae is a distinctive group of about 700 described species, most of which have relatively long and sometimes strikingly stilt-like mid and hind legs. These common flies are often seen standing prominently on leaves or similar surfaces, apparently emboldened by their resemblance to well-defended Hymenoptera such as ants, spider wasps and ichneumonid wasps. Many Neotropical micropezids in particular seem to mimic ichneumonids in both morphology and movement. The forelegs of micropezids in the largest subfamily, Taeniapterinae, are shorter and narrower than the mid and hind legs and are often colored and moved like wasp antennae. Ant mimicry has also evolved repeatedly in the family; it is spectacularly exhibited by the only species in the Australian genus *Badisis* (subfamily Eurybatinae), some species in the southern South American genus *Cryogonus* (Micropezinae) and many species in the Neotropical genus *Plocoscelus* (Taeniapterinae). *Badisis ambulans*, a wingless species that develops only in the cup leaves of Western Australian pitcher plants (Albany Pitcher Plant, *Cephalotus follicularis*), not only looks exactly like nearby dolichoderine ants but also behaves the same way, jumping when disturbed.

Micropezids often engage in elaborate courtship behaviors that involve male displays, transfer of regurgitated fluids from male to female and other complex behavior such as posturing, stroking and "kissing." Male Taeniapterinae often have prominent eversible pouches in the membranous sides of the abdomen; these are usually inflated during courtship and probably serve to disperse pheromones.

Larval habitats are little known; Taeniapterinae and Eurybatinae often develop in decaying wood or stems such as banana stumps, ginger roots, cut sugarcane or the rotting roots and stem bases of cattails (*Typha*). Some *Mimegralla* species (Taeniapterinae) apparently damage otherwise healthy roots, suggesting that not all taeniapterine species are strictly saprophagous. Larvae of the mostly Holarctic subfamily Calobatinae have been found under the bark of dead trees but are more often found in decaying vegetation such as piles of grass clippings or accumulations of plant debris in wet areas. Calobatines have also been reared from sewage tanks and pig dung, although adult calobatines are not regular dung visitors. Most Micropezinae have never been reared or collected

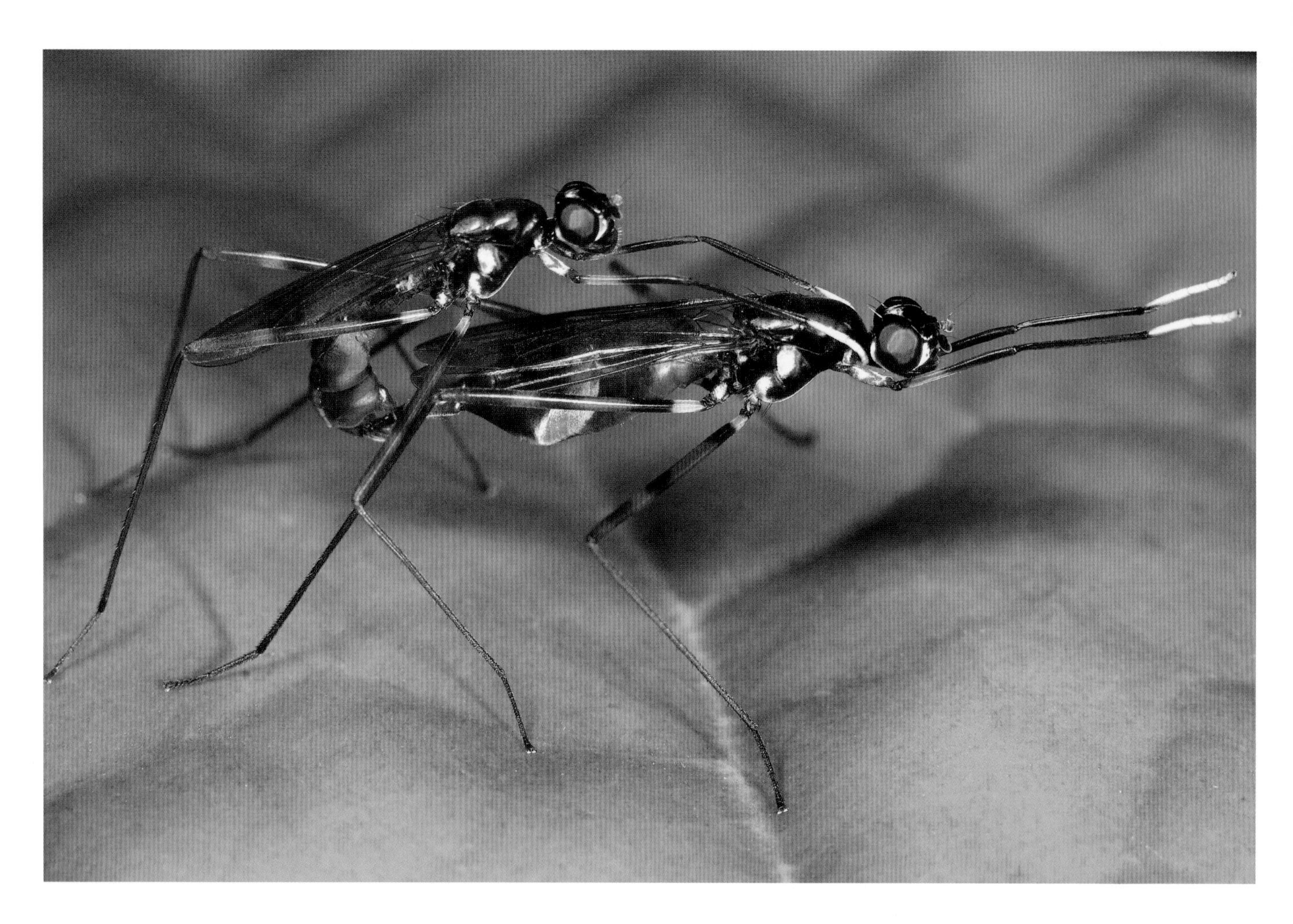

These Southeast Asian taeniapterine micropezids have white-tipped forelegs waved like the antennae of similar ichneumonid wasps. The male of this pair has gripped the tip of the female's abdomen with a fork-like process on his fifth abdominal sternite, also typical for this large and widespread subfamily.

as larvae, but the secondarily widespread species *Micropeza corrigiolata* develops in fresh, healthy root nodules of several leguminous plants, hollowing them out before leaving the empty nodule shells and pupariating in the soil. Other *Micropeza* species, which are commonly collected on legumes, are likely to have similar habits. The subfamily Calycopteryginae is restricted to the subantarctic islands, where the larvae of the only species in the subfamily, *Calycopteryx moseleyi*, develop in both the living and dead parts of endemic plants, and possibly in other habitats such as decaying seaweed.

Adults of many Taeniapterinae and some Eurybatinae are attracted to dung; Neotropical species can often be seen in great numbers on leaves spattered with fecal material from monkeys in the trees above. Micropezinae are occasionally seen feeding on honeydew but are not normally attracted to dung or other baits. Calobatinae, unlike other Micropezidae, are predators as adults and are often seen feeding on small flies, usually nematocerans. Some Micropezidae are associated with edge habitats or even highly disturbed areas, but most are associated with native forests and a few, such as the Neotropical taeniapterine genus *Mesoconius*, are exceptionally restricted in habitat. Most *Mesoconius* species do not seem to occur outside pristine or almost pristine high-elevation forests, and patches of high-altitude forest throughout the neotropics are home to endemic *Mesoconius* species (mostly undescribed).

The family Micropezidae is currently divided into five subfamilies, of which one, Calycopteryginae, is restricted to the subantarctic islands; two, Micropezinae and Taeniapterinae, have widespread distributions; one, Calobatinae, is a mostly northern group; and one, Eurybatinae, is mostly Pacific and Oriental. Three of the subfamilies were treated as the separate families Calobatidae, Taeniapteridae and Micropezidae by Hennig (1973).

Eurybatinae The subfamily Eurybatinae is primarily Pacific-Oriental and includes only a single New World species (a rare Costa Rican species placed in the endemic Central American genus *Notenthes*) and a single Afrotropical species (a rare species of *Cothornobata* known only from the Mascarene islands of Réunion and Mauritius). Most of the Australasian species of Micropezidae are among the Eurybatinae, as is the remarkable Indonesian species *Anaeropsis guttipennis*, the only member of the entire Nerioidea with stalked eyes. The Eurybatinae and the Taeniapterinae are closely related; both subfamilies are often associated with fallen, cut or damaged trees.

Taeniapterinae The overwhelming majority of micropezids belong to the diverse worldwide subfamily Taeniapterinae. Despite the relatively large size and conspicuous behavior of most Taeniapterinae, this mostly tropical group remains poorly known, with many undescribed species and an unstable generic classification. The subfamily Taeniapterinae as a whole is a well-defined group characterized by a number of distinctive features, such as the absence of surstyli (male claspers) and the presence of a vertical row of bristles on the side of the thorax (katepisternum), but many of the included genera as currently defined are artificial and can be difficult to identify. Most Taeniapterinae occur in the neotropics, but the subfamily also includes some common North American species and a number of Old World genera. The 35 or so currently recognized genera of Taeniapterinae include the incredibly long-legged Madagascar genus *Stiltissima*, one of two genera of micropezids entirely restricted to Madagascar; the endemic Micronesian genus *Steyskalia*, known only from Ponape; the Mascarene endemic *Courtoisia*, known only from Réunion; the endemic African genus *Glyphodera*, with its long and distinctive thoracic "neck"; and many striking Neotropical groups such as the ant-like genus *Plocoscelus* and the black-bodied, red-headed *Scipopus* species.

Calycopteryginae *Calycopteryx moseleyi*, the only member of the subfamily Calycopteryginae, is restricted to the windswept subantarctic islands of Kerguelen and Heard, where it is among the dominant insects. *Calycopteryx moseleyi* is flightless, and otherwise so modified for its extreme environment that it looks very different from all other micropezids.

Micropezinae Species of the mostly Neotropical and Holarctic subfamily Micropezinae have relatively delicate bodies that seem barely able to balance their large, often elongate heads. With the exception of a handful of species in the southern South American genus *Cryogonus*, all members of the subfamily are currently treated in the genus *Micropeza*, but the taxonomy of the group is long overdue for revision. Micropezines are most diverse in the neotropics, but the subfamily occurs throughout Europe and Asia, and it was recently recorded from South Africa as well. The originally European but now widespread *M. corrigiolata* develops inside the root nodules of legumes, a habit that may turn out to be widespread in the genus or even the subfamily.

Calobatinae The subfamily Calobatinae is a small Holarctic group with so little in common with other micropezids — other than a stilt-legged appearance — there remains some question about whether it should be treated as a separate family, Calobatidae. Larvae are known from various kinds of decomposing vegetable material, including dung and compost, and adults are at least facultatively predaceous.

Neriidae

Cactus Flies and Their Relatives

The 120 or so species of Neriidae are sometimes called "cactus flies" because the best-known species, *Odontoloxozus longicornis*, develops in the necrotic tissue of wounded cacti in the southern United States, Mexico and Central America. Most neriid species, however, frequent damaged tree trunks, fallen trees and similar localized pockets of decomposing plant material such as cut banana stems. These attractive, medium-sized (usually around 10 or 11 mm) flies have a porrect, heads-up stance made all the more characteristic by a prominent, often pointed antenna with an apical or almost apical (dorsoapical) arista. Look on wounded tree trunks or sappy fallen trees almost anywhere in the New World or Old World tropics to see female neriids extending their telescoping ovipositors into pockets of moist decay while males stand by or protectively straddle the ovipositing females. Oozing beetle borings and moist bark edges on recently cut or broken stems are particularly attractive to neriids; they constitute discrete oviposition sites often held by males defending their exclusive right to mate with females arriving to lay eggs.

Males of more than one species of neriid sometimes stake out the same patch of attractive habitat, where the different species ignore one another but vigorously interact with members of their own species. Eberhard (1998) observed the Neotropical species *Glyphidops flavifrons* and *Nerius plurivittatus* on a fallen tree in Panama, where males of each species aggressively rebuffed conspecific male rivals. Male–male interactions involved "stilting" — the males trying to look as tall as possible to swiftly deter smaller rivals — as well as more energetic head-to-head clashes and chest-butting matches between similarly sized males. Males also slashed and whacked each other with impressively armed front legs. Male–female interactions were found to include complex leg waving and species-specific genitalic fondling, vibration, stroking, tugging and tapping during copulation.

Preston-Mafham (2001) studied an analogous pair of Old World species, *Gymnonerius fuscus* and an unidentified *Telostylinus* species, on a fallen mango tree in Sulawesi. These flies were attracted to oozing beetle borings that constituted relatively small, discrete oviposition sites; Preston-Mafham found that both species energetically defended their patches of attractive turf from conspecific males whether females were present or not. Arriving females were allowed to oviposit only after mating with the territory-holding males, which vigorously defended their attractive real estate from other males with ritualized displays such as wing flicking and stilting to look as tall as possible, or by pitched battles involving wrestling and head pushing between stilting contestants. Similar behavior has been recorded for North American populations of the Cactus Fly, *Odontoloxozus longicornis* (Mangan 1979).

Despite the conspicuous appearance and interesting deportment of these attractive long-legged flies, and despite the thorough studies of mating behavior mentioned above, the Neriidae remain a poorly known family. One of the reasons for this is that the male genitalic structures, which usually provide reliable evidence for species and genus characterizations in other Diptera, are extraordinarily uniform in Neriidae (and most other Nerioidea), probably because of the complex behavioral sequences that characterize courting or mating neriids. Genitalic morphology has apparently remained relatively static through speciation events while female choice has driven rapid evolution in other characteristics, such as behavior, involved with mate selection. Whatever the reason, taxonomy of nerioids is made especially challenging by the lack of useful morphological characters in the male genitalia. The only two species now included in the cactus fly genus *Odontoloxozus*, for example, have identical male and female genitalia. The second species in the genus was discovered only when Robert Mangan and David Baldwin noticed that Mexican (Baja California) and southwest American (Arizona) individuals of what they thought were *O. longicornis* had difficulty mating. They went on to document differences in behavior, external morphology and chromosomes in support of their recognition of the Mexican populations as a separate species, *O. pachycericola* (Mangan and Baldwin 1986).

The Neriidae are divided into two subfamilies, the Neriinae and the Telostylinae. The largest of these, the subfamily **Neriinae**, seems to be a natural (monophyletic) group characterized by an unusual development of the upper face, which creates a shelf-like "antennal base" just below the ptilinal fissure. The Neriinae are most diverse in the neotropics but range as far north as the southern United States and also include some Oriental and Australian species, mostly in the genus *Telostylinus*. All named New World Neriidae belong to the Neriinae. The mostly tropical Old World and Pacific subfamily **Telostylinae**, on the other hand, seems to be characterized only by a lack of the derived characters that define the Neriinae, and is thus likely to be an artificial (paraphyletic) assemblage of basal lineages.

Cypselosomatidae and Pseudopomyzidae

Although the bright black and yellow species in the Neotropical genus *Latheticomyia* are attractive flies that look a bit like miniature cactus flies (Neriidae), other members of the small families Cypselosomatidae and Pseudopomyzidae are generally small, dark flies rarely noticed even by dipterists. The two families are sometimes treated together as the single family Cypselosomatidae, including genera that have at one time or another been treated as Sphaeroceridae, Micropezidae, Sepsidae, Heleomyzidae and Clusiidae. These are poorly known flies, with many species still awaiting description and with little known about their combined 35 or so named species. Adults can sometimes be collected by using small dung baits, while both of the Nearctic species in the group were first discovered by using traps baited with rotting banana. Larvae are probably microbial grazers in decomposing material.

The family **Pseudopomyzidae** is a small but widespread group known from every zoogeographic region except the Afrotropical. One of the four pseudopomyzid genera found in the New World, *Heloclusia*, occurs only in Chile; the others occur throughout Central and South America and (except for *Pseudopomyzella*) the Caribbean. With the exception of the two species of *Latheticomyia* that occur in the southwestern United States, pseudopomyzids are strangely absent from North America, but they reappear in Europe and Asia. The most common genus in both South America and Europe is *Pseudopomyza*, although the Neotropical members of the genus are in a distinct subgenus, *Rhinopomyzella*. Other *Pseudopomyza* subgenera occur in Europe, Asia, Australia and New Zealand, and further genera of Pseudopomyzidae are found in Asia (*Polypathomyia*) and the Oriental region (*Tenuia*). An extinct genus, *Eopseudopomyza*, was described from Baltic amber. Larvae of an Old World species have been found under bark but larvae of New World species remain unknown.

The **Cypselosomatidae**, excluding the Pseudopomyzidae, are absent from the Holarctic, Afrotropical and Neotropical regions, with all extant species occurring in the Australasian and Oriental regions. An extinct Palaearctic species is known from Baltic amber, but the family placement of that fossil, *Cypselosomatites succini*, has been questioned. One Australian cypselosomatid species has been reared from guano in the caves of bent-wing bats and has also been collected in large numbers flying out of pit latrines.

Diopsoidea

Stalk-eyed Flies, Syringogastrids and Their Relatives

Psilidae

Psilid Flies

The Psilidae are sometimes called "rust flies" because larvae of one species in the family — the Carrot Rust Fly, *Psila rosae* — feed on developing parsnip, turnip or carrot roots, causing rust-colored etching and patches of decay. The other 320 or so species in this group of small (3–11 mm), bare, generally shiny acalyptrates are mostly stem or root borers or occur under the bark of coniferous or broadleaf trees. The greatest known diversity is Holarctic and Afrotropical; adults of the large genus *Chyliza* are common on low foliage in much of the world. *Chyliza* is usually recognized as distinct from other Psilidae and treated in a separate subfamily **Chylizinae**. Most psilids are in the subfamily **Psilinae**, which is made up mostly of the large genus *Psila* and the distinctive genus *Loxocera*. *Loxocera* species, conspicuous for their remarkably long antennae, are mostly Holarctic or Old World tropical, and the genus was unknown from the neotropics until 2006. A third subfamily, **Belobackenbardiinae**, was recently described for three South African species.

Diopsidae (including Centrioncidae)

Stalk-eyed Flies

Although the common name "stalk-eyed flies" refers only to the small (just under 200 species) family Diopsidae, many families of flies include hypercephalic (broad-headed) species, often with the eyes stalked or elongated far to the sides of the head. Hypercephaly is found in several acalyptrate families, including the Micropezidae, Tephritidae, Periscelididae, Platystomatidae, Richardiidae, Ulidiidae, Drosophilidae and Diopsidae, but it is only in the last family that it is an almost universal feature, absent in only the small subfamily Centrioncinae. Furthermore, it is only in the Diopsidae that both the eyes and antennae are at the ends of stalks. Both male and female diopsids have stalked eyes, while in most other broad-headed groups only the males are hypercephalic. In some diopsid species females prefer males with the longest stalks; this has driven conspicuous sexual dimorphism. These same relatively long-stalked males use their broad heads in ritualized male–male battles that leave the best-endowed in possession of mating sites. In these species it appears that size really matters as an indicator of male fitness or fertility.

North Americans and Europeans can be excused for being less than impressed by diopsid displays, since the one European and two North American species of stalk-eyed flies are only moderately hypercephalic. South Americans and Australians are even more out of luck, since the family Diopsidae is absent from both the Neotropical region and Australia. Most species occur in the Old World tropics, especially Africa and Southeast Asia, where stalk-eyed flies are extremely common insects on almost any low foliage, especially shrubs and herbaceous vegetation near streams. Adults can be seen sparring or grazing on microflora on leaf surfaces; larvae most often develop in dead vegetation, although several species are strictly phytophagous. Adult diopsids are often armed with spikes, spines or other conspicuous processes, and some have armatures and shapes that closely resemble co-occurring ant species.

The family Diopsidae is divided into three subfamilies, the Centrioncinae (sometimes treated as a separate family, Centrioncidae), Sphyracephalinae and Diopsinae. Although true Diopsidae do not occur in the neotropics, the probable sister group to the Diopsidae is the entirely Neotropical family Syringogastridae.

Centrioncinae The subfamily Centrioncinae includes only the very closely related African genera *Centrioncus* and *Teloglabrus*, which lack the eyestalks that characterize other Diopsidae. Feijen (1983) created the family Centrioncidae for these two genera and suggested that they are related to the Neotropical family Syringogastridae rather than to other Diopsidae. Newer molecular and morphological data have not supported this idea and most dipterists currently treat *Centrioncus* and *Teloglabrus* as part of the Diopsidae. Larvae of Centrioncinae remain unknown.

Sphyracephalinae The subfamily Sphyracephalinae is the only subfamily of the Diopsidae to reach north temperate regions, and the only one to extend from the Old World to the New World. The genus *Sphyracephala*, with one recently discovered European species and two North American species, is a group of moderately broad-headed saprophagous diopsids. The most common Nearctic species, *S. brevicornis*, is found in wetlands, and adults are often seen on skunk cabbage. Larvae have been reared from decaying material among sphagnum moss.

Diopsinae The subfamily Diopsinae is home to most of the species in the family, including several spectacularly broad-headed groups and some well-known phytophagous species. *Diopsis macrophthalma*, for example, feeds only on rice plants, and *Diopsis indica* larvae consume maize (corn) plant tissue. Other diopsines are saprophagous or facultatively phytophagous, developing at least partly on decomposing plant tissue and associated microorganisms. Some oviposit in plants already damaged by stem-boring caterpillars, developing in the frass and decomposing tissue in the borings.

Syringogastridae

Syringogastrid Flies

The Neotropical fly family Syringogastridae is one of those remarkable little fly groups that vividly illustrate the significant gaps that remain in our knowledge of fly diversity. Although syringogastrids are distinctive and striking in shape, size and color, the family was described

only in 1969, most of the species in the family were first named in 2009, and we still have no idea where the larvae live or what they do. The adults are similar in size (around 5 mm) and shape to stinging ants in the genus *Pseudomyrmex* ("twig ants"). Their wing pigmentation augments the petiolate abdomen to render the similarity so remarkable that even experts can mistake them for ants until the moment they take wing. All 23 species in the family are placed in the genus *Syringogaster*; the 21 extant species are found from Mexico to Argentina and two extinct species occur in Miocene amber from the Caribbean. Although rare in collections, aggregations of *Syringogaster* are not uncommon on broad leaves in the lowland neotropics, and they sometimes occur in fairly dense aggregations of more than one species. It is not hard to find mating pairs or gravid females, so it is remarkable that the biology of the family remains a mystery.

The family is a very well-defined and easily recognized group, as suggested by the inclusion of all species in a single genus, and most specialists agree that the Syringogastridae are closely related to the stalk-eyed flies, Diopsidae. There is, however, some disagreement about whether this relationship is between the Syringogastridae and

Stalk-eyed flies, like this *Eurydiopsis* from the Philippines, have the sides of the head extended out as stalks, with both the antennae and eyes at the ends of the stalks. Both male and female diopsids have broadened heads.

Diopsidae as a whole or between Syringogastridae and the group of African flies variously treated as the family Centrioncidae or as a subfamily, Centrioncinae, in the Diopsidae. It is most likely that Diopsidae as a whole — including Centrioncinae — is the sister group (most closely related) to the Syringogastridae.

Tanypezidae and Strongylophthalmyiidae

Tanypezid and Strongylophthalmyiid Flies

This small lineage of around 75 long-legged flies is sometimes divided into two families, Tanypezidae and Strongylophthalmyiidae, and sometimes treated as the single family Tanypezidae, with two subfamilies, Tanypezinae and Strongylophthalmyiinae.

Tanypezidae The extremely skittish and speedy adults of the family Tanypezidae are common on foliage throughout the neotropics, where they appear to graze on microflora, honeydew and the occasional bird dropping. The larvae are poorly known, although one Nearctic species has been reared from eggs placed on decaying watermelon rind. Only two species in this family occur in North America, and one of these, *Tanypeza longimana*, is the only member of this essentially Neotropical group that occurs in the Palaearctic region. All 25 species of Neotropical Tanypezidae, 10 of which are just about to be named by Canadian dipterist Owen Lonsdale, belong to the genus *Neotanypeza*.

Strongylophthalmyiidae The family Strongylophthalmyiidae is an essentially Old World group that occurs mostly in Southeast Asia. The group is not known from the neotropics, and only two members of the family occur in North America. One of these, *Strongylophthalmyia pengellyi*, was only recently described; it was named in honor of a great teacher of entomology, the late D.H. Pengelly, my PhD advisor. *Strongylophthalmyia pengellyi* was discovered under the bark of damaged old-growth poplar trees in northern Ontario and named by another of Pengelly's former students, Kevin Barber. Most of the 50 or so species in the family are in the genus *Strongylophthalmyia*, although there is a second genus, *Nartshukia*, with a single Oriental species.

Megamerinidae

Megamerinid Flies

The 15 or so species in the small Oriental and Palaearctic family Megamerinidae are distinctive elongate flies with swollen and colorful hind femora. Immature stages are poorly known, but larvae of a European species have been found under bark, where they were probably eating other subcortical Diptera. Some megamerinids, especially Oriental species in the genus *Texara*, are strikingly similar to the Neotropical Syringogastridae, which were originally treated as part of the Megamerinidae. However, the family Megamerinidae is related to a larger diopsoid lineage combining Diopsidae and Syringogastridae rather than to the superficially similar *Syringogaster*. The superfamily placement of this little family has a contentious history, having been previously grouped with the Nerioidea (D.K. McAlpine 1997b) and the Sciomyzoidea (Griffiths 1972).

Nothybidae

Nothybid Flies

The small family Nothybidae is the only family of flies entirely restricted to the Oriental region. All nine named species of Nothybidae are in the genus *Nothybus*, and most occur in wet rainforests of Indonesia. Nothing is known about the biology of this group and larval habitats remain unknown, although at least some species are macrolarviparous, a conclusion based on the presence of large larvae in the abdomens of preserved specimens. Nothybids are medium-sized (7–15 mm) flies, usually with patterned wings and a strikingly elongate thorax.

Somatiidae

Somatiid Flies

Somatiidae are small, squat, black and yellow flies with feathery aristae and a dome-like abdomen. The seven species in the family are all Neotropical, and the adult flies are most often found underneath the leaves of leguminaceous plants, often strutting about with their wings held outstretched. Essentially nothing is known about their biology beyond what is reflected in one photograph, of an adult feeding on a dead insect in Costa Rica. These rarely seen little (3–5 mm) flies are so different from other fly families that it has been difficult to ascertain their relationships. Different authors have treated them as close to Periscelididae, Richardiidae, Psilidae or Heleomyzidae.

Gobryidae

Gobryids or Hinge Flies

The family Gobryidae is a rarely collected group with one genus and only five described species, mostly distributed in the Oriental region but with records as far east as New Guinea. Almost nothing is known about the biology of this recently described family. Larvae remain unknown but adults characteristically perch on the undersides of broad leaves. It is unusual for such an obscure group to have a common name, but the term "hinge fly" was coined in the original paper describing the family (D.K. McAlpine 1997a). The so-called hinge is a pinched area at the base of the abdomen that gives it a broad, hinge-like articulation with the thorax. Gobryids are markedly similar to Megamerinidae, although they are a bit smaller and characterized by the distinctive hinge.

Ephydroidea (Drosophiloidea)

Shore Flies, Laboratory Fruit Flies, Quasimodo Flies, Bee Lice and Their Relatives

Drosophilidae

Vinegar Flies and Their Relatives

The family Drosophilidae is best known for the Laboratory Fruit Fly, a species so familiar that its scientific name, *Drosophila melanogaster*, is probably the most widely recognized formal name of any animal other than *Homo sapiens*. Less widely appreciated is the size (about 4,000 species) and diversity of the Drosophilidae as a whole, a family that includes predators and plant feeders as well as yeast-grazing species such as *D. melanogaster*.

The genus *Drosophila* itself, with some 1,500 species, is larger and more diverse than most families of living things. *Drosophila* species are generally saprophagous filter-feeders attracted to yeast and other microorganisms in various oozing substrates, including fungi, flowers and fruit; but some of the most specialized lineages within the genus are predators or commensals associated with a narrow range of other invertebrates. The range of forms and life histories in the genus is especially striking in the Hawaiian Islands, where the arrival of an ancestral *Drosophila* species 26 million years ago was followed by explosive adaptive radiation into almost 1,000 Hawaiian drosophilid species. Many of the Hawaiian species are large, colorful insects with boldly patterned wings, in contrast with the familiar little (3 mm) yellowish *Drosophila melanogaster* and its mainland relatives. *D. melanogaster* itself is a secondarily widespread species that probably originated in Africa, but many similar species are native to North America. The other native genus of Drosophilidae in the Hawaiian Islands is *Scaptomyza*, which is thought to have originated on the islands from an ancestor held in common with the Hawaiian *Drosophila* (O'Grady and Desalle 2008). That, of course, would mean that *Drosophila* (without *Scaptomyza*) is not a natural group, and seems to suggest that the large number of mainland *Scaptomyza* are all descended from a species that "escaped" Hawaii, perhaps crossing the 4,000 kilometers to the nearest mainland as a larva or puparium in a fruit stuck to a migrating bird.

Nothybidae, the only entirely Oriental fly family, is a poorly known group found in the wet forests of Southeast Asia.

The genus *Drosophila*, with some 1,500 species, is larger and more diverse than most families of living things.

Given the size and heterogeneity of the genus *Drosophila* one might wonder why it has not been broken into several, more manageable genera, especially given that the phylogeny of the group is well known and a number of subgroups are already recognized as subgenera. There have been proposals to divide the genus, putting the species closely related to the type species, *D. funebris*, into a more narrowly defined *Drosophila* and recognizing other groups under other generic names (mostly the same groups currently treated as subgenera). However, because the famous *Drosophila melanogaster* is in one of those other groups — it is not closely related to the type species — such a division would land this most famous of flies in another genus, *Sophophora* (currently a subgenus). A reclassification that changed the name of this iconic insect to *Sophophora melanogaster* would be unlikely to gain widespread acceptance, so I think most people will continue to recognize a megagenus *Drosophila* that includes *D. melanogaster*. Similar problems occur with enormous genera in other families, such as the genus *Megaselia* in the Phoridae and the genus *Aedes* in the Culicidae, in which communication and taxonomic stability are arguably best served by keeping the huge genera.

The Drosophilidae belong to the well-defined superfamily Ephydroidea, along with Ephydridae and several smaller families. Some evidence suggests that Curtonotidae and Drosophilidae are closely related. The Drosophilidae are divided into two subfamilies: the enormous subfamily Drosophilinae and the relatively small Steganinae.

Drosophilinae The subfamily Drosophilinae includes the big, closely related genera *Scaptomyza* and *Drosophila* discussed above, and much of the diversity in the subfamily is reflected in these two genera. *Scaptomyza* species normally live in rotting fruit or similar wet, decaying material, but the 150 or so Hawaiian species in the genus include some exceptional insects. *Scaptomyza cyrtandrae* larvae, for example, live under the leaves of one kind of shrub, *Cyrtandra*, where they subsist on the secretions of leaf glands. Larvae of another group of Hawaiian *Scaptomyza*, the subgenus *Tantalia*, scavenge exposed on the surfaces of fallen leaves, protecting themselves by gluing bits of debris to their bodies. Yet another subgenus of *Scaptomyza*, *Titanochaeta*, is made up of species that live as predators in the egg sacs of endemic Hawaiian crab spiders.

Among the many subgenera of *Drosophila* we find species that prey on black fly and midge larvae, devour dragonfly eggs, specialize in slime fluxes, breed only in specific living flower hosts, depend on decaying cacti, live between the upper and lower surfaces of living or dead leaves, coexist as commensals on crabs, consume living embryos in the egg masses of Neotropical frogs and, of course, specialize in all kinds of living and dead fruits, fungi and other microbially rich substrates. At least one *Drosophila* species has been reared from the spittle masses of Cercopidae (spittlebug) nymphs, a habitat more widely exploited by species in the large New World, mostly Neotropical genus *Cladochaeta*. Spittlebug nymphs in the genus *Clastoptera* often share their spittle masses with *Cladochaeta* larvae, which either feed on the host nymphs without harming them or feed on the spittle mass itself. Similar habits seem to have arisen independently in some African species of the large, widespread genus *Leucophenga*, in the subfamily Steganinae.

Several of the *Drosophila* species associated with decaying fruit can become domestic nuisance pests, but only a very few *Drosophila* species are primary pests that develop in fresh fruit. *Drosophila suzukii* — which North Americans call the Spotted Wing Drosophila, or simply SWD, because males have distinctively spotted wings — is an originally Oriental pest of small fruit crops that was first detected as an invasive species in Hawaii in the 1980s. It was then detected in California some 28 years later and in Europe at almost the same time. By 2010 *D. suzukii* had become a serious pest in western Canada and the Pacific states, and as I write this (2011) it seems poised to become a problem in eastern North America as well. The Spotted Wing Drosophila can quickly build up large populations because its short life cycle allows multiple generations per season; its unusual ability to saw into fresh fruit to lay eggs has resulted in heavy economic damage in infested areas, as larval feeding encourages decay and destroys the fruit. A great deal of energy and funding is now devoted to monitoring and researching this major new pest in North America. We are lucky that the Spotted Wing Drosophila happens to be distinctively marked and relatively easy to recognize, since there are no easy tools for identification of most other *Drosophila* species, and a less obvious invasive species would likely remain unidentified and undetected for a long time. Incidentally, *D. suzukii* has long been known under the common name Cherry Drosophila in its native range, where its closest relatives also have spotted wings. There are no formal international rules governing common names, but the International Code of Zoological Nomenclature strictly forbids this sort of arbitrary replacement of a scientific name such as *D. suzukii*.

Steganinae Species of *Stegana*, the nominate genus of the relatively small subfamily Steganinae, have a markedly beetle-like appearance because of the way the wings are plastered to the top of the body (several acalyptrate flies in other families, including Lauxaniidae and Sphaeroceridae, are similarly beetle-like). These striking little flies are often common on freshly fallen logs, where the larvae probably feed on fungus-infested sap. Many other steganines are associated with other invertebrates as predators or inquilines. *Acletoxenus* species are predators of whiteflies; some species of *Cacoxenus*, *Pseudiastata* and *Rhinoleucophenga* eat scale insects; and some *Cacoxenus* are recorded as inquilines in the nests of solitary bees. *Leucophenga*, a large and widespread genus of mostly fungus-breeding species, has a few Afrotropical species that develop inside the spittle masses of spittlebugs (Cercopidae), much like the genus *Clastoptera* in the Drosophilinae.

Ephydridae (including Risidae)

Shore Flies

The Ephydridae make up a diverse family of almost 2,000 species, most of which develop in aquatic or semiaquatic habitats. Adults, usually distinctive for their bulging faces and correspondingly massive mouth openings, are among the dominant dipterans of damp places, especially the moist margins of ponds, streams, lakes and oceans. Although they occur in all sorts of wet habitats, ephydrids, or "shore flies," can be spectacularly abundant in saline lakes and other extreme environments such as hot springs. One western American and Caribbean species, *Helaeomyia petrolei*, is called the "petroleum fly" because it develops in seepages of crude oil. Larvae have their posterior spiracles on long stalks, allowing them to stay largely submerged while feeding on the remains of other insects trapped in the oil.

Although most larval shore flies are microbial browsers or filter feeders in aquatic or semiaquatic environments, there is an enormous morphological and biological diversity in this group of tiny (around 1 mm) to large (over 10 mm) flies. Some develop in carrion or dung, one species develops as a predator in the foam nests of leptodactyline frogs, some species develop as parasitoids in spider egg sacs, and several are leaf or stem miners. Adult diets are a little less varied than those of the larvae — most species graze on algae or microorganisms

— but both adults and larvae in the widespread and common genus *Ochthera* are predators. The habitat range in the family is as impressive as the range in size, morphology and food habits, with species in the extremes of aquatic habitats ranging from hot springs through to bone-chilling mountain waterfalls.

The Ephydridae have traditionally been divided into four subfamilies: Ephydrinae, Hydrelliinae, Parydrinae and Psilopinae. However, most dipterists now recognize five subfamilies: Ephydrinae, Hydrelliinae, Discomyzinae, Gymnomyzinae and Ilytheinae, on the basis of a phylogenetic analysis by Zatwarnicki (1992).

Hydrelliinae Almost half of the 500 or so species in the subfamily Hydrelliinae belong to the large and widespread genus *Hydrellia*; its phytophagous larvae are normally found mining in the leaves or stems of aquatic plants, including rice. *Notiphila*, another large (about 150 species) and widespread genus in this subfamily, has distinctive aquatic larvae with posterior spiracles on long processes. Some *Notiphila* larvae puncture and breathe through the roots of aquatic plants, while others use their spiracular processes to breathe at the surface of the water or mud in which they feed as microbial grazers. Adults of *Notiphila* are common on emergent vegetation in and around ponds and lakes the world over.

Discomyzinae The Discomyzinae form a relatively small (212 species) subfamily best known for a number of terrestrial species that breed in decomposing material. *Discomyza*, for example, breeds in dead snails, while *Leptopsilopa* has been reared from rotting lettuce, and the larvae of some *Psilopa* species are miners in sugar-beet leaves. *Helaeomyia petrolei* (the Petroleum Fly), formerly included in *Psilopa*, is also in this subfamily, as are the *Trimerina* species that develop in spider egg sacs. The controversial genus *Risa*, a group of five little milky-winged flies found in arid parts of the Old World, is usually included in this subfamily, although some dipterists previously treated it as a separate family, Risidae. Known *Risa* larvae are leaf miners, like *Psilopa* and some other genera of Discomyzinae.

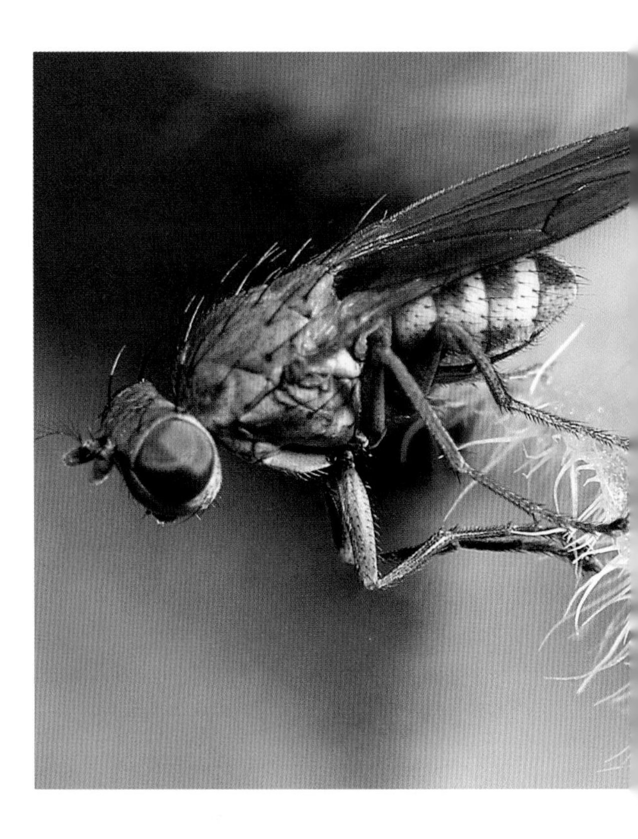

Shore flies in the genus *Notiphila* are common on emergent vegetation in and around ponds and lakes the world over. Larvae are aquatic, with posterior spiracles on long processes.

Gymnomyzinae The subfamily Gymnomyzinae is a widespread group of more than 400 mostly aquatic or semiaquatic species. *Athyroglossa*, for example, is a cosmopolitan genus of more than 30 species that includes *A. granulosa*, a Nearctic species that has been reared from decaying skunk cabbage. The closely related *Platygymnopa helicis* (the only species in its genus) has been reared from dead aquatic snails, as have some species in the widespread genus *Allotrichoma*. Most other members of this large (more than 50 species) genus and other members of the subfamily develop in wet, microbially rich microhabitats. Ferrar (1987) cites records of *Hecamede* species, for example, from such savory situations as "foul sand beneath a human corpse," dead horseshoe crabs and putrescent marine mollusks. Members of the widespread and distinctive genus *Ochthera* are exceptional because they are predaceous as both adults and larvae. Adult *Ochthera* species are common flies along shorelines, conspicuous for their huge and often silvery reflective raptorial forelegs. Other remarkable habits in the Gymnomyzinae include those of species in the Neotropical *Gastrops*, which develop in the foam nests of frogs.

Ephydrinae The subfamily Ephydrinae is a group of almost 500 species, of which the most common genera are probably the dark-winged species of the widespread genus *Scatella* and the chunky, relatively bare members of the mostly north temperate genus *Parydra*. Larvae and adults of *Parydra* abound on muddy pond margins, where the larvae consume diatoms; *Scatella* are also associated with freshwater margins but some (mostly the Holarctic *S. stagnalis*) can be greenhouse pests when they build up large populations in damp, alga-covered materials. Adults of the small, widespread genera *Brachydeutera* and *Setacera* are common on algal mats and the water surfaces of still, sometimes scummy pools and backwaters, where their larvae consume decomposing vegetable matter. The flattened larvae of the small Neotropical genus *Diedrops* graze diatoms from the surfaces of rocks in rapids or waterfalls, and many other ephydrine larvae are aquatic filter-feeders in a variety of aquatic and semiaquatic habitats, including extreme environments such as hot springs and alkaline or saline water. Most of the species that form hordes along salt-lake margins belong to the genera *Ephydra* and *Cirrula*, while other *Cirrula* and *Dimecoenia* live in salt marshes. Larvae of *Dimecoenia* use salt-marsh plants like living snorkels, inserting spinelike spiracles into the roots to breathe.

Ilytheinae The 225 or so species in the subfamily Ilytheinae include several small specialists on certain blue-green algae, or cyanobacteria, found along the edges of ponds, ditches or marshes. Foote (1977) found that seven North American *Lytogaster* and *Hyadina* species appear to be specific to the blue-green alga genus *Cylindrospermum*.

Curtonotidae

Hunchbacked or "Quasimodo" Flies

The family Curtonotidae is a small group (about 80 species) of slightly to spectacularly humpbacked flies that often have a spiked wing margin (spinose costa) like that of most Heleomyzidae, and usually have the general appearance and feathery arista of an extra-large fruit fly (Drosophilidae). Of the four named genera in the family, two — *Cyrtona* and *Tigrisomyia* — are restricted to Africa, and one, *Axinota*, is mostly Oriental-Australasian. The largest genus, *Curtonotum*, is currently found almost everywhere in the world except the Australasian region, but the genus will probably be eventually redefined to include only Neotropical species. The family remains poorly documented; a recent thesis by University of Guelph student John Klymko led to the discovery of 13 new species of *Curtonotum*, almost doubling the known New World fauna to 32 species (Klymko and Marshall 2011).

The most common of the three North American species currently treated in the genus *Curtonotum*, *C. helvum*, occurs on sand dunes. Larvae probably develop on grasshopper egg pods, as do the related Afrotropical and Palaearctic species *Curtonotum simile* and the Afrotropical *C. saheliense*. Otherwise, not much is known about the biology of curtonotids. Adults are occasionally attracted to dung but most curtonotids probably develop on invertebrate carrion, such as damaged egg pods or the bodies of prey insects in the nests of predaceous wasps. Adults of some Afrotropical species have been observed sheltering in mammal burrows or under the overhangs along riverbanks, and their numbers sometimes reach into the hundreds in warthog burrows. According to Ashley Kirk-Spriggs, the expert on African Curtonotidae, Zimbabwean *Cyrtona* adults are found under overhangs along rivers in the dry season and in adjacent vegetation in the rainy season.

Most curtonotids are active in the late afternoon and evening, often appearing right at dusk and mating on sand or low vegetation as the light disappears. They are most often seen perched on the ground (all North American species) or on branch tips (many South American species) or making characteristic short, hovering flights over their perches. Similar flight behavior — a vertical dash followed by slow descent back to a perch — is seen in some Syringogastridae and Asteiidae. Egg-laying habits remain unknown, but at least some species of *Cyrtona* are viviparous. The best-known species are associated with sand, but one recent paper reported tens of thousands of

This North American species of Curtonotidae, currently treated as *Curtonotum helvum*, occurs on sand dunes, where it probably develops as a predator of grasshopper egg pods.

curtonotids collected in banana-baited traps hung from trees in mangrove forests in southern Brazil (this interesting publication is unfortunately based on specimens unidentified beyond the family level and unavailable to specialists for further identification or confirmation).

Although dipterists now agree that the Curtonotidae is a distinct family of Ephydroidea, within that superfamily the group has in the past been treated variously as a subfamily of the Drosophilidae, a subfamily of Ephydridae and as part of the family Diastatidae.

Diastatidae

Diastatid Flies

The family Diastatidae is a small group of attractive little drosophilid-like flies generally associated with cooler altitudes and latitudes but completely absent from the Australasian region. Most have intricately pigmented wings and are further distinguished from similar ephydroids by a row of spines along the costa, somewhat like those of most Heleomyzidae and many Curtonotidae. The biology of the group remains little known, but adults are most often found on shrubbery along the margins of wetlands, and larvae of one North American species have been reared to adults on the feces of small rodents.

Despite being a small and relatively obscure group (only 47 named species worldwide), the diastatids have attracted considerable taxonomic attention because of disputes over their relationships and classification. The type genus, *Diastata*, has at various times been treated as a family by itself or as a member of the Ephydridae, while the other genera are sometimes treated as another family, the Campichoetidae. Most specialists today treat *Diastata* and two related genera (plus an extinct genus) as a family closely related to the Ephydridae and Camillidae, but some recent studies suggest a close relationship between the Curtonotidae, Camillidae and Diastatidae. Within the Diastatidae the relatively large and widespread genus *Diastata* is treated as one subfamily, **Diastatinae**, leaving the dozen or so species of *Campichoeta*, along with the single Palaearctic species of *Euthychaeta* and the fossil genus *Pareuthychaeta*, in the subfamily **Campichoetinae**.

Camillidae

Camillid Flies

Despite a name that suggests the glamor of a prince's consort, the genus *Camilla* is a nondescript group of nondescript flies. The other three genera of Camillidae include some interestingly ant-like species, but most of the 40 or so species in the family are small (usually 2–3 mm) and dark, with a superficial resemblance to shore flies (Ephydridae) or vinegar flies (Drosophilidae), although the long hairs of the arista are shorter below than above and the body usually has a characteristic submetallic appearance. Camillids are known mostly from Europe and Africa, but these inconspicuous little flies are easily overlooked; the relatively few records from the Nearctic and northern Neotropical regions may reflect lack of appropriate collecting efforts rather than a real absence. A few rearing records from Africa show that camillid larvae develop in guano, bat dung and other fecal material in caves and burrows, so careful sampling of the tiny flies in these relatively poorly collected habitats is likely to turn up many new species and new records in this generally overlooked family.

The relationships of the Camillidae remain a matter of dispute, although there is general agreement that the family belongs in the Ephydroidea.

Cryptochetidae

Cryptochetid Flies

The small, shiny black or blue-black acalyptrate flies in the family Cryptochetidae are named for their unusually reduced arista — *crypto* means "hidden" and *cheta* means "hair." They are generally uncommon insects, although adults occasionally occur in adequate abundance to become nuisance "eye gnats" in Japan as well as some parts of far eastern Russia and Southeast Asia. Such misbehavior on the part of adult cryptochetids is more than compensated for by the role this entirely parasitic group plays in the biological control of scales and mealy bugs, since all known species in the family develop as internal parasitoids of coccids. In fact, one of the earliest and best-known success stories in applied biological control of insect pests involves the importation of *Cryptochetum iceryae* from Australia to California in the late 1880s to help control Cottony Cushion Scale, *Icerya purchasi*. These obscure little flies were brought in at the same time as a somewhat more charismatic lady beetle (Vedalia Beetle, *Rodolia cardinalis*), with which they share credit for the ongoing suppression of an otherwise economically devastating pest of citrus fruit.

Flies in the small family Diastatidae are distinctive in the field for their characteristic wings-up posture, as well as the spines along the leading edge of their pigmented wings.

The Cryptochetidae are distributed mostly in the Old World tropics, although a few species have been moved around the world as biological control agents. The genus *Cryptochetum*, characterized by total loss of the arista, includes all but one of the 30 species in the family. The other species, an odd Australian insect placed in its own genus, *Librella*, is of unknown biology and only tentatively placed in this presumably entirely parasitic family. The genus *Cryptochetum* itself has a tortured taxonomic history involving previous placements in the Agromyzidae, Milichiidae, Chloropidae, Drosophilidae and Chamaemyiidae. Dipterists now generally treat the Cryptochetidae as a distinct family closely related to the Chloropidae and Milichiidae (in the Carnoidea), although a recent phylogenetic analysis (Weigmann et al. 2011) suggests that both the Cryptochetidae and the odd little family Braulidae are very close to — and maybe even part of — the Drosophilidae.

Braulidae

Bee Lice

As the common name "bee lice" suggests, braulid adults are wingless insects able to cling, louse-like, to their host's hairs. Not only do they lack wings but they also lack other normal fly bits and pieces such as halters, ocelli and a scutellum. The net effect is of a strange-looking ovoid, convex animal that looks more like a mite than a fly, with only dots where the eyes should be and the tarsal claws uniquely modified into combs to cling to their hosts' branched hairs. The hosts are bees in the Honey Bee genus *Apis*, and the relationship between bee lice and their hosts is not at all louse-like: instead of parasitizing the adult bees, the adult braulids simply go along for the ride, stealing a bit of food from the bee's mouth from time to time. Larvae live in honeycombs in the hosts' nest, causing some damage as they tunnel through the wax and feed on pollen.

Bee lice, like this *Megabraula*, are wingless flies with reduced eyes. Their tarsal claws are uniquely modified to cling to their host bee's branched hairs.

The family Braulidae includes only eight species, divided between the two genera *Braula* and *Megabraula*. *Braula* includes most of the species in the family, including a couple of pests of commercial Honey Bee colonies. Of these, *Braula coeca* is the most widespread, having been moved to most parts of the world with the European Honey Bee, *Apis mellifera*. The other genus is called *Megabraula* because of its relatively large body size (about 3 mm, while *Braula* species are less than 2 mm long). The two species of *Megabraula* occur in the nests of the world's largest Honey Bee, the Himalayan Honey Bee, *Apis laboriosa*, in Nepal.

Extremely specialized lineages such as the Braulidae are difficult to place phylogenetically. This unusual group of flies is sometimes treated as part of the Carnoidea, although a recent analysis (Weigmann et al. 2011) treats the Braulidae and Cryptochetidae as closely related to the Drosophilidae, in the Ephydroidea.

Nannodastiidae

As the *nanno* prefix suggests, Nannodastiidae are minute flies (often less than 1 mm long). These tiny flies occur on widely separated coastal areas of the Mediterranean as well as shorelines of the Pacific, Atlantic and Indian Oceans, but they are rarely collected. The two genera in the family were once placed in different families — Ephydridae and Asteiidae — but they have since been grouped together and then bounced around from family to family, usually either the Ephydridae or the Chyromyidae; the separate family Nannodastiidae was named in 1994. They seem to be closely related to the Curtonotidae, but given the lack of solid evidence for relationships, this odd little group of just five species is only tentatively placed in the Ephydroidea. The authors of a recent review of the nannodastiids (Papp and Mathis 2001) point out that they lack the defining features of the Ephydroidea but suggest no alternative superfamily placement. Nothing is known of the biology of Nannodastiidae, but the larvae are probably microbial grazers in shoreline debris.

Lauxanioidea

Aphid Flies and Their Relatives

Lauxaniidae

Lauxaniid Flies

The family Lauxaniidae is one of the most diverse lineages of acalyptrate flies and can be found on every continent except Antarctica, although most of the 2,000 or so species currently described in the family occur in either the neotropics or Southeast Asia. Lauxaniid flies, formerly known as Sapromyzidae, come in almost every imaginable shape, color and size. Even though the "typical" lauxaniid is a small yellow or gray fly, the morphological diversity of adult lauxaniids is such that if an unrecognized fly has no vibrissae and a complete subcosta, my first guess (usually correct) is that it is a lauxaniid. Some species with mottled wings look very much like species of Tephritidae and Ulidiidae, smaller species often look like *Drosophila*, and some genera are apparent mimics of beetles and leafhoppers. The beetle mimics bend their wings flat against the body so their backs look like beetle backs; the leafhopper mimics even have snouts like similar leafhoppers. Other Lauxaniidae, especially those in the subfamily Eurychoromyiinae, are outrageous in appearance, with broadened heads and elongated appendages.

Lauxaniid larvae frequently develop in decay ing leaves, often as miners between the upper and lower leaf surfaces; some feed on mycelia in rotting vegetation and some occur only in birds' nests, but most are unknown. Mature larvae of some species coat themselves with excrement, which hardens into what amounts to an extra shell over the puparium — itself the hardened skin of the last larval stage; at least one species pupariates inside a distinct cocoon. Adults feed on fungal hyphae and spores, scraped from leaves with a specialized labellum bearing prongs and scoops for cutting and raking the food toward the oral opening. Few adult higher flies are able to feed on solid particles in this fashion; these special labial structures appear to be unique to some (but not all) Lauxaniidae.

Lauxaniidae have traditionally been divided between two subfamilies, the Lauxaniinae and Homoneurinae. A third subfamily, Eurychoromyiinae, was recently added when the family Eurychoromyiidae was combined with the Lauxaniidae (Gaimari and Silva 2010).

Eurychoromyiinae Members of the subfamily Eurychoromyiinae are odd-looking flies, so much so that they have been variously treated as a separate family, as part of the Sepsidae, as part of the Ropalomeridae and as part of the Chamaemyiidae. The type genus, *Eurychoromyia*, remains known only from four specimens collected near a small Bolivian village in 1903, and two of the six genera described since then remain known only from one specimen each. Most of the recent specimens of this subfamily were collected by fogging the rainforest canopy with insecticides, suggesting that the apparent rarity of the group might simply reflect a relatively inaccessible treetop habitat.

Homoneurinae The subfamily Homoneurinae, which is characterized by short black bristles on the costa, is dominated by the huge genus *Homoneura*. These usually yellowish flies seem to be common everywhere but the Neotropical and Afrotropical regions, with more than 100 species in the Holarctic region and more than 200 in the Australasian region. They are often associated with fallen leaves in wetlands and moist forests, although some species can be found in sand dunes and other dry environments.

Lauxaniinae The subfamily Lauxaniinae is dominated by the large genus *Sapromyza*, a huge artificial (paraphyletic) group of hundreds of species worldwide. Like other Lauxaniidae, *Sapromyza* species are saprophagous, sometimes feeding between the upper and lower surfaces of fallen leaves and sometimes in other decomposing plant material. Other Lauxaniinae larvae have similar habits; for example, those of the large genus *Minettia* graze on fungal hyphae and spores. The subfamily Lauxaniinae is a sort of leftover legion of Lauxaniidae that lack the distinctive features of the other two subfamilies; some dipterists consider it to be an artificial assemblage rather than a real lineage.

Chamaemyiidae

Aphid Flies

Chamaemyiidae, sometimes called "silver flies" because of their frequently silvery gray vestiture and sometimes "aphid flies" because of their importance as predators of aphids, are infrequently collected despite the diversity (about 350 species) and worldwide distribution of the group. All members of the family with known larvae are predators of soft-bodied plant-sucking true bugs such as aphids and scale insects (Sternorrhyncha) and can often be found in pockets of appropriate prey, especially in grasslands. Some larvae, including those of the large genus *Leucopis* — which includes most aphid fly species — are active feeders in aphid colonies, much like the larvae of syrphine syrphids. Others are associated with sessile hosts such as mealy bugs or scales and sometimes develop in a single host gall or egg sac. Some species are important predators of pests and have been moved around for use as biological control agents. Among these are a couple of western North American *Leucopis* species and a European *Cremifania* species introduced to the east coast of North America to help control the Hemlock Wooly Adelgid, *Adelges tsugae*.

The family Chamaemyiidae is divided into the three subfamilies Chamaemyiinae, Leucopinae and Cremifaniinae.

The subfamily **Cremifaniinae** is a small Holarctic group often treated as the separate family Cremifaniidae, even though it contains only one genus and three species. The larvae of *Cremifania* are similar in habits to those of other Chamaemyiidae, but some adult morphological characters suggest a closer relationship to other families. **Leucopinae** is the largest subfamily, made up mostly of species in the widespread genus *Leucopis*. Look for *Leucopis* adults scuttling sideways, backwards and forward on grasses near the populations of mealy bugs, scale insects or aphids that serve as lunch for their larvae. The subfamily **Chamaemyiinae** includes the distinctively black-spotted species of *Chamaemyia*, which develop as predators of soft-bodied bugs in the leaf sheaths of grasses; similar habits are found in *Pseudodinia* and related genera with known larvae.

The family Chamaemyiidae includes the distinctively black-spotted species of *Chamaemyia*, which develop as predators of soft-bodied bugs in the leaf sheaths of grasses.

Paraleucopidae

Paraleucopid Flies

Much of what we know about the small, shining flies in the family Paraleucopidae is summarized in Robert Smith's 1981 paper called "The Trouble with 'Bobos.'" "Bobo" is a local name for *Paraleucopis mexicana*, a fly from the Gulf of California discovered by Smith and formally described by George Steyskal in the same journal issue, but it might easily be applied to all four (three named and at least one unnamed) species in the western Nearctic genus *Paraleucopis*. *Paraleucopis mexicana* adults are abundant on the west coast of Sonora, Mexico, from March until June, when they annoy visitors to shorelines with their attraction to blood, sweat and tears. Similar habits are known for other bobos, including an unnamed species recently discovered by Robert Smith in the Sonoran Desert near his house on the outskirts of Tuscon, Arizona. When I visited Dr. Smith in May 2010, his backyard bobos ("Arizona bobos") obligingly swarmed around us, alighting in our ears and eyes and on scratches on our legs, much like the Mexican species he found some 30 years earlier. Larvae of bobos remain unknown, but the type species of the genus, from New Mexico, was named *Paraleucopis corvina* because some of the type specimens were reared from the nest of a raven (genus *Corvus*).

Other than the few bobo (*Paraleucopis*) species, only another dozen or so species in three genera are included in the family Paraleucopidae. The only described paraleucopid species outside of *Paraleucopis* — all but one being in the genus *Mallochianamyia* — are from Chile and Argentina, but some undescribed Australian species (in an undescribed genus) probably also belong in this group. Until recently the genera now included in the family Paraleucopidae were usually treated as "unplaced" or of "uncertain family placement," or as awkward members of the aphid fly family Chamaemyiidae. J.F. McAlpine (1989) argued that *Paraleucopis* and related genera are not closely related to Chaemaemyiidae, but instead form a distinct group (thus an unnamed family) in the superfamily Opomyzoidea. *Paraleucopis* and related genera were first formally treated as a named family in the recent *Manual of Central American Diptera* (Wheeler, 2010) but there is still no consensus among fly specialists about how to solve the taxonomic troubles with bobos. *Paraleucopis* and its relatives are included here next to the Chamaemyiidae only because of their traditional association with that family.

Celyphidae

Beetle Flies

"Beetle flies" are so named because of the way part of the thorax — the scutellum — is usually spectacularly inflated to cover the wings and most of the abdomen with a shiny, beetle-like shell. Thanks to that hard shell, which is frequently combined with a flattened, leaflike antenna, the Celyphidae are among the most easily recognized flies. Exceptionally distinctive groups such as the beetle flies often turn out to be oddball subgroups (autapomorphic lineages) that really belong inside larger families; it has been suggested that Celyphidae are either just very unusual members of Lauxaniidae or are at least very closely related to the Lauxaniidae. Known beetle fly larvae have similar habits to some Lauxaniidae, feeding on decaying grass or skeletonizing the upper surfaces of dead leaves.

Beetle flies are usually found in wet areas, often among grasses near bodies of water. Most of the 120 or so species in this small family are Oriental, several species occur in Africa, some range as far north as Afghanistan and Nepal, and others are found southeast to New Guinea and the Solomon Islands. These attractive insects are relatively well known despite their mostly tropical range, and there is even a good key to the Oriental species, by Tenorio (1972).

Opomyzoidea

Upside-down Flies, Druid Flies, Leaf-mining Flies and Others

The superfamily Opomyzoidea is a loose assemblage of small, slender acalyptrate flies, with more than three-quarters of its 4,000 or so included species placed in one enormous, important family and the rest spread over about a dozen relatively minor to intriguingly obscure families. The giant family of the group is Agromyzidae, one of the largest and most economically important of all acalyptrate families. At the other end of the spectrum, the opomyzoid families Marginidae and Teratomyzidae, with less than a dozen species between them, are among the smallest and least-known families of flies. The superfamily Opomyzoidea includes four subgroups of similar (related?) families sometimes formally named as "suprafamilies": Clusioinea for Clusiidae; Opomyzinae for Opomyzidae and Anthomyzidae; Agromyzoinea for Fergusoninidae, Agromyzidae and Odiniidae; and Asteioinea for all the rest. The family discussions here are not grouped by suprafamilies but instead are ordered from largest to smallest families within the superfamily as a whole.

Agromyzidae

Leaf-mining Flies

Relatively few lineages of Diptera have made the transition from saprophagy to phytophagy, but some of the groups that have adopted an entirely plant diet have been conspicuously successful. The two largest phytophagous lineages of flies, Tephritidae and Agromyzidae, are also the first and third largest acalyptrate families, and the mostly plant-associated family Chloropidae is the fourth largest group of acalyptrates. The family Agromyzidae is the only one of these large groups that is entirely phytophagous, with all species developing on living plant tissue. The common name "leaf-mining flies" is accurate for most species of Agromyzidae, which spend their larval lives between the upper and lower surfaces of host leaves; however, at least a quarter of the more than 3,000 species in the family develop in fruits, roots, stems, seeds, woody tissues or galls rather than leaves.

Unsurprisingly for such a large and successful family, agromyzids are found in a wide range of habitats in every zoogeographic region and all latitudes, from the Canadian Arctic in the north to the subantarctic islands in the south. These small (usually 2–3 mm) acalyptrates are most diverse in north temperate regions and usually attack angiosperms, although they are recorded from more than 140 plant families. Most of the small, non-shiny black or black and yellow acalyptrates routinely seen on plant stems or foliage belong to this huge family. Females are distinctive for their stout oviscape (the conical visible tip of the abdomen), used to puncture plants to insert eggs or to start sap flows on which they feed; males also feed on plant juices, often taking advantage of fluids exuding from feeding punctures made by the better-armed females.

As is true for the gall midges (Cecidomyiidae), agromyzid species are often more easily identified by their host plants and characteristic feeding damage than by visible features of the small, often nondescript adult flies. The leaf-mining larvae lie sideways as they feed within the leaf, leaving a characteristic trail or "mine" as they feed. Mines of different species differ in shape (straight, wiggly, blotchy) and how the frass is deposited (in a single mass or dropped in regular piles along the trail). Some species can be identified by their specific host plants, but others develop on a wide range of plants; *Liriomyza trifolii*, for example, attacks tomatoes, cucumbers, asters and a variety of legumes.

The Agromyzidae are closely related to the Fergusoninidae and Odiniidae; these three families are sometimes grouped together as the suprafamily Agromyzoinea within the superfamily Opomyzoidea. The family Agromyzidae is divided into two subfamilies, Phytomyzinae and Agromyzinae, each widely recognized as a natural group and confirmed as monophyletic by recent phylogenetic work (Scheffer et al. 2007).

Agromyzinae The subfamily Agromyzinae is made up largely of dull black species most easily identified by their leaf mines, although a few genera mine in plant stems. Both sexes and all species in the typical genus *Agromyza* stridulate by rubbing a file on the first abdominal segment against a scraper on the hind leg, presumably for communication between the sexes. Similar structures also occur in some species of the genera *Liriomyza* and *Cerodontha* (subfamily Phytomyzinae). The genus *Agromyza* includes several widespread and familiar species, such as *A. youngi*, which mines the petioles, leaves and flower stalks of dandelions; *A. laterella*, which produces galls on young iris leaves and mines on older leaves; *A. schineri*, which mines poplar leaves; and *A. simplex*, which bores in stems of asparagus.

Phytomyzinae The subfamily Phytomyzinae includes the huge worldwide genus *Phytomyza*, a probably paraphyletic genus with 200 species in North America alone. The more distinctive genus *Liriomyza* also includes hundreds of species worldwide, some of which are secondarily widespread pest species. *Liriomyza huidobrensis* and *L. trifolii*, for example, are highly polyphagous species and serious agricultural pests.

Clusiidae

Druid Flies

The small (3–6 mm), slender, attractively pigmented acalyptrate flies in the medium-sized (400 or so species) family Clusiidae have been called "druid flies" because, like the druids of 2,000 years ago, they are best known for a range of male-only rituals carried out deep in ancient forests. The male rituals of druid flies are carried out in defense of mating territories, often on large, bare logs at some distance from the beetle burrows and bits of bark-covered decomposing wood in which female druid flies lay their eggs. Since the mating territories or arenas are separate from any resource used as adult or larval food, they are referred to as leks.

Lekking male druid flies employ an astonishing array of morphological and behavioral ploys to hold their turf without resorting to simple physical combat with rival males. Wing color patterns, leg modifications, broadened heads and elongated antennae have been shown to serve in ritual battles in which territory-defending males assess interlopers and vanquish inferior individuals. Evenly matched males may go from mutual measuring to pitched battles involving head butting and grappling, sometimes with unlikely grappling tools such as the bedspring-like coiled vibrissae sported by some broad-headed Australian *Hendelia*. The Australian *Hendelia armiger* boosts its apparent advantage by flipping out its long antennae at right angles to exaggerate head width in head-measuring matches, and the common North American *Heteromeringia nitida* has brilliantly bicolored front legs that can be folded back at the tibial-femoral joint and used in ritual boxing matches between rival males. Some tropical druid flies, such as *Hendelia kinetrolicros* and *Procerosoma alni*, have improbable-looking growths that extend far out of their cheeks. The spectacular cheek processes probably play a role in male–male interactions, although these two species, like most clusiids, have never been observed "doing their thing" in nature.

Strikingly modified lekking males also occur in some Tephritidae, Ulidiidae, Richardiidae and Drosophilidae, but lekking is less common in the Diptera than resource-defense or mate-defense strategies such as those found in Neriidae and most Tephritidae. Even within the Clusiidae, lekking has been properly documented in only one of the three subfamilies, the Clusiodinae, although members of the other two subfamilies, Sobarocephalinae and Clusiinae, can be found on the same exposed logs, where they patrol broad areas and sometimes chase off all other flies. Some clusiids lek and lurk under living leaves; at least one Australian species defends territories on fallen palm fronds.

Unlike related (other opomyzoid) families, which generally inhabit living — albeit sometimes wounded — plant tissue, druid flies develop in dead wood. Larvae have a reduced cephaloskeleton and lack the pharyngeal filter that characterizes saprophagous (microbe-grazing) maggots, so they probably prey on other insect larvae.

Clusiodinae The subfamily Clusiodinae occurs in most parts of the world and includes the large, mostly Oriental genera *Czernyola*, *Hendelia* and *Heteromeringia*. *Heteromeringia* occurs throughout the world, including North America, Australia and Africa, but most of the 67 named species are Oriental. Most temperate species are split between the primarily northern hemisphere genus *Clusiodes* and the primarily southern hemisphere *Hendelia*, both of which include well-documented lekking species with associated modifications such as genal

Druid flies, like this *Clusia occidentalis* perching on a bare log in western United States, occur commonly in wooded areas worldwide. Larvae are subcortical predators.

processes. *Hendelia* is a mostly Old World and Pacific group of about 50 species, with one species group in the neotropics and a few species in temperate Australia. *Clusiodes* is the most abundant group of north temperate druid flies, with 31 species generally characteristic of North American and European forests.

Clusiinae The subfamily Clusiinae is a relatively small group, including 13 mostly north temperate *Clusia* species, some of which are among the largest and most conspicuous druid flies. Other members of the subfamily are mostly tropical and south temperate in distribution, with most species occurring in Southeast Asia.

Sobarocephalinae The subfamily Sobarocephalinae, the largest clusiid subfamily, is dominated by the genus *Sobarocephala*, one of the least-known and most speciose genera of acalyptrates. This group of small, attractively patterned but mostly yellowish flies is extremely diverse in South and Central America; one recent paper includes descriptions of 171 new Neotropical *Sobarocephala* species (Lonsdale and Marshall 2012). The southern South American genus *Apiochaeta* is among the most common and conspicuous acalyptrates in Chile, but sobarocephaline genera other than *Sobarocephala* and *Apiochaeta* are relatively uncommon.

Asteiidae

Asteiid Flies

The small (1–3 mm), delicate flies in the family Asteiidae are more often encountered on windows inside buildings than in the field. They usually have a distinctive appearance, thanks to their long (much longer than the body) and clear, slightly iridescent wings. The most common species are distinctively colored, with a dark top and pale underside or with a distinctive spotted or banded pattern. At least some species have a characteristic response to disturbance, flying straight up when disturbed and then slowly hovering back down. Sometimes these rarely collected flies can be found under the tips of leaves along forest trails.

Despite a worldwide distribution, with about 140 known species spread over every zoogeographic region, not much is known about asteiid biology. Some have been reared from fungi and others are reported from trees, flower buds, dried reed stems, debris in tree hollows and even the stalks of marijuana plants. It is likely that in most cases the larvae are saprophagous and typical microbial grazers, possibly in the waste products of other insects. Some species of *Leiomyza* occasionally occur in large numbers on fresh mushrooms (Agaricales), where they engage in elaborate courtship displays before mating and depositing a few very large eggs. Males of the European *Asteia elegantula* aggregate in leks on leaves, where they court arriving females not only with wing waving and swaying but also with a nuptial gift in the form of regurgitated fluid (Freidberg 1984).

The family is divided into ten genera, but half the species in the family are in the cosmopolitan genus *Asteia*. As is true for most small acalyptrate Diptera, the family is probably much more diverse than the numbers of described species indicate. By way of illustration, a recent study of the Asteiidae and related flies (suprafamily "Asteioinea") of Fiji recorded five *Asteia* species, four of which were new to science (Grimaldi 2009). Some of the asteiids recorded from Fiji were collected in rolled leaves along with the Stenomicridae more typical of such habitats.

Anthomyzidae

Anthomyzid Flies

Anthomyzids are distinctively slender acalyptrate flies most often found among low woodland or wetland vegetation, where they develop as microbial grazers between the leaf sheaths of herbaceous plants or, in the case of the recently described Holarctic genus *Fungomyza*, within the tissues of soft fungi. Although some are recorded as stem miners it is likely that all larvae in the family are saprophagous, feeding on decaying tissue and associated microorganisms rather than living plant tissue. Most are associated with specific communities of grasses or sedges.

The 90 or so described species in the family Anthomyzidae are mostly Holarctic and seem to be especially diverse in Europe, where the anthomyzid fauna was recently treated in a beautifully illustrated monograph by Jindřich Roháček (2006, 2009). The Pacific anthomyzid fauna is small, with only an endemic genus in New Zealand, a couple of introduced species in Hawaii and a species of the mostly Afrotropical/Oriental genus *Amygdalops* in Australia. Many species remain to be described in both the New World and Old World tropics.

All extant species are in the subfamily **Anthomyzinae**, as is a Dominican amber fossil, *Grimalantha*, but a Baltic amber genus, *Protanthomyza*, is treated as the separate subfamily **Protanthomyzinae**. Anthomyzid flies are closely related to another mostly grass-associated family, the Opomyzidae. Both groups include species in which the wings have become narrow and useless for flight, and some are functionally wingless. Several high-altitude Neotropical species such as *Mumetopia messor*, recently described by Roháček and Barber (2008) from the Ecuadorian *páramo*, have extremely narrow straplike or needlelike wings similar to those found in some southern South American Sphaeroceridae. The specific epithet *messor*, from the Latin for "scythe," nicely reflects the sharp, narrow wings of *M. messor*.

Stenomicridae

Stenomicrid Flies

One of the best ways to gain an immediate, firsthand appreciation of the enormous task facing scientists trying to sort out the diversity of Neotropical flies is to visit any rainforest in South or Central America and peer down into the partially unfurled leaves of a broad-leaved plant in the ginger or banana families. Almost inevitably you will see dozens of tiny, somewhat flattened stenomicrid flies smoothly and swiftly gliding backwards, forward or sideways down the leafy funnel as you approach. These are flies in the enormous genus *Stenomicra*, and after you have peered at half a dozen species in half a dozen furled plants, you will start noticing these distinctive little acalyptrates on the upper and lower surfaces of expanded broad leaves as well, although the only known larvae in the genus develop in the fluid organic matter in the funnel-like bases of the furled leaves. The obvious diversity of this group is especially staggering in the neotropics, but enormous numbers of these little flies await discovery throughout the tropical world. As a case in point, a recent review of the Asteioinea (Periscelididae and related families) of Fiji reported 15 species of stenomicrids, of which 14 were new to science. Published estimates suggest that at least 100 species of *Stenomicra* await description, but my guess is that the genus includes several hundred undescribed species in addition to the 40 or so currently named.

Stenomicra, *Stenocyamops*, *Cyamops* and possibly *Planinasus* together form the group here treated as the Stenomicridae. Of these, only the cosmopolitan *Stenomicra* and the very closely related Asian and Pacific genus *Stenocyamops* hang out in rolled leaves. *Planinasus* adults occur reliably on wet rocks or splashed logs in and over streams in Central and South America, but the larvae of this small (three named species) Neotropical genus remain unknown and its inclusion in the Stenomicridae is tentative. The 30 described species in the cosmopolitan genus *Cyamops* are generally larger and darker and are most often found on rocks or leaves near rivers and streams. Larvae of *Cyamops* are also unknown, although there is one record of a puparium of an American species from a wetland.

While dipterists generally agree that *Cyamops* and *Stenomicra* are closely related, their relationships to other acalyptrates remain a bit contentious; various authors have put them in the families Drosophilidae, Asteiidae, Geomyzidae, Periscelididae, Anthomyzidae or Aulacigastridae. At the moment the debate seems to be whether to treat *Stenomicra* and its relatives as part of the Periscelididae or to place them in a separate family, Stenomicridae. I'm opting for the latter classification here because the group is so distinct, and because there is no convincing evidence that a Periscelididae including *Stenomicra* would be a monophyletic group. *Planinasus* and *Stenomicra* are both known from Dominican amber fossils; another putative stenomicrid genus, *Procyamops*, is known only from Baltic amber.

Periscelididae

Dwarf Flies

The Periscelididae comprise a well-defined group that includes some of the most beautiful and least often collected acalyptrate flies. The dozen distinctive, often brightly colored described species of the very closely related Neotropical genera *Scutops*, *Neoscutops*, *Marbenia* and *Parascutops* are occasionally seen sitting on rainforest and cloud-forest foliage, and they are rare enough to be exciting finds every time they are encountered. The only other Neotropical periscelidid genus, *Diopsosoma*, is a spectacular genus of one species known only from the type series of three specimens collected in Peru in 1931. *Diopsosoma primum*, first described by John Malloch in 1932, has extremely long eyestalks that render the head about as wide as the body is long, much like the male *Plagiocephalus latifrons* (Ulidiidae). Like *Plagiocephalus* but unlike the spectacularly stalk-eyed flies in the family Diopsidae, the antennae remain in their normal positions while the eyes are way out at the ends of the stalks. Like Diopsidae but unlike *Plagiocephalus*, both the males and females have long eyestalks. Eyestalks or broadened heads occur in more than 20 Diptera families, where they often play a role in male–male mutual assessments or battles for mates or mating territories. Obviously, such interactions have yet to be observed in *Diopsosoma*.

Nothing is known of the biology of the Neotropical dwarf fly genera, but the bright blue to violet eggs and flattened larvae of the widespread (every zoogeographic region but the Afrotropical) genus *Periscelus* have been found in the sap of bleeding deciduous trees. They are probably microbial grazers, although Mathis and Papp (1998) reported that some European species prefer fresher, more fluid sap than flies in the family Aulacigastridae (adults of North American *Periscelus* sometimes co-occur with *Aulacigaster*). *Periscelis* includes 14 extant species and four amber fossil species. The Periscelididae are probably most closely related to the Neurochaetidae, Stenomicridae or Aulacigastridae. Some evidence suggests that *Planinasus*, a poorly known Neotropical genus currently in the Stenomicridae, belongs in the Periscelididae.

Odiniidae

Ribbon Flies

Odiniidae are larger (3–6 mm), stouter and more attractive than most other opomyzoids, usually gray with contrasting colors, spots or bands on the wings, eyes and bodies. Although fairly rare in general collections, adults can be common around the right kinds of sappy wood (or sometimes woody fungus), especially at fermenting sap around the burrows of wood-boring insects such as long-horned beetles and carpenter moths. Although one Palaearctic species has been reared from mealy bug egg masses, odiniid larvae are normally found in tree sap or the tunnels of wood-burrowing insects, where they feed at least partly on microorganisms in the decomposing sap and frass (older larvae may be facultative predators). The biologies of most species, including all 24 Neotropical ribbon fly species, remain unknown, but about a dozen species spanning both subfamilies have been reared from tree sap or gum, often associated with insect borings, in the Nearctic, Palaearctic and Afrotropical regions.

The vernacular name "ribbon flies" was coined by von Tschirnhaus (2008) based on the original generic name, *Odinia*, applied to this group since 1830. He thought that it was from the Greek for "sadness" or "mourning" and referred to the ribbon-like markings on the wings of many Odiniidae. I think that is a stretch, but "ribbon flies" has a nice ring to it anyway, and the only alternative name I know of is the less imaginative "odiniid fly."

The Odiniidae are closely related to Agromyzidae and Fergusoninidae but lack the stiff, tubelike abdominal tip (oviscape) that those phytophagous groups use to inject eggs into plant tissue. The family is divided into two subfamilies, Odiniinae and Traginopinae. The subfamily **Odiniinae** is a mostly Old World group of four genera, although the small genus *Neoalticomerus* is Holarctic and the large genus *Odinia* occurs in every zoogeographic region except the Australasian. The subfamily **Traginopinae** is mostly Neotropical, with three species in Europe, a couple of species in North America and four species in Africa. The two North American species, both in the genus *Traginops*, are among the most distinctive sap-loving flies because of their brightly spotted wings and body, banded eyes and prominently inflated ocellar plate. Although the total number of named species in the family is currently about 60, many further species await description. New species of this and other "rare" families are often collected during mass trapping programs using Malaise traps to sample previously poorly known areas such as Madagascar.

Opomyzidae

Opomyzid Flies

Opomyzids occur among the grasses in which their phytophagous larvae develop, and these attractive little (2–5 mm) flies can be locally abundant in moist, grassy areas along seacoasts and rivers. Most species are yellow or brown with distinctively patterned or dark-spotted wings. The group is mostly Palaearctic, with a few species in North America, Africa and the Oriental region but none in the neotropics or the Australasian region. All but two of the world's 60 or so opomyzid species are in the common genera *Opomyza* and *Geomyza*. *Opomyza* is an Old World genus with ten species native to Europe and three species native to South Africa (Barraclough 1999). Two *Opomyza* species also occur in North America, but both were probably introduced from Europe,

Flies in the family Opomyzidae, like this *Opomyza petrei*, are often locally abundant in moist, grassy areas along seacoasts and rivers. Larvae feed in the stems of grasses.

as were two of the 11 Nearctic *Geomyza* species. Opomyzids are thought to be closely related to another family of mostly grass-associated flies, the Anthomyzidae.

Opomyzid larvae and puparia inconspicuously occupy grass stems, probably predisposing them to accidental movement from continent to continent with straw, animal bedding and similar materials. A few species of opomyzid develop in the stems of economically important grasses, causing damage known as "deadheart." In particular, the Palaearctic species *Opomyza florum* attacks a wide range of crops, including wheat, rye, oats and barley, while the secondarily Holarctic *Geomyza tripunctata* inhabits stems of a wide variety of graminoids and has the potential to become a serious pest of corn or maize (*Zea mays*).

Fergusoninidae

Myrtle Gall Flies

The family Fergusoninidae is made up entirely of one genus, *Fergusonina*, of about 30 described species that probably all develop only in galls induced by a specialized and codependent group of nematodes, *Fergusobia*. The *Fergusonina–Fergusobia* symbiotic relationships are intricate and specific, with each species of fly associated with one species of nematode that induces characteristic galls on only a single kind of host plant (*Eucalyptus* or a closely related tree or shrub in the Myrtaceae or Myrtle family). Adult female *Fergusonina* flies are always infected with nematodes that they carry to host plants, where the nematodes induce galls in which the fly larvae feed and develop. Mated female nematodes invade female fly larvae in the gall and later release eggs into the fly's haemolymph, where they hatch into juvenile nematodes to be in turn delivered to the host plant when the adult fly lays eggs in buds or other tissue.

Some *Fergusonina–Fergusobia* pairs are pests of eucalyptus because the galls sometimes reduce seed set and flower production; others are potentially beneficial because they attack Australian plants that have become invasive aliens elsewhere. The Australian broad-leaved paperbark, *Melaleuca quinquenervia*, for example, has become a serious invasive weed in the southeastern United States, and some research is now being directed to the possibility of importing a host-specific fly–nematode pair — *Fergusonina turneri* and the nematode *Fergusobia quinquenerviae* — for use in biological control.

Many species of Fergusoninidae remain to be described, almost all of them from Australia, although the genus is now known from New Zealand (one species), New Guinea (one undescribed species) and India (one species). *Fergusonina* species are usually yellowish or black and yellow flies and, according to David McAlpine (1998), they are weak fliers. Fergusoninidae strongly resemble Agromyzidae and are very closely related to (if not actually part of) that large and diverse family.

Neurochaetidae

Upside-down Flies

Best known from Down Under, the "upside-down flies" comprise a small (around 25 species) family of small, slender flies known from Australia, Southeast Asia, Madagascar and Africa. Australian dipterist David McAlpine coined the common name for these rarely collected flies in the same paper (1978) in which he formally named the family, because he noted that they always maintain a head-down position on a vertical surface, no matter which way they run. Like the widespread genus *Stenomicra* (Stenomicridae), upside-down flies have a remarkable ability to move backwards and sideways in a swift, smooth, crab-like scuttle, all the while keeping the wings folded over the body.

Flies in the related family Stenomicridae usually perch heads-up on vertical surfaces within the confines of still-furled leaves such as those of *Heliconia*, and the few species of *Stenomicra* for which larvae are known develop in the wet material trapped in the leaf axils. Upside down flies are found in similar places. Biologies of species other than the eastern Australian *Neurochaeta inversa* remain unknown, but *N. inversa* lives on or in a broad-leaved rainforest plant, *Alocasia brisbanensis* (Araceae), where adult flies feed on pollen and lay eggs on the female flowers. Larval flies live within the spathe — a sort of sheath that envelops developing fruit in the Araceae — grazing on microorganisms in the watery environment inside the protected chamber. Some Southeast Asian species are associated with other broad-leaved plants, presumably ones that offer a similar moist, microbe-rich pocket for the larvae of this special family of flies. The Western Australian upside-down fly *Nothoasteia clausa*, described by David McAlpine (1988), lives among the leaves of tall grass trees (Xanthorrhoeaceae).

The Eastern Australian Upside-down Fly, *Neurochaeta inversa*, lives on or in the broad-leaved rainforest plant *Alocasia brisbanensis* (Araceae), where adult flies feed on pollen and lay eggs on the female flowers.

Aulacigastridae

Sap Flies

The small family Aulacigastridae has suffered a great deal of confusing taxonomic volatility since it was established to hold the single European species *Aulacigaster leucopeza* in 1925. Most of the genera historically included in the family are of uncertain relationship and have been shuffled back and forth between the Aulacigastridae and the closely related family Periscelididae. Alessandra Rung and coauthors (2005) simplified the issue by limiting the Aulacigastridae to the type genus,

Aulacigaster, plus three obviously very closely related Southeast Asian species, for which they created a new genus curiously called *Curiosimusca*. Most of the other extant genera previously treated as Aulacigastridae are now placed in the Stenomicridae (*Cyamops*, *Stenomicra*, *Planinasus*) or Neminidae (*Ningulus*, *Nemo*), although the problematic genus *Echidnocephalodes* — with one species, from Madagascar and Seychelles, tentatively moved from the Anthomyzidae to Aulacigastridae by Roháček in 1998 — remains "unplaced."

"Sap flies," so called because the best-known species of *Aulacigaster* are routinely found on tree-trunk sap fluxes, occur in the Afrotropical, Holarctic, Neotropical and Oriental regions. Most of the diversity, however, is in the neotropics, and 37 of the 55 species in the family were just recently named on the basis of Central and South American specimens (Rung 2011). Most *Aulacigaster*, including the relatively few north temperate species, are found around fermenting tree wounds on deciduous trees (beetle burrows that penetrate the bark of live trees often provide particularly good habitat for *Aulacigaster*). Female flies deposit their eggs in moist decaying sap, where the larvae develop as microbial grazers, breathing through a long, tail-like respiratory tube and long, thin spiracular lobes on the thorax. Neotropical sap flies have more varied habits; they include long-legged species that aggregate in bromeliads and elongate species found along the ribs of heliconia leaves.

Neminidae

Nobody Flies

Finding *Nemo* is always exciting for an acalyptrate fly enthusiast, since *Nemo* is the type genus of a small, rarely seen family of flies with species in Australia, New Guinea, Africa and Madagascar. "Nobody flies" were discovered by the Australian dipterist David McAlpine, who named the genus *Nemo* in 1983, placing it in a new subfamily of the Aulacigastridae. The Israeli acalyptrate specialist Amnon Freidberg added a Madagascar genus to the group in 1994, but he argued that it could not belong in the Aulacigastridae and elevated the subfamily (now with three genera) to family status. Larvae of this obscure group remain unknown, but adults have been collected on the very same broad-leaved eastern Australian plants (*Alocasia*) as adults and larvae of another of McAlpine's unusual acalyptrate discoveries, the upside-down flies. *Nemo* adults have also been collected on the smooth bark of eucalyptus trees. It is likely that adults have been overlooked by most entomologists because of their minute size (0.8–2.5 mm), short activity period and unusual (for acalyptrate flies) habit of being active in the rain.

Xenasteiidae

Xenasteiid Flies

Although only formally named in 1980, Xenasteiidae have been recognized as a distinct group for the better part of a century, starting with a handwritten label that dipterist John Malloch attached to a tiny (less than 2 mm) specimen in the United States National Museum in the 1930s. Malloch's label — *Xenasteia*, from the Greek *xenos*, meaning "strange" or "foreign," combined with the name of the genus *Asteia* — suggests that he intended to describe it as a genus in the small family Asteiidae, but he never got around to it. When Curtis Sabrosky, a fly specialist also at the United States National Museum, published a synopsis of the New World Asteiidae in 1957, he made no mention of Malloch's unpublished genus, perhaps because *Xenasteia* was known only from Pacific islands and perhaps because he did not think it was related to *Asteia*. John Malloch's odd little fly was not formally dealt with until Sabrosky brought it to the attention of Hawaiian entomologist Elmo Hardy, who agreed with Malloch that it shared many characters with *Asteia* but was nonetheless such a "strange" fly that it represented an entirely new family. So, about 50 years after Malloch recognized the genus and 17 years after his death, *Xenasteia* was formally named and given its own brand-new family.

Hardy's case for describing the new family Xenasteiidae pivoted on differences between *Xenasteia* and the genera then in Asteiidae, emphasizing differences in key characters such as the costal breaks (Asteiidae none, *Xenasteia* two) and the phallus (long and slender in Asteiidae, short in *Xenasteia*). He also considered suggestions from other leading dipterists that *Xenasteia* was related to the Carnidae or the Anthomyzidae, but rejected those suggestions for similar reasons. *Xenasteia*, then, became a family simply because it did not fit well into existing families.

Shortly after Hardy published the family name Xenasteiidae, Hungarian dipterist Laszlo Papp described the family Tunisimyiidae for a single new species, *Tunisimyia excellens*, from Tunis (Papp 1980). *Tunisimyia* turned out to be the same as *Xenasteia*, so Tunisimyiidae was the same as Xenasteiidae. Taxonomic synonyms like this are resolved using the rule of priority — the older name is correct. So, if we are to recognize this small group at the family level, it is Xenasteiidae and not Tunisimyiidae. Another small new family, Risidae, was named in the same 1980 paper in which Tunisimyiidae was proposed. The Risidae are now treated as part of the Ephydridae.

Most records of *Xenasteia* are from islands in the Pacific and Indian Oceans, although since 1980 species have been described from Spain, Israel and Taiwan as well as several Pacific islands. The biology of *Xenasteia* remains unknown but the dozen species currently included in this family probably develop as microbial grazers on seashore debris. One specimen was reared from a dead fish on an Indian Ocean island, and I have collected several specimens from a succulent seashore plant, *Sesuvium portulacastrum*, on the east coast of Australia.

Teratomyzidae

Fern Flies

Fern flies are elegantly elongate little flies very similar in general appearance to flies in the larger and more familiar family Anthomyzidae, from which they differ in significant wing, bristle and abdominal features. Most species in the family Teratomyzidae, including all of those for which any biological information is available, occur in Australia, where they can be found on fern foliage in moist forest. This fern association is known to be a close one for at least one Australian species, *Auster pteridii*, the larvae of which appear to be microbial grazers on the surfaces of fern fronds in eastern Australia. The generic classification of Teratomyzidae was reviewed by McAlpine and Keyzer (1994) — with descriptions of the nicely named genera *Auster*, *Lips*, *Camur*, *Stepta* and *Pous* — but they point out that most of the species in the family remain undescribed, and that unnamed species were known from Ecuador, southeast Asia and Japan. Several of those species have since been described, including a dozen new Oriental species and a new genus just named by Papp (2011).

Marginidae

Margin Flies

It seems remarkable that one of the world's rarest insect families should have a common name, but when David McAlpine named this family in 1991, he came up with the common name "margin flies" because of the distinctively dark leading edge of the wing on these truly unusual little (about 2 mm) acalyptrate flies. The entire family — which includes only the two *Margo* species named by McAlpine when he named the family — is known from only four specimens, probably fewer than any other widely accepted family of insects other than the nematoceran family Valeseguyidae. Three of the specimens, all males of one species, were collected in the 1950s in Zimbabwe; one, a female of a second species, was collected in 1987 in Madagascar. Both species were collected in wet forest, a threatened habitat everywhere, and McAlpine speculated that the Marginidae, along with the Mormotomyiidae and Eurychoromyiidae, are "among the fly families most at risk of extinction." The Eurychoromyiidae are now treated as a subfamily of Lauxaniidae and known from many new collections, leaving Mormotomyiidae, Valeseguyidae and Marginidae as the families known from the fewest localities.

The two *Margo* species seem to warrant their own family because they show a strange mixture of characters and cannot be reliably put even in a recognized superfamily, let alone a family. McAlpine (1991) only tentatively placed them in the Opomyzoidea, drawing attention to the very nerioid-like head and antennae, especially on the male specimens. The round, shiny thorax, unbroken costa and tachinid-like bulging postscutellum are among the most distinctive features of this mysterious little group. Of course, nothing is known of their biology.

Carnoidea

Frit Flies, Filth Flies, Grass Flies, INBio Flies, Beach Flies and Others

The superfamily Carnoidea is a loose assemblage of acalyptrate families reviewed by Buck in 2006; he tentatively considered it to be a natural (monophyletic) group mostly because males in all the included families have a long, soft, hairy penis — or, to be more precise, "phallus microtrichose, flexible, unsclerotized, simple and elongate." Although I may seem to be poking a bit of fun at taxonomy by using a vernacular rephrasing of part of a significant published paper, it is important to recognize that many of the useful and predictive taxa we recognize in the Diptera are defined on the basis of shared characters in the very complex male genitalic structures. These structures tend to be remarkably constant within species but often undergo drastic change coincident with speciation events, usually leaving a phylogenetic trail that can be followed through careful morphological study. Once the composition of natural groups such as the Carnoidea is established on the basis of complex internal characters (for example, the phallus and associated structures), those groups often turn out to be recognizable on the basis of biology or external features. However, Carnoidea is an unusually heterogeneous group on both counts.

Chloropidae

Frit Flies, Grass Flies, Eye Flies and Their Relatives

The Chloropidae, with just shy of 3,000 species, is one of the largest families of acalyptrate Diptera. Most members of the group are instantly recognizable as chloropids by a conspicuous, generally shiny triangular plate, or ocellar triangle, covering much of the front of the head, and even those that lack a charismatic shining triangle are reliably recognized as members of this family by a signature carina, or sharp ridge, crossing the propleuron. Chloropids are diverse in size and shape, generally small (2–3 mm), relatively bare and inconspicuous, although several are brightly patterned or striped in yellows, greens or even reds and a few exceed 5 millimeters in length. The most familiar Chloropidae are those that abound in lawns, fields and meadows, where the larvae develop in grass and cereal stems, and those that become nuisances because of their attraction to moist orifices or exudates of mammals and reptiles. However, chloropids occur in a wide variety of habitats. Larvae of different groups are saprophagous, predaceous, phytophagous or even occasionally parasitic. Some of the phytophagous species are pests of cereal crops and some of the predaceous species are beneficial predators of aphids, scale insects and grasshopper eggs. The family Chloropidae is usually divided into three worldwide subfamilies, the common and diverse Chloropinae and Oscinellinae, and the enigmatic small subfamily Siphonellopsinae.

Oscinellinae Most larvae of the subfamily Oscinellinae are phytophagous or saprophagous, with many species developing in living stems, seedpods or flower heads and others occurring only in decaying plant material. Some of the primary plant invaders, such as the Frit Fly, *Oscinella frit*, are significant pests of grasses in Europe and North America, but in many cases oscinellines are secondary invaders of plant material already attacked by other Chloropidae or other insects, sometimes feeding on the frass of their predecessors. Saprophagy in this subfamily reaches an extreme in an American species, *Aphanotrigonum darlingtoniae*, that is one of the specialized fly species that develop on decomposing insects inside the California pitcher plant, *Darlingtonia californica*.

Other unusual members of Oscinellinae include the remarkable Australian (and New Guinea) genus *Batrachomyia*, the species of which parasitize frogs in much the same way as rodent bots (Oestridae, Cuterebrinae) attack their hosts. Larval *Batrachomyia* apparently seek out amphibian hosts (frogs in the families Hylidae, Myobatrachidae and Ranidae), penetrate the

Some species in the large chloropid genus *Olcella* are kleptoparasites that aggregate and mate while feeding on the juices flowing from some kinds of insects captured by, and being consumed by, spiders or other invertebrate predators.

skin and develop in swellings under the skin. This subfamily also includes those that predate or parasitize spider egg sacs, a habit that seems to have arisen independently in a number of basally saprophagous lineages. *Gaurax*, for example, is a relatively large, colorful genus that includes predators of spider egg masses as well as several saprophagous and fungivorous species. Another genus, *Pseudogaurax*, is made up mostly of predators of spider egg sacs — including *P. signatus*, an egg predator of black widow spiders in North America — but also includes a couple of species that have been reared from African mantid egg masses. Several members of this subfamily parasitize or predate other arthropods or their eggs; some, such as species in the genus *Fiebrigella*, are specialized predators of grasshopper egg pods.

The adult habits of some oscinellines are of special interest, especially those of the New World *Liohippelates* species known as "eye flies," "sore flies" or (at least among dipterists who own male dogs) "pecker gnats." Their localized abundance and exudate-seeking habits make *Liohippelates* species potentially dangerous mechanical carriers of diseases such as pinkeye (conjunctivitis) and yaws, a serious disease caused by a spirochete, from person to person. Some Old World *Siphunculina* species have similarly annoying habits (as do some Drosophilidae). A few other adult oscinellines are attracted to invertebrate rather than vertebrate exudates; included among these are the kleptoparasitic species of *Olcella*, which aggregate, mate and feed on the juices flowing from some kinds of insect prey impaled by robber flies, spiders or other predaceous invertebrates. Similar habits are found in some Milichiidae.

Chloropinae The subfamily Chloropinae is best known for its many phytophagous species, including pest species such as the Gout Fly, *Chlorops pumilionis*, which causes gall-like swellings called "gouts" on barley, wheat and rye stems in the Old World, and the Wheat Stem Maggot, *Meromyza americana*, which is a pest of wheat in North America. As is true for the Oscinellinae, however, a number of lineages have apparently independently switched from a vegetarian diet to a carnivorous one. The Old World genus *Camarota*, for example, includes one species that is a pest of wheat and barley and another that preys on larvae of a different fly (a stalk-eyed fly, Diopsidae) that is itself a pest in rice.

The large (42 species) and widespread genus *Thaumatomyia* is also recorded as a grass stem borer, although the genus is primarily predaceous and includes an important predator of the Sugarbeet Root Aphid (*Pemphigus populivenae*). One common Old World species of *Thaumatomyia*, *Th. notata*, develops as a predator of root aphids. It periodically builds up enormous populations that appear as gigantic swarms, sometimes invading houses in huge masses and sometimes swarming at sufficient densities to be mistaken for the smoke of burning buildings. *Thaumatomyia notata* swarms are common in Europe but the only recorded mass occurrence of a Nearctic species, *Th. annulata*, was noted in Michigan in 1940. Unlike the Oscinellinae, Chloropinae are rarely saprophagous, although some only develop in plants that have already been damaged by other insects.

Siphonellopsinae The small subfamily Siphonellopsinae includes a few Palaearctic and Oriental genera but is dominated by the large cosmopolitan genus *Apotropina*. This is a particularly common genus in Australia, where one species scavenges in the nests of sand wasps (*Bembix*), others abound on beaches and another has been reared from mammalian carrion and collected in carrion-baited pitfall traps. This odd little subfamily fits uneasily into Chloropidae, so much so that some dipterists have treated it as a separate family, Siphonellopsidae.

Milichiidae

Freeloader Flies and Their Relatives

The family Milichiidae, sometimes called "freeloader flies" because of the kleptoparasitic (food-stealing) habits of a few species, consists mostly of nondescript little black acalyptrate flies, although a few stand out because of their silvery reflective structures, conspicuous swarms or unusual behaviors. This family of about 300 species occurs throughout the world. Larvae are generally saprophagous, usually developing in decaying plants but sometimes found in dung, carrion or more unusual deposits of organic material inside the nests of ants and other social insects. Adults of some ant-associated species hitch rides on foraging ants, clinging to the ants or their loads en route back to their nests; some solicit regurgitated food from associated ants, while others feed from the ant's anus. Adult females of some milichiid species are kleptoparasites, stealing food from spiders, robber flies, predaceous bugs and other large insect predators, in some cases specializing on particular kinds of prey such as Honey Bees. Females of several milichiid genera are common flower visitors.

Milichiidae are probably most closely related to the Chloropidae. The family is here divided into three widespread subfamilies, Milichiinae, Madizinae and Phyllomyzinae, following Brake (2000), although the latter two groups are often treated as the one subfamily Madizinae.

The subfamily **Milichiinae** is known from a wide range of microbe-rich habitats, including farmyards, bat dung, dead fish, guano, latrines and ant nests. Larvae of several species of *Milichiella* are reported as scavengers in ant nests; one species has even been described as covering itself with excreta on which attending ants feed. Larvae of some species of *Pholeomyia* develop in the nests of leafcutter ants, and some *Eusiphona* develop in the nests of leafcutter bees. The subfamily **Madizinae** includes the large, widespread genus *Desmometopa*, with more than 50 described species known from a variety of decomposing material, including rotting cacti, chicken dung, snails, rotting fruit, sewage and silage. Adults of *Desmometopa* are often seen on flowers but are perhaps most often noticed on recently killed Honey Bees being consumed by invertebrate predators such as spiders or robber flies. Some other Madizinae develop in ant nests. The subfamily **Phyllomyzinae** is dominated by the large genus *Phyllomyza*, known larvae of which occur in ant nests and adult females of which are frequently noted as kleptoparasites of

Many Milichiidae develop in the nests of social insects. This *Eusiphona* was reared from a leafcutter bee (*Megachile*) nest where its larvae developed on stored pollen.

spiders and other predaceous invertebrates. The worldwide genus *Paramyia* is another common group of phyllomyzines. *Paramyia*, especially the widespread New World *P. nitens*, often occur in crowds on flowers or on dead insects (especially stink bugs) being consumed by spiders.

Canacidae (including Tethinidae)

Beach Flies, Surf Flies and Surge Flies

Seashores are fantastic places to find highly habitat-restricted flies, including several specialized shoreline families such as Coelopidae, Heterocheilidae and Helcomyzidae. The most speciose and widespread flies of the ocean's edge, however, are the little (usually 2–3 mm) flies in the large (more than 300 species) family Canacidae. Although some canacid lineages have shifted from marine to inland habitats, this group is essentially a seashore family, ubiquitous on beaches and wave-splashed intertidal rocks, where their larvae feed on marine algae.

The small silvery gray "beach flies" invariably present on beaches and similar salty sediments the world over were until recently placed in the family Tethinidae, separately from the "surf fly" family Canacidae, but when it was discovered that the Canacidae were really just a special lineage of Tethinidae, the two families were combined (Buck 2006; D.K. McAlpine 2007). The Canacidae are probably related to the Chloropidae and Milichiidae.

Canacinae The Canacidae are divided into six subfamilies, of which the worldwide subfamily Canacinae, or surf flies, includes the 122 species that constituted the family Canacidae before it was combined with the Tethinidae. The surf flies, so called because they often occur on surf-splashed intertidal rocks, are similar to shore flies, although this Ephydridae-like appearance is only superficial: these groups are in different superfamilies that differ widely in details such as wing venation and genitalia. One of the more obvious distinguishing features of the Canacinae is a pair of conspicuously large spines that curve up from the tip of the female abdomen.

The worldwide subfamily **Tethininae** includes most but not all of the species formerly treated as the family Tethinidae, or beach flies. These are typically silvery or grayish flies frequently found in high densities at beaches, salt marshes, mangrove swamps and similar environments, sometimes including inland habitats such as desert oases. The mostly New World and Holarctic subfamily **Pelomyiinae** includes about 50 species formerly included in the Tethinidae, although most species of this subfamily are not beach flies but instead occur in inland sites such as meadows.

The subfamily **Zaleinae**, which was given the common name "surge flies" in the original paper describing it (D.K. McAlpine 2007), includes two genera and some 16 species known mostly from the water-whipped surge zone of intertidal rocks in Australia and New Zealand. A few species of these minute (usually less than 1.5 mm) flies occur elsewhere in the Australasian region, and (remarkably!) two species are known from Palaearctic shores in Israel, Egypt and Oman. The four species in the small subfamily **Horaismopterinae** occupy beaches in New Zealand, the Indian Ocean, the Middle East and southern Africa. The small (four species) subfamily **Apetaeninae** is endemic to the subantarctic islands, where they occur on seabird guano.

Carnidae

Bird Flies and Filth Flies

Many, if not most, of the 90 or so described species of small (1–2 mm) flies in the family Carnidae develop as saprophagous larvae in birds' nests and similar habitats. Adults normally feed on pollen, honeydew or fluids on the surfaces of decomposing material. *Carnus hemapterus*, however, times its transformation from nest-dwelling larva to two-winged adult to coincide with the appearance of new nestlings in spring. This Holarctic species overwinters as a puparium in bird nests, emerging in spring as a winged adult able to fly to other nests containing nestlings. Adult flies then shed their wings along shear lines near the base and spend the rest of their lives feeding on blood, skin particles and other bits on their young bird hosts. *Carnus hemapterus* is a well-studied species, common in the nests of tree-nesting birds (especially raptors), but other species of *Carnus* are poorly known. Three of the five species in the genus are each known from only a single collection in a single nest.

Over 80 percent of the species in the family Carnidae are in the filth fly genus *Meoneura*, a mostly Holarctic genus of tiny, dull (downright ugly!) flies normally associated with carrion, dung or decaying material in nests and other "filthy" microhabitats. Many North American species of *Meoneura* still await description.

The Carnidae have been traditionally treated as close to or even part of the Milichiidae, but it is now thought that they are more closely related to the small families Australimyzidae and Inbiomyiidae.

Inbiomyiidae

INBio Flies

The family Inbiomyiidae is a small but distinctive Neotropical group of 11 described (and many undescribed) species that has a fascinating history of discovery and description. The existence of the family first came to light when a hymenopterist, William Mason, collected an odd female fly specimen in Monteverde, Costa Rica, in 1980. He passed it on to a fellow Ottawa-based entomologist, dipterist Frank (J.F.) McAlpine, who recognized it as an undescribed genus and suspected that it might belong in the Aulacigastridae. The specimen was examined shortly thereafter by the Australian dipterist David McAlpine, who thought it was probably a new family, although it was hard to be sure on the basis of a single female specimen; he nonetheless arranged for head and wing drawings to be made.

Those drawings languished for about 20 years, until they were circulated at a meeting of dipterists in Monteverde with the comment that we should all watch for this mysterious Monteverde fly while in Costa Rica. One of the scientists at the meeting was Matthias Buck, then curator of the University of Guelph's insect collection. He immediately recognized the drawing as similar to specimens collected in Peru and Guatemala that he had earlier noted in the Guelph collection.

Buck went on to carefully study these remarkable flies, first putting them in a new genus, *Inbiomyia*, and a new family, Inbiomyiidae, based on the species originally collected at Monteverde (Buck 2006) and later adding another ten species to the genus *Inbiomyia* (Buck and Marshall 2006). The name for this group was chosen to honor the Costa Rican National Biodiversity Institute (INBio) because of its world leadership in documenting biodiversity and because of its support of Neotropical dipterology.

The family Inbiomyiidae is extraordinarily distinctive, with a large number of unique features such as an extremely short head with a nonfunctional ptilinum and few bristles, a short first flagellomere (third antennal segment) with a very elongate arista, a proboscis with largely separate labellar lobes that point in different directions, and several unusual features in the male and female genitalia. The larval habitat of *Inbiomyia* remains unknown, but some of the adult female specimens examined were found to contain large and uniquely flattened eggs, suggesting some sort of specialized biology.

It is remarkable that the family remained undiscovered for so long, because *Inbiomyia* species are not uncommon, especially around green decaying foliage near recently fallen trees throughout the tropical part of the Neotropical region (the family seems to be absent from the Caribbean). Inbiomyiidae occupy rainforests and cloud forests from sea level to 2,000 meters, but most species occur at lower elevations. The small size of these flies (less than 2 mm) is probably the main reason they escaped detection for so long.

Buck (2006) placed the Inbiomyiidae as the sister group of the monogeneric Australasian family Australimyzidae, within the superfamily Carnoidea.

Australimyzidae

Australimyzid Flies

The small (nine species) family Australimyzidae has an austral distribution, as the family name suggests: its one genus, *Australimyza*, is found in Australia, New Zealand and associated islands. Although larvae are known for only one species in the genus (and thus in the family), *Australimyza* seems to be an entirely coastal group normally associated with decomposing material along shorelines, including kelp, thatch among tussock grasses and seabird nests. Australimyzids can be common insects and are easily collected in shoreline pan traps, but — as one might expect for small (around 2 mm), dull-colored acalyptrates with specialized habitats — they are infrequently collected. Brake and Mathis (2007) revised the family based on all known specimens.

The Australimyzidae are probably closely related to the Inbiomyiidae, even though these groups are superficially dissimilar and occur in widely separated parts of the world. Dipterists have generally treated *Australimyza* in the superfamily Carnoidea since the genus was described (in the Milichiidae) by New Zealand dipterist Roy Harrison in 1953. An explicit analysis showing the relationship between Inbiomyiidae and Australimyzidae within this superfamily was provided as part of Matthias Buck's 2006 review of the Carnoidea, published with his description of the new family Inbiomyiidae.

Acartophthalmidae

Shaved-eye Flies

The family Acartophthalmidae includes only four extant species, all similar small, grayish flies in the Holarctic genus *Acartophthalmus*. Although adults have been collected from fungi, dung, tree wounds and carrion, usually in mesic woodlands, the larvae remain unknown. Since one author reports females laying eggs in a dead snake and another reports rearing adult acartophthalmids from decaying wood, it seems likely that the larvae are saprophagous.

The relationships between this uniform little group and other acalyptrates is far from clear, although recent studies at least place it in the superfamily Carnoidea (Buck 2006) and perhaps closest to the Chloropidae and Milichiidae (Brake 2000). It is a pity to have to recognize an entire family for such a small and relatively nondescript group of flies, especially when the acalyptrate flies as a whole are split into so many families that communication about diversity in this group is impeded. But there seems to be no logical home for this little cluster of leftover acalyptrates in any other named family.

The Calyptrates

Although calyptrates vary widely in size, shape and habit, they form a generally distinctive group represented by that most familiar of all flies, the common House Fly, *Musca domestica*, and including many other well-known robust higher Diptera such as blow flies and cluster flies. Calyptrates are traditionally divided into three groups. The blood-sucking superfamily Hippoboscoidea forms a single lineage that is probably the sister group to the rest of the Calyptratae. The superfamily Muscoidea is an artificial assemblage of calyptrate lineages similar to House Flies and root maggots, and the superfamily Oestroidea is a distinctive and natural lineage comprising the blow flies, screwworms, cluster flies, flesh flies, satellite flies, tachinids, bots and warbles.

Oestroidea

Blow Flies, Screwworms, Cluster Flies, Flesh Flies, Satellite Flies, Tachinids, Bots and Warbles

The superfamily Oestroidea is a well-defined lineage that includes most species of calyptrate flies. Members are easily recognized by the row of stout bristles on the meron (the side of the thorax just in front of and above the base of the hind leg) and further characterized by enough morphological and molecular evidence to render it one of the most widely accepted natural groups in the Diptera.

Calliphoridae

Blow Flies, Screwworms and Cluster Flies

Although the family Calliphoridae is best known for the ubiquitous and conspicuous metallic bluebottle and greenbottle flies, the 1,600 or so world species in this group span a wide range of parasitic, predaceous, flesh-eating and carrion-feeding habits. The common name "blow fly" refers to the effect of synanthropic (human-associated) sarcosaprophagous (carrion-feeding) species on exposed meat, which is referred to as "fly-blown" once it is contaminated by the eggs and larvae of Calliphoridae.

Although many species of Calliphoridae, especially species that attack other invertebrates, are larviparous or ovoviviparous and therefore bypass or effectively bypass the egg stage, the most common meat-associated species lay hundreds of eggs — or thousands, depending on the species — over their short adult lives, often swiftly covering exposed meat with clusters of rice-like whitish eggs. The eggs hatch into larvae (known to fishermen as "gentles") that rasp at the substrate with their mouthhooks while consuming a mixture of carrion and bacteria, usually feeding as a mass of maggots that collectively condition and raise the temperature of the blown meat. Development is temperature dependent; the most common species complete their larval development in about a week before migrating away from the larval substrate to pupate (pupariate) in the soil. Adults typically emerge another week or two later. Precise development times for each stage under different conditions have been calculated for the most common species, and these can be used by forensic entomologists to reconstruct the history of dead bodies on the basis of their resident maggots.

There is some disagreement over the treatment of recognizable calliphorid subgroups as tribes, subfamilies or even separate families, but all the important carrion-associated blow flies and most of the flesh-eating (myiasis-causing) species are found in the familiar large, widespread groups treated here as the subfamilies Calliphorinae, Luciliinae and Chrysomyinae. These three important subfamilies are thought to form a single lineage, or clade, along with the smaller, more localized and less familiar subfamilies Toxotarsinae and Melanomyinae. The relationships, and thus the classification, of the other calliphorid subfamilies remain a matter of debate.

Calliphorinae The subfamily Calliphorinae includes the big (6–14 mm) blue-black flies that appear quickly on carrion or cadavers, especially

Calliphora species are often the most conspicuous, if not the most abundant, flies attracted to carrion, and species in this genus are most important in forensic entomology work. This is a South American species, *C. nigribasis*.

in shaded or indoor sites. The largest and most important genus in the group, *Calliphora*, is most diverse at the northern and southern ends of the world — northern Europe and Asia, northern North America and Australia — with relatively few endemic species in between, although some secondarily widespread synanthropic species are common almost everywhere. *Calliphora vomitoria* and *C. vicina* are among the most predictable visitors to dead bodies throughout the Holarctic region; the latter species is now spread over much of the world and is frequently the most common insect species on human corpses, especially in urban areas.

Although sarcosaprophagy is the rule for the subfamily Calliphorinae, some calliphorines develop in the flesh of living vertebrates, several attack other invertebrates, and a few invade unusual pockets of decay such as those found deep in the pitchers of Southeast Asian pitcher plants. For example, the mostly Palaearctic genus *Bellardia*, the mostly Pacific genus *Onesia* and some *Calliphora* species are parasites or predators of earthworms, as are the generally more common species in the subfamily Polleniinae. The originally Australian species *Calliphora stygia* (the Eastern Golden-haired Blow Fly) is a common myiasis-causing species that often follows *Lucilia cuprina* (see subfamily Luciliinae, below) in causing "sheep strike" by invading the flesh of livestock in Australia, Polynesia and New Zealand.

Luciliinae The subfamily Luciliinae (treated by some specialists as a tribe in the Calliphorinae) is best known for the familiar greenbottle flies, those extremely common metallic green flies frequently seen on flowers and sunny foliage as well as on garbage, dung and carrion. Greenbottle (*Lucilia*) species are smaller (4–10 mm) and shinier than *Calliphora* and much less likely to be found indoors or in shaded places. Several synanthropic species of *Lucilia* can be found on a range of smelly substrates such as dung and carrion over much of the world; the most cosmopolitan, best-known and most abundant species in the group is probably *Lucilia sericata* (sometimes treated as *Phaenicia sericata*). This is a fly that makes rural dumps buzz, but it is also a species with the potential to penetrate flesh, living or dead. Some strains of *Lucilia sericata* are the major "medicinal maggots" maintained in culture for use in maggot therapy. These strains feed only on decaying material, and maggots raised under sterile conditions are routinely used to clean necrotic tissue out of deep and difficult wounds, a medical value enhanced by the bacteria-killing compounds produced by the larvae of many calliphorid species. In the wild, however, greenbottle flies are not necessarily limited to necrotic tissue, and many feed on both the living and dead flesh of animals (and humans). *Lucilia sericata* and some similar species, especially *L. illustris* and *L. cuprina*, are serious pests when they lay eggs in the wounds or fouled fleece of sheep, where their larvae can infest the animal's living flesh. This is known as "sheep strike," and it can be costly; *L. cuprina* alone is thought to cost the Australian sheep industry more than $170 million per year.

Some Luciliinae routinely develop on living amphibians. The common greenbottle flies *Lucilia silvarum* (Holarctic) and *Lucilia bufonivora* (Palaearctic) preferentially infest the flesh of frogs and toads, often killing their hosts. Several other lineages of blow flies, especially the true screwworms and bird blow flies in the subfamily Chrysomyinae, have similarly crossed the line between immersing themselves in bacteria-laden dead meat and feeding on living flesh. When fly larvae attack living vertebrates it is referred to as myiasis, a phenomenon more broadly discussed in Chapter 3.

Chrysomyinae The subfamily Chrysomyinae is infamous for the flesh-eating habits of the genera *Cochliomyia* and *Chrysomya*. *Cochliomyia* is a New World genus and *Chrysomya* was originally distributed in the Old World (mostly in the Afrotropical, Oriental and Australasian regions), but several species are now widespread. Both genera include primary screwworms, which are obligate feeders in living flesh, as well as secondary screwworms, which are facultative feeders on living flesh. Other species normally develop in the default calliphorid pabulum of dead tissue.

Chrysomya bezziana, the Old World Primary Screwworm, and *Cochliomyia hominivorax*, the New World Primary Screwworm, are significant pests of livestock in warmer parts of the Old World and the Americas, respectively; the latter once ranged well into the United States but was eradicated from the Nearctic region by the mass release of sterilized males. Screwworm eggs are usually laid near wounds such as a tick bite or a scratch from a thorn or barbed-wire fence, and larvae develop swiftly on the animal's flesh before dropping to the ground to pupate. Hosts are literally eaten alive, sometimes with fatal consequences.

Secondary screwworms include the widespread and common *Chrysomya megacephala* and *C. rufifacies*, both of which were originally Old World species but are now established in much of the world, including North and South America. The latter species, known as the Hairy Maggot because of its hirsute larvae, now ranges all the way north to Canada. Each Hairy Maggot female lays only male or only female eggs that hatch to facultatively predaceous larvae, able to shift from eating carrion to consuming competing maggots.

One of the largest genera of Chrysomyinae is the bird blow fly genus *Protocalliphora*, a group of obligate parasites of nestling birds. *Protocalliphora* maggots live in the nests of particular host species (most are host-specific), where they suck blood from the nestlings and hide in nest debris between sanguinary snacks. Maggots of the closely related genus *Trypocalliphora* take their association with nestlings a little further, burrowing under the host's skin to feed. Not all Chrysomyinae are parasites or screwworms, however; one of the most abundant of all carrion-feeding blow flies, the Black Blow Fly, *Phormia regina*, belongs to this group. Black Blow Flies are among the most common carrion-associated insects in the Holarctic region.

Toxotarsinae The subfamily Toxotarsinae is a small South American group with habits similar to the Calliphorinae. Although they now compete with widespread species such as *Calliphora vicina* and *Lucilia sericata*, some toxotarsine

Although blow flies in the subfamily Chrysomyinae are infamous for the flesh-eating habits of their "screwworm" larvae, the adults are often strikingly beautiful insects. This is a *Compsomyiops* species from Peru.

species remain very common within their limited ranges. In much of Chile, for example, the brilliantly metallic blue-green *Sarconesia magellanica* is one of the most common species attracted to dung and carrion, along with *Sarconesia chlorogaster* and *Sarconesiopsis chilensis*. Most species of Toxotarsinae, however, are rarely collected and little-known flies, perhaps because some have not been as successful in holding their own against species introduced from more northern latitudes, and perhaps because some have always occurred only in small areas and limited microhabitats.

Melanomyinae Although the subfamily Melanomyinae is thought to be closely related to the mostly carrion- and flesh-feeding blow flies discussed above, this small subfamily is probably entirely parasitic. The life cycle of the nominate genus, *Melanomya*, remains unknown, but some other Melanomyinae are parasitoids of snails, developing inside their hosts and ultimately killing them. Species of these uncommon nonmetallic, muscid-like calliphorids occur throughout Europe, North America and the northern neotropics, and have been recently recorded from the Oriental and Afrotropical regions as well. The limits of the subfamily remain uncertain.

Polleniinae The subfamily Polleniinae, the largest and most common parasitic lineage of Calliphoridae, is made up mostly of dull, House Fly–sized calyptrates adorned with conspicuous fine yellow hairs like crinkly corn silk. Most Polleniinae belong to the large worldwide genus *Pollenia*, some species of which are called "cluster flies" because the adults cluster in huge numbers on walls prior to taking shelter for the winter in attics and similar places. When they become active again in early spring, they can become nuisance flies in houses, even though these sluggish flies neither bite nor contaminate food. Some cluster flies lay their eggs on soil inhabited by earthworms, from whence the newly hatched larvae make their way through the soil in search of earthworm hosts to consume as parasitoids (usually) or predators, but the biology of this group is relatively poorly known and a few have been recorded from other hosts, including caterpillars and bees. Cluster fly adults can be among the most abundant insects visiting flowers and are often extremely common on flowers in urban and agricultural settings.

Like some other Calliphoridae, the Polleniinae are diverse at both ends of the planet, with lots of species restricted to either the northern or southern hemisphere. Australia has seven *Pollenia* species, all endemic, while New Zealand alone has around 32 endemic species and at least two species, *P. pediculata* and *P. rudis*, apparently accidentally brought in from North America in the 1980s. The latter, along with North America's five other *Pollenia* species, probably first came to North America from Europe along with the earthworms on which they feed. *Pollenia pediculata* (also known as *P. pseudorudis*) is now a nuisance pest in New Zealand, just as introduced cluster flies are nuisance pests in North America. All of the widespread nuisance *Pollenia* species appear to be parasitoids of earthworms; however, none of the native New Zealand members of the genus are known as larvae. New Zealand has a remarkable earthworm fauna that includes some 173 native species, but none have been recorded as hosts for *Pollenia* species.

Helicoboscinae Cluster flies are widespread and familiar insects, but other parasitic Calliphoridae are generally more localized or uncommon insects in genera scattered over a range of subfamilies. Many of these parasitic or predaceous lineages deposit larvae rather than eggs. Members of the small subfamily Helicoboscinae, for example, deposit relatively large (late first-instar) larvae directly on their snail hosts — usually dead or dying Helicidae. This group includes only two genera, the monotypic Himalayan genus *Gulmargia* and the western Palaearctic genus *Eurychaeta*, which includes species previously treated as *Helicobosca*. Helicoboscines used to be treated as part of the Sarcophagidae, and treatment of this group as part of the Calliphoridae remains controversial.

Ameniinae As far as is known, the relatively robust, brightly colored Ameniinae are parasitoids that deposit large (second- or third-instar) larvae on snail hosts, but relatively few species have been reared. Ameniinae are most characteristic of the Australasian region and also occur in the Oriental region. Several endemic Australian *Amenia* species look much like larger Tachinidae, so it is not surprising that the Ameniinae were previously included in that entirely parasitic family.

Phumosiinae and Aphyssurinae Known species in the small Old World tropical and Pacific subfamilies Phumosiinae and Aphyssurinae are similarly macrolarviparous, but hosts remain unknown for the strictly Australian subfamily Aphyssurinae. The Phumosiinae include an Oriental species, *Phumosia coomani*, that develops as a predator inside the foam egg masses of the brown tree frog, *Polypedates megacephalus*. Macrolarvipary also crops up in the odd Pacific genus *Dyscritomyia* (Luciliinae), of which several endemic Hawaiian species have been reared from land snails.

Bengaliinae Adults of the Afrotropical and Oriental blow fly subfamily Bengaliinae are typically brown or yellowish flies that depart from the usual adult calliphorid habits of lapping up liquids from carrion and dung; some *Bengalia* even actively hunt and consume other insects, including termite workers, ant pupae and prey stolen from ants. Larval habits of most Bengaliinae are poorly known, but one Oriental species, *Verticia fasciventris,* was recently recorded as an internal parasitoid that develops in the head capsule of the termite *Macrotermes barneyi*. Larvae leave the host's head through the neck foramen and migrate through the termite's body before popping out of the nether regions to pupariate (Sze et al. 2008). The classification of the *Bengalia* and related genera is highly controversial, in part because of disagreements about a monograph by Lehrer (2005) in which the genus *Bengalia* was elevated to family level and divided into a dozen genera in four subfamilies. Lehrer's classification has not been widely accepted, and his genera of "Bengaliidae" are still usually treated as one genus, *Bengalia*.

According to Rognes (2011) most Bengaliinae other than *Bengalia* belong to a lineage that includes saprophagous species, species associated with termite or ant nests, and the myiasis-causing species previously treated as the subfamily Auchmeromyiinae (these were moved to Bengaliinae by Rognes in 1998). *Auchmeromyia senegalensis*, known as the Congo Floor Maggot even though it is found throughout sub-Saharan Africa and the Cape Verde Islands, lays eggs on dry earth, sometimes the earthen floors of huts but normally in nests or burrows of animal hosts such as wild pigs, warthogs, aardvarks or hyenas. The larvae visit hosts, including people, during the night, sucking blood for about 20 minutes at a time before dropping off and hiding. Some blow flies in the Chrysomyinae attack nestling birds in a similar fashion, but *Auchmeromyia* is the only known genus of blood-sucking maggot to attack humans.

The Tumbu Fly, *Cordylobia anthropophaga*, is a related African species that attacks a range of mammals, also including *Homo sapiens*. Tumbu Flies lay their eggs on substrates contaminated with urine or feces, usually in sand or soil but often on diapers or other soiled clothing. The eggs hatch upon contact with the skin of a potential host and the maggots burrow under the skin, where they develop in boil-like warbles not unlike those created by some bot and warble flies in the family Oestridae. Development under the host's skin takes a week to ten days, after which the mature maggot pops out and puypariates among debris or soil on the ground. Other *Cordylobia*, *Auchmeromyia* and related genera attack a variety of mammals. Some are highly host-specific, as reflected in names such as Elephant Skin Maggot (*Booponus indicus*, known only from the skin of Asian elephants) and deer and water-buffalo skin maggots (four other *Booponus* from eastern Europe and Asia).

Rognes (2011) makes a strong case for treating the Bengaliinae (including the former Auchmeromyiinae) as a separate family Bengaliidae.

Although this family is one of the most studied groups of flies, much remains to be discovered about the Calliphoridae. The Neotropical and Afrotropical subfamily **Prosthetosomatinae**, for example, remains known only from larvae found in the nest mounds of termites. Adults are entirely unknown; in fact Prosthetosomatinae were until recently treated as part of the Muscidae and might yet turn out to be Rhiniidae when the adults are found. Even the very legitimacy of the family Calliphoridae in the current sense is in question. One group traditionally placed in the Calliphoridae, as the subfamily Rhiniinae, is now thought to be more closely related to other calyptrate families than to blow flies and it is treated below as the separate family Rhiniidae. On the other hand, some authors argue that a couple of calyptrate groups widely treated as separate families — the Mesembrinellidae, Mystacinobiidae and Rhinophoridae — actually belong in the Calliphoridae. These groups are also treated as separate families below.

Adult flies in the blow fly genus *Bengalia* are predators of other insects, an unusual habit for the higher Diptera. This Vietnamese fly is eating a cockroach nymph.

Mesembrinellidae

Mesembrinellid Flies

The robust, partly metallic and partly brown, long-legged adults of the family Mesembrinellidae

Mesembrinellids are large flies, common and conspicuous throughout the neotropics where they can usually be found at carrion or dung. The biology of the group remains almost unknown, but the female abdomen typically holds only a single large egg at a time. This group of flies is sometimes included in the Calliphoridae.

are strikingly conspicuous insects in native Neotropical forests from southern Mexico to northern Argentina, and they are often the largest flies present on dung, carrion or other decomposing materials. Females produce a single large egg at a time; it hatches inside the uterus, where it is apparently nourished by secretions from the enlarged spermathecae (structures used for sperm storage in other flies). Although one mesembrinellid species has been partially reared on dung and carrion, the natural host or larval food remains unknown for the whole group.

This distinctive group of fewer than 50 named species is often treated as a subfamily of Calliphoridae, but the status of *Mesembrinella* and related genera as a subfamily of Calliphoridae has long been controversial; the most significant paper on the group to date, by Guimarães (1977), treats the group as the family Mesembrinellidae. More recently a major study of calyptrate phylogeny (Kutty et al. 2010) provided some evidence that the Mesembrinellidae are more closely related to the Tachinidae than to the Calliphoridae.

Rhiniidae

Rhiniid Flies

The family Rhiniidae is a group of some 400 species traditionally treated as a subfamily of Calliphoridae, although they are not associated with carrion and dung as are typical blow flies. This mostly eastern hemisphere group is diverse in the Old World tropics and the Australasian region; only a few species are found in Europe and Asia, where the syrphid-like adults of the genus *Stomorhina* are often seen on flowers. Rhiniidae are completely absent from the New World, with the exception of a single introduced species in Bermuda. Some rhiniid species have been observed laying eggs in freshly excavated termite mounds, but the best-known species in the genus, *S. lunata*, is an important predator of grasshopper egg pods. *Stomorhina lunata* females oviposit on the soil above the buried egg pods, and hatching larvae burrow down and destroy the egg pods before pupating in the soil.

Some African and Australian rhiniid maggots feed on termites in nests that have been recently broken open by other animals, and adults of a few *Rhinia* species have been reared from the nests of sand wasps, where their larvae probably developed on dead or paralyzed insects. Parasitism, predation or scavenging on immobile below-ground arthropods such as termite brood, grasshopper eggs or wasp prey seems to be general in this group, although most Rhiniidae remain unknown as larvae. This is perhaps the least-known of the large oestroid families.

Rhinophoridae and Axiniidae

Woodlouse Flies and Axe Flies

The **Rhinophoridae** are best known to North American entomologists only on the basis of the common Holarctic Woodlouse Fly, *Melanophora roralis*, a distinctive little dark-winged fly with prominent white dots on the female's wing tips. However, this is a family of a couple of hundred species of parasitoids, possibly all associated with isopod hosts (older records of rhinophorids with other hosts are of species since transferred to other families). The Woodlouse Fly is most common in disturbed areas and even inside buildings such as greenhouses, where it is an endoparasitoid of "sowbugs" (isopods in the family Oniscidae). Both the fly and its hosts are synanthropic species probably accidentally introduced from Europe to North America; *Melanophora roralis* has now been recorded as far south as Chile. Rhinophorids are most diverse in southern Europe and Africa; there are only a few Neotropical and Oriental species. The Rhinophoridae are probably closely related to the Calliphoridae, and some authors treat them as the subfamily Rhinophorinae in the family Calliphoridae.

Some dipterists consider the Rhinophoridae to be absent from the Australasian region, but that is a matter of definition, since the small Australian family **Axiniidae**, or "axe flies," is now considered by some dipterists to be part of the family Rhinophoridae. The axe flies — so named because of their axe-shaped antennae — include 16 species in four genera in Australia and New Guinea. All are unknown as larvae but it is a good guess that axe flies, like the related Rhinophoridae, are parasitoids of isopods or related crustaceans.

Sarcophagidae

Flesh Flies, Satellite Flies and Their Relatives

Most members of the large and common family Sarcophagidae are readily recognizable by the pattern of three — not four, as in many Tachinidae — longitudinal black stripes on the thorax combined with a tessellated pattern on top of the abdomen. Such typical Sarcophagidae can be found almost everywhere in the world, even in city parks and backyards, where some breed abundantly in mundane debris such as dog dung. The common name "flesh fly" reflects the fact that several abundant species of Sarcophagidae develop in cadavers and carrion; some scarce species actually consume the flesh of living animals, although most of the more than 3,000 world species of Sarcophagidae probably develop on invertebrate tissue as scavengers, predators, parasitoids or kleptoparasites. The family also has its full share of unusual specialists in such habitats as the soup inside insectivorous pitcher plants, the dung of bats and the nests of social wasps. Sarcophagids are almost all ovoviviparous (their eggs hatch at the moment they are expelled) and are thus probably pre-adapted to capitalize on ephemeral food sources or evasive hosts by depositing active larvae rather than eggs.

The family Sarcophagidae is divided into three subfamilies, the typical Sarcophaginae (*Sarcophaga* and relatives), the Miltogramminae (satellite flies and their relatives) and the relatively small subfamily Paramacronychiinae. Because of the remarkably uniform appearance of most sarcophagid species, species identification is often difficult and largely dependent on examination of the male genitalia.

Sarcophaginae The subfamily Sarcophaginae is dominated by the ubiquitous genus *Sarcophaga*, a huge genus with about 800 species, most of which are superficially similar medium-sized

Flesh flies throughout the world have a remarkably uniform grey-striped appearance. This Peruvian *Ravinia* probably developed in dung, a typical habitat for species in this large genus.

gray and black striped flies associated with dead invertebrates, although many species develop in vertebrate carrion, dung or other decomposing material, sometimes in specialized habitats such as nests of paper wasps. This huge genus also includes predators or parasitoids that attack tent caterpillars, snails, earthworms, spider egg sacs and grasshopper egg pods. Several sarcophagine species develop in the cuplike leaves of New World or Old World pitcher plants; the large grey and black adults of pitcher-plant flesh flies in the genus *Fletcherimyia* are very common in eastern North American peatlands, where each pitcher plant (*Sarracenia* spp.) typically houses one large *Fletcherimyia* larva. Other sarcophagine genera inhabit a variety of larval media such as dung, carrion and the bodies of living or dead invertebrates.

Blaesoxipha, the second largest genus in the family, is largely parasitic, usually attacking acridid grasshoppers and tenebrionid beetles, although there are *Blaesoxipha* species associated with a wide range of invertebrates. Some members of this diverse genus are significant parasitoids of locusts and other acridid grasshopper adults, and a few develop in carrion or the flesh of reptiles and amphibians. Some sarcophagines are parasitoids of cicadas and are among the few parasitoids that locate their hosts by sound. Like tachinids that locate cricket hosts by song, the sarcophagine cicada parasitoids have an enlarged prosternum that functions like an ear. Members of the sarcophagine genus *Tripanurga* are commonly found feeding on eggs in the underground nests of various turtle species and were until recently considered specialized predators and major pests of rare turtles. Recent work, however, suggests that *Tripanurga* are just specialized scavengers, normally invading only eggs that are already damaged. Larvae of another turtle-associated sarcophagine, the North American *Cistudinomyia cistudinis*, are myiasis-causing maggots that invade adult turtles through tick bites or similar wounds.

Miltogramminae The subfamily Miltogramminae includes the familiar "satellite flies," so called because of their habit of flying behind solitary bees or hunting wasps at a fixed distance, like little satellites, as they await an opportunity to place a larva where it can steal the wasp's paralyzed prey or the bee's stored pollen. Some such kleptoparasitic miltogrammines larviposit directly on paralyzed prey as it is being transported, some larviposit into nest entrances and some go right into nests to larviposit on the provisions. Although egg laying is rare in the Sarcophagidae, some members of this group glue incubated eggs onto the host wasp or its paralyzed prey. Kleptoparasitic miltogrammines are almost invariably found wherever solitary bees or wasps are found, and they are major mortality factors for many if not most aculeate (stinging) wasp species.

Other members of the subfamily have converged on habits also found in the Sarcophaginae. Some *Eumacronychia* species, for example, are opportunistic scavengers or predators on buried turtle nests, as are *Tripanurga* species (Sarcophaginae); recent studies have shown *Eumacronychia sternalis* to develop in sea turtle nests in much the same way as *Tripanurga* species develop in damaged clutches of several freshwater turtle species. Other members of the genus *Eumacronychia* have been reared from dead or paralyzed insects and from vertebrate carrion. Several miltogrammines are parasitoids, including some species that tackle termites and one pest species, *Senotainia tricuspis*, that develops inside adult Honey Bees and bumble bees. Other members of the same genus are kleptoparasitic on the nest provisions (prey) of aculeate wasps or are predators of grasshopper egg pods — habits found in several other genera in the subfamily.

Paramacronychiinae The subfamily Paramacronychiinae has fewer species and a more restricted distribution — almost entirely Holarctic, although a few species occur in southern Africa and Central America — than the Sarcophaginae (worldwide but mostly New World) and Miltogramminae (worldwide but mostly Old World). Paramacronychiinae is a diverse group that includes parasitoids of grasshoppers and snails, parasites that cause myiasis in humans, scavengers, and predators of a variety of insects such as bumble bee brood and grasshopper egg pods. Perhaps the best-known of the group are the parasitic members of the genus *Wohlfahrtia.* Although the genus *Wohlfahrtia* itself is diverse and includes saprophagous and predaceous species, it is infamous because of the parasitic habits of the grandly named *Wohlfahrtia magnifica.* Also known as the Spotted Flesh Fly because *Wohlfahrtia* species have a spotted abdomen, *W. magnifica* is an Old World species known for causing myiasis in mammals, including humans. Females larviposit in wounds or orifices such as ears, eyes, mouth or genitalia; the larvae develop quickly and generally destructively, although *Wohlfahrtia* maggots have been used for cleaning human wounds. New World *Wohlfahrtia* — currently treated as the two similar species *W. opaca* and *W. vigil* — have similar habits but rarely cause myiasis in adult humans.

Mystacinobiidae

New Zealand Bat Flies (and McAlpine's Fly?)

The New Zealand Bat Fly, *Mystacinobia zelandica*, has traditionally been treated as its own single-species family because of its unusual morphology and its unique phoretic association with the endemic New Zealand short-tailed bat, *Mystacina tuberculata*, itself the only extant member in its family. (A second species in the bat family Mystacinidae is thought to have become extinct in the 1960s, before the fly family Mystacinobiidae was described in 1976, and further extinct species are known from Australian fossils.) New Zealand Bat Flies are bizarre little (3 mm) wingless flies with long, bristly legs that give them a spidery appearance similar to unrelated bat flies in other families. Larvae live in the bat guano while adults cling to the host bats, laying eggs collectively in a sort of "nursery" of eggs and larvae. Remarkably, there seems to be some sort of communal care of the larvae: female flies groom the larvae (and each other) and some males act as guards, producing a high-frequency buzz to deter bats from disturbing the fly colony. Recent molecular and morphological work suggests that despite the unusual appearance and biology of Mystacinobiidae, this group belongs in the superfamily Oestroidea.

Kutty and colleagues (2010) suggest that the Mystacinobiidae plus "McAlpine's fly" form the sister group to the Sarcophagidae. McAlpine's fly is an enigmatic Australian species that has puzzled dipterists for many years and remains unnamed, even though it has been discussed in the literature for decades under its informal name — given because Dr. D.K. McAlpine collected the first specimens to be studied. Ferrar (1979) described the immature stages and biology of McAlpine's fly, showing that it is a macrolarviparous species in which the female develops a single larva inside the uterus before larvipositing the large first-instar larva on dung. Larvae develop as microbial grazers in dung and pupariate in the nearby soil. Ferrar and later authors point out that McAlpine's fly is an enigma, superficially similar to Anthomyiidae but with some important characters more like those of Calliphoridae. Current work suggests that McAlpine's Fly and *Mystacinobia* are either separate basal lineages of the Oestroidea or together form the sister group to other oestroids. Now that we are coming closer to understanding how these enigmatic species fit into calyptrate classification, McAlpine's fly will soon either need its own family name or will be subsumed into the Mystacinobiidae.

Tachinidae

Parasitic Flies or Tachinid Flies

The Tachinidae is in many ways the ultimate fly family. With almost 10,000 named species and thousands more still awaiting description, this ubiquitous group is rivaled in diversity only by the Tipulidae (with around 15,000 species) and exhibits an unparalleled variety of sizes, shapes and colors. The range of life history strategies is equally amazing, at least within the constraint that every known species in the group is a parasitoid that develops inside another insect — usually a caterpillar, sawfly larva, beetle or true bug — or related arthropod. Despite the ubiquity and impressive diversity of the Tachinidae, flies in this family are almost invariably recognizable as tachinids by their prominent subscutellum (a distinct bulge under the scutellum).

More than half of all parasitoids in the order Diptera are tachinids, and because most tachinids are parasitoids of phytophagous insects, they are enormously beneficial in the regulation and control of insect pests of plants. A few tachinid species, all in the subfamily Tachininae, attack arthropods other than insects, and there are a few scattered records of species that develop as true parasites — parasites don't kill their hosts, while parasitoids do — but such records are exceptional. Although adult tachinid flies vary widely in morphology, often with a beauty belying their categorization as "killer flies," the typical tachinid is House Fly–like in general size and shape, with its abdomen wrapped in armored tergites and conspicuously studded with long, stout bristles. This well-defended abdomen, entirely encased as it is in spiked armor, sometimes serves to deliver eggs to a potentially resistant recipient, although tachinids employ a spectrum of attack strategies that don't always involve direct contact between the female fly and the host.

The most basic mode of attack — now abandoned by most lineages of tachinids other than the Phasiinae and some Exoristinae — is simply to nail the host with an undeveloped egg. When the fly larva emerges from the egg, it burrows into the host, sometimes directly through the relatively thin bottom of an egg earlier stuck conspicuously to the host's integument. This system has some obvious disadvantages for the tachinid. For example, if the eggs are left exposed on the outside of the host they could be bitten or scraped off. Furthermore, the adult female tachinid has to come in direct contact with the host in order to lay eggs, something the host could avoid by hiding, by coming out only at night or by developing a protective coat of hairs or scales. The strategies that tachinid flies have evolved for circumventing such defenses are partly responsible for the great morphological and behavioral diversity of this group of "killer flies." Most tachinids incubate their eggs so that they are ready to hatch immediately upon oviposition, spending little or no time exposed as a vulnerable unhatched egg. Some female tachinids use a piercing

"Hedgehog flies" like this large and obviously well-defended Peruvian *Adejeania* are parasitoids of caterpillars. This one is not smoking a cigar; the big black process is the pair of enlarged maxillary palpi.

ovipositor to inject their eggs, releasing them safely inside the host. Others lay eggs that hatch immediately into planidial (active) larvae, while females in one group lay thousands of minute eggs that hatch only when ingested by a host.

Tachinid larvae are specialized endoparasitoids with reduced mouthhooks, the labrum fused to the cephaloskeleton in the first instar, and a variety of mechanisms for dealing with the hostile environment inside another animal's body. If some sort of parasitoid were to insert eggs directly into you and me, our bodies would react to this effrontery by healing over the surface wound and attacking the inserted egg, just as any foreign object is attacked by our body's defensive system. Insect hosts react to tachinid attacks in much the same way, but, surprisingly, many tachinid larvae turn these responses to their own advantage. In species that adopt this strategy, the wound-healing process of the host is distorted so that the "scab" does not close the entry hole but instead fuses with the capsule formed around the invader by the host's internal defenses. The ultimate effect is a tube, called a respiratory funnel, that runs from the tachinid larva to the surface of the host. Since growing tachinid larvae have to breathe air through their posterior spiracles, the tube serves as a sort of snorkel, allowing the parasitoid to breathe at the surface of the host while dining on the host's entrails. Other tachinids have different breathing strategies. Many tap the internal air supply of their host, which is analogous to a parasitoid sticking a breathing tube into your lung while feeding on your other internal organs.

The family Tachinidae is not only one of the biggest fly families but is also widely considered one of the most taxonomically difficult. Enormous numbers of species remain undescribed, most described species are effectively unidentifiable except by a handful of specialists, and hosts are known for only a relatively small proportion of the family. Many related species with different hosts are morphologically indistinguishable, or almost so, and the higher classification of the family is at best poorly supported. Despite these problems, most tachinid researchers divide the family into the same four or five subfamilies. The biology of the family is discussed below in the context of the four most generally recognized and biologically disparate groups: Phasiinae, Dexiinae, Tachininae and Exoristinae. Of these, only the Dexiinae is strongly supported as a monophyletic group and only the small subfamilies Phasiinae and Dexiinae are readily recognizable by non-specialists.

Phasiinae Most members of the small subfamily Phasiinae are relatively bare-bodied parasitoids of Hemiptera (true bugs) that they parasitize using the basic or primitive attack strategy of depositing a large, unincubated egg on the host, although some are equipped to inject eggs into the host's body cavity. Almost all phasiines attack adult insects, most often larger true bugs such as stink bugs (Pentatomidae) and leaf-footed bugs (Coreidae), but there are exceptions: some records exist of phasiines attacking Lepidoptera, Coleoptera, Orthoptera and Mantodea. Phasiine females are often attracted to the odors of stink bugs and other large Hemiptera; baits based on bug pheromones can attract large numbers of some phasiine species.

Many Phasiines are distinctive, common and familiar flies. Species in the type genus are familiar flower visitors and include some strikingly broad and brightly colored species, while other genera show a wide variation in shape and size that belies the relatively low diversity of the subfamily. The small, squat, bare-bodied flies in the genus *Gymnosoma*, for example, could hardly be more different from the common slender *Cylindromyia* or the species of *Trichopoda* with spectacularly broad, flat scales on their hind legs. Hosts of all these common genera are adult bugs, and members of the latter genus have even been pressed into service for biological control of pest species of Coreidae. The generally low fecundity of phasiines, however, renders this subfamily less important in biocontrol than the larger subfamilies.

Dexiinae Members of the subfamily Dexiinae, the second smallest and best defined subfamily of Tachinidae, could hardly be more different in appearance and behavior than species in the subfamily Phasiinae. Instead of being small, generally bare flies that lay unincubated eggs on or in true bugs, Dexiinae are relatively large, bristly and conspicuously long-legged flies that lay ready-to-hatch eggs. Eggs are incubated in a special ovisac or uterus prior to deposition, so they hatch immediately into larvae that usually assume the task of approaching and penetrating suitable hosts. Their hosts are usually beetles (especially Scarabaeidae) or caterpillars found in concealed places accessible to the active, armored first-instar larvae of dexiine tachinids but not to the fragile adult flies. Caterpillars burrowing in plant stems usually leave excrement (frass) along the burrow, thus providing a cue for adult tachinids to lay their eggs in that area. The active tachinid larvae burrow through the frass to reach concealed hosts.

Dexiines tend to be much more fecund than phasiines, scattering hundreds or thousands of eggs near host habitat. The large, generally inconspicuously colored, long-legged tachinids common on tree trunks are typical members of the subfamily Dexiinae, but the group varies widely in size and color. Probably the most spectacular of the many large, metallic, bristly tachinids are the Oriental and Australian dexiines in the tribe ***Rutiliini***, which rival the Neotropical "hedgehog flies" from the subfamily Tachininae for spectacular size, setosity and colors.

Exoristinae The subfamily Exoristinae (often treated under the name **Goniinae**) is a huge and biologically diverse group with close to half the species in Tachinidae and most of the species of economic value as biological control agents. Although some recent molecular studies have confirmed the validity of the subfamily, Exoristinae is not well defined with respect to the large tachinid subfamily Tachininae, nor is it easy to make the sorts of generalizations about life history that were possible for the smaller subfamilies Dexiinae and Phasiinae; indeed, the Exoristinae exhibit the whole range of oviposition strategies found in

Tachinidae in the Oriental-Australian tribe Rutiliini rival the Neotropical "hedgehog flies" for spectacular size, setosity and colors. This eastern Australian *Rutilia* (*Grapholostylum*) *variegatum* is carrying a pollinium, probably from an orchid flower, on its head.

the family. Some lay exposed eggs on their hosts; others insert eggs right into the body of the host, using a female abdomen modified into a piercing tool; and some employ "attack larvae" like those of the Dexiinae. Members of one subgroup produce thousands of tiny eggs that they deposit on the foodstuff of the host. These tiny (microtype) eggs survive consumption by the host and then hatch into larvae that proceed to eat that host, most often a caterpillar, from the inside out.

The Exoristinae include many of the tachinids that have been moved around the globe as biological control agents of (usually) introduced pest insects. One classic success story is that of the introduction of *Cyzenis albicans* from Europe to North America for the control of a pest geometrid moth, the Winter Moth, *Operophtera brumata*. However, another exoristine brought to North America for control of pest Lepidoptera offers a cautionary tale about the introduction of tachinids for biocontrol. *Compsilura concinnata*, originally introduced to help control Gypsy Moths (*Lymantria dispar*) and other pest species, has become a common parasitoid of a wide range of native moths and has been held responsible for precipitous declines in various species, including the Luna Moth, *Actias luna*, and the Cecropia Moth, *Hyalophora cecropia*. One study showed that this polyphagous parasitoid was killing over 80 percent of the Cecropia Moths in an area in Massachusetts, and other studies suggest that depredations of other Nearctic giant silkworm moths are equally catastrophic. *Compsilura concinnata*, with more than 200 known hosts, is an unusually polyphagous species; many tachinid species attack only a single host or a narrow range of hosts. Recent work in the neotropics has suggested that some of the species previously considered

Very few higher flies are nocturnal, but *Ormia* species fly by night and use their inflated prosternum as an "ear" to pick up the nocturnal stridulations of their cricket hosts. This one (in a group of relatively dark species sometimes put in *Ormiophasia*) was attracted to a light in Costa Rica. Most *Ormia* are relatively pale, usually yellow or orange in color.

generalists are actually complexes of morphologically similar specialists with different hosts (Smith et al. 2007).

Tachininae The subfamily Tachininae is more diverse in size, shape and color than the other Tachinidae subfamilies combined, with sizes ranging from 2 to 20 millimeters. Although some tachinines deposit eggs directly on their hosts, most are like the Dexiinae in depositing large numbers of eggs in their host's habitat — often on foliage damaged by host feeding — and not directly on the hosts. Newly hatched larvae are typically heavily armored and sufficiently desiccation-resistant to wait some time for potential hosts to pass by, often taking an upright host-seeking stance as they try to detect an opportunity to attack and burrow into their doomed victim. This sort of indirect attack strategy not only bypasses the fly's vulnerable egg stage but also allows access to hosts hidden in webs, nests, burrows or other refuges difficult for adult flies to penetrate.

As is true for most Tachinidae, tachinines usually attack caterpillars (Lepidoptera), but some members of the subfamily are parasitoids of other arthropods, including other flies as well as beetles, grasshoppers, sawflies, earwigs and even centipedes and scorpions, although very few tachinids attack non-insect arthropods. One of the more unusual attack strategies in the family is found in the genus *Ormia*, a distinctive group of pale-colored flies with a conspicuous "breast," or inflated prosternum. Unlike most parasitic flies, which fly by day and locate hosts by the scent of their pheromones, frass or body odor, *Ormia* species fly by night and use their inflated prosternum as an "ear" to pick up the nocturnal stridulations of their cricket hosts. One species of this widespread genus has been moved from Brazil to Florida for use in biocontrol of pest mole crickets.

Oestridae

Bots and Warbles

The family Oestridae is a small group of about 150 species that develop under the skin of various mammals, in the digestive tracts of horses, rhinos and elephants, or in the sinus cavities of deer, sheep, elephants and many other herbivorous mammals. All larvae in the family are parasites in the bodies or tissues of mammals, and all develop into robust, often beelike flies that lack functional mouthparts — a few species imbibe fluids but most cannot feed at all. The adults are infrequently encountered except on hilltops and at similar elevated lek sites, even though some species are common enough for their larvae to be serious pests of livestock or annoying pests of humankind.

Oestrids are neatly divided into four subfamilies, each biologically and morphologically distinct enough to have been treated as its own family in the past. Oestridae now occur worldwide, but the Cuterebrinae (New World skin bot flies) are typical of North and South America while the other subfamilies, Hyperdermatinae (Old World skin bot flies or warble flies), Oestrinae (nose bot flies) and Gasterophilinae (stomach bot flies), probably originated in the Old World.

Cuterebrinae The subfamily Cuterebrinae, or "New World skin bot flies," is a strictly western hemisphere group consisting almost entirely of large (often 2–3 cm), robust, bumble bee–like flies in the genus *Cuterebra*. The eastern North American Emasculating Bot Fly, *Cuterebra emasculator*, is typical of this group. Female flies lay eggs in areas frequented by eastern chipmunks (*Tamias striatus*), where the warmth of a passing animal stimulates them to hatch. The first-instar larva makes its way inside the host's body through a convenient orifice such as the mouth or nose, then develops as an excruciatingly large subcutaneous parasite before popping out through the chipmunk's belly — through a breathing hole already maintained by the larva — and pupating in the soil. Despite their common name, Emasculating Bot Flies attack chipmunks of both sexes and rarely emasculate their victims. Another *Cuterebra* of special interest, the Howler Monkey Bot Fly, *Cuterebra baeri* (until recently treated as the genus *Alouattamyia*), is the only bot fly restricted to a primate host; it is a common parasite of howler monkeys in Central and South America. Other *Cuterebra* species are generally associated with rodents and lagomorphs. Although they often reach impressive densities, infesting a third or more of all individuals in local populations of mice, chipmunks, squirrels and rabbits, they are of little or no economic importance and are rarely noticed by people.

The only species of Cuterebrinae not currently placed in the genus *Cuterebra* is the Human Bot Fly, *Dermatobia hominis*, a Neotropical species that is unfortunately much more likely to be encountered than other bot flies. The first evidence that you have encountered a *Dermatobia hominis* — also known as a Torsalo — is likely to be a boil-like swelling somewhere on your body that opens to a small craterlike opening, through which you might be able to see the glistening spiracles on the tapered hind end of a subcutaneous maggot. Left alone, the fully mature Torsalo maggot would pop out about two months later (usually five to ten weeks), dropping to the ground to form a puparium. The fat, shining blue adult would likely emerge about three weeks later. Torsalo adults, like other Oestridae, are rarely seen, and they are never seen laying eggs on people. *Dermatobia* females have the remarkable habit of attaching their eggs not to their hosts but to the bodies of mosquitoes, muscids and other flies likely to visit a vertebrate host. When an egg-encumbered mosquito or Stable Fly lands on your skin, or the skin of almost any warm-blooded animal, body heat causes the bot fly egg to instantly hatch into a little larva that promptly burrows under the skin, where it feeds on the spot until maturity. Despite its reputation as the Human Bot Fly, the Torsalo is a frequent parasite of livestock, pets and a few wild mammals (and, perhaps very occasionally, big birds).

Hypodermatinae The subfamily Hypodermatinae, the "Old World skin bot flies," is a small group of about 30 species mostly restricted to Europe, Asia and the Oriental region, with only a couple of secondarily widespread pest species — both *Hypoderma* species — now found in much of the world. Unlike their New World counterparts in the Cuterebrinae, all Old World skin bots lay eggs directly on the host. Although oestrid flies have neither functional mouthparts nor stingers, the arrival of ovipositing adult *Hypoderma* seems to strike fear into the hearts of livestock.

When adult warble flies or cattle grubs, especially *Hypoderma bovis*, arrive to painlessly attach their eggs to cattle hairs, the cows gallop wildly and often injure themselves in an attempt to escape. This behavior is well enough known to have its own special term: "gadding," and occurs in hosts of some other oestrids as well. It is almost as if the cattle know those eggs will hatch into spiny larvae that will burrow under their skin to torment them with subcutaneous migrations before coming to rest in boil-like swellings (warbles), each containing a massive maggot that breathes through a hole in the cow's hide.

The two widespread cattle grub species that are now pests throughout the Holarctic region have slightly different biologies. *Hypoderma bovis*, the Northern Cattle Grub, and *H. lineatum*, the Common Cattle Grub, both glue their eggs to the body hairs of cattle, but only *H. bovis* attacks active hosts. *Hypoderma lineatum* normally oviposits on standing or resting livestock. Larvae of both species crawl to the base of the hair and penetrate the host's hide by using mouthhooks and secretions of proteolytic enzymes. Both spend months burrowing through the host, but *H. bovis* maggots cluster around the spinal column during the winter and *H. lineatum* larvae cluster around the esophagus. Come spring, the larvae move up to the animal's back to form boil-like warbles, each with an open breathing hole. As with the Human Bot Fly, the maggot ultimately pops out of its warble to pupariate in the ground, damaging the cow's hide in the process. The exit holes take a while to heal, sometimes providing an avenue for further attacks by screwworm flies (Calliphoridae). Although the pest *Hypoderma* species are the best-known hypodermatines, other members of the family are common parasites of reindeer, rodents and rabbits in Europe and Asia. A handful of hypodermatine species occur in Africa, but the only hypodermatines found in Australia and South America are introduced populations of *Hypoderma bovis*.

Oestrinae The subfamily Oestrinae, the "nose bot flies," includes about 35 species that parasitize a diversity of hosts ranging from kangaroos to elephants and giraffes, as well as a few domestic animals. Unlike oestrids in other subfamilies, oestrines are ovolarviparous, producing eggs that hatch into squirming larvae as they are ejected from the abdomen of a female fly hovering near her host's face. A range of wild animals, along with horses, sheep and the occasional unlucky shepherd, are subject to attack by Oestrinae that share the habit of squirting a stream of live larvae up the noses or into the eyes of their hosts in swift aerial attacks. The larvae work their way into the throat or sinuses, where they develop until they are sneezed out at maturity.

Eastern North America's most common nasal bot fly maggots, the native *Cephenemyia phobifera*, or "snot bot," develop (probably painfully!) as parasites in the throat and sinus cavities of deer, ultimately leaving the nose with a snort or sneeze when they are ready to pupate. Other *Cephenemyia* include the Holarctic *C. trompe*, which is called the Reindeer Throat Bot although it also attacks deer, moose and caribou, and several Nearctic species. All native New World nasal bot flies belong to the Holarctic genus *Cephenemyia* and attack cervids such as caribou, deer and moose, but an introduced species called the Sheep Nose Bot, *Oestrus ovis*, attacks sheep.

The subfamily Oestrinae is more diverse in Eurasia and Africa, where most species attack large herbivorous wild mammals such as elephants, gazelles, antelope and warthogs, but some attack domestic animals, including camels and horses. Horses in Eurasia, Africa and the Far East are attacked by the Horse Nose Bot, *Rhinoestrus purpureus*, a species that can reach high and damaging population densities. This species occasionally larviposits in the eyes of people who handle horses, causing ocular myiasis. Given the practice of moving horses around the world and the propensity of most insects associated with domestic animals to become widespread, it is remarkable that Horse Nose Bots have not reached the New World or Australia. The only native nose bot fly in Australia (in fact, the only native Australian bot fly) is a remarkable endemic species that develops in the trachea of the Red Kangaroo (*Macropus rufus*). A couple of accidentally introduced

Oestrinae are parasites with the horrifying habit of squirting a stream of live larvae up the noses or into the eyes of their hosts in swift aerial attacks. The larvae work their way into the throat or sinuses, where they develop until they are sneezed out at maturity. This is the eastern North American nasal bot fly *Cephenemyia phobifera*, which develops in the throat and sinus cavities of deer.

oestrine species, the Sheep Nose Bot and the Camel Nose Bot, *Cephalopina titillator*, also occur in Australia.

Gasterophilinae The subfamily Gasterophilinae comprises about a dozen stomach bot species that develop in horses, elephants and rhinos, plus a couple of anomalous elephant bot flies that develop under the thick skin of their hosts, forming warbles much like those caused by the Cuterebrinae. Most authors treat the latter two species, *Ruttenia loxodontis* and *Neocuterebra squamosa*, as basal lineages of the Gasterophilinae, indicating that the stomach-invading habit originated later in the evolution of the subfamily.

The best-known members of Gasterophilinae are the horse stomach bot flies, which are beelike flies that develop as large, spiny larvae in horse stomachs. Some horse stomach bot species, including the cosmopolitan species *Gasterophilus haemorrhoidalis* and *G. nasalis*, lay their eggs in or near the horse's mouth, but the most common species in North America, Australia and Europe, *G. intestinalis*, sticks its eggs to the horse's foreleg hairs or other places where the horse will likely lick them off. The eggs hatch in response to the warmth of the tongue; larvae burrow into the tongue and ultimately end up in the stomach, where they attach themselves to the stomach lining to feed. When ready to pupariate, the mature larvae pass out with the feces.

All three common cosmopolitan pest species of *Gasterophilus* are European or Asian in origin, but a fourth pest species of *Gasterophilus* has failed to spread around the world with its domestic hosts and remains a regional pest bot in parts of Eurasia and Africa. *Gasterophilus pecorum*, the Dark-winged Horse Bot, causes more damage to its host than its cosmopolitan congeners, sometimes killing the host horse by damaging and blocking its esophagus. This species is also unique among the Gasterophilinae for its habit of producing an unusually large number (thousands) of eggs and sticking them on grasses rather than on the host's body. The eggs hatch when ingested by a horse. Further species of *Gasterophilus* are associated with wild equids such as zebras, which characteristically sustain a very high rate of infestation.

The hosts of Gasterophilinae include some of Africa's most threatened large mammals, such as the black rhinoceros and the white rhinoceros; the Rhinoceros Stomach Bot Fly, *Gyrostigma rhinocerontis*, is thus one of Africa's rarest insects. It also happens to be a contender for the bulkiest fly on the continent, with an adult body length that reaches a whopping 40 millimeters. The genus *Gyrostigma* includes two other rare rhino-parasitizing species, one of which is known only from a single specimen reared from a captive Sumatran rhinoceros in the late 1800s. The African Rhinoceros Stomach Bot Fly is not extinct yet, but stomach bots have disappeared along with their hosts in the past. The Mammoth Bot Fly, *Cobboldia russanovi*, for example, was discovered in 1973 as a subfossil larva in the stomach of a wooly mammoth preserved for the past 100,000 years or so in the Siberian permafrost. Two of the living species of *Cobboldia* are stomach parasites of African elephants; a third parasitizes the Indian Elephant.

Scathophaga stercoraria (Scathophagidae) is a widespread predator, especially common around pasture dung in Europe and North America. This one is feeding on a flesh fly through its oral cavity.

Muscoidea

Much More Than House Flies

The superfamily Muscoidea is easily distinguished from other calyptrates because muscoid families lack the distinctive characters that readily define the other two calyptrate superfamilies, Oestroidea and Hippoboscoidea. Oestroidea are most easily diagnosed and defined on the basis of a prominent row of meral bristles, while Hippoboscoidea are unmistakable for their suite of characters associated with ectoparasitism and "pupiparous" development. The four families that comprise the superfamily Muscoidea are apparently united by little more than a bare meron — the absence of the meral bristles that define the Oestroidea — but despite this weak foundation the muscoid families are generally similar to one another and the superfamily is almost universally recognized at least as a convenient grouping, if not a natural group. Some molecular analyses (e.g., Kutty et al. 2008, 2010; Wiegmann et al. 2011) suggest that Muscoidea is not a monophyletic group, a possibility previously raised on the basis of morphology (Griffiths 1972).

Many muscoids, including the most common species, are dull-colored and somewhat nondescript. The most common calyptrate species almost everywhere in the world are among those morphologically modest muscoids that generally develop as scavengers/predators in decomposing material (most Muscidae) or as plant feeders/saprophages in plants (many Anthomyiidae). These generalizations, however, give short shrift to the enormous variety of forms and lifestyles in this large and abundant superfamily.

Scathophagidae

Scathophagid Flies

The Scathophagidae form one of the most familiar yet one of the most unfamiliar of all calyptrate families — familiar because of the synanthropic and easily recognizable Pilose Yellow or Golden Dung Fly, *Scathophaga stercoraria*, and unfamiliar because the other 400 or so species are typically inconspicuous and poorly known. Most are elongate flies with the lower calypter much smaller than that of other calyptrates, which explains the frequent misidentification of some scathophagids, especially the smaller, relatively bare species, as acalyptrates. Scathophagids are essentially northern flies, absent from the Pacific region and represented in South America, Africa and Southeast Asia only by a few high-altitude species and a few populations of the secondarily widespread Pilose Yellow Dung Fly in South Africa, the Caribbean and Brazil. Not only is the family Scathophagidae largely restricted to the Holarctic region, it also includes a higher percentage of arctic, subarctic and high-alpine species than any other fly family.

Despite the family stereotype created by *S. stercoraria* and similar synanthropic dung-breeding *Scathophaga* species, Scathophagidae vary widely in appearance and habits. Adults are generally predators of other insects but larvae live in a variety of habitats and can be either saprophagous or predaceous on other invertebrates. Some abound in decaying seaweed and others are stem borers, leaf miners or aquatic predators of insect eggs. The family is usually divided into two subfamilies, the Scathophaginae and the small subfamily Delininae.

The relatively few species in **Delininae** with known habits deposit their eggs on the leaf surfaces of plants in the orchid, lily and spiderwort families; hatching larvae burrow into the leaves and develop as miners.

Scathophaginae Leaf-mining habits also show up in a few species of the subfamily Scathophaginae, but scathophagine females are adapted to insert their eggs into a variety of substrates ranging from dung to living plants. Members of the genus *Scathophaga* develop in decomposing material, usually dung, although some live in decaying seaweed and form part of the specialized wrack community (see Chapter 2). *Scathophaga stercoraria* is a widespread rural species that is one of the most common predators in dung (the larvae) and around it (the adults), especially cattle dung. Best known as the Pilose Yellow Dung Fly, *S. stercoraria* occupies a wide range and undergoes multiple generations per year; this easily found, easily reared species has been used to investigate several questions in animal ecology and evolution.

Predaceous larvae are also found in the aquatic genera *Spaziphora* and *Acanthocnema*, but otherwise most scathophagids are phytophagous as larvae. Members of the large Holarctic genus *Cordilura* usually bore in sedges (Cyperaceae) or similar plants, and the two species of *Hydromyza*, *H. confluens* in North America and *H. livens* in Europe, are common species that bore in the stems of water lilies (*Nymphaea* and *Nuphar*). One unusual North American species, *Orthachaeta hirtipes*, invades sedges already bored by *Cordilura*; the European scathophagid *Cleigastra apicalis* has a superficially similar habit of invading rush (*Phragmites*) stems already bored by other insects, but it feeds mainly on frass left by the primary borers. Other scathophagids with noteworthy known biologies include several species of *Nanna* that damage the flowering heads or ears of grasses, and some *Gimnomera* species that develop in the seeds and flower parts of louseworts (Scrophulariaceae).

The Scathophagidae have long been considered close relatives of the Anthomyiidae. Older works treat both groups as a single family (Anthomyiidae), which was probably a better reflection of relationships than recognition of the two families Anthomyiidae (probably paraphyletic) and Scathophagidae (monophyletic). Although some studies (e.g., Michelsen 1991) suggest that the Anthomyiidae are closer to other Muscoidea (Muscidae and Fanniidae) than to the Scathophagidae, Kutty et al. (2008, 2010) show that the Anthomyiidae and Scathophagidae are closely related to one another and together form a group that is more closely related to the Oestroidea than to other muscoids.

Anthomyiidae

Root Maggot Flies and Their Relatives

The Anthomyiidae are best known for the many species of *Delia* that damage economically valuable plants by developing as maggots inside roots, shoots, seeds or leaves, but members of this important family of about 1,900 species show a wide range of habits, including parasitism, kleptoparasitism, saprophagy and predation.

The family Anthomyiidae, like the closely related Scathophagidae, is an essentially Holarctic group, with only about a dozen species in the Pacific region and only a handful in the tropics.

Fucellia (Anthomyiidae) is one of the specialized genera associated with masses of decomposing algae tossed up on ocean shores. *Fucellia tergina*, shown here, occurs on seaweed from Australia to Sweden.

Less than 100 anthomyiid species are known from Africa and slightly over 100 occur in the Neotropical region, where they are mostly confined to the cooler high altitudes and low latitudes. The family was until recently generally divided into two subfamilies, Anthomyiinae and Fucelliinae. The great majority of species were traditionally placed in the large subfamily Anthomyiinae, almost all species of which have distinctive downward-pointing hairs on the underside of the scutellum. The species traditionally placed in the other, much smaller subfamily, Fucelliinae, lack the ventral scutellar hairs (as do the aberrant genus *Coenosopsia* and a few small groups of Anthomyiinae, such as the grasshopper parasitoids in the genus *Acridomyia*). Most species traditionally treated as Fucelliinae are in the large genus *Fucellia*, one of the specialized genera associated with wrack, the masses of decomposing algae tossed up on ocean shores. The southern hemisphere wrack lacks endemic *Fucellia* species, but one secondarily widespread species, *F. tergina*, occurs on seaweed from Australia to Sweden.

The previously widely accepted division of the anthomyiids into the subfamilies Anthomyiinae and Fucelliinae has recently fallen out of favor among muscoid taxonomists, but it has yet to be replaced by a generally agreed-upon alternative subfamily classification. Michelsen (2000) divided most anthomyiids into the subfamilies Myopininae, Pegomyinae and Anthomyiinae but left a couple of mostly lowland Neotropical/southern Nearctic genera, *Coenosopsia* and *Phaonantho*, outside these subfamilies as the probable sister group of the remaining Anthomyiidae. *Coenosopsia* is now included in the Anthomyiidae largely because the females have well-developed seventh abdominal spiracles (lost in all Muscidae) and because the frons has a pair of interfrontal setulae as in many other anthomyiids (and a few muscids!); this genus was until recently placed in the Muscidae. Opinions still differ as to where (and if) *Coenosopsia* fits into the family Anthomyiidae. The taxonomy of Anthomyiidae in general is difficult, with species-level identification usually requiring examination of the male genitalia.

The Anthomyiidae are very closely related to the Scathophagidae. Recent phylogenetic analyses (Kutty et al. 2010) suggest that the family Anthomyiidae is not a natural group because some members of the family are more closely related to the Scathophagidae than to other Anthomyiidae.

Anthomyiinae The typical subfamily Anthomyiinae is now home to most of the species and genera in the family, including the wrack flies (*Fucellia* spp.) and the typical root maggots, such as the several pest species in the genus *Delia*. *Delia antiqua* (Onion Maggot), *D. radicum* (Cabbage Maggot) and *D. platura* (Seed Corn Maggot) are some of the economically important *Delia* species that damage host plant roots and shoots by tunneling in the tissue and inducing decay that facilitates larval feeding. Some other *Delia* species develop as miners in leaves and stems, a couple of

Phorbia species are wheat shoot pests, and larvae of species in the genus *Strobilomyia* are cone maggots that damage various conifers. Relatively few other Anthomyiinae are pests; many more are beneficial pollinators or parasitoids. *Acyglossa* and *Acridomyia* species, for example, develop as internal parasitoids of grasshoppers, including pests such as the Migratory Locust, *Locusta migratoria*. Female *Acridomyia* use rasping mouthparts to puncture the host's cuticle before inserting the ovipositor to lay eggs in the grasshopper's body. Many other larval Anthomyiinae, including the type genus *Anthomyia* and numerous species in the important genus *Delia*, are saprophagous in dung, fungi, carrion or rotting plant material.

Several anthomyiid genera have highly specific relationships with host plants. *Chiastocheta* species, for example, have an obligate mutualism with *Trollius* flowers, in which the larvae feed on the developing seeds but the adult flies are essential pollinators. *Egle* species develop in poplar and willow catkins, *Heterostylodes* species develop in the flower heads of host composites, and *Chirosia* species develop inside fern fronds. A couple of common anthomyiine genera, *Eustalomyia* and *Leucophora*, are kleptoparasites that develop on the paralyzed prey of solitary wasps or the pollen stored by solitary bees. *Botanophila* is one of the more diverse genera of Anthomyiidae, with species that develop in dung, damaged plants, leaves, flowers, seed capsules and fungi. Several *Botanophila* species develop only in *Epichloe* fungus, which is itself an encrusting parasite of grasses.

Myopininae and Pegomyinae The subfamilies Myopininae and Pegomyinae — often combined into the single subfamily Pegomyinae — are smaller groups with a correspondingly narrower range of biologies, although the large genus *Pegomya* includes species that live in leaves, flowers, seeds, cones, fungi and birds' nests. Some, such as the Holarctic *P. hyoscyami* (the Beet Leafminer or Spinach Leafminer), are pests, and some develop in unusual hosts, such as horsetail (*Equisitum*). Other Pegomyinae occur in dung, mammal nests and roots.

Fanniidae

Little House Flies and Their Relatives

The most familiar fanniids are the relatively small (4–8 mm), House Fly–like *Fannia* species often seen endlessly circling around light fixtures in unsanitary buildings or around the ceilings of outhouses and latrines. They develop from strongly flattened larvae with feathery lateral body processes, much like those of some Phoridae and Platypezidae, that are probably adaptations to larval life in liquefying organic matter such as wet feces, rotting fungi or well-rotted carrion, although some species are also found in relatively dry habitats such as birds' nests. Several *Fannia* species, such as *F. scalaris* (Latrine Fly), *F. canicularis* (Lesser House Fly) and *F. pusio* (Chicken Dung Fly) are synanthropic, or human-associated, insects now found all over the world. A few, including the Latrine Fly and Lesser House Fly, have been reported as causing myiasis, with their larvae occasionally developing in the human urethra and digestive tract. Species in the small genus *Piezura* develop in fungi; some *Euryomma* species have been found in army ant refuse piles.

About 285 of the 300 or so species of fanniids are in the genus *Fannia*, with several secondarily widespread species as well as numerous native New World and Palaearctic species. The remaining species are in the genera *Australofannia* (with only a single, Australian species), *Euryomma* (New World, mostly Neotropical) and *Piezura* (Holarctic, mostly Palaearctic). The Fanniidae were until recently treated as a subfamily (Fanniinae) of the Muscidae; one of the most important treatments of the group remains Chillcott's (1961) "Revision of the Nearctic Species of Fanniinae."

The Fanniidae are thought to be one of the "basal" calyptrate lineages, along with the Hippoboscoidea. Kutty et al. (2010) suggest that the Fanniidae form the sister group to all calyptrates other than Hippoboscoidea.

Muscidae

House Flies, and Thousands More

When most people refer to flies, they are probably thinking of one species in the genus *Musca*, the type genus of the family Muscidae. The common House Fly, *Musca domestica*, is *the* familiar pest fly, ubiquitous on feces and other bacteria-rich waste the world over. Many other *Musca* species have similarly filthy habits, and a few, such as Australia's notorious Bush Fly, *Musca vetustissima*; the tropical Old World Bazaar Fly, *Musca sorbens*; and the now widespread Face Fly, *Musca autumnalis*, are abundant regional pests, dangerous for their unsanitary habits and association with people. However, the House Fly is the most important of the 60 or so *Musca* species. This originally Old World species, now found on every continent except Antarctica, has followed mankind around the planet, breeding in bacteria-rich human and domestic animal waste, garbage dumps and other effluvia. Adult House Flies routinely move from filth to food, bringing with them a bestiary of pathogens and associated enteric diseases. Diarrhea, often caused by fecal-borne pathogens spread by adult House Flies and related *Musca* species, is a serious worldwide problem that causes more than a million deaths annually (most, but by no means all, in developing countries). *Musca*, like *Anopheles*, should be recognized as one of the most deadly genera of animals. Some other synanthropic muscids, such as the widespread tropical and subtropical *Synthesiomyia nudiseta* and the cosmopolitan *Muscina stabulans* (False Stable Fly), similarly breed in refuse and can also be significant pests.

Most of the world's 5,200 or so species of Muscidae are innocuous saprophages or predators in a wide variety of natural habitats. Muscid biology and larval morphology were thoroughly reviewed by Skidmore (1985). As Skidmore points out, many are facultative predators as larvae, shifting to a high-protein diet late in their development. Very few muscid lineages are phytophagous or parasitic on other invertebrates, but either the larvae or the adults of several genera feed on vertebrate blood. The most familiar biting adults in this family look much like House Flies: if you think a House Fly has bitten you, the culprit was probably a Stable Fly, *Stomoxys calcitrans*, a cosmopolitan biting muscid with a forward-directed "lance" held under its chin. That characteristic lance is actually a stiffened lower lip (labium) that ends in thorny lobes (labella), used by both sexes to rasp painfully through your skin, even through thin shirts and socks. Female Stable Flies need a blood meal in order to develop eggs, which are laid in moist decaying material such as urine-soaked stable straw or rotting vegetation along lakeshores.

Some other muscids have similar adult biting habits, some feeding preferentially on livestock and others restricted to wildlife hosts. A few muscids do their blood feeding in the larval stage. Larvae in the large (50 species) New World genus *Philornis* and the smaller (five species) Old World genus *Passeromyia* are mostly blood feeders that attack nestling birds. A few *Philornis* and *Passeromyia* species are scavengers in nests, suggesting that larval blood feeding appeared independently in these two genera. It does not occur elsewhere in the Muscidae, but similar habits are found in the bird blow flies of the family Calliphoridae and in one species of Piophilidae.

Adult muscids are among the most common predators of other flies, and glades and gardens the world over are among the haunts of the large predaceous genus *Coenosia*. *Coenosia* and other members of the subfamily Coenosiinae are commonly seen perched conspicuously on foliage while consuming recently captured flies, leafhoppers or other insects.

There has been, to say the least, considerable disagreement about the composition and limits of the subfamilies of Muscidae. The family has thus suffered from a confusingly unstable higher classification, and most of the subfamilies are hard to characterize. I here follow the most widely used subfamily classification, dividing the muscids among the subfamilies Achanthipterinae, Muscinae, Azeliinae, Cyrtoneurininae, Coenosiinae, Mydaeinae, Phaoniinae and Atherigoninae, even though there is considerable disagreement about which genera belong in which subfamilies, and even though recent studies of muscid relationships (Kutty et al. 2010; Schuehli et al. 2007) suggest that many of the traditional muscid subfamilies are artificial (para- or polyphyletic) groups. In the absence of general agreement about alternative

Although best known for a few pests such as the House Fly, Bush Fly, Bazaar Fly and Face Fly, the genus *Musca* is a diverse group of some 60 species.

arrangements, the subfamily classification of the Muscidae is best considered as "tentative."

Although the subfamily classification of the Muscidae is controversial, the family itself is a natural group that has long been treated as closely related to the Fanniidae, Anthomyiidae and Scathophagidae. Recent molecular analyses suggest that the Fanniidae, Muscidae and the Anthomyiidae/Scathophagidae lineage are "basal" calyptrate clades, and that the sister group to the Muscidae is a lineage comprising the Scathophagidae, Anthomyiidae and Oestroidea (Kutty et al. 2010).

Muscinae The subfamily Muscinae includes the "typical" muscids in the House Fly genus *Musca* and related genera such as *Morellia*. As larvae, most members of this group are microbial grazers in dung, and they are among the more abundant insects associated with livestock waste as well as the dung of native mammals. Some of the more familiar dung-breeding muscines include the robust *Mesembrina* species, with their distinctive yellow wing bases, and the metallic green *Eudasyphora* and *Dasyphora* species.

Current classifications put the Stable Flies and their relatives in the subfamily Muscinae as the tribe ***Stomoxyini***, although this lineage is sometimes treated as a separate subfamily, Stomoxyinae. The nine genera in this group are all bloodsuckers as adults and most are dung feeders as larvae, but *Stomoxys ochrosoma* is an exception. Females of this species hover over columns of African driver ants before releasing a clutch of eggs in front of a foraging ant destined to carry them back to the colony's bivouac, where they will hatch into scavenging larvae. Some of the better-known Stomoxyini that develop in dung include the Horn Fly, *Haematobia irritans*, and the Buffalo Fly, *H. exigua*, both of which are major pests of cattle. The Moose Fly, *Haemotobosca alcis*, has similar habits but is less familiar, since it feeds only on moose (*Alces alces*).

Azeliinae The subfamily Azeliinae is largely made up of the cosmopolitan genus *Hydrotaea*, with around 130 species including the familiar little blue-black flies that abound on almost any kind of organic material exposed in temperate forests. *Hydrotaea* adults seem to appear instantly on any kind of sweaty scratch or wound, which is worrisome, given that this genus includes several synanthropic species that breed and feed in dangerously pathogen-rich decomposing materials, including animal waste and carrion. *Muscina*, another cosmopolitan genus widely treated as part of the Azeliinae, is often found with *Hydrotaea* in dung and carrion. Both genera are of public health concern and are of interest to forensic entomologists because of their roles in carrion insect communities.

The 27 species of *Muscina* include several secondarily widespread species such as *M. stabulans*, the False Stable Fly — which, unlike a true Stable Fly, does not bite. As is true for many "saprophagous" muscids, older larvae of *Hydrotaea* and *Muscina* shift from grazing microorganisms to preying on other insect larvae. Similar predaceous habits are found in late instars of the carrion-breeding *Synthesiomyia nudiseta*, one of the most common Azeliinae in tropical countries and an unusually large muscid at almost 10 millimeters, almost double the size of most *Hydrotaea* species. The subfamily Azeliinae, as usually defined, also includes the only muscid genera with blood-feeding larvae, the Old World and Pacific genus *Passeromyia* and the New World genus *Philornis*. However, some specialists consider these genera to belong in entirely different subfamilies, placing *Passeromyia* in the Reinwardtiinae (along with *Muscina*, *Synthesiomyia* and some other genera) and *Philornis* in the Dichaetomyiinae along with *Dichaetomyia* and related species.

Mydaeinae The subfamily Mydaeinae includes the enormous and widely distributed genus *Mydaea*, larvae of which live in dung, tree holes and fungi. Other Mydaeinae have similar habits, mostly developing in dung or similar decomposing material and probably preying on other larvae, at least in later instars. Perhaps the most frequently noticed members of the group are the boldly striped species of *Graphomya*, which are among the more common and easily recognized flower-visiting muscids. *Graphomya* larvae are found in wet to very wet rotting material — they are described as "subaquatic to aquatic" — and are variously characterized as saprophagous or predaceous. The genus *Graphomya* seems to be difficult for muscid specialists to classify, as it has regularly been moved around from subfamily to subfamily.

Cyrtoneurininae Although most of the subfamilies and larger genera of Muscidae are widespread, the subfamily Cyrtoneurininae as currently defined is a strictly Neotropical group. None of the genera of this small group are common or familiar flies, and few are known as larvae. At least one genus with known breeding habits, *Charadrella*, develops in dead invertebrates, but others, such as *Cyrtoneurina* and *Neomuscina*, are dung breeders.

Phaoniinae The subfamily Phaoniinae is a large and heterogeneous group that includes some of the most common "wild" (non-synanthropic) muscids, such as the widespread and superficially similar genera *Helina* and *Phaonia*. *Helina* and *Phaonia* both hatch from the egg as a relatively large larva, ready to prey on other larvae in decaying matter such as dung, nest material, fungi and rot holes. In most classifications this subfamily also includes *Dichaetomyia*, a large but relatively poorly known group and one of the most common muscid genera of the Old World tropics, where its many species develop as predators in a range of decomposing material. As noted above, *Dichaetomyia* and related species are sometimes treated as the subfamily Dichaetomyiinae.

The genus *Eginia* and some close relatives comprise a "problem" lineage, tentatively included in the Phaoniinae but sometimes treated as a separate subfamily, a separate family or even part of the Calliphoridae or Tachinidae. Members of this Old World group are thought to be parasitoids of millipedes, based on a record of a Japanese species, *Syngamoptera flavipes*, from a millipede host and some 80-year-old rearing records of the rare European species *Eginia ocypterata*. Recent records of *E. ocypterata* parasitizing millipedes appear to be based on misidentification of a different species of

Lispe are common shoreline Muscidae often seen poised over the water's edge waiting to pounce on vulnerable newly emerged aquatic insects, like this mayfly. Both adults and larvae of this genus are predators.

parasitoid commonly found in Portuguese millipedes. Once reared, these millipede-parasitizing maggots turned out to be acalyptrates (*Pelidnoptera nigripennis*, Phaeomyiidae) (Baily 1989). A recent review of *Eginia ocypterata* (Michelsen 2006) argued that not all records of this species as a millipede parasitoid are in error, and provided a description of *Eginia* eggs as "helmet-shaped" and adapted for adherence to the host's smooth exoskeleton. The flat underside of the egg has a couple of circular, soft "windows," possibly to help the egg stick to the millipede or to facilitate larval emergence into the host.

Coenosiinae The subfamily Coenosiinae is a huge, cosmopolitan group found in a variety of habitats. Several genera of Coenosiinae are aquatic or semiaquatic predators. *Limnophora*, *Lispoides* and *Lispe*, for example, are common shoreline predators as adults and aquatic or semiaquatic predators as larvae. *Lispe* adults, which are characterized by broad, conspicuous palpi, are often seen poised over the water's edge waiting to pounce on vulnerable newly emerged aquatic insect adults; *Lispe* larvae develop in wet soil along the water's edge. *Lispoides* and *Limnophora* adults are commonly seen engaging in elaborate courtship displays and male–male territorial disputes along the margins of streams, where the larvae of some of the better-known species develop as predators of black flies and other aquatic insect larvae. These common flies deposit unusually large eggs that hatch to almost mature third-instar larvae, ready to attack prey the size of black fly larvae.

Spilogona larvae range from aquatic to terrestrial, but most are inhabitants of semiaquatic substrates such as sphagnum moss, which may in part explain why this is one of the most speciose and abundant groups of higher flies in arctic, subarctic and alpine environments. Most of the hundreds of species of *Spilogona* occur in the northern Holarctic region, but the genus is also diverse in temperate southern South America, Australia and New Zealand. *Lispocephala* is a cosmopolitan genus with most of its 150 or so species endemic to the Hawaiian Islands, where some are highly specialized predators of particular species of endemic *Drosophila*. The most frequently encountered terrestrial members of Coenosiinae belong to the typical genus *Coenosia*, an enormous, cosmopolitan group of small predaceous muscids common on low foliage. The best-known species of *Coenosia* is the common Holarctic *C. tigrina*, a familiar black-spotted gray predator abundant in rural and urban areas throughout North America and Europe.

Atherigoninae Atherigoninae is a small subfamily dominated by the type genus *Atherigona*, an important group that includes phytophagous species with stem-boring larvae, some of which are economically important pests of rice and other cereals in the Old World tropics. The Sorghum Shoot Fly, *Atherigona soccata*, for example, is one of the most serious pests of sorghum in Africa, Asia and the Mediterranean. The genus *Atherigona* also includes species with more typical muscid biologies — saprophagous or predaceous larvae in decaying material — and possibly even species that develop as parasitoids of other insect larvae in stems. The Atherigoninae are sometimes treated as part of the Phaoniinae.

Achanthipterinae The subfamily Achanthipterinae contains only a single Old World species, *Achanthiptera rohrelliformis*, that develops in the debris in yellowjacket and hornet (Vespidae) nests.

Hippoboscoidea

Louse Flies, Bat Flies and Tsetse Flies

The superfamily Hippoboscoidea comprises two important families of bloodsucking flies that completely bypass the free-living larval stage, giving birth instead to single, completely developed larvae. Female Hippoboscidae (louse and bat flies) and Glossinidae (tsetse flies) have internal "milk glands" that sustain their progeny almost to maturity before they deposit (larviposit) each one as a large, ready-to-pupate third-instar larva. Tsetse flies, bat flies and louse flies are recognized as a distinct lineage, sometimes called the "Pupipara," because this remarkable life cycle — called pseudo-placental viviparity — is markedly different from the more common ovoviviparity, in which the developing larva is not provided with food to supplement that available from the egg yolk. The only major lineage outside the Hippoboscoidea that is characterized by pseudo-placental viviparity is the poorly known family Mesembrinellidae, in which developing larvae are nourished by secretions of the spermathecae (normally sperm-storage organs) rather than uterine accessory glands as in the Hippoboscoidea.

The flies treated here as the louse fly family Hippoboscidae are sometimes divided into three families, but the bat fly groups often treated as the separate families Nycteribiidae and Streblidae are now widely considered to be part of the Hippoboscidae. The superfamily Hippoboscoidea probably forms the sister group to the rest of the Calyptratae.

Glossinidae

Tsetse Flies

The family Glossinidae includes only the single genus *Glossina*, with some 23 extant species almost entirely restricted to sub-Saharan Africa (there are also some populations in small areas of southern Saudi Arabia). Thirty million years ago the group was much more widespread, as indicated by enormous extinct species known from North American and German Oligocene fossils. Extant *Glossina* species, known as "tsetse flies," are the infamous biting flies responsible for the African scourges of sleeping sickness and nagana,

Tsetse fly adults (*Glossina*, family Glossinidae) of both sexes are bloodsucking flies armed with conspicuous bayonet-like mouthparts. This one is on the back of a dipterist's T-shirt in Tanzania.

a livestock disease related to sleeping sickness. Both of these enormously important diseases are caused by trypanosomes (*Trypanosoma* species) carried by tsetse flies.

Tsetse fly adults of both sexes are bloodsuckers armed with conspicuous bayonet-like mouthparts superficially similar to the probosces of the more familiar Stable Flies (Muscidae). But, like the closely related Hippoboscidae, tsetse flies are somewhat flattened and hold their wings tightly overlapping the abdomen like scissor blades. As is true for all bloodsucking calyptrate flies, both sexes feed on blood. Tsetse fly females use a long, bulbous-based proboscis to puncture a variety of mammal and reptile hosts, imbibing blood by contracting robust muscles inserted on the pharynx so that it acts like a pump. The female's blood meal, encased in a membrane (peritrophic membrane) secreted at the junction of the foregut and midgut, has to be digested with the aid of symbiotic bacteria before she is able to nurture and deposit, one at a time, a very few large larvae. A single mature larva is produced every ten days or so over an adult's life, which ranges from about three weeks to three or four months. Larvae burrow into the soil upon deposition and pupariate almost immediately, emerging as adults about a month later. Unlike the generally host-bound adults of other Hippoboscoidea, tsetse fly adults visit their vertebrate hosts only for brief bouts of blood feeding.

The genus *Glossina* is routinely divided into three species groups (sometimes treated as subgenera) with distinct habitat preferences, host preferences and distributions. The *G. palpalis* group is found mostly along rivers, lakes and streams, where the normal hosts are reptiles, although mammals (including humans) are also attacked. Members of this group are the most important vectors of sleeping sickness in West Africa, where the flies often pick up the trypanosomes from human carriers and then infect other people. Most species of the *G. morsitans* group are found in open savanna, where they feed on giraffes, rhinos and buffalo and other mammals of the African plains. Species in this group are the major vectors of sleeping sickness in East Africa, where visitors are at risk of being bitten by a tsetse fly that has previously picked up the disease-causing trypanosome from a large mammal. Tourists on safari are sometimes bitten when they are mobbed by flies attracted to the warmth, movement or profile of vehicles on the open savanna. The *G. fusca* group, in contrast, is largely associated with forest animals and, although members of this group are widespread in the great tropical forests of West Africa, they are relatively unimportant in transmission of human disease.

Hippoboscidae

Louse Flies and Bat Flies

The only Hippoboscoidea found outside Africa are the louse and bat flies now grouped into the family Hippoboscidae, different species of which suck blood from a variety of mammals and birds. About three quarters of the 800 or so hippoboscid species are associated with bats and most of the remainder are found on birds. The most commonly encountered louse flies are flattened flies with a somewhat crab-like appearance as they rapidly scuttle through fur or feathers. Some species are wingless or lose their wings — and sometimes other appendages — after finding a host. Although they occasionally alight on humans, you are most likely to see hippoboscids smoothly sliding along the feathered surface of a recently killed bird or perhaps scuttling about the wool of a sheep. Two of the three subfamilies are strictly bat parasites, and therefore rarely encountered by most people.

Nycteribiinae and Streblinae The bat-parasitizing subfamilies Nycteribiinae and Streblinae — sometimes treated as the families Nycteribiidae and Streblidae — are mostly confined to the warmer parts of the world, with the former group diverse in the Old World tropics and the latter more characteristic of the neotropics. Adults of both groups have small heads with the eyes reduced or absent, and flattened bodies with the wings often reduced (some Streblinae) or entirely lost (all Nycteribiinae). While adults of both subfamilies are strongly modified for life on their bat hosts, the wingless Nycteribiinae carry this to the extreme, with a dorsal shifting of the head and legs that gives members of this well-defined subfamily a characteristically bizarre spiderlike appearance.

Perhaps the most extreme bat flies are the 20 or so species of the Old World genus *Ascodipteron*, an unusual group in the heterogeneous subfamily Streblinae. Males and newly emerged females are normal-looking fully winged bat parasites, but the females undergo extraordinary transformations once they find their hosts. Upon arrival on an appropriate bat, the *Ascodipteron* female burrows almost entirely under her host's skin, loses her wings and legs, and undergoes extensive bloating of the abdomen so that it envelops the head and thorax. Once encysted under the bat's skin, the female fly becomes little more than a blood-sucking bag that contains a developing larva; it is hardly recognizable as an insect, let alone a fly. Only the posterior tip of the abdomen remains free of the host's tissue, allowing the female to push out single large larvae nurtured to maturity in her uterus (as in all Hippoboscoidea). Other bat flies normally leave their hosts to place their single larvae on surfaces such as cave walls near their hosts' roosts. Most are fussy about which hosts they attack and where they attack them, often homing in on particular body parts of particular kinds of bats. More than one bat fly species, and sometimes more than one subfamily, can often be found on a single bat.

Hippoboscinae Most of the 200 or so species of the subfamily Hippoboscinae are fully winged, flattened "louse flies" found among the feathers of birds. The best-known member of this subfamily, however, is the Sheep Ked, *Melophagus ovinus*, a wingless species that spends its entire life cycle in the wool of sheep. Even the puparia remain on the host, glued to the wool, in contrast to other Hippoboscidae, which normally drop to the ground before pupariating. Sheep Keds have spread to almost every corner of the world along with their domestic sheep hosts and were regarded as the most damaging ectoparasitic pests of sheep in North America until very recently. They have quietly disappeared from much of United States and Canada in the past

few years, presumably because of insecticide use and quarantine measures, although they remain common in many other parts of the world. Where they still occur, they are often mistakenly called "sheep ticks" because of their faint similarity to ticks. Another *Melophagus*, found on northwestern North American Dall sheep, shares the wool-restricted lifestyle of its cosmopolitan congener.

Several louse flies in the genus *Hippobosca* are associated with domestic mammals, including dogs, horses, donkeys and cattle in the Old World, but *Hippobosca* is known in North America only from transient infestations of Dog Flies, *H. longipennis*, reported from zoos in the United States. The Dog Fly does not appear to be established in North America, in contrast with the common European Deer Ked, *Lipoptena cervi*, which was apparently introduced from Europe to North America about 100 years ago. European Deer Keds are now common on introduced deer in North America; other species of *Lipoptena* (and *Neolipoptena*) are found on native North American deer. The louse flies most frequently encountered by people are *Lipoptena* species that accidentally alight on humans as they fly from host to host.

The largest subgroup of Hippoboscinae, the tribe ***Ornithomyiini***, is normally associated with wild birds, although a couple of endemic Australian genera parasitize wallabies. Pigeons are parasitized by the widespread Pigeon Fly, *Pseudolynchia canariensis*, but otherwise domestic birds are remarkably free of hippoboscid parasites. One of the most common louse fly species throughout the New World is *Icosta americana*, a species that has been recorded from a wide variety of birds, both raptors and passerines, and which has been implicated in the transmission of West Nile virus.

The Sheep Ked, *Melophagus ovinus*, is a wingless biting fly in the family Hippoboscidae. This species spends its entire life cycle in the wool of sheep.

A Photographic Guide to the Cyclorrhapha

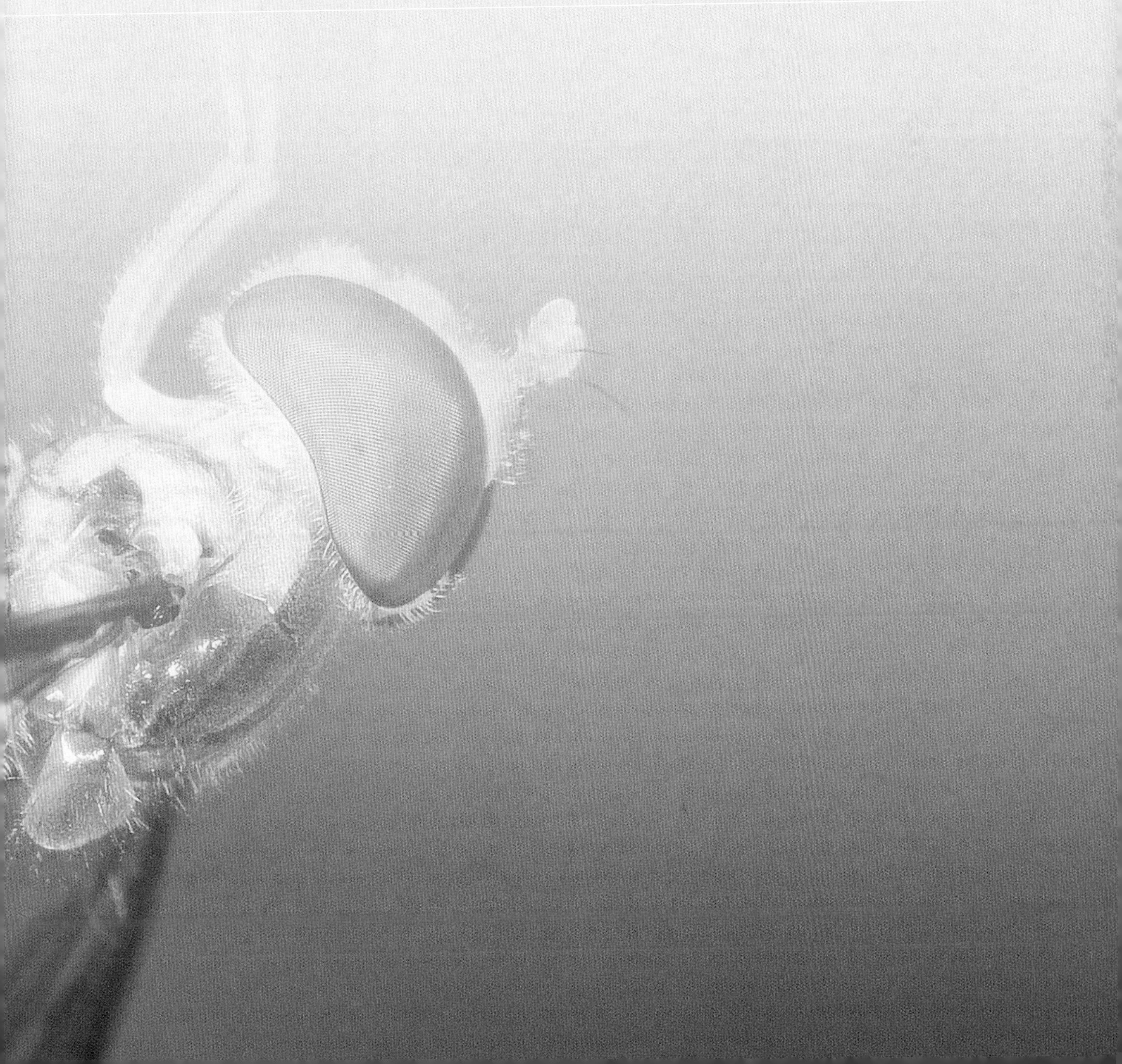

SUPERFAMILY PHOROIDEA

FAMILY PLATYPEZIDAE (FLAT-FOOTED FLIES) Subfamily Callomyiinae ❶ Members of the subfamily Callomyiinae, like this Canadian ***Agathomyia***, are distinguished by the distinctive row of bristles (acrostichals) in the middle of the thorax. *Agathomyia* occurs in every zoogeographic region except the Afrotropical and is the largest genus of Platypezidae, with more than 50 species. Known larvae develop in polypore fungi. ❷ ***Bertamyia notata*** is a remarkably widespread species, found throughout the New World from southern South America to Canada. The only other species in the genus is from South Africa. Larvae of *B. notata* pupate inside cocoons made of frass and bits of host polypore fungus, much like the platypezine *Polyporivora*. ❸ ❹ Males and females of the Holarctic genus ***Callomyia*** often look very different, so much so that the males and females of some species have been erroneously described as different species. This male (big eyes) and female (small eyes) from Greece belong to the ***C. dives*** species group. Known *Callomyia* larvae feed on wood-encrusting fungi. ❺ ❻ These strikingly different flies from southern Europe are a male and female of the same species, an undescribed taxon of the ***Callomyia speciosa*** group. ❼ ❽ ***Grossoseta*** is a Nearctic genus with only two species, the western ***G. pacifica*** (the male shown here is from western Canada) and the eastern ***G. johnsoni*** (the mating pair here is from eastern Canada). Male platypezids have big eyes that meet on top of the head, and the facets of the top part are distinctively enlarged. **Subfamily Microsaniinae** ❾ ***Microsania***, the only genus in the smoke fly subfamily Microsaniinae, is a cosmopolitan genus of flies routinely attracted to smoke, like that of the Appalachian campfire that brought in this mite-covered male. Unlike other Platypezidae, *Microsania* often carry a heavy load of presumably phoretic mites. Larval *Microsania* remain unknown. They probably breed in living fungi like other platypezids, even though some of the associated mites have been identified as dung- and compost-breeding *Macrocheles* species.

FAMILY PLATYPEZIDAE (continued) Subfamily Platypezinae ❶ ***Calotarsa*** is a small Nearctic genus with five of its six species found in western North America or Mexico. This spectacularly ornamented male is a specimen of ***C. pallipes***, an eastern species that develops in gilled fungi, including common honey mushrooms (*Armillaria mellea* and related species). ❷ This ***Lindneromyia*** male was one of dozens on an exposed rock in the middle of an Arizona stream. The cosmopolitan genus *Lindneromyia*, named after the great German dipterist Erwin Lindner, now includes species previously in *Symmetricella* (like this one), *Plesioclythia* and *Grossovena*. Known larvae are mostly associated with the domestic mushroom genus *Agaricus*. ❸ ❹ Most of the tropical and subtropical Platypezidae are now treated as part of the genus ***Lindneromyia***. This male (with large, contiguous eyes, like most male platypezids) and this silver-patterned female are from Tanzania. ❺ ❻ Larvae of *Paraplatypeza* and the closely related *Platypeza* are found in gilled mushrooms; this ***Paraplatypeza velutina*** larva was one of a large number in an oyster mushroom (a soft gilled fungus). ❼ This ***Platypeza*** female has her ovipositor still extruded after laying her eggs in a honey mushroom (*Armillaria mellea*). *Platypeza* is a mostly Holarctic genus of about 15 species, all with similar host associations. ❽ As the name suggests, members of the small Holarctic-Oriental genus ***Polyporivora*** develop in polypore fungi, where larvae pupate in a cocoon embedded in the host fungus. This is the Nearctic ***P. polypori***. **FAMILY OPETIIDAE (OPETIID FLIES)** ❾ A point-mounted specimen of ***Opetia nigra*** from the U.S. National Museum. Opetiidae is a strictly Palaearctic family of only one genus.

FAMILY LONCHOPTERIDAE (SPEAR-WINGED FLIES) ❶ Most spear-winged flies, like this unidentified ***Lonchoptera*** species from Vietnam, occur in the Oriental and eastern Palaearctic regions. Many species await description. ❷ ❸ Spear-winged flies are distinctively slender flies with somewhat spear-shaped wings. ***Lonchoptera bifurcata*** occurs throughout the world; the fly seen here in lateral view is from Chile, and the one seen from the top is from Canada. **FAMILY IRONOMYIIDAE (IRONIC FLIES)** ❹ The Ironomyiidae now occur only in Australia, where two of the three named species of ***Ironomyia*** were just discovered in 2008. The only other species in this family are northern hemisphere fossils. **FAMILY PHORIDAE (SCUTTLE FLIES) Subfamily Metopininae** ❺ ❻ This ***Apocephalus*** is darting in to oviposit on a carpenter ant in northern Canada. The hatching phorid larva will develop inside the ant, ultimately decapitating it prior to pupariating in the detached head capsule. Although members of this huge genus are best known as ant-decapitating flies, some *Apocephalus* species attack other arthropods, including bees and beetles. ❼ ❽ Ant-decapitating flies in the genus *Apocephalus* and ***Eibesfeldtphora*** are major enemies of leafcutter ants, and the tiny guard ants often seen riding on leaf fragments carried by the ants serve to fight off these parasitic phorid flies. This *Eibesfeldtphora* has a perch over a column of ants in Bolivia, from which she can dart down to insert an egg in her doomed host. ❾ The formidably huge Neotropical ants known as bullet ants or three-sting ants (*Paraponera clavata*) are frequently injured in intercolony clashes. The odors released by injured ants invariably attract the specialized parasitic phorid fly ***Apocephalus paraponerae***.

FAMILY PHORIDAE (continued) ❶ ❷ These two images, taken less than a minute apart, show a *Labidus praedator* (a Neotropical army ant) finding a rove beetle and then swarming it. Look closely, however, and you can see a phorid fly (***Dacnophora***) hovering over the action. ❸ Flies in the widespread genus ***Gymnophora***, like this European species, are scavengers, and some species develop in dead or dying snails. The prefix *gymno* means "bare," referring to the short pelage of these relatively large phorids. Members of this group are often common on low foliage in moist forests. ❹ The genus ***Megaselia***, with thousands of species, is one of the largest, most biologically diverse and taxonomically difficult genera in the entire animal kingdom. This is a *Megaselia* female from Costa Rica. ❺ Despite the taxonomic intractability of the mega-genus ***Megaselia***, phorid expert Brian Brown has identified the fly standing on the abdomen of this dead Bolivian monkey grasshopper as ***M. aurea***, a species commonly attracted to dead grasshoppers. ❻ These ***Megaselia*** maggots and puparium (enlarged as an inset) were found in a mass of grasshopper eggs in Canada, probably feeding on damaged eggs. ❼ This ***Megaselia aurea*** was attracted to a dead scarab beetle in an Ecuadorian cloud forest. ❽ This Costa Rican ***Melaloncha elongata*** has a sharp ovipositor for delivering eggs to her stingless bee host. *Melaloncha* is a large genus of more than 150 species, found only in the neotropics. All species with known hosts are parasitoids of bees (Apidae). ❾ This ***Myriophora*** was attracted to an injured millipede in an eastern North American forest. Most *Myriophora* species with known biologies are parasitoids of myriapods, and are often associated with dead or dying millipedes.

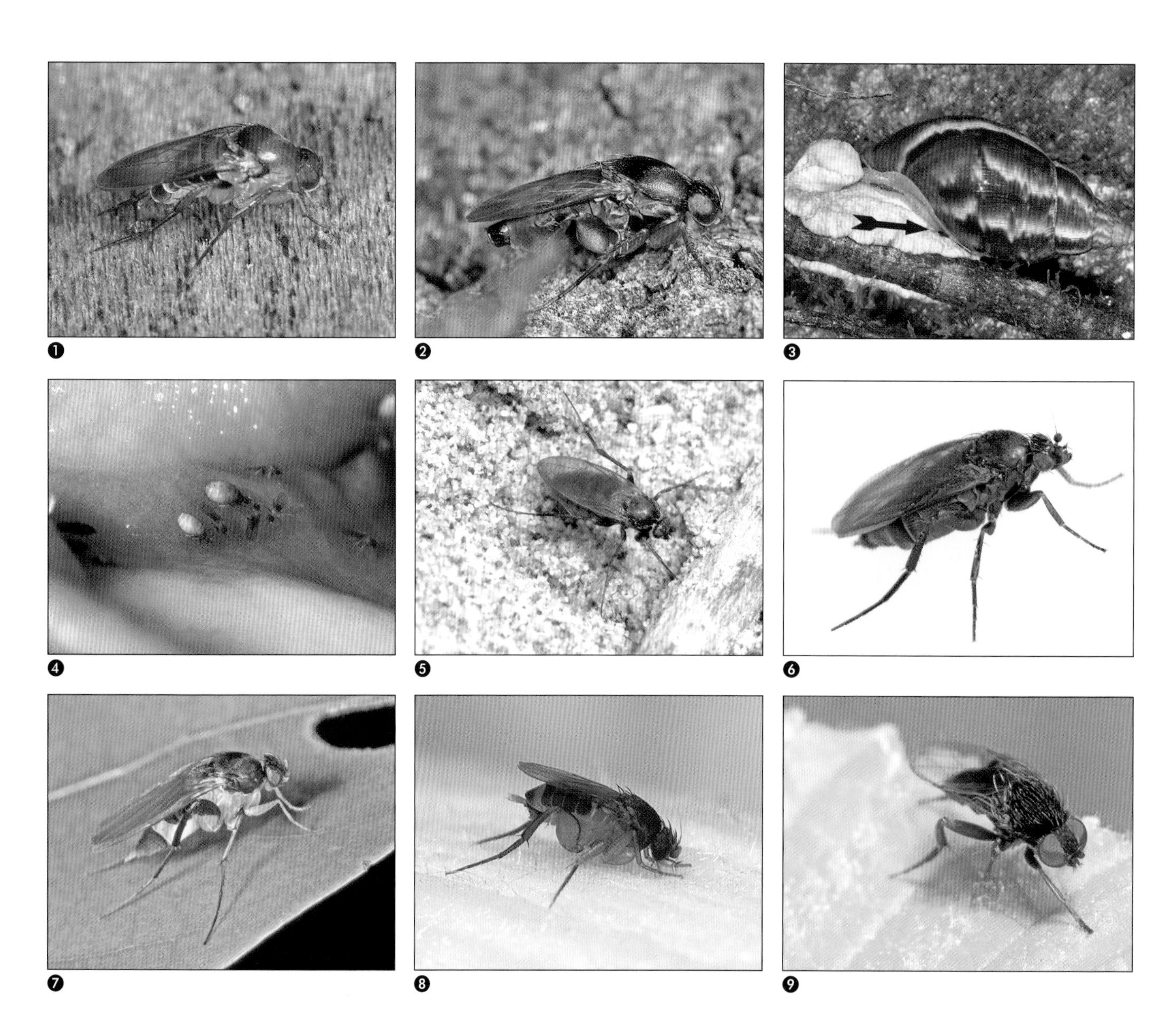

FAMILY PHORIDAE (continued) ❶ ❷ This orange ***Phalacrotophora longifrons*** is on a rotting log in Canada, and this darker African ***Phalacrotophora*** is on a tree trunk. Several *Phalacrotophora* species parasitize lady beetle (Coccinellidae) pupae; others prey on spider egg masses or scavenge in the nests of bees. ❸ ❹ The mantles of giant African achatinid snails provide the only known habitat for the minute phorid fly ***Wandolleckia achatinae***; the magnified portion of this Tanzanian snail shows winged male and wingless female flies, including physogastric females (with swollen abdomens). Larvae apparently develop in the snail's feces while adults are phoretic on the snail's mantle. **Subfamily Phorinae** ❺ This ***Anevrina*** emerged from a rabbit burrow, where it presumably developed on nest debris; other species in the genus have been collected on corpses. This small genus is mostly Holarctic, but species also occur in the Neotropical and Oriental regions. ❻ ***Diplonevra*** is a large, widely distributed genus of about 80 species, including the Holarctic ***D. nitidula*** shown here. *Diplonevra* species, which are routinely reared from dead arthropods, were formerly included in the Aenigmatinae, along with the related genus *Dohrniphora*. ❼ ***Diplonevra bifasciata***, an unusually large and colorful scuttle fly, is a widespread Oriental species, shown here from Vietnam. ❽ ***Dohrniphora*** is the largest genus of Phorinae, with some 167 species worldwide (most Neotropical). This is ***D. cornuta***, a cosmopolitan and synanthropic species, found in my backyard compost heap. Other species have been reared from termite nests, ant nests and a variety of decomposing materials. ❾ Phorid adults are attracted to a wide variety of decomposing materials, and velvety black adults in the widely distributed genus ***Phora*** are commonly seen on bird droppings. This one is feeding on honeydew.

FAMILY PHORIDAE (continued) ❶ This eastern North American fly is one of about 30 species in the globally distributed genus ***Stichillus***. Larval biology remains unknown for the whole group, which was formerly treated as part of the subfamily Hypocerinae. **Subfamily Termitoxeniinae** ❷ ❸ These remarkable photographs of the termite flies ***Cliteroxenia formosana*** (long proboscis, head and thorax exposed) and ***Pseudotermitoxenia nitobei*** (short proboscis, thorax and part of head covered by hood) were taken in the nests of fungus-gardening termites by Japanese entomologist Komatsu Takashi. Termite fly females shed their wings upon entering the nests of Old World termites, where they are kleptoparasites. The pale mass that looks like a termite nymph on the fly's back is the abdomen, which becomes swollen after the female sheds her wings and enters the termite nest. **Subfamily Sciadocerinae** ❹ Other than this Australia/New Zealand species (***Sciadocera rufomaculata***) the only extant species in the subfamily Sciadocerinae is *Archiphora patagonica*, from southern Chile. Sciadocerines are usually found on dead animals.

SUPERFAMILY SYRPHOIDEA

FAMILY SYRPHIDAE (FLOWER OR HOVER FLIES) Subfamily Eristalinae ❺ ❻ ❼ Members of the Holarctic (and Oriental) genus ***Blera***, such as these Nearctic species (the orange-tipped ***B. analis*** and the yellow-sided ***B. umbratillis*** on leaves; ***B. confusa*** at a raspberry flower), are often black with characteristic paler patches. Larvae of some *Blera* develop in tree sap. ❽ Species in the Holarctic (and Oriental) genus ***Brachyopa*** are generally orangish flies that look superficially more like Sciomyzidae (marsh flies) than Syrphidae. Larvae develop in tree sap, and this undescribed Holarctic species, near ***B. (Hammerschmidtia) ferruginea***, is associated with old poplar trees. It has become rare in Europe but remains relatively common in eastern North America. ❾ ***Cacoceria*** is a small (two species) Neotropical genus; this Ecuadorian species (probably ***C. willistoni***) was found on the foliage of a newly fallen tree. Larvae of this genus remain unknown.

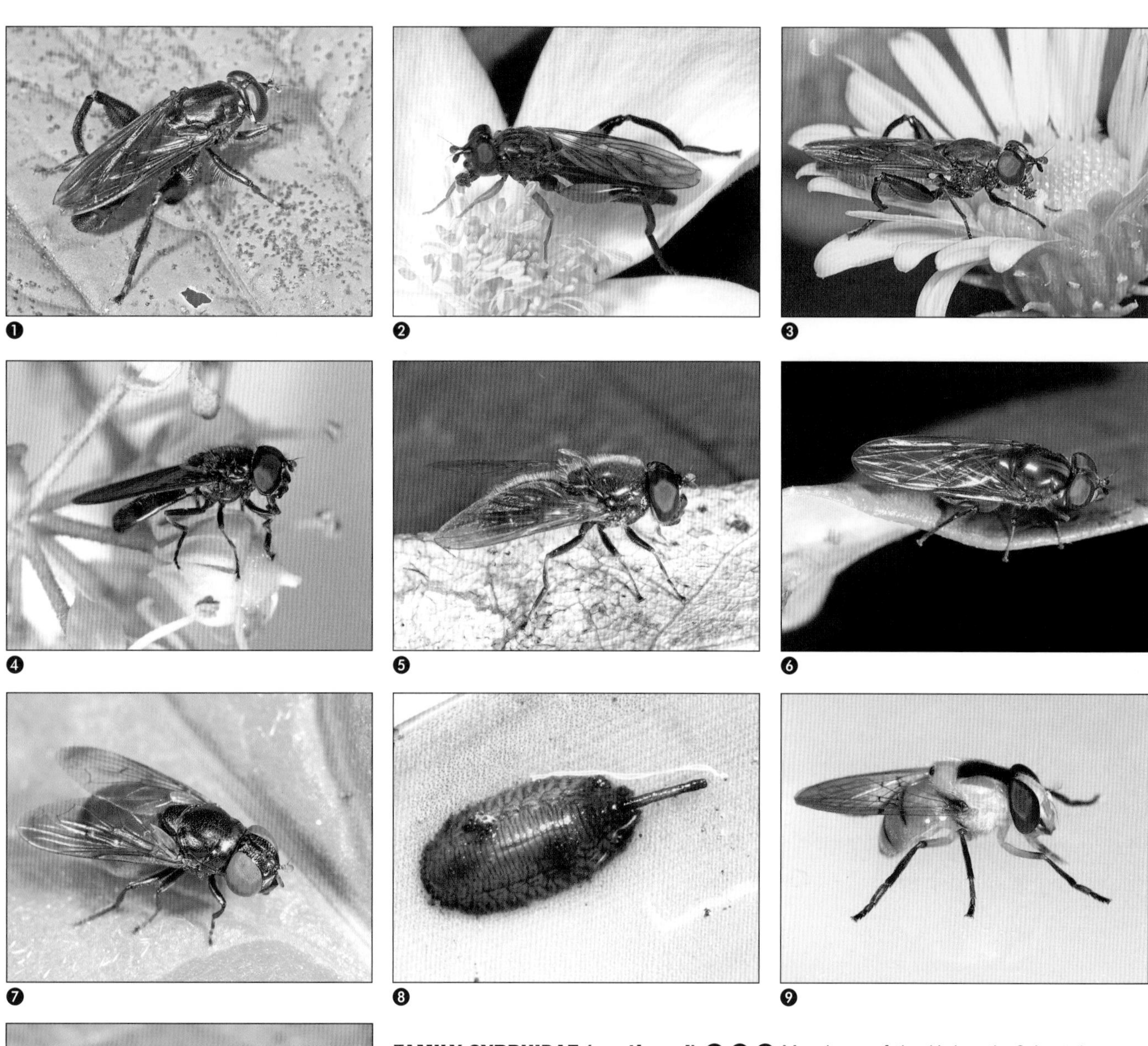

❶ ❷ ❸ ❹ ❺ ❻ ❼ ❽ ❾

❿

FAMILY SYRPHIDAE (continued) ❶ ❷ ❸ Members of the Holarctic-Oriental genus ***Chalcosyrphus*** are usually distinctive flies with swollen hind legs (similar to *Xylota*). Larvae normally occur under bark. The North American species shown here include the blue-black mud-dauber wasp mimic ***C. chalybius***, the orange-legged ***C. vecors***, and the black-legged ***C. piger***. ❹ *Cheilosia* is a large Holarctic-Oriental genus of generally dull black flower flies, often with unusual (for a syrphid) larval habits such as plant feeding or fungus feeding. ***Cheilosia thessala***, here visiting an ivy flower in Greece, develops in fungi. ❺ With almost 500 species, ***Cheilosia*** is currently the largest syrphid genus, but these dark, often nondescript flies are among the least-known flower flies. This ***C. prima*** is perched in a sunny patch on leaf litter in eastern Canada. ❻ ***Chromocheilosia pubescens*** is one of the many remarkable flower flies found only in southern South America. All three species of *Chromocheilosia* are from Chile and Argentina. ❼ Some members of the Holarctic (and African) genus ***Chrysogaster*** have aquatic larvae with posterior spiracles modified into a spine-like tap, used to breathe through the roots of aquatic plants. ❽ This ***Copestylum*** larva has its posterior spiracles fused into a rigid tube to help it breathe while immersed in pockets of water among decomposing plant material. ❾ ❿ These two Bolivian flies are among the more than 300 species in the huge genus ***Copestylum***. Their larvae probably developed in detritus in water trapped in *Heliconia* bracts or similar pockets of rotting plant material.

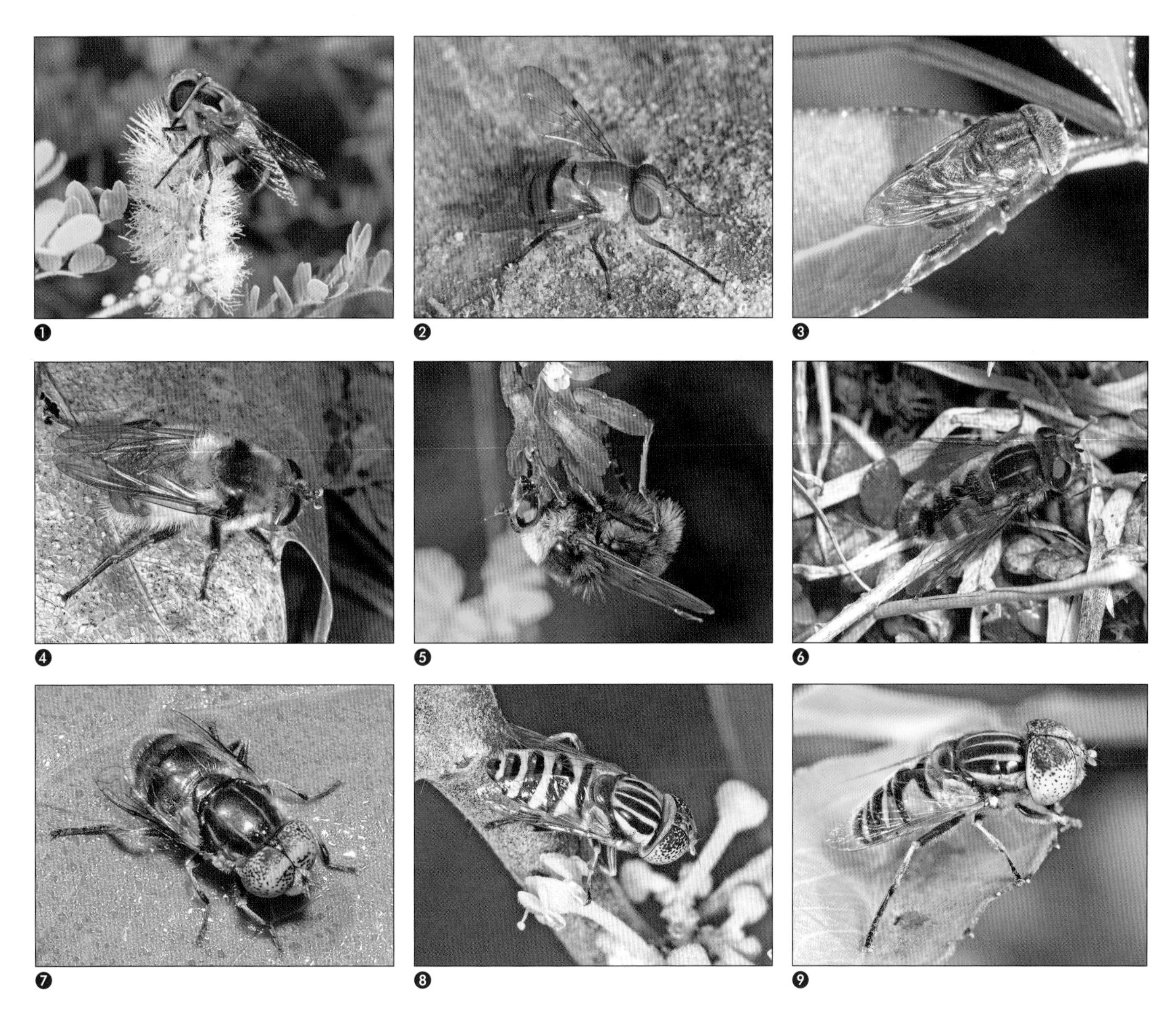

FAMILY SYRPHIDAE (continued) ❶ Almost all ***Copestylum*** are Neotropical, but a few species are North American and a few have been introduced to the Pacific region. This western North American species, ***C. satur***, ranges north to Canada. *Copestylum* is the largest genus of Syrphidae in the Neotropical region, followed closely by the syrphine genus *Ocyptamus*. ❷ ❸ These Ecuadorian ***Copestylum*** (an orange female in the ***C. vagum group***, attracted to a cut log, and a small, beelike male ***C. viridis***) illustrate the impressive diversity of form in this large genus, which has about 300 named Neotropical species. Many, if not most, Neotropical *Copestylum* have not yet been named. ❹ ❺ The large, furry flies in the genus ***Criorhina*** develop in dead wood and are usually active as adults only in early spring. Shown here are ***C. verbosa***, on dead leaves in an eastern Canadian hardwood bush, and a western North American ***C. aurea*** at a flower. These spectacular bumble bee mimics are mostly north temperate, but the genus occurs as far south as Central America in the New World and India in the Old World. ❻ This ***Dolichogyna chilensis***, a south temperate species in a Neotropical genus, was active on a near-freezing day near the southernmost tip of South America. ❼ The distinctively spotted eyes make ***Eristalinus aeneus*** distinctive. *Eristalinus* is an Old World genus but *E. aeneus* was introduced from Europe to North America. ❽ ❾ ***Eristalinus megacephalus*** occurs in much of the Oriental region, as well as parts of Africa and the Pacific. This female fly is visiting jujube (*Ziziphus ziziphus*) flowers in Vietnam, and this male is spattered with pollen from the same flowers.

FAMILY SYRPHIDAE (continued) ❶ ❷ ***Eristalis*** is an enormous worldwide genus of generally robust, often bee-like flies. Their larvae are known as rat-tailed maggots because of their long, tail-like posterior breathing tube that lets them develop in shallow water habitats such as tree holes, puddles and sewage ponds. Similar rat-tails are found in other genera of Eristalinae, but the most common rat-tailed maggots are *Eristalis*, especially the widely distributed and synanthropic ***E. tenax*** **(Drone Fly)**. ❸ ❹ The speciose, widespread genus ***Eristalis*** includes several bumble bee mimics, such as these Canadian ***E. flavipes*** on a *Eupatorium* flower and a willow flower. ❺ *Eristalis* includes many vaguely wasp-like species such as this Canadian ***Eristalis transversa***. ❻ Pollen-covered ***Eristalis*** adults, like this ***E. dimidiata*** on a Canadian willow flower, are common sights on many flowers. ❼ Although the genus ***Eumerus*** now occurs worldwide, the three North American species in the genus are all introduced, probably having arrived in shipments of bulbs (such as onion or narcissus) from Europe. The **Lesser Bulb Fly (*E. funeralis*)**, like this one on a spurge flower, is considered a pest of bulbs worldwide, but it probably develops only in already damaged bulbs. ❽ The genus ***Eumerus*** is a large one, with many species in the Old World tropics. These three (a mating pair and an incoming male) were among many at a small patch of flowers in a Tanzanian mountain forest. ❾ Larvae of the mostly Holarctic genus ***Ferdinandea***, including this southwestern American ***F. aeneicolor***, develop in sappy tree wounds such as those around the exit holes of wood-boring moths and beetles.

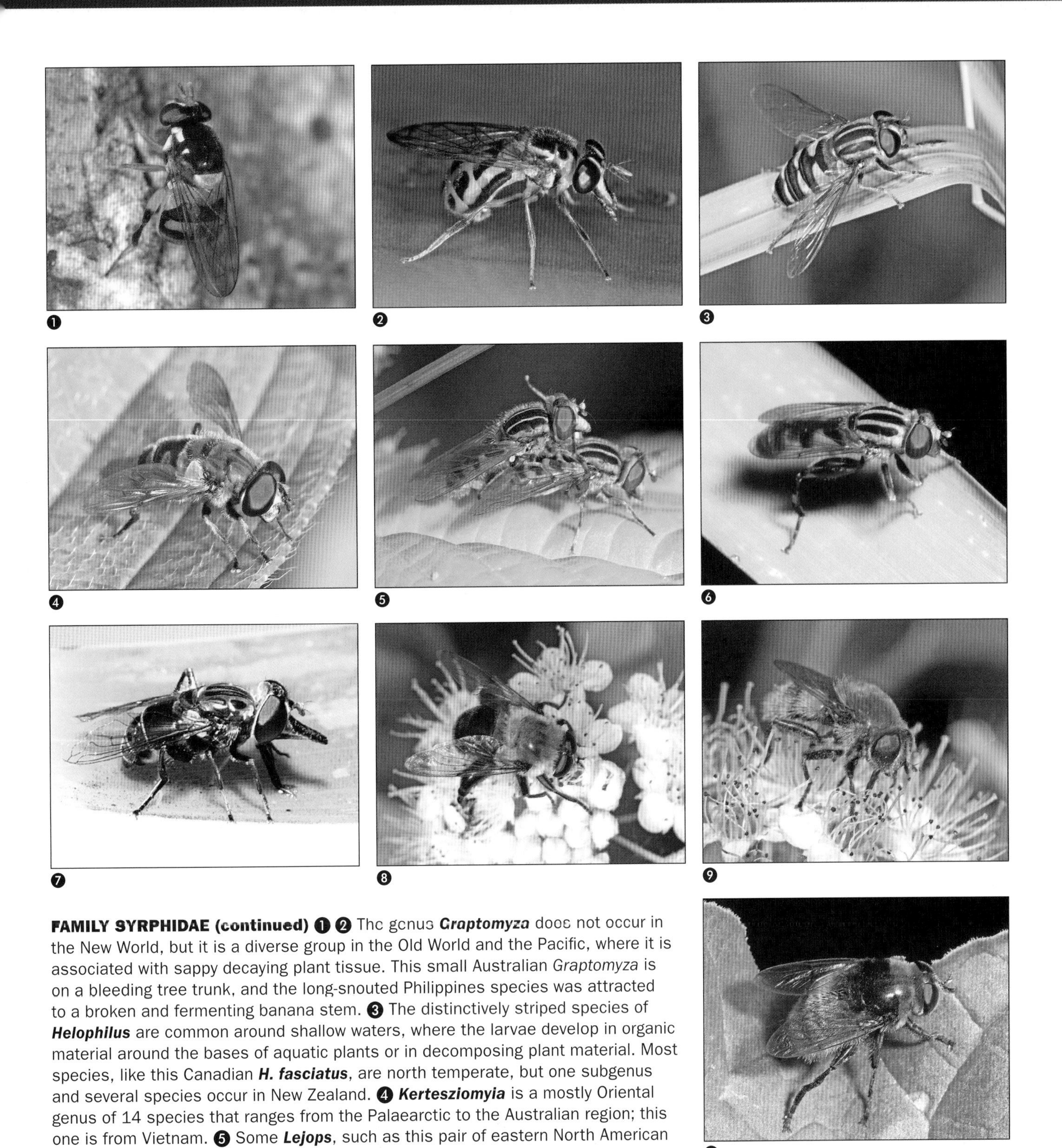

FAMILY SYRPHIDAE (continued) ❶ ❷ The genus ***Graptomyza*** does not occur in the New World, but it is a diverse group in the Old World and the Pacific, where it is associated with sappy decaying plant tissue. This small Australian *Graptomyza* is on a bleeding tree trunk, and the long-snouted Philippines species was attracted to a broken and fermenting banana stem. ❸ The distinctively striped species of ***Helophilus*** are common around shallow waters, where the larvae develop in organic material around the bases of aquatic plants or in decomposing plant material. Most species, like this Canadian ***H. fasciatus***, are north temperate, but one subgenus and several species occur in New Zealand. ❹ ***Kertesziomyia*** is a mostly Oriental genus of 14 species that ranges from the Palaearctic to the Australian region; this one is from Vietnam. ❺ Some ***Lejops***, such as this pair of eastern North American ***L. lineatus***, are striped like the much more robust *Helophilus* species. They occur in the same marshy areas where their saprophagous larvae develop. ❻ ***Lejops*** is a mostly north temperate genus but it also occurs in Africa, and one New World species ranges south to Central America. This is the Nearctic ***L. vittatus***. ❼ ***Lycastrirhyncha nitens***, the only species in its genus, is a strikingly long-snouted species found throughout much of the neotropics. ❽ *Mallota* is a widespread genus often associated with rot holes in old trees; this is ***Mallota bautias***, a common species in eastern North America. ❾ ❿ ***Merodon equestris* (Bulb Fly)** is a common, secondarily widespread fly that develops in bulbs of the lily family. This is a well-known polymorphic species, with different color forms as shown here.

FAMILY SYRPHIDAE (continued) ❶ Although represented in the New World only by the introduced Bulb Fly, the genus ***Merodon*** is very diverse in the Old World. This pollen-covered *Merodon* (an undescribed species in the *geniculatus*-group) is one of many Mediterranean *Merodon*, some of which represent species complexes still being unraveled by taxonomists. ❷ This ***Merodon pulveris*** on the Greek island of Samos seems to be deliberately dangling a blob of egested fluid from the tip of her abdomen, perhaps to attract a mate. ❸ Forty of the 43 species of ***Meromacrus*** are Neotropical; this ***M. basiger*** is from Ecuador. ❹ Robust, striped members of the Old World genus ***Mesembrius*** are commonly encountered in the Old World tropics; this one is from the Philippines. ❺ As the name suggests, this southwestern American ***Milesia bella*** is a beautiful fly; larvae of this genus of large wasplike flies are filter feeders in rot-holes in trees and similar habitats. ❻ ❼ The European ***Milesia crabroniformis*** (seen here on an ivy flower) and the North American ***M. virginiensis*** are large, colorful wasp mimics, and among the most spectacular flies of southern Europe and eastern North America respectively. ❽ The large, wasplike flower flies in the widespread genus ***Milesia*** develop in wet rotting vegetation in tree stumps and similar habitats. This one is laying eggs in the muddy sludge between rotting logs in a Philippines rainforest. ❾ ***Myathropa florea*** is a striking European flower fly that was first given its species name by Linnaeus in his tenth edition of *Systema Naturae* (1758), the official starting point for all animal taxonomy. The genus *Myathropa*, named almost a hundred years later, is a small Palaearctic-Oriental group of only three species; larvae are filter feeders in rot holes or similar habitats.

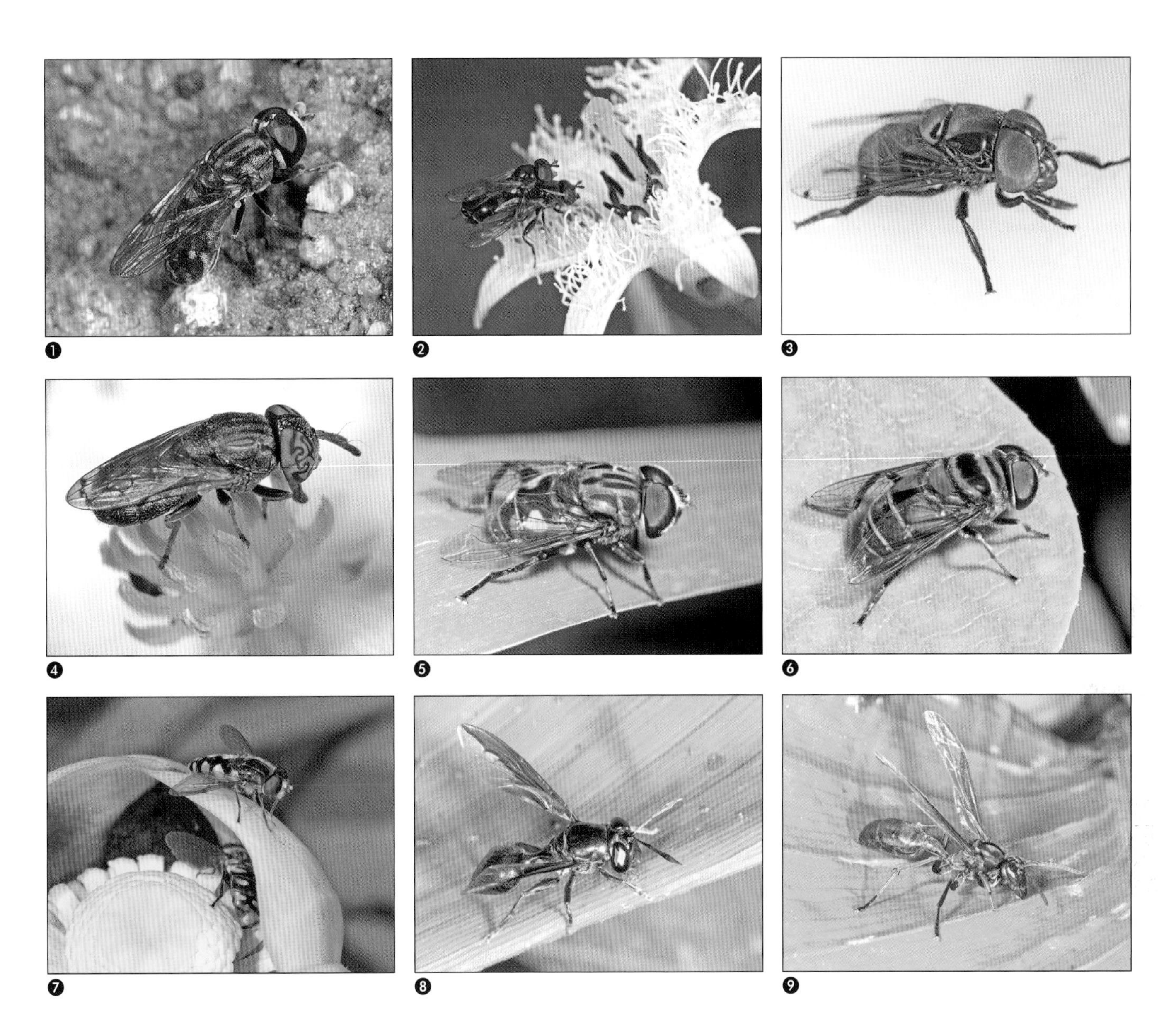

FAMILY SYRPHIDAE (continued) ❶ ***Nausigaster*** is a diverse New World (mostly Neotropical) genus that includes several species, such as this one, in the American southwest. Larvae have been reared from standing water in decaying plants. ❷ These ***Neoascia metallica*** were among many visiting Buckbean (*Menyanthes trifoliata*) flowers in a Canadian wetland. The larvae are semiaquatic or aquatic scavengers, and members of this Holarctic-Oriental genus are often among the most abundant flies at wetland flowers such as Marsh Marigold (*Caltha palustris*). ❸ The genus ***Ornidia*** is a Neotropical genus of five species, one of which (*O. obesa*) is a secondarily widespread, common synanthropic species in tropical and subtropical areas throughout the world. Larvae of ***O. obesa*** develop in sewage, pit toilets and similar putrescent soupy substrates. ❹ ***Orthonevra nitida*** is among the most readily recognizable North American flies because of its decorated eyes; most other species in this widespread but mostly Holarctic genus have more subtle eye pigments. ❺ ❻ *Palpada* is a large New World (mostly Neotropical) genus of more than a hundred species of *Eristalis*-like flies with *Eristalis*-like rat-tailed, semiaquatic or aquatic filter-feeding larvae. These two Costa Rican flies are ***Palpada furcata*** and ***Palpada vinetorum*** (more orange). ❼ ***Parhelophilus rex*** occurs across Canada and the northern United States, where it is almost invariably abundant in Yellow Water-lily (*Nuphar lutea*) flowers, often along with similar *Lejops* species. Flies in the Holarctic genus *Parhelophilus* are common wetland insects that develop in aquatic or semiaquatic habitats such as the leaf sheaths of emergent plants like *Typha*. ❽ This Ecuadorian ***Polybiomyia bergothi*** is strikingly similar to co-occurring "polybiine" vespid wasps. *Polybiomyia* is sometimes treated as a subgenus of *Ceriana*. ❾ This vespid wasp (*Agelaia* spp.), which is sharing a leaf with the *Polybiomyia* in the previous image, shows the distinctive wing, narrowed waist and long antennae apparently copied by its defenseless syrphid neighbor.

FAMILY SYRPHIDAE (continued) ❶ ***Quichuana*** is a genus of 23 species that develop in wet, decaying pockets on Neotropical plants, such as heliconia bracts. This one is from the Ecuadorian Amazon. ❷ Although the genus *Rhingia* is best known for a Palaearctic species that lays eggs above cow dung, into which the hatching larvae drop, little is known about the biology of other species in this widespread genus. This is the common Nearctic ***Rhingia nasica***. ❸ Long-snouted ***Rhingia*** species are common in forests of the Afrotropical, Oriental and Neotropical regions, where they have a characteristic habit of alighting underneath leaves. ❹ This ***Sericomyia chrysotoxoides*** has extruded her telescoping ovipositor to deposit eggs among leaf litter in a Canadian hardwood swamp, where her rat-tailed larvae will develop as semiaquatic scavengers. This Holarctic-Oriental genus includes several robust and distinctively colored species. ❺ ***Somula*** is a small Nearctic genus of two scoliid wasp-like species, of which the most widespread is the distinctive ***S. decora***. Like many eristalines, it develops in tree rot holes. ❻ ***Sphecomyia*** is a small Holarctic genus of distinctively large wasp mimics; this North American ***S. vittata*** is feeding at a Canada Anemone (*Anemone canadensis*) flower near a wetland, where it probably developed as a filter-feeding semiaquatic larva. ❼ ❽ ❾ ***Sphegina*** is a diverse Holarctic-Oriental genus of distinctively wasp-waisted flies with a superficial similarity to the acalyptrate family Syringogastridae. The flies shown here hovering and visiting flowers are western North American species not identifiable at this time. ❿ This eastern North American ***Sphegina brachygaster*** is feeding on pollen at a Marsh Marigold (*Caltha palustris*) flower.

FAMILY SYRPHIDAE (continued) ❶ ❷ ❸ The mostly Holarctic genus *Spilomyia* includes several precise mimics of sympatric stinging wasps, such as the three North American species shown here. ***Spilomyia longicornis*** looks like a yellowjacket (*Vespula* spp.), ***S. fusca*** is a dead ringer for a Bald-faced Hornet (*Dolichovespula maculata*), and the southwestern ***S. kahli*** is a close match to a potter wasp (*Euodynerus* spp.). ❹ *Sterphus* is a diverse genus occurring throughout the neotropics. This is ***Sterphus coeruleus***, known only from Chile and Argentina. ❺ ***Syritta pipiens*** is a common backyard species that breeds in compost heaps and similar habitats throughout the Holarctic region and beyond. The genus *Syritta* is an Old World and Pacific group; both species now found in the New World are accidental introductions from Europe. ❻ ❼ ❽ Like the superficially similar genus *Spilomyia*, *Temnostoma* is a mostly Holarctic genus of remarkably precise wasp mimics. Seen here are the North American yellowjacket wasp mimics ***Temnostoma aequale*** (previously known as *T. vespiforme*) (*in copula*) and ***T. alternans***, and the North American potter-wasp mimic ***T. balyras***. Larvae are found in dead and rotting wood. ❾ ***Tropidia*** larvae, such as those of this North American ***T. quadrata***, are found in dung and other rotting material. *Tropidea* species occur in the Neotropical and Holarctic regions.

FAMILY SYRPHIDAE (continued) ❶ ***Valdiviomyia*** is a small genus of only half a dozen species in the extreme south of South America. Most species, like this ***V. ruficauda***, are restricted to Chile. ❷ This Holarctic bumble bee mimic (***Volucella bombylans***) is one of several *Volucella* species that develop in the nests of social Hymenoptera, either as scavengers or as brood predators. *Volucella bombylans* is a polymorphic species that resembles different bees in different areas, but it develops in yellowjacket wasp nests. ❸ ❹ ***Xylota***, including this unidentified Tanzanian species and this southern European ***X. tarda***, is a large genus found throughout the Holarctic region and the Old World tropics. ❺ ❻ ***Xylota*** species are often seen picking up pollen from leaves, but they are also frequent flower visitors. These are among the many similar North American *Xylota* species. **Subfamily Syrphinae** ❼ ❽ Almost all members of the subfamily Syrphinae are predators of aphids and related Sternorrhyncha. Adult syrphines can often be seen hovering over aphid colonies, periodically darting in to deposit an egg. ❾ Members of the subfamily Syrphinae, such as this ***Allobaccha*** from the Philippines, are characteristically bright yellow and black flies. *Allobacha* occurs in the Old World and Australia.

FAMILY SYRPHIDAE (continued) ❶ ❷ *Allograpta* is a large (more than 100 species) and diverse worldwide genus that includes very common species in the Holarctic region as well as more exotic species such as this ***Allograpta (Allograpta) hortensis*** on a yellow *Ranunculus* flower in southern Chile, and this long-nosed ***Allograpta (Rhinoprosopa) lucifer*** from Costa Rica. Syrphine larvae are almost all predators, but some exceptional species of *Allograpta* have plant-eating (leaf-mining) larvae. ❸ ❹ ***Anu una*** is a unique New Zealand flower fly first named in a 2008 paper, in which a male is illustrated with its abdomen strangely arched up and forward. As shown in these pictures of dead and living flies taken on a 1999 trip to New Zealand, this species looks quite ordinary in life; the abdomen apparently curls forward only in dried specimens. ❺ Hover flies are so named for their common habit of pulling up their landing gear and hovering motionless in one spot, like this Tanzanian ***Asarkina***. *Asarkina* is a *Syrphus*-like genus widely distributed in the Old World as well as the Oriental and Australasian regions. ❻ This delicate ***Baccha elongata*** is hovering over a honeydew-spattered leaf in eastern Canada. *Baccha* is a huge genus with hundreds of species, mostly Neotropical. ❼ ❽ ***Chrysotoxum*** species, found in most parts of the world other than Australia and Oceania, are distinctively large, brightly colored flies with long antennae. These adult flies superficially resemble some of the large, wasp-mimicking Eristalinae; larvae remain unknown but probably prey on root aphids. This eastern Canadian species seems to be picking up honeydew from a scale-covered twig. ❾ Most of the 25 or so species of ***Citrogramma***, like this one from Vietnam, occur in the Oriental region. ❿ This ***Dasysyrphus pauxillus*** from western North America is covered with pollen. Larval *Dasysyrphus* are aphid predators, often on conifer foliage. The genus is widespread, although not present in Africa or Australia.

FAMILY SYRPHIDAE (continued) ❶ The white lines visible through the transparent abdominal tergites of this eastern North American ***Didea fuscipes*** are tracheae (air tubes) leading to the spiracles in the sides of the abdomen. *Didea* is a Holarctic-Oriental genus. ❷ This eastern North American ***Doros aequalis*** looks much like common potter wasps (Vespidae: Eumeninae) both in flight and appearance. For this reason, and because the adult flight period is short, species in this small Holarctic genus are easily overlooked and not often collected or photographed. ❸ Members of the large and widespread genus ***Epistrophe*** are common flower visitors, often seen hovering over blossoms before touching down to feed on pollen and nectar. ❹ ***Episyrphus balteatus*** is a widespread Old World and Pacific species common in agroecosystems, where adult flies are important pollinators and larvae are important predators. ❺ ❻ *Eupeodes*, *Epistrophe* and *Syrphus* are among the most common "typical" syrphines at flowers in Europe and North America. ***Eupeodes***, such as this pollen-feeding ***E. pallipes***, have a distinctive margin (a ridge or groove) along the edge of the abdomen. The *Eupeodes* larva is munching through aphids on an introduced weed (Himalayan Impatiens) in my backyard in Canada. ❼ ***Leucopodella*** is one of several genera of delicate, petiolate (the base of the abdomen narrows like a stem) syrphids common in the New World tropics. It resembles *Baccha*, but *Baccha* species have a prominent knob in the middle of the face. ❽ ❾ ***Melangyna*** is a Holarctic and Oriental genus with several small species, like this ***M. mentalis***, that visit spring flowers.

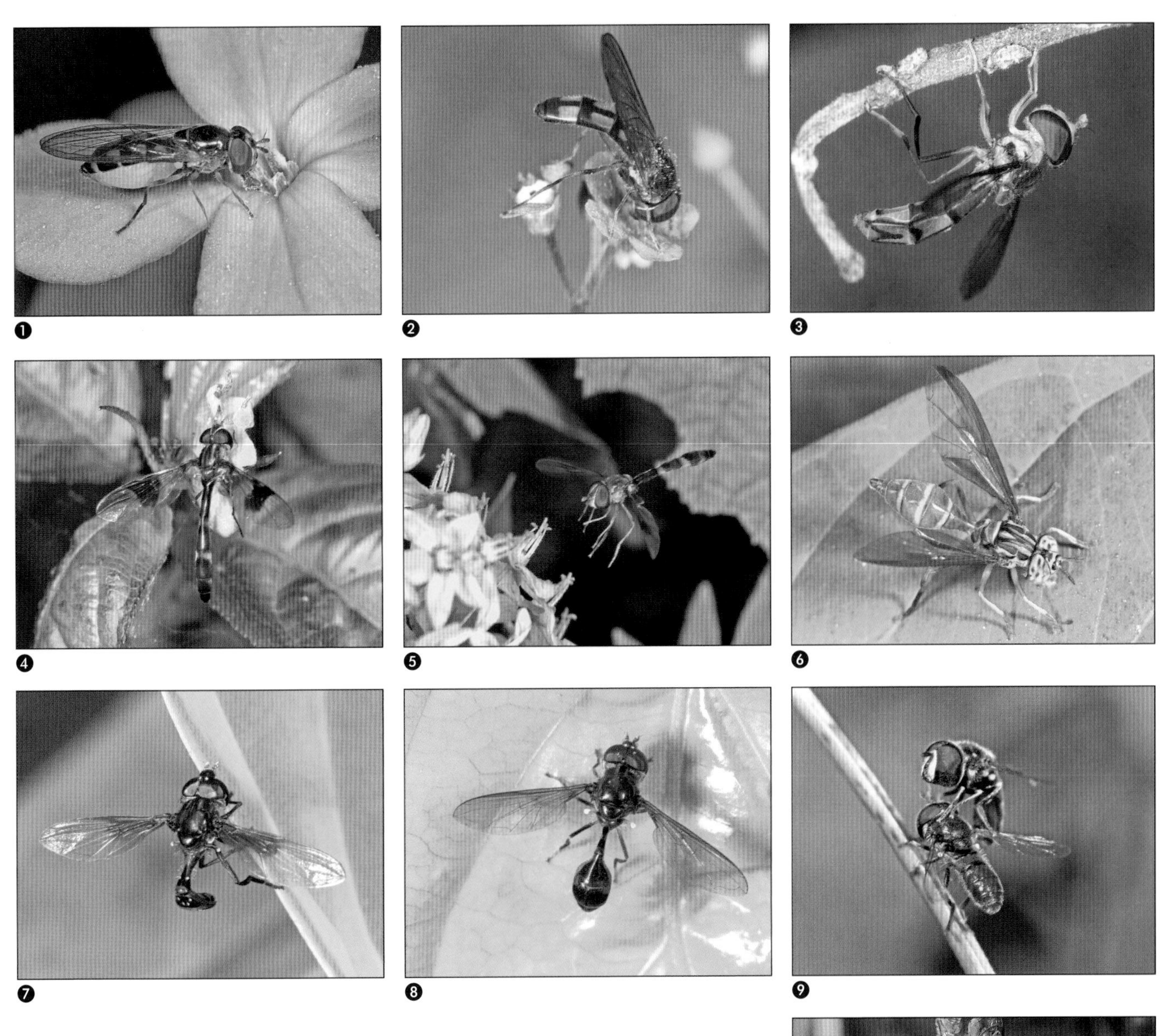

FAMILY SYRPHIDAE (continued) ❶ ❷ ***Melanostoma mellinum***, one of the fly species described by Linnaeus in 1758, is a common flower visitor throughout much of the Holarctic region. Variable in color, this species is similar to the more diverse genus *Platycheirus*. The female on the purple flower has bright green-blue halters and an abdomen bursting with eggs. The halters are normally pale, but in some *Melanostoma* they turn bright green in females ready to lay eggs. ❸ ***Meliscaeva***, seen here in Vietnam, is a Holarctic-Oriental genus most diverse in the Oriental region. ❹ ❺ ***Ocyptamus fascipennis*** is a widespread Nearctic species and one of 22 Nearctic species in the large New World genus *Ocyptamus*. ❻ This Ecuadorian ***Ocyptamus conjunctus*** is a striking mimic of vespid wasps in the Neotropical genus *Angiopolybia*. ❼ ❽ Three hundred or so of the 322 ***Ocyptamus*** species occur in the neotropics, where almost half of all Syrphini belong to this genus. Shown here are a yellow-spotted ***O. adspersus*** and the strikingly wasp-waisted ***O. zeteki***. ❾ ❿ The small flower flies in the widespread genus *Paragus* are aphid predators (like other Syrphinae). This pair and these larvae belong to the common Holarctic species ***Paragus haemorrhous*** (a species name that presumably refers to its red posterior). This species seems to be a specialized predator of aphids defended by ants, which tend the aphids for their sweet honeydew. The body projections on the *Paragus* larva seem to stave off ants and give these relatively small (about 4 mm) flower flies an advantage over other syrphids.

FAMILY SYRPHIDAE (continued) ❶ Most of the 90 or so *Paragus* species, such as this ***Paragus quadrifasciatus*** from Greece, occur in the Old World, although a couple of species are secondarily widespread and now occur in North America, Central America and Australia. ❷ This ***Pelecocera (Chamaesyrphus) pruinosomaculata*** was among many visiting wild cyclamen (*C. hederifolium*) in Greece. The genus and subgenus also occur in North America. ❸ ***Pipiza*** and related genera (the tribe Pipizini) used to be included in the Eristalinae because of adult morphology, but their predaceous larvae, chromosomal evidence and newer molecular evidence suggest that they belong in the Syrphinae. *Pipiza* is a large genus, diverse in the Holarctic region and also found in temperate South America. ❹ ❺ ❻ Members of the large, widespread genus ***Platycheirus***, such as this ***P. nearcticus*** on a geranium flower, feed on pollen at a variety of flowers. Many species are characteristically associated with grasses and can be found in abundance on grass flowers in the early morning or evening. This larva was found on half-submerged foliage in a Canadian fen. ❼ Male ***Platycheirus*** species can often be identified by their characteristically modified forelegs; this Canadian ***P. quadratus*** has a twisted tibia and broadened fore tarsomeres. ❽ ***Salpingogaster*** species, such as this one from the high Andes of Peru, look remarkably like some stinging ichneumonid wasps, right down to an abdominal process that looks like a wasp's sharp ovipositor. Larvae of some species of this common New World genus have the unusual habit of penetrating the spittle masses of spittlebugs (Cercopidae) to prey on the nymphs within. ❾ This ***Salpingogaster nigra*** from near Quito, Ecuador, has a mite attached to an abdominal tergite.

FAMILY SYRPHIDAE (continued) ❶ ❷ ***Scaeva***, like this hovering individual from southern Chile and this ***S. pyrastri*** lapping up honeydew from a stem in eastern Canada, is a widespread genus of relatively large, distinctively marked syrphines. The larvae, as with most syrphines, are aphid-eaters. ❸ ***Simosyrphus grandicornis***, the only member of the genus *Simosyrphus*, occurs in Oceania, including Australia and New Zealand. ❹ These eastern North American ***Sphaerophoria contigua*** show conspicuous sexual dimorphism (the male, above, has much larger eyes). *Sphaerophoria* is a common, widespread genus found almost everywhere except the neotropics. ❺ ❻ Some of the most common flower flies in both the New World and the Old World are in the genus ***Syrphus***. Predaceous *Syrphus* larvae are easily found among groups of aphids on a wide variety of hosts. ❼ This ***Syrphus*** larva is eating aphids on a rosebush; larvae such as this can each kill hundreds of aphids. ❽ This bright green ***Syrphus*** puparium is stuck to a rosebush among the aphids that served as larval prey. ❾ ***Syrphus torvus*** is a common and widespread Nearctic flower fly that ranges from Alaska across to Greenland and south to New Mexico.

FAMILY SYRPHIDAE (continued) ❶ ❷ ***Syrphus*** is an enormous genus found almost everywhere except the Australasian region. These are both *Syrphus* species from Chile; the relatively bright male is ***S. octomaculatus*** and the drab female is an undescribed species. ❸ ❹ Probably the most common small flower flies worldwide belong to the large genus ***Toxomerus***, most of which are bright black and yellow flies like the Central American pair here (flying united). The duller species shown here is ***T. vertebratus***, from southern Chile. Almost 90% of the 150 or so species in the genus are Neotropical, but some of the Nearctic species are very common. ❺ High-altitude species of many genera of flies have a characteristically long-winged appearance. These two flies, from high in the Ecuadorian Andes, are a yellow-banded ***Allograpta fasciata*** and a more intricately patterned ***Toxomerus nasatus***. ❻ ***Tuberculanostoma*** is a small genus of distinctively snouted but otherwise *Platycheirus*-like syrphids found in the Neotropical and Palaearctic regions (but not in between). This one is from the high Andes of Peru. ❼ ❽ Members of the Holarctic and Oriental genus ***Xanthogramma***, such as this North American ***X. flavipes*** and this European ***X. pedissequum***, are strikingly colored wasp mimics. **Subfamily Microdontinae** ❾ This is an unnamed species of Ecuadorian ***Hypselosyrphus*** that showed up at a honeydew-spattered leaf along with several almost identical stingless bees. The fly even flew with its hind legs dangling below, just like the bees.

FAMILY SYRPHIDAE (continued) ❶ The mollusk-like dome beside this Canadian carpenter ant is a ***Microdon*** larva that developed in the ant's nest. ❷ The widespread Southeast Asian species ***Microdon auroscutatus*** usually has a black abdomen, but some individuals (like this female from Vietnam) have a partly red abdomen and are referred to as "variety *variventris*." ❸ The orange fly on this arboreal Ecuadorian ant nest is a female ***Microdon tenuicaudus***, probably looking for an opportunity to leave eggs in the nest. ❹ ***Microdon*** species, such as this Tanzanian ***M. obesus***, develop in ant nests. ❺ *Microdon* species like this Tanzanian ***Microdon obesus*** are generally uncommon, and are most likely to be found near their ant hosts. ❻ Although ***Microdon*** species are not often encountered, this ***M. ruficrus*** was one of many found at the same time around some Canadian *Formica* ant nests. ❼ This southwestern American species in the *Microdon* subgenus ***Omegasyrphus*** lacks the extremely long antennae that mark most *Microdon*. All known *Microdon* are ant-associated, and this one probably developed in nearby *Monomorium* nests. ❽ ❾ This black Ecuadorian ***Mixogaster*** is on a leaf just visited by an almost identical vespid wasp; the pair of more brightly colored ***M. rarior*** are under a leaf in Costa Rica. The 17 species in this New World genus, like other microdontines, probably all develop in ant nests.

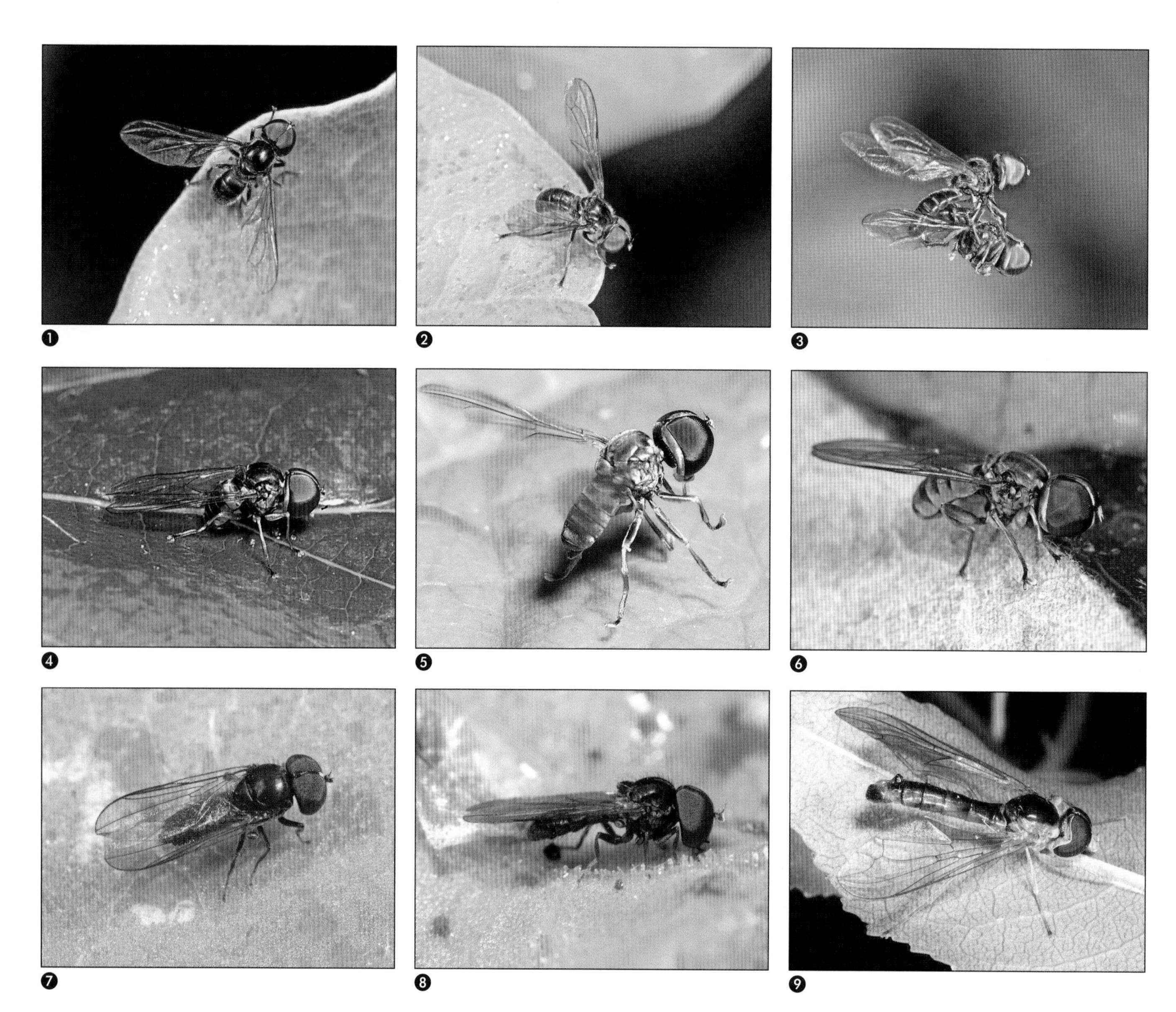

FAMILY PIPUNCULIDAE (BIG-HEADED FLIES) Subfamily Pipunculinae ❶ ❷ This male (with contiguous eyes) and female (with separated eyes) of ***Cephalops conjunctivus*** were attracted to a honey-sprayed leaf in Greece. ❸ Big-headed flies are the ultimate hoverers, able to hover in confined spaces and even remaining apparently motionless while mating, as shown by this pair of ***Cephalops conjunctivus*** in Greece. ❹ The New World genus ***Elmohardyia*** was named in honor of the late Elmo Hardy, one of the great Pipunculidae specialists. Hosts remain unknown for all 51 species in the genus, including this Nearctic ***E. atlantica***, but it is a safe assumption that they are parasitoids of leafhoppers or planthoppers. ❺ ***Eudorylas*** is the largest genus of big-headed flies, with more than 400 species worldwide. This Canadian species is taking off from a honeydew-spattered leaf. ❻ Intermittent waterways are excellent places to find a variety of flies, including this southwestern American ***Eudorylas*** that has landed for a drink. **Subfamily Chalarinae ❼** Each genus of big-headed fly (except for *Nephrocerus*) is associated with a single subfamily of Auchenorrhyncha, and all species of ***Chalarus*** with known hosts are parasitoids of the leafhopper (Cicadellidae) subfamily Typhlocybinae. The 25 or so species in this genus are mostly north temperate, although *Chalarus* does occur in the Neotropical and Oriental regions. ❽ ***Jassidophaga*** is a small, widespread genus associated with the leafhopper subfamily Macropsinae. **Subfamily Nephrocerinae ❾** ***Nephrocerus*** species, unlike other Pipunculidae, are parasitoids of crane flies. Other members of the family are parasitoids of leafhoppers and related Auchenorrhyncha. This ***Nephrocerus acanthostylus*** photograph was taken by Jeff Skevington, the entomologist who named the species in 2005.

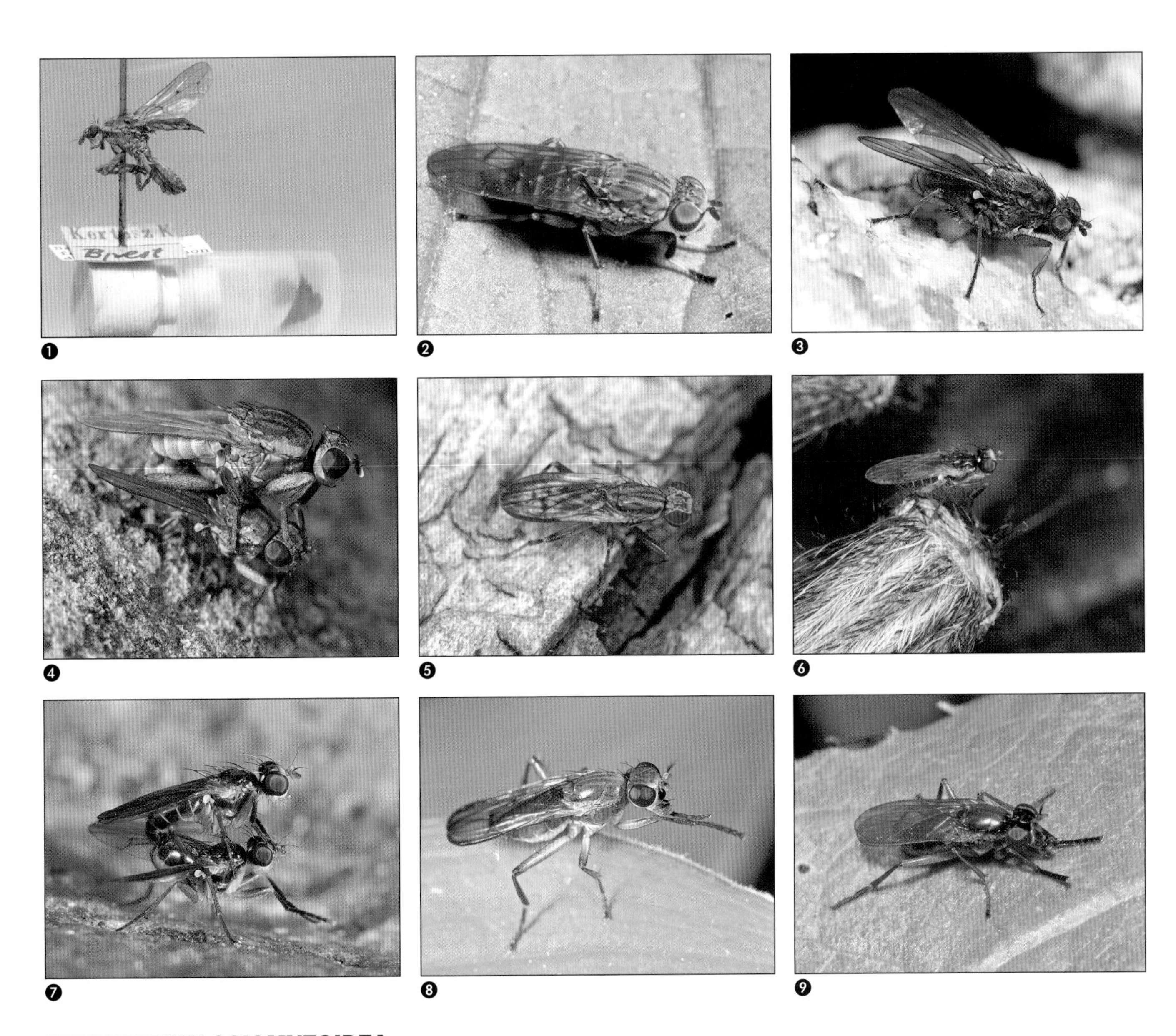

SUPERFAMILY SCIOMYZOIDEA

FAMILY SCIOMYZIDAE (MARSH OR SNAIL-KILLING FLIES) Subfamily Salticellinae ❶ This specimen of the European species ***Salticella fasciata*** is a museum specimen; part of its abdomen, removed for study by a taxonomist, is stored in a tube of glycerin pinned under the fly. **Subfamily Sciomyzinae: Tribe Sciomyzini** ❷ ***Atrichomelina*** includes only a single species, the widespread Nearctic ***A. pubera***. Look for it along shorelines, where it develops in living or dead stranded snails. ❸ ***Calliscia*** is a Chilean genus of one species, ***C. calliscales***, known only from a couple of collections on the rocky coast of central Chile. Larvae presumably attack intertidal or supralittoral mollusks. ❹ The three species of ***Ditaeniella***, sometimes treated as part of *Pherbellia*, occur in Europe, North America and Chile. Species with known larvae are parasitoids of a variety of snails, both aquatic and terrestrial, but larvae of the species shown here, ***D. patagonensis***, from south Chile, are unknown. ❺ ***Pherbellia*** is a large, cosmopolitan and taxonomically difficult genus. Known larvae develop as parasitoids of both aquatic and terrestrial pulmonate snails. ❻ ❼ ***Parectinocera*** is a small Neotropical genus ranging from Ecuador to Chile. These ***P. inaequalis*** were found on or near a tarantula near Santiago, Chile; the single specimen here is on the knee of the spider, and mating pairs were found on the ground between the spider's legs. Larvae are unknown. ❽ Species of the Holarctic genus ***Sciomyza***, like this Canadian ***S. aristalis***, are parasitoids of snails, attaching their eggs directly to the host snails. **Tribe Tetanocerini** ❾ Although most Tetanocerini oviposit away from their hosts, ***Anticheta*** species such as this Nearctic ***A. melanosoma*** lay eggs on exposed egg masses of snails in the genera *Lymnaea* and *Succinea*.

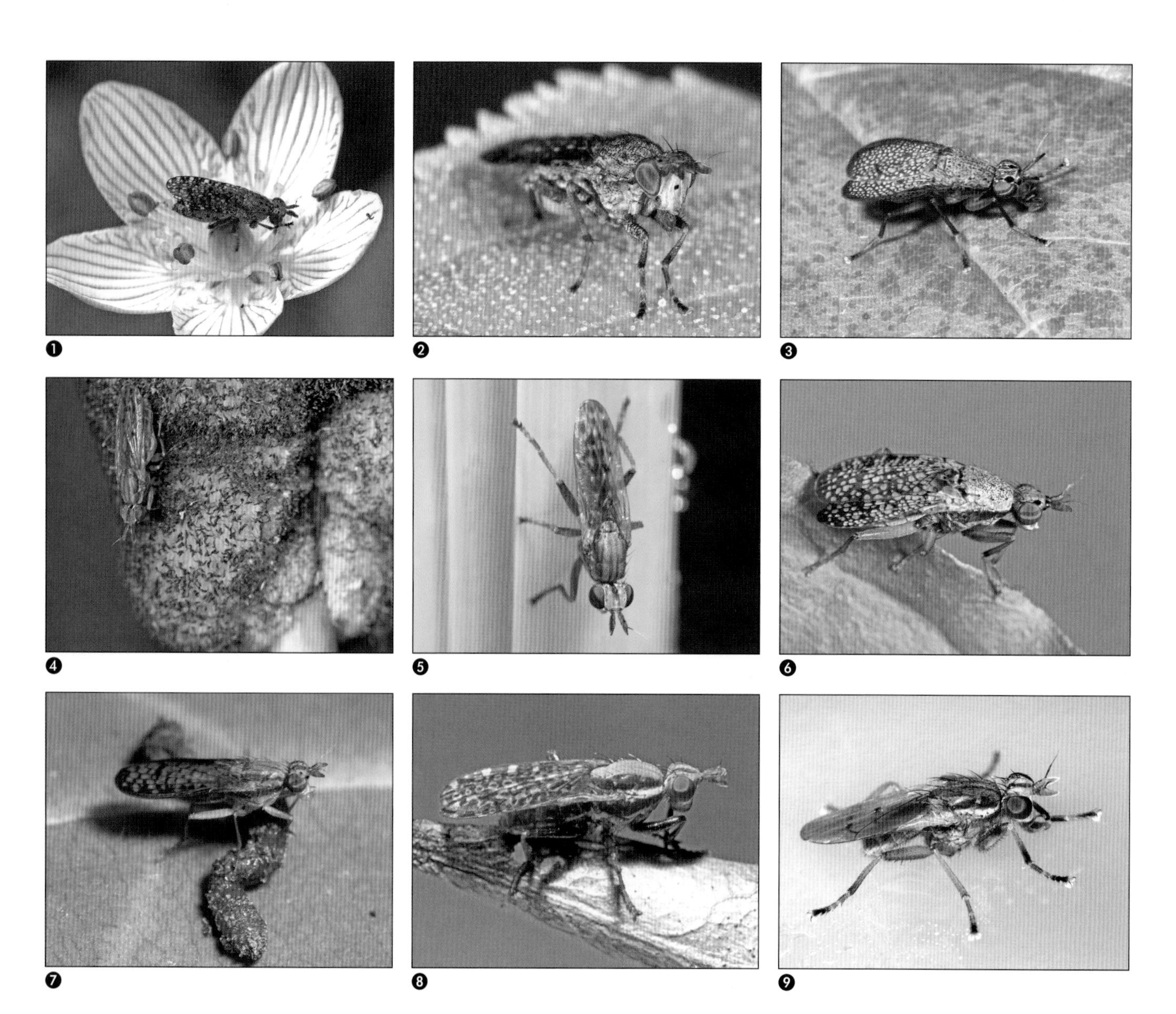

FAMILY SCIOMYZIDAE (continued) ❶ ❷ ***Dictya*** is a commonly encountered group in the New World, where most of the 29 species are associated with wetlands and prey on aquatic pulmonate snails; a single species occurs in the Old World. A signature black dot on the middle of the face makes these common flies easy to recognize. ❸ This is ***Dictyacium firmum***, one of two species in this small northern Nearctic genus. These rarely seen flies are associated with fens (alkaline peatlands), where the as yet unknown larvae probably develop in peatland snails. ❹ ❺ Most Tetanocerini (other than *Anticheta*) lay eggs away from their hosts; this Canadian ***Elgiva solicita*** is ovipositing in a cattail head (*Typha*) over a pond in which her larvae will develop as aquatic predators. This small Holarctic genus has two species in North America and three in Europe. ❻ *Euthycera* is a mostly European genus, with 20 of its 22 species occurring only in the Palaearctic region. This is the North American ***Euthycera arcuata***, one of two New World species (the other occurs in Mexico). Larvae feed on terrestrial snails. ❼ ***Limnia*** is a common Holarctic group of species generally distinguished from similar genera by antennae that end in contrasting white aristae. *Limnia* species are associated with both aquatic and terrestrial snails; this North American species, ***L. boscii***, probably develops in terrestrial snails. ❽ The Australian region has a small sciomyzid fauna dominated by the New Zealand genus ***Neolimnia***. The 14 *Neolimnia* species occupy a variety of habitats, with hosts that include aquatic and terrestrial pulmonate and operculate snails. ❾ Known ***Perilimnia*** larvae develop in stranded snails along shorelines. Both described species in this genus are restricted to southern South America; this one is from southern Chile.

FAMILY SCIOMYZIDAE (continued) ❶ The genus ***Pherbecta*** includes only ***P. limenitis***, a rare fly found in fens of northeastern North America. Larval habits remain unknown. ❷ ***Poecilographa decorum***, the only member of its genus and probably the prettiest of all Sciomyzidae, is one of the least-known Nearctic snail-killing flies: its natural hosts remain a mystery. Formerly widespread in eastern North America, it is now rare in much of its former range. ❸ Known larvae of the small Holarctic genus ***Renocera*** are predators of small bivalve mollusks ("fingernail clams," family Sphaeriidae), an unusual habit otherwise found only in the New Zealand genus *Eulimnia*. ❹ Members of the large, cosmopolitan genus ***Sepedon*** are almost unique among higher Diptera in having lost a functional ptilinum and the associated ptilinal fissure. This is ***S. convergens***, an African species. ❺ ***Sepedon*** species, such as this North American ***S. armipes***, are aquatic predators of pulmonate snails. The deep notch under the hind femur is characteristic of males of this species, and the porrect (long and projecting) antennae are characteristic of the genus. ❻ ***Sepedon*** species are distinctive the world over for their prominent antennae and reduced ptilinal suture. Shown here is a Vietnamese species, ***S. aenescens***. ❼ ***Shannonia*** is a southern South American genus of only two species found in Chile and Argentina; this is ***S. costalis*** from southernmost Chile. Larvae develop in pulmonate snails stranded on shorelines. ❽ ***Tetanocera*** is a mostly Holarctic genus that includes many common species that develop as aquatic snail predators in ponds and marshes in Europe and North America. This widespread Nearctic species, ***T. plumosula***, develops as a predator on snails along shorelines or in semiaquatic habitats. ❾ Although the Sciomyzidae include a wide range of terrestrial and aquatic mollusk scavengers, mollusk parasitoids and mollusk predators, some of the most common larvae are free-living aquatic predators that hang from the water surface as they hunt snails. This is a ***Tetanocera***, but similar habits are found in *Elgiva* and *Sepedon*.

FAMILY SCIOMYZIDAE (continued) ❶ Marsh fly adults are often seen scavenging on dead insects, bird dung and other organic material, but they also sometimes eat tabanid (horse fly) egg masses. This pair of flies striking a characteristic mating pose (male feet on the inner margins of the female's eyes, typical of most sciomyzids other than *Sepedon*) is ***Tetanocera plebeja***, a species that develops in terrestrial slugs. ❷ This North American species, ***Tetanocera valida***, attacks slugs, laying eggs on the slugs' invaginated eyestalks. Larval habits vary widely in this diverse genus of more than 40 species. ❸ ***Tetanoceroides*** is a small southern South American genus of half a dozen poorly known species. This one is on driftwood on a river floodplain in the extreme south of Chile, where it presumably developed on pulmonate snails in shallow pools. ❹ ❺ ***Thecomyia***, the typical snail-killing fly of South and Central American rainforests, is an entirely Neotropical group. Known larvae are predators of aquatic snails. **FAMILY SEPSIDAE (ANT-LIKE SCAVENGER FLIES) Subfamily Orygmatinae** ❻ ❼ ***Orygma luctuosum***, the only species in this Holarctic genus, occurs only in piles of seaweed (or "wrack") along the northern shores of the Atlantic Ocean. **Subfamily Sepsinae** ❽ Sepsidae such as these Cuban ***Archisepsis*** often occur in huge aggregations near fresh dung. This small genus of about a dozen species is mostly Neotropical, but one species has been introduced to the Azores. ❾ Sepsids are often seen in dense aggregations near larval habitat; these ***Dicranosepsis*** were among hundreds on foliage near some water buffalo dung in the Philippines. *Dicranosepsis* is a common genus in parts of the Old World tropics, although only two uncommon species occur in Africa.

FAMILY SEPSIDAE (continued) ❶ ***Lasionemopoda hirsuta*** is an endemic New Zealand sepsid, and the only New Zealand member of the family. ❷ Sepsidae often have elaborate mating behaviors. Males of several genera have forelegs modified to grasp the base of the female's wing, as can be seen in this headstanding couple of Peruvian ***Lateosepsis***. ❸ The ant-like silhouette of typical sepsids, including this member of the small African genus ***Leptomerosepsis***, justifies their common name, "ant-like scavenger flies," and possibly gives them some protection from ant-shy predators. ❹ ***Meropliosepsis sexsetosa*** is a widespread Neotropical species, and the only species in the genus *Meropliosepsis*. ❺ ***Nemopoda*** is a mostly Old World genus, but ***N. nitidula*** is a Holarctic species, often associated with animal dung in urban and agricultural environments. ❻ ***Paratoxopoda***, such as this one on an elephant dropping in Tanzania, is a common African genus. Note the small sphaerocerid on the dung surface in the foreground. ❼ Sepsids, like many other flies, often "bubble." This ***Archisepsis excavata*** in the Amazon lowlands is fanning its wings rapidly, apparently to increase the rate of evaporation of its extruded fluid bubble. ❽ The inner surfaces of the front femur and tibia of this male ***Palaeosepsis*** are characteristically modified to clamp the margin of the female's wing during mating. Similar but specifically distinct foreleg modifications are found in most sepsids. *Palaeosepsis* is an entirely Neotropical genus; many of the species formerly in *Palaeosepsis* are now in the more recently described genus *Archisepsis*. ❾ This Ecuadorian ***Archisepsis*** has joined a number of small tachinids feeding on exudate or honeydew on a plant stem.

FAMILY SEPSIDAE (continued) ❶ ***Saltella sphondylii*** is one of several Holarctic species of Sepsidae common in disturbed environments. All other species in this genus are Old World, so *S. sphondylii* was probably introduced from Europe to North America. ❷ ***Sepsis*** is a large (78 species), mostly Old World, genus with many secondarily widespread species. Most North American species also occur in Europe. ❸ Many ***Sepsis*** species are found only in Africa; this one (***S. pronodosa***) is known only from Tanzania. ❹ ***Themira*** is another mostly Old World genus associated with dung or sewage. This is ***T. (Encita) annulipes***, one of several widespread Holarctic species. ❺ This Holarctic species, ***Themira nigricornis***, emerges in early spring, and males can be seen displaying in sunny spots while shadier areas remain snow-covered. This male has used elaborate processes on his fore femur and tibia to clamp the female's wing. ❻ ***Toxopoda*** is an Old World Pacific genus with many species endemic to parts of the Oriental, African and Australasian regions (one species ranges into the Palaearctic). This is an East African species.
FAMILY STYLOGASTRIDAE (STYLOGASTRID FLIES) ❼ ❽ Most of the 92 known ***Stylogaster*** species occur in the neotropics, but 14 species are known from Africa. This Tanzanian fly hovering over a driver ant raid is probably a new species close to ***S. westwoodi***. ❾ This African ***Stylogaster*** was hovering over the column of driver ants, waiting for the ants to flush out host insects (composite image).

FAMILY STYLOGASTRIDAE (continued) ❶ ***Stylogaster***, the only genus of Stylogastridae, is a mostly Neotropical group almost always seen hovering over the swarm fronts of New World army ants, awaiting an opportunity to dart harpoon-like eggs into roaches and crickets flushed out by the ants. This was one of many hovering over a *Labidus praedator* raid in Ecuador. ❷ This male Costa Rican ***Stylogaster*** has knocked down a hovering female and is trying to force a mating, but she appears to be preventing that by bending the long tip of her abdomen up over her body and out of the way. ❸ Only two species of the mostly Neotropical genus ***Stylogaster*** reach the Nearctic region, ***S. neglecta*** (shown here in eastern Canada) and ***S. biannulata***. **FAMILY CONOPIDAE (THICK-HEADED FLIES) Subfamily Myopinae** ❹ Like other conopids, ***Myopa*** species attack adult bees and wasps; they can often be found on flowers as they await potential hosts. This is one of several eastern North American species. ❺ ***Myopa*** species are found in every region except the Afrotropical. This is the Australian species ***M. ornata***, photographed in Tasmania. ❻ This small male of ***Myopa peplexa*** is struggling with a larger female on a roadside flower in California. ❼ ***Thecophora occidensis*** is one of the most common thick-headed flies throughout the Nearctic region. Similar congeners occur in every zoogeographic region. ❽ ❾ ***Thecophora*** species can be seen on flowers worldwide. The ones shown here were awaiting Hymenoptera hosts in an eastern American meadow.

FAMILY CONOPIDAE (continued) Subfamily Dalmanniinae ❶ ❷ ***Dalmannia*** species often look strikingly like small solitary bees, including their potential hosts. Shown here are the eastern North American ***D. nigriceps*** (on a pink flower) and the western North American ***D. vitiosa***. This genus occurs in the Holarctic and Oriental regions. **Subfamily Conopinae ❸ ❹ ❺** ***Physocephala*** is a cosmopolitan genus and perhaps the most frequently seen member of the family, in part because of its large size and wasplike appearance, but also because members of this genus are frequent visitors to common flowers (especially composites and umbellifers). They attack adult wasps and bees, often bumble bees and sometimes Honey Bees. Shown here are three North American species, ***P. clavata*** (dark wings), ***P. furcillata*** (mating) and ***P. sagittaria***. **❻** ***Physocephala burgessi*** is a western North American species. **❼ ❽** ***Physoconops***, a widespread genus in the New World and in the Old World tropics, is superficially similar to the common and widely distributed genus *Physocephala*, but *Physoconops* lack the basally thickened hind femur that characterizes *Physocephala* species. Shown here are the Nearctic *Physoconops* species ***P. excisus*** (on a white flower) and ***P. obscuripennis***. The *P. excisus* is a female; you can see her large abdominal structures (theca) used to grip a struggling host bee or wasp to insert an egg. **❾** Many wasp-mimicking flies have their wings pigmented to resemble the longitudinally folded wings of vespid wasps, but this Costa Rican ***Physoconops nigromarginatus*** is the first fly I have seen to physically fold its wings like a vespid.

FAMILY CONOPIDAE (continued) Subfamily Zodioninae ❶ ❷ ❸ Conopids, like these North American ***Zodion***, are important both as parasitoids of hymenopteran pollinators and as pollinators themselves. *Zodion*, a large, cosmopolitan genus and the most common conopid genus with a beak that is not bent at the middle, is sometimes treated in the subfamily Myopinae. **FAMILY COELOPIDAE (SEAWEED FLIES) ❹** This ***Coelopa (Fucomyia) alluaudi*** was photographed on the island of Mindoro in the Philippines. This species is found from Madagascar to Australia, but Coelopidae are otherwise mostly distributed on southern and northern temperate shores, with the conspicuous exception of southern South America. **❺ *Coelopa (Fucomyia) frigida*** is a species of North Atlantic shores that abounds in the wrack piles of Europe and northeastern North America. **❻ ❼ *Coelopa (Neocoelopa) vanduzeei*** is a Pacific coast species found along North American shorelines from Alaska to California. **❽ *Dasycoelopa***, with only the one species, ***D. australis***, occurs only on the east coast of Australia. **❾** These piles of wrack on the American west coast support impressively dense ***Coelopa*** populations, and are literally crawling with coelopids (but swarming with sphaerocerids).

FAMILY ROPALOMERIDAE ❶ ***Apophorhynchus*** is a South American genus of four named species; this Bolivian fly is probably an undescribed species. ❷ ***Ropalomera*** is a relatively large genus, with about 15 species ranging throughout the Neotropical region. This ***R. femorata*** was one of many attracted to a newly fallen tree in the Ecuadorian Amazon. ❸ ❹ ***Willistoniella*** is a small genus of four described species found in Central and South America. Like other Ropalomeridae, they are attracted to tree wounds and similar fermentation. Shown here are an undescribed species from Bolivia and ***W. ulyssesi*** from Ecuador. **FAMILY DRYOMYZIDAE** ❺ ❻ ❼ ***Dryomyza anilis*** (sometimes treated as ***Neuroctena anilis***) is an extremely common species throughout North America and Europe, where it is most common on dung but also occurs on flowers and low foliage. Shown here are a male guarding a smaller female, a mating pair and a male spattered with pollen. ❽ ❾ The female of this mated pair of ***Oedoparena glauca*** is laying eggs in a barnacle while guarded by the male. This is the more common of the two *Oedoparena* species on the Pacific coast of North America; a third species occurs in Japan.

FAMILY HELOSCIOMYZIDAE (COMB-WINGED FLIES) ❶ The genus ***Helosciomyza*** is found in both Australia and New Zealand; this is an Australian species. ❷ All three species in the small genus ***Napaeosciomyza*** occur only in New Zealand. **FAMILY HELCOMYZIDAE** ❸ ***Helcomyza mirabilis***, the only Nearctic helcomyzid, occurs on the Pacific beaches of the northwestern United States. ❹ ❺ Species of ***Paractora***, such as these Chilean flies, occur mostly in southern South America, where they seem to replace Coelopidae as the dominant large wrack flies. **FAMILY HETEROCHEILIDAE** ❻ This ***Heterocheila hannai*** is standing on rotting seaweed along the shoreline in Vancouver, British Columbia. The only other species in the genus (and the family) is the European *H. buccata*. **FAMILY HUTTONINIDAE** ❼ The family Huttoninidae occurs only in New Zealand; this specimen of ***Huttonina glabra*** was photographed in the Australian Museum. **FAMILY PHAEOMYIIDAE (MILLIPEDE-KILLING FLIES)** ❽ All four species in the Phaeomyiidae are endemic to the Palaearctic region, although this species, the millipede parasitoid ***Pelidnoptera nigripennis***, was deliberately introduced to Australia for biological control of pest millipedes. Dmitry Gavryushin took this photo in the Moscow area of Russia. **FAMILY NATALIMYZIDAE** ❾ This specimen, in the United States National Museum, is one of the type specimens (paratypes) used to name the South African species ***Natalimyza milleri***, upon which the family Natalimyzidae is based. The specific name honors Ray Miller, who was the first to formally recognize the family, in 1984.

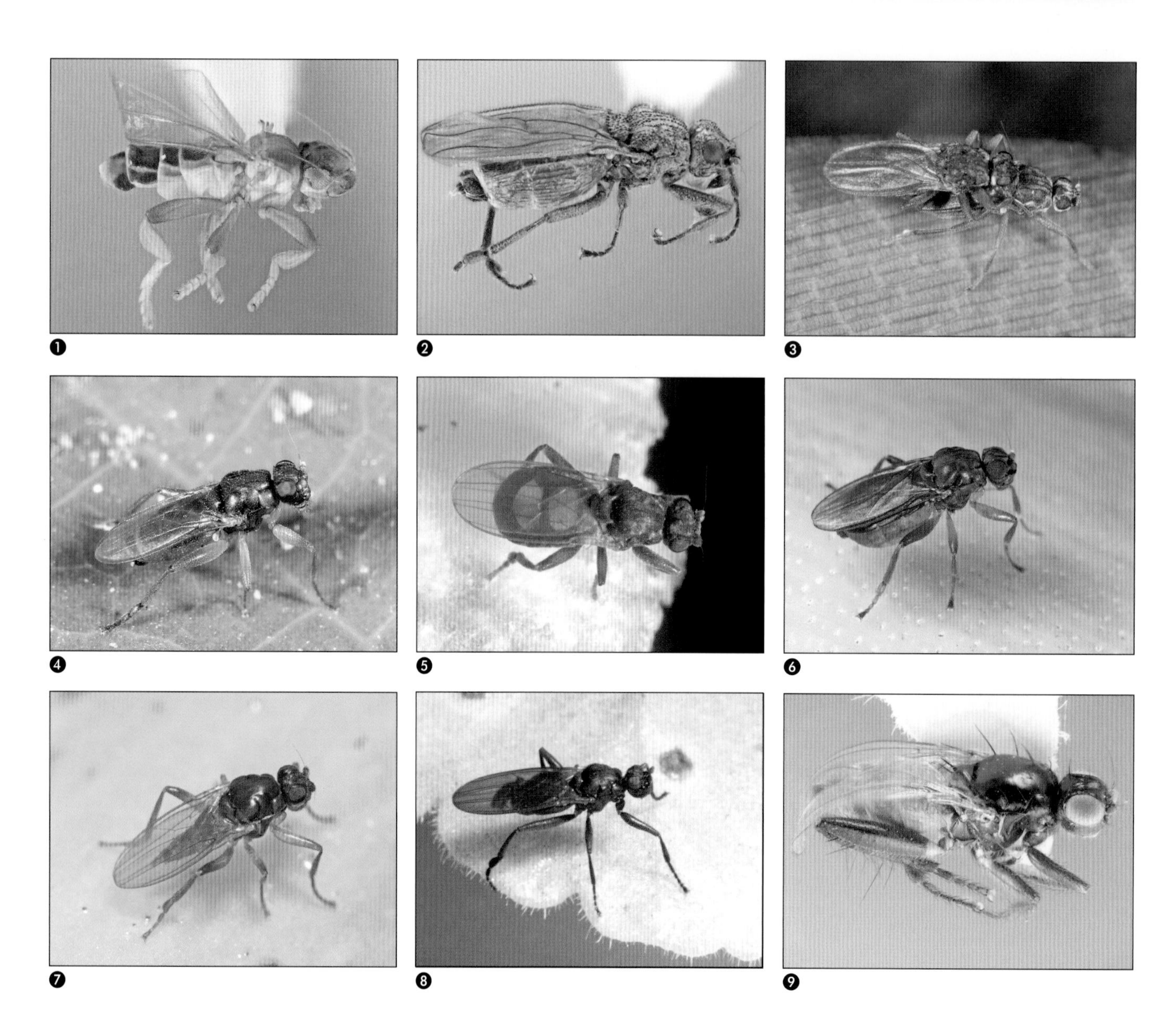

SUPERFAMILY SPHAEROCEROIDEA

FAMILY SPHAEROCERIDAE Subfamily Homalomitrinae ❶ This is the type specimen (the holotype) of the Peruvian ***Sphaeromitra inepta***, the only species of *Sphaeromitra* and one of only half a dozen species in the rare and unusual subfamily Homalomitrinae. To my knowledge, nobody has seen a homalomitrine alive. **Subfamily Sphaerocerinae** ❷ ❸ Flies in the widespread genus ***Ischiolepta***, such as this point-mounted ***I. denticulata*** from Europe and this mating pair from east Africa, are characterized by a row of 6 to 9 warts along the margin of the scutellum. ❹ This Tanzanian fly belongs in ***Lotobia***, a mostly African genus characterized by a row of 12 or more warts along the margin of the scutellum. ❺ ❻ Flies in the New World (almost entirely Neotropical) genus ***Parasphaerocera*** are characterized by fused abdominal tergites that form a plate usually punctuated by conspicuous "windows," which are clearly visible on this Costa Rican fly. ❼ ❽ ***Neosphaerocera*** is a Neotropical genus closely related to *Parasphaerocera*. **Subfamily Tucminae** ❾ ***Tucma tucumana*** is one of only two species in the subfamily, both restricted to northern Argentina. *Tucma* (and thus the subfamily Tucminae) is the "basal lineage" of Sphaeroceridae, the sister group to the rest of the family.

FAMILY SPHAEROCERIDAE (continued) Subfamily Archiborborinae ❶ ❷ The Neotropical subfamily Archiborborinae is the dominant group of Sphaeroceridae at the higher altitudes and lower latitudes of South and Central America. Most species in the subfamily were until recently treated in the single large genus *Archiborborus*. Shown here are ***Antrops annulatus*** (feeding on honeydew) and ***Maculantrops hirtipes*** (at the edge of a leaf) from Chile. ❸ These ***Boreantrops talamanca*** were attracted to some dung in a high-elevation Costa Rican cloud forest. ❹ ***Antrops femoralis*** is one of the most common acalyptrates at the extreme southern tip of South America. ❺ The 14 species of ***Frutillaria*** are flightless flies mostly restricted to fragments of Valdivian forest in Chile. This species, ***F. abdita***, was discovered in a small ravine completely surrounded by pasture. As far as is known it occurs only in that one ravine, and only in parts that are too steep for cattle to access. **Subfamily Copromyzinae** ❻ The bare, shining thorax of this African fly is reflected in its generic name ***Achaetothorax***. This genus of 16 species is mostly African, but one species occurs in Malaysia. ❼ Members of the Holarctic genus ***Crumomyia*** are often associated with cold conditions and are often collected in northern regions, high altitudes and cooler months. ❽ ***Lotophila atra*** is a common Holarctic species that often abounds on dog droppings and other sorts of dung in disturbed environments. *Lotophila* is mostly an Old World genus, although one species (*L. confusa*) occurs in the Nearctic, Palaearctic and Oriental regions. ❾ ***Norrbomia*** is a mostly Holarctic and Old World tropical genus that includes widespread species such as the nearly cosmopolitan *N. sordida* and localized species found only in the Nearctic, Palaearctic or Afrotropical regions. This African *Norrbomia* is sharing an elephant dropping with a much smaller member of the subfamily Limosininae.

FAMILY SPHAEROCERIDAE (continued) ❶ ***Norrbomia frigipennis*** is a white-winged southeastern Nearctic species with the interesting habit of riding on dung-rolling scarabs and laying eggs in their dung balls as they are buried. Those shown here are riding the back of a *Phaeneus vindex* in Florida. Similar habits are found in some Limosininae (genus *Ceroptera*). **Subfamily Limosininae** ❷ ❸ Loss of flight is common in the Sphaeroceridae, especially in the subfamily Limosininae. These two point-mounted specimens, an ant-like ***Myrmolimosina andersoni*** from Mexico and a more robust ***Anatalanta aptera*** from the subantarctic Crozet archipelago, have lost their wings entirely. ❹ ❺ Wing loss or reduction is common in the Limosininae, and tiny flightless limosinines are often abundant in leaf litter, especially on islands and at higher altitudes where there are few ants. These are a ***Howickia*** species from New Zealand (on the edge of a leaf) and an ***Aptilotella*** species from Costa Rica (on the gills of a fungus). ❻ ❼ Although many genera of Sphaeroceridae are completely wingless, the genus ***Aptilotus*** includes apterous (wingless), brachypterous (short-winged) and macropterous (fully winged) species. These are a brachypterous ***A. nigriscapus*** from the American southwest and an apterous ***A. politus*** from the American northwest. ❽ ❾ The genus ***Ceroptera*** occurs in Europe, Africa and the southeastern United States, where the adults cling to adult scarab beetles, awaiting the opportunity to oviposit in the scarab's dung ball or other food cache. These ***C. longicauda*** are clinging to the tail end of a *Mycotrupes gagei* in Florida.

FAMILY SPHAEROCERIDAE (continued) ❶ ***Coproica*** is an enormous and ubiquitous group of coprophagous sphaerocerids invariably present on meadow muffins. Given these habits, it is not surprising that 6 of the 43 species in the genus are now cosmopolitan, but this genus is still rich in local endemics. The undescribed species all over this Ecuadorian cow flop occurs throughout the neotropics. ❷ This distinctive African species, first described by British dipterist O.W. Richards in 1938, was put in its own genus by Hungarian dipterist L. Papp in 2008. It is now known as ***Giraffimyiella giraffa***. ❸ This ***Leptocera hexadike*** from Costa Rica became part of the type series when the species was described as new in 2009. *Leptocera* is a very common worldwide group, with 56 described species, and is one of the most frequent genera in general collections made with nets and Malaise traps. ❹ The 56 world species of ***Poecilosomella***, generally recognizable by the pigmented wings and peculiar wing venation, occur mostly in the Old World tropics, although some species are widespread. The species shown here, ***P. angulata*** (photographed in Cuba), now occurs in warmer areas throughout the world. ❺ Although the great majority of species in the New World genus ***Pterogramma*** remain undescribed, this is one of the most common genera of flies in the Neotropical region, occurring abundantly anywhere there is decaying vegetation. Most members of this genus are very small (1–2 mm) flies with unusual wing venation and often with banded wings. ❻ ***Rachispoda***, a large worldwide genus of 158 described species, is easily recognized by the long inclinate bristles at the front of the thorax and by the somewhat pointed head. These are ubiquitous and generally abundant flies around shorelines, wet meadows and other habitats with wet exposed soil and decaying vegetation. ❼ The common New World (mostly Neotropical) genus ***Robustagramma*** includes 51 named species, all but two of which were newly described in 2005. ❽ ❾ Although ***Sclerocoelus*** is one of the most common groups of insects in Neotropical cloud forests, most species in this large genus still await formal description. Shown here are species from Costa Rica and Ecuador.

FAMILY SPHAEROCERIDAE (continued) ❶ ***Spelobia*** is one of the larger genera of Sphaeroceridae, with 80 named species, including some of the most common sphaerocerids in North America and Europe. Although the genus is mostly Holarctic, the fly shown here is an undescribed Neotropical species in a fungivorous species group, the ***S. aciculata*** group. ❷ ***Eulimosina ochripes*** is a common Holarctic species until recently known as *Spelobia ochripes*. Adults of this species are often common in open areas in early spring; this one was seen on a patch of remaining snow. ❸ ***Thoracochaeta*** is a large coastal genus found in huge numbers in seaweed along shorelines the world over. The long inclinate bristles on the wing margins mark these western American flies as females of ***T. johnsoni***, a species that ranges from Canada to Chile. ❹ This Neotropical genus of Limosininae, first found in 2011 and still unnamed, has been collected only deep in rolled *Heliconia* leaves. Countless other small Limosininae remain to be discovered in other specialized microhabitats. **FAMILY HELEOMYZIDAE (WOMBAT FLIES, CAVE FLIES, ETC.) Subfamily Heteromyzinae** ❺ Members of the widespread Old World and Holarctic genus ***Tephrochlamys*** breed in a wide variety of decomposing materials, including birds' nests, fungi and dung. This is a Holarctic species, ***T. rufiventris***. **Subfamily Diaciinae** ❻ ❼ The subfamily Diaciinae includes only three very rare species: one each in the South American genera *Diacia* and *Dichromya*, and one in the Western Australian genus *Amphidysis*. Shown here are specimens of ***Amphidysis hesperia*** (Australian Museum, on a paper point) and ***Diacia diadema*** (United States National Museum, pinned). **Subfamily Cnemospathidinae** ❽ ❾ All 23 species of ***Borboroides*** occur in Australia, where several species, such as this one, seem to occur only on wombat dung.

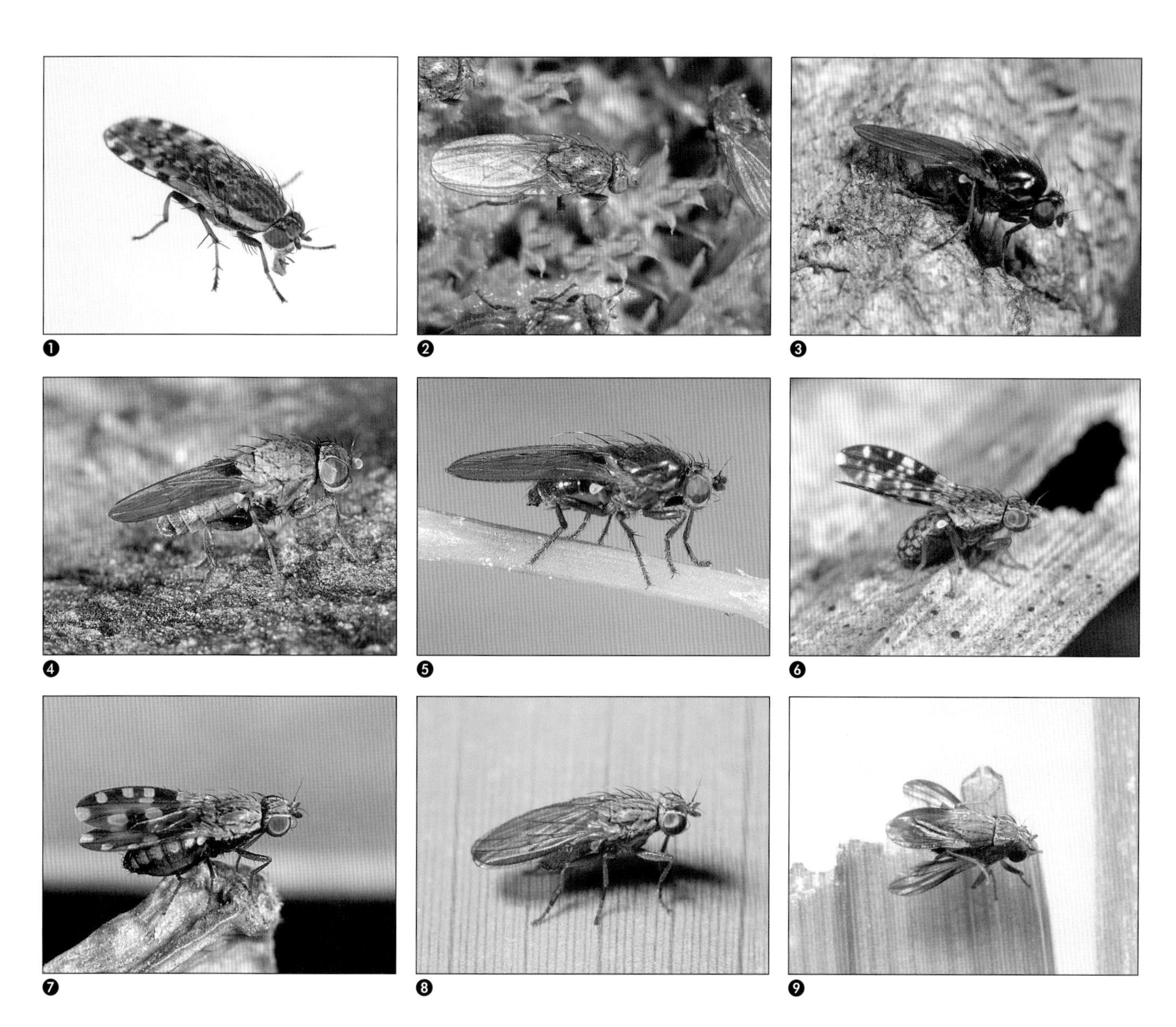

FAMILY HELEOMYZIDAE (continued) ❶ ***Diplogeomyza*** is one of the more diverse and distinctive Australian heleomyzid genera; almost all of its 16 species were described by D.K. McAlpine. This is ***D. hardyi*** from New South Wales. ❷ ***Epistomyia*** is one of the heleomyzid genera that blur the boundary between the Heleomyzidae and Sphaeroceridae and (but for the unmodified first hind tarsomere) looks much like a "primitive" sphaerocerid. There is only one named species of *Epistomyia*, *E. aurifrons* from Argentina. ❸ ***Blaesochaetophora***, with the single species ***B. picticornis***, is one of several small, sphaerocerid-like genera of Heleomyzidae restricted to southern South America. ❹ ***Cephodapedon*** is a small southern South American genus of only three species, of which two are known only from Chile. ❺ ***Notomyza*** is a southern South American genus sometimes treated as a separate family because of its unusual (secondarily symmetrical) male genitalia. **Subfamily Trixoscelidinae** ❻ ❼ ***Spilochroa*** is a small genus closely related to the large genus *Trixoscelis*; shown here are ***S. ornatus*** from Florida and ***S. guttata*** from Chile. ❽ ***Trixoscelis*** is a large, widespread genus sometimes reared from birds' nests. This one is from California. Trixoscelidinae were until recently usually treated as a separate family. ❾ ***Trixoscelis*** is diverse in western North America, but ***T. fumipennis*** is the only described member of the genus found in eastern North America. It is most easily found in sparsely grassed areas such as older cemeteries and airfields.

FAMILY HELEOMYZIDAE (continued) Subfamily Rhinotorinae ❶ ❷ Rhinotorines are distinctive because of their broadened heads and characteristic wings, and the group was until recently treated as a separate family. Shown here are pinned specimens of the Australian ***Zinza grandis*** and the South American ***Rhinotora diversipennis***. **Subfamily Tapeigastrinae ❸** ***Tapeigaster*** is an entirely Australian genus of uncertain placement, sometimes treated as a separate family. These relatively large orange flies are often conspicuous on mushrooms. **Subfamily Suilliinae ❹** ***Allophyla*** is a small North American genus closely related to the more widespread *Suillia*, from which it differs in having a humeral ("shoulder") bristle. Larvae develop in fungi, like most members of the subfamily. ❺ ❻ ❼ Most Suilliinae belong to the large, widespread (except Australia) genus ***Suillia***. Shown here are a mating pair of east African ***S. ingens***, a top view of a Costa Rican ***S. inens*** and a member of the ***S. variegata*** species group from Greece. ❽ This African ***Suillia vockerothi*** (named after Canadian dipterist Richard Vockeroth) shows the characteristic oblique orbital plates (the bristle-bearing patches just inside the eyes) of the Suilliinae, but it lacks the strong costal bristles (along the wing margin) found in most Heleomyzidae. ❾ ***Suillia plumata*** is one of the most common North American heleomyzids.

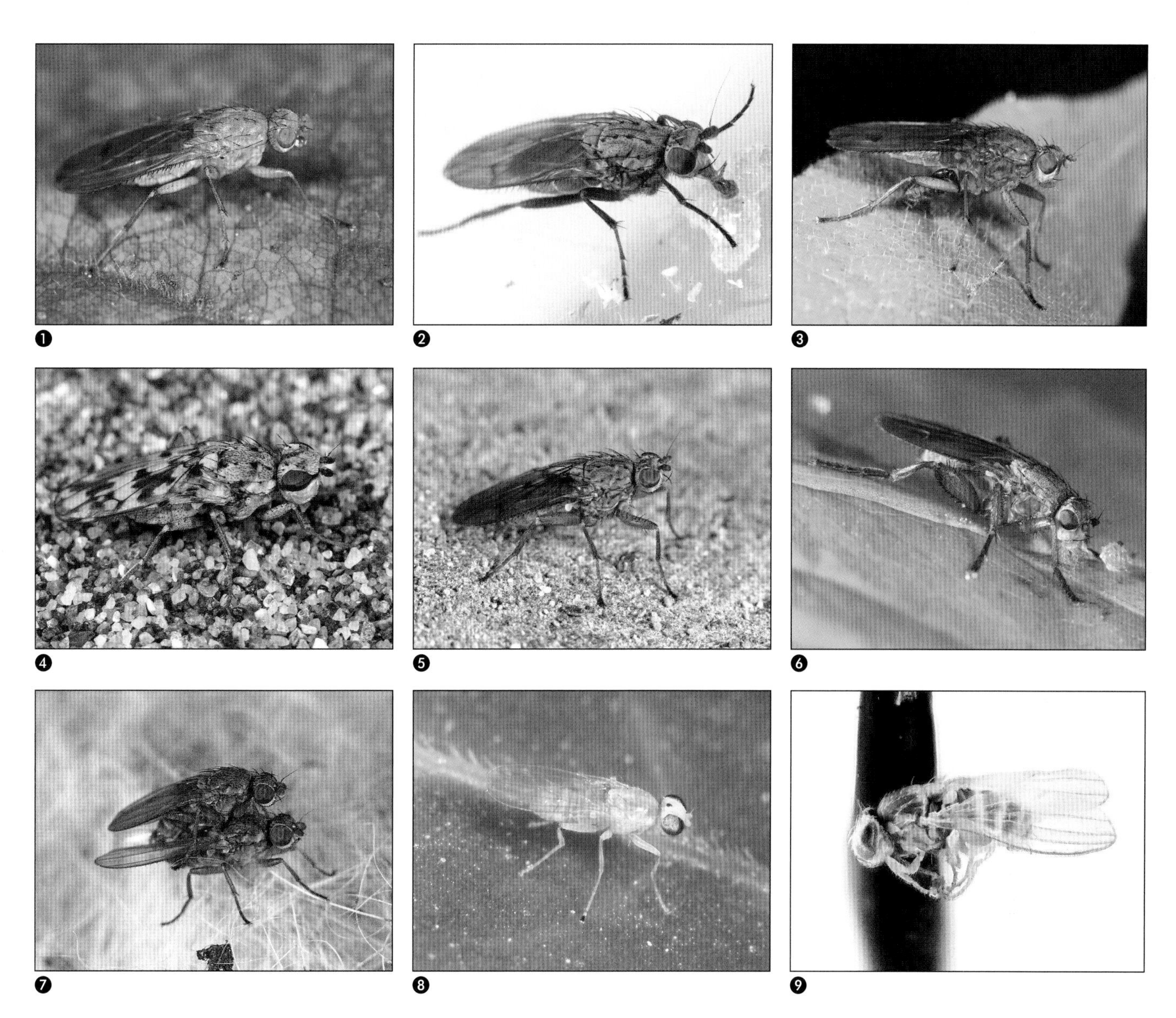

FAMILY HELEOMYZIDAE (continued) Subfamily Heleomyzinae ❶ This eastern North American ***Scoliocentra helvola*** (one of about a dozen Nearctic species until recently treated in the genus *Amoebaleria*) is bubbling out a sphere of liquid that it will later re-ingest. Many, if not most, acalyptrate flies bubble, probably in most cases to concentrate the fluid contents of the crop. ❷ A wide variety of acalyptrates, such as this ***Scoliocentra sackeni***, can be attracted to honey baits. ❸ Heleomyzids, such as this North American species in the Holarctic genus ***Scoliocentra***, often carry pseudoscorpion hitchhikers like this one. Pseudoscorpions are predators of small invertebrates. ❹ ❺ *Anorostoma* is a New World genus of which most species, including this camouflaged seashore species ***Anorostoma maculatum*** and this bicolored ***A. alternans***, occur in western North America. ❻ ***Lutomyia*** is a small North American genus of four species. This eastern species (***L. spurca***), like other *Lutomyia*, usually has its wings mutilated to a knife-like remnant, with most of the hind part broken off. ❼ Members of the Holarctic genus ***Neoleria*** are common carrion visitors, almost invariably found on recently dead small animals. This eastern Canadian fly is probably ***N. lutea***. **FAMILY CHYROMYIDAE (GOLDEN-EYED FLIES) Subfamily Chyromyinae** ❽ This tiny ***Gymnochiromyia nigrimana***, a Nearctic member of a widespread genus, was attracted to honeydew on low vegetation in Arizona. The glittering green eyes and pale color are characteristic of the family. **Subfamily Aphaniosominae** ❾ ***Aphaniosoma*** is a large and widespread genus, but these minute flies are easily overlooked and most have been discovered only recently. This specimen is glued directly to the side of an insect pin, a mounting method preferred by many acalyptrate specialists. Others glue specimens to paper points, and some use tiny "minuten" pins inserted into small foam blocks impaled by a larger main pin.

FAMILY MORMOTOMYIIDAE (HOBGOBLIN FLIES) ❶ This specimen is part of the famous collection of "hobgoblin flies" made by V.G.L. van Someren in 1948. Note the strap-like wings and minute eyes. ❷ ❸ These are the first live images of Hairy Hobgoblin Flies, ***Mormotomyia hirsuta***, taken by Robert Copeland in 2010. ❹ This rock, close to a major highway in Kenya, is the only known locality for the family Mormotomyiidae (photo by Robert Copeland).

SUPERFAMILY TEPHRITOIDEA

FAMILY TEPHRITIDAE (TRUE FRUIT FLIES, ETC.) Subfamily Phytalmiinae ❺ ❻ ❼ ❽ Species in the genus *Phytalmia* are the famous "antler flies," males of which use elaborate cheek processes, or "antlers," to defend territories on fallen trees. Shown here are ***Phytalmia mouldsi***, ***P. cervicornis***, ***P. alcicornis***, and ***P. antilocapra***. With the exception of *P. mouldsi*, which occurs in north Queensland, Australia, all seven species in this spectacular genus are found only on the island of New Guinea. ❾ This ***Rioxa lucifer*** was one of several on the same fallen log in the Philippines. Like most of the subfamily, *Rioxa* is an Oriental-Australasian group. None of the 331 species in this subfamily occur in the New World, and only a few live in the Afrotropical and Palaearctic regions. ❿ ***Rioxoptilona formosana***, shown here in Vietnam, is a pest of bamboo shoots. Maggots bore into the shoots, damaging them and accelerating decay.

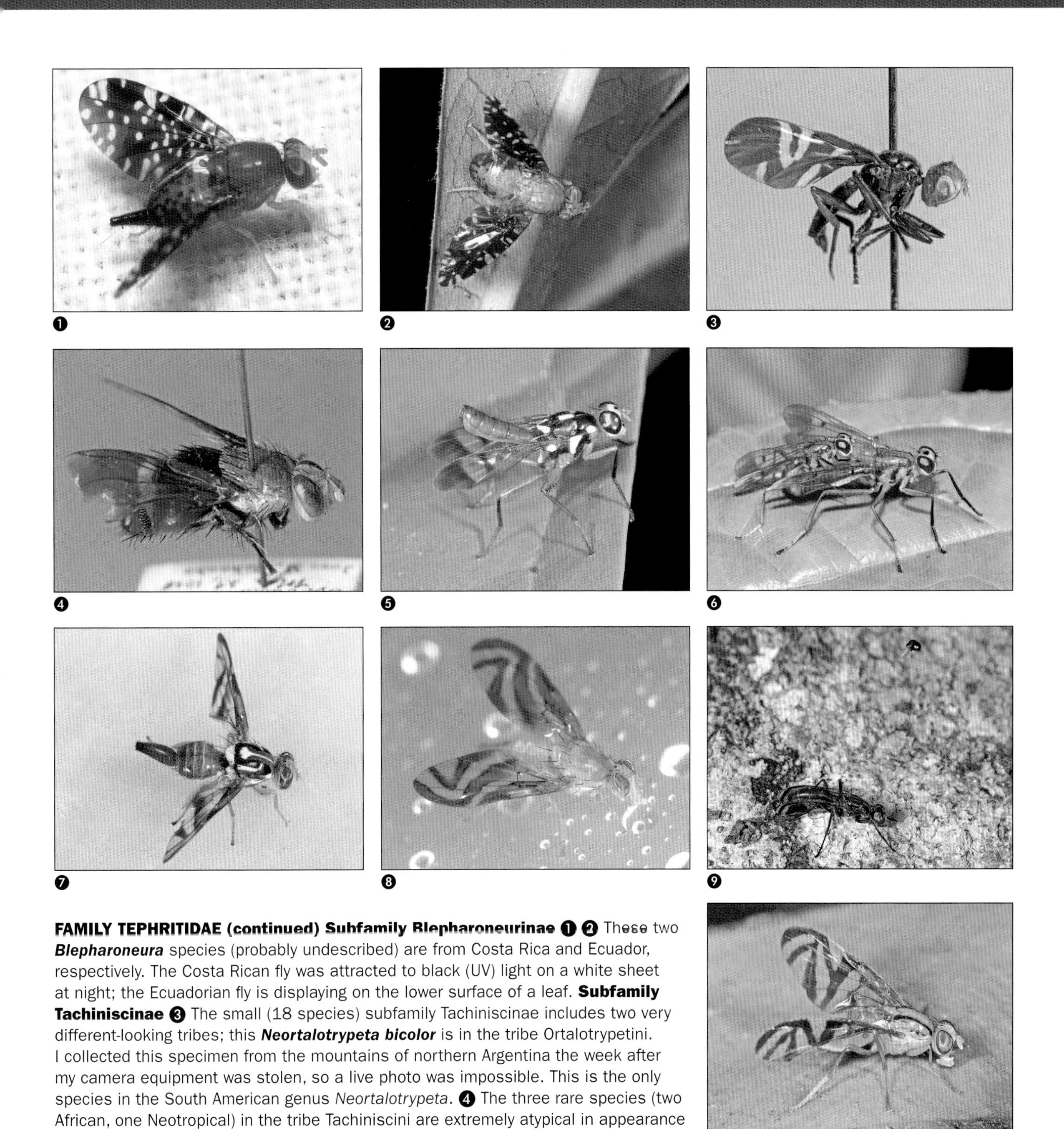

FAMILY TEPHRITIDAE (continued) Subfamily Blepharoneurinae ❶ ❷ These two ***Blepharoneura*** species (probably undescribed) are from Costa Rica and Ecuador, respectively. The Costa Rican fly was attracted to black (UV) light on a white sheet at night; the Ecuadorian fly is displaying on the lower surface of a leaf. **Subfamily Tachiniscinae** ❸ The small (18 species) subfamily Tachiniscinae includes two very different-looking tribes; this ***Neortalotrypeta bicolor*** is in the tribe Ortalotrypetini. I collected this specimen from the mountains of northern Argentina the week after my camera equipment was stolen, so a live photo was impossible. This is the only species in the South American genus *Neortalotrypeta*. ❹ The three rare species (two African, one Neotropical) in the tribe Tachiniscini are extremely atypical in appearance and look superficially more like Tachinidae than Tephritidae. This is a specimen of the Neotropical ***Tachinisca cyaneiventris*** in the U.S. National Museum of Natural History. **Subfamily Trypetinae** ❺ ***Adrama*** is an Oriental-Australasian genus, and this species (***A. rufiventris***) ranges fom Thailand to New Guinea. ❻ These mating flies from northern Vietnam were captured to become the holotype and paratype of a new genus and species, recently published as ***Sapadrama citrina***. ❼ ❽ ***Anastrepha*** is a huge genus in the Neotropics, with over 250 species, including over a dozen serious pests. Shown here are a brightly banded female **Guava Fruit Fly** (***A. striata***) in Costa Rica and a paler Ecuadorian male *Anastrepha*. ❾ ***Euphranta macularis*** is one of many Southeast Asian fruit flies that develop in shoots of the ubiquitous bamboo trees of the region. This one is resting on a tree trunk in Vietnam; note the tiny acalyptrate fly (probably Celyphidae) hovering just above it. ❿ This Bolivian fly is one of 25 species in the Neotropical genus ***Pseudophorellia***, all known only from adults.

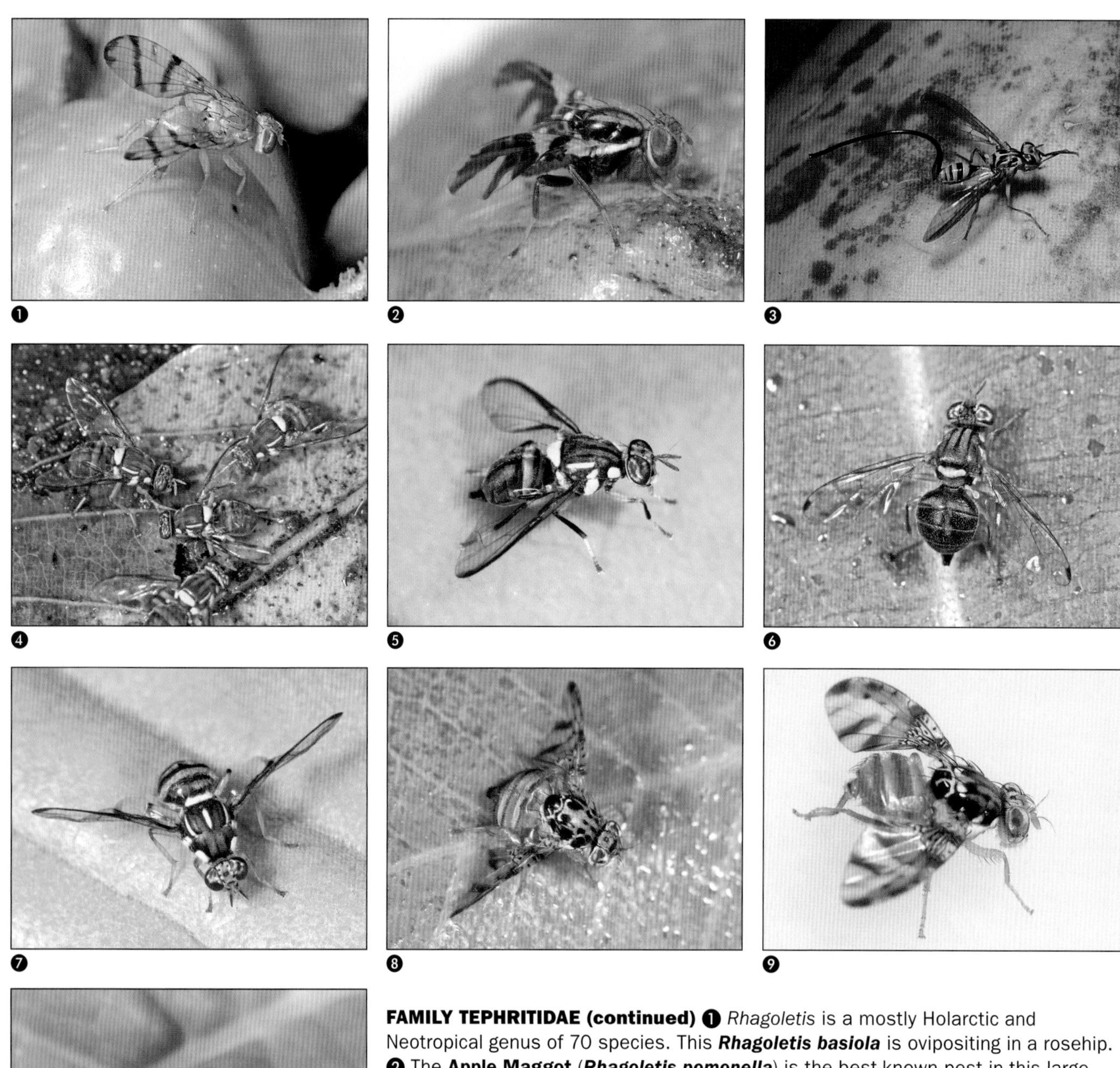

FAMILY TEPHRITIDAE (continued) ❶ *Rhagoletis* is a mostly Holarctic and Neotropical genus of 70 species. This ***Rhagoletis basiola*** is ovipositing in a rosehip. ❷ The **Apple Maggot** (***Rhagoletis pomonella***) is the best-known pest in this large genus of fruit-infesting flies. ❸ The **Papaya Fruit Fly** (***Toxotrypana curvicauda***) is a striking pest of papaya in the New World. Other members of this small genus are all Neotropical. **Subfamily Dacinae** ❹ ***Bactrocera (Bactrocera) dorsalis***, the **Oriental Fruit Fly**, can develop in many kinds of fruits. The flies shown here are aggregating on fallen vegetation in Vietnam, which is within the native range of the species, but Oriental Fruit Flies are serious introduced pests in Hawaii, Tahiti and elsewhere. ❺ ***Bactrocera (Bactrocera) occipitalis***, here seen on a fallen mango in the Philippines, is a member of the *B. dorsalis* complex and a pest of mango and guava. ❻ The **Olive Fruit Fly**, ***Bactrocera (Daculus) oleae***, is a native pest of olives in the Old World but only appeared in the New World for the first time in 1998. It now infests olive fruits throughout California. ❼ ***Bactrocera (Zeugodacus) scutellata***, the **Striped Fruit Fly**, is a widespread fruit fly throughout Southeast Asia, where it develops in the fruits, buds and stems of Cucurbitaceae (including squash, pumpkin and related plants) and Solanaceae (including eggplant). ❽ The **Medfly** (***Ceratitis capitata***) is a major (or potential) fruit pest throughout the warmer parts of the world. ❾ Most of the 78 or so species of ***Ceratitis*** are restricted to Africa, where some, like this one, have not even been formally described and named. ❿ This ***Gastrozona proterva*** is "paddling" its wings in typical tephritid fashion, as you can see by the twisted wing bases. *Gastrozona* is a large Oriental-Palaearctic genus that includes many species, such as this one from Vietnam, that breed in bamboo shoots.

FAMILY TEPHRITIDAE (continued) Subfamily Tephritinae ❶ The cosmopolitan genus ***Campiglossa*** includes some 180 species, of which most develop in the flowerheads of Asteraceae. This is the widespread Nearctic species ***C. albiceps***. ❷ ***Terellia ruficauda*** is an originally European species; larvae develop in the heads of thistles. ❸ Tephritids have a distinctive right-angled bend at the end of the subcostal vein, nicely visible in the wing of this Peruvian ***Epochrinopsis***. *Epochrinopsis* is a little-known genus of two species described from Bolivia. ❹ The New World genus ***Euaresta*** ranges from Canada to Argentina, and this species (***E. bellula***) ranges from Canada to Mexico. Larvae develop in ragweeds and related plants. ❺ Flies in the small New World genus ***Euarestoides*** are typical tephritines and develop in flowers of Asteraceae. This is the widespread North American ***E. acutangulus***. ❻ ❼ ❽ ❾ The familiar ball galls on goldenrod (*Solidago* spp.) stems are induced by larvae of the North American **Goldenrod Gall Fly** (***Eurosta solidaginis***), which develops in the gall, then tunnels to near its surface before pupating in early spring. Seen here are a gall, a puparium in a cut-open gall, an adult emerging from the gall with its ptilinum still fully inflated, and an adult female. ❿ ***Eutreta novaeboracensis***, like *Eurosta solidaginis*, is a common northeastern North American goldenrod (*Solidago*) specialist. Unlike the gall-forming *E. solidaginis*, however, this species develops inconspicuously in the host plant's rhizomes.

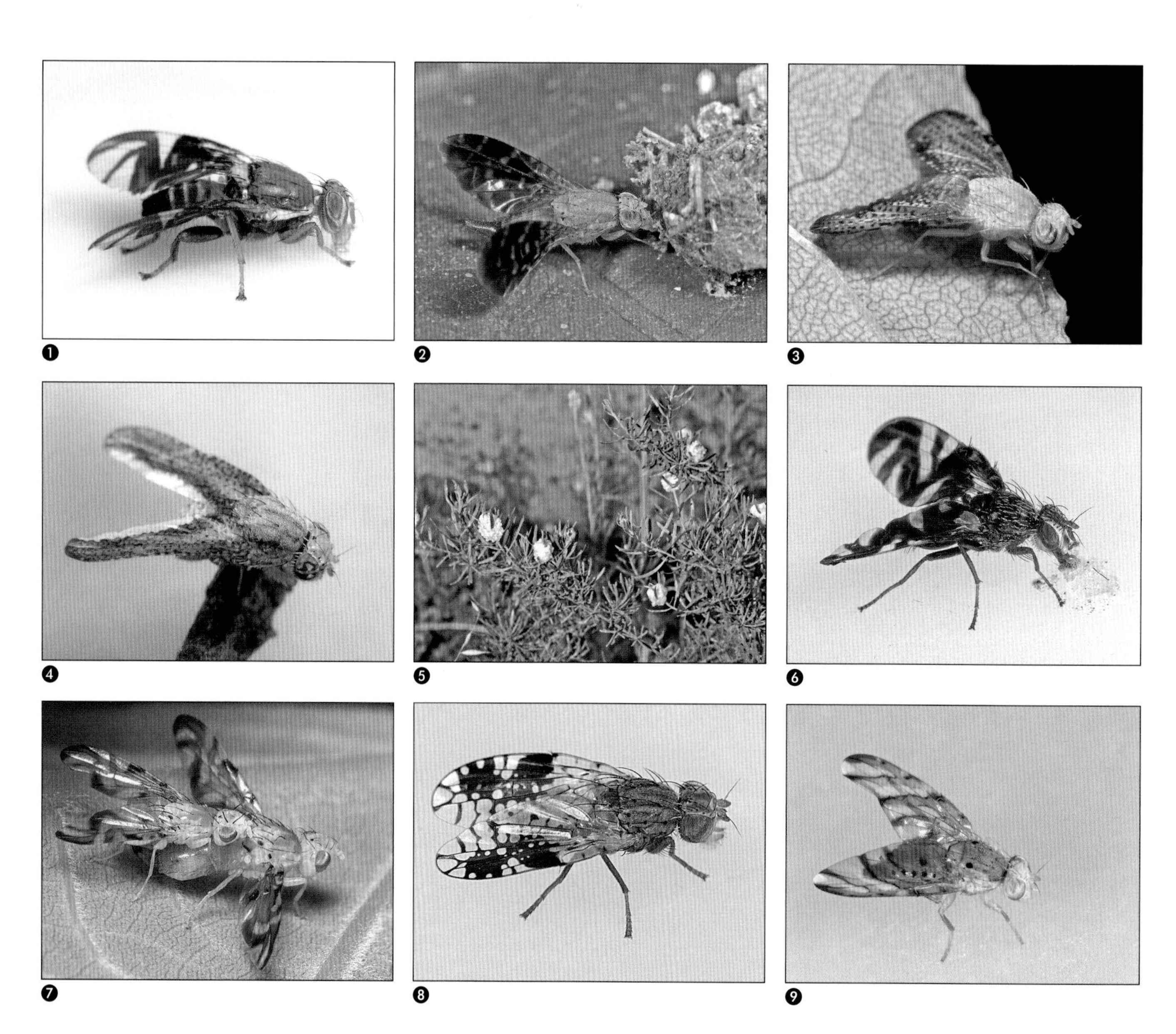

FAMILY TEPHRITIDAE (continued) ❶ ***Hexachaeta*** is a Neotropical group of 31 species sometimes associated with fruits of Moraceae (mulberries). ❷ ***Hexacinia pellucens*** is known only from the Philippines, but other members of the genus occur in the Oriental, Pacific and southern Palaearctic regions. ❸ ***Icterica seriata*** is an abundant North American species wherever its host (*Bidens*) occurs. ❹ ❺ The southern South American tephritid ***Rachiptera limbata*** is responsible for conspicuous white galls on the common shrub *Baccharis linearis*. All 11 *Rachiptera* species are southern South American; two are restricted to Chile. ❻ ***Stenopa vulnerata*** is a widespread North American species that makes stem galls on ragwort (*Senecio aureus*). The only other *Stenopa*, *S. affinis*, occurs in the western United States. ❼ ***Strauzia longipennis*** (**Sunflower Maggot**) is a common garden fly that develops in sunflowers and related plants. Most of the 18 or so species in this New World genus were described relatively recently, between 1985 and 1990. ❽ The type genus of the Tephritinae is the genus ***Tephritis***, a mainly Holarctic group of some 150 species of "typical" tephritines assocated with composites, usually developing in the flowers. This is the widespread Nearctic species ***T. pura***. ❾ This is a southwestern American species in the large (60-plus species) New World genus ***Tomoplagia***, a group usually associated with composite flowers (Asteraceae).

FAMILY TEPHRITIDAE (continued) ❶ ***Trupanea*** is a huge (220 species and counting), cosmopolitan genus of typical tephritines that usually develop in composite flowerheads. ❷ ❸ The **Thistle Gall Fly** (***Urophora cardui***) was introduced from Europe to North America for biological control of invasive thistles; it is now a common species in parts of Canada and the United States. Other species in the large, widespread genus *Urophora* have been moved around for similar purposes; *U. quadrifasciata*, for example, was introduced in North America to help control knapweed. ❹ Most ***Ensina*** species occur in the Old World tropics but some occur in continental South America. This one is on a windswept flower high in the Peruvian Andes. ❺ ***Xanthaciura tetraspina*** ranges from Canada south to Brazil; the other 16 named species in the genus occur throughout the New World. **FAMILY PLATYSTOMATIDAE (SIGNAL FLIES) Subfamily Platystomatinae** ❻ Males and females of some species in the large Oriental-Australasian genus ***Achias*** are strikingly dimorphic: the males have tremendously widened heads with eyes at the ends of long stalks. These specimens of the New Guinea species ***A. rothschildi*** are in the Australian Museum. In contrast to the similarly stalk-eyed Diopsidae, the antennae of stalk-eyed Platystomatidae remain in the middle of the head. ❼ The large, conspicuous adults of the eastern African genus ***Bromophila*** are sluggish flies that react to disturbance by regurgitating a yellow liquid that probably contains defensive compounds sequestered by the root-feeding larvae. Although ***B. caffra*** is the only named species of *Bromophila*, according to D.K. McAlpine the genus includes a number of undescribed species. ❽ This is an Australian species of the mostly Oriental and Afrotropical genus ***Elassogaster***. Larvae of this poorly-known genus are reported both as predators of locust egg pods and scavengers in rotting vegetables. ❾ This Tanzanian fly is in the genus ***Engistoneura***, a small Afrotropical group.

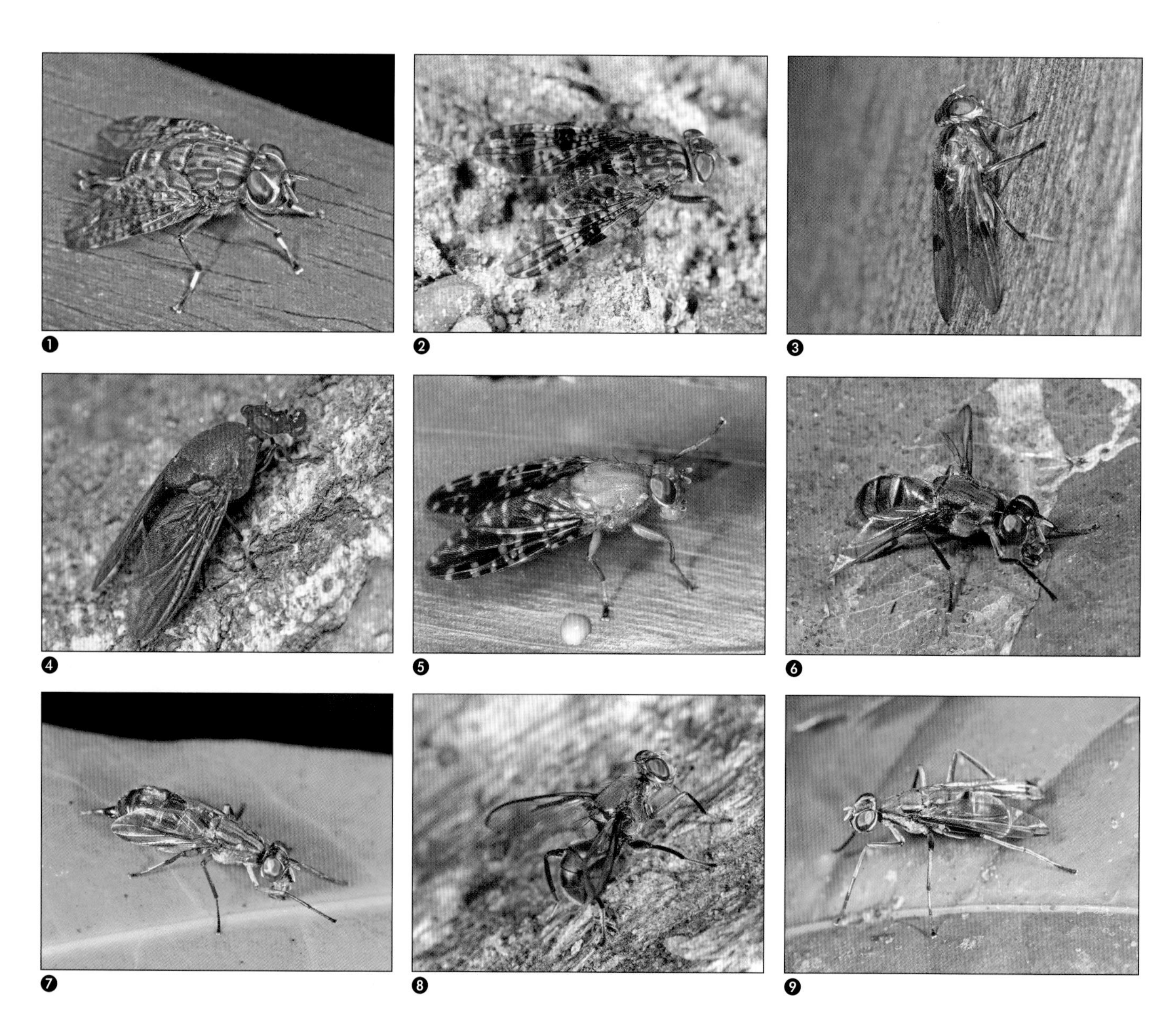

FAMILY PLATYSTOMATIDAE (continued) ❶ ❷ ***Euprosopia***, a mostly Australasian genus with some Oriental and a few Palaearctic species, includes 122 Australasian species and is the most common platystomatid genus in Australia. ❸ ***Lamprogaster*** is a mostly Australasian genus, with an estimated 92 species in the region, although most have not yet been formally described. This one was spotted on a New South Wales tree trunk late at night. ❹ ❺ ***Peltacanthina*** is a large Afrotropical group; shown here are ***P. fumipennis*** (with uniformly smoky wings) and an unidentified species from Tanzania. ❻ ❼ ***Plagiostenopterina*** occurs throughout the Old World tropics and ranges through Australasia to northern Australia. Shown here is ***P. submetallica***, a widely distributed African species, feeding over a brown leaf mine, and an unidentified Vietnamese species on a green leaf. ❽ ❾ ***Pogonortalis*** is native to the Oriental and Australasian regions, although one species has been introduced to California. Most *Pogonortalis* species are endemic to single islands, including Flores, Lord Howe Island, Norfolk Island, New Caledonia, and Guam. The genus has not been recorded from the Philippines, but the flies shown here, one patrolling a rotting log and the other on a leaf, are from the island of Luzon.

FAMILY PLATYSTOMATIDAE (continued) ❶ ❷ ❸ ***Rivellia*** is the largest and most widely distributed genus of Platystomatidae and can be found in most countries of the world. Shown here are the orange Nearctic species ***R. inaequata***, a similarly colored Philippines species, and a more typically colored black African species. Known larvae of this genus develop in the root nodules of legumes. ❹ Although the Platystomatidae are largely an Old World and Australian family, species of the Neotropical genus ***Senopterina*** occur throughout Central and South America, where adults are often common visitors to dung and carrion. **Subfamily Scholastinae** ❺ Five of the six named species in the Australian genus ***Lenophila*** are associated with the endemic Australian grass trees (*Xanthorrhoea* spp.). Larvae presumably develop in damaged plant tissue or organic material trapped at the base of the plant. Adults of this Western Australian species appear on the thin grass tree leaves at dusk, when males and females engage in an exchange of oral fluids that starts head-to-head and then continues as the male moves into mating position. ❻ ***Scholastes bimaculatus*** (**Coconut Fly**) occurs on several Pacific islands, including the islands of Hawaii and Fiji. ❼ ***Scholastes*** species are found throughout tropical parts of the Australasian and Oriental regions; this one is from the Philippines. ❽ ❾ These two related platystomatids (species of ***Pterogenia***) are both from Vietnam.

FAMILY PLATYSTOMATIDAE (continued) Subfamily Plastotephritinae ❶ ***Oeciotypa*** is a small genus of half a dozen Afrotropical species; this is ***O. hendeli*** from Tanzania. ❷ ❸ ❹ Males of many members of this subfamily have bizarrely shaped heads, as seen in this Tanzanian ***Conopariella tibialis*** with its hyperinflated lower cheek. **FAMILY ULIDIIDAE (PICTURE-WINGED FLIES) Subfamily Otitinae** ❺ ***Callopistromyia strigula***, sometimes called the "peacock fly" for its habit of strutting around with its wings at right angles to its body, is a common and flamboyant associate of fallen trees in eastern North America. The only other species in the genus, *C. annulipes*, has a similar distribution. ❻ The eastern North American genus ***Delphinia*** (with only one species, ***D. picta***) is readily recognizable by its boldly patterned wings and conspicuous habit of twisting and "paddling" its wings. Larvae develop in decaying plant material. ❼ The only species of ***Idana***, the northeastern North American ***I. marginata***, is one of the largest picture-winged flies (about 10 mm). Larvae develop in compost. ❽ ***Otites michiganus*** occurs among the shoreline reeds of the Great Lakes, but other Nearctic species of this Holarctic genus occur in the western part of the continent. ❾ ***Pseudotephritis***, like this eastern North American ***P. approximata***, develop under bark and can be common on fallen trees. *Pseudotephritis* is a mostly Nearctic genus with a couple of species in the eastern Palaearctic region.

FAMILY ULIDIIDAE (continued) ❶ ❷ ❸ Most larvae in the subfamily Otitinae are saprophagous inhabitants of decomposing plant material such as compost, rotting vegetable matter or fermenting tissue under the bark of fallen trees, but the genera *Tetanops* and *Tritoxa* are phytophagous. This dull-colored ***Tetanops integer*** develops in root crops, while the more distinctive ***Tritoxa flexa*** (**Black Onion Fly**) and ***Tritoxa incurva*** are usually associated with wild onions. **Subfamily Ulidiinae** ❹ Most of the 13 species of the New World genus ***Chaetopsis*** are Nearctic, but some species occur as far south as Argentina. One species of these common flies was recently documented as a pest that feeds and develops in corn ears in the southeastern United States, another attacks seedling corn in Central America, and two species have been recorded as developing in the stems of Saltmarsh Cordgrass (*Spartina alterniflora*). ❺ Although there are no native ***Euxesta*** in the Old World, a few species have been introduced to Europe and Asia. One of them is shown here feeding on a bleeding banana stem in the Philippines. ❻ ***Euxesta*** is a diverse group of about 90 New World and Pacific species, including this one from a salt marsh in the southeastern United States. ❼ This male fly from Ecuador with its flashy genal processes is an undescribed species in the poorly known Neotropical genus ***Megalaemyia***. Specialists recognize about four times as many species in this genus as are yet named. ❽ Flies in the entirely Neotropical genus ***Paragorgopis***, like this Ecuadorian ***P. amoena***, range from Mexico south to Brazil. ❾ These four ***Paragorgopis*** are sharing a scat with an undescribed *Parophthalmoptera* species in Ecuador.

FAMILY ULIDIIDAE (continued) ❶ ***Physiphora*** is an almost cosmopolitan genus, including a couple of species that have established themselves almost everywhere there are people. This one is from the Philippines. ❷ ❸ ❹ These spectacularly broad-headed flies are males of ***Plagiocephalus latifrons*** from Ecuador and Bolivia; females (such as this one from Ecuador) have a normal head. There are two other species of *Plagiocephalus*, both Neotropical. ❺ ❻ ❼ ***Pterocalla*** species are very common flies in the neotropics and often seen waving and paddling their wings on wood or foliage. Shown here are a fly from Bolivia and two from Ecuador, one of which is laying eggs in a crack in the bark of a fallen tree. ❽ The five species of ***Pterotaenia*** all occur in southern South America, those with known larvae developing in rotting fruits and vegetables. ❾ This is ***Seioptera vibrans***, a common Holarctic species. Other *Seioptera* species occur in China, Chile and North America. ❿ This unnamed Peruvian ulidiid (close to the genus *Apterocina*) has a distorted head, with hornlike processes on top and cheek processes extending up into the eye.

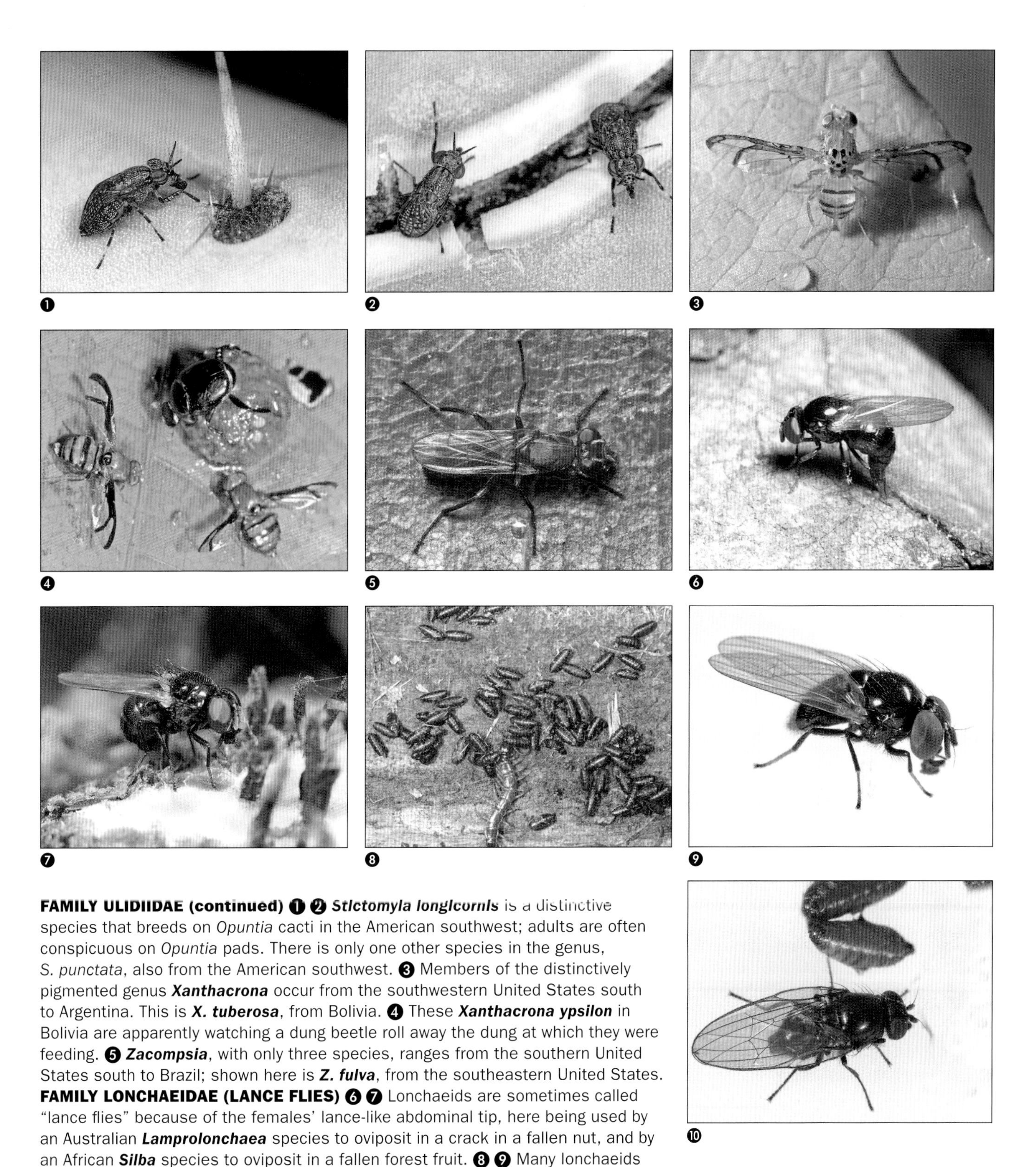

❶ ❷ ❸ ❹ ❺ ❻ ❼ ❽ ❾ ❿

FAMILY ULIDIIDAE (continued) ❶ ❷ ***Stictomyia longicornis*** is a distinctive species that breeds on *Opuntia* cacti in the American southwest; adults are often conspicuous on *Opuntia* pads. There is only one other species in the genus, *S. punctata*, also from the American southwest. ❸ Members of the distinctively pigmented genus ***Xanthacrona*** occur from the southwestern United States south to Argentina. This is ***X. tuberosa***, from Bolivia. ❹ These ***Xanthacrona ypsilon*** in Bolivia are apparently watching a dung beetle roll away the dung at which they were feeding. ❺ ***Zacompsia***, with only three species, ranges from the southern United States south to Brazil; shown here is ***Z. fulva***, from the southeastern United States. **FAMILY LONCHAEIDAE (LANCE FLIES)** ❻ ❼ Lonchaeids are sometimes called "lance flies" because of the females' lance-like abdominal tip, here being used by an Australian ***Lamprolonchaea*** species to oviposit in a crack in a fallen nut, and by an African ***Silba*** species to oviposit in a fallen forest fruit. ❽ ❾ Many lonchaeids develop under the bark of dead trees, and their puparia are often exposed in numbers by peeling bark. Shown here are puparia and a reared adult of a Canadian ***Lonchaea*** species. ❿ Lonchaeids are common flies throughout the world, except in New Zealand. This ***Protearomyia*** was one of many reared from puparia under the bark of a fallen tree in Chile.

FAMILY PYRGOTIDAE (SCARAB-KILLING FLIES) Subfamily Pyrgotinae ❶ ***Boreothrinax*** is a small genus of three Nearctic species. This is ***B. dichaetus***, from New Mexico. ❷ ***Leptopyrgota*** is a rarely encountered group of about 21 described species. Most of the named species are known from only a single specimen each, and several species still await description. ❸ ***Pyrgota undata*** is probably the most common pyrgotid fly in eastern North America, where it develops as a parasitoid of June bugs or May beetles (Scarabaeidae, *Phyllophaga* and related genera). *Pyrgota* includes three Neotropical and two Nearctic species. ❹ ***Pyrgotella chagnoni*** is a common eastern North American fly that is a parasitoid of chafers (Scarabaeidae, *Dichelonyx*); female flies often pursue hosts while they are on the wing at dusk. This is the only *Pyrgotella* species. ❺ This new species from Costa Rica is recognized as an undescribed genus, that is, one that has not yet been formally described or named. **Subfamily Teretrurinae** ❻ This Chilean ***Teretrura flaveola***, like most pyrgotids, is probably a parasitoid of melolonthine scarabs. *Teretrura* species, unlike most other Pyrgotidae, have ocelli. **FAMILY RICHARDIIDAE Subfamily Richardiinae** ❼ ❽ Species of the Neotropical genus ***Beebeomyia*** have been reared from the water-holding bracts and flowers of *Heliconia* and similar plants. This female, sliding the tip of her ovipositor between layers of plant tissue, is one of several undescribed species from Costa Rica. ❾ This ***Coilometopia trimaculata*** is taking a characteristic adult feeding position with its abdomen pointed up, perhaps mimicking the head of a potentially threatening insect or spider. There are six species in this Neotropical genus. The other fly in this photograph is a micropezid (*Poecilotylus*).

FAMILY RICHARDIIDAE (continued) ❶ ❷ ***Hemixantha*** is a large, diverse genus from Central and South America with about 20 described species. Shown here are species from Bolivia and Ecuador. ❸ Species of ***Melanoloma*** are widespread throughout the neotropics, where they are common visitors to dung. This Ecuadorian species is feeding on a leaf spattered with monkey dung. ❹ Extrafloral nectaries, like this pool of sweet exudate on a Bolivian *Inga* leaf, supposedly serve to attract ants that in turn protect the plant from foliage-eating pests. But many flies, such as this ***Melanoloma***, are also attracted. ❺ ***Odontomera*** is one of the few richardiid genera that range north to the Nearctic region, but most of the 18 named species in the genus are South American, like the Ecuadorian fly shown here. ❻ ***Richardia***, the largest genus of Richardiidae, includes 29 named species in Central and South America, all characterized by enlarged hind femora. Shown here are a male and a female feeding on a bird dropping in Costa Rica. ❼ ❽ ❾ ***Richardia*** males sometimes have broadened heads or elaborate genal processes, probably used in male–male battles for mating territories.

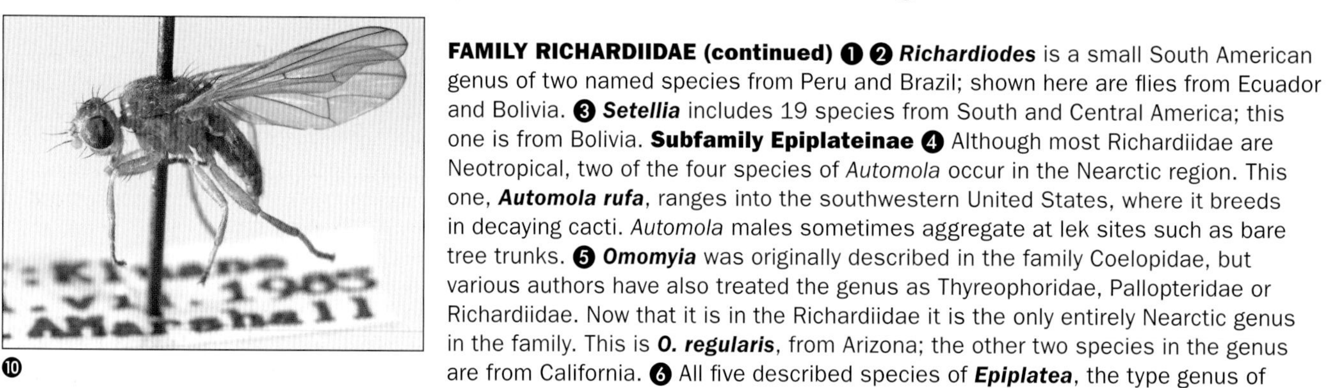

FAMILY RICHARDIIDAE (continued) ❶ ❷ ***Richardiodes*** is a small South American genus of two named species from Peru and Brazil; shown here are flies from Ecuador and Bolivia. ❸ ***Setellia*** includes 19 species from South and Central America; this one is from Bolivia. **Subfamily Epiplateinae** ❹ Although most Richardiidae are Neotropical, two of the four species of *Automola* occur in the Nearctic region. This one, ***Automola rufa***, ranges into the southwestern United States, where it breeds in decaying cacti. *Automola* males sometimes aggregate at lek sites such as bare tree trunks. ❺ ***Omomyia*** was originally described in the family Coelopidae, but various authors have also treated the genus as Thyreophoridae, Pallopteridae or Richardiidae. Now that it is in the Richardiidae it is the only entirely Nearctic genus in the family. This is ***O. regularis***, from Arizona; the other two species in the genus are from California. ❻ All five described species of ***Epiplatea***, the type genus of the subfamily, are Neotropical. This one is from Ecuador. **FAMILY PALLOPTERIDAE (FLUTTER FLIES) Subfamily Pallopterinae** ❼ Larvae of this European species, ***Toxonevra muliebris***, live under the bark of trees, where they prey on other insects, especially larval gall midges. *Toxonevra* is a Holarctic-Oriental genus frequently misspelled as *Toxoneura*. ❽ ❾ This is the Nearctic ***Toxonevra superba***, seen from the front in a characteristic wing display and from the side with a cluster of mites around the tip of its abdomen. **FAMILY PIOPHILIDAE (SKIPPER FLIES) Subfamily Neottiophilinae** ❿ ***Actenoptera*** is a small northern Holarctic genus in the small subfamily Neottiophilinae. This ***A. hilarella*** is from Canada's Yukon Territory.

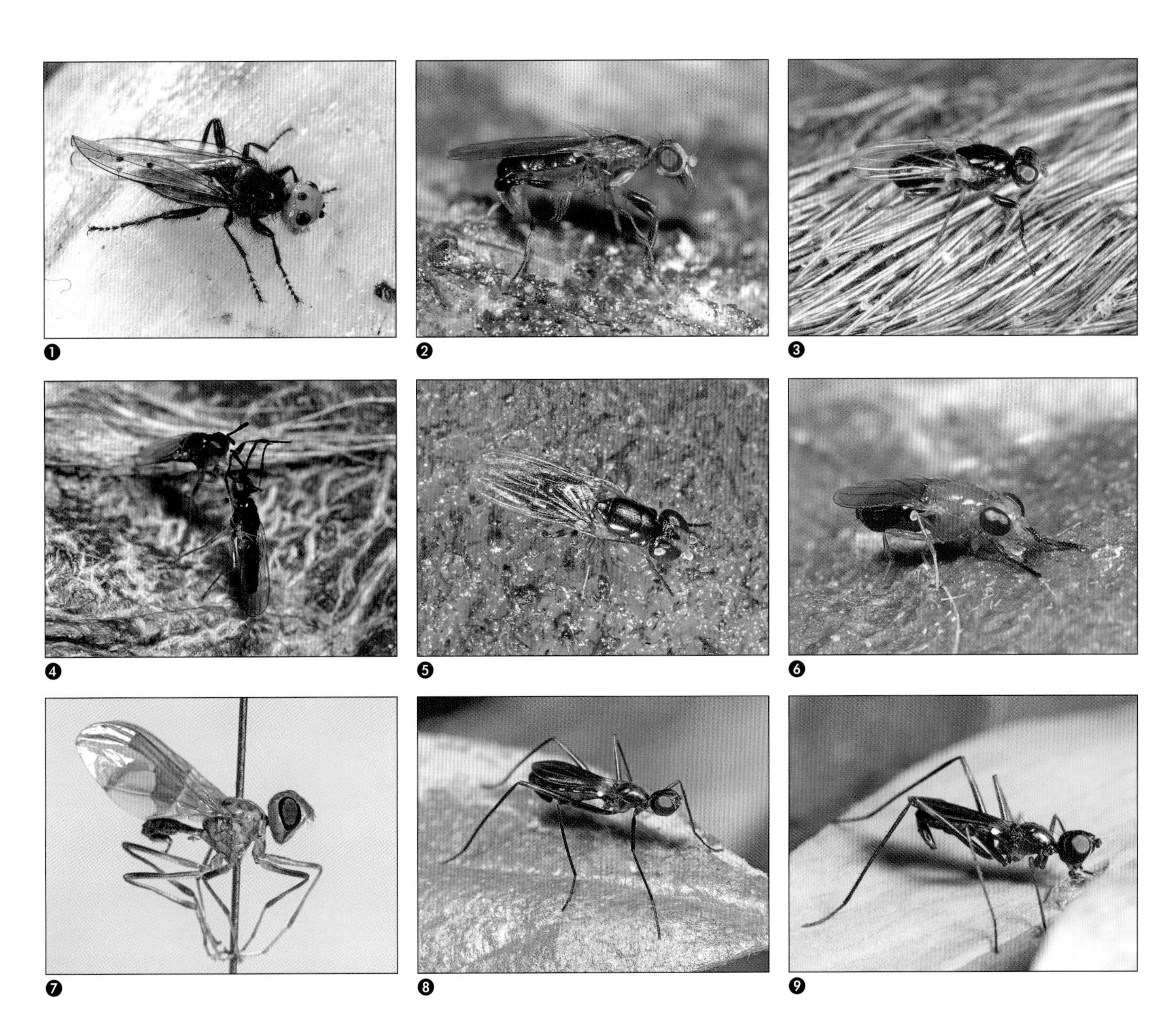

FAMILY PIOPHILIDAE (continued) ❶ The two species of bone skippers in the genus ***Thyreophora*** are rarely seen winter-active Palaearctic flies associated with large carcasses. This is ***T. cynophila***, widely considered extinct until it was "rediscovered" in 2010 (photo by P. Rodríguez). **Subfamily Piophilinae** ❷ The northern Holarctic genus ***Amphipogon*** ("two-bearded skipper flies") includes a northwestern Nearctic species and a Palaearctic species. This is ***A. spectrum***, photographed by J. Sevcik in the Czech Republic. ❸ The **Cheese Skipper**, ***Piophila casei***, is a common cosmopolitan species on carcasses of all sorts. Other species of *Piophila* are mostly Old World and Australian. ❹ Males of ***Prochyliza xanthostoma*** have the front of the head extended and the antennae elongated to facilitate male–male battles for mating territories, such as the bit of dead cat serving as an arena for the battle shown here (similar height-measuring competitions are seen in some Neriidae). The other seven *Prochyliza* species are Holarctic and Neotropical. ❺ ❻ These two species of *Protopiophila* were found on the same dead sloth in Bolivia; if they are described species, the black one is ***Protopiophila atrichosa*** and the orange one is ***Protopiophila nigriventris***. There are only three named South American species in this cosmopolitan genus. Although the Neotropical *Protopiophila* species are poorly known, the Nearctic species *P. litigata* breeds in discarded antlers. **FAMILY CTENOSTYLIDAE** ❼ This male ***Lochmostylia borgmeieri*** was collected at lights during a Costa Rican night.

SUPERFAMILY NERIOIDEA

FAMILY MICROPEZIDAE (STILT-LEGGED FLIES) Subfamily Micropezinae ❽ ❾ ***Cryogonus formicarius***, one of three named species in this southern South American genus, is an ant mimic known only from Chile.

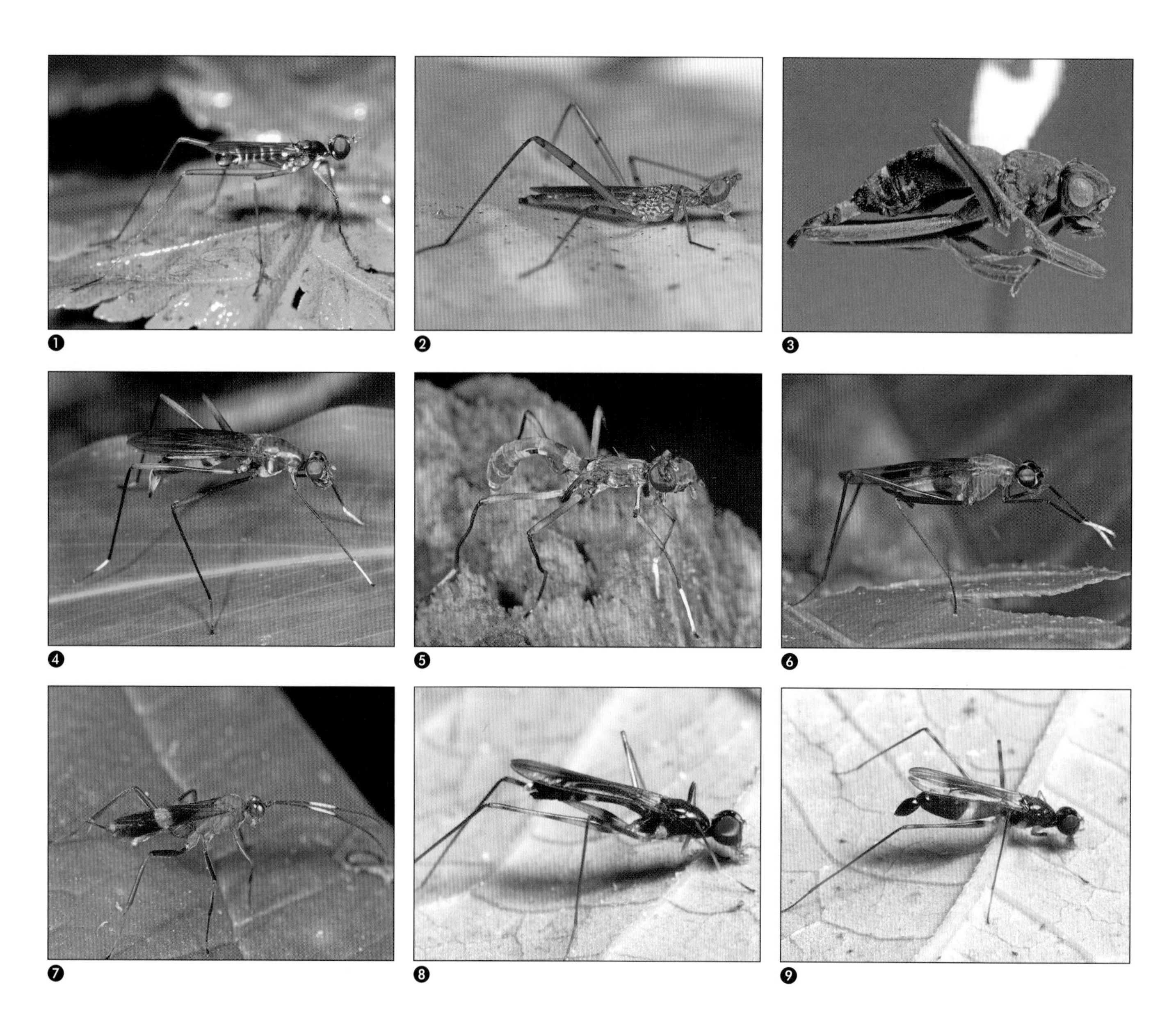

FAMILY MICROPEZIDAE (continued) ❶ ❷ The large, worldwide genus *Micropeza* is sometimes divided into two subgenera. This male from Ecuador is in the subgenus ***Micropeza***; the relatively long-headed female here feeding on honeydew on a Costa Rican leaf is in the subgenus ***Neriocephalus***. **Subfamily Calycopteryginae ❸** ***Calycopteryx mosleyi***, the sole member of the Calycopteryginae, is one of many flightless fly species found on the subantarctic islands of Kerguelen and Heard. **Subfamily Taeniapterinae ❹ ❺** ***Aristobatina*** is a small east African genus of two species. This mature female ***A. rufa*** from Tanzania probably emerged from some decaying wood, like this immature female with her wings not yet inflated and some larval habitat still stuck to her head. **❻ ❼** This orange ***Calosphen*** (an undescribed species) was photographed in Ecuador, meters away from this similar ichneumonid wasp. Most members of the subfamily Taeniapterinae have partially white forelegs that they wave in front of the head like wasp antennae, and many occur with similarly colored wasp models. **❽** When this Ecuadorian micropezid was photographed, it was an undescribed genus and species, but it was later pinned and formally described as the type specimen of the genus *Globopeza* and of the species ***Globopeza ecuadoriensis***. **❾** This ***Globopeza ecuadoriensis*** female was used in the original description but was not designated as the type; it therefore became a paratype. (The holotype male is shown in the previous image.)

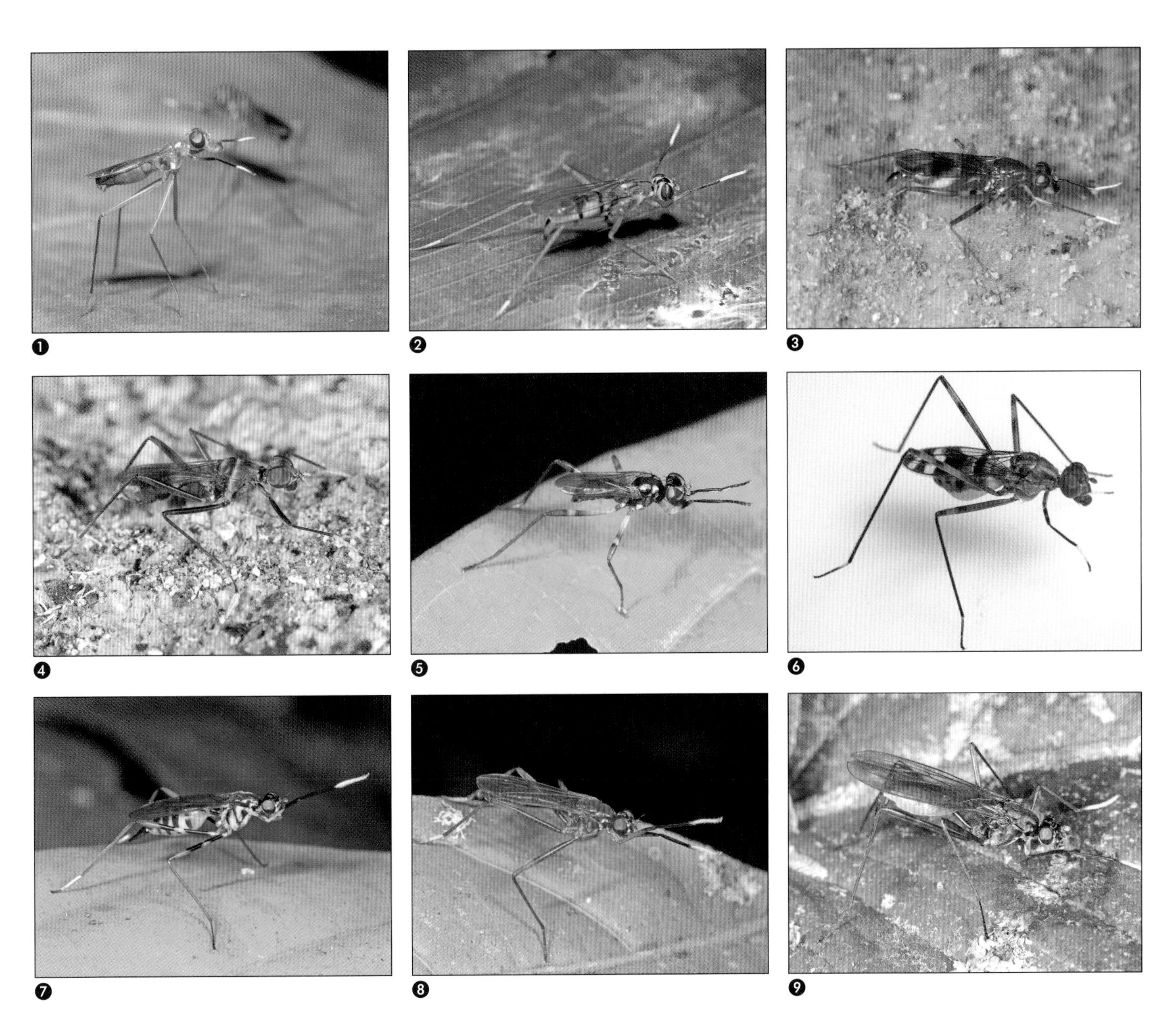

FAMILY MICROPEZIDAE (continued) ❶ ❷ ***Grallipeza*** is a common Neotropical genus with about 35 named species and many times as many still awaiting description. Most are orange flies like the Cuban male ***G. placida*** displaying here, but the genus also includes black species, as well as garishly colored species such as this undescribed female from Costa Rica. ❸ Although ***Grallipeza*** adults are easily collected at baits, almost nothing is known about their larval habitats. This recent photograph of a ***G. gracilis*** female ovipositing in a freshly fallen but still living tree in Ecuador is a first indication that micropezids can develop in fresh or newly dead wood. The object projecting from her left side is a phoretic pseudoscorpion; pseudoscorpions are common on other micropezid genera (especially *Scipopus*) associated with older dead wood. ❹ This male ***Grammicomyia*** from the Philippines is holding a patch of habitat on a fallen log, vigorously defending it from a stream of speedy incoming males while awaiting the arrival of a female. Such behavior is unknown among other micropezids, most of which are slow and stately in their movements. ❺ The small Neotropical genus ***Hemichaeta*** is sometimes treated as a subgenus of the large genus *Poecilotylus*, but these distinctive squat little flies are not closely related to *Poecilotylus*. This is an undescribed Central American species. ❻ ***Hoplocheiloma*** is a Caribbean genus with one species that ranges to Florida; several species are found only on single islands. This ***H. maculosum*** was photographed in a Cuban botanical garden, but the species also occurs on Hispaniola. ❼ ❽ ***Mesoconius*** is a marvelous genus of large (often 15–20 mm), colorful flies characteristic of cool mountain habitats in the neotropics, where many occur only in small, isolated patches of high-altitude cloud forest. Shown here are an undescribed species and ***M. ujhelianus***, both at Monteverde Biological Station in Costa Rica. ❾ This undescribed species of ***Mesoconius*** from the high Andes of Ecuador is one of many new species of this distinctive genus recognized in an unpublished revision.

FAMILY MICROPEZIDAE (continued) ❶ ***Mimegralla*** species, such as these members of the ***M. albimana*** complex from the Phillipines, are common flies throughout the Old World tropics, with about 50 species divided among the Oriental, Pacific and Afrotropical regions. Larvae of this group develop in rhizomes and can be significant pests of turmeric and ginger. ❷ ***Plocoscelus*** is a common Neotropical genus of about 17 ant-mimicking species, some of which are familiar insects of forest edges and other disturbed areas. Several species form conspicuous mating aggregations on low foliage, where mating pairs display complex rituals of dancing, stroking and kissing. ❸ Courtship rituals that include an exchange of oral fluid ("kissing") occur in many Taeniapterinae. This pair of Ecuadorian ***Poecilotylus*** was alternating brief smooches with stilting and stroking. *Poecilotylus* is an extremely diverse Neotropical genus, with 15 named species and at least as many awaiting description. ❹ ***Pseudoeurybata*** is a small Central American genus of large, long-legged flies closely related to the large Neotropical genus *Scipopus*. This Costa Rican female is laying eggs in a beetle burrow in a mossy log. ❺ Females of ***Ptilosphen tetrastigma*** have extensive white banding on the abdomen (unlike the males) and display to males using an odd position of the front legs, as demonstrated by this Costa Rican female. ❻ ***Rainieria*** is a mostly Neotropical genus with a couple of species in the Holarctic region and one mysterious species found in the Bonin Islands, south of Japan. This is an undescribed Ecuadorian species. ❼ The large, black-bodied, red-headed species of ***Scipopus*** are among the most easily recognized of all Neotropical flies. Many of the 22 named species are common insects in rainforests of both Central and South America. This Ecuadorian female is ovipositing in a moss-covered log. ❽ ❾ ***Scipopus*** species oviposit in beetle borings in older fallen trees. Some have the remarkable habit of "peering" down the boring first, probably using their antennae to assess the current occupancy of the burrow.

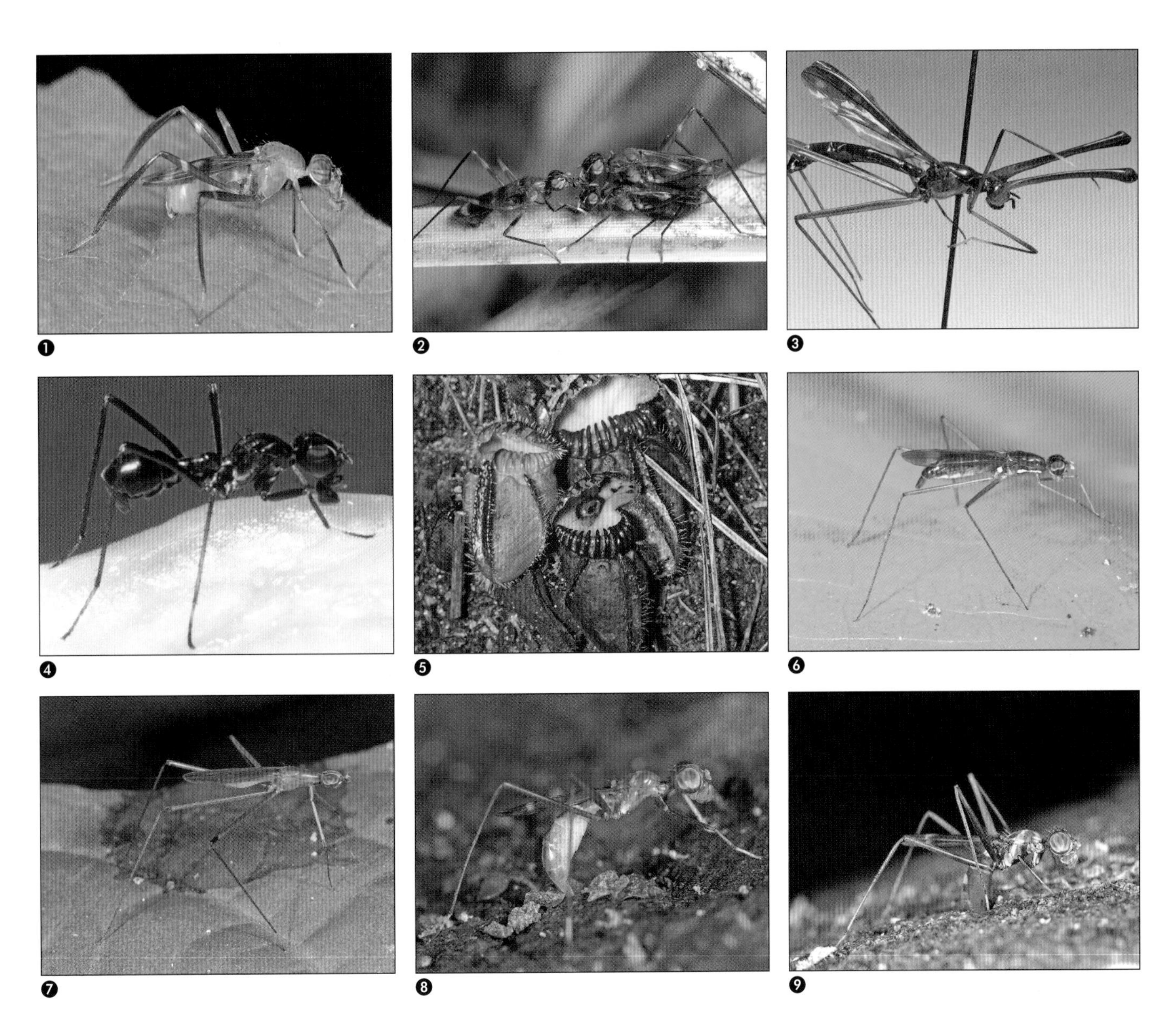

FAMILY MICROPEZIDAE (continued) ❶ ***Paragrallomyia*** (until recently treated as part of *Taeniaptera*) is a large Neotropical group with dozens of species. Like the closely related *Taeniaptera*, members of this genus have a broad and characteristically shaped palpus, clearly visible in this Costa Rican male ***P. platycnema***. ❷ Oral exchange between male and female *Taeniaptera* is common, but these Canadian males seem to have locked lips while one is still mating with the female below. ***Taeniaptera trivittata*** is the only species of this small, mostly Neotropical, genus to range north to Canada. **Subfamily Eurybatinae** ❸ The only known ***Anaeropsis***, this Indonesian ***A. guttipennis***, is also the only known stalk-eyed micropezid — and one of the strangest-looking flies in the world. ❹ ❺ The rare genus ***Badisis*** includes only one species, the entirely wingless (even lacking halters), remarkably ant-like ***B. ambulans*** from southwestern Australia. Larvae develop only inside the pitchers of the rare Albany Pitcher Plant (*Cephalotus follicularis*). This adult female was reared from a larva found in the gunk at the bottom of one of these plants. ❻ Most of the dozen or so named species of ***Cothornobata***, such as this darkly striped ***C. aczeli*** from New South Wales, occur in Australia, but others occur throughout Southeast Asia. ❼ ***Trepidarioides territus***, the only member of its genus, is a very rare Southeast Asian species. This is a male from the Philippines. ❽ This ***Crepidochetus argentifasciata*** is laying eggs in an old log riddled with the burrows of wood-boring beetles. *Crepidochetus* is a small Southeast Asian–Australian genus. ❾ This undescribed species of ***Crepidochetus*** from the Philippines is inserting much of her abdomen into a beetle burrow as she lays eggs in a rotting log.

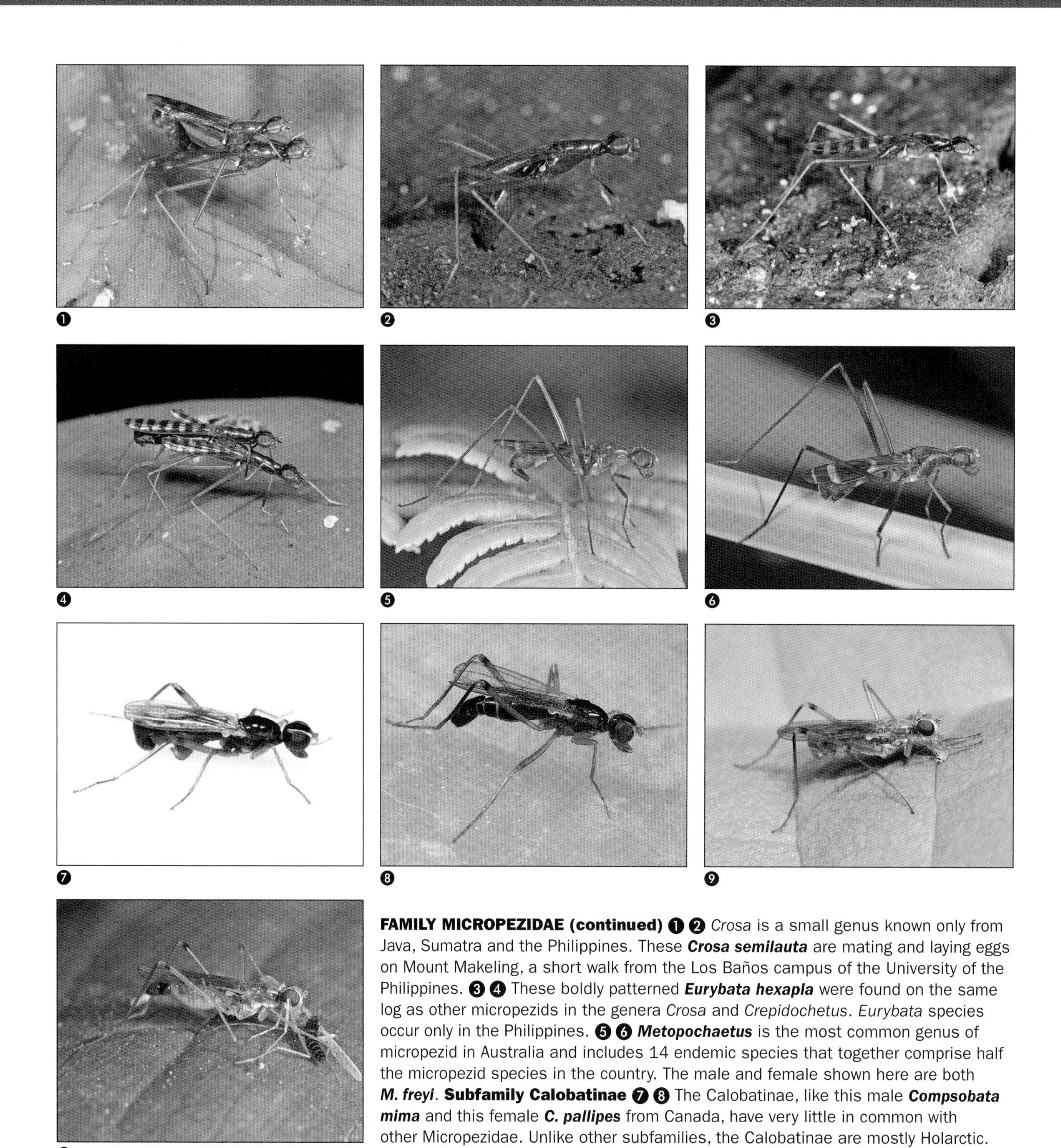

FAMILY MICROPEZIDAE (continued) ❶ ❷ *Crosa* is a small genus known only from Java, Sumatra and the Philippines. These ***Crosa semilauta*** are mating and laying eggs on Mount Makeling, a short walk from the Los Baños campus of the University of the Philippines. ❸ ❹ These boldly patterned ***Eurybata hexapla*** were found on the same log as other micropezids in the genera *Crosa* and *Crepidochetus*. *Eurybata* species occur only in the Philippines. ❺ ❻ ***Metopochaetus*** is the most common genus of micropezid in Australia and includes 14 endemic species that together comprise half the micropezid species in the country. The male and female shown here are both ***M. freyi***. **Subfamily Calobatinae** ❼ ❽ The Calobatinae, like this male ***Compsobata mima*** and this female ***C. pallipes*** from Canada, have very little in common with other Micropezidae. Unlike other subfamilies, the Calobatinae are mostly Holarctic. ❾ ❿ Calobatinae, like these ***Compsobata univittata*** from Canada, are predators as adults and are often seen feeding on small flies (usually nematocerans).

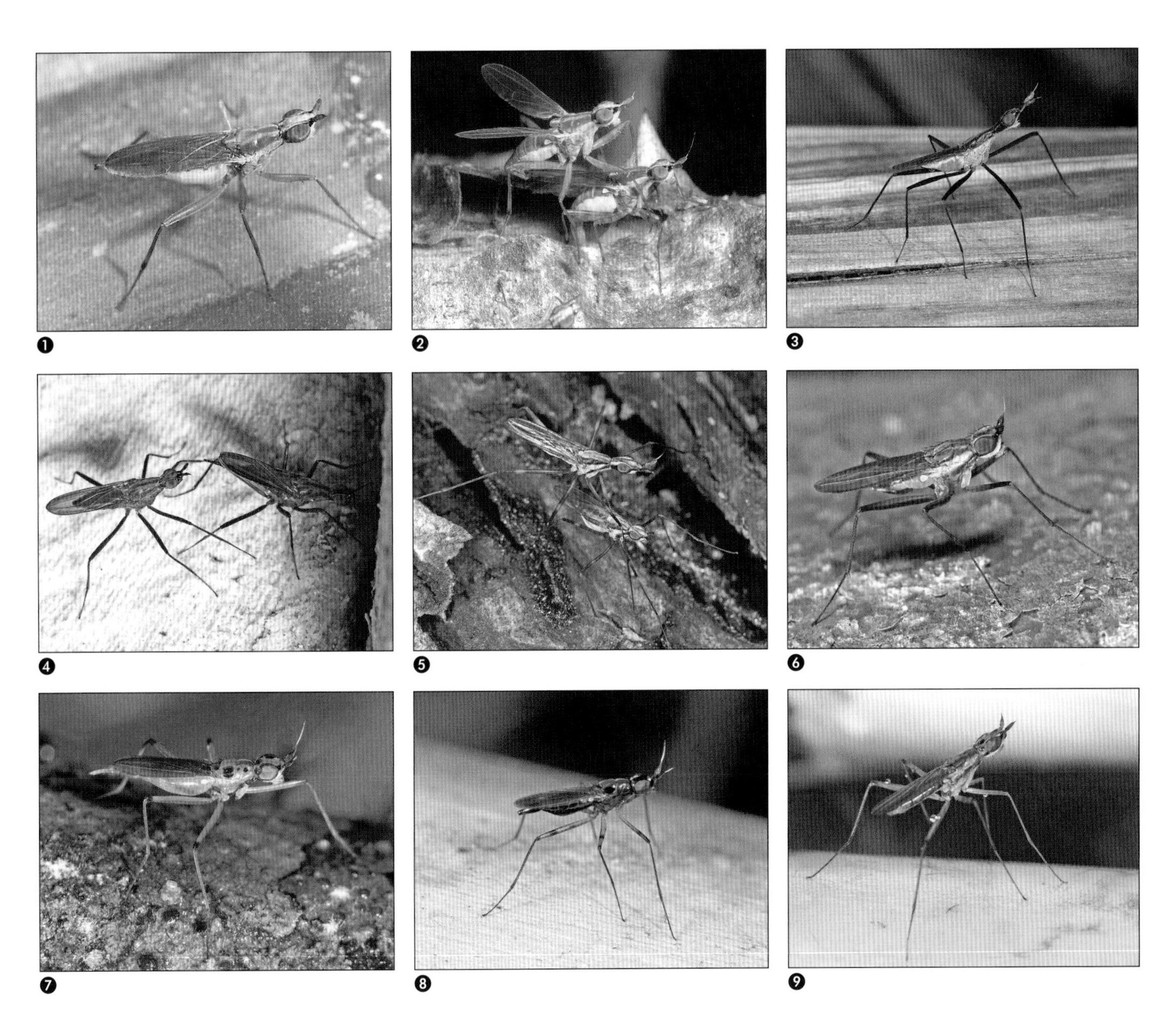

FAMILY NERIIDAE (CACTUS FLIES, ETC.) Subfamily Neriinae ❶ The common Neotropical genus ***Glyphidops***, with two subgenera, develops in decaying plant tissue, usually in living but damaged branches or stems. This is an Ecuadorian species in the larger subgenus ***Oncopsia***, which has seven species. ❷ ***Glyphidops*** is a mostly Neotropical genus of 11 named species, of which one occurs in the Nearctic region. Shown here is ***G. (Oncopsia) flavifrons***, a common species found in damaged twigs and stems from Central America to the southwestern United States. Male Neriidae often guard their mates, straddling to protect them from competing males as they lay eggs. ❸ This elegant fly attracted to a cut banana stem in the Philippines is ***Gymnonerius dimidiatus***, known only from the island of Luzon. Other species in this small genus occur throughout the Oriental region. ❹ This Bolivian ***Nerius*** male is guarding his mate as she lays eggs in a crack in the bark of a damaged tree. All five species of *Nerius* are Neotropical. **Subfamily Telostylinae** ❺ Although most Neriidae are tropical, two species of ***Telostylinus*** occur in Australia. This one, ***T. angusticollis*** (sometimes treated as *Derocephalus angusticollis*), is common in temperate New South Wales. ❻ All African Neriidae, such as this one on the trunk of a tree in Tanzania, belong to the genus ***Chaetonerius***. ❼ ❽ ***Telostylus*** is the most widespread and diverse Oriental genus of Neriidae. Shown here are two species from the Philippines. ❾ ***Telostylus*** species are usually common on decaying vegetation throughout Southeast Asia; this was among many on some cut stems in Vietnam.

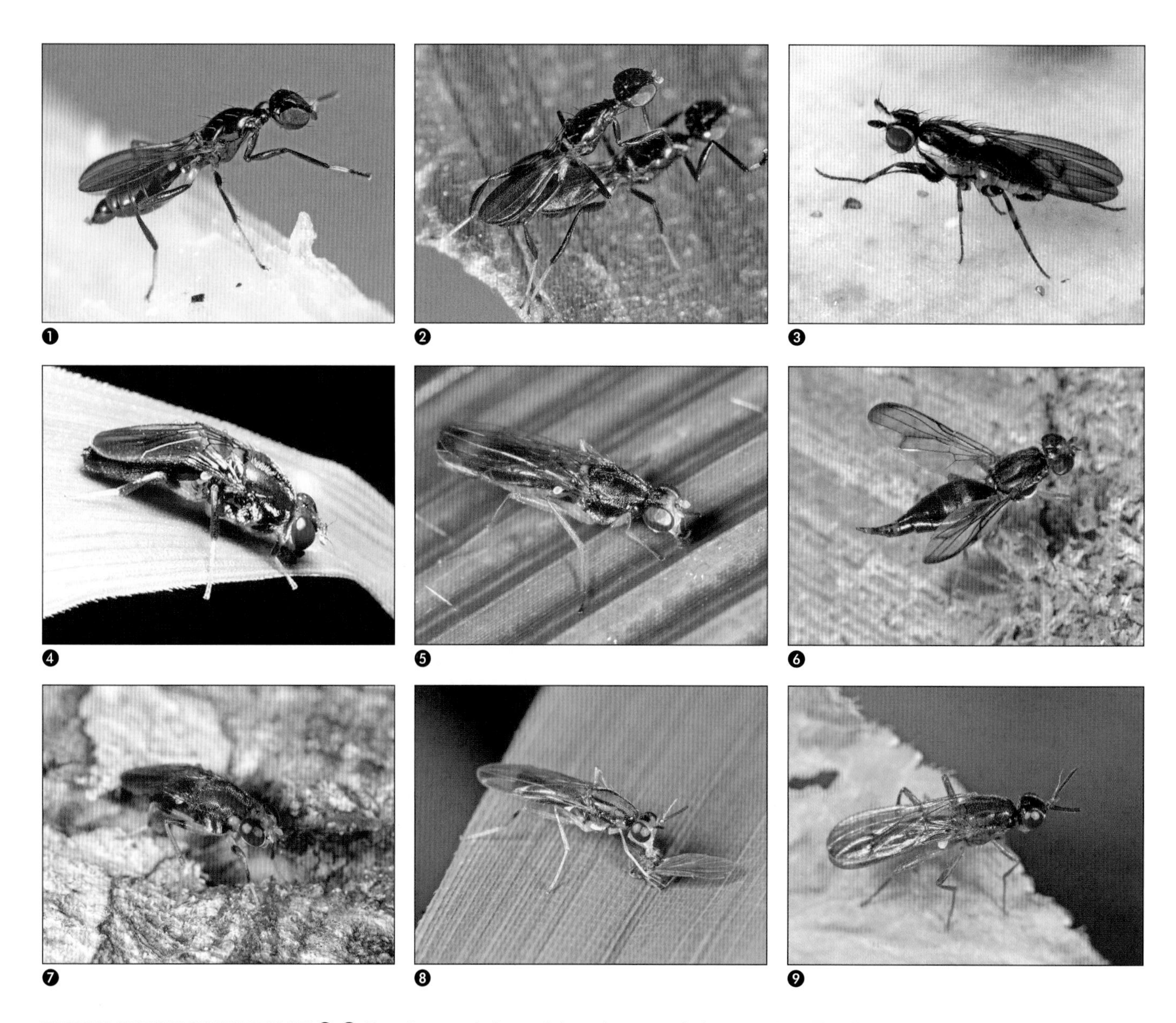

FAMILY CYPSELOSOMATIDAE ❶ ❷ ***Formicosepsis*** is an Oriental genus of nine species. The flies shown here, probably ***F. tinctipennis***, were attracted to a cut banana stem in the Philippines. **FAMILY PSEUDOPOMYZIDAE ❸** ***Latheticomyia*** is a small genus of five named species ranging from the western United States to southern South America. Several other Neotropical species are recognized but not yet named.

SUPERFAMILY DIOPSOIDEA

FAMILY PSILIDAE Subfamily Chylizinae ❹ ❺ ***Chyliza*** is a large, cosmopolitan genus of flies often seen patrolling leaves on shrubs or lower tree branches. Many, like this Costa Rican species, have the wing abruptly bent to follow the line of the abdomen. The other species shown here is from Tanzania. **❻ ❼** The larvae of several ***Chyliza*** species occur under bark, while others are stem borers. Shown here are a Canadian female ***C. notata***, about to lay eggs in the end of a cut log, and an Ecuadorian female laying eggs in a tree wound. **Subfamily Psilinae ❽** Adult acalyptrate flies rarely feed on other insects, but this Vietnamese ***Loxocera*** has either captured or (more likely) scavenged a small lower dipteran. **❾** This is the widespread Nearctic ***Loxocera cylindrica***, a common species with phytophagous larvae that develop in sedge (*Carex*) shoots.

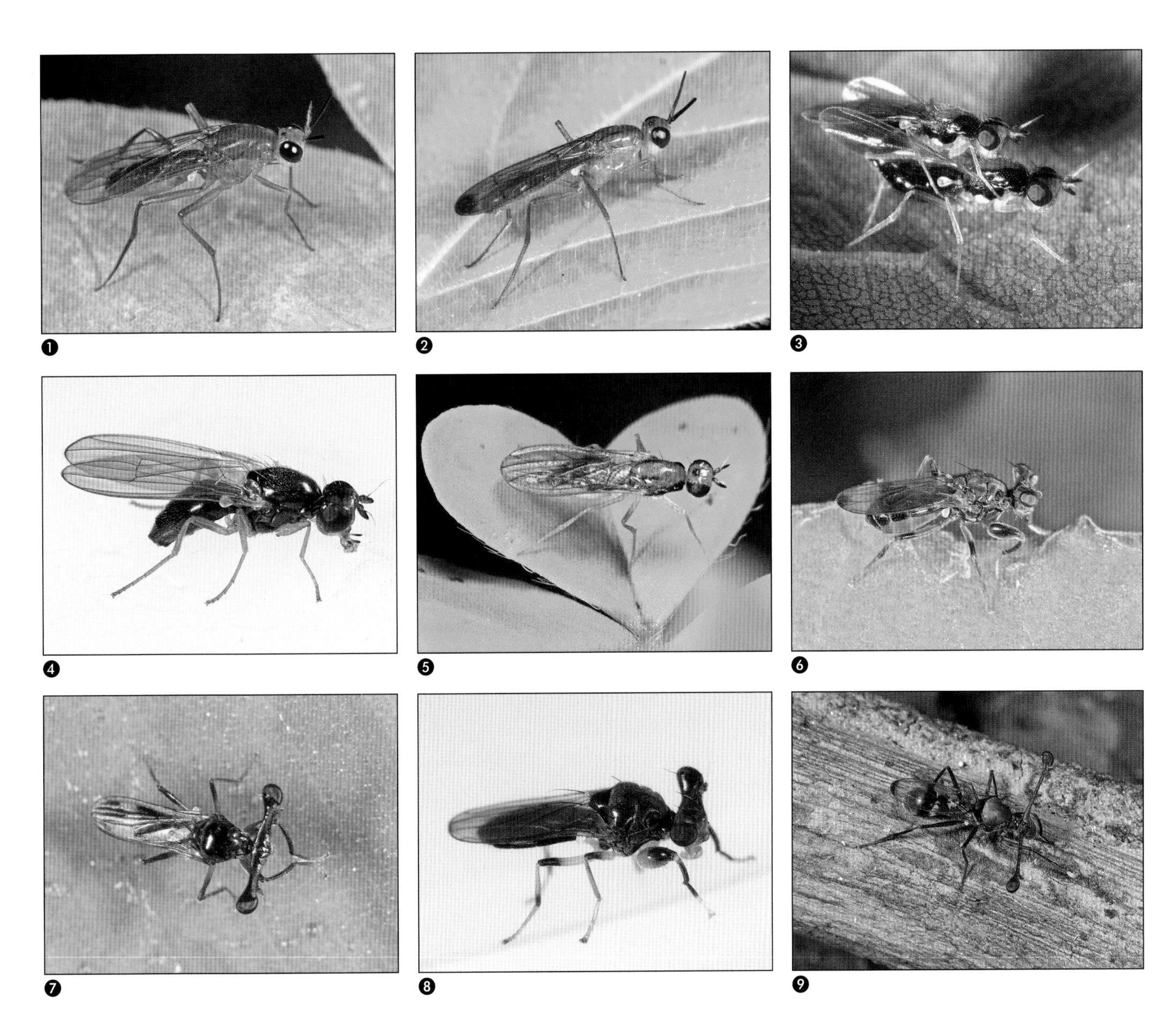

FAMILY PSILIDAE (continued) ❶ ❷ ***Loxocera*** species, conspicuous for their remarkably long antennae with at least the first flagellomere strikingly elongate, are mostly Holarctic or Old World tropical but occur in all zoogeographic regions. Shown here are two East African species. ❸ Larvae of ***Psila*** species usually develop in roots or stems of plants. Shown here is a mating pair of ***P. collaris*** from eastern Canada. ❹ ***Psila rosae***, the **Carrot Rust Fly**, is a widespread pest with larvae that bore into carrot roots, disfiguring and damaging the crop. *Psila* is divided into subgenera, with the Carrot Rust Fly in the subgenus *Chamaepsila*, which some treat as a separate genus. This species was known for a while as *P. hennigi*, but this name was declared invalid by the ICZN in 1999. ❺ The large genus *Psila* occurs in every zoogeographic region but is most diverse in the Holarctic; this is ***Psila (Chamaepsila) setalba*** from Greece. **FAMILY DIOPSIDAE (STALK-EYED FLIES) Subfamily Sphyracephalinae** ❻ Diopsidae is a mostly Old World tropical family, but one species of ***Sphyracephala*** occurs in Europe and another two are found in North America. This is the Nearctic ***S. brevicornis***, a species associated with decaying plant material in wet areas. ❼ ***Sphyracephala*** is the most northerly genus of stalk-eyed flies, but it is still most diverse in the Old World tropics. This is a species near ***S. detrahens*** in the Philippines (*detrahens* is included by some authors in *Pseudodiopsis*). ❽ Most members of the genus ***Sphyracephala***, such as this ***S. beccarii*** from Tanzania, have only moderately broadened heads. **Subfamily Diopsinae** ❾ The eyes — and the antennae — of this African stalk-eyed fly are at the ends of improbable-looking stalks. This is a ***Chaetodiopsis***, a genus treated by some as a synonym of *Diasemopsis*.

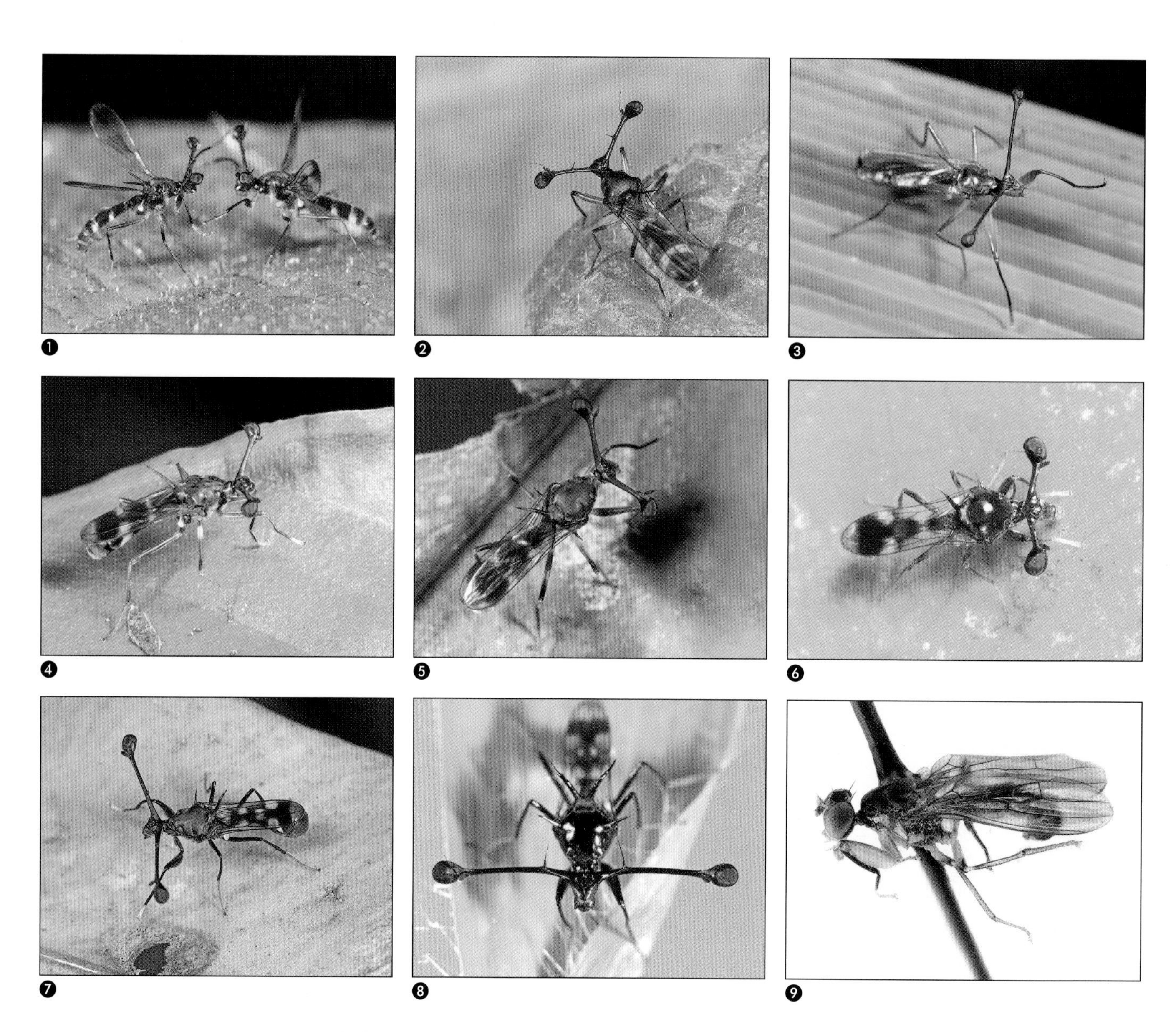

FAMILY DIOPSIDAE (continued) ❶ ❷ ***Diasemopsis*** is a large (more than 50 species), mostly African genus. Shown here is ***D. nebulosa***, from Tanzania. Some *Diasemopsis* species are saprophagous, developing in decaying plant tissue or in the frass of stem-boring caterpillars. ❸ Stalk-eyed flies in the large genus ***Diopsis*** are the best known in the family, partly because this genus includes several pests of rice. Most species, such as this Tanzanian ***D. surcoufi***, are Afrotropical. ❹ ***Eurydiopsis*** is an Oriental genus of nine described species; this fly from Vietnam is an undescribed species soon to be named by diopsid specialist Hans Feijen. ❺ ***Eurydiopsis*** is a small Oriental genus; this species (***E. subnotata***) is widespread in the region. ❻ This stalk-eyed fly from Vietnam is an undescribed species soon to be named as a species of ***Megalabops*** by diopsid specialist Hans Feijen. *Megalabops* is sometimes treated as part of *Teleopsis*. ❼ This is ***Teleopsis pharao***, a stalk-eyed fly known only from the Philippines; other *Teleopsis* species occur elsewhere in the Oriental and Afrotropical regions. Species with males that have a much greater eyespan than females have been used extensively as experimental organisms for studying the evolution of ornamental sexual traits. ❽ In the Diopsidae, not only the eyes but also the antennae are widely separated by head stalks. This head-on view shows the antennae close to the eyes. **Subfamily Centrioncinae** ❾ Species in the African genus ***Centrioncus*** lack the broad head of other diopsids, but they are recognizable as members of the family by their long scutellar processes.

FAMILY SYRINGOGASTRIDAE ❶ ❷ ❸ ❹ All Syringogastridae are Neotropical, and all belong to the genus ***Syringogaster***. These are ***S. atricalyx*** from Peru (mating), ***S. brunneina*** from Costa Rica (bubbling), and ***S. dactylopleura*** from Ecuador (two images). All these species were undescribed until 2009. **FAMILY TANYPEZIDAE** ❺ ❻ The neotropical genus ***Neotanypeza*** includes 15 named species and another 10 recognized but not as yet formally named. Shown here are as yet undescribed species from Costa Rica (sharing a bird dropping with a blow fly) and Ecuador (feeding on a bird dropping). ❼ Tanypezidae, such as this ***Neotanypeza*** from Costa Rica, frequently carry phoretic pseudoscorpions. The pseudoscorpions are predators presumably hitching a ride to wherever the *Neotanypeza* develop, but the larval habitat of this genus remains unknown. ❽ ***Tanypeza longimana*** is a Holarctic species in this otherwise mostly Neotropical family. Little is known about its habits but larvae are presumably saprophagous. **FAMILY STRONGYLOPHTHALMYIIDAE** ❾ ***Strongylophthalmyia*** is a big name for a relatively small genus of 48 species, mostly Old World tropical. This is ***S. angustipennis***, one of two Nearctic species in the genus. Both are associated with fallen or damaged poplar trees. The only other genus of Strongylophthalmyiidae, *Nartshukia*, is known from only one Vietnamese specimen.

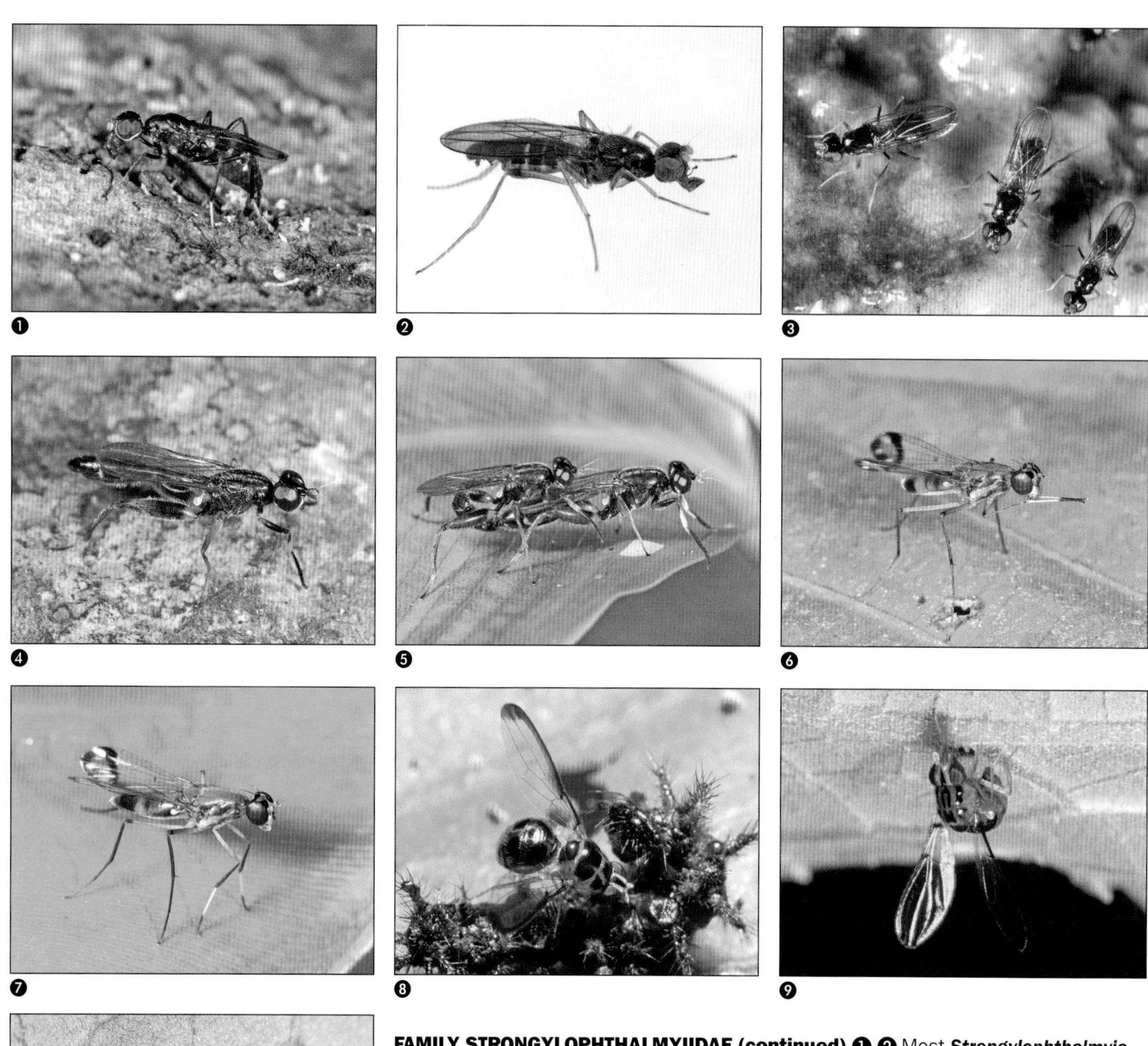

FAMILY STRONGYLOPHTHALMYIIDAE (continued) ❶ ❷ Most ***Strongylophthalmyia*** species are found in Southeast Asia. Shown here are a female laying eggs in a fallen log in the Philippines, and a male from Vietnam. ❸ This group of ***Strongylophthalmyia*** was photographed at a tree wound in coastal New South Wales. **FAMILY MEGAMERINIDAE** ❹ ❺ Most megamerinids are in the genus ***Texara***, an Oriental and eastern Palaearctic group of about a dozen species. These ***T. annulifera*** were frequenting a fallen tree in the Philippines or mating on nearby foliage. **FAMILY NOTHYBIDAE** ❻ ❼ The family Nothybidae, with only the one genus ***Nothybus***, is a poorly known group of nine described species. Shown here are a male and a female of two unnamed species from Vietnam, distinctive not only for morphology but also for their characteristic posture, with the pictured wingtips angled and prominently displayed. Nothing is known of the biology of this group. **FAMILY SOMATIIDAE** ❽ ❾ All species of Somatiidae are similar, and all are in the genus *Somatia*. Shown here are the Central American ***S. schildi*** (feeding on a dead caterpillar) and the widespread Neotropical ***S. aestiva*** (on the lower surface of a leaf on a Bolivian leguminaceous shrub). **FAMILY GOBRYIDAE (HINGE FLIES)** ❿ The small family Gobryidae has only a few species, scattered from Taiwan to New Guinea, all in the genus ***Gobrya***. This species, found under leaves in the forests of Mount Makeling in the Philippines, has not yet been formally named. It has a name in an unpublished manuscript, but if that name were printed here it would comprise what is called a *nomen nudum* — a formal scientific name first published without a proper description. *Nomina nuda* (the plural term) can cause considerable confusion; they are avoided throughout this book by simply referring to taxa as "unnamed" or "undescribed," even if they do have names in unpublished works.

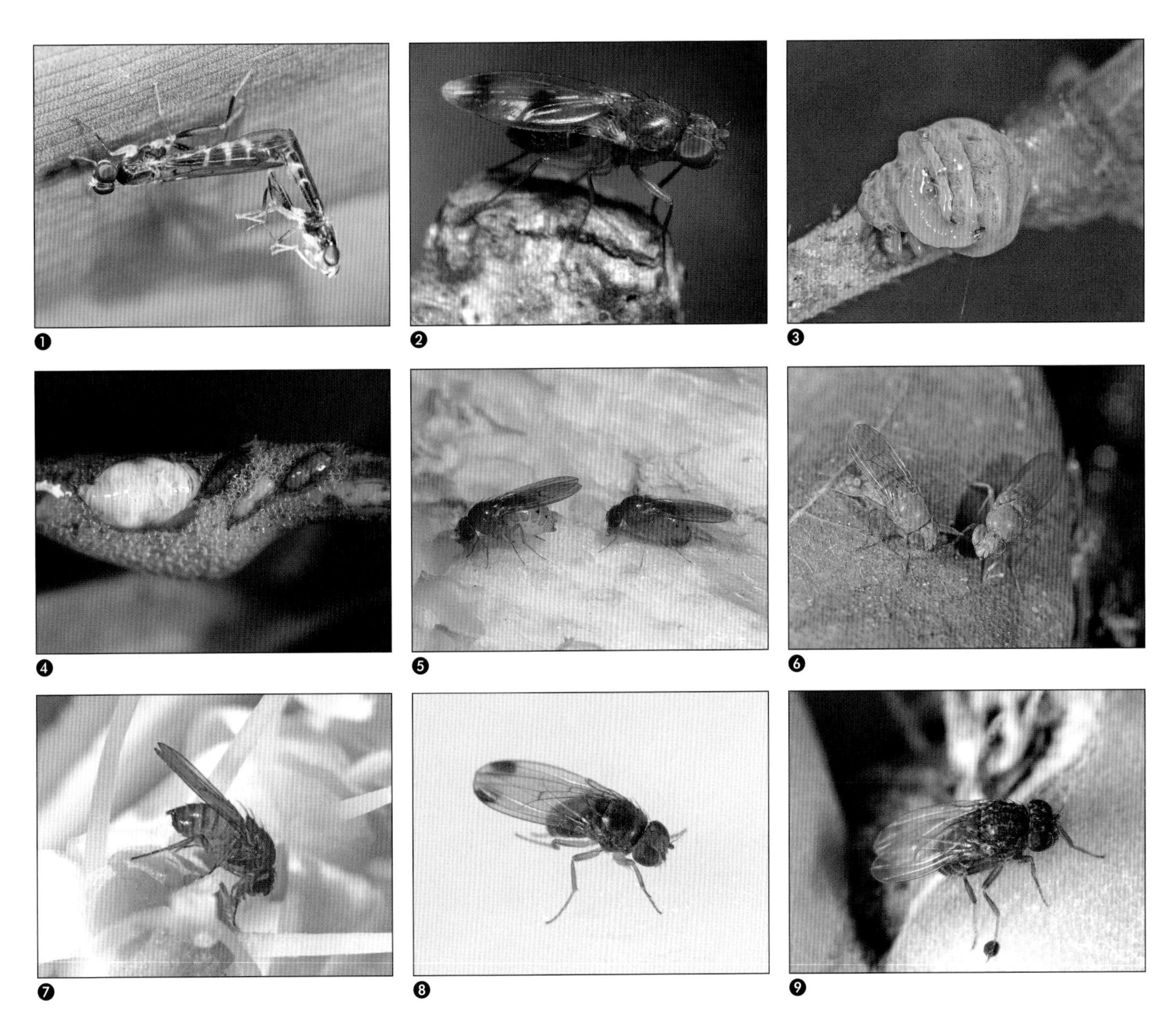

FAMILY GOBRYIDAE (HINGE FLIES) (continued) ❶ Gobryids are always found on the undersides of leaves; shown here is a mating pair of an undescribed species from Vietnam, with the male suspended from the female at a 90-degree angle.

SUPERFAMILY EPHYDROIDEA

FAMILY DROSOPHILIDAE (VINEGAR FLIES, ETC.) Subfamily Drosophilinae ❷ ***Chymomyza*** is a large, cosmopolitan genus almost invariably present on fallen or wounded trees. This is the common and widespread Holarctic species ***C. amoena***. ❸ ❹ Most known larvae of the large, mostly Neotropical, genus ***Cladochaeta*** occur in spittle masses of *Clastoptera* (Cercopidae), either feeding on the host nymphs without harming them or feeding on the spittle mass itself. ❺ ❻ Over half the species in the family Drosophilidae are in the huge, diverse genus ***Drosophila***. These flies on the yellow flesh of an *Amanita* fungus in Canada and this pair of flies on a dead leaf in the Philippines are typical for this huge worldwide genus of about 1,500 species. ❼ ❽ ***Drosophila suzukii***, called the **Spotted Wing Drosophila** because of the male's appearance, is a new pest of small fruits in North America. Females of this originally Oriental species have a sawlike ovipositor used to cut into fresh fruit, unlike most other fruit-associated *Drosophila*, which lay their eggs in soft, rotting or damaged fruit. ❾ ***Drosophila nigrospiracula***, one of several *Drosophila* species associated with cacti in the American southwest, breeds in saguaro cacti.

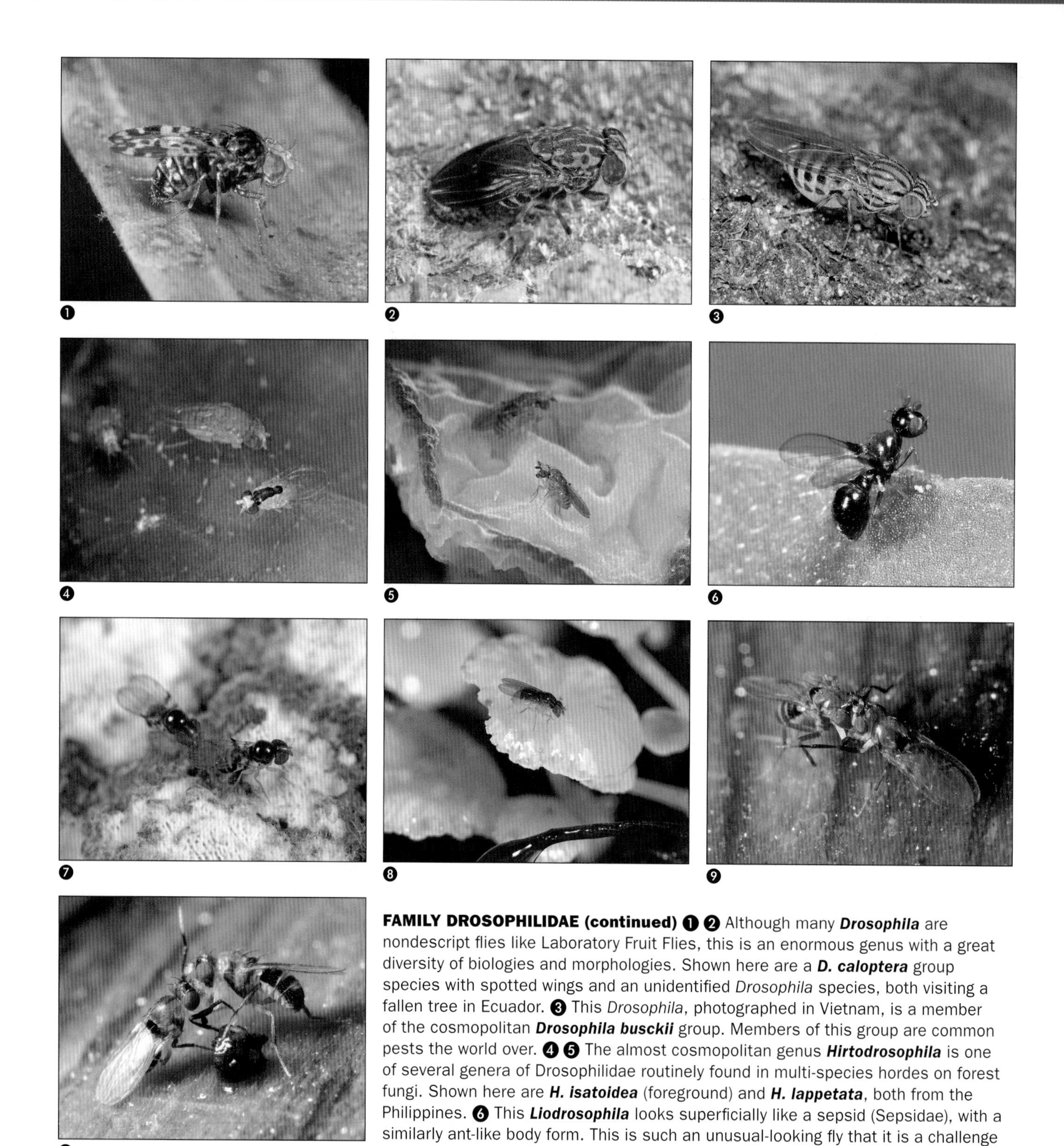

FAMILY DROSOPHILIDAE (continued) ❶ ❷ Although many ***Drosophila*** are nondescript flies like Laboratory Fruit Flies, this is an enormous genus with a great diversity of biologies and morphologies. Shown here are a ***D. caloptera*** group species with spotted wings and an unidentified *Drosophila* species, both visiting a fallen tree in Ecuador. ❸ This *Drosophila*, photographed in Vietnam, is a member of the cosmopolitan ***Drosophila busckii*** group. Members of this group are common pests the world over. ❹ ❺ The almost cosmopolitan genus ***Hirtodrosophila*** is one of several genera of Drosophilidae routinely found in multi-species hordes on forest fungi. Shown here are ***H. isatoidea*** (foreground) and ***H. lappetata***, both from the Philippines. ❻ This ***Liodrosophila*** looks superficially like a sepsid (Sepsidae), with a similarly ant-like body form. This is such an unusual-looking fly that it is a challenge to recognize it as a drosophilid, despite the typical pectinate arista. ❼ ❽ Fleshy fungi, especially in tropical forest habitats, are almost invariably attended by the same few genera of drosophilids. Shown here are Bolivian and Ecuadorian species of the widespread genus ***Mycodrosophila***. ❾ ❿ These Ecuadorian ***Neotanygastrella*** males are butting heads in an apparent battle for prime turf on a sappy log that is attracting potential mates. Similar behavior is common in the closely related genus *Chymomyza*.

FAMILY DROSOPHILIDAE (continued) ❶ ***Scaptomyza*** is a large group of around 300 species, including common temperate species such as the widespread Holarctic ***S. pallida***. Larvae usually mine stems and leaves. ❷ This distinctively striped fly belongs to the large Old World and Australian genus ***Zaprionus***. Most of the 56 species in the genus are Afrotropical, but one originally African species, ***Z. indianus***, was accidentally introduced to the New World and now occurs in Florida and parts of South America. ❸ ❹ Deep flowers often house numerous drosophilids, and this South American angel's trumpet (*Brugmansia* spp.) holds hundreds of ***Zapriothrica*** individuals. The larvae of some *Zapriotheca* species feed on flower buds. ❺ ❻ ❼ ***Zygothrica*** is one of the widespread drosophilid genera characteristically abundant on forest fungi. Many Neotropical species have broadened heads and engage in elaborate territorial displays; one of those shown here has a female flipping over his back. Shown are species from Costa Rica and Ecuador. **Subfamily Steganininae** ❽ ***Acletoxenus formosus*** is an originally Palaearctic predator of whiteflies that has been moved around for biological control of whitefly pests. It is now established in Australia. ❾ This Vietnamese ***Leucophenga*** is sharing a bird dropping with a couple of small scuttle flies (Phoridae, probably *Megaselia*).

FAMILY DROSOPHILIDAE (continued) ❶ ❷ ❸ Members of the large, cosmopolitan genus ***Leucophenga*** are often found around fungi, but their life histories are poorly known. Shown here are species from Bolivia (feeding on a small dropping), the Philippines (on fungus) and Tanzania (investigating a dead leafhopper). ❹ The genus ***Amiota*** is a large, cosmopolitan group of drosophilids with the irritating habit of hovering in front of people's faces and getting into their eyes. Most of the almost 100 species in the genus occur in the Old World tropics, but this large Costa Rican species is of the small Central American subgenus ***Sinophthalmus***. ❺ ❻ ❼ ***Stegana*** is a cosmopolitan genus of distinctively shaped beetle-like flies commonly found on tree trunks. Shown here are species from Bolivia (black), Ecuador (orange) and the Philippines (brown). **FAMILY EPHYDRIDAE (SHORE FLIES) Subfamily Hydrelliinae** ❽ This tiny ***Atissa*** is a silvery white coastal fly, almost invisible against the sand of Florida's Gulf Coast. ❾ The nine species of ***Dryxo***, such as this Tanzanian ***D. ornata***, occur in the African, Oriental and Australasian regions.

FAMILY EPHYDRIDAE (continued) ❶ ❷ ***Hydrellia*** species often have elaborate courtships, like this wing-twisting Cuban species and these "kissing" Canadian *Hydrellia*. Larvae in this widespread genus are usually miners in the leaves of grasses and aquatic plants, and some are pests of rice. ❸ The small (usually around 2 mm) flies in the large (more than 200 species), cosmopolitan genus ***Hydrellia*** are often abundant. The Pacific ***H. tritici*** is one of the most common flies in Australia; larvae mine in grasses (*Poa*) in damp areas. ❹ ***Hydrellia*** adults are commonly seen feeding on drowned insects such as this mayfly in a Canadian pond. ❺ The minute floating leaves of duckweed (*Lemna* spp.) are mined by larvae of ***Hydrellia*** and ***Lemnaphila***. ❻ ❼ ***Notiphila***, with about 150 species worldwide, is a common genus on or around shallow waters with emergent vegetation. Shown here are species from Canada and Tanzania. ❽ Larvae of the aquatic genus ***Notiphila*** lack anterior spiracles but have posterior spiracles at the end of a flexible snorkel-like air tube; some species puncture the roots of aquatic plants with the posterior spiracles, using the plant as a living snorkel. This newly emerged adult is standing beside a puparium with a nearly developed adult visible inside. ❾ Robust adults of the large (85 species) genus ***Paralimna*** are common on water margins throughout the world; this one is on an ocean beach in the Philippines.

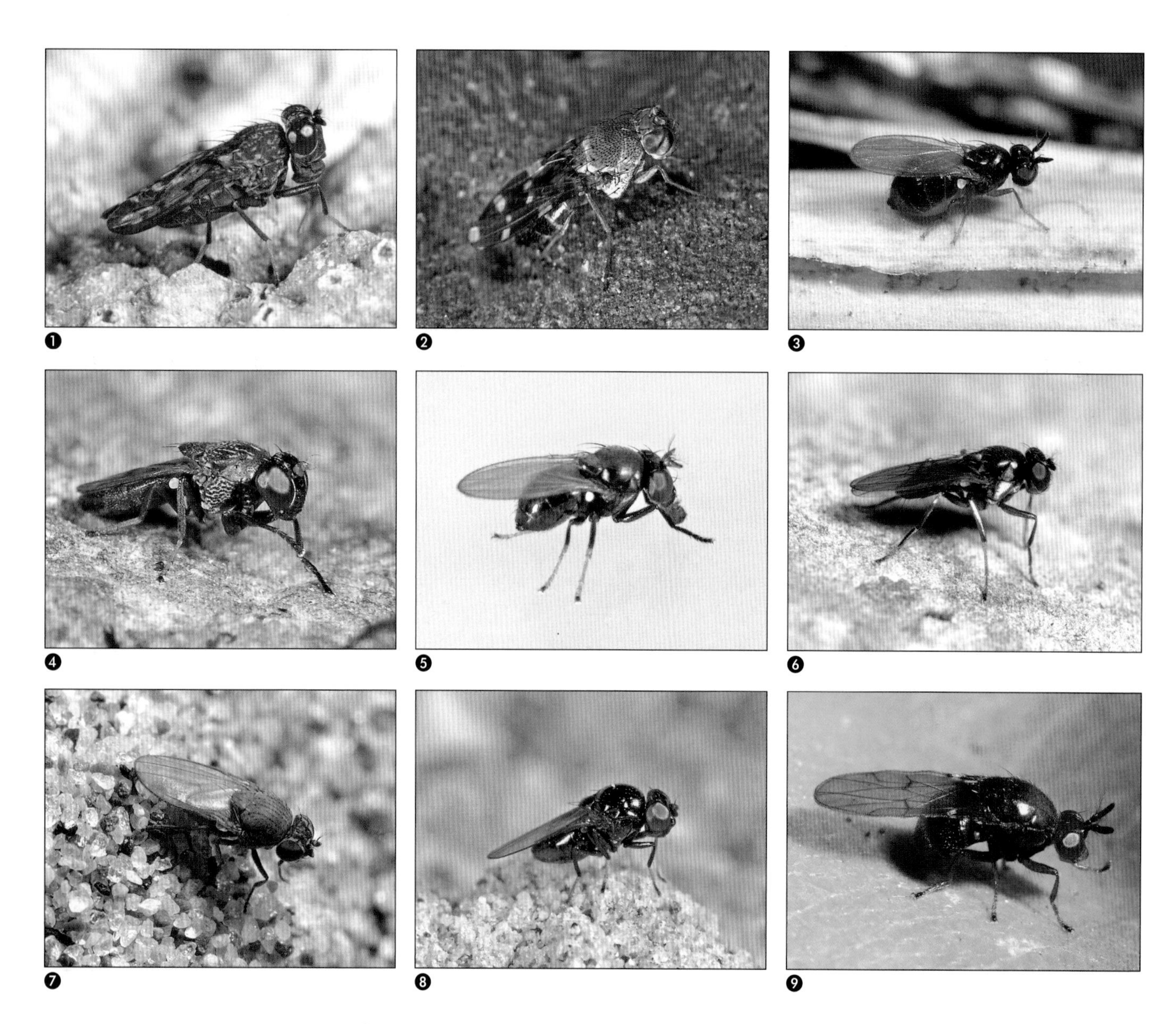

FAMILY EPHYDRIDAE (continued) ❶ This distinctively patterned Ecuadorian ***Paralimna*** is an unnamed species. Known larvae of this widespread genus develop in wet decomposing material. **Subfamily Discomyzinae** ❷ ***Actocetor*** is a small Old World tropical, mostly African group. Shown here is a species from the Philippines. ❸ ***Ceropsilopa*** such as this one from Florida are characteristic of low vegetation near beaches. The genus includes 18 species worldwide. ❹ ***Discomyza*** species usually develop on dead mollusks but are also reported as parasitoids of living snails. This species (***D. maculipennis***) is from the Philippines; most other species of this widespread genus also occur in the Old World. ❺ ***Psilopa*** is a large (70 species), widespread but mostly Old World genus; this one is from the Philippines. Known larvae are leaf miners. ❻ ***Rhysophora*** is a small New World genus of five species; this one was photographed on a rock in a Costa Rican river. **Subfamily Gymnomyzinae** ❼ ***Allotrichoma*** species are often abundant on sand beaches such as this Great Lakes dune in Canada. ❽ ***Athyroglossa*** is a cosmopolitan genus that includes ***A. granulosa***, a Nearctic species that has been reared from decaying skunk cabbage. ❾ ***Beckeriella*** species are characteristic flies of tropical forests in both the New and Old Worlds. This one is from Costa Rica.

FAMILY EPHYDRIDAE (continued) ❶ This ***Discocerina*** was among many on rocks emerging from a swift mountain river in Costa Rica. The other 27 species in the genus are widely distributed and normally associated with water margins. ❷ Known larvae of the small New World genus ***Gastrops*** develop in the foam nests of frogs. Shown here is a Costa Rican species. ❸ The dozen tiny ***Hecamede*** species are found on Pacific and Old World shorelines such as this Philippines beach. ❹ ***Hecamede granifera*** is a widespread species found on seashores around the Old World tropics and the Australasian region. Most of the other 11 species in the genus have similar habits and distributions; some occur in numbers on dead fish and similar shoreline debris. ❺ ❻ Most members of the shore fly family (Ephydridae) eat algae, and many species abound along the moist margins of lakes and ponds. Members of the worldwide genus ***Ochthera***, however, are predators that eat other insects such as midges and mayflies. This Cuban ***Ochthera loreta*** has flexed its raptorial forelegs to turn a smaller fly into its next meal; the other photo is an *Ochthera* from the Philippines. **Subfamily Ephydrinae** ❼ This ***Brachydeutera*** is walking on the surface of a Costa Rican tide pool, feeding on particles trapped by the surface tension. *Brachydeutera* is a widespread group that is common on still water surfaces.
❽ ***Calocoenia platypelta***, one of only two species in this small genus, is a widespread shore fly in the western United States and Canada. ❾ This mass of ***Cirrula hians*** is feeding on some algae on the Great Salt Lake, Utah. This species, formerly treated as part of *Ephydra*, is the most abundant of several shore flies found on and around saline lakes in western North America.

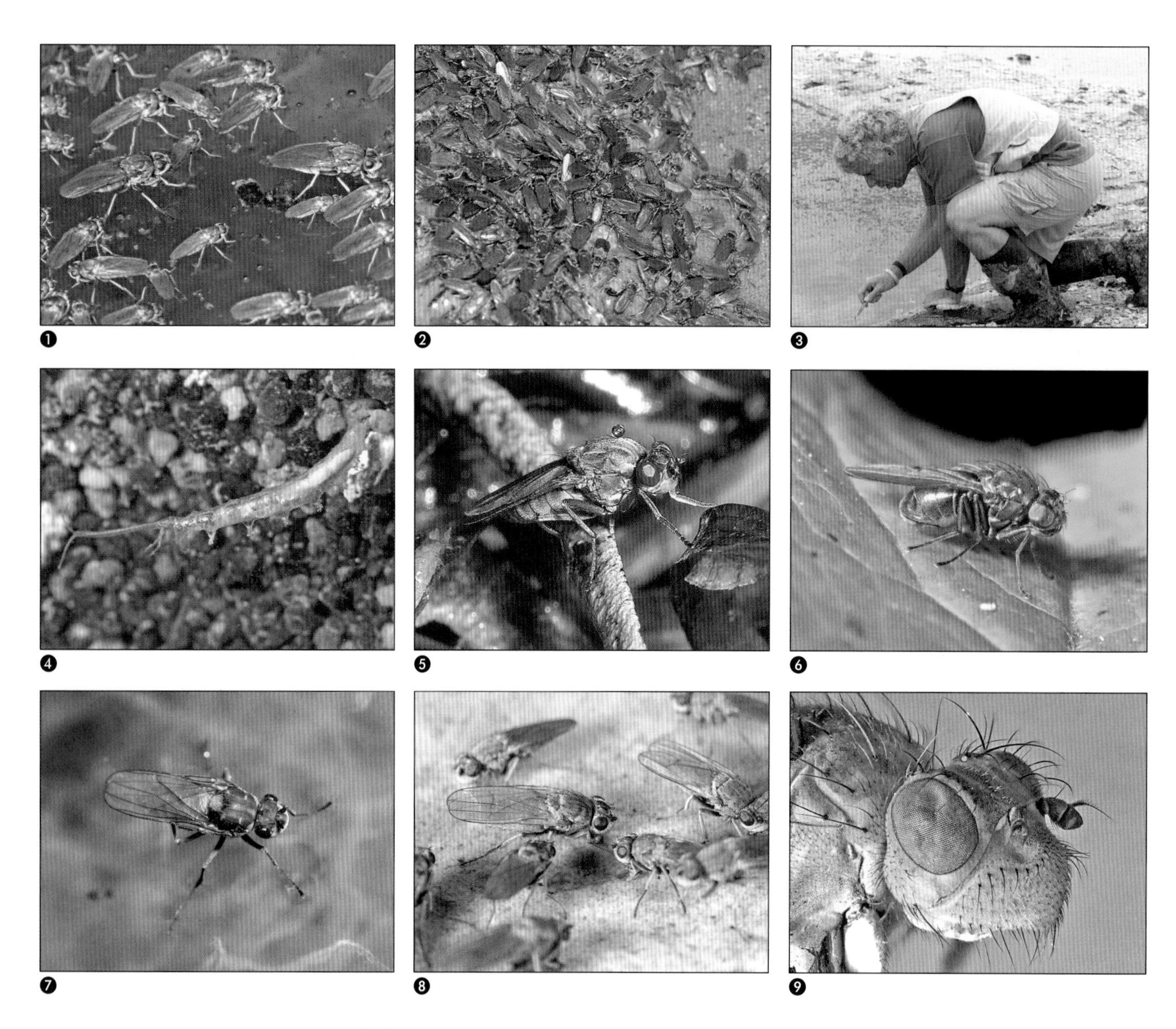

FAMILY EPHYDRIDAE (continued) ❶ ❷ The famous shore flies of the Great Salt Lake in Utah include a number of species, of which the most common are ***Cirrula hians*** and *Ephydra gracilis*. The larger flies shown here are *C. hians*. The shore flies of the Great Salt Lake reach enormous densities, often forming conspicuous windrows that contain as many as a billion flies per kilometer of shoreline. ❸ ❹ This entomologist is picking larval shore flies from the Great Salt Lake as the surrounding surface buzzes with adult shore flies. The larvae, which feed on algae in the lake, have a long, bifurcate breathing "tail," long prolegs and typical maggot mouthparts. ❺ ***Cirrula gigantea*** develops as a filter feeder in salt marshes of the North American east coast. ❻ ***Coenia curvicauda*** is a common North American species in cattail (*Typha*) marshes in North America, where it feeds on algae and detritus; most other species in this small genus are Palaearctic. ❼ This Chilean ***Dimecoenia*** is walking on the surface of a calm tidal pool, surrounded by the algal mats in which it developed. Other members of this mostly Neotropical genus occur in similar habitats. ❽ Species of *Ephydra*, such as these ***Ephydra gracilis*** along a small puddle on the edge of the Bonneville Salt Flats, are often abundant in saline environments. ❾ ***Ephydra*** and related genera are characterized by a strongly produced, bristly face.

FAMILY EPHYDRIDAE (continued) ❶ ❷ The chunky, relatively bare members of the large (73 species) genus ***Parydra*** often abound on pond margins. ❸ Some of the 135 species of the widespread genus ***Scatella*** are common indoors and in greenhouses, where they feed on algae on wet growing media; others are found on a variety of wet, alga-rich substrates. This ***S. triseta*** is on a rock seep in Oregon. ❹ ***Setacera*** species are common on algal mats such as the surface of this Canadian ditch. ❺ This ***Teichomyza fusca*** is on a wave-splashed rock along the coast of Chile, but the species also occurs in Europe. Its semiaquatic larvae often occur in high densities on wet, microbe-rich substrates such as urinals or wet materials around cadavers. *Teichomyza* is closely related to *Scatella*. **Subfamily Ilytheinae** ❻ Known larvae of the genus ***Nostima*** feed on blue-green algae. Shown here is ***N. semialata*** from Canada, but most members of this widespread genus are Neotropical or Afrotropical. **FAMILY CURTONOTIDAE (HUMPBACKED FLIES)** ❼ ❽ ❾ Although most Curtonotidae are brown humpbacked flies, these three South American ***Curtonotum*** (a striped ***C. bivittatum***, a dark ***C. impunctatum***, and a spot-winged ***C. tryptetipenne***) are relatively distinctive.

FAMILY CURTONOTIDAE (continued) ❶ ***Curtonotum helvum*** can be abundant on eastern North American sand dunes but it becomes active only around dusk. This copulating pair was among many that appeared on blades of American Beach Grass just as it got dark. ❷ ❸ Many Neotropical Curtonotidae belong to two distinct clusters of described and undescribed species (species complexes) that have not yet been revised. This humpbacked fly on a Bolivian twig belongs to the ***Curtonotum murinum*** complex, and this yellowish fly feeding on a gelatinous Ecuadorian seed represents the ***C. vulpinum*** complex. ❹ Although this undescribed species from Tanzania could be called a ***Cyrtona*** for the moment, it belongs to a group due to be described as a new genus in the near future. ❺ ***Tigrisomyia amnoni*** is the type species of the recently described East African genus *Tigrisomyia*, and this Tanzanian fly was used in the original description of the species. **FAMILY DIASTATIDAE Subfamily Diastatinae** ❻ Most members of the small family Diastatidae are in the genus ***Diastata***, which has 35 named species. This eastern North American species is similar to ***D. pulchra***. ❼ Most ***Diastata*** species occur in the Holarctic region, but the genus also occurs in South America, Africa and the Oriental region. This Tanzanian fly is probably ***D. freidbergi***. **Subfamily Campichoetinae** ❽ Species in the small genus ***Campichoeta***, found in the Holarctic, Afrotropical and Oriental regions, are sometimes treated as the family Campichoetidae. This is ***C. griseola***, a widespread Holarctic species. **FAMILY CAMILLIDAE** ❾ Most species of the small family Camillidae are in the genus ***Camilla***, a poorly known group of small flies apparently associated with mammal burrows in the Holarctic and Afrotropical regions.

FAMILY CRYPTOCHETIDAE ❶ This Indian ***Cryptochetum*** shows the characteristic compact appearance and unusual antennae (with no arista) of the small flies in the family Cryptochetidae. **FAMILY BRAULIDAE (BEE LICE) ❷** ***Braula coeca*** ("bee louse") is a widespread associate of the European Honey Bee (*Apis mellifera*). ❸ The two species of ***Megabraula*** are associates of the Himalayan Honey Bee (*Apis laboriosa*). **FAMILY NANNODASTIIDAE ❹** This ***Nannodastia horni*** specimen (indicated by the arrow) seems dwarfed beside a *Pantophthalmus* specimen (Pantophthalmidae, see page 256). Nannodastiids are found on Mediterranean, Pacific, Atlantic and Indian Ocean shores; this one (in the U.S. National Museum of Natural History) is from a remote Pacific island.

SUPERFAMILY LAUXANIOIDEA

FAMILY LAUXANIIDAE Subfamily Eurychoromyiinae ❺ Eurychoromyiinae are among the rarest of flies, possibly because they are normally restricted to the forst canopy. This is ***Physegeniopsis hadrocara***, from a high Ecuadorian cloud forest. **Subfamily Homoneurinae ❻** The Oriental and Afrotropical genus ***Cestrotus*** includes about a dozen African species. This couple is on a fallen tree in the Udzungwa Mountains of Tanzania. ❼ Although most members of the large, widespread genus ***Homoneura*** are yellow or orange, this Vietnamese ***H. euaresta*** is a conspicuous exception. ❽ ***Homoneura*** is one of the largest and often the most common genera of Lauxaniidae. Although the group is absent from the Neotropics it includes some abundant Nearctic species. Males of some, such as this Canadian ***H. melanderi***, have ornamented hind tarsi. ❾ ❿ ***Homoneura*** is a large, common group with hundreds of species in the Old World tropics, including these two colorful flies from Tanzania.

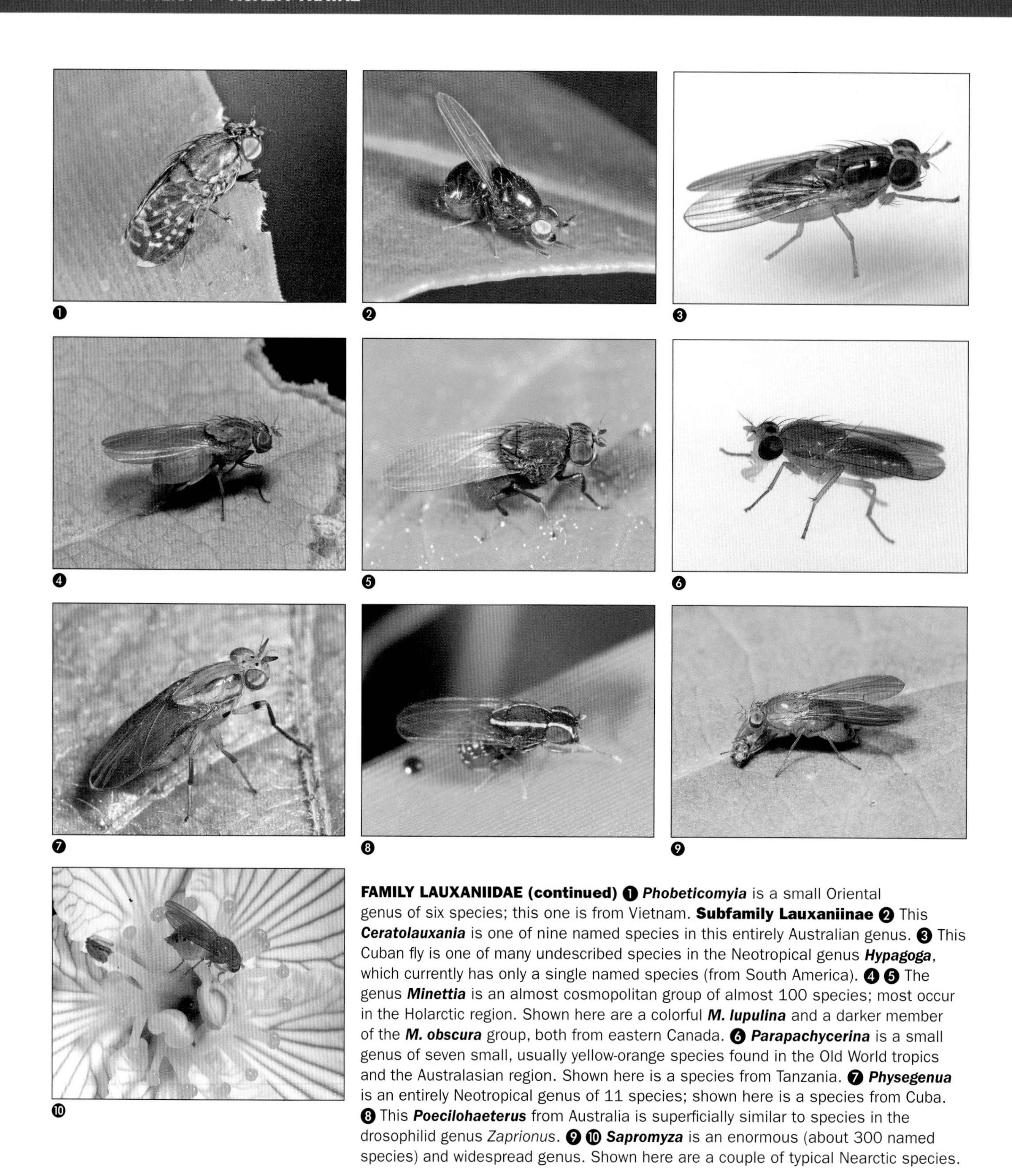

FAMILY LAUXANIIDAE (continued) ❶ ***Phobeticomyia*** is a small Oriental genus of six species; this one is from Vietnam. **Subfamily Lauxaniinae** ❷ This ***Ceratolauxania*** is one of nine named species in this entirely Australian genus. ❸ This Cuban fly is one of many undescribed species in the Neotropical genus ***Hypagoga***, which currently has only a single named species (from South America). ❹ ❺ The genus ***Minettia*** is an almost cosmopolitan group of almost 100 species; most occur in the Holarctic region. Shown here are a colorful ***M. lupulina*** and a darker member of the ***M. obscura*** group, both from eastern Canada. ❻ ***Parapachycerina*** is a small genus of seven small, usually yellow-orange species found in the Old World tropics and the Australasian region. Shown here is a species from Tanzania. ❼ ***Physegenua*** is an entirely Neotropical genus of 11 species; shown here is a species from Cuba. ❽ This ***Poecilohaeterus*** from Australia is superficially similar to species in the drosophilid genus *Zaprionus*. ❾ ❿ ***Sapromyza*** is an enormous (about 300 named species) and widespread genus. Shown here are a couple of typical Nearctic species.

FAMILY LAUXANIIDAE (continued) ❶ ❷ ***Sapromyza*** is a huge genus, especially diverse in Australia, as shown by this ***S. mallachomyia*** (striped thorax) and ***S. pictigera*** (spotted thorax) from New South Wales. ❸ ***Siphonophysa*** is a Neotropical genus with only two described species, but, as is so often the case with tropical acalyptrate flies, many species remain to be described. ❹ ***Trivialia*** is a small New World genus with seven named species, but this Cuban species is undescribed. According to lauxaniid expert Stephen Gaimari, at least five Caribbean and Central American *Trivialia* species still await description.
FAMILY CHAMAEMYIIDAE (APHID FLIES) Subfamily Chamaemyiinae ❺ Species in the large, almost cosmopolitan genus ***Chamaemyia*** develop as predators of mealy bugs in leaf sheaths of grasses. This one is from Canada. ❻ Most Chamaemyiidae are relatively inconspicuous clear-winged flies, but this undescribed chamaemyiine from Ecuador is an exception. ❼ Most species in the New World genus ***Ortalidina***, such as this Chilean species, are from southern South America, but the genus ranges as far north as the southeastern United States. ❽ This eastern Canadian fly belongs to the genus ***Plunomia***, an entirely Nearctic genus of only four species. ❾ ***Pseudodinia*** is a mainly Nearctic genus, usually associated with grasses, where the larvae feed on mealy bugs in the leaf sheaths. This is ***P. antennalis***.

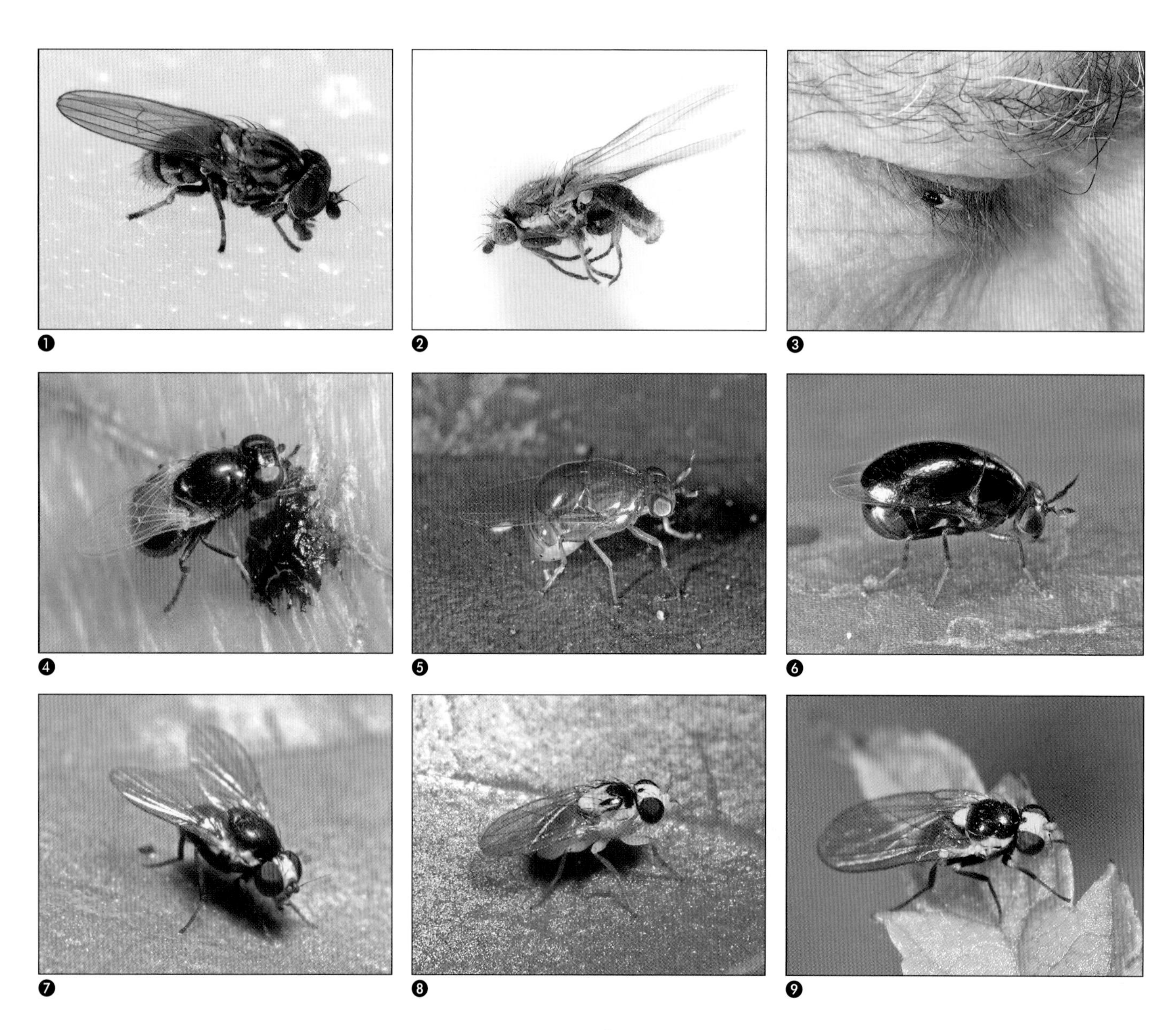

FAMILY CHAMAEMYIIDAE (continued) Subfamily Leucopinae ❶ ***Pseudoleucopis*** is a genus of about 15 Australian species that prey on soft-bodied bugs such as scales and mealy bugs; they are sometimes important in the control of pest mealy bugs on eucalyptus trees. **Subfamily Cremifaniinae** ❷ ***Cremifania*** includes only four species — one in the American southwest and three in Europe (one of which has been introduced to North America) — but they are sometimes treated as a separate family, Cremifaniidae. Larvae are parasitoids of Balsam Wooly Adelgids (*Adelges picea*) and related hemipterans. **FAMILY PARALEUCOPIDAE (BOBOS)** ❸ An undescribed "bobo" (***Paraleucopis*** spp.) on bobo expert Robert Smith's eye. ❹ This undescribed species of ***Paraleucopis***, here attracted to a scratch on an entomologist's leg, is known only from Sonoran Arizona; other species occur in California, New Mexico and Sonoran Mexico. **FAMILY CELYPHIDAE (BEETLE FLIES)** ❺ The nine species in the Oriental genus ***Idiocelyphus*** are characterized by a relatively short scutellum. This is a female of one of the eight *Idiocelyphus* species described from the Philippines. ❻ This ***Spaniocelyphus levis*** from the Philippines is a typical beetle fly, with part of the thorax (the scutellum) spectacularly inflated to cover the wings and most of the abdomen with a shining, beetle-like shell. *Spaniocelyphus* is a large Old World tropical genus that includes five species in the Philippines.

SUPERFAMILY OPOMYZOIDEA

FAMILY AGROMYZIDAE (LEAF-MINING FLIES) Subfamily Phytomyzinae ❼ ***Calcomyza*** is a large, widespread genus. One species mines the leaves of lantana; it has been moved around the world for use as a biological control agent in areas where lantana has become a weed. ❽ ***Liriomyza*** is a huge genus with many superficially similar species. This is probably ***L. marginalis***, a common species that mines in the leaves of corn and other grasses. ❾ ***Liriomyza septentrionalis*** is an extremely common grass-feeding species along the west coast of North America.

FAMILY AGROMYZIDAE (continued) ❶ ❷ This is a **Milkweed Leaf Miner** (***Liriomyza asclepiadis***) puncturing a milkweed leaf with its sharp abdominal tip. The patches of white sap are puncture points. Larvae of this eastern North American species make splotch mines in milkweed leaves. ❸ This yellow ***Liriomyza schmidti*** larva is visible as it feeds on the upper part of the leaf from inside its mine in a Cuban botanical garden. According to Cuban agromyzid expert Gabriel Garces Gonzalez, the mines of this species turn silver once the larva has left to pupate. ❹ This Ecuadorian ***Liriomyza*** is "bubbling," presumably concentrating its gut contents by regurgitating bubbles of fluid for evaporation. ❺ This Canadian ***Napomyza*** is piercing a leaf with her abdominal tip, either to oviposit or to liberate sap on which to feed. *Napomyza* is a large, widespread genus that includes several pests, such as the Allium Leaf Miner and the Carrot Fly. ❻ ***Phytobia*** is a cosmopolitan genus of about 65 species that apparently all develop as cambium borers in the twigs or trunks of trees, sometimes causing economic damage (for example, the Birch Cambium Miner and Ash Cambium Miner). Shown here is an unidentified Bolivian species. ❼ ***Phytoliriomyza*** includes about 70 species worldwide, but the host plants are known for only a few. Shown here is a minute Canadian ***P. pacifica***. ❽ ❾ ***Phytomyza***, with more than 560 species, is the largest genus of leaf-mining fly. This adult female on a buttercup flower is ***P. bicolor***; the leaf mines shown here were made by larvae of ***P. aquilegiana*** (**Columbine Leafminer**).

FAMILY AGROMYZIDAE (continued) Subfamily Agromyzinae ❶ ***Agromyza*** is a large, widespread genus of almost 200 species that is most diverse in the Holarctic region, although the fly shown here is from Tanzania. *Agromyza* species mine in the leaves of a wide range of plants. **❷** ***Hexomyza schineri*** (**Poplar Twiggall Fly** or **Sallow Stem Galler**) is a widespread Holarctic "leaf-mining" fly that causes galls on twigs of poplar and aspen trees. Shown here are galls on *Populus tremuloides* in the western United States. **❸** ***Melanagromyza*** is a mostly tropical genus with about 300 described species. Larvae of most species are found in stems (sometimes of crop plants); a few develop in roots, seeds or flowerheads, but only a couple of species are leaf miners. **FAMILY CLUSIIDAE (DRUID FLIES) Subfamily Clusiodinae ❹** ***Allometopon*** is a small, mostly Southeast Asian genus, but a few species also occur in Japan, Australia and the Seychelles. This unnamed female is extending her abdomen to insert eggs in a wet rotting log on a mountain trail in the Philippines. **❺** ***Clusiodes ater***, like other members of the mostly Holarctic genus *Clusiodes*, can often be found perching conspicuously on exposed wood. **❻** ***Clusiodes*** species, such as this Canadian female from the large ***C. melanostoma*** group, often lay eggs in beetle borings in fallen or standing dead trees. **❼** This unnamed ***Hendelia*** female has extended her abdomen to lay eggs in a rotting log in the Philippines. **❽** This Vietnamese ***Hendelia (H. extensicornis)*** is one of many species with enlarged antennae, probably used in male–male "head-measuring" battles. **❾** ***Hendelia kinetrolicros*** is one of two Neotropical *Hendelia* species in which the males have spectacular processes arising from their cheeks. To my knowledge, the specimen that posed for this photo is the only one anyone has observed alive, but since it was alone we can only speculate that the cheek processes serve a role in male–male combat or territoriality.

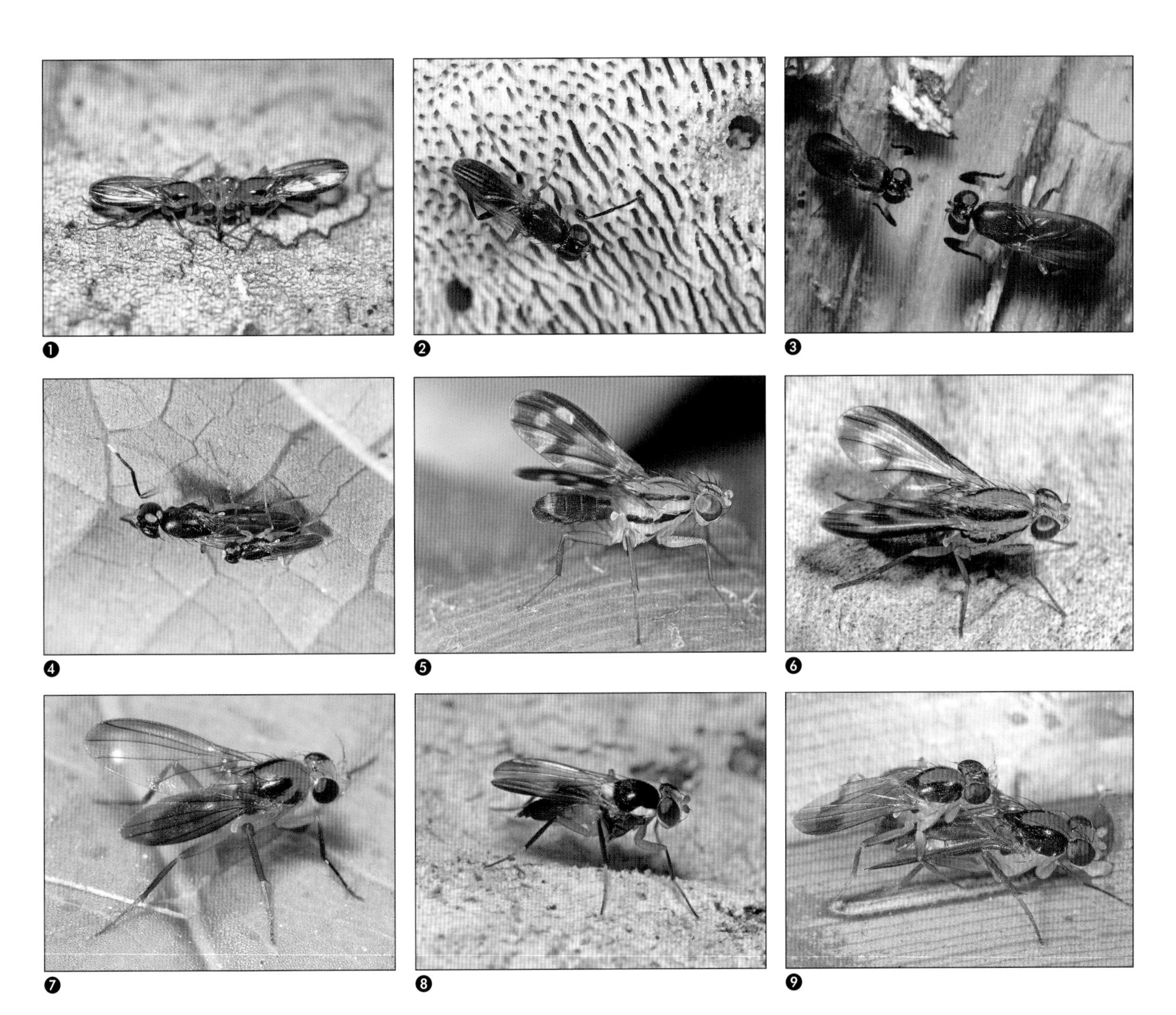

FAMILY CLUSIIDAE (continued) ❶ Male–male territorial battles are common in the Clusiodinae, and these Australian ***Hendelia armiger*** males have pivoted their antennae out 90 degrees to make their already widened heads seem even wider. This is more like a head-measuring contest than a head-bashing battle, since the less well-endowed male will abandon the territory to his rival. ❷ ***Heteromeringia*** occurs throughout the world, including North America, Australia and Africa, but most of the 80 named species are Oriental. This Costa Rican species is on the underside of a bracket fungus, where it probably developed in a beetle boring. ❸ Males of the North American ***Heteromeringia nitida*** face off and defend territory from other males by "boxing" with their black "elbows." ❹ ***Heteromeringia*** such as this Philippines pair are often seen on the undersides of leaves. **Subfamily Sobarocephalinae** ❺ ❻ The southern South American genus ***Apiochaeta*** is among the most common and conspicuous acalyptrates in Chile. These are ***A. similis*** (on a leaf) and ***A. vitticollis*** (on a log), both in southern Chile. ❼ ❽ ***Sobarocephala***, a mostly Neotropical genus, includes over 250 species, most of which were just described in 2012. Although *Sobarocephala* species are usually orange and black flies such as this Canadian ***S. setipes***, the genus includes several black (or black and yellow) species like this Central American ***S. diversa***. ❾ These mating Central American ***Sobarocephala*** were captured to become type specimens of the recently described species ***S. schildimyia***.

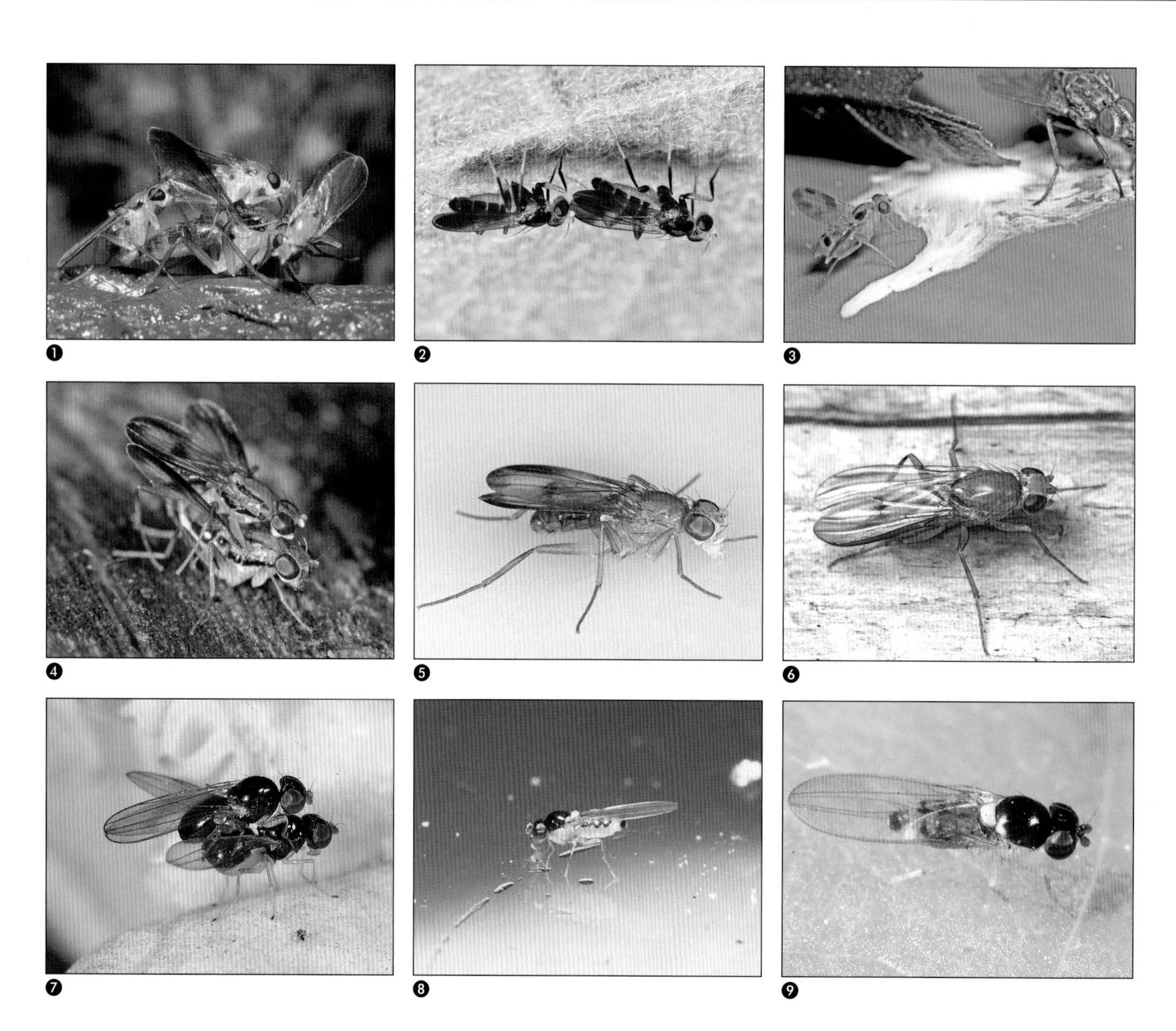

FAMILY CLUSIIDAE (continued) ❶ ❷ ***Sobarocephala*** males often aggregate on specific substrates (lek sites). This cluster of four pale Canadian ***S. latifrons*** includes three males that were waiting on a small animal dropping and which mobbed the female as soon as she arrived. These two dark ***S. daidaleos*** were among many on the lower surface of a broad leaf in Bolivia. ❸ Clusiidae, like this Central American ***Sobarocephala rubsaamenii*** tentatively approaching a bird dropping already attended by a larger calyptrate fly, are often attracted to the droppings of small animals and birds. **Subfamily Clusiinae** ❹ ❺ ❻ Members of the mostly Holarctic genus ***Clusia***, such as this mating pair of eastern North American ***C. czernyi*** and these western North American ***C. occidentalis***, are among the largest and most conspicuous clusiids. **FAMILY ASTEIIDAE** ❼ These tiny ***Leiomyza scatophagina*** are mating on a fresh mushroom, where the female will later lay a few relatively large eggs. This is a Holarctic species found in both Europe and North America. ❽ This ***Sigaloessa*** was one of many on the inner side of windows in a lodge in Costa Rica. Adult flies in this otherwise uncommon New World (and Australian) genus are often found on windows. ❾ The tiny (less than 2 mm) ***Asteia beata*** occurs across North America, where it is the most common member of the Asteiidae but still a rarely noticed fly.

FAMILY ANTHOMYZIDAE ❶ ***Amygdalops*** is a large, mostly Afrotropical/Oriental genus often associated with decaying stems, such as this banana stem in the Philippines. ❷ This African ***Amygdalops*** is "bubbling." Bubbling (regurgitation and reingestion of a droplet of fluid) is common throughout the Brachycera, possibly serving to evaporate water and concentrate the content of the fly's crop. ❸ Species in the Holarctic genus ***Anthomyza*** are "typical" anthomyzids usually found in grassy areas. ❹ Most anthomyzids develop in plant tissue, but larvae of the recently described Holarctic genus ***Fungomyza*** develop within the tissues of soft fungi. This southeastern American fly was captured and used in the original description of the only Nearctic species in the genus, ***F. buccata***. ❺ ***Ischnomyia spinosa*** is an eastern Nearctic species characteristic of open forest. This small Holarctic genus includes only three species. ❻ ***Mumetopia*** is a small New World genus, but one species has been accidentally introduced to Hawaii. This eastern North American species has normal wings, but a recently described congener from high *páramo* in Ecuador has its wings narrowed to needle-like strips. **FAMILY STENOMICRIDAE** ❼ The 30 described species in the cosmopolitan genus ***Cyamops*** often occur near rivers and streams, but this Australian fly was found in a coastal wetland. ❽ The Neotropical genus ***Planinasus*** includes only three named extant species; many more await description. These inconspicuous little flies are common in riparian habitats throughout Central and South America, sometimes along shorelines but usually on moist fallen branches or trunks in, over or alongside flowing waters. This Ecuadorian fly is on a waterfall-splashed log. ❾ ***Stenomicra*** species can be found on broad leaves throughout the world's tropics. This one, from Vietnam, is an undescribed species.

FAMILY STENOMICRIDAE (continued) ❶ ❷ ❸ Known larvae of the enormous cosmopolitan genus ***Stenomicra*** develop in fluid and organic matter in the funnel-like bases of some kinds of leaves. Adults of Neotropical species often hang out by the hundreds in young, partially unfurled leaves of *Heliconia* and similar plants (the flies on the upper part of the leaf shown here are fungus gnats, but most of those in the unfurled part are *Stenomicra*). Most species, probably including these Ecuadorian species, remain undescribed. ❹ ❺ ***Stenomicra*** species are common in Australia, although very poorly known. Shown here are a couple of undescribed species from New South Wales. **FAMILY PERISCELIDIDAE (DWARF FLIES)** ❻ ***Periscelis*** species, like this ***P. annulata*** on a beetle-bored aspen in Canada, are found on bleeding deciduous trees in every region except the Afrotropical. ❼ The strikingly colored and shaped species of ***Scutops***, such as this Costa Rican fly, occur only in the neotropics. These poorly known flies are not commonly encountered. **FAMILY ODINIIDAE (RIBBON FLIES) Subfamily Traginopinae** ❽ ***Traginops***, including this pair of Canadian ***T. irrorata***, are among the most distinctive sap-loving flies because of their brightly spotted wings and body, banded eyes and prominently inflated ocellar plate. *Traginops* has only a couple of Nearctic species; it is more diverse in the Old World. **Subfamily Odiniinae** ❾ ***Neoalticomerus*** is a Holarctic genus with one species in Europe and one in North America. This is the Nearctic species, ***N. seamansi***.

FAMILY ODINIIDAE (continued) ❶ ❷ ***Odinia*** is a widespread genus, at least in the northern hemisphere, but it is relatively rarely encountered except at bleeding tree wounds. Shown here are a relatively dull ***O. boletina*** from Canada and a more colorful Ecuadorian species (possibly ***O. surumuana***). ❸ Most ***Schildomyia***, like this female ***S. reticulata*** ovipositing over a sap flow on a Costa Rican tree, are Neotropical, but (remarkably) a couple of species are also known from the eastern Palaearctic. **FAMILY OPOMYZIDAE** ❹ ❺ Although there are many native Nearctic species of the large genus ***Geomyza*** (which includes most species of Opomyzidae), the most common species in North America are Holarctic species that were probably introduced from Europe. Shown here are an individual of the long-established species ***G. apicalis*** and a mating pair of the recently introduced ***G. tripunctata***. ❻ ❼ ***Opomyza*** is an Old World genus, but this species, ***O. petrei***, was accidentally introduced to North America, where it is now a common west coast species. Opomyzid flies usually develop as stem miners in grasses and are thus predisposed to accidental transport around the world. **FAMILY FERGUSONINIDAE (MYRTLE GALL FLIES)** ❽ Flies in the small family Fergusoninidae develop only in galls induced by an equally specialized and co-dependent group of nematodes (*Fergusobia*). Shown here is a ***Fergusonina*** species under consideration as a biological control agent for use against Australian broad-leaved paperbark (*Melaleuca quinquenervia*), now an introduced weed in parts of southeastern United States (photo by Susan Wineriter, USDA). ❾ The antennae of fergusoninids are deeply recessed into the front of the head, rendering this family as morphologically distinctive as it is biologically unique.

FAMILY NEUROCHAETIDAE (UPSIDE-DOWN FLIES) ❶ ❷ ❸ ***Neurochaeta inversa***, the **Eastern Australian Upside-down Fly**, is found on the broad leaves of *Alocasia macrorrhiza* (Araceae) in New South Wales. Adult flies feed on pollen and lay eggs on the female flowers; larvae graze on microorganisms in the watery environment inside the spathe. Related species are found in Southeast Asia and Africa. ❹ ***Nothoasteia clausa***, the **Western Australian Upside-down Fly**, lives among the leaves of tall grass trees (Xanthorrhoeaceae) in Western Australia; adults can be found smoothly scuttling along the leaves (head down, of course). The only other species in this genus is described from Queensland. **FAMILY AULACIGASTRIDAE (SAP FLIES)** ❺ These ***Aulacigaster korneyevi*** were among many in a Costa Rican bromeliad (tank plant), walking up and down the inner leaf surface and waving their legs. This is an older (1999) photo that documents the first discovery of the habitat of this species, which was not formally named until 2011. ❻ This elongate ***Aulacigaster ecuadoriensis*** (formerly in *Schizochroa*) is on a broad *Heliconia* leaf in Ecuador. This species runs along with its abdomen raised in a good imitation of a rove beetle (Staphylinidae) when disturbed. ❼ ***Aulacigaster neoleucopeza*** is a widespread and common species in North America and also known from a few records in Europe. Adults can be found in abundance on fermenting sap runs on many kinds of trees; larvae develop in the sap. ❽ ***Aulacigaster*** is a diverse genus in the neotropics, with some 37 species described so far. Shown here is an Ecuadorian species. **FAMILY NEMINIDAE (NOBODY FLIES)** ❾ ***Nemo*** is a small genus of eight species known from Australia, South Africa and New Guinea. These tiny flies are rarely collected but are most often found on broad-leaved plants.

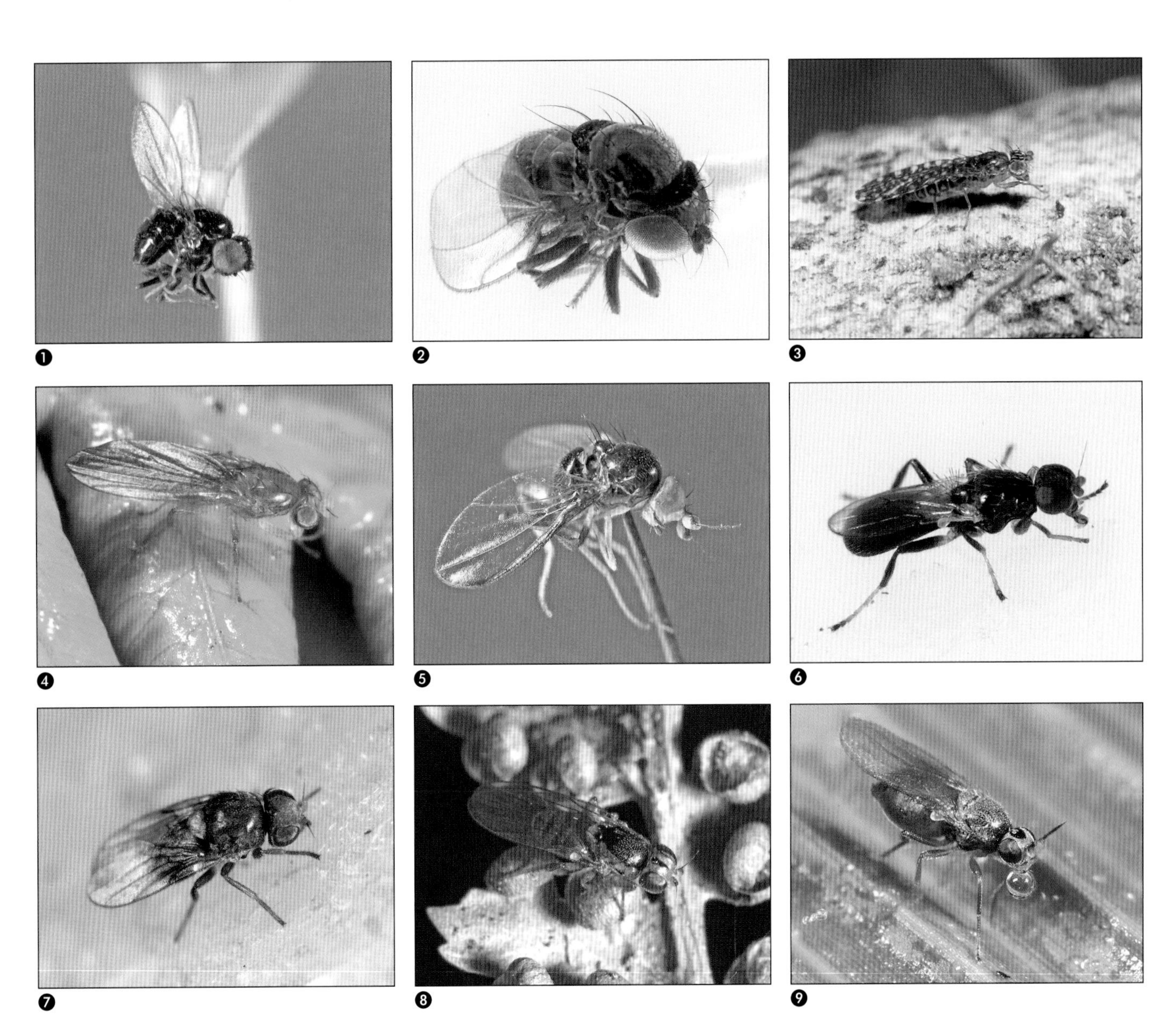

FAMILY XENASTEIIDAE ❶ Xenasteiidae (all in the genus ***Xenasteia***) are minute flies that occur (generally unnoticed) on seashores along the Pacific and Indian Oceans. This ***X. sabroskyi***, from Hawaii, is in the U.S. National Museum of Natural History. ❷ This undescribed species of ***Xenasteia*** was collected among seashore plants (*Sesuvium portulacastrum*) on the east coast of Australia, far south of the previously known range of this small family of 12 named species. **FAMILY TERATOMYZIDAE (FERN FLIES)** ❸ This distinctively pigmented ***Teratomyza undulata*** is on its host fern in the Blue Mountains above Sydney, Australia. Other members of this small genus occur in New Zealand and the Oriental region. ❹ This unnamed species of ***Auster***, like most Teratomyzidae, occurs on fern foliage in moist Australian forests. **FAMILY MARGINIDAE (MARGIN FLIES)** ❺ This is the holotype of ***Margo aperta*** in the Australian Museum, and one of only four known specimens of Marginidae. It is also the type of the family Marginidae, since *M. aperta* is the type species of this mysterious family.

SUPERFAMILY CARNOIDEA

FAMILY CHLOROPIDAE (FRIT, GRASS AND EYE FLIES, ETC.) Subfamily Oscinellinae ❻ The small Afrotropical genus ***Anatrichus*** has only three species, of which ***A. erinaceus*** is the most widespread. ❼ Members of the large, cosmopolitan genus ***Conioscinella*** are associated with a variety of habitats and have been recorded as scavengers in mines, galls, dead invertebrates and grasses already damaged by other Diptera. This one was attracted to a cut stem in the Philippines. ❽ This Australian fly probably belongs in ***Gaurax***, a huge, ill-defined, cosmopolitan genus with a wide variety of described and undescribed species, including more than 80 Australian species (there is some question about whether Australian "*Gaurax*" species should be in *Gaurax, Lioscinella* or undescribed genera). Nearctic larvae assigned to this genus are scavengers in fungi, frass and other decaying material, but Australian species with known habits develop in spider egg masses. ❾ This Tanzanian ***Melanochaeta*** is "bubbling," blowing a sphere of liquid from its mouth, then reingesting the bubble. *Melanochaeta* is a mostly Old World group although it does include two North American species.

❶ ❷ ❸ ❹ ❺ ❻ ❼ ❽ ❾

❿

FAMILY CHLOROPIDAE (continued) ❶ Members of the common New World genus ***Olcella*** are often kleptoparasitic, sharing prey captured by arthropod predators such as spiders, robber flies and mantids. This North American ***O. cinerea*** is feeding on an ant captured by a crab spider. Similar habits are seen in some Australian Chloropidae. ❷ This is the Holarctic species ***Siphonella oscinina***, known both as a predator of scale insects and a parasitoid of spider egg cases. Other members of this Holarctic-Afrotropical genus are recorded as predators of grasshopper egg masses. ❸ This Tanzanian fly belongs in the widespread genus ***Steleocerellus***, a mostly Old World tropical group also represented by a few species in the neotropics. ❹ This Australian fly belongs to the large, widespread genus ***Tricimba***. American species of *Tricimba* have been reared from decaying plants, but Australian rearing records of the genus are from spider egg sacs. **Subfamily Chloropinae** ❺ ***Bathyparia praeclara*** is common on the sand beach at Dar es Salaam, Tanzania. *Bathyparia*, which includes only one species, is also known from South Africa and Madagascar. ❻ ❼ ***Chlorops*** is a large genus, diverse in temperate parts of both northern and southern hemispheres. Known larvae are phytophagous on a variety of plants, especially grasses and sedges. The orange *Chlorops* is from Canada and the yellow one is from Australia. ❽ ***Gampsocera*** is a widespread Old World and Pacific (mostly Oriental) genus associated with decomposing plant material such as rotting banana stems. This one is from Vietnam. ❾ ***Ectecephala*** is a diverse, mostly Neotropical group, but this one (***E. albistylum***) ranges north to Canada. ❿ Known larvae of the Holarctic genus ***Epichlorops***, such as these ***E. exilis*** from Canada, are phytophagous in sedges.

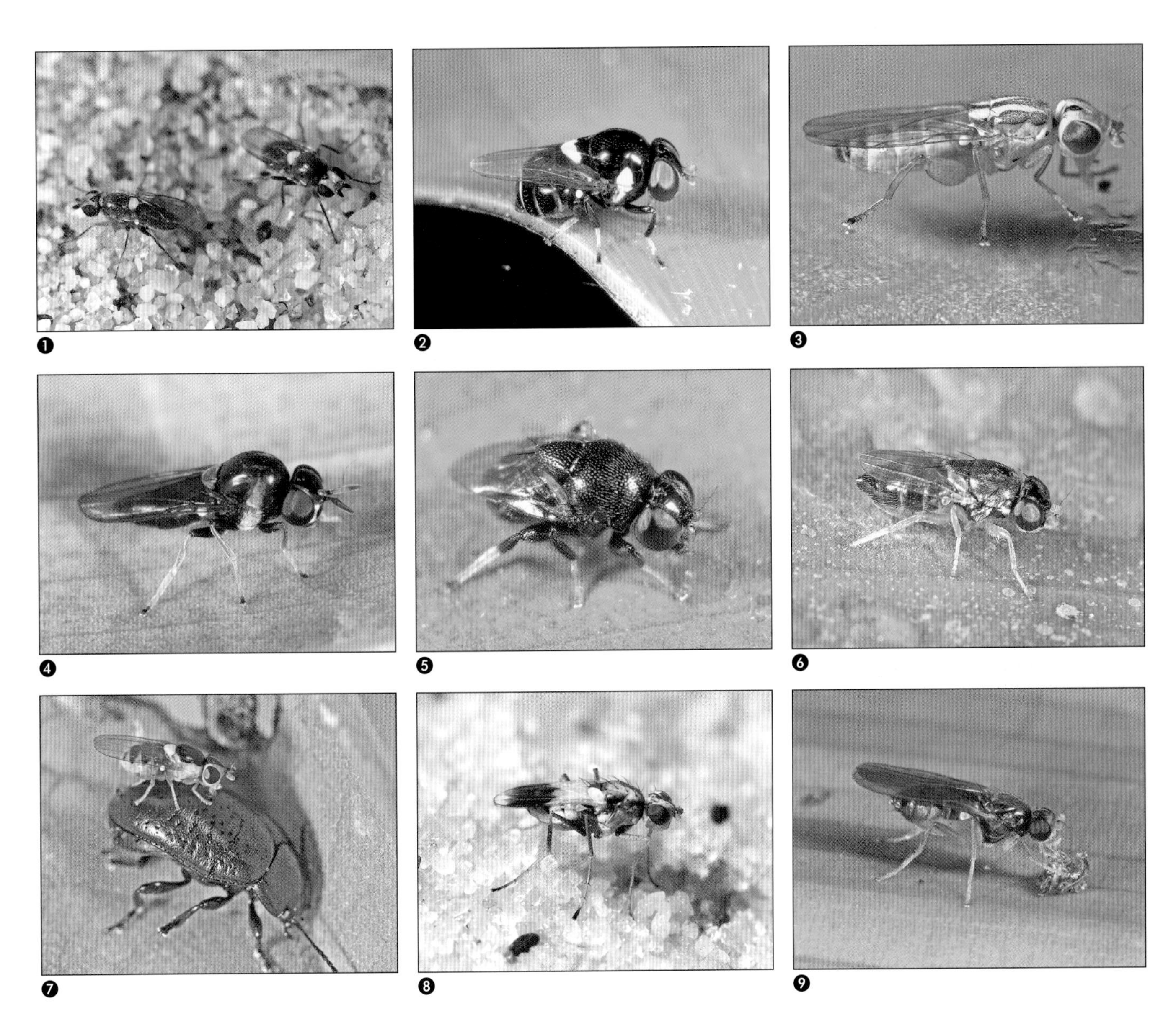

FAMILY CHLOROPIDAE (continued) ❶ This ***Eutropha*** species was abundant among the sparse vegetation on a Philippines beach. The only known larvae of this Old World genus were found in dead animals washed up on a Namibian beach. ❷ The distinctively robust species of ***Merochlorops***, such as this ***M. lucens*** from the Philippines, occur mostly in Australia and the Old World tropics. Larvae are phytophagous and one species is a shoot fly pest of large cardamom (*Amomum subulatum*). ❸ ***Meromyza*** is a well-known genus of phytophagous chloropines that includes the Wheat Stem Maggot (*M. americana*). Shown here is the Holarctic species ***M. pratorum*** (**Grass Fly**). ❹ This large, almost syrphid-like ***Neoloxotaenia luzonicus*** was attracted to a cut banana stem on the island of Luzon in the Philippines. *Neoloxotaenia* is a small Oriental–eastern Palaearctic genus with only five described species. ❺ ❻ ***Rhodesiella*** is a common and diverse group in the Old World tropics with just one described species confirmed from the New World. Shown here are a robust African species and a more gracile fly from the Philippines. ❼ ***Thaumatomyia*** is a diverse, widespread genus of more than 40 species, but little is known of their biology. Some develop as predators of aphids and others periodically appear as adult swarms. This Canadian fly is feeding on exudates from an injured but mobile leaf beetle. **Subfamily Siphonellopsinae** ❽ ***Apotropina ornatipennis*** is an ornately colored fly found on the beaches of eastern Australia. A similar species, *A. exquisita*, is found on the beaches of Western Australia. ❾ This Ecuadorian ***Apotropina*** is feeding on what appears to be insect fragments.

FAMILY CHLOROPIDAE (continued) ❶ ❷ These ***Apotropina*** were among hundreds massed under broad leaves along a stream in the Philippines. ❸ ***Apotropina*** is one of the most common acalyptrate genera in Australia and is almost as abundant in many other parts of the world. Some species develop as scavengers in wasp nests, but most species of this huge genus remain unknown as larvae. **FAMILY MILICHIIDAE (FREELOADER FLIES, ETC.) Subfamily Phyllomyzinae** ❹ ***Paramyia nitens*** is a common fly on composite flowers in much of the New World from Chile to Canada. This poorly known genus is diverse in both the New World and Old World tropics. ❺ ***Paramyia nitens*** is one of the kleptoparasitic species routinely attracted to some kinds of prey in spider webs. This stink bug trapped in a Canadian orb-weaving spider's web is almost concealed by about 30 of these small flies. ❻ Several ***Phyllomyza*** species in both the New and Old Worlds are associated with ants. This one (in the Philippines) disappeared underground with its ant associate right after the picture was taken. Larvae are probably scavengers in the ant nest. ❼ ***Phyllomyza***, like this female from the American southwest, is a cosmopolitan genus but is especially diverse in the Old World tropics. ❽ ***Stomosis vittata*** is one of two Australian species in its genus. The two other described *Stomosis* species occur in the New World. **Subfamily Madizinae** ❾ Members of the large, widespread genus ***Desmometopa*** seem to appear instantly around recently killed Honey Bees, even in parts of the world where Honey Bees are not native. Half a dozen flies can be seen on this pollen-spattered bee captured by a robber fly (*Laphria*).

FAMILY MILICHIIDAE (continued) ❶ ❷ These two Honey Bees, one dead on a leaf in the United States and one captured by a lynx spider in Cuba, have both attracted numerous ***Desmometopa***. These flies have also been reported as stealing pollen from living Honey Bees. Larvae of the genus occur in all sorts of decomposing material. ❸ This Vietnamese ***Desmometopa*** is bubbling, probably to evaporate water and concentrate crop contents. ❹ ***Leptometopa*** is a widespread milichiid genus associated with foul substrates such as pit toilets, cesspools and animal stalls. This adult is on a flowerhead in the Arizona desert. ❺ ***Madiza*** adults are often common on flowers but known larvae are found in decomposing material, including pig dung. This is a mostly Old World genus; the only North American species (***M. glabra***, shown here) is probably introduced. **Subfamily Milichiinae** ❻ This was one of several ***Eusiphona*** individuals reared from a leafcutter bee (*Megachile*) nest in Ontario, Canada. Larvae feed on pollen in the bee's nest. ❼ ❽ Males of some species of the large, widespread genus ***Milichiella*** have reflective silvery abdomens and form conspicuous mating swarms that flash in the sunlight. Similar habits are found in a few other milichiid genera. This individual fly is from Chile and the mating swarm is in Tanzania. ❾ ***Milichiella lacteipennis*** is a cosmopolitan species associated with a variety of filthy materials, including manure heaps, chicken dung and rotting vegetables. This individual was pulled from a huge swarm that extended from ground level almost to my seventh-floor hotel room in the Tanzanian city of Dar es Salaam.

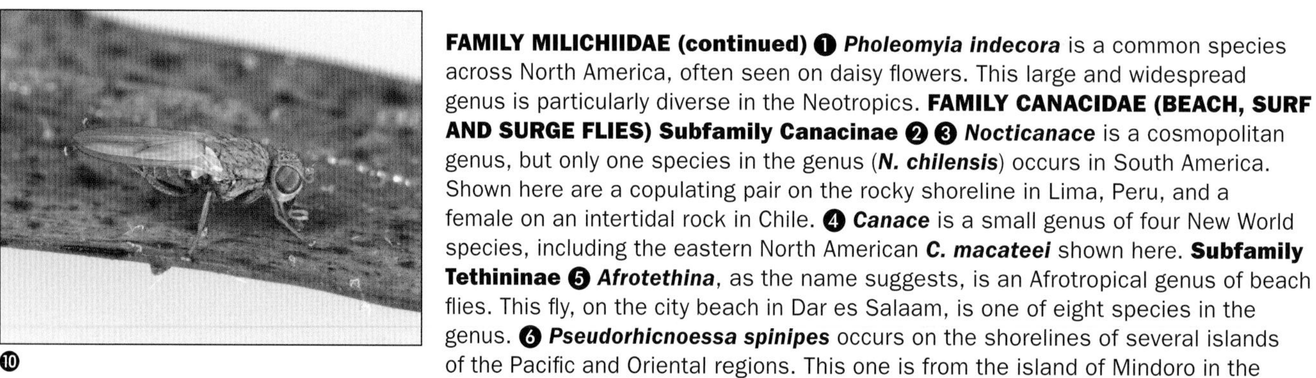

FAMILY MILICHIIDAE (continued) ❶ ***Pholeomyia indecora*** is a common species across North America, often seen on daisy flowers. This large and widespread genus is particularly diverse in the Neotropics. **FAMILY CANACIDAE (BEACH, SURF AND SURGE FLIES) Subfamily Canacinae** ❷ ❸ ***Nocticanace*** is a cosmopolitan genus, but only one species in the genus (***N. chilensis***) occurs in South America. Shown here are a copulating pair on the rocky shoreline in Lima, Peru, and a female on an intertidal rock in Chile. ❹ ***Canace*** is a small genus of four New World species, including the eastern North American ***C. macateei*** shown here. **Subfamily Tethininae** ❺ ***Afrotethina***, as the name suggests, is an Afrotropical genus of beach flies. This fly, on the city beach in Dar es Salaam, is one of eight species in the genus. ❻ ***Pseudorhicnoessa spinipes*** occurs on the shorelines of several islands of the Pacific and Oriental regions. This one is from the island of Mindoro in the Philippines. ❼ ❽ These pairs of ***Tethina albula*** were among many on crab carcasses on an American east coast beach. A second male is trying to horn in on one of the pairs shown here. ❾ This Australian ***Tethina nigriseta*** is carrying a heavy load of mites under its abdomen. ❿ This is ***Tethina grisea***, a Mediterranean member of this worldwide genus.

FAMILY CANACIDAE (continued) Subfamily Pelomyiinae ❶ Species in the small subfamily Pelomyiinae are unusual in the Canacidae for their inland distribution. This is a ***Pelomyia*** specimen collected in Arizona. **Subfamily Horaismopterinae** ❷ The small subfamily Horaismopterinae includes only four species, all in the genus ***Horaismoptera*** and all found on southern hemisphere beaches such as this one in Tanzania. **FAMILY CARNIDAE (BIRD AND FILTH FLIES)** ❸ The Holarctic bird fly species ***Carnus hemapterus*** shed their wings before settling in to feed on their young bird hosts. Shown here are four males (about 1 mm) and four fat females. ❹ Species in the large genus ***Meoneura*** are known as "filth flies" because some species breed in dung and carrion. These tiny flies are poorly known, but different species have distinctive and diagnostic male genitalia, as demonstrated by the surstyli ("claspers") and phallus visible on this pointed specimen. **FAMILY INBIOMYIIDAE (INBIO FLIES)** ❺ This tiny ***Inbiomyia*** was photographed on a fallen Costa Rican leaf in 2005, before the genus or the family Inbiomyiidae were described. Nothing is known about the biology of the flies in this recently discovered family. ❻ ❼ These ***Inbiomyia*** (probably ***I. empheres***) are only a couple of millimeters long but are nonetheless recognizable by their broad heads and long aristae. *Inbiomyia* can be common in fresh leaf litter, such as that around the Ecuadorian treefall where these were photographed. **FAMILY AUSTRALIMYZIDAE** ❽ This is a specimen of ***Australimyza kaikoura*** that I collected along the shoreline of the Banks Peninsula in New Zealand. All nine species of this little-known group are tiny flies found in coastal New Zealand or nearby islands. **FAMILY ACARTOPHTHALMIDAE (SHAVED-EYE FLIES)** ❾ The entire family Acartophthalmidae is made up of four rarely encountered species of ***Acartophthalmus***. This is the Holarctic ***A. nigrinus***.

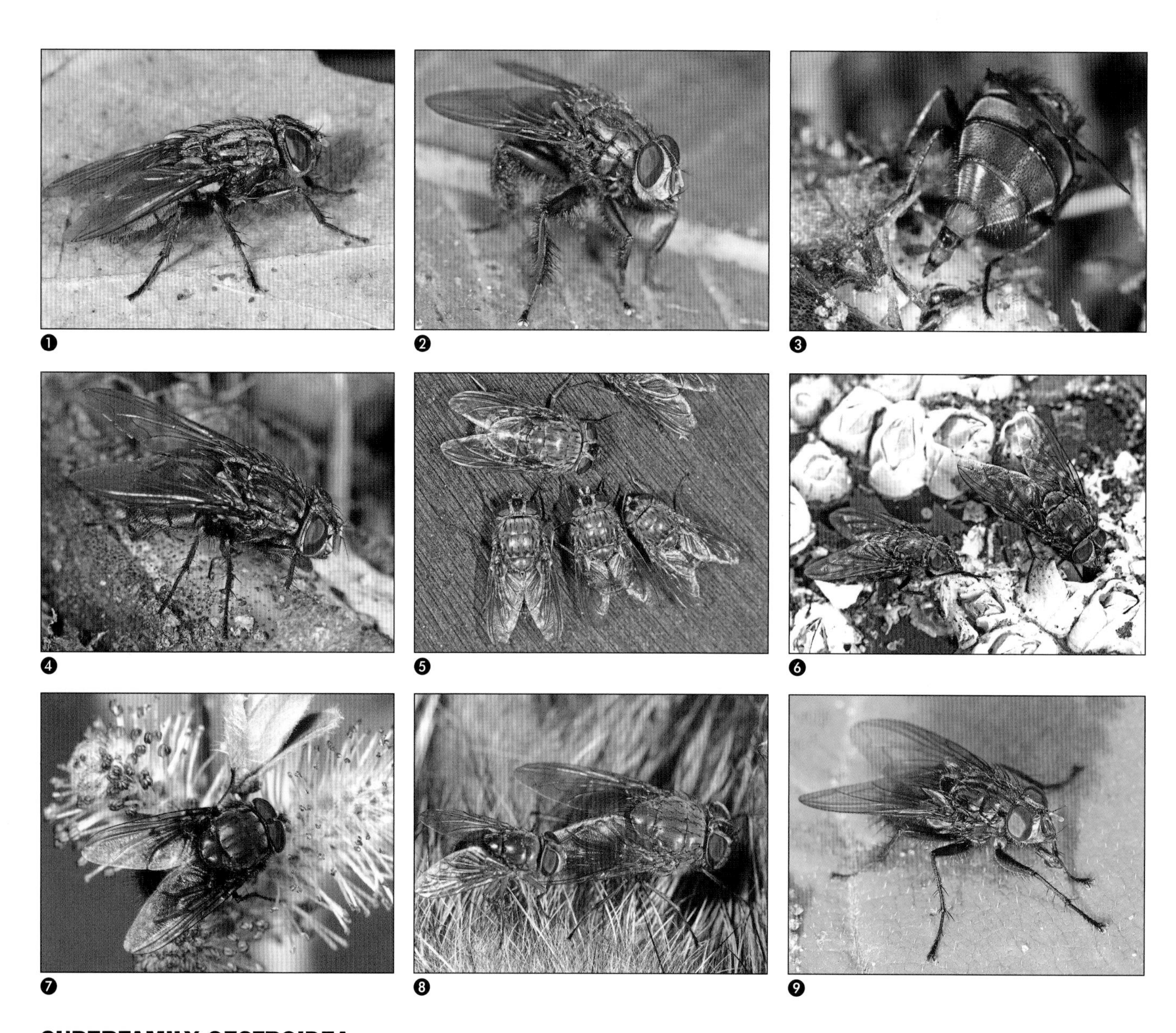

SUPERFAMILY OESTROIDEA

FAMILY CALLIPHORIDAE (BLOW FLIES, SCREWWORMS AND CLUSTER FLIES) Subfamily Calliphorinae ❶ ***Bellardia*** is a common European genus with several species that probably attack earthworms. ***B. vulgaris*** also occurs in eastern North America, where it has apparently been accidentally introduced. ❷ ***Blephericnema*** is a spectacular Andean genus containing the single large, hairy-legged species ***B. splendens***, seen here in a high-altitude cloud forest in Ecuador. ❸ ❹ This Canadian ***Calliphora livida*** is laying eggs on a dead fish, using her soft telescoping abdominal tip to deliver eggs to pockets of damaged tissue. ❺ ***Calliphora stygia*** (**Eastern Golden-haired Blow Fly**) is a common nuisance fly in Australia, Polynesia and New Zealand, where it can abound on carrion and sometimes invades the living flesh of wounded animals. ❻ ***Calliphora vicina*** is sometimes called the **Urban Bluebottle Fly** and sometimes the **European Bluebottle**, reflecting the origin and synanthropic habits of this common species, seen here investigating some stranded barnacles on the Canadian west coast. *Calliphora vicina* now occurs in every zoogeographic region but is most common in the Holarctic region. ❼ ***Calliphora vomitoria***, distinctive for its reddish "cheek" hairs, is a common carrion fly throughout the Holarctic region and also found in Hawaii and South Africa. ❽ This relatively large ***Calliphora ochracea*** (**Reddishbrown Blow Fly**) is dwarfing a small *Calliphora* (probably ***C. macleayi***) on a dead rabbit by an Australian roadside. Australia has a fascinating *Calliphora* fauna, including more than 40 named species in half a dozen subgenera. ❾ The ominously named ***Cynomya cadaverina*** is indeed a common visitor to cadavers and other decomposing objects in northern North America and Europe. Both species in the genus *Cynomya* have a Holarctic distribution.

FAMILY CALLIPHORIDAE (continued) ❶ This Tanzanian ***Pericallimyia*** was attracted to honeydew on a leaf; the other dozen or so species in this genus are also found in Africa. **Subfamily Phumosiinae** ❷ The small subfamily Phumosiinae is an Old World tropical group, some species of which are predators of frog eggs. The biology of this African species, ***Phumosia congensis*** (specimens in Chicago's Field Museum), is unknown. **Subfamily Luciliinae** ❸ The eastern North American ***Lucilia coeruleiviridis*** commonly visits both flowers and decomposing materials. Like many *Lucilia* species, this species can vary from iridescent green to brilliant bronze. ❹ ***Lucilia sericata*** (**Common Greenbottle**) originated in the Holarctic region but is now a common pest throughout the world. This is a pest species often responsible for sheep strike, but it is also a beneficial species used in maggot therapy to treat difficult infections. ❺ This Costa Rican ***Lucilia*** is probably ***L. purpurescens***, but there are many similar undescribed Neotropical species. **Subfamily Chrysomyinae** ❻ I bumped into this half-submerged dog skull while sampling insects in an urban pond. The larger of the two flies and most of the writhing mass of maggots surrounding the skull are **Black Blow Flies**, ***Phormia regina***. ❼ Stinkhorn fungi (aptly placed in the genus *Phallus*) expose their spores in a stinky, sticky mass that attracts a variety of flies, in this case several Black Blow Flies (***Phormia regina***). This very widespread fly is the only species in the genus. ❽ The tropical **African Latrine Blow Fly**, ***Chrysomya putoria***, does not breed in carrion like most *Chrysomya* but instead favors human waste. This abundant synanthropic fly is a common inhabitant of pit latrines. ❾ This ***Chrysomya megacephala*** is recognizable as a male by its large eyes with the upper facets enlarged. Now common in the New World tropics, *C. megacephala* was originally an Old World tropical species; it has only recently (in the 1970s) established itself in the neotropics.

FAMILY CALLIPHORIDAE (continued) ❶ ***Chrysomya rufifacies***, known as the **Hairy Maggot** because of its unusually hirsute larva, is another blow fly that has recently arrived in the New World. Originally Southeast Asian and Australian, this species appeared in North America around 1980. ❷ ❸ ***Cochliomyia macellaria*** (**Secondary Screwworm**) is a widespread New World species ranging from Canada to southern South America. Shown here are a male on a flower and a female sharing a bit of dung with a *Lucilia*. This small New World genus also includes the Primary Screwworm, *C. hominivorax*. ❹ ***Compsomyiops*** species, like this one (probably ***C. fulvicrura***) from Peru, look much like *Cochliomyia*, except for a slightly broader palpus. This New World (almost entirely Neotropical) genus includes half a dozen species. **Subfamily Aphyssurinae** ❺ ❻ ***Protocalliphora*** larvae suck blood from nestling birds, and abandoned nests are often full of *Protocalliphora* puparia. Both of these flies were reared from puparia in birds nests in eastern Canada. **Subfamily Toxotarsinae** ❼ Most Toxotarsinae are southern South American, but this hairy-eyed species (***Sarconesia roraima***) seems to range from the Andes in Peru (where this picture was taken) to the *tepuis* of northeastern South America (where the species was first discovered). ❽ Most ***Sarconesia*** species occur in southern South America; this species, ***S. magellanica***, is one of the more common Chilean blow flies. **Subfamily Melanomyinae** ❾ ❿ The small flies in the rarely encountered subfamily Melanomyinae occur in the Holarctic and Neotropical regions; these are North American specimens of the genera ***Angioneura*** (dark legs) and ***Opsodexia*** (yellow legs). Known larvae are parasitoids of snails.

FAMILY CALLIPHORIDAE (continued) Subfamily Polleniinae ❶ Cluster flies (***Pollenia*** spp.) are frequently attacked by parasitic fungi (Entomophthorales) that predispose the fly to hang from a prominent spot where its body can most effectively disseminate the fungal spores. ❷ ***Pollenia*** are common flower visitors and important pollinators. This is ***Pollenia rudis***, a common species in both Europe and North America. All six *Pollenia* species found in North America were until recently treated as *P. rudis*. ❸ ❹ The difference in eye size between male and female is marked in these pairs of ***Pollenia***, a male and female of ***P. pediculata*** side by side, and a mating pair of ***P. angustigena***. Both species have a Holarctic distribution. ❺ Just as some Sarcophagidae develop in New World pitcher plants (*Sarracenia*), some Calliphoridae live in Old World pitcher plants. These specimens of ***Wilhelmina nepenthicola*** were reared from *Nepenthes* pitchers in Indonesia; the puparia from which they emerged are glued above the data labels. Although originally treated as part of the Polleniinae, Rognes (2011) suggests that it belongs in the Phumosiinae. **Subfamily Ameniinae** ❻ ❼ Larvae of the endemic Australian genus ***Amenia*** are parasitoids of snails; this is ***A. leonina***. ❽ ***Amenia*** species, such as this ***A. nigromaculata***, are superficially similar to Tachinidae and, like Tachinidae, known species are parasitoids. But these Australian flies attack snails while tachinids attack arthropods. ❾ The very tachinid-like species of Ameniinae occur mostly in Australia, but this Vietnamese fly belongs to the large Oriental ameniine genus ***Silbomyia***.

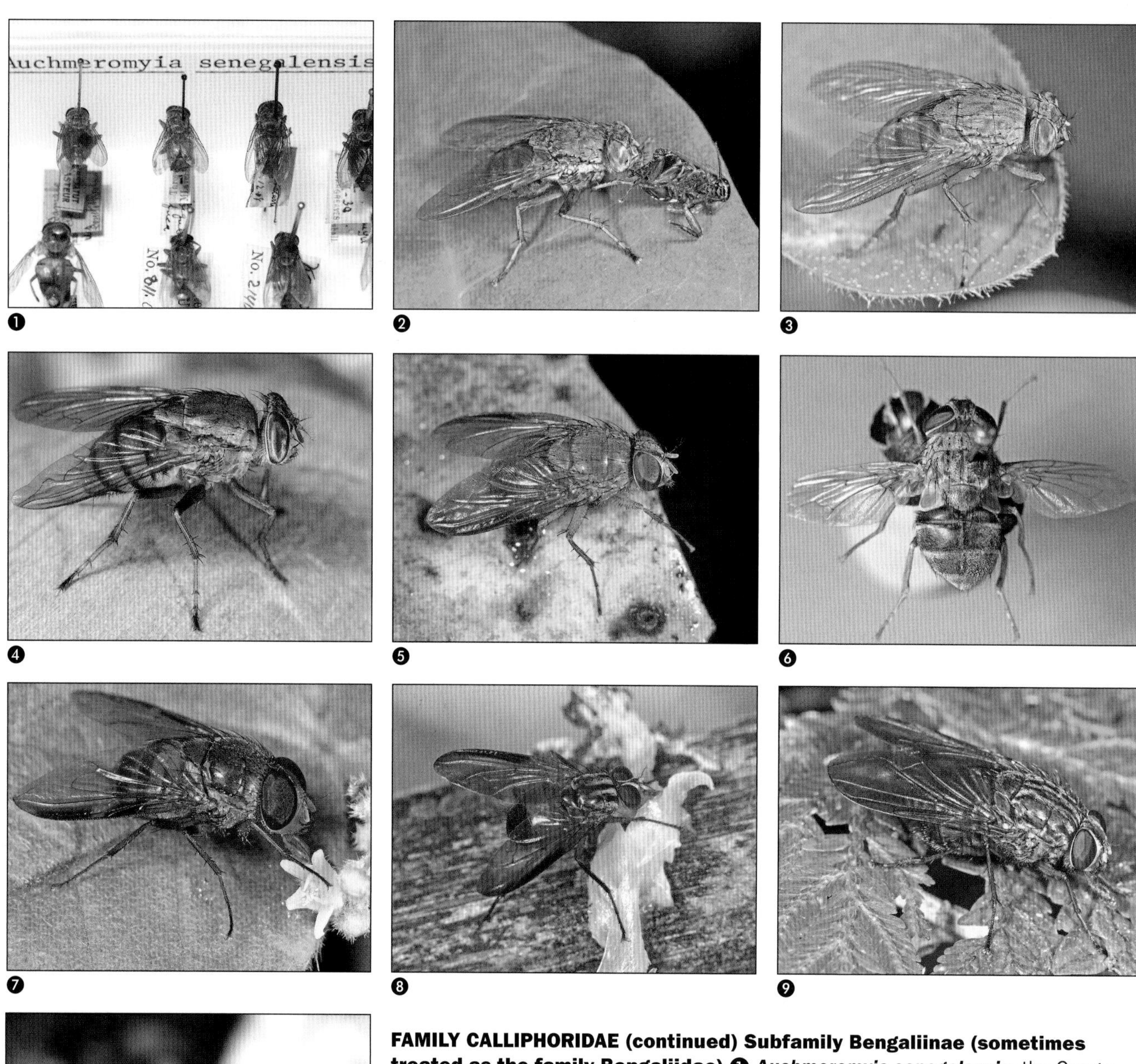

❶ ❷ ❸ ❹ ❺ ❻ ❼ ❽ ❾ ❿

FAMILY CALLIPHORIDAE (continued) Subfamily Bengaliinae (sometimes treated as the family Bengaliidae) ❶ ***Auchmeromyia senegalensis***, the Congo Floor Maggot, is an African species with blood-sucking larvae. This is a tray of *Auchmeromyia* specimens in the Smithsonian Institution in Washington, DC. ❷ ***Bengalia*** is one of the very few calyptrate groups with predaceous adults; this Vietnamese species has captured a small cockroach. ❸ ❹ ***Bengalia*** adults, such as these from Tanzania, are robust brownish predators found throughout the Old World tropics. Larvae are thought to be associated with termites. ❺ The taxonomy of this group of Calliphoridae is controversial, and the taxon here referred to as ***Bengalia*** is sometimes treated as a separate family (Bengaliidae) and divided into 11 genera in four subfamilies. This one is from the Philippines. ❻ Larvae of the African Tumbu Fly, ***Cordylobia anthropophaga***, burrow under vertebrate skin, where they develop in boil-like warbles. This is an African Tumbu Fly specimen in the Smithsonian Institution in Washington, DC. ❼ The large genus ***Tricyclea*** includes many of the most common Calliphoridae in Africa; this flower-visiting example is from Tanzania. Larvae of this genus occur in the nests of ants or termites, where they probably develop as scavengers. **FAMILY MESEMBRINELLIDAE** ❽ The large and colorful ***Eumesembrinella randa*** is a common fly of the Amazonian rainforest. ❾ ❿ Most Mesembrinellidae, including these two Costa Rican species, are in the genus ***Mesembrinella***. Females carry a single large egg at a time; the larval habitat for these very common Neotropical flies remains unknown.

FAMILY RHINIIDAE ❶ ***Fainia*** is a small group of five named Afrotropical species. This is one of at least three *Fainia* species to occur in Tanzania. ❷ ***Isomyia*** is an Old World tropical genus. This brilliantly metallic species is from the Philippines. ❸ ***Rhyncomya*** is a large Old World genus that includes species that have been reared from "termite-infested cow dung." Shown here is a species from Greece. ❹ ***Strongyloneura*** is a mostly Oriental-African genus. This small species, ***S. prasina***, ranges from Japan and Taiwan south to the Philippines. ❺ ***Stomorhina*** is a widespread genus with species in Europe, Australia and throughout the Old World tropics; this one is from Greece. Larvae of some species consume grasshopper egg pods, while others are predators of termites. ❻ ***Sumatria*** is a small Southeast Asian genus of four described species, unusual among the normally dark and metallic rhiniids for their pale pigmentation. Shown here is an unidentified species from Vietnam. **FAMILY AXINIIDAE (AXE FLIES)** ❼ ***Barrinea cantrelli*** belongs to the small Australian family Axiniidae, often called axe flies because of their distinctively shaped antenna. **FAMILY RHINOPHORIDAE (ISOPOD KILLERS)** ❽ ***Melanophora roralis*** is a cosmopolitan and common fly with strikingly different males and females; only the females have a white spot on the wingtip and only the males have a bottlebrush-like arista. Like other members of the family, they are isopod (woodlice) parasitoids. ❾ Neotropical Rhinophoridae are poorly known; this Ecuadorian fly is a new (undescribed) species of ***Trypetidomima***. The only named species in the genus is from Brazil and has partly yellow legs. The tongue-like lower calypter, just visible on the right side of this fly, is characteristic of the family. ❿ ***Ventrops*** is an Afrotropical genus of four species; shown here is a species from Tanzania.

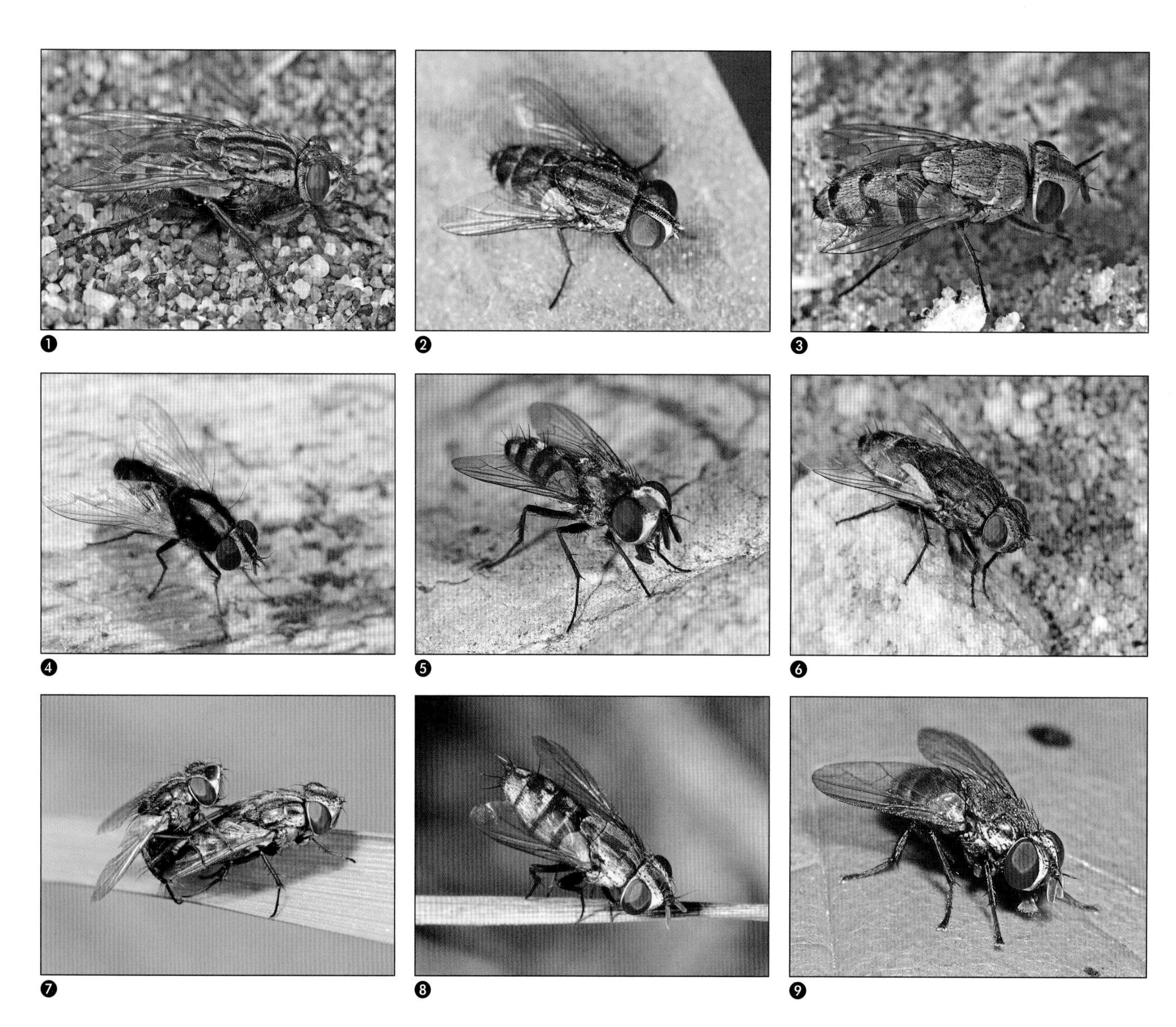

FAMILY SARCOPHAGIDAE (FLESH AND SATELLITE FLIES, ETC.) Subfamily Paramacronychiinae ❶ ***Wohlfahrtia*** species usually have a characteristically spotted abdomen in contrast to the tessellated or checkered abdomen of most flesh flies. Some members of this genus are parasitic, causing myiasis in vertebrate flesh. **Subfamily Miltogramminae** ❷ This ***Amobia aurifrons*** from the southwestern United States is one of 13 species in the widespread genus *Amobia*, larvae of which are usually kleptoparasites in the nests of twig-nesting or mud-daubing sphecid and crabronid wasps. ❸ This ***Craticulina seriata*** was one of many on the beach right in the city of Dar es Salaam, Tanzania, where it undoubtedly developed as a kleptoparasitic larva in a sand wasp nest. ❹ Seven of the eight species in the genus ***Dolichotachina***, like this ***D. cuthbertsoni*** from Tanzania, are Afrotropical; the other species is in Sri Lanka. ❺ This ***Sphenometopa tergata*** from New Mexico is associated with the nests of spider wasps and thread-waisted wasps, where it presumably develops as a kleptoparasite on the paralyzed prey. ❻ Flies in the Old World tropical genus ***Hoplacephala***, like this one from Tanzania, are larger than most miltogrammines and are apparently associated with termites, although few details of their life history have been confirmed. This is one of about 28 species in the genus. ❼ ❽ This mating pair of ***Metopia argyrocephala*** from Canada show the typical sexual dimorphism of this genus. The lone female ***Metopia*** shown here is from the Philippines. These common flies larviposit in nests of solitary bees and wasps, where the larvae develop on stored pollen or paralyzed prey. ❾ ***Oebalia*** species, such as this ***O. minora*** from Canada, are kleptoparasites that develop on stored paralyzed prey in the twig nests of crabronid wasps. This small genus has a dozen species worldwide.

FAMILY SARCOPHAGIDAE (continued) ❶ This ***Opsidia intonsa*** perching on the end of a Chilean twig is one of 10 species in this New World genus, known larvae of which are associated with the nests of solitary bees and sphecid wasps. ❷ ***Phrosinella*** is a mostly Holarctic genus of 23 species. This one was tracking sand wasps on a Great Lakes beach, where it probably developed as a kleptoparasite in a wasp nest. ❸ ❹ ***Senotainia*** species are common flies often seen trailing sphecid or crabronid wasps at a fixed distance, almost as though they are tethered to the wasp. Shown here are a female ***S. trilineata*** watching for wasps from a twig-tip vantage point, and another one in the act of larvipositing on a paralyzed stink bug being carried by a Canadian sand wasp (*Astata bicolor*). *Senotainia* is a widespread genus of some 64 species. ❺ This Nearctic ***Sphenometopa tergata*** is probably watching for a spider wasp or other solitary hunting wasp, since larvae of this species are kleptoparasites that consume paralyzed spiders or other prey in wasps' nests. *Sphenometopa* is a mostly Holarctic genus of some 47 species. **Subfamily Sarcophaginae** ❻ Flesh flies deposit active first-instar larvae rather than eggs. This female is larvipositing on some fresh dung in the southeastern United States. ❼ ***Blaesoxipha*** is a large, worldwide genus of about 250 species. This ***B. angustifrons*** from Nevada belongs to the subgenus ***Acridiophaga***, a group of grasshopper-parasitizing species. ❽ Most genera of the subfamily Sarcophaginae conform to a remarkably uniform appearance, with a gray and black striped thorax and grey and black patterned abdomen. This North American species, in the New World genus ***Boettcheria***, probably developed as a parasitoid inside another insect. ❾ This ***Fletcherimyia*** has extended its fleshy labellum onto the outer surface of a yellow pitcher plant. Larvae of *Fletcherimyia* develop inside the tanklike leaves of these insectivorous plants.

FAMILY SARCOPHAGIDAE (continued) ❶ ❷ ***Fletcherimyia*** is a genus of seven Nearctic species associated with pitcher plants (*Sarracenia* spp.), in which their larvae develop. Shown here are an adult exploring the lip of a yellow pitcher plant (*S. flava*) and a larva of ***F. fletcheri*** exposed by cutting open the pitcher of a Northern Pitcher Plant (*S. purpurea*). ❸ ***Helicobia rapax*** is a common species that has been reared out of dead snails and other dead invertebrates. This species is very widespread in the Nearctic region; the other 32 *Helicobia* species occur throughout the New World. ❹ ❺ ***Macronychia*** is a widespread group of 16 species, including some associated with wasp nests. Shown here are an adult with a puparium and an associated larva among the paralyzed prey of a twig-nesting wasp (*Ectemnius*, Crabronidae) in eastern Canada. ❻ ***Oxysarcodexia*** is a large (79 species) New World genus of typical dung-breeding sarcophagines. This pair is from Canada, but the genus ranges south to Argentina. ❼ This ***Peckia gulo*** from the Pacific coast of Costa Rica is an impressively large flesh fly that seems to specialize in scavenging on dead coastal crabs. *Peckia* is a large genus that contains another 69 New World species. ❽ ***Ravinia*** is a large, mostly New World genus common in urban areas, where the larvae develop in dog dung. Larvae of the genus have been reared from various kinds of dung and some other materials, such as dead crabs. ❾ ***Sarcophaga*** is an enormous worldwide genus of some 800 species organized into several subgenera. This is a member of the subgenus ***Robineauella*** nectaring on American Bittersweet flowers in Canada.

FAMILY SARCOPHAGIDAE (continued) ❶ ❷ *Sarcophaga* species occur throughout the world. Shown here are ***Sarcophaga (Liosarcophaga) rohdendorfi*** from coastal Tanzania and ***Sarcophaga (Mehria)***, probably ***S. travassosi***, from southern Chile. ❸ ❹ ***Sarcophaga*** species are often associated with carrion; these were attracted to a dead snake and to the carrion-like aroma of an *Aristolochia* flower. ❺ ❻ Species in the small New World (mostly Nearctic) genus ***Spirobolomyia*** are associated with large millipedes in the order Spirobolida. The larvae seen here emerged from a dead millipede; one later developed into this adult. ❼ Known larvae of the New World genus ***Tripanurga*** develop on buried turtle eggs, possibly as obligate predators, although recent studies suggest that they are opportunistic and develop only in damaged eggs. This is ***T. aurea*** from California, one of 14 species in the genus. **FAMILY MYSTACINOBIIDAE (NEW ZEALAND BAT FLIES)** ❽ The only known member of the Mystacinobiidae is the **New Zealand Bat Fly**, ***Mystacinobia zelandica***, which is restricted to colonies of the New Zealand short-tailed bat, *Mystacina tuberculata*. Larvae live in the bat guano while adults cling to the host bats, laying eggs collectively in a sort of "nursery" of eggs and larvae. This photo was taken by Rod Morris. **FAMILY TACHINIDAE (PARASITIC FLIES) Subfamily Dexiinae** ❾ This ***Amphibolia (Amphibolia) vidua*** from eastern Australia belongs to the tribe Rutilini, a mostly Australian group of scarab beetle parasitoids that includes some of the biggest and shiniest of all tachinids. This one was among many aggregating on top of Mount Kaputar in New South Wales.

FAMILY TACHINIDAE (continued) ❶ ***Calodexia*** females are almost invariably seen perched over the swarm fronts of Neotropical army ant raids, awaiting the opportunity to parasitize cockroaches flushed out by the marauding ants. This female is watching an *Eciton burchelii* raid in Bolivia. ❷ ❸ ***Calodexia*** is a common Neotropical genus of about 40 species. Females, such as this one from Ecuador (top view), are regular army ant associates. Males, like this one from Ecuador (side view), are harder to find. ❹ Flies in the cosmopolitan genus ***Campylocheta*** are parasitoids of cutworms, inchworms and related caterpillars. ❺ Species in the distinctive New World genus ***Cordiligaster***, such as this Costa Rican ***C. petiolata***, are known to parasitize pyralid and crambid moths. ❻ ***Dexia***, the type genus of the subfamily Dexiinae, is an originally Old World group of beetle parasitoids. This one is from Vietnam. ❼ ***Euchaetogyne roederi***, the only species in the genus, occurs in Mexico and the southwestern United States. ❽ ***Euthera*** is a cosmopolitan genus of bug parasitoids. This North American species, ***E. tentatrix***, is a parasitoid of stink bugs (Pentatomidae). ❾ This ***Morphodexia barrosi*** from Chile is on a tree trunk, a typical haunt for many similar dexiine parasitoids of scarab beetles.

FAMILY TACHINIDAE (continued) ❶ ***Neosophia*** is a rare Neotropical genus with only a couple of named species; this long-legged fly is from primary rainforest in Bolivia. ❷ ***Oestrophasia*** is a small New World genus of scarab beetle parasitoids; this ***O. signifera*** was photographed in Canada, but the same species occurs south to Costa Rica. ❸ Small flies in the mostly Holarctic and northern genus ***Periscepsia*** are often seen frenetically searching the leaf litter in search of caterpillar hosts, flicking their wings in a characteristic fashion. ❹ ***Prosena*** species, such as this Australian species, can be seen taking up a typical dexiine pose on tree trunks in most zoogeographic regions except the Neotropical. Like most similar dexiines, they are parasitoids of scarab beetles. ❺ ***Ptilodexia*** species are common tachinids around dead or fallen trees where their host beetles occur. ❻ ❼ Some of the 50 or so species in the Australian genus ***Rutilia*** are among the biggest and shiniest flies anywhere, but despite their attractive appearance many species remain to be named, and the named species are often difficult to identify. Like most other Dexiinae, known larvae are beetle parasitoids. ❽ This eastern Australian ***Rutilia (Grapholostylum) variegatum*** is carrying a pollinium, probably from an orchid flower, on its head. ❾ This striking Costa Rican fly is a member of the small, poorly known Neotropical genus ***Scotiptera***.

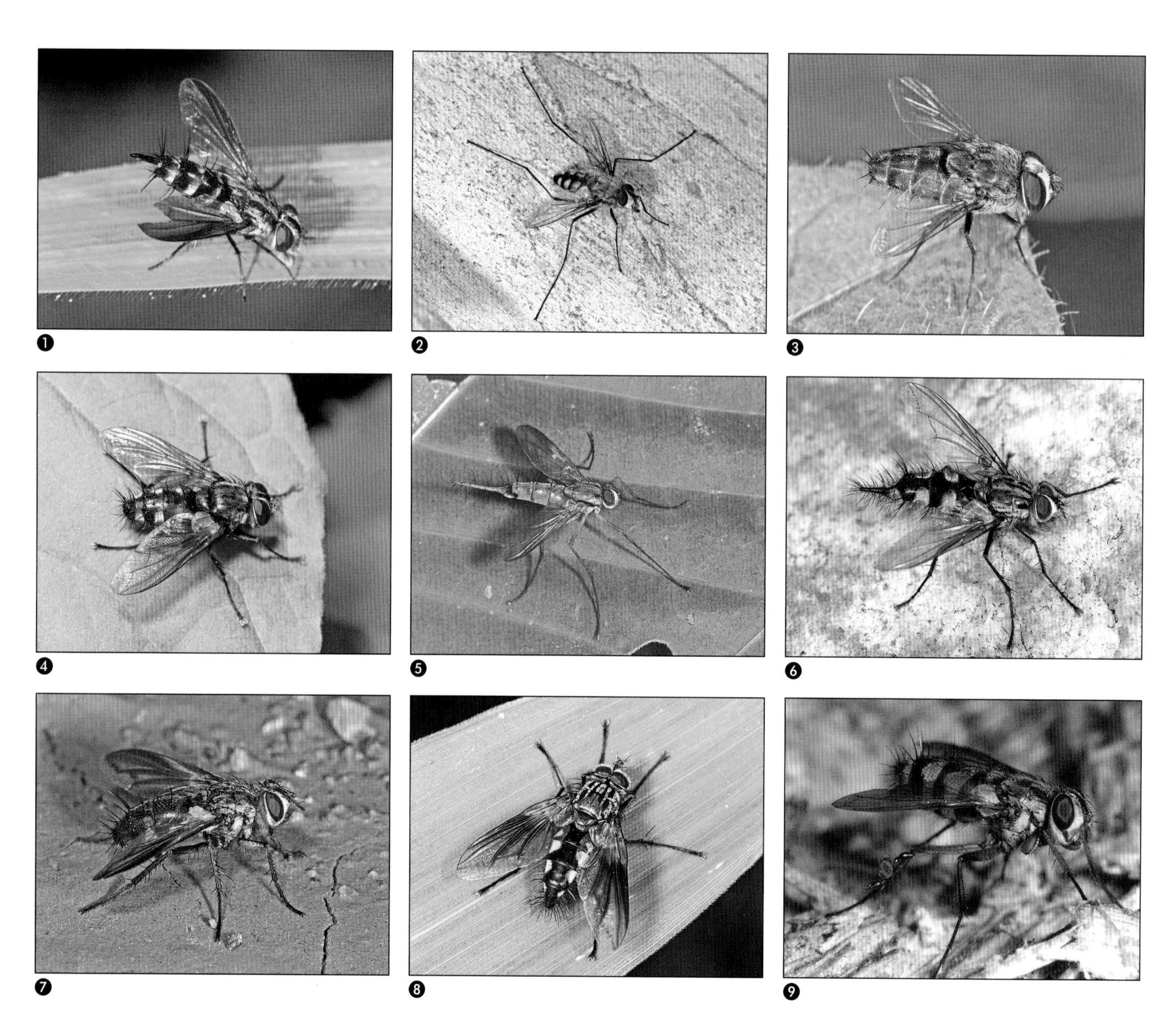

FAMILY TACHINIDAE (continued) ❶ The spathe-like abdominal tip on this Central American ***Spathidexia*** is distinctive, but its function is unknown. Some species in this poorly known Neotropical genus have been reared from skipper caterpillars (Hesperiidae). ❷ This strikingly long-legged ***Senostoma longipes*** standing on a eucalyptus trunk in New South Wales is probably a parasitoid of scarabs. ❸ This ***Telothyria*** was among many perching over a Costa Rican stream, presumably in search of a mate. *Telothyria* is a small New World (mostly Neotropical) group of caterpillar parasitoids. ❹ This ***Thelaira americana*** probably developed inside a Woolly Bear Caterpillar (Arctiinae). The genus *Thelaira* is almost cosmopolitan and *T. americana* is a widespread Nearctic species. ❺ The remarkably long, pointed abdomen of this Ecuadorian ***Trichodura*** has nothing to do with attacking its hosts and is in fact strongly developed only in males. Eggs of this genus hatch upon deposition, leaving the active first-instar larvae to seek hosts (as yet unknown) in typical dexiine fashion. ❻ Like males of the Neotropical genus *Trichodura*, males of ***Uramya halisidotae*** have a strikingly ovipositor-like tubular abdominal tip of unknown function. The first-instar larvae of this western North American species actively pursue their hosts, tent caterpillars and related Lepidoptera. ❼ ❽ ***Uramya*** species range through the Americas, but ***U. indita***, seen here on a mudflat in Arizona, is the only species common to both the Nearctic and Neotropical regions. The other species shown here, on a leaf in Peru, is one of a dozen South American species in the genus (possibly ***U. quadrimaculata***). ❾ This eastern North American ***Zelia vertebrata*** has a pseudoscorpion attached to its hind leg. Pseudoscorpions are predators that often hitch rides on flies that visit prey-rich habitats such as rotting wood. *Zelia vertebrata* is a parasitoid of large scarab larvae found in rotting wood.

FAMILY TACHINIDAE (continued) Subfamily Exoristinae ❶ ❷ Males of the Neotropical genus ***Actinodoria***, like this Ecuadorian species perched on a tree trunk, have an unusually wide frons. The other fly (feeding on a wet leaf surface) is a Costa Rican *Actinodoria*. Some species of this genus have been reared from skipper caterpillars. ❸ Species of the small New World genus ***Ametadora*** are, like most exoristines, parasitoids of caterpillars. Costa Rican species such as this one have been reared from various butterfly and moth larvae. One species (*A. harrisinae*) has been used for biocontrol of the Western Grapeleaf Skeletonizer (*Harrisina brillians*) in California. ❹ Tachinid specialist D.M. Wood has observed this eastern Nearctic species (***Anisia flaveola***) depositing eggs on material likely to be eaten by camel crickets (*Ceuthophilus*). The eggs presumably hatch upon ingestion by the host. Most other species in this New World genus are Neotropical. ❺ This Costa Rican ***Anoxynops aurifrons*** is one of only a few species in this New World genus. ❻ ***Aplomya*** is a large, cosmopolitan genus of caterpillar parasitoids. This one is from Vietnam. ❼ Female ***Belvosia*** deposit large numbers of small microtype eggs on foliage where they are likely to be consumed by host caterpillars (a typical life cycle for the tribe Goniini). This is a Bolivian species, but *Belvosia* is a large genus that ranges throughout the New World. ❽ This is an east African species of ***Blepharella***, a large Old World–Pacific genus. ❾ Although one species ranges north to Canada, most species of ***Calolydella*** are parasitoids of Neotropical Lepidoptera.

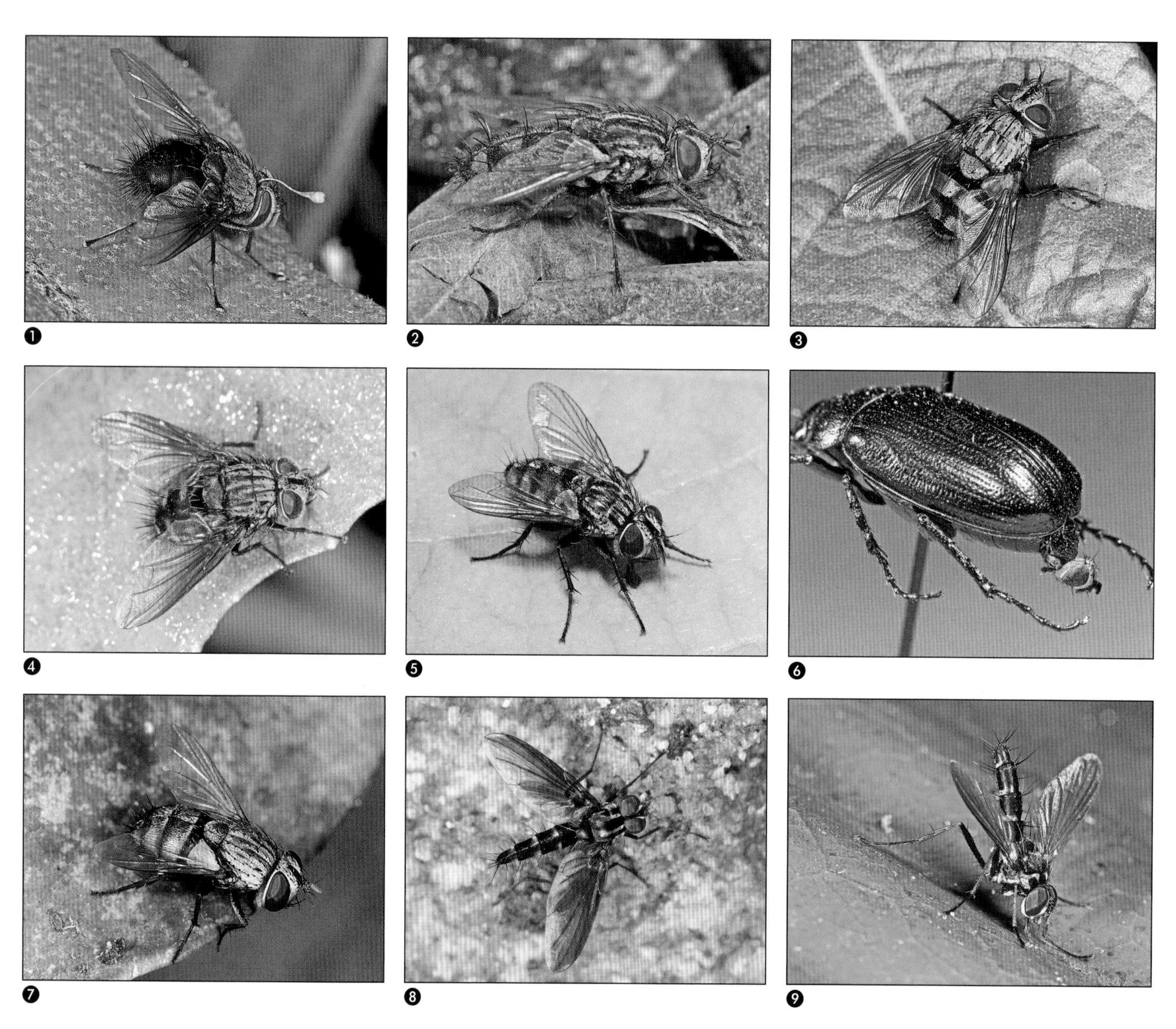

FAMILY TACHINIDAE (continued) ❶ Members of the large genus ***Carcelia*** are found naturally around the world and some species have been moved around as biological control agents of pest Lepidoptera. Females of this genus attach their eggs to hosts by stalks; larvae burrow into the host upon hatching. This Bolivian ***C. montana*** has picked up a conspicuous pollinium, probably from an orchid flower. ❷ This ***Chetogena*** was one of many landing on a small sunny spot on the Chilean forest floor. Hosts of this large, cosmopolitan genus are usually larger moth larvae such as Noctuidae. ❸ ❹ The brilliant green metallic dusting of species in the New World genus ***Chrysoexorista*** — such as the golden green of this Ecuadorian fly and the lime green of this southwestern American species — fades to brown in pinned specimens. Like related flies in the exoristine tribe Goniini, females lay microtype eggs that hatch upon ingestion by caterpillar hosts. ❺ This is the notorious ***Compsilura concinnata***, a highly polyphagous species introduced from Europe to North America in 1906 in an effort to halt the spread of Gypsy Moths (*Lymantria dispar*). Females of this fly grip and rip into caterpillar hosts to inject their parasitic larvae. Their hosts include more than 200 species, and they are sometimes blamed for declining populations of native moths. ❻ Females in the New World exoristine genus ***Cryptomeigenia*** deposit hard-shelled eggs on adult scarab beetle hosts such as this Canadian June beetle. Hatching larvae penetrate the host's integument to develop inside. The teneral fly seen trapped halfway out the anus of this beetle pupariated inside the host and died before fully emerging. ❼ ***Drino*** is a taxonomically difficult worldwide group. This Tanzanian species is one of many in the Old World tropics and probably developed inside a sphinx caterpillar (Sphingidae). ❽ ***Eophyllophila*** is a small genus of four Old World tropical species; this one is from Vietnam. ❾ This tiny Tanzanian ***Eophyllophila*** is lapping up honeydew from a leaf.

FAMILY TACHINIDAE (continued) ❶ Despite the specific name, this ***Eurygastropsis tasmaniae*** is a widespread Australian species. This is the only species in its genus and one of several tachinids with a superficial similarity to co-occurring blow flies (*Calliphora*). ❷ Although the dazzling diversity of Tachinidae is concentrated in the Neotropical region, the type genus of the Exoristinae (genus ***Exorista***) occurs in every zoogeographic region except the Neotropical. This is one of the few North American species (possibly ***E. larvarum***, introduced from Europe). Hosts include a wide range of caterpillars and sawflies. ❸ ***Frontina*** is a mostly Palaearctic genus, but the fly shown here is from the mountains of northern Vietnam, where several "Palaearctic" taxa penetrate the Oriental region. ❹ ***Gonia*** species are distinctively fat-headed flies common on flowers throughout the Holarctic region; they lay microtype eggs meant for ingestion by host caterpillars in several families. ❺ ***Kaiseriola*** is a monotypic genus (with only a single African species, ***K. aperta***) named by the European tachinid expert L.P. Mesnil. ❻ ***Lixophaga*** is a huge, mostly New World genus that includes a species that has been used for control of Sugarcane Borer (*Diatraea saccharalis*, a small moth with larvae that feed inside the stems of sugarcane). ***Lixophaga diatraeae*** females are cued by the odor of the moth's frass to lay eggs near the entrance burrow, and the hatching larvae penetrate the frass to parasitize the host. Other *Lixophaga* species use the same approach to parasitizing other hidden hosts. The fly shown here is a female from Ecuador. ❼ ❽ This Chilean ***Myiopharus pirioni*** is apparently kissing a larval leaf beetle (Chrysomelidae, Alticinae, *Procalus*), taking up a fluid regurgitated by the larva. All known species in the large genus *Myiopharus* are parasitoids of larval or adult Chrysomelidae; it is not known if this "kiss" was a prelude to parasitism. ❾ This Canadian ***Nilea*** has been attracted to the honeydew produced by some *Kermes* scales on an oak branch. Larvae of this Holarctic genus are parasitoids of caterpillars.

FAMILY TACHINIDAE (continued) ❶ Its bright orange calypters render this monotypic African genus (***Ossidingia***, which includes only ***O. cruciata***) distinctive. ❷ This African fly belongs to the large and abundant Old World and Pacific genus ***Pales***. *Pales* species lay eggs on the undersides of leaves and parasitize various caterpillars that ingest the eggs along with leaf tissue. ❸ Species of ***Patelloa***, such as this Peruvian ***P. andina***, parasitize caterpillars throughout the New World. ❹ Species in the Holarctic genus ***Phebellia***, like this ***P. cerurae*** eyeing a notodontid caterpillar, deposit eggs on caterpillar hosts. This caterpillar already has one egg just behind its head. ❺ ***Phorinia*** is an Old World and Pacific group of caterpillar parasitoids. ❻ ***Phyllophilopsis*** species develop as parasitoids inside the larvae of fungus beetles (Endomychidae). This is ***P. nitens***, the more common of the two *Phyllophilopsis* species found in Canada; the remainder of the genus is Neotropical. ❼ Members of the Holarctic-Neotropical genus ***Spallazania***, such as this Greek ***S. rectistylum***, parasitize cutworm moths and related Noctuoidea. ❽ ***Sturmia*** species, including this ***S. bella*** from Greece, occur throughout the Old World and the Pacific. Like most Exoristinae, they parasitize Lepidoptera. ❾ Some of the 10 species in the Nearctic genus ***Tachinomyia*** are large, conspicuous and common flies and are thus well represented in North American insect collections. Known hosts are mostly larger caterpillars such as cutworms, tent caterpillars, tussock moths and tiger moths, but sawflies are sometimes attacked as well. This is ***T. apicata***.

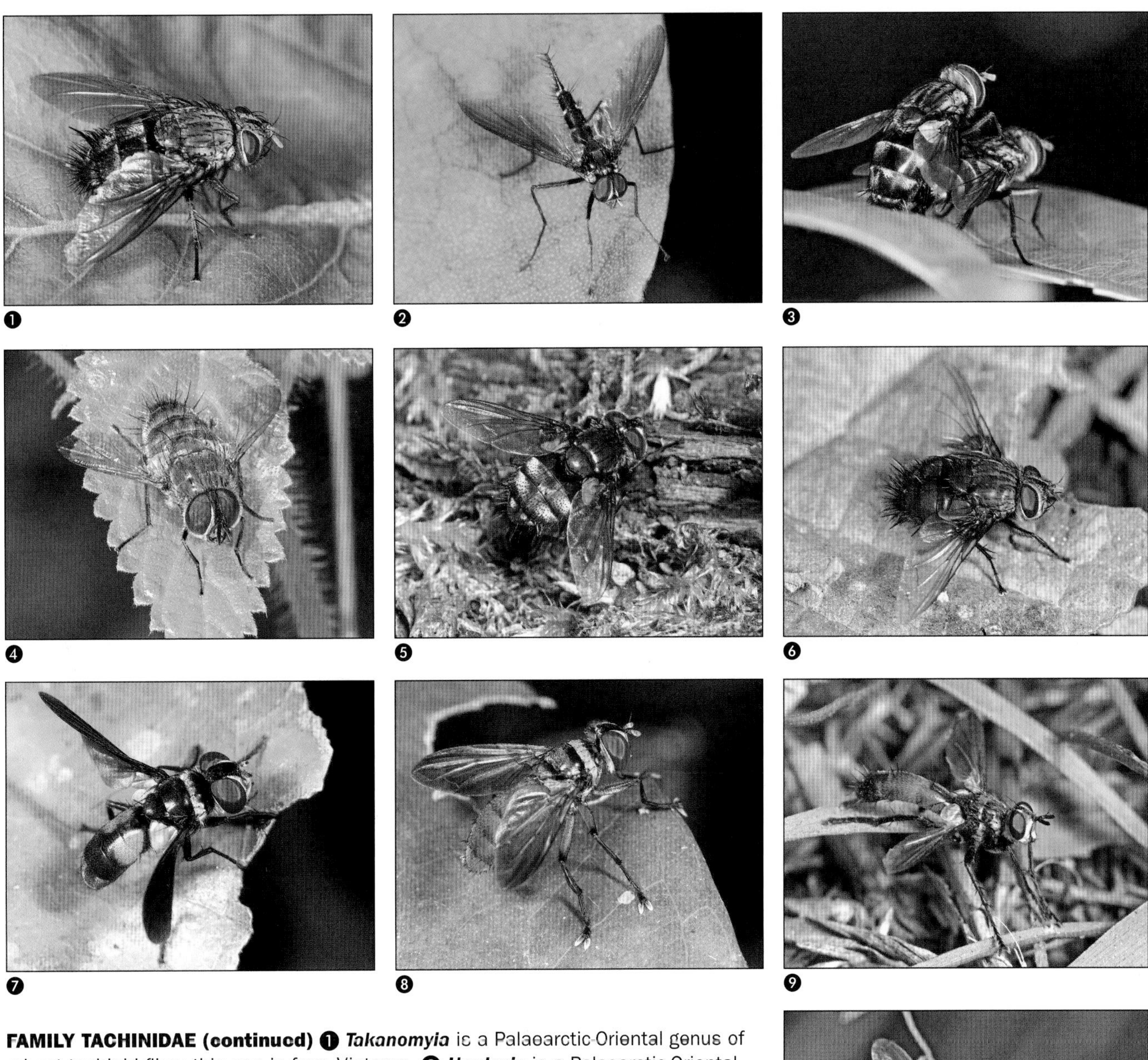

FAMILY TACHINIDAE (continued) ❶ ***Takanomyia*** is a Palaearctic-Oriental genus of robust tachinid flies; this one is from Vietnam. ❷ ***Urodexia*** is a Palaearctic-Oriental genus of slender tachinid flies; this one is from Vietnam. ❸ ***Winthemia*** is a huge, cosmopolitan genus; shown here is a mating pair from Vietnam. At this angle it is easy to see the male's subscutellum — the bulge under the scutellum that is a "signature" of the Tachinidae. ❹ ❺ ***Winthemia*** is a common cosmopolitan genus of more than 100 species and is well known for depositing eggs on a wide range of caterpillar hosts, where emerging larvae later penetrate the host's skin. This relatively big-eyed male is perched on a leaf in Tanzania; this female ***W. occidentis*** is searching for a host on the ground in Canada. ❻ ***Leschenaultia*** is a large New World genus of caterpillar parasitoids, including some that attack tent caterpillars in North America. Shown here is a Peruvian species. **Subfamily Phasiinae** ❼ ❽ Flies in the Afrotropical genus ***Bogosia***, like these two from Tanzania, have been used for the biological control of pest Pentatomidae on coffee plantations. ❾ ❿ ***Cylindromyia*** is a cosmopolitan genus, and these distinctive cylindrical flies are common in most countries. Shown here are a Canadian and an Australian species, both probably parasitic on stink bugs.

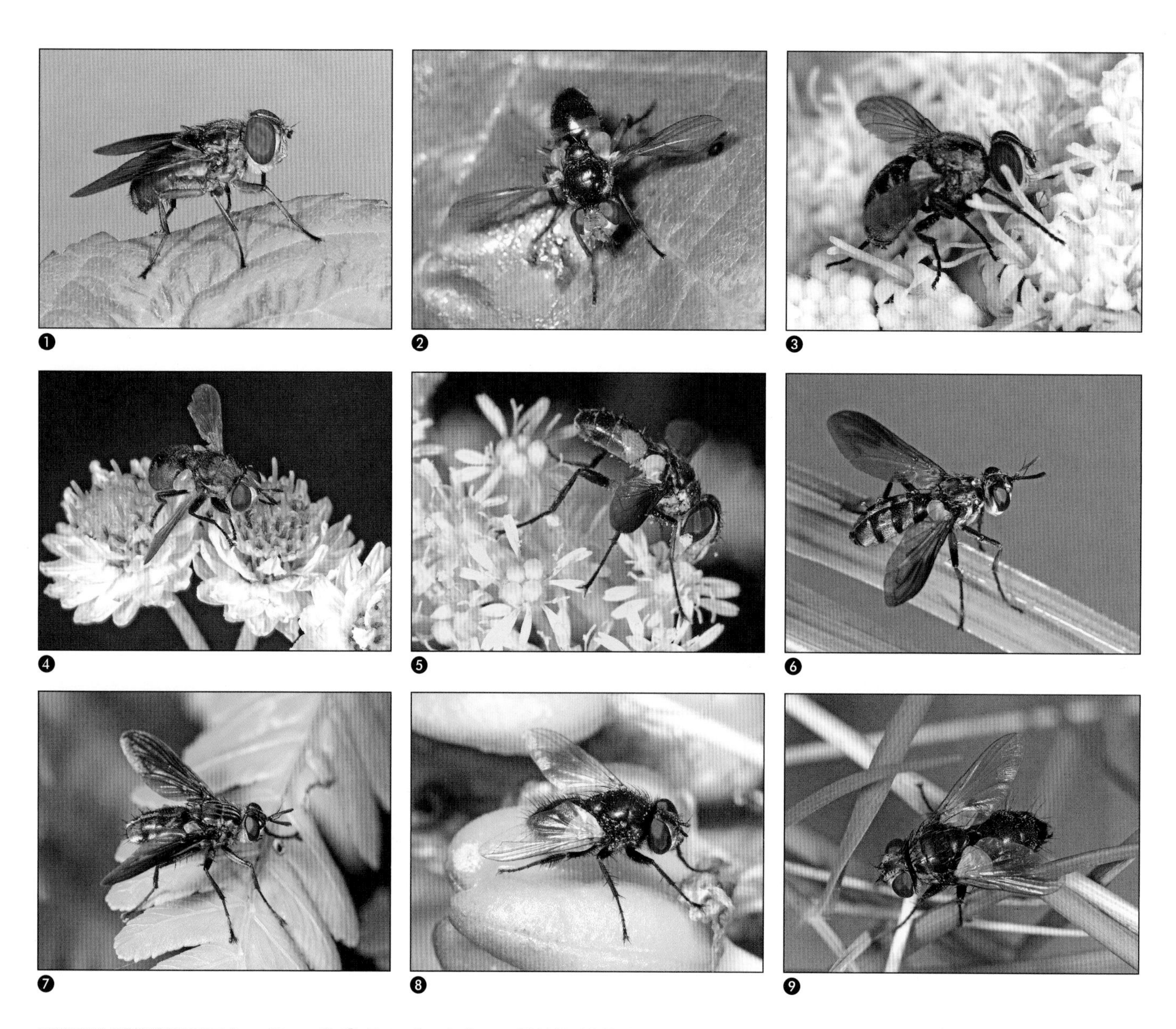

FAMILY TACHINIDAE (continued) ❶ ***Ectophasia*** is an Old World (Palaearctic and Oriental) genus. This ***E. leucoptera*** was photographed in Greece. ❷ The North American genus ***Euclytia*** includes a single described species, ***E. flava***, a parasitoid of stink bugs. ❸ Members of the New World genus ***Gymnoclytia*** are common flower visitors. Most host records of *Gymnoclytia* are stink bugs (as is true for many other Phasiinae). ❹ Members of the almost worldwide genus ***Gymnosoma*** are common flower visitors and common parasitoids of stink bugs and related Heteroptera. Their conspicuous white eggs are laid on the host; larvae emerge much later to penetrate the host. ❺ ***Hemyda aurata***, the only North American species in this New World and Palaearctic genus, has been reared from stink bugs (Pentatomidae). ❻ ❼ ***Hermya*** is a mostly Old World tropical group of flies that develop in larger true bugs such as Coreidae. Shown here are species from Vietnam. ❽ ❾ Tachinids in the distinctive genus ***Leucostoma*** are parasitoids of seed bugs, which the female flies attack by using specialized structures (pincer-like cerci) to grasp their hosts while they inject eggs with a sword-like piercing structure. *Leucostoma* occurs almost everywhere in the world; this male and female were photographed in Greece.

FAMILY TACHINIDAE (continued) ❶ ❷ ***Phasia*** is a cosmopolitan genus of 75 species of bug parasitoids, usually associated with Pentatomidae or Pyrrhocoridae. Shown here are two North American species, ***P. aurulans*** (with a golden thoracic mark) and ***P. fenestrata*** (with darkly marked wings). ❸ ***Trichopoda*** species, which parasitize Coreidae and related bugs, are distinctive for the broad combs of bristles on their hind legs. ❹ ***Trichopoda pennipes*** is a North American species that has been transported out of its native eastern Nearctic range for biological control of pest bugs (Hemiptera, including stink bugs and Squash Bugs) and now occurs in western North America and Europe. ❺ These eastern North American ***Xanthomelanodes flavipes*** probably developed in an assassin bug (Reduviidae) host. *Xanthomelanodes* is a New World taxon that is most diverse in the neotropics. **Subfamily Tachininae** ❻ ***Adejeania vexatrix*** is a spectacular "hedgehog fly" that ranges from Mexico to western Canada. Larvae are parasitoids of caterpillars. Most similar hedgehog flies (tribe Tachinini), including 37 of the 38 species in *Adejeania*, occur in the Neotropical region. ❼ "Hedgehog flies" like this large and obviously well-defended Peruvian ***Adejeania*** are especially common and conspicuous at higher elevations in the Andes. ❽ ***Archytas apicifer*** is one of several common North American species in this large New World genus, but most of the hundred or so species in the genus are Neotropical. Hosts include a wide range of moth and butterfly caterpillars. Related species have been used for biocontrol of Corn Earworm (*Helicoverpa zea*). ❾ ***Chaetophthalmus*** is a large endemic Australian genus and one of the relatively few Australian genera of Tachininae. This one is remarkably similar to more common Australian *Calliphora* species found at the same time and place.

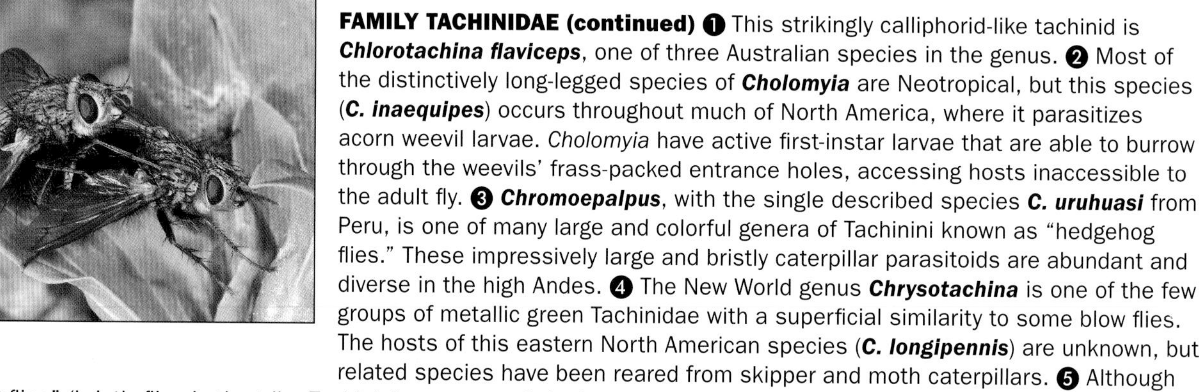

FAMILY TACHINIDAE (continued) ❶ This strikingly calliphorid-like tachinid is ***Chlorotachina flaviceps***, one of three Australian species in the genus. ❷ Most of the distinctively long-legged species of ***Cholomyia*** are Neotropical, but this species (***C. inaequipes***) occurs throughout much of North America, where it parasitizes acorn weevil larvae. *Cholomyia* have active first-instar larvae that are able to burrow through the weevils' frass-packed entrance holes, accessing hosts inaccessible to the adult fly. ❸ ***Chromoepalpus***, with the single described species ***C. uruhuasi*** from Peru, is one of many large and colorful genera of Tachinini known as "hedgehog flies." These impressively large and bristly caterpillar parasitoids are abundant and diverse in the high Andes. ❹ The New World genus ***Chrysotachina*** is one of the few groups of metallic green Tachinidae with a superficial similarity to some blow flies. The hosts of this eastern North American species (***C. longipennis***) are unknown, but related species have been reared from skipper and moth caterpillars. ❺ Although "hedgehog flies" (bristly flies in the tribe Tachinini) occur mostly in the neotropics, some, like this Tanzanian ***Dejeania***, are found in Africa. All four *Dejeania* species are Afrotropical. ❻ ***Deopalpus*** ranges from Canada to Chile; this one is visiting a flower high in the mountains above Santiago, Chile. ❼ ❽ These two very different-looking ***Epalpus*** species, a red ***E. constans*** and a black and white unnamed species, are among the 30 or so "morphospecies" of Costa Rican *Epalpus* recognized by tachinid expert D.M. Wood. ❾ ❿ Although mostly a Neotropical group, ***Epalpus*** includes a few species in North America. This is ***E. signifer***, a common fly throughout the United States and southern Canada.

FAMILY TACHINIDAE (continued) ❶ ❷ ❸ ❹ The abundance and diversity of the hedgehog fly genus ***Epalpus*** is illustrated by these four flies, all photographed on the same day at the same site in the Peruvian Andes. ❺ This massive black ***Eudejeania pallipes*** dwarfs the other Tachinidae attracted to a leaf sprayed with sugar in the high Andes of Peru. ❻ This ***Eusaundersiops ornatus*** from Peru is the only species in its genus. ❼ ***Fasslomyia*** includes only one named species, the nicely named Andean species ***F. fantastica***. ❽ ***Ginglymia*** is a New World group of about half a dozen species, including the Nearctic ***G. johnsoni*** shown here. *Ginglymia* with known hosts parasitize snout moth (Pyralidae) caterpillars. ❾ ***Strongygaster robusta*** occurs across the Nearctic region; known hosts of this genus include various beetles and even adult moths.

FAMILY TACHINIDAE (continued) ❶ ***Hystricia abrupta*** is a widespread and common fly across temperate North America, where it has been reared from caterpillars of tiger moths (Arctiinae) and giant silkworm moths (Saturniidae). Like the superficially similar hedgehog flies (which are in a different tribe), *Hystricia* is a mostly Neotropical group and is common in the Andes. ❷ ***Jurinia*** is a mostly Neotropical genus, but this species (***J. pompalis***) ranges north to Canada. ❸ ***Leskia*** species occur in every zoogeographic region, but this species (***L. depilis***) is an eastern North American parasitoid of small moths. ❹ ***Lindigepalpus*** is one of many superficially similar Neotropical "hedgehog fly" genera; this one is from the Peruvian Andes. ❺ ***Linnaemya*** species such as this Tanzanian fly are found in every zoogeographic region but are particularly diverse and colorful in Africa. ❻ This ***Macromya crocata*** has stopped for a drink of water along an Arizona stream. Most species of this genus are Neotropical, but *M. crocata* ranges from Brazil to the southwestern United States. ❼ ***Mintho*** is an Old World (Palaearctic and Afrotropical) genus of small moth parasitoids. This distinctive Greek fly with an unusually compressed abdomen is the aptly named ***M. compressa***. ❽ This Tanzanian tachinid is in the tribe Minthoini, along with similar ***Mintho*** species and a handful of similar small genera. ❾ The spectacular yellow-rimmed lower calypters on this Tanzanian ***Ocypteromima*** give it a striking appearance. *Ocypteromima* is a small Afrotropical-Oriental genus in the tribe Leskiini.

FAMILY TACHINIDAE (continued) ❶ ❷ The Orthoptera parasitoids in the New World genus ***Ormia*** are especially diverse (34 species) in the neotropics. Unusually for higher Diptera, *Ormia* species are nocturnal (this one was attracted to a light in Costa Rica); females home in on chirping male hosts using an inflated prosternum that serves as a sensitive "ear." The prosternal "ear" is indicated by an arrow on this close-up photograph of an orange North American *Ormia* species, ***O. brevicornis***. ❸ ***Paradidyma*** is a small New World genus of 16 species that develop as parasitoids of small caterpillars. This one is in a Canadian garden. ❹ ***Pararchytas*** is a small Nearctic genus with two species that range north to Canada and one (***P. apache***, shown here in Arizona) known only from the southwest United States. ❺ ***Peleteria*** is a huge Holarctic and Neotropical genus of caterpillar parasitoids often reared out of cutworm or owlet moths (Noctuidae). This is the Nearctic ***P. anaxias***. ❻ ***Protodejeania*** is a mostly Mexican and Central American genus of "hedgehog flies," but two of the half-dozen or so species in the genus range north to western Canada. ❼ Most of the 20 named species in the New World genus ***Rhachoepalpus***, like this one from Peru, occur at high elevations in the Andes. ❽ Most Australian Tachinidae are in the subfamily Dexiinae, but this ***Semisuturia*** (Tachininae) from New South Wales is an exception. ❾ Small ***Siphona*** species can be found the world over and are often seen using their long, elbowed beaks to feed in flowers. Most are parasitoids of Lepidoptera, but a few species have been reared from other Diptera (Tipulidae). ❿ ***Tachina***, the type genus for the Tachininae and the Tachinidae, is a large genus found in most of the world but not in Australia or Africa. This is a ***T. casta*** from Greece.

FAMILY TACHINIDAE (continued) ❶ *Xanthoepalpus* is a small New World genus with one species ranging north to Canada and another half-dozen strictly Neotropical species like this Costa Rican fly. **FAMILY OESTRIDAE (BOTS AND WARBLES) Subfamily Gasterophilinae** ❷ ❸ ***Gasterophilus intestinalis***, the most common horse stomach bot fly in most areas, places eggs on horses' forelegs, where they are likely to get licked off and hatch into larvae that burrow into the horse's tongue *en route* to the stomach. **Subfamily Oestrinae** ❹ The North American ***Cephenemyia phobifera*** ("snot bot") develops as a parasite in the throat and sinus cavities of deer. ❺ The **Sheep Nasal Bot Fly**, ***Oestris ovis***, normally deposits newly hatched larvae in the nasal cavities of sheep, rarely parasitizing humans as well. Here photographed in Australia by Paul Zaborowski, this originally Old World species now occurs worldwide. **Subfamily Hypodermatinae** ❻ This ***Hypoderma tarandi*** (**Reindeer Warble Fly**) was photographed in Norway by Arne C. Nilssen. Typical for the subfamily, this species lays eggs on its reindeer host and develops as a warble under the skin. **Subfamily Cuterebrinae** ❼ ***Cuterebra austeni*** develops under the skin of White-throated Wood Rats (*Neotoma albigula*) in the southwestern United States. This adult was "hilltopping" in New Mexico. ❽ ❾ ❿ ***Cuterebra fontinella*** is a common parasite of the White-footed Mouse (*Peromyscus leucopus*) in eastern North America. Adult bot flies are non-feeding and have greatly reduced mouthparts. Larvae are covered with spines.

FAMILY OESTRIDAE (continued) ❶ The so-called **Human Bot Fly** or **Torsalo** (***Dermatobia hominis***) is a common Neotropical fly that more often parasitizes livestock than people. Eggs are laid on another fly, such as a mosquito or House Fly, and hatch when that fly comes in contact with the warmth of a host's skin. Larvae develop in a boil-like swelling. This teneral specimen developed in an entomologist host, and has just emerged from its puparium.

SUPERFAMILY MUSCOIDEA

FAMILY SCATHOPHAGIDAE Subfamily Delininae ❷ Larvae of the North American ***Neochirosia nuda*** are unknown, but the related *N. atrifrons* is a leaf miner specific to false hellebore or corn lily (*Veratrum*). All three *Neochirosia* species are Nearctic. **Subfamily Scathophaginae** ❸ This eastern North American ***Acanthocnema (Clinoceroides)*** probably developed as an aquatic larva that fed on egg masses of other aquatic insects; related species are specialized predators of caddisfly eggs. ❹ ❺ ❻ ❼ *Cordilura* species are borers in sedges and rushes, mostly *Carex* and *Scirpus*, although *Cordilura (Snyderia) praeusta* is associated with *Juncus*. Despite its uniform habitats, the genus is morphologically diverse and is divided into several subgenera. Shown here are the North American ***Cordilura (Achaetella) varipes*** mating, ***Cordilura (Cordilura) angustifrons*** lapping up honeydew beside an aphid, ***Cordilura (Cordilurina) gracilipes*** on coniferous foliage and ***Cordilura (Snyderia) praeusta*** perching on the edge of a leaf. ❽ Larvae of ***Hydromyza confluens***, the only North American member of the small Holarctic genus *Hydromyza*, burrow into the underwater stems of water lilies (*Nuphar* and *Nymphaea*). The European *H. livens* is a leaf miner on the same water lily genera. ❾ Although the Scathophagidae is a characteristically Holarctic family, this one (an undescribed species of ***Cordilura***) was found in the mountains of northern Vietnam.

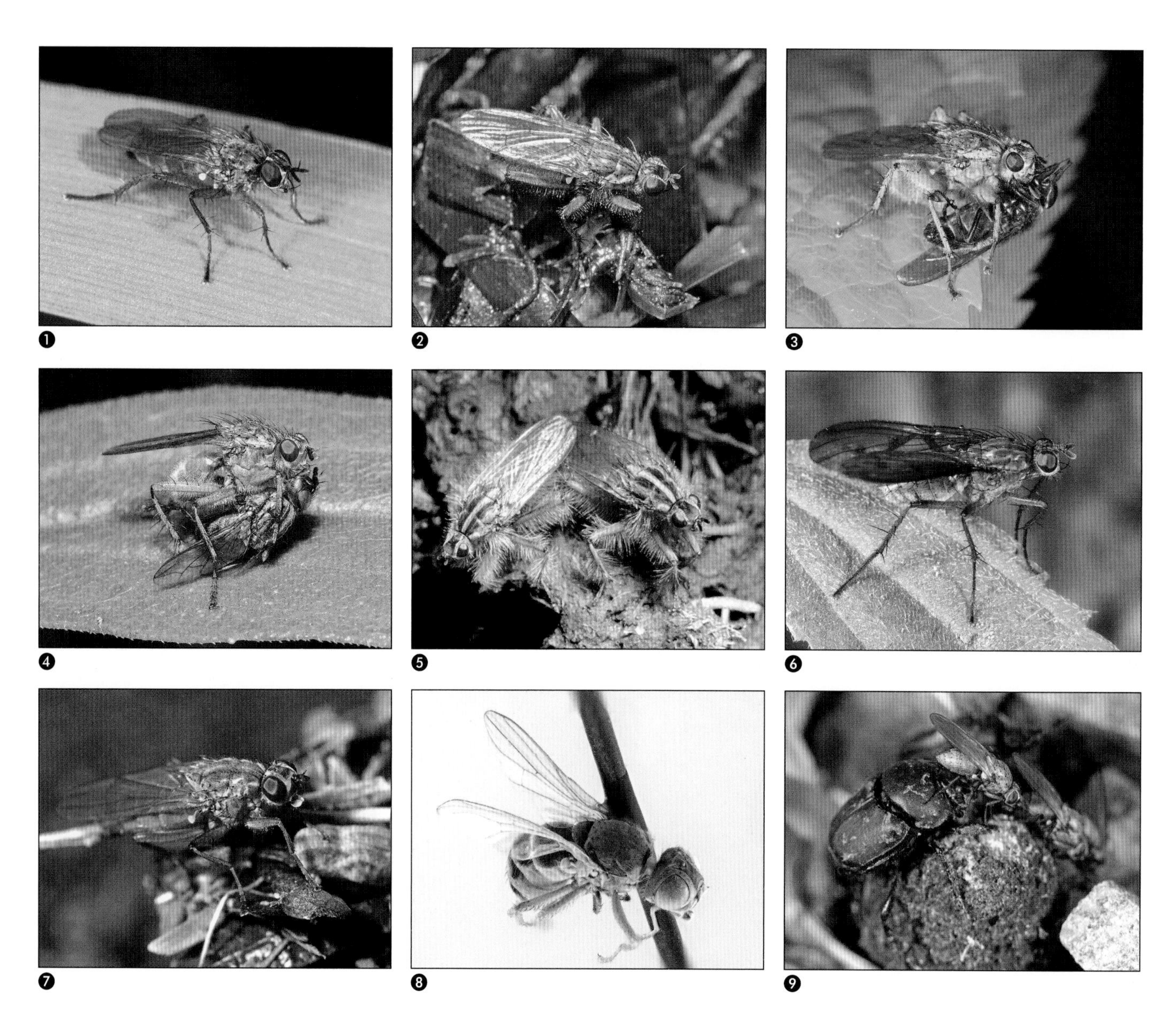

FAMILY SCATHOPHAGIDAE (continued) ❶ ***Neorthacheta dissimilis* (Iris Bud Fly)** develops in the buds of iris flowers, often resulting in disfigured blooms. Although it is a native North American fly, this species can be a pest of introduced iris cultivars. ❷ ***Scathophaga intermedia***, the characteristic supralittoral (marine shoreline) scathophagid species of northeastern North America, often occurs in high densities in decaying seaweed. ❸ ❹ ***Scathophaga*** is a diverse genus, especially in the Holarctic region, where most areas support several similar species, but the most widespread and abundant species in the genus is ***S. stercoraria***. This species abounds on cow-flops and other animal droppings, and the predaceous adults are common in a wide range of habitats. ❺ Despite its name, ***Scathophaga tropicalis*** is a south temperate species that lives in the *altiplano* of Peru, Bolivia and Argentina. This pair was among thousands on the shore of Lake Titicaca, on the Peru–Boliva border. ❻ Scathophagidae are rare insects in tropical countries, but the Andean genus ***Scatogera*** is an interesting exception. The only described species in the genus (***S. primogenita***) can be abundant in the mountains of Ecuador. ❼ ***Spaziphora*** is a small Holarctic genus of shoreline predators, with one species in the Old World and one in North America. This is ***S. cincta*** hunting prey on the shores of the Great Lakes in Canada. **FAMILY ANTHOMYIIDAE (ROOT MAGGOT FLIES, ETC.) Subfamily Anthomyiinae** ❽ The small (three species) Holarctic genus ***Acridomyia*** is unusual both in appearance and biology. Females attack grasshoppers, chewing a hole in the hopper's body wall through which to lay eggs. Larvae are internal parasitoids of grasshoppers. ❾ These ***Adia cinerella*** were persistently following and riding this dung-rolling scarab (*Canthon simplex*) and its pungent prize. This Holarctic coprophagous anthomyiid species is not known to be a kleptoparasite (food thief) like some scarab-associated flies in the Sphaeroceridae, but it seems likely that larvae develop in the scarab's burrow.

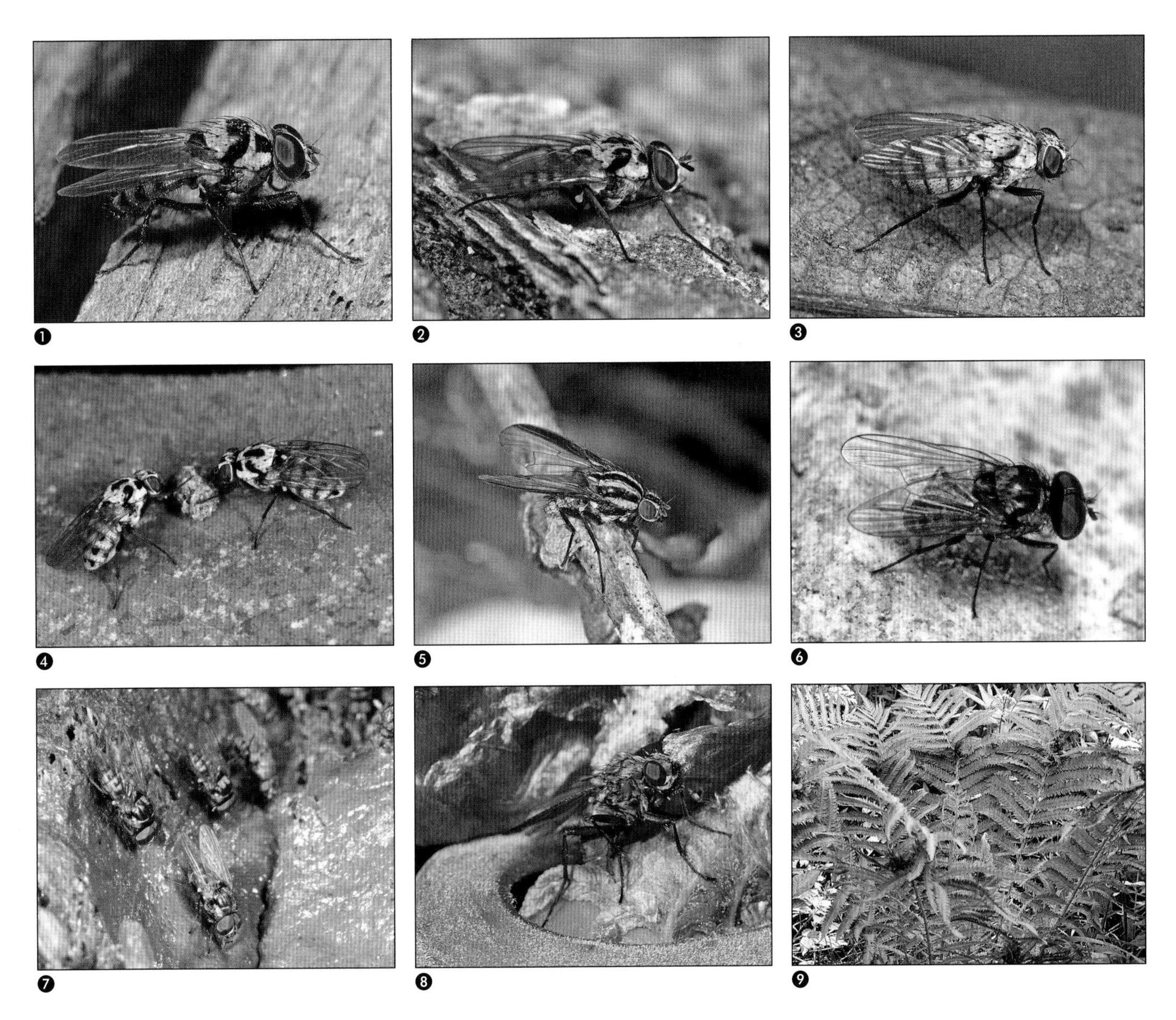

FAMILY ANTHOMYIIDAE (continued) ❶ ❷ ❸ ❹ ***Anthomyia*** is a large, widespread and diverse genus with species developing in a range of decomposing materials. Shown here are ***A. oculifera*** from eastern North America (on wood), ***A. procellaris*** from western North America, ***A. illocata*** from the Philippines (on a dead leaf), and a pair of ***A. pluvialis*** sharing some debris on a Greek leaf. ❺ This is one of several undescribed ***Anthomyia*** species in the Andes of Peru. ❻ ❼ Species in the widespread genus ***Calythea*** often develop in cattle dung and sometimes occur in immense numbers in the rangeland of the American southwest. ❽ ❾ All members of the large genus ***Chirosia***, at least those with known larvae, are leaf miners, stem borers or gall makers on ferns. This ***C. betuleti***, a species recently introduced from Europe to North America, is ovipositing on a fern at the fiddlehead stage. Larvae develop in conspicuous swellings on the developing fronds.

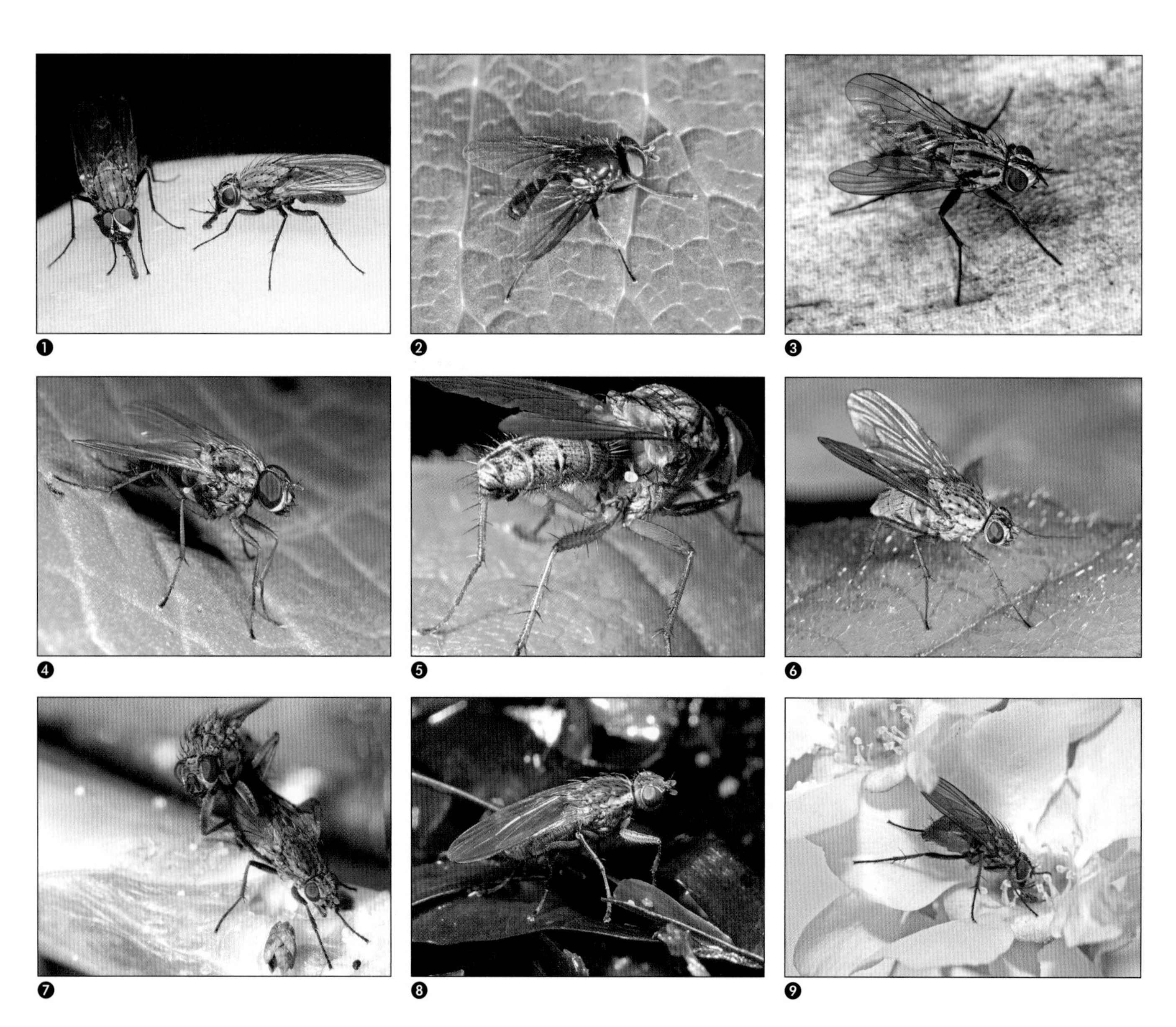

FAMILY ANTHOMYIIDAE (continued) ❶ The **Onion Maggot** (***Delia antiqua***) is one of many pest root maggots in the genus *Delia*, but this genus also includes some common flower visitors and pollinators. The male of this pair of Onion Maggots is the one with the big eyes. ❷ Larvae of the large Holarctic genus ***Egle*** develop in the female catkins of poplar and willow. ❸ ***Eustalomyia*** species, such as the widespread Holarctic ***E. festiva***, are kleptoparasites that develop in the nests of wasps (Crabronidae) that nest in twigs or dead wood. The fly larvae consume the paralyzed prey stored by the wasps. This small genus ranges throughout the Holarctic region and the Old World tropics. ❹ ❺ ❻ ***Eutrichota*** is a large, mostly Holarctic genus often found in mammal burrows, presumably developing as scavengers in dung. Shown here are a female from California and two males of the widespread Nearctic ***E. lipsia***. The ventral scutellar hairs diagnostic of most Anthomyiidae are clearly visible on one of the males. ❼ ❽ ***Fucellia tergina*** (widely known under the synonym ***F. intermedia***) is a widespread member of the wrack community, developing in piles of seaweed on rocky coastlines worldwide. This cosmopolitan genus includes more than 30 similar species. ❾ ***Hydrophoria*** is a large, common, mostly Holarctic genus associated with a variety of larval habitats, including dung.

FAMILY ANTHOMYIIDAE (continued) ❶ Anthomyiids, as the name suggest, are often found on flowers. This ***Lasiomma*** was one of many on a willow flower in eastern Canada. *Lasiomma* is a big, mostly Holarctic-Oriental genus that includes several species that develop in the cones of coniferous trees. ❷ ***Leucophora*** species with known habits are kleptoparasites, developing on pollen in the nests of solitary bees. Shown here is a Canadian species, but the genus is widespread and found in every region but the Australasian. They are sometimes called "satellite" flies because their bee-following behavior is like that of true satellite flies (miltogrammine sarcophagids). ❸ ***Taeniomyia*** is a New World, mostly Neotropical, group of species that lack the sexual dimorphism of most anthomyiids (the male head is like that of the female). Shown here is a colorful but probably undescribed species from Costa Rica. **Subfamilies Myopininae and Pegomyinae** ❹ ❺ Larvae of the large, mostly Holarctic genus ***Pegomya*** are often found in mushrooms, but some species develop in leaves (as leaf miners), flowers, seeds, cones, and birds' nests. This male scavenging on a dead caterpillar and this female taking up honeydew are from eastern Canada. ❻ Three leaf-mining larvae of a Canadian ***Pegomya*** are seen here inside a burdock leaf, thanks to light projected from beneath the leaf. The parasitoid wasp ovipositing through the leaf into the larva is an *Opius* (Braconidae, Opiinae). ❼ Most ***Pegomya*** occur in north temperate countries, but this male, with a red mite attached to his leg, is from a high-altitude cloud forest in Ecuador. **FAMILY FANNIIDAE (LITTLE HOUSE FLIES, ETC.)** ❽ ***Fannia canicularis***, a species first named by Linnaeus in 1761, is a common domestic fly associated with all kinds of wet decomposing materials. ❾ ***Fannia*** larvae, like these ***F. canicularis*** picked out of some poultry manure, are distinctive for their long body processes and the two broad tubes sticking up from the back, each ending in three spiracular lobes.

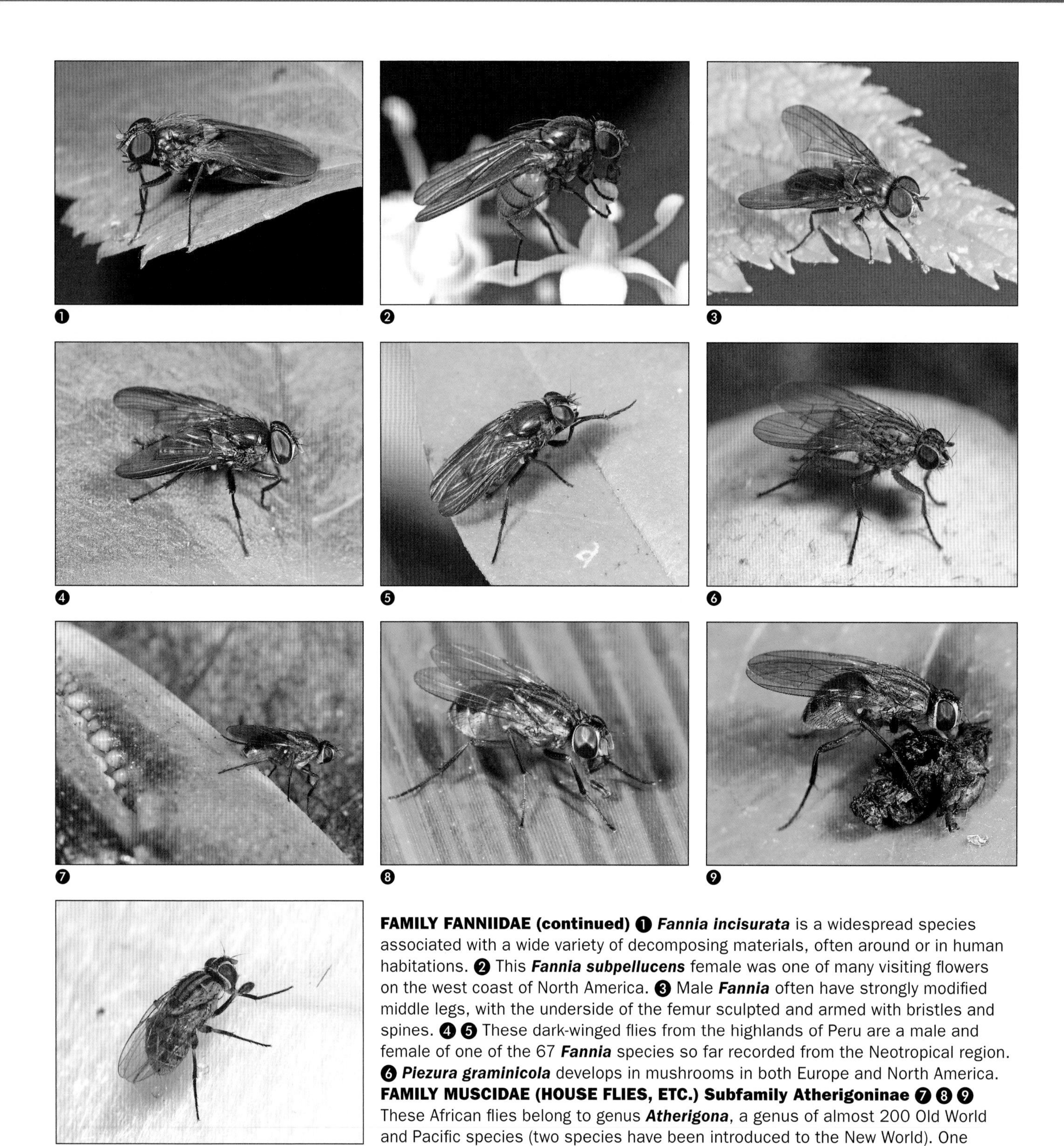

FAMILY FANNIIDAE (continued) ❶ ***Fannia incisurata*** is a widespread species associated with a wide variety of decomposing materials, often around or in human habitations. ❷ This ***Fannia subpellucens*** female was one of many visiting flowers on the west coast of North America. ❸ Male ***Fannia*** often have strongly modified middle legs, with the underside of the femur sculpted and armed with bristles and spines. ❹ ❺ These dark-winged flies from the highlands of Peru are a male and female of one of the 67 ***Fannia*** species so far recorded from the Neotropical region. ❻ ***Piezura graminicola*** develops in mushrooms in both Europe and North America. **FAMILY MUSCIDAE (HOUSE FLIES, ETC.) Subfamily Atherigoninae** ❼ ❽ ❾ These African flies belong to genus ***Atherigona***, a genus of almost 200 Old World and Pacific species (two species have been introduced to the New World). One lineage in this genus is made up of species with stem-boring larvae, some of which are significant pests. ❿ ***Atherigona*** is a tremendously diverse group in the Old World tropics; this Vietnamese fly is in the subgenus ***Acritochaeta***.

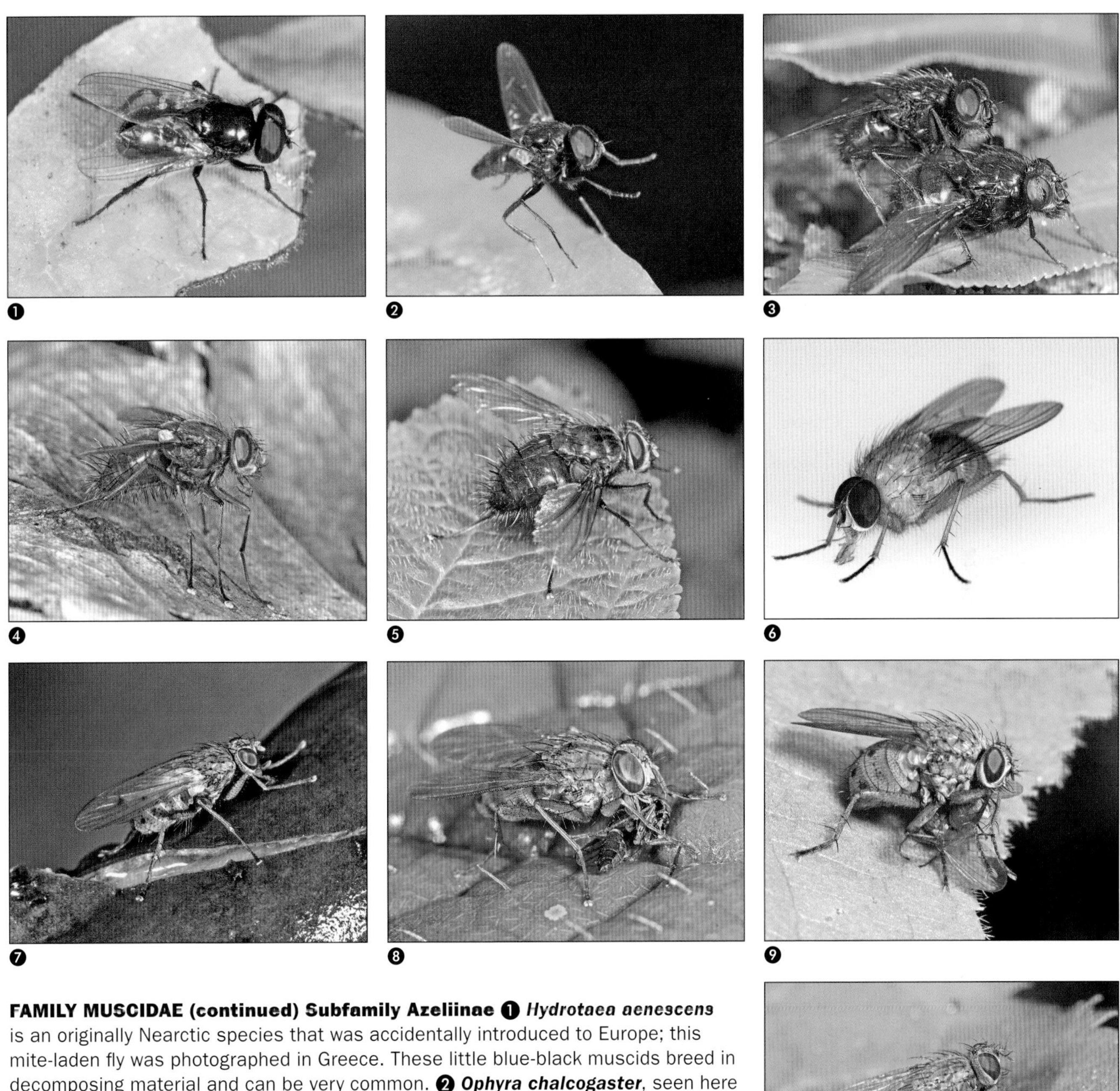

FAMILY MUSCIDAE (continued) Subfamily Azeliinae ❶ ***Hydrotaea aenescens*** is an originally Nearctic species that was accidentally introduced to Europe; this mite-laden fly was photographed in Greece. These little blue-black muscids breed in decomposing material and can be very common. **❷** ***Ophyra chalcogaster***, seen here launching itself in response to my camera flash in Vietnam, is a widespread species found in most parts of the world. **❸** Muscids in several subfamilies have converged on the metallic blue appearance more typical of Calliphoridae (and some tachinids). Species in the South American genus ***Brachygasterina*** (the pair here *in copula* are probably ***B. humboldti***) have traditionally been treated as part of the Phaoniinae, but are now considered part of the Azeliinae. **❹ ❺** This large, robust, tachinid-like Peruvian fly is in an undescribed species in the genus ***Reinwardtia***. *Reinwardtia* and its relatives are sometimes treated as a tribe (Reinwardtiini) in the Azeliinae, and sometimes as a separate subfamily (Reinwardtiinae). **❻** This bright orange Canadian ***Thricops diaphanus*** is one of the most distinctively colored Nearctic muscids. **Subfamily Coenosiinae ❼** This Canadian ***Lispocephala*** (previously treated as *Caricea*) was attracted to the sap flowing from a cut twig in early spring. *Lispocephala* is a large, widespread genus particularly diverse in Hawaii. **❽** Coenosiine muscids, like this ***Cephalispa*** in Vietnam, are often seen consuming other flies. *Cephalispa* species are mostly Oriental, but the genus occurs in Madagascar and the Australasian region. **❾** The distinctively spotted ***Coenosia tigrina*** is a common Holarctic species sometimes used for biological control in greenhouses; larvae and adults of this genus are predators. **❿** This Canadian ***Coenosia***, one of over 360 species in this cosmopolitan genus, captured a chironomid midge on the wing and then returned to its perch to consume its catch.

FAMILY MUSCIDAE (continued) ❶ ❷ These two Costa Rican ***Cordiluroides***, one stuck into a mycetophiloid meal, are among the half a dozen species in this Neotropical genus. ❸ ❹ ***Limnophora*** is a large and widespread genus of aquatic predators; the male and female (eating) shown here are from Tanzania. ❺ ❻ ***Lispe*** is one of the most common flies found along fresh waters the world over, and these big-palped flies are often seen feeding on newly emerged aquatic insects. These North American *Lispe* are feeding on a dead mayfly and mating. ❼ ❽ This American ***Lispe tentaculata*** larva developed into this adult, seen beside its recently vacated puparium. ❾ ***Lispe*** species are common along waterways in Africa, as in most other parts of the world. ❿ ***Macrorchis ausoba*** is a common and widespread Nearctic fly, but little is known of its biology beyond some observations of larvae occurring in soil around bulbs.

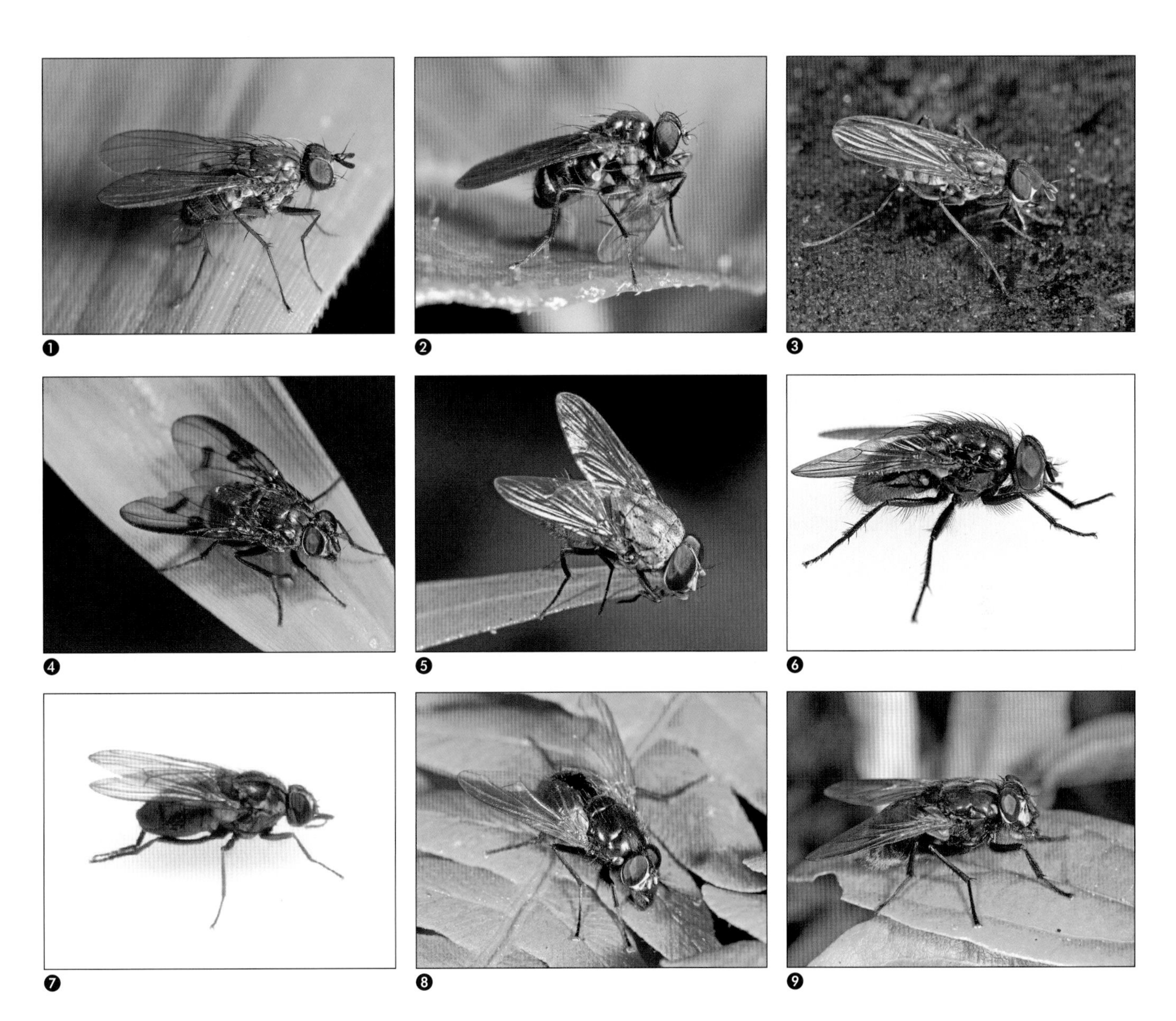

FAMILY MUSCIDAE (continued) ❶ ❷ ***Neodexiopsis***, seen here in Peru and Costa Rica (with prey), is a large New World genus with most of its hundred or so species occuring in the neotropics. ❸ This ***Xenomyia*** was among hundreds hunting invertebrate prey on the alga-covered rocks in a shallow east African stream. *Xenomyia* is a distinctive African genus of five species associated with swift streams, where their larvae prey on black fly (Simuliidae) larvae. **Subfamily Cyrtoneurininae** ❹ ❺ ***Cyrtoneuropsis*** is a Neotropical genus of around 30 named species; this black female and golden male are both from Ecuador. **Subfamily Muscinae** ❻ Members of the small, mostly Holarctic genus ***Eudasyphora***, such as this ***E. canadiana***, are metallic muscids that develop as microbial grazers in dung. ❼ ***Haematobia irritans***, the **Horn Fly**, is a biting fly originally found in the Old World but now a common pest of cattle and related livestock almost everywhere animal agriculture is practiced. The species is sometimes split into two subspecies; the southern Asian–Pacific subspecies (*H. irritans exigua*) is called the Buffalo Fly. ❽ The distinctively orange-winged ***Mesembrina*** species are mostly Holarctic-Oriental and mostly found in forests. The larvae develop in dung, feeding on both microbes and other larvae. ❾ ***Mesembrina*** species are among the most striking of all muscids, and this Vietnamese fly (probably ***M. aurocaudata***, previously known only from a couple of specimens from Myanmar and Nepal) is surely the most attractive of all *Mesembrina*.

FAMILY MUSCIDAE (continued) ❶ These ***Morellia (Parapyrellia)*** are seeking sugar on a honeydew-spattered leaf in Ecuador. ❷ ***Morellia*** is a large and widespread genus in the neotropics; this Ecuadorian female appears to be eating part of a dead arthropod. ❸ ❹ ***Musca domestica***, the **House Fly**, originated in the Old World (probably the eastern hemisphere) but is now one of the most widespread and dangerous of all insects because of its association with humans and their associated filth. Shown here are a female recently emerged from her puparium, with her ptilinum still partly visible above her antennae, and a female feeding on some honey bait. ❺ ❻ ***Musca autumnalis***, the **Face Fly**, is best known as a nuisance pest of cattle but is also a common domestic fly frequently seen on spring and fall flowers. Seen here are a male on a yellow flower and a female on a barn wall. Face Flies differ from House Flies in having a much narrower frontal vitta (the dark band up the middle of the front of the head). *Musca autumnalis* is an originally Old World species now a pest in North America. ❼ ***Musca sorbens***, the **Bazaar Fly**, replaces *Musca domestica* as the main winged bundle of bacteria in much of the Old World tropics and tropical Pacific, even though Bazaar Flies are less inclined to come indoors than House Flies. Those shown here were photographed on a meat scrap as I was waiting for lunch to arrive in an open-air restaurant in Tanzania. ❽ ***Musca vetustissima***, the infamous **Bush Fly** of Australia, is the insect responsible for the so-called "Aussie salute," the constant swish of the hand required to keep these abundant flies off your face while in the outback. The male (big eyes) and female shown here were attracted to a scrape on my kneecap. Larvae develop in dung. ❾ ***Musca*** larvae are typical maggots, with the head reduced to a pair of stout hooks in a tapered anterior end.

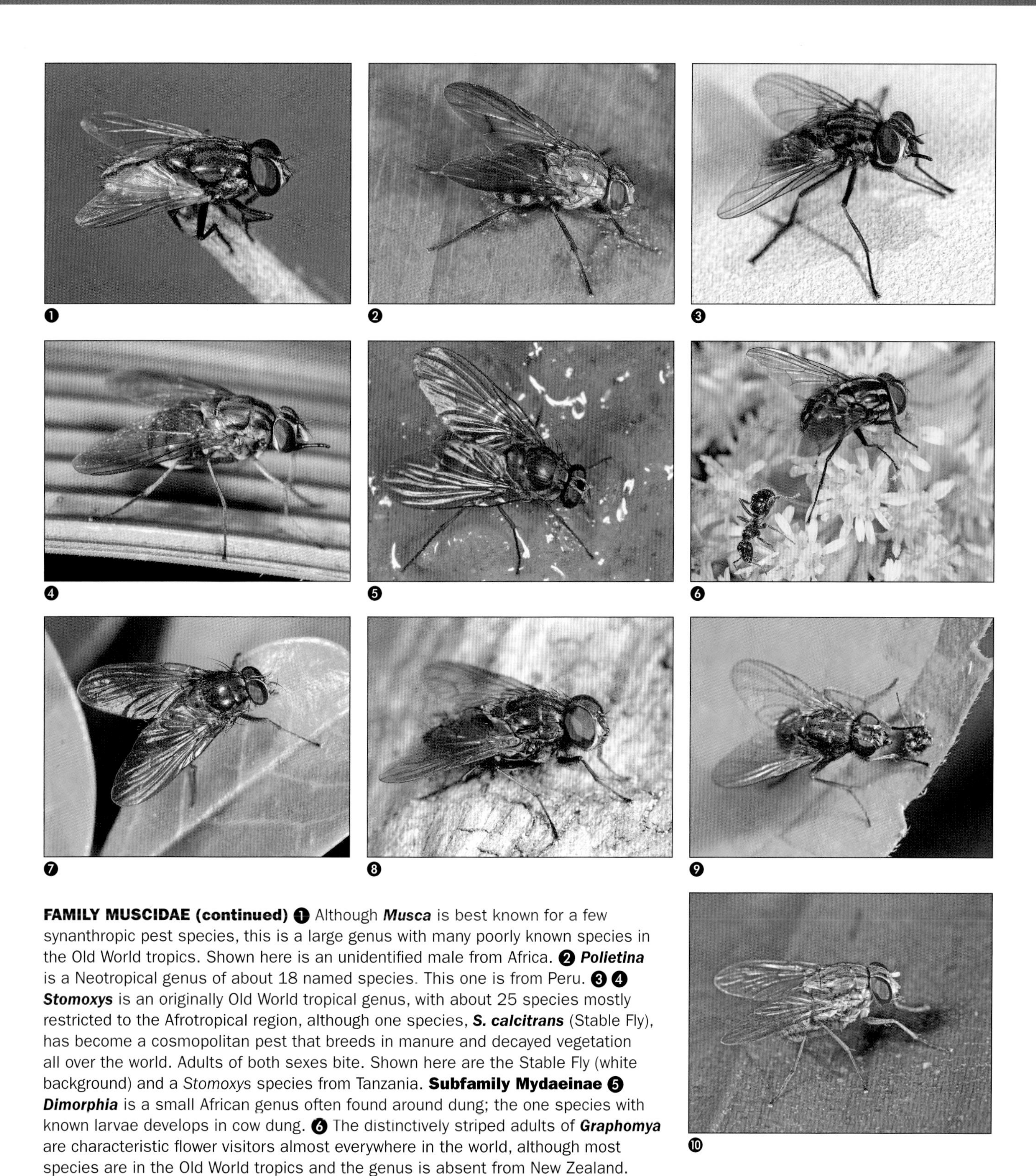

FAMILY MUSCIDAE (continued) ❶ Although ***Musca*** is best known for a few synanthropic pest species, this is a large genus with many poorly known species in the Old World tropics. Shown here is an unidentified male from Africa. ❷ ***Polietina*** is a Neotropical genus of about 18 named species. This one is from Peru. ❸ ❹ ***Stomoxys*** is an originally Old World tropical genus, with about 25 species mostly restricted to the Afrotropical region, although one species, ***S. calcitrans*** (Stable Fly), has become a cosmopolitan pest that breeds in manure and decayed vegetation all over the world. Adults of both sexes bite. Shown here are the Stable Fly (white background) and a *Stomoxys* species from Tanzania. **Subfamily Mydaeinae** ❺ ***Dimorphia*** is a small African genus often found around dung; the one species with known larvae develops in cow dung. ❻ The distinctively striped adults of ***Graphomya*** are characteristic flower visitors almost everywhere in the world, although most species are in the Old World tropics and the genus is absent from New Zealand. Larvae are semiaquatic predators. ❼ ***Hebecnema*** is a Holarctic, Old World tropical and Australian genus usually found around ungulate dung, where the larvae develop as predators. ❽ ***Myospila meditabunda*** is a common Holarctic species that develops in dung, pupariating in a cocoon made from sand or soil and sticky secretions. ❾ ***Mydaea*** species, which often have the characteristic yellowish wing bases of this Canadian species, are commonly seen as adults scavenging on dead insects or similar debris; larvae are usually in fungi or rotting vegetation. This large genus is common in the Holarctic, Oriental and Neotropical regions. ❿ Members of the small African genus ***Pseudohelina*** have been reared from a variety of rotting fruits and vegetables.

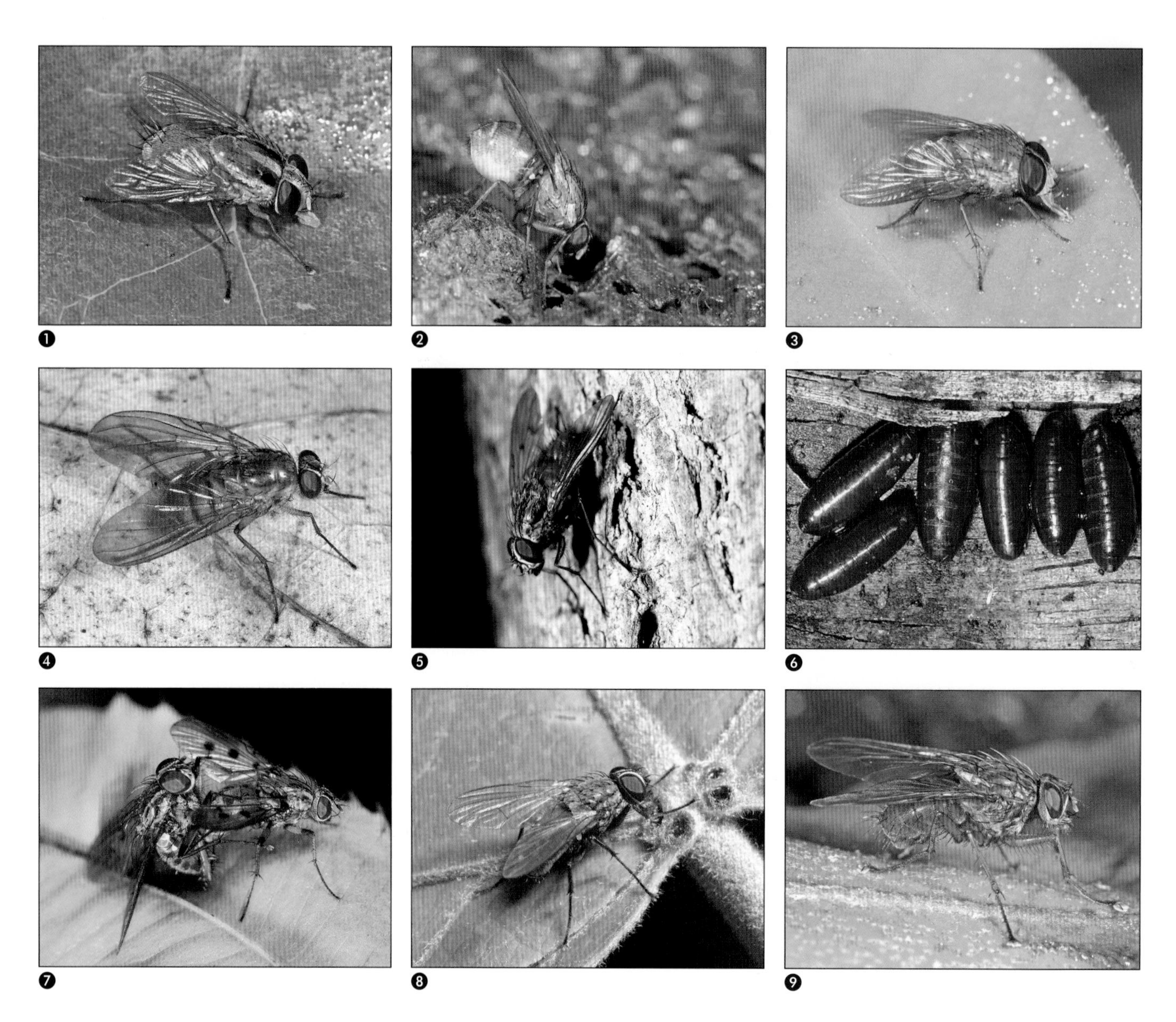

FAMILY MUSCIDAE (continued) Subfamily Phaoniinae ❶ Members of the small (half a dozen species) Afrotropical genus ***Alluaudinella*** feed on dead snails. ❷ This female ***Dichaetomyia*** is sharing an elephant dung pat with a relatively minute sphaerocerid fly. Few species of this large genus have been reared, but larvae are probably predators in dung and other decomposing materials. ❸ ❹ ***Dichaetomyia*** species, such as this male (on a leaf) and female (on leaf litter) from Tanzania, are common flies throughout the Pacific and Old World tropics. ❺ This ***Helina troene*** is taking a characteristic pose on a Canadian tree trunk, possibly the same tree where its larvae developed as a predator of other subcortical (under-bark) larvae. ❻ These ***Helina troene*** puparia were found under the bark of a dead tree. ❼ ***Helina*** is a large, ill-defined, cosmopolitan genus. This pair was photographed in Tanzania. ❽ Although extrafloral nectaries are usually thought of as a tropical phenomenon, some temperate plants also have them, presumably to attract ants that will in turn protect the young foliage from phytophagous insects, but of course many flies are attracted too. Shown here is what appears to be a ***Helina*** attracted to a young poplar leaf. ❾ This large Peruvian muscid is in the genus ***Helina***, but inclusion of a species in this widespread and heterogeneous genus tells us little about its relationships or biology. Like most other Phaoniinae, it probably developed as a predator in decomposing organic matter.

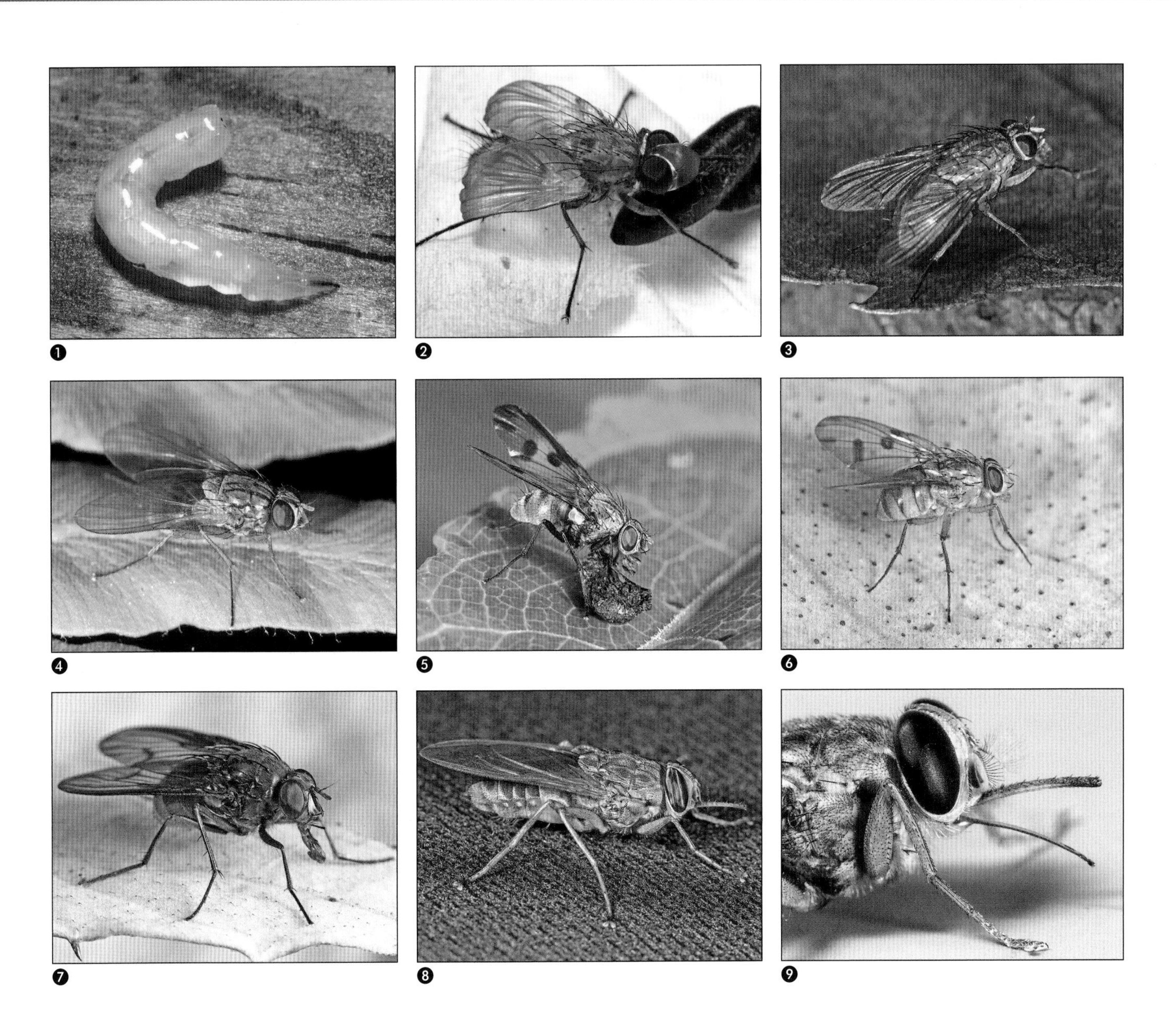

FAMILY MUSCIDAE (continued) ❶ ❷ ***Phaonia*** larvae, like those of the similar and related genus *Helina*, are predators of other insect larvae in rotting wood as well as a wide range of other moist and bacteria-rich media. This adult has not yet fully retracted its ptilinum (the bulge on the front of the head) or fully inflated its wings. ❸ ❹ ***Phaonia*** is a large (almost 800 species so far described), cosmopolitan and poorly defined genus. Shown here are females of the European ***P. pallida*** (pale orange, as the name suggests) and an eastern North American species (probably ***P. winnemanae***). ❺ Many muscids, like this unidentified species from Peru, are attracted to organic debris or fragments of dead insects on foliage. ❻ ❼ These Peruvian ***Phaonia*** are among several co-occurring high Andean species with pointed or bladelike thoracic bristles. The species currently relegated to the large genus *Phaonia* form a heterogeneous assemblage that needs to be reviewed on a worldwide basis.

SUPERFAMILY HIPPOBOSCOIDEA

FAMILY GLOSSINIDAE (TSETSE FLIES) ❽ All species in the family Glossinidae are in the genus ***Glossina***, which has some 23 extant species almost entirely restricted to sub-Saharan Africa. These are the infamous "tsetse flies," notorious as vectors of the trypanosome that causes African sleeping sickness. ❾ Both male and female tsetse flies have a prominent piercing proboscis with an elongate labrum and hypopharynx, sheathed in a long lower lip (labium) that is tipped by cutting teeth. The proboscis, unlike that of Stable Flies, has a bulbous base; it is exposed here beneath the elongate palpi.

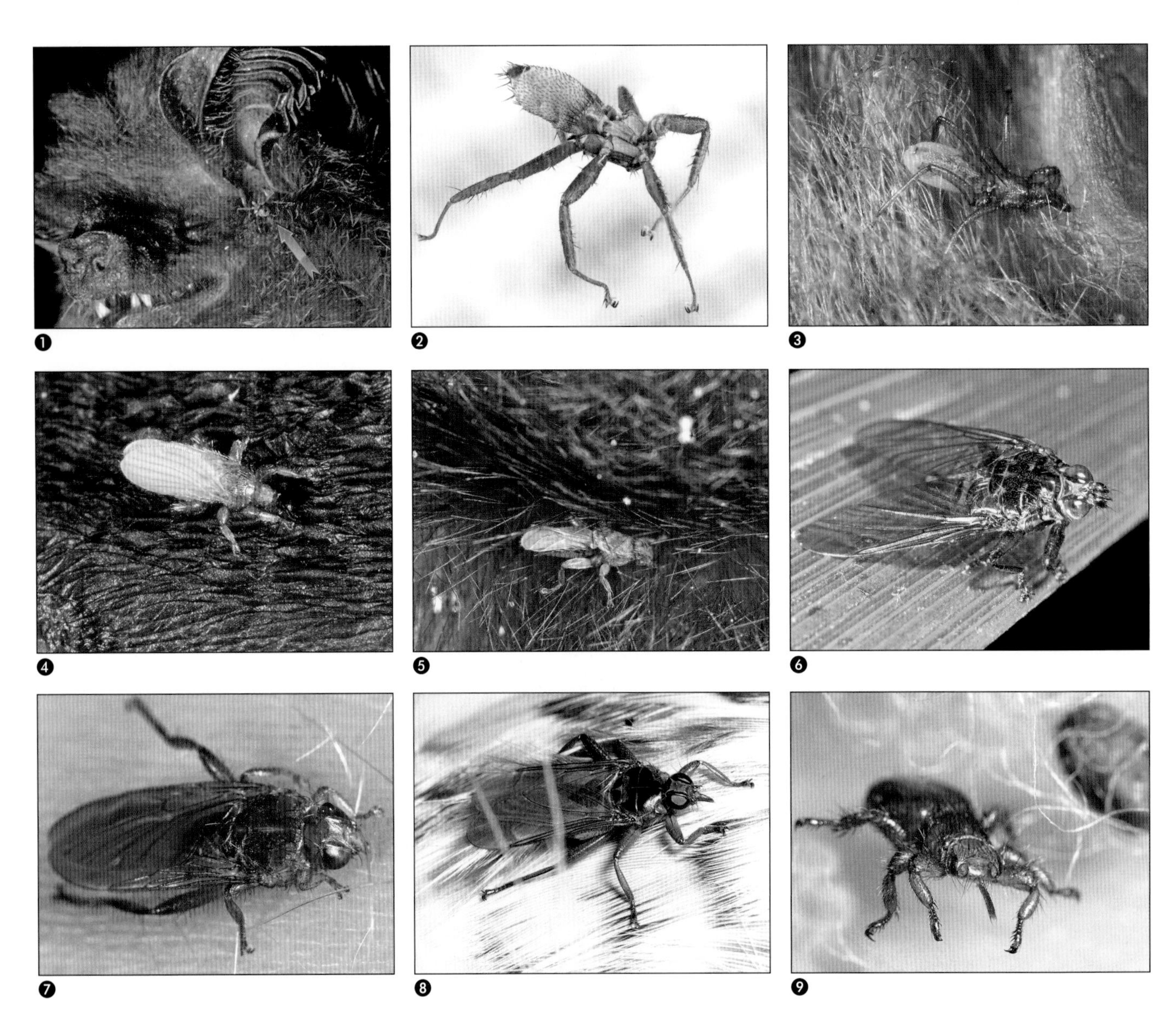

FAMILY HIPPOBOSCIDAE (LOUSE AND BAT FLIES) Subfamily Nycteribiinae ❶ ❷ The arrow under this Australian bat's ear points to a tiny wingless fly in the family Nycteribiidae. The second image shows a specimen of another nycteribiid species; note the tiny head positioned dorsally on the thorax. **Subfamily Streblinae ❸** This tiny ***Neotrichobius*** is just behind the ear of a phyllostomid bat in the Ecuadorian Amazon. The wings of this small (four species) genus are narrow and strap-like. ❹ Most Streblinae, like this Ecuadorian ***Trichobius***, are fully winged and quick to fly off their hosts when disturbed. This one on a captured bat took off seconds after the photo was taken. ❺ This ***Raymondia***, on a bat host in Vietnam, has curled its wings as it zips through its host's fur. **Subfamily Hippoboscinae ❻ ❼** ***Ornithoica*** is a cosmopolitan genus of louse flies found on every continent except Antarctica. Shown here are species from Costa Rica and Australia. Although this Australian fly landed on my son's arm, *Ornithoica* species are normally associated with birds. ❽ ***Icosta americana*** is a widespread and common louse fly that occurs on both raptors and passerine birds. ❾ This **Sheep Ked** (***Melophagus ovinus***) is a wingless louse fly that spends its entire life cycle in the wool of sheep. The oval object in the background is a puparium glued to the host's wool.

PART 3

CHAPTER 9

Identifying and Studying Flies

CHAPTER 10

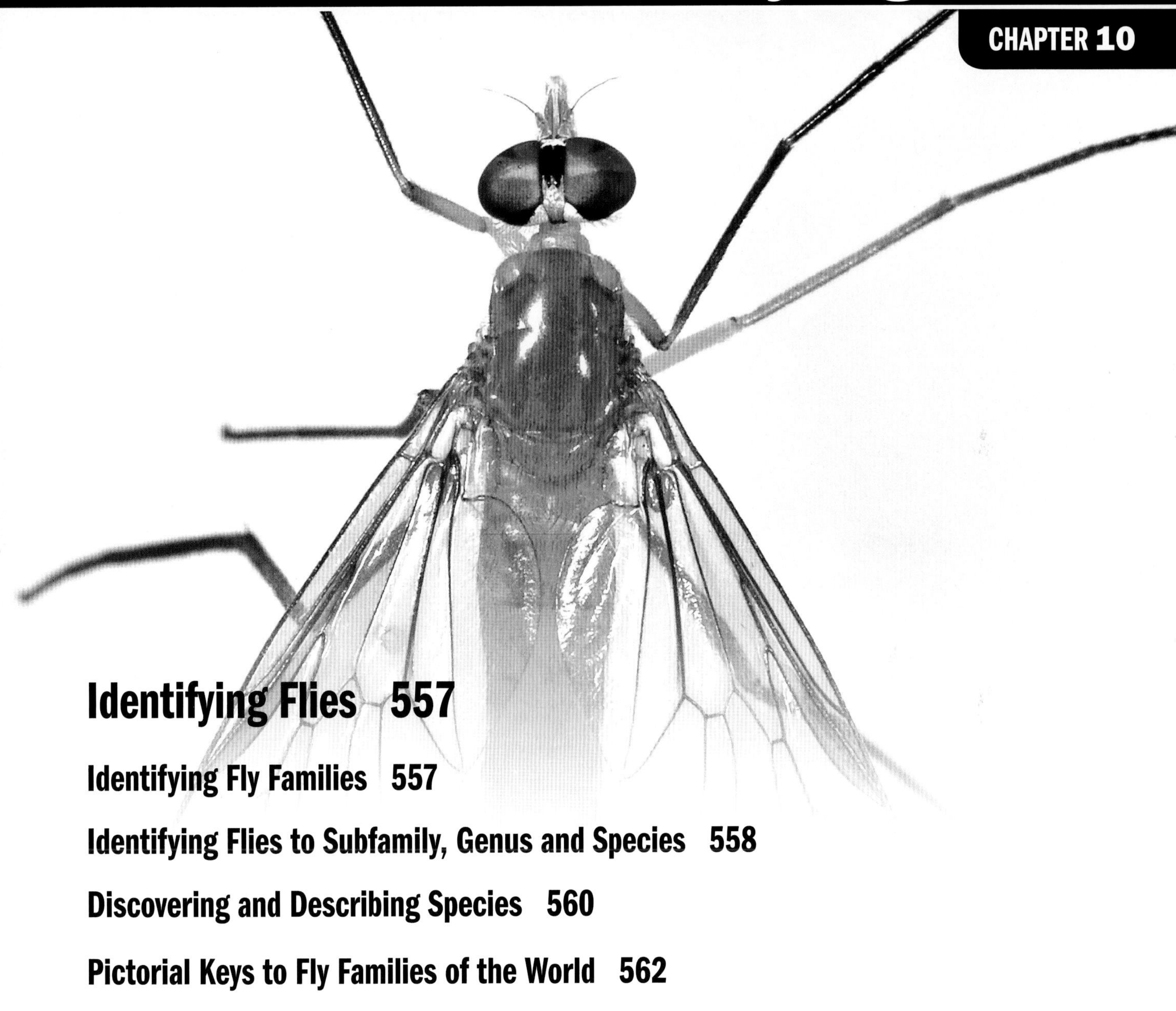

9

Collecting and Preserving Flies

WITH A BIT OF EXPERIENCE (AND A GOOD book!) it is easy to recognize most fly families in the field and to become familiar with their habits and habitats. But the greatest excitement in getting to know diverse groups such as the Diptera comes more from discovering the new than from observing the familiar. There is satisfaction, of course, in recognizing the flower flies that regularly visit your garden, but it is the flies you don't recognize that will intrigue, challenge and tantalize with the possibility of new discoveries. Wherever you live, you are within easy range of flies with unknown larvae or undocumented behaviors, and if you live anywhere but Europe or eastern North America the odds are high that undescribed species are close at hand. Species new to science still await us even in these relatively well-known areas, and species that are new to you are almost certain to enliven your every outing, especially if you are willing to explore unusual habitats and novel techniques. What, for example, are the kinds of flies associated with your favorite flowers, local ponds or perhaps the top of that mountain you climb twice a year? Discovery and identification are just the critical first steps, and interesting questions about natural history and distribution are sure to follow. But everything flows from those first steps, so let's look at some of the tools and techniques needed to find and identify flies.

Photographing Flies

Most beginning dipterists today will document their first fly specimens not by using the traditional nets, bottles and boxes but with a digital camera. The relatively recent appearance of affordable and easy-to-use digital cameras with adequate magnification and resolution to capture, identify and share insect images is currently revolutionizing insect study. Insect photography is the new bird-watching, as you can see for yourself by typing "insect photographs" or "fly images" into an Internet search engine. Literally millions of insect images and associated discussions have been posted to major photo-sharing websites, and an increasing proportion of them belong to the most ubiquitous insect order — Diptera, of course! Images of adequate quality to post on the Web can be taken with all sorts of digital cameras, ranging from tiny shirt-pocket models to the heavy professional models,

OPPOSITE PAGE
This small flower fly (*Platycheirus nearcticus*) picking up pollen on a geranium flower was photographed at f20 to ensure adequate depth of field (Nikon D2X, 105 macro lens, ISO 100, shutter speed 1/250s, hand-held flash).

These photos, taken by Stephen Marshall Jr, show the author photographing a gobryid under a leaf (see page 476) and preparing to shoot a cypselosomatid on a cut banana stem (see page 472). Note the hand-held, corded flash.

but somewhat higher-quality images are needed for print.

The images used in this book were captured over a thirty-year period using a wide variety of SLR (single-lens reflex) cameras with macro lenses (55, 60 and 105 mm), but most were taken in the past five years using digital SLR cameras. Older images taken using slide film were used only for taxa I have been unable to reshoot digitally, and for images from areas (Tasmania, Brazil, New Zealand, Western Australia, etc.) that I have been unable to revisit since moving from film to digital, because the digital images are generally superior to scanned versions of the older slides. This is in part because scanned images never seem as sharp as the originals, but also because it is relatively easy to get high-quality insect images with a digital camera. Any kind of photography is like painting with light, but with a good digital camera the amount and distribution of light reaching your "canvas" is readily controlled to give the desired depth of field, sharpness and shadowing.

The biggest challenge in photographing a tiny animal like a fly is to get enough of your little target in focus, because the greater your magnification, the shallower the depth of field at a given aperture size. A close-up shot of a small fly taken at f8, for example, will be largely out of focus because the depth of field at that aperture will give you only one "slice" — perhaps one eye or a wing — in focus. There are two possible solutions to that problem. One is to take multiple "slices" and then stitch them together digitally, using one of many commercially available software programs — for example, Combine Z or Helicon Focus — to make a composite image. This approach, called focus stacking, is great for laboratory photographs of mounted specimens, but generally impractical for field photography of active animals such as flies. The other approach is to increase the depth of field by decreasing the aperture size.

Changing aperture size is the easiest way to control depth of field, and understanding the effect of different aperture sizes is key to getting decent fly photos at different magnifications and in different light conditions. A low-magnification shot of a flower fly on a daisy at f11 is likely to look good because the fly and the flower will be in focus and the background will probably be pleasingly blurred out. The same fly shot full frame (at higher magnification) at the same aperture size is likely to be largely out of focus. Narrowing the aperture, perhaps to f16, can address that issue by increasing the depth of field, but that creates a new set of problems. To ensure that enough light gets through that narrowed aperture to provide a properly exposed image, it is necessary to use one or more of the following tactics: (1) use a very high ISO (film speed equivalent), (2) hold the shutter open for a longer time or (3) boost the light with a flash.

High ISO settings come at the cost of "noise" and lower picture quality; the very high settings needed for natural-light photography of smaller flies are not yet practical. This is changing with the new generation of digital cameras, but until a useful ISO of several thousand is within reach, macrophotography will demand either slow shutter speeds or the use of artificial light. Slow shutter speeds are impractical for handheld photographs, because the shutter is open long enough for the camera to shake and cause a blurry photograph. Natural-light photographs of small objects usually require a tripod, which is fine for flowers but generally unworkable for small, mobile insects such as flies. It is much easier and more effective to use a fast shutter speed while ensuring that there is enough light to make the combination of a fast shutter speed and a small aperture work. In other words, fly photography is generally flash photography.

Fortunately, flash photography has been simplified by new flash systems that communicate with the camera, firing at just the right intensity to give you the needed extra light. I use a handheld corded flash, but wireless units and flash brackets are available and work well. Sometimes two flashes are needed to detail both the upper and lower surfaces of a fly, and sometimes the flash intensity needs to be manually adjusted to get the desired effect. One of the undesirable properties of high-depth-of-field flash photography is the creation of a black background, as the powerful flash units swamp the natural-light background. Sometimes this can be avoided by striking the right balance between depth of field and lighting (that is, don't stop down further than you have to); mitigated by moving a solid background such as a large leaf into position behind the subject; or solved by shooting at a higher ISO.

Fly photography is great fun, and an efficient way to develop a collection that can be used to document your dipteran discoveries. In many ways a digital image collection is superior to a traditional collection of pinned or preserved specimens, since it bypasses the thorny issue of collecting permits, is easily organized and shared and can be readily geo-referenced — many cameras now include geographic positioning systems that record latitude and longitude in the image metadata. But image collections have some serious limitations, especially when it comes to naming digital "specimens" that are not linked to actual dead bodies in a physical collection. And don't forget that those collections of dead bodies still form the foundation for the science of Diptera taxonomy, an active science indeed, given the number of fly species still awaiting discovery. Taxonomy based on the description and study of specimens gives us the framework of names and inferred relationships on which all biology depends.

Catching and Killing Flies

Even though many naturalists may be satisfied with a collection of images alone, the task of catching, killing and preserving fly specimens remains an important and rewarding activity. It is also, believe it or not, an illegal activity in most countries if it is carried out without the correct permits. Increasingly stringent and unfortunately inclusive laws, put in place largely to protect vertebrates, mean that in many places you must apply months or even years in advance for permits to collect any organisms, including flies, and for the most part you must be an accredited scientist to get permits. Such regulations apply to parks and protected areas everywhere, and in many cases apply to entire countries. If you try to leave almost any tropical country with specimens but without a collecting permit (and an export permit), you could end up in jail. So think twice before casually collecting any insects while on an exotic holiday.

Snap-cap plastic tubes ("pill bottles") can be used to capture and individually store most kinds of flies. Most flies will fly up into a bottle carefully lowered towards them.

OPPOSITE PAGE (TOP) Canadian dipterist Kevin Barber is captured here as he examines acalyptrates in his straight-tube aspirator.

(BOTTOM) The widespread use of Malaise traps has had an enormous impact on dipterology, and many of the fly specimens making their way into museums today were taken in these simple traps. Shown here are two raised emergence traps (foreground) and a Malaise trap.

Fly photography, on the other hand, is unregulated in most (but not all!) countries. But even fly photography should be linked to some collecting if possible, since photos are often unidentifiable without specimens, and new species that have been photographed without collection of "voucher" specimens usually cannot be given names. Fly collecting is no threat to habitats or populations. Instead of being restricted it should be encouraged as a beneficial educational activity and recognized as a critical contribution to our understanding of biodiversitiy.

Collecting Equipment

The minimum equipment for collecting flies is simply a place to put the captured specimens. I usually catch only a single insect at a time, placing each one separately in a different "pill bottle" (a small snap-top bottle). Any small container will do, but clear plastic bottles with an opening of 2 to 3 centimeters work well for "tubing" flies — slowly lowering the container over the target and capturing it as it flies up — and they fit well into pockets or packs. My field kit includes a few dozen such bottles and a few Ziploc bags in which to put full bottles, along with appropriate tags, from each habitat. The Ziploc bags go into the freezer at the end of the day and the specimens can be thawed and mounted later on.

Most fly collectors carry somewhat more elaborate kits, normally including some sort of "killing bottle." The ideal killing bottle is a plastic centrifuge tube (unbreakable soft plastic, slips nicely into a pocket, does not react to chemicals) containing an appropriate poison in the bottom of the tube. A layer of potassium cyanide covered with a layer of plaster of Paris is the traditional favorite in this role, as it is effective and lasts a long time, but cyanide has a nasty reputation and is now rarely used. Ethyl acetate is currently the most widely used killing agent; it is easy to obtain and relatively safe. When I need a killing bottle for bulk collecting (or when I have no access to a freezer), I soak some absorbent paper with acetone-free nail polish remover, which is largely ethyl acetate and is available from most shops that sell women's cosmetics. That soaked paper goes in the bottom of the bottle, separated from the catch by layers of dry tissue — the result is a killing bottle that will last for a few days. Some collectors skip the chore of collecting dry specimens for later mounting, instead collecting directly into vials of alcohol. This works well for minute, delicate or soft-bodied specimens and is standard practice for collecting immature specimens. However, it is rarely appropriate for larger flies unless they are being collected for special purposes, such as molecular analysis.

Although a great deal can be accomplished with nothing more than a plastic bottle or a vial of alcohol, most fly collectors carry some other equipment, almost invariably including nets and aspirators. Nets come in a variety of shapes, sizes and weights. I prefer a short-handled general-purpose net about 27 centimeters in diameter, with a net bag made from the tough material used in tent flies. My net bags are custom-made with a tapered tip the shape of my killing bottle, allowing me to slip in the bottle and efficiently harvest the catch. Dipterists who like to look into their nets to examine specimens usually prefer white mesh; those who like to look through the net material at their catch often prefer the more transparent black mesh. Nets can be purchased in a variety of shapes and sizes, ranging from heavy nets for sweeping vegetation through to ultralight aerial nets and folding pocket nets, or they can easily be made from a loop of number 8 wire.

After the bottle and the net, the most common bit of equipment found in a dipterist's kit is the mouth aspirator, or pooter. At its simplest, the aspirator is a soft, flexible tube (surgical tubing is ideal) about 80 centimeters long, with a hard plastic tube (a clear pen barrel is perfect) inserted in one end; a fine screen covers the end of the plastic tube where it is inserted into the soft tube. Flies are sucked up into the hard plastic tube, prevented from entering the soft tube (and thus the collector's lungs!) by the screen, and then blown back out into a vial or other container. Some aspirator designs incorporate a bottle between the plastic tube and the screen, allowing retention of the catch in the aspirator bottle. When aspirating from potentially dangerous substrates such as carrion, it is best to use a modified car vacuum cleaner to provide the suction, rather than the traditional mouth tube. I rarely use an aspirator, preferring instead to simply lower bottles over my target flies, which almost invariably — with the noteworthy exception of some tachydromiine Hybotidae — fly up into the containers.

A good fly collector's field kit will include larger containers to gather live material and habitats such as galls for rearing or observation. It is also likely to include baits (such as honey) to attract target taxa. If I am doing "serious" collecting rather than just taking photo vouchers, I also carry a good supply of small yellow bowls for use as pan traps. A few dozen bowls full of soapy water can out-collect even a good dipterist in some habitats, and catches can be boosted by stretching a vertical mesh panel over the bowls; small flying insects intercepted by the panel usually drop into the trap fluid. Pan traps can also be baited by hanging a mesh bag of mushrooms (or dung, carrion or whatever) over the fluid, and preservatives — propylene glycol, salt or sodium benzoate — can be added to the fluid if the traps are to be left out for more than a day.

Other traps that intercept or attract flying insects include Malaise traps, raised emergence traps and light traps. A Malaise trap is much like a slope-roofed pup tent with the sides open but with a panel in the centre. Flies hit the panel and either drop (into pan traps below) or move up and toward the light, where they are collected in a transparent trap head at the apex of the "tent." A raised emergence trap is a bit like a short Malaise trap without the centre panel; it is installed over mammal burrows, fungi, rotting wood or other microhabitats, from which it collects arriving and emerging flies. Light traps of various designs, especially those using ultraviolet bulbs, can be effective for collecting some kinds of flies. Other kinds of traps are as varied as the taxa they are designed to collect. Frog midges (Corethrellidae), for example, are collected in traps "baited" with recordings of their tree frog hosts, and smoke flies (Platypezidae) can be brought in with billows of smoke.

Preserving Flies

It is easy to collect a diversity of flies but somewhat harder to deal with myriads of specimens in such a way that they end up properly preserved, organized and linked to useful data. Dipterists who collect into alcohol have it easiest in the short run, since they need only insert a permanent label in the vial to have a labeled preserved specimen. They pay for this convenience in the long run, however, because archival-quality vials are expensive (alcohol will eventually evaporate from inexpensive screw-top vials), they are difficult to organize unless there is only one species per vial, and specimens in alcohol are generally much more difficult to identify than dry specimens. When I collect into alcohol I usually end up drying and mounting the specimens for study and storage, but drying specimens directly from alcohol will result in shriveled and useless specimens without some time-consuming intermediate steps such as soaking in ethyl acetate or other solvents. Professional dipterists routinely dry small specimens from alcohol using a critical point dryer — essentially a brass bomb that enables flushing of the original alcohol with liquid carbon dioxide at a high temperature and pressure. This results in beautiful specimens, but for most collectors it is easier to keep the specimens dry in the first place. So let's say you have a freezer full of nicely labeled bags, each containing a dozen tubes of dry flies. What next?

RIGHT
Large collections of flies are usually organized in movable unit trays in a larger drawer, like this drawer of deer flies from the University of Guelph Insect Collection.

Mounting

Larger flies, such as most syrphids and many calyptrates, are mounted by simply impaling them through the right side of the mesonotum (the top of the thorax) with an appropriate-sized insect pin. Insect pins come in sizes from 000 (very flimsy) to 6 (almost a nail), but sizes 00 to 3 will cover the necessary range for most fly collectors. Specimens too small to mount with a 00 pin should be glued to a point instead (but see below regarding alternatives to pointing). Flies should be pinned at a uniform height that allows just enough room for a large thumb and forefinger to safely grip the pin well above protruding bristles and wings.

Smaller flies (*Drosophila*-sized and smaller) should not be impaled but instead should either be glued to the end of a paper triangle (called a point, and usually punched out to a standard size using a "point punch") or glued directly to the side of a pin. I prefer point-mounted specimens, since it is easier to view them from every angle. White glue (carpenter's glue or school glue) works fine for point-mounting, but shellac is also widely used for this purpose. Some dipterists prefer a system of double mounting very small specimens, using

These acrocerid specimens demonstrate alternative mounting techniques. Two are pinned, one is glued to a paper point, and one is glued to the side of a pin. An alternative method, called double mounting, is illustrated on page 487.

a tiny pin called a "minuten" to pin the specimen on a soft block that is in turn impaled by the main pin. I dislike double mounts, but that is a matter of personal preference.

Labeling Specimens

The most common error made by beginning insect collectors is to laboriously mount dozens of specimens and then procrastinate about labeling them. Trust me on this one — if you don't label your specimens right after you mount them, you will forget where they came from and your efforts will have been wasted. I mount material from each collecting "event" in a cluster, which I normally immediately label with a handwritten label. Occasionally I'm in a rush and put off writing the label until "tomorrow" — those specimens almost invariably end up in the garbage, which is where specimens without data belong. Handwritten labels should provide enough information so that you can later make printed labels that include at least the place and date, and it is always useful to include further information.

I often use Google Earth to make my labels now, marking the collecting localities on the map and copying the latitude and longitude into a text document to make the labels. After "blocking and copying" repeatedly to make the number of labels I need, I put the whole document into the smallest readable font (4 point Arial is good) and print it using a laser printer and heavy rag paper (cover stock). Labels are pinned at a uniform height under each specimen following the long axis of the specimen — minimum space use, maximum specimen protection — and the labeled collection is then stored in a sealed container to prevent invasion by museum pests such as dermestid beetles, which would otherwise ultimately convert the pinned specimens into dust.

Once you have a box or two of labeled specimens, you have a fly collection ready for organization and study. Organization is easiest if you either use several small boxes or divide up a large box, such as a glass-topped insect drawer, with small moveable trays (unit trays) lined with soft high-density foam. But proper organization depends on proper identification to at least the family level, which is what Chapter 10 is all about.

10

Identifying Flies

Identifying Fly Families

THE EMPHASIS IN THIS BOOK IS ON THE FAMILY level, since this is generally the minimum identification level necessary to communicate about flies. If you know what family a fly belongs to it is usually possible to make meaningful predictions about habits and habitats, and a family name also provides the handle needed to seek further information on the Web or elsewhere. Most flies belong to instantly recognizable families such as robber flies (Asilidae), flower flies (Syrphidae) or bee flies (Bombyliidae), but even these familiar families include a few atypical members. Furthermore, a few fly families — Lauxaniidae, for example — are so variable in general appearance they defy easy characterization, and many flies are too small and nondescript for easy "eyeballing" (sight recognition). For these reasons we frequently need identification aids — called keys — to identify flies to the family level.

Keys are simply schematic decision trees that guide you through a series of questions, progressively narrowing the choices to arrive at a family (or other taxon) name. The two main types of keys are *matrix keys*, which present a long list of attributes to choose from in any order, and *dichotomous keys*, such as those in this book, which guide the user along a road that repeatedly divides — each such junction is signposted with contrasting pairs or groups of attributes, or "character states."

Keys have traditionally been considered difficult to use ("written by people who don't need them for people who can't use them"), but the user-unfriendliness of many keys stems mostly from three problems: (1) inadequate illustration of character states or taxa, (2) inappropriate selection of characters (for example, choosing phylogenetically significant but cryptic characters rather than easily observed ones, or choosing non-discrete characters rather than characters with clear gaps between states), and (3) excessively broad zoogeographic coverage. The most important of these is the illustration of characters, a problem addressed here with character images that guide the user through most of the key decisions, and by representative sets of color photographs that can be consulted to confirm or question identifications. Characters used in the keys have been kept as simple as possible and, because of space constraints, are often limited to the easiest one or two out of several possible diagnostic characters.

The keys could have been further simplified by dividing them up according to area (the smaller the area, the easier it is to develop a comprehensive

OPPOSITE PAGE Images are often difficult to identify without associated specimens, but odds of a successful identification are increased if care is taken to capture "key" characters. This African athericid was photographed in direct dorsal view to capture the detail of wing venation.

key), but they are instead presented as a single set of condensed world keys that represent a compromise. Many couplets were complicated by the unusual or atypical species that invariably make key-writing for a large area difficult, and there are still some uncommon species that will not key out properly, especially among the Acalyptratae. Most users, however, will find that all of their winged fly specimens are identifiable to family. Wingless flies are not covered in the keys, so the distinctive family Braulidae (bee lice) and the small, highly localized and very unusual families Mystacinobiidae and Mormotomyiidae are excluded, as are most flightless species of otherwise fully winged families. The entirely wingless families are easily recognized from the photos, and flightless species of other families usually resemble their winged relatives in non-wing characters.

With the exception of the entirely wingless families, all widely recognized extant families of flies are included in the keys. The designation "widely recognized" is, of course, subjective; as discussed in Part Two of this book, some authors recognize families that are not included in the keys — for example, by splitting the Heleomyzidae and Hippoboscidae into multiple families — and other authors don't accept some of the families that are included here — for example, Mesembrinellidae, Stylogastridae and Mythicomyiidae.

All the regional Diptera manuals have more comprehensive keys, specifically written for restricted regional faunas. Those interested in a matrix key to fly families may wish to check out the key to Australian Diptera families by Hamilton et al. (2006), now available at http://anic.ento.csiro.au/insectfamilies/.

Identifying Flies to Subfamily, Genus and Species

Subfamilies

Fly subfamilies, like fly families, are sometimes distinctive and easily recognized groups that are part of the routine language of dipterology, but some subfamilies serve more to reflect perceptions of phylogeny than to provide useful taxonomic units. For this reason, providing a key to all fly subfamilies would be impractical. I have instead provided sample photographs and comments for most subfamilies in Part Two. Many distinctive subfamilies can be recognized from the photographs and notes, but others require careful study, and sometimes even genitalic dissection, to recognize.

Genera

Since there are more than 10,000 genera of flies, it is impossible for any one person to recognize all or most of them, and impractical to provide a key to genera in a book like this. However, it is often extremely useful to know what genus a fly belongs to, because generic identification can unlock a wealth of literature about biology, distribution, relationships and diversity. Keys to genera are available for most, but not all, zoogeographic regions and can be found in the Diptera manuals covering the Nearctic, Central American and Palaearctic fly faunas (although not all families are covered in the Palaearctic manual). A Diptera manual is in preparation for the Afrotropical region, but generic keys for the Oriental and Australasian regions are either lacking or scattered widely in the primary literature.

Paper keys such as those in the manuals are currently the most important resources for the identification of fly genera, but there are growing numbers of alternatives. More and more digital keys are appearing on websites such as that of the *Canadian Journal of Arthropod Identification* (http://www.biology.ualberta.ca/bsc/ejournal/ejournal.html). There you can find regional keys covering the Tabanidae, Bombyliidae, Calliphoridae, Tephritidae, Syrphidae, Culicidae and other groups, along with a growing number of keys with broader coverage. One such digital key, for example, covers the world genera of Clusiidae.

Even in the absence of accessible and easily used keys, many genera can still be identified by using reference collections of identified specimens or identified photographs. Many such collections of photographs are appearing on the Web, often on sites supported by professional fly taxonomists who contribute identifications and advice. Two of the most useful sites for identifying fly genera are Diptera.info (http://www.diptera.info/news.php) and BugGuide (http://bugguide.net), which post extensive image libraries and support dialogue between taxonomists and individuals looking for fly identifications (BugGuide is North American and Diptera.info is mostly European). Other sites, including Encyclopedia of Life (http://eol.org), also host extensive image libraries, but most are still of limited use for generic identification.

The images in this book cover less than 10 percent of the world's fly genera, but many of the larger and more common genera are illustrated; of these a significant proportion can be recognized on the basis of the photographs in the text. This approach to identification must be used with caution, though, because a user unfamiliar with the diagnostic characters of a genus could easily misidentify specimens on the basis of superficial resemblance to a picture of a different genus.

Species

Given that there are more than 160,000 named species of flies, routine identification of fly species is far beyond the scope of a book like this. But species are the working units of biodiversity, and species-level identification provides the link between specimens and the knowledge and information linked to those species names. Complete and usable species keys are readily available for some groups of flies, especially medically or agriculturally important taxa and especially for European fauna, but fly species keys for most areas and most groups either do not yet exist or are widely scattered in the primary literature.

Paper keys, which are often hard to find and harder to use, still provide the main route by which people identify fly species, but websites such as the *Canadian Journal of Arthropod Identification* include a growing number of guides to the regional species of relatively well-known families. These richly illustrated keys provide a model by which species could be made more widely identifiable, but they remain available only for a select few fly groups.

Even in the absence of accessible and easily used keys, fly species can often be identified by using reference collections of identified specimens or photographs mentioned above. Many of the photographs in this book are identified to the species level and frequently represent distinctive species that can be recognized from the photographs. Again, however, this approach must be used with caution because of the possibility of misidentifying a superficially similar species.

The main obstacle to fly identification is the dearth of good recent taxonomic revisions and reviews that synthesize information on the regional species of a genus or family. Many, if not most, fly species remain unnamed, and few fly genera have been revised on a world basis. This means that flies are often unidentifiable to the species level and will remain so until the taxonomic work is done. But there are many flies and few Diptera taxonomists, so this will take time. In the meantime, you may be wondering about alternative ways to identify flies, at least those species that have been named and properly classified (which must be done first). In particular, are there alternatives to using keys, photographs and morphology to identify flies?

For the most part the answer to that question remains no, although there has been a great deal of buzz lately about the use of DNA sequences for species identification. This idea has been around for a long time, and has most recently manifested itself as "DNA barcoding," an ambitious program to develop a central library of standardized DNA sequences that can be used for subsequent identification of species represented in that library. If you have access to the expertise, equipment and chemicals necessary to amplify, sequence and analyze DNA from an unknown fly, you can try matching your specimen's "barcode" sequence — a standardized fragment of cytochrome c oxidase subunit 1 (commonly called cox1 or COI) — to the barcode library. If your fly's sequence matches something in the library, it *might* be the same species. Some different species have identical barcodes, many species have multiple barcodes, and barcoding apparently does not work for some groups (such as blow flies), but for most taxa a good match is treated as a probably correct identification. A more serious limitation of this sort of approach to fly identification is the scope and accuracy of the barcode "library," because it depends on a set of reference sequences from authoritatively identified and preferably freshly collected specimens. Since that set of sequences currently comprises only a small fraction (around 5 percent) of fly species, barcoding is not yet a useful approach to fly identification except for a few small, intensively studied groups.

It remains possible that these obstacles will be overcome and the necessary technology will eventually be within reach of the average biologist (perhaps even the average naturalist), but I'm not holding my breath in anticipation of routinely identifying fly species using a (currently imaginary) handheld barcoder or even a desktop barcode system. At least for the foreseeable future, molecular diagnostic tools are likely to play a significant role only for professional identification of fragmentary specimens or larvae, and then only for a relative few economically or medically important species. Digital keys, on the other hand, are likely to provide the most widely useful tools for fly identification for many years to come, and they certainly represent the most accessible and cost-effective approach. And, of course, basic taxonomic work — discovering, describing and classifying species — remains a prerequisite for any kind of identification.

Discovering and Describing Species

Throughout this book I have emphasized how much remains to be discovered about flies, and the text and captions are replete with reminders that the order Diptera is rich in undescribed taxa. There has, however, been little discussion of how those undescribed species are discovered and what follows such a discovery. What happens, for example, if the keys in this book tell you that the little fly you found in your Minnesota woodlot belongs to the family Clusiidae, and if perusal of the pictures in this book then suggests that it seems to be a *Heteromeringia* with distinctively bicolored forelegs? Let's say this piques your curiosity, so you do an Internet search and find that there is an open-access digital key to the world genera and North American species of Clusiidae (Lonsdale and Marshall 2011). You open the file and find a couple of kinds of generic keys: a matrix key that allows you to sequentially pick from a long list of illustrated characters until only one genus remains, and a dichotomous key, like the ones in this book, in which you are guided through a series of questions. You try them both and swiftly find that your specimen keys to *Heteromeringia*, a genus with only four species in North America.

You then open the species key and find that your specimen keys to *Heteromeringia nitida*, but the forelegs are a different color and it does not quite fit the description and illustrations on the species page. Something is wrong — have you made a mistake? Fascinated, you dig deeper and find that Lonsdale and Marshall revised the New World species of *Heteromeringia* in 2007. That is great luck, since modern revisions are available for relatively few fly genera. You download the paper from the Internet and discover that all the species of *Heteromeringia*, like so many flies, have distinctively different male genitalia. Now obsessed, you follow the steps in the "Materials and Methods" section of the paper to examine the abdomen of your male specimen, comparing the genitalic structures piece by piece with those illustrated in the paper. If they are different, you have probably discovered a new species! Of course, you will still need to eliminate the possibility that you only have a new record of a species previously named from Europe or Asia, but for now, pat yourself on the back for your discovery — and then consider the real work of turning your find into published science.

Discovering new species is a routine activity for experienced dipterists, especially in groups that have not been recently revised. It is easy for a fly taxonomist to walk into a major insect collection, pull out a drawer full of species in an unrevised genus and make a pronouncement such as "This genus has three described species. I see six different morphologically distinct species in this drawer — therefore, at least three must be new."

To move from simply discovering a new species to formally naming ("describing") it is a significant task, especially to do it properly so that the new species is placed in the correct phylogenetic context and is rendered easily recognizable by others. Describing demands a thorough familiarity with every named species in the genus, as well as every named species that might belong in the genus. There must be no doubt that the species to be described is indeed a new one before a scientific name can be coined, following the rules laid out in the International Code for Zoological Nomenclature (http://www.nhm.ac.uk/hosted-sites/iczn/code/). The Code states that a new name is valid only once it has been formally published with a description and designation of a type specimen that serves as a sort of standard for the species. Modern descriptions are usually illustrated well enough to allow subsequent recognition of the species without checking the type specimen, but older descriptions are often so vague it is impossible to be sure what the species is without obtaining and examining "types." Familiarity with all the type specimens in a group is often a practical prerequisite for describing a new species in that group.

To really be useful, a new species description needs to be published in the context of a treatment of all species (or at least all regional species) in the genus, with a new key covering both old and new species. Isolated descriptions of new taxa in unrevised groups are of little or no use and often turn out to be mistakes, as authors who describe species without revising the whole group often give new names — called synonyms — to previously named species; the newer name is then a junior synonym and is invalid. It is not unusual for a taxonomist to track down all the type specimens in a genus for a first comprehensive revision, only to find that some are identical despite their different names. Only the oldest name is correct; the other names get sunk as junior synonyms. It is this sort of ongoing task of cleaning up names, combined with the shuffling of species from genus to genus as we clarify our ideas about who is most closely related to whom, that causes the periodic name changes that so often confuse beginning students of diverse groups such as the Diptera (sorry!).

In scientific publications the name of the describer of a species normally follows the species name, in parentheses if the species has been moved to a genus other than that in which it was first described. Author names were left off the species names throughout this book to improve flow; for example, what I refer to as *Syrphus torvus* is more correctly referred to as *Syrphus torvus* Osten Sacken, since this species was first named (in the genus *Syrphus*) by Baron von Osten Sacken in 1875. Authorship, distribution and other information on almost every taxon (species, genus, family) mentioned in this book can be found on the Systema Dipterorum website at http://www.diptera.org.

The many mentions in this book of undescribed species or undescribed genera refer to species or genera that are recognized as new — they have been "discovered" — but for which the long process of formal description and associated group revision has not been completed. In most cases they are flagged as new because someone has done the work of examining all the relevant literature and type specimens, has recognized the species as new and is planning to name it formally in the near future in a paper covering the larger group. In many cases the species have been named in unpublished manuscripts, but those names cannot be released until the papers have been formally published. You may also have noticed that some photographs are captioned as "probably" or "possibly" a given species. This is usually because there is no adequate key or because there is no specimen to go with the picture, and a specimen is needed for a confident identification. Such names (indeed all names based only on images) are best guesses, based on distribution and general appearance.

Pictorial Keys to Fly Families of the World

How to Use the Keys

The keys to fly families that follow are like illustrated roadmaps. Each junction in the road leads to two alternatives; pick the alternative that best fits your specimen and keep moving on to the next junction until you reach a family name. The alternatives are morphological attributes such as size, color, wing venation or bristles. Sometimes it is easy to decide which way to turn, as when one road leads to "black" and one leads to "white" and you have a white fly in hand. But often the choice is not so easy, for example, when the characters compared are unfamiliar structures referred to in necessarily technical terms. Thumbnail illustrations are provided at most junctions to remind you where to look for these structures, but you will find that you can work through the keys more reliably and quickly if you spend a bit of time learning the main parts of a fly's body, the main wing veins and some of the named bristles often used in keys.

Even if you know exactly what you are looking for and where to look, some key characters are difficult to interpret on some specimens and some species. You may, for example, be well aware that the subcosta is a vein just above R_1 (the first branch of the radial vein), and you might recognize that some species have a complete subcosta (reaching the costa as a separate vein) while others have an incomplete subcosta (petering out or running into R_1 before reaching the costa). But every now and then you will encounter a specimen on which this character is difficult to interpret; perhaps the subcosta is very pale or you are not sure whether it ends in the costa or not. When you run into this sort of problem when using the key, the best strategy is to go down both roads and end up at two different destinations. Compare your two answers to the sets of photographs in the book and it should then be fairly obvious which answer is wrong and which is right.

And don't be discouraged if some specimens just don't key. It is impossible to take every exception into account in a short key covering the whole world. The key is just a guide, to be used in conjunction with photos and other information.

The following descriptions and associated illustrations of fly bits and pieces are not meant to be comprehensive treatments of fly morphology but rather are brief introductions to the main structures, or characters, used in the keys. A more complete morphological glossary can be found in the recent *Manual of Central American Diptera.* The terms and abbreviations used here follow the manuals of Central American and Nearctic Diptera.

Characters Used in the Lower Diptera Keys (Keys 1 and 2)

Head Characters

The antenna of lower Diptera includes a small basal scape (first antennal segment), a variously enlarged pedicel (second antennal segment) and a flagellum (subdivided into multiple segments called flagellomeres). The shape and size of the pedicel and the number of flagellomeres are important key characters for lower Diptera.

Ocelli, or simple eyes, are often present at the top of the head. The presence or absence of ocelli and the number (usually three, sometimes two) are important key characters. The ocelli can be difficult to see on pinned dried specimens but are usually obvious on fresh material. Most species of lower Diptera have large compound eyes, sometimes fused over the antennae to form an eye bridge, sometimes divided into lower and upper parts, with different-sized facets. Males of many groups are holoptic, with large eyes that meet on top of the head.

Wing Characters

The six main longitudinal wing veins — costa, subcosta, radius, media, cubitus and anal veins — present a diversity of character states created by branching, fusion, loss or interconnection by crossveins. The costa (C) is the leading edge of the

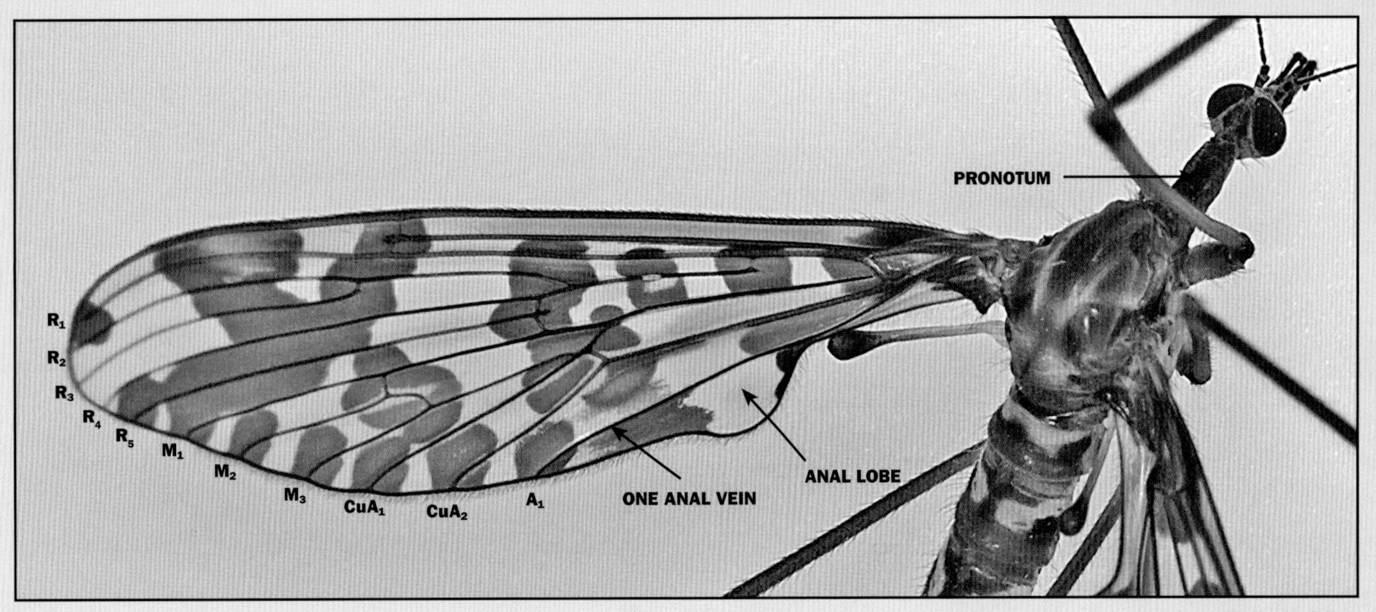

Tanyderidae

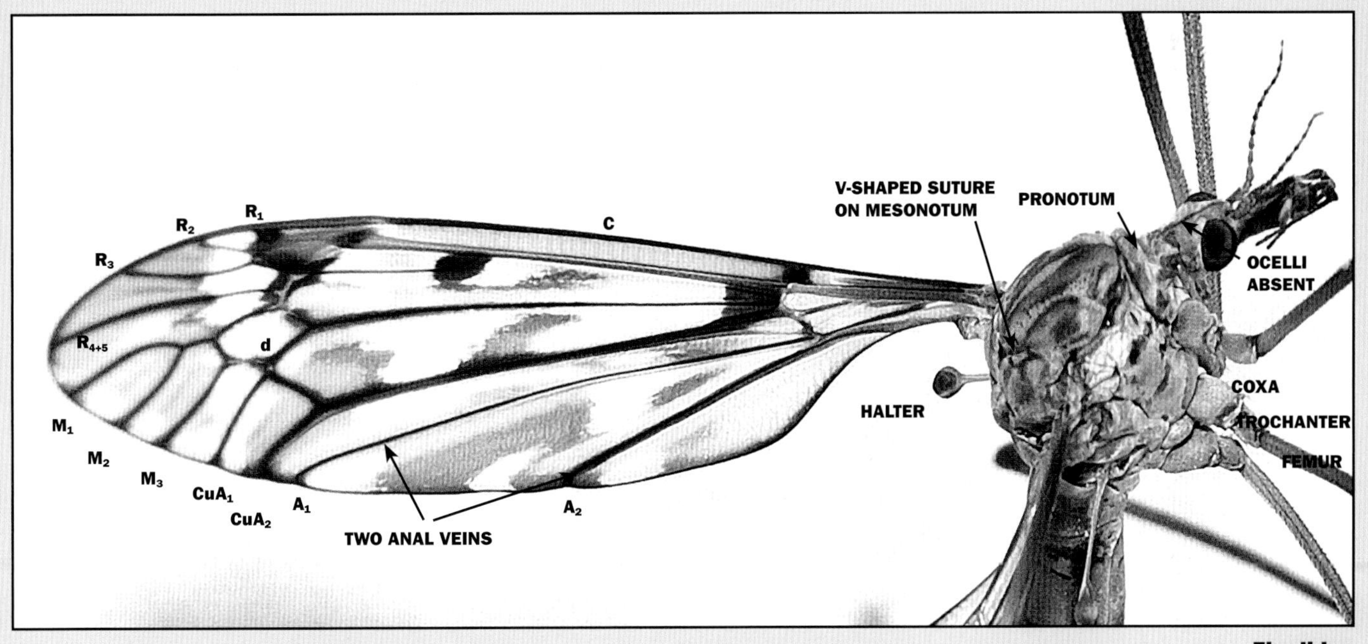

Tipulidae

wing and sometimes extends around the tip, but the hind margin of the wing is often membranous. The radius (R) is divided near the base into R_1 (first vein) and Rs (radial sector), with Rs dividing into upper (R_{2+3}) and lower (R_{4+5}) branches, one or both of which can be further forked (producing R_2, R_3, R_4 and R_5). The radial veins in lower Diptera usually bracket the wing tip.

The wing vein following R is the medial, or "middle," vein (M), which is usually forked into M_{1+2} and M_3. M_{1+2} can be further forked into M_1 and M_2. The branches of M are sometimes connected by a crossvein (m-m), creating a closed cell (d, or discal, cell) in the base of the fork. The wing vein following M is the cubital vein (Cu), which forks into anterior (CuA) and posterior (CuP) branches. The anterior branch can be further forked into CuA_1 and CuA_2, while the posterior branch is usually indistinct and incomplete. The last two veins of the wing are the anal veins

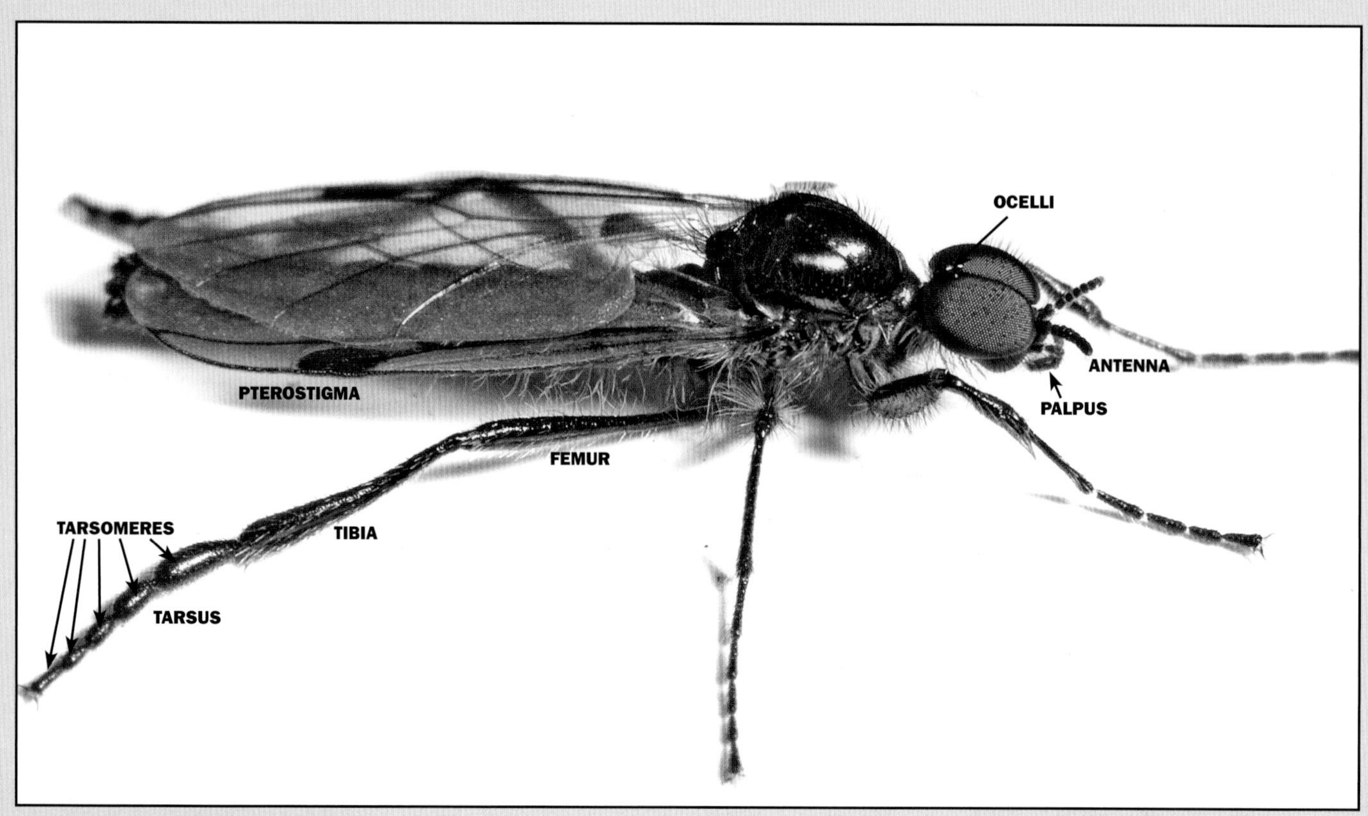

A male March fly (Bibionidae)

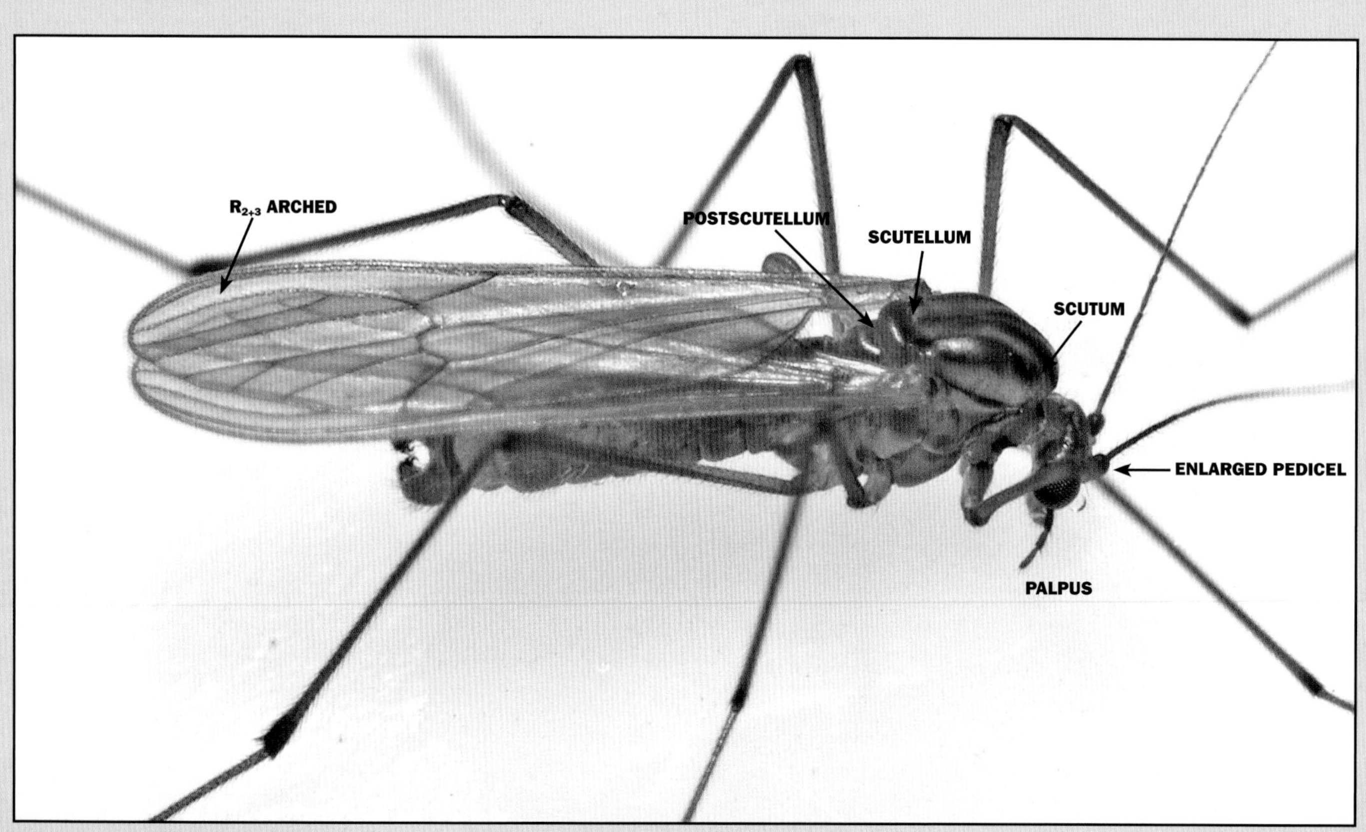

A male dixid midge (Dixidae)

(A_1 and A_2); both sometimes reach the wing edge independently, as in Tipulidae, but A_2 is often lost or reduced.

Leg Characters

Each leg consists of a basal coxa (usually short, but elongate in the Mycetophiloidea), a small trochanter, a long femur, a long tibia and a tarsus. The tarsus is divided into tarsomeres and ends in a small terminal sclerite that supports a pair of tarsal claws and a median, usually bristle-like empodium. Each tarsal claw is sometimes subtended by a pad-like pulvillus. The tibiae sometimes have spurs or bristles on the ventral surface right at the apex. The presence or absence of these spurs, the number and length of tarsomeres and the shape of the pulvilli and empodium are the main leg characters used in the lower Diptera family keys.

Thoracic Characters

The thorax is made up mostly of the wing-bearing second thoracic segment, the mesothorax. The first and third thoracic segments (prothorax and metathorax) are greatly reduced; the metathorax supports the third pair of legs and a pair of knob-like halters, and the prothorax supports the first pair of legs. Thoracic characters used in identifying lower Diptera families include the presence or absence of a prehalter (a prominent lobe sticking out in front of the halter of phantom crane flies) and the presence or absence of a V-shaped groove on the scutum (characteristic of crane flies). The scutum is the main part of the top of the mesothorax (mesonotum) and it is followed by a shelf-like posterior part called the scutellum. Behind the scutellum, the thorax forms a bulging postnotum. The presence of a longitudinal suture or seam running up the middle part of the postnotum is a useful feature for recognizing most members of the family Chironomidae; the presence of long postnotal setae is characteristic of the Ditomyiidae.

Abdominal Characters

Abdominal characters are critical to most species keys for lower Diptera but are not usually used in family keys.

Characters Used in the Lower Brachycera, Empidoidea and Aschiza Keys (Keys 3 and 4)

Head Characters

The antenna is made up of a small basal scape (first antennal segment), a variously enlarged pedicel (second antennal segment) and the flagellum, which is divided into multiple segments called flagellomeres. In the suborder Brachycera the flagellum is relatively compact, in contrast to the long, multi-segmented flagellum of lower Diptera. It normally includes eight or fewer flagellomeres, with the distal ones coalescing into an arista (in the Aschiza and other Cyclorrhapha) or a stylus (in most lower Brachycera and Empidoidea). The stylus, or style, of lower Brachycera can be thin and hairlike, similar to the arista of Cyclorrhapha, or broad and indistinctly divided into two to eight flagellomeres.

The front of the head below the antennae is called the face, and the part just below the face is the clypeus. In some Tabanomorpha the face is reduced and the clypeus takes up most of the area between the antennal bases and mouthparts; in the Asilomorpha it is the clypeus that is reduced. The size and shape of the clypeus can be useful in separating lower brachyceran families; it is sometimes bulbous and evenly convex — as in the Rhagionidae — and sometimes flat, concave or unevenly convex. The face is sometimes characteristically shaped or adorned, as in the family Asilidae, which is characterized by a tuft of forward-facing facial bristles called a mystax. Males of many lower Brachycera groups are holoptic, with large eyes that meet on top of the head.

Wing Characters

Many of the characters used in the keys pertain to the wing veins. The six longitudinal veins are the costa, subcosta, radius, media, cubitus and anal veins. The costa (C) is the leading edge of the wing, sometimes ending well before the wing tip — as in most Stratiomyidae — and sometimes extending around the tip. The subcosta (Sc) is unbranched. The radius (R) is the main vein of the anterior half of the wing; it is divided near

Stratiomyidae

the base into R_1 (first vein) and Rs (radial sector), with Rs dividing into upper (R_{2+3}) and lower (R_{4+5}) branches, and R_{4+5} sometimes further forked into R_4 and R_5. The wing vein following R is the medial, or middle, vein (M), which is often forked into M_1 and M_2 and occasionally has a third branch (M_3) running from a cell (cell d, or discal cell) made up of a patch of wing membrane enclosed between the bases of the branches of M. M_1 sometimes curves forward to join R_{4+5}, enclosing cell r_{4+5}.

The wing vein following M is the cubital vein (Cu), which forks into anterior (CuA) and posterior (CuP) branches. The posterior branch is indistinct and can be ignored for the purposes of this key, but the anterior branch is distinct and further forked into CuA_1 and CuA_2. The last veins of the wing are the anal veins (A_1 and A_2). A_1 sometimes ends independently in the wing margin and sometimes joins CuA_2 to form A_1+CuA_2. Whether and where these veins join is an important key character: some major groups are characterized by a long, pointed anal cell (cell cu*p*) formed when these veins join near the wing margin, while other groups have a short anal cell created when A_1 and CuA_2 join near the base of the wing. A_2 is usually inconspicuous and fades out long before the wing margin.

Areas of wing membrane completely enclosed by veins are called cells, like the anal (cu*p*) cell and discal (d) cell mentioned above. Cells can be closed distally by crossveins, which are short veins that run between the main longitudinal veins, especially between R and M and between M and CuA_1. Each cell is named (using lowercase) for the vein that forms the anterior boundary of the cell. Thus the cell between the bases of R and M is the basal radial cell, or br, closed by crossvein r-m. M and CuA_1 are sometimes connected by two crossveins, a basal one (bm-cu) and a distal one (dm-cu), thus enclosing two cells, the bm (basal medial) and dm (distal medial).

At the very base of the wing, between the wing base and the side of the thorax, are two membranous flaps called the upper and lower calypters. The lower calypter is greatly enlarged in some families — markedly so in the Acroceridae — and both calypters are enlarged in some other families, such as the Tabanidae.

Leg Characters

The basal segment (coxa) of each leg is followed by the small trochanter, the femur, the tibia and the five tarsomeres that make up the tarsus. The foretibia (tibia of the first leg) sometimes has a spur or bristle on the ventral surface right at the apex; although sometimes small and easily overlooked, it is important in the key. The most important leg character used in this key is the apex of the tarsus, where a small terminal sclerite — the acropod — supports a pair of tarsal claws usually subtended by pad-like pulvilli and a median empodium. The empodium can be bristle-like, in which case the tarsus ends in two pads, or similar to the pulvilli, in which case the tarsus ends in three pads. A few species that key to the "two tarsal pads" section have a slightly broadened empodium, but it is always narrower and more tapered than the pulvilli.

Thoracic Characters

Thoracic characters used in identifying lower Brachycera families include the presence or absence of a postspiracular scale, which is a scale-like process behind the posterior thoracic spiracle, and the presence or absence of a prominent subscutellum behind the scutellum. The scutellum is the shelflike piece on the posterior margin of the scutum, which in turn makes up most of the top of the thorax.

Abdominal Characters

Abdominal characters are widely used in species keys but less so in family keys. The dorsal sclerites of the abdomen are called tergites; the ventral ones are sternites. In some families the first abdominal tergite is notched or has a median (middle) suture.

Characters Used in the Schizophora Keys: Calyptratae and Acalyptratae (Keys 5–9)

Head Characters

The antenna of higher flies is divided into three basal segments and a hairlike distal arista. The basal segments are the scape (first antennal segment), pedicel (second antennal segment) and first flagellomere (third antennal segment). The scape is usually inconspicuous but the pedicel is prominent and often characteristically shaped. The pedicel of calyptrates and some acalyptrate families has a dorsal seam, and that of some acalyptrate families has a distinctive lobe on the inner or outer surface. The arista, usually made up of three flagellomeres, is inserted in the third antennal segment, sometimes basally or dorsobasally but sometimes apically (at the tip), as in Neriidae. The arista can be bare, pubescent (with inconspicuous hair), plumose (like a feather) or pectinate (with a comb-like series of hairs). The long hairs on pectinate aristae are sometimes restricted to the upper surface of the arista, as in the Ephydridae.

The front of the head is divided into the frons above the antenna and the face below; this division is sharply delineated by the ptilinal fissure. The face may be characteristically shaped (either concave or convex) and usually has a prominent ridge — the facial ridge — along its lateral margin. The clypeus is usually an inconspicuous U-shaped sclerite below the lower margin of the face.

Bristles of the frons are important key characters, especially those closest to the eye (orbital bristles) and the next row in (frontal bristles, usually anterior to the orbitals), which are sometimes referred to collectively as the fronto-orbitals. Some fly families have their fronto-orbitals characteristically inclined or directed backward (reclinate), forward (proclinate), inward (inclinate) or out over the eye (lateroclinate). The two rows of bristles

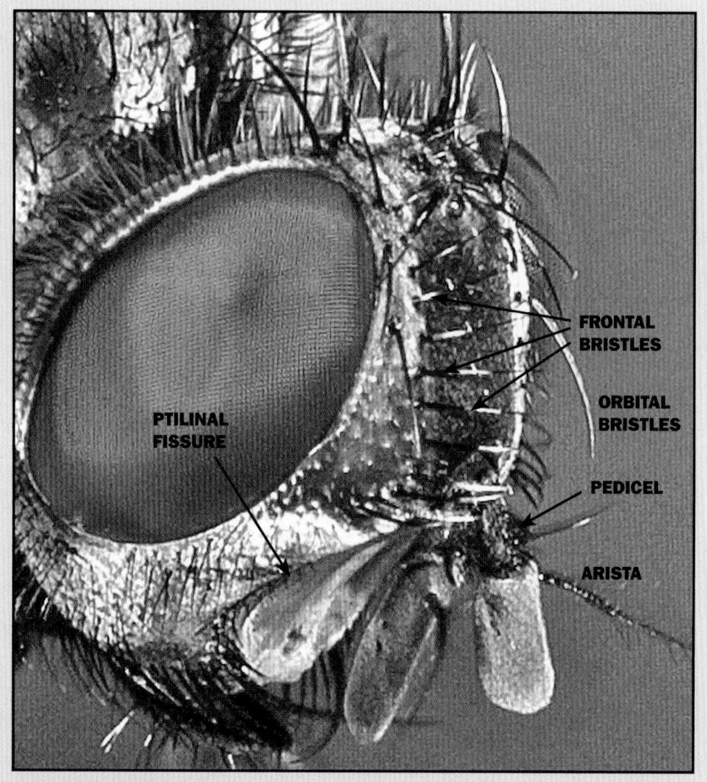

Sarcophagidae

Syrphidae

sometimes present near the middle of the frons are called the interfrontal bristles.

The top of the head, or vertex, bears the simple eyes, or ocelli — usually three, arranged in a triangle — and several bristles, almost always including a pair of ocellar bristles between the ocelli and a pair of postocellar bristles immediately behind. The postocellar bristles can be divergent, parallel or convergent in different families, and thus provide a useful key character.

The side of the head is dominated by the compound eye; the area below the eye is called the gena, or cheek. Where the front of the gena meets the face there is usually a prominent angle at the bottom of the facial ridge called the vibrissal angle, which may or may not support a large, anteromedially directed bristle called the vibrissa. The generally inconspicuous palpi usually project just below the vibrissal angle. The mouthparts of higher Diptera are most often inconspicuous and spongelike but are sometimes extended into structures of importance to family identification, such as a long proboscis that is sometimes geniculate (with a sharp bend in the middle).

Wing Characters

In the Schizophora the costa (the leading edge of the wing) often has breaks or gaps — called the humeral and subcostal breaks — near the base. The presence or absence of these gaps, especially the one near the end of the subcosta, can often help distinguish between otherwise similar acalyptrate families.

The subcosta or subcostal vein (Sc) is a short vein between the costa and the radius (R). It can be difficult to see but is an important character in the Schizophora keys, since the acalyptrate families are traditionally split into those with a complete subcosta (ending in C) and those with an incomplete subcosta (ending in R or petering out in the membrane before reaching C). Projecting light through the wing from underneath will help bring this and other venational characters into view. The radius, or radial vein, branches into R_1 (often called the first vein or first longitudinal vein) which usually runs to the costa very close to the wing base, and the

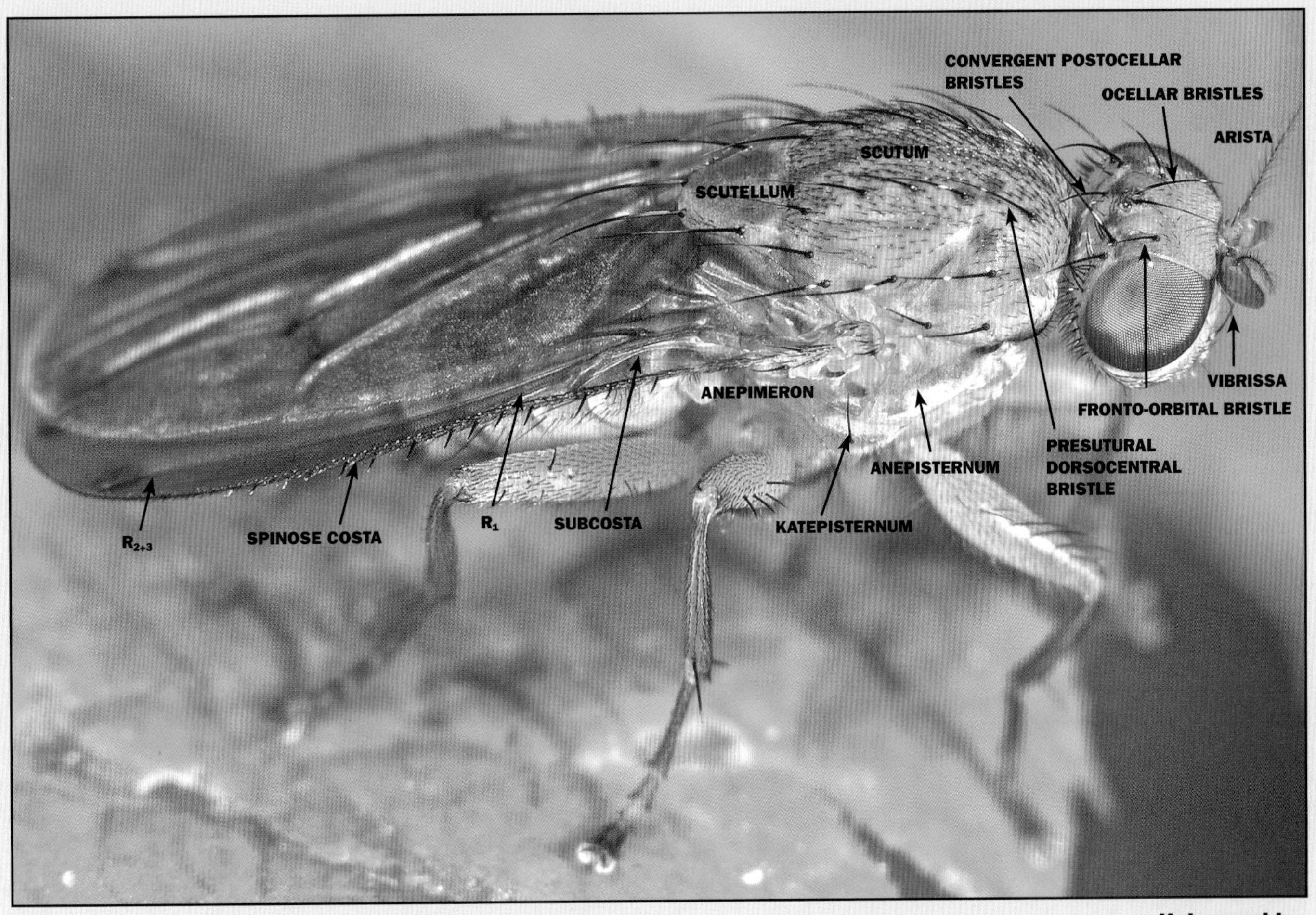

Heleomyzidae

Calliphoridae

Sciomyzidae

radial sector (Rs), which in turn branches into R_{2+3} (second longitudinal vein) and R_{4+5} (third longitudinal vein).

The next vein to reach the wing margin is appropriately called the media or medial vein — literally, the "middle vein," although it usually reaches the wing margin beyond the middle of the wing. The medial vein (M, or fourth longitudinal vein) is unbranched in the Schizophora but is occasionally reduced and sometimes curves up and fuses with R_{4+5} before reaching the wing margin. The next long vein after M is the cubital vein or cubitus (Cu; fifth longitudinal vein). Cu is divided into anterior and posterior branches (CuA and CuP), although the posterior branch is inconspicuous and restricted to the base of the wing. CuA has two long branches, CuA_1 and CuA_2, with CuA_1 (treated by some authors as M_4) often reaching the wing margin. CuA_2 in the Schizophora fuses with the first anal vein (A_1) to become A_1+CuA_2 and is variously shortened or reduced in different families. The last vein, the second anal vein (A_2), is often weak or indistinct, but its extent and shape can be important in separating families such as the Fanniidae, Anthomyiidae and Muscidae.

The main wing veins are often connected by crossveins, usually named for the veins they connect, although the short crossvein linking the subcosta to the costa near the base of the wing is called the humeral crossvein (h). Veins R and M are connected by crossvein r-m and veins M and Cu are connected by two crossveins, distinguished by the prefixes *b* for basal and *d* for distal — thus

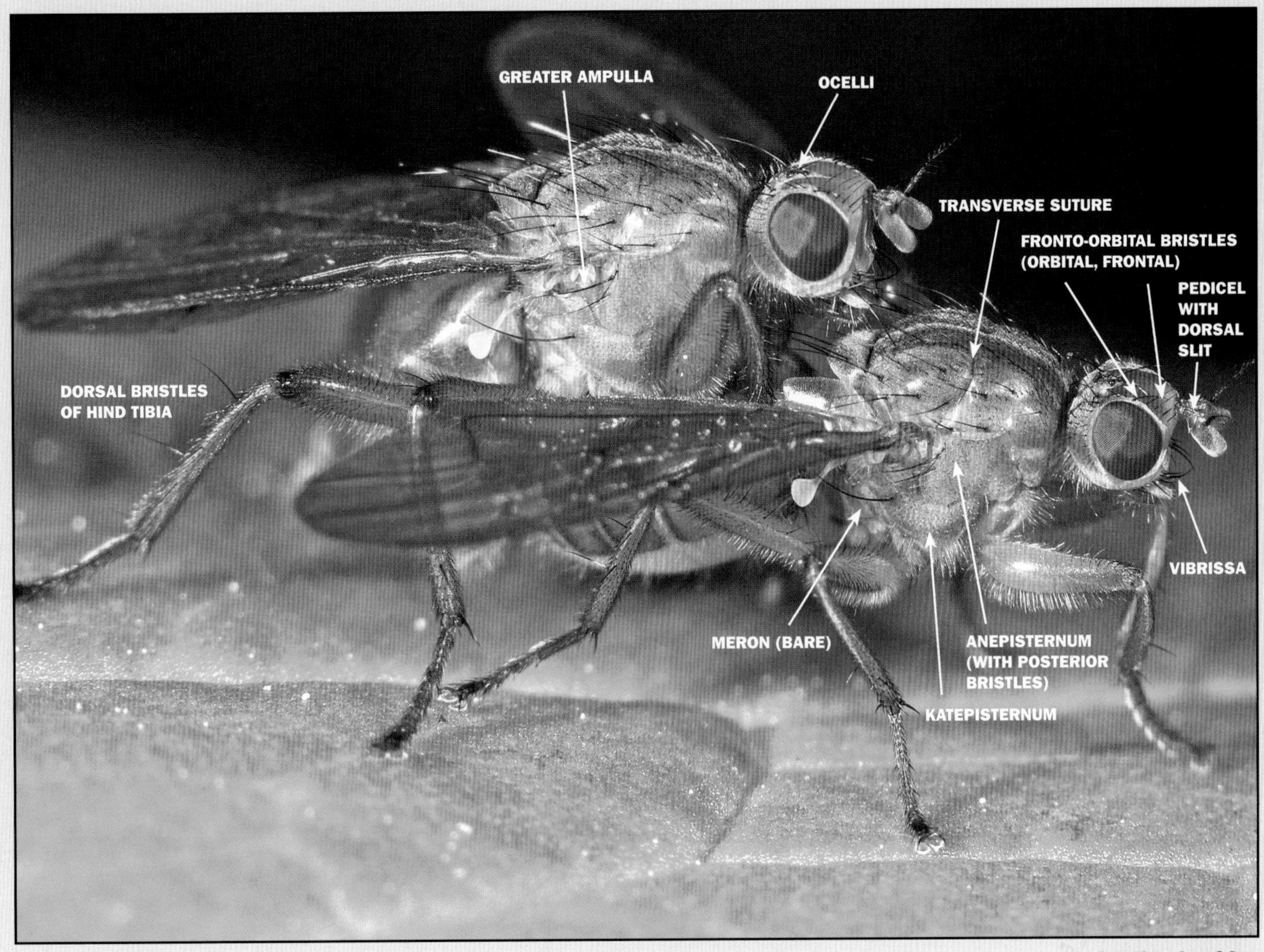

Scathophagidae

bm-cu near the base of the wing and dm-cu distally. The areas bounded by veins and crossveins are called cells, and each cell is usually named for the vein that precedes it. For example, the cell behind M and closed by bm-cu is cell bm, or the basal medial cell, and the cell behind M and closed by dm-cu is called dm, or the distal medial cell. Crossveins and cells are always abbreviated in lowercase to distinguish their abbreviations from vein abbreviations.

The most important cell in keying Schizophora families is the last one, at the base of the anal area of the wing. I prefer the term *anal cell* for this cell, although it is often called cell cu*p* because it is at least partly bounded anteriorly by the posterior cubital vein (CuP). Some families lack a closed anal cell, and in others it is closed by the crossvein-like basal part of CuA_2 to form a closed cell. When the anal cell is closed, its distal margin can be short or extended into a longer process. Determining whether a closed anal cell is present or absent is often critical to identifying acalyptrate families.

The basal hind part of the wing (with the anal veins) is called the anal lobe, and basal to the anal lobe there is a more distinct lobe called the alula. At the very base of the wing — between the wing and the side of the thorax — are two flaps called the upper and lower calypters. The lower calypter is often large, forming a hood over the halter. The size and shape of the lower calypter is often a useful character, as in the Rhinophoridae, where it sticks out like a tongue at right angles to the body.

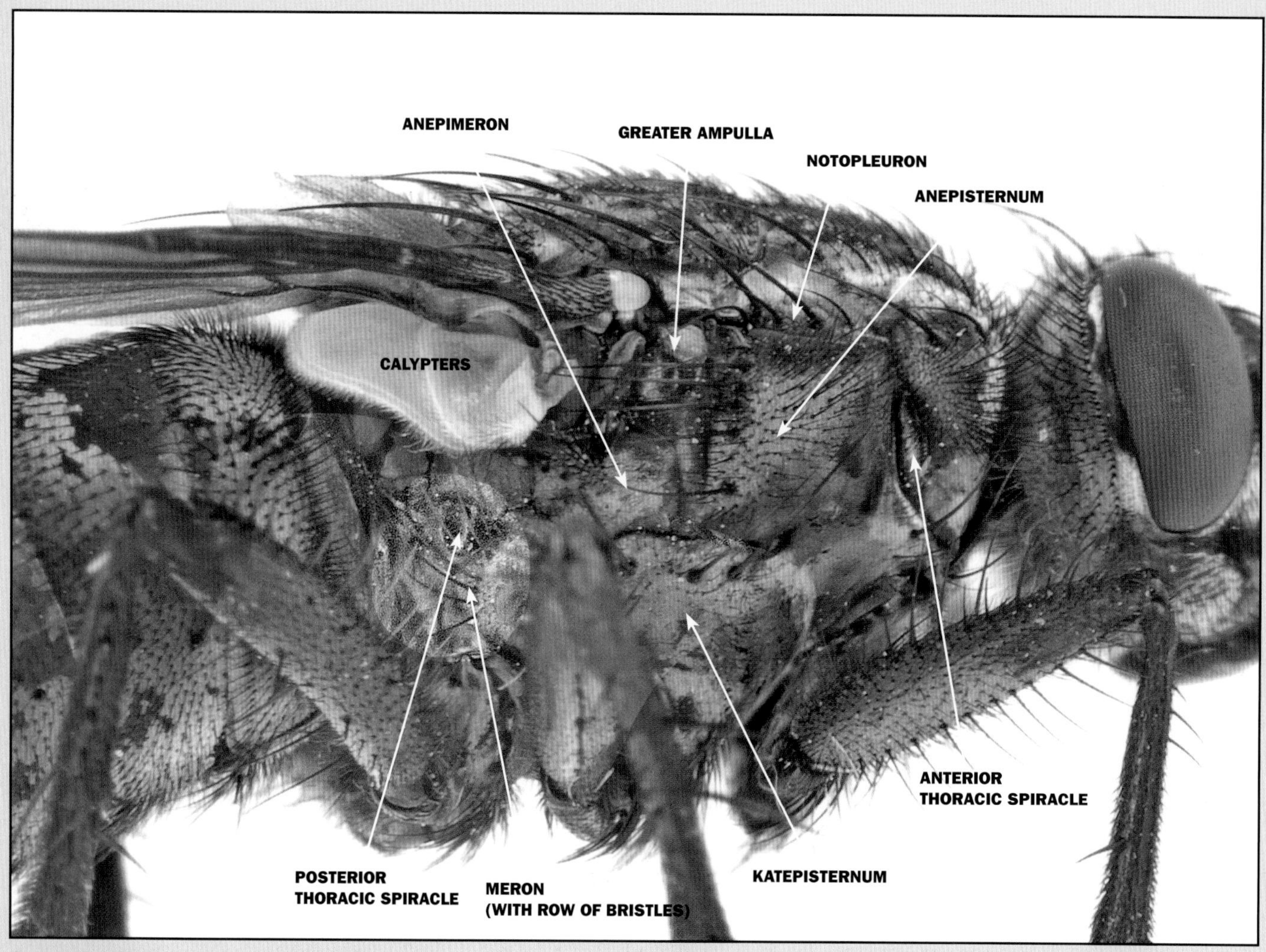

Sarcophagidae

Leg Characters

Leg characters are of great importance in generic and specific keys, but only a few are used in higher dipteran family keys. Of the main leg parts — coxa, trochanter, femur, tibia and tarsus — only the last three are widely used in family keys, and the arrangement of bristles on the tibiae is most important. Leg bristles are described by position: proximal (near the base) versus distal (nearer the tip); dorsal (upper surface) versus ventral (lower surface); apical (at the tip) versus preapical (just before the tip); anterior (front surface) versus posterior (back surface); and so on. One or more of the femora may be swollen and may have a row of spines or stout bristles ventrally. The tarsus is divided into five tarsomeres, which may be characteristically shaped, and the last tarsomere ends in a pair of claws. The basal (first) tarsomere of the hind leg is characteristically shortened and swollen in the Sphaeroceridae.

Thoracic Characters

The fly thorax is made up almost entirely of the muscle-packed mesothorax, which drives the single pair of wings and has useful key characters dorsally (mesonotum) and laterally (mesopleuron). The prothorax is normally small and inconspicuous but is sometimes extended to form a neck, as in the Syringogastridae. The propleuron sometimes bears a distinctive vertical ridge called a carina, which is characteristic of the Chloropidae in which the area before the carina is often bare and shiny. The mesonotum is made up mostly of the scutum and a smaller, shelflike posterior scutellum. Some families have a bulging subscutellum, or postscutellum, behind the scutellum — between it and the postnotum. The bristles of the scutum, especially the two longitudinal rows of large bristles called dorsocentral bristles, are often used as taxonomic characters. The scutum can be divided into anterior and posterior parts by a furrow called the transverse suture, which is distinct in calyptrates and indistinct in acalyptrates. The area in front of the transverse suture is called presutural and the area behind postsutural.

The mesopleuron is divided into a front part, the episternum, and a back part, the epimeron, by a furrow called the pleural suture, which runs from the wing base to the base of the mid-leg; this suture is straight in the lower Diptera but zigzag in the Schizophora. The episternum and epimeron are in turn divided into upper parts (the anepisternum and anepimeron) and lower parts (the katepisternum and katepimeron). Of these, the anepisternum and katepisternum are most frequently used in family keys, and both sclerites are easily located. The anepisternum is the large plate right in front of the wing base; the katepisternum is the conspicuous triangular part right in front of the base of the middle leg. The smaller plate right behind the katepisternum — between the middle leg and the posterior spiracle — is the meron. The meron is an important character in the calyptrate key, since the first couplet in the key asks if there is a vertical row of bristles on the meron or not.

The two thoracic spiracles are referred to as the anterior and posterior spiracles, although they are really the mesothoracic and metathoracic spiracles (there are no prothoracic spiracles in the Diptera). The anterior spiracle is near the front corner of the pleuron and the posterior spiracle opens just below the halter. Some flies, including all calyptrates and a few acalyptrates, have a globular bulge, called a greater ampulla, just below the wing base on the upper part of the epimeron.

Tachinidae

Abdominal Characters

Abdominal characters, especially the complicated claspers and other structures of the male genitalia, are very important in identifying species but are rarely used in family keys. The female abdomen is usually simple, with the apical segments telescoping into one another to make a soft, extendible ovipositor. However, some acalyptrates have the tip of the abdomen modified into a prominent, hard tube, or oviscape, and this is a useful key character for several families, especially those related to Tephritidae. Some families have unusual abdominal shapes and some have a narrowed abdominal base, described as petiolate. Occasional reference is made to the dorsal sclerites (tergites) or ventral sclerites (sternites) of the abdomen; the abdominal pleuron is unsclerotized, or soft and membranous.

All of the above bits and pieces will be easiest to examine on fresh, well-mounted flies, as some key characters will be hard to interpret on damaged, dirty, shriveled or poorly mounted specimens. Most characters used in the key can be distinguished using a good hand lens in good light, but a stereomicroscope will make the job easier. In a pinch, a camera with a macro lens and close-up rings or a teleconverter can serve as a substitute microscope. Lighting is critical, with transmitted light often necessary to see subtle wing characters and bright but diffused light best for picking out smaller bristles. LED ring lights work well. And remember, if you can't interpret a key character just follow both alternatives and then check your results against the photos and text.

KEY ONE

Start here to identify a winged adult fly unless you know your specimen is a terrestrial lower dipteran with ocelli (key two); lower Brachycera, Empidoidea or Aschiza (keys three and four); or Schizophora (calyptrates and acalyptrates, keys five to ten).

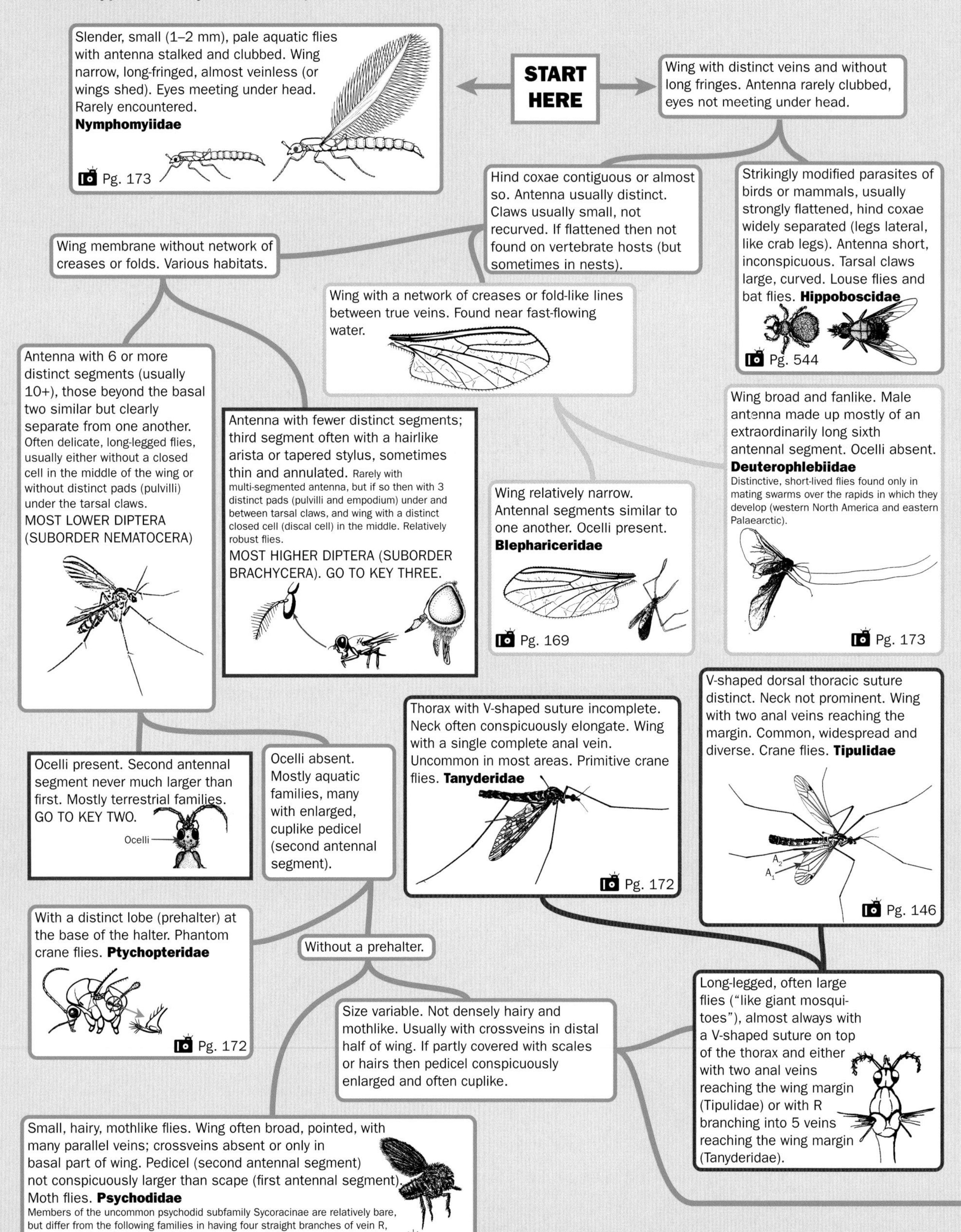

Postscutellum usually with a distinctive longitudinal groove (absent in the small and uncommon subfamily Podonominae). Radial veins usually ending close to wing tip. Pulvilli (pads under claws) often present. Vein M not forked. Non-biting midges. **Chironomidae**

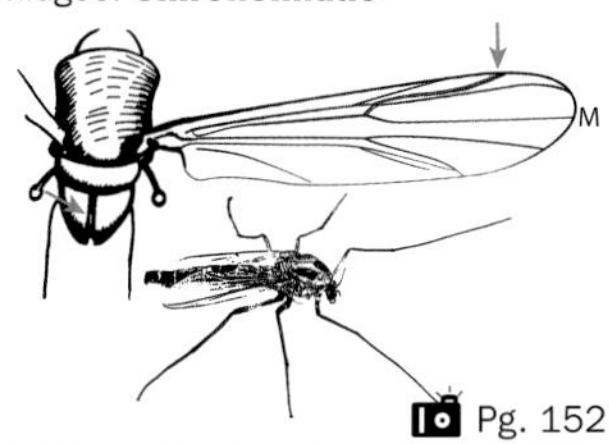

Pg. 152

Postscutellum not grooved. Radial veins often shortened and ending far before wing tip. Female mouthparts modified for biting. Vein M usually forked, but fork often indistinct. Pulvilli not developed. Biting midges. **Ceratopogonidae**

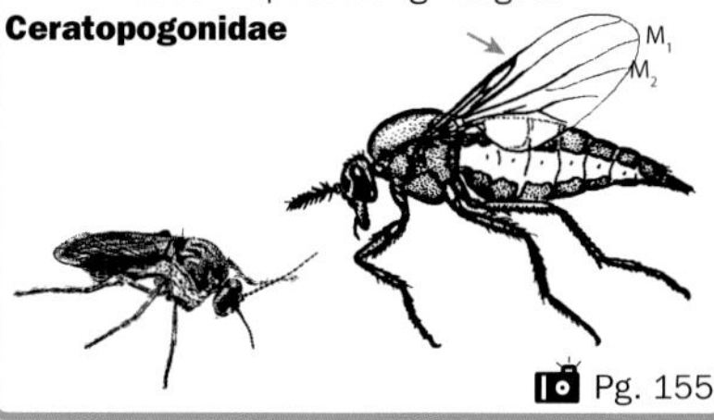

Pg. 155

Proboscis elongate and covered with scales (flattened hairs); scales also present on body, legs, wing margin and wing veins. Mosquitoes. **Culicidae**

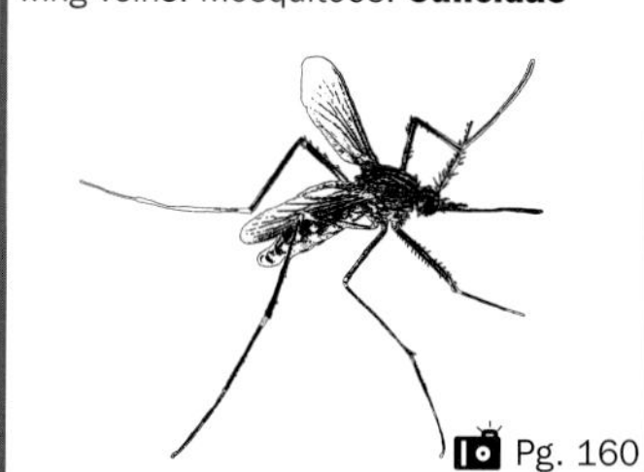

Pg. 160

Eight or fewer longitudinal wing veins reaching wing margin. Wings usually relatively bare.
Some uncommon Mycetophilidae that lack ocelli would key to this point but they have long coxae and long tibial spurs, as is typical of fungus gnats (key Two).

Nine or more longitudinal wing veins reaching wing margin. Wings and body with conspicuous scales or hairs.

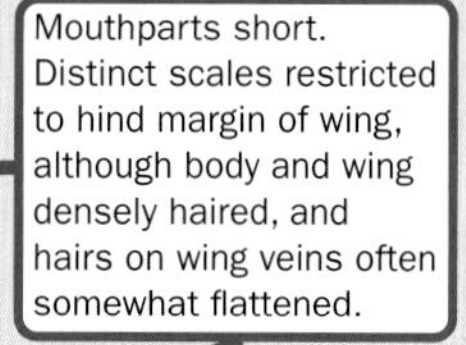

Mouthparts short. Distinct scales restricted to hind margin of wing, although body and wing densely haired, and hairs on wing veins often somewhat flattened.

Wing without an arched vein (R_{2+3} straight). Usually not similar to small crane flies.

Wing with a forwardly convex, forked vein (R_{2+3}) ending near wing tip. Long-legged, delicate, non-biting flies similar to small crane flies. Meniscus midges. **Dixidae**

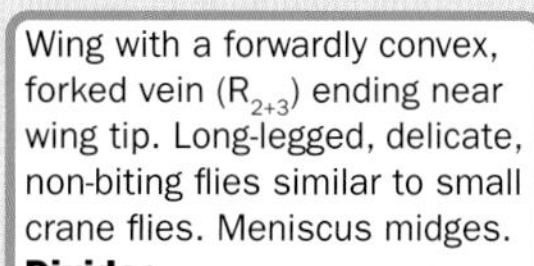

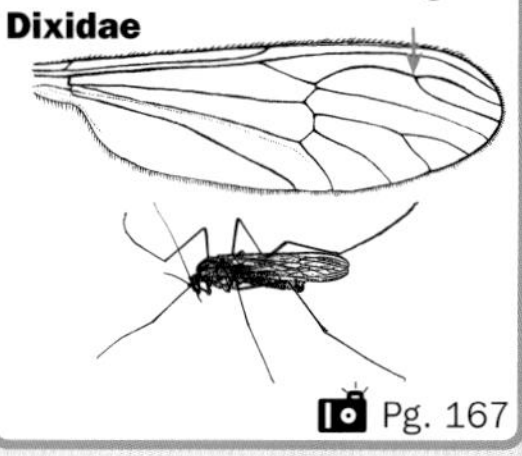

Pg. 167

Small to minute flies (usually 1–2 mm), females with short biting mouthparts. Vein R_1 short, ending closer to subcosta (Sc) than to R_2. Frog-biting midges. **Corethrellidae**

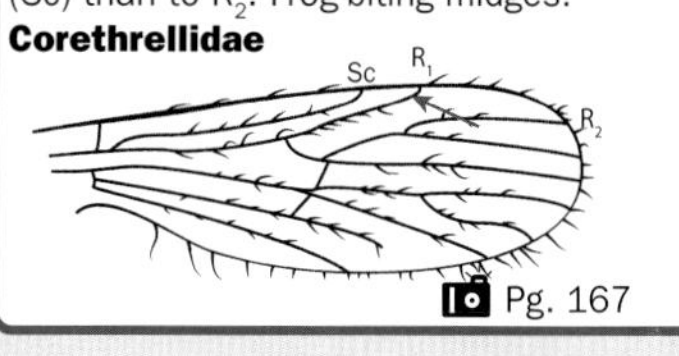

Pg. 167

Usually larger (1–10 mm), non-biting flies. Vein R_1 elongate, ending closer to R_2 than to Sc. Phantom midges. **Chaoboridae**

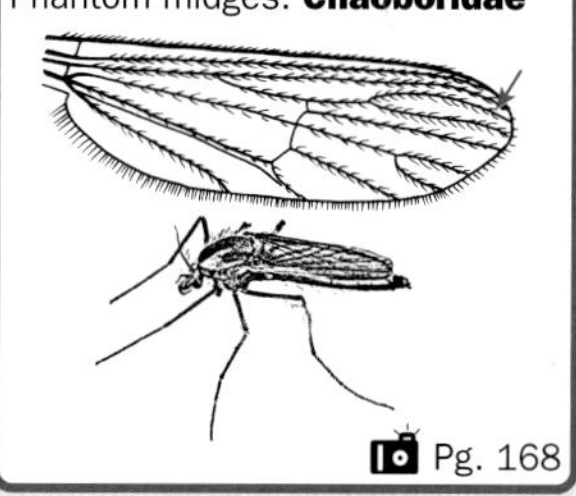

Pg. 168

Antenna usually much longer than head, usually with pedicel cuplike, conspicuously larger than other segments (more so in males). Relatively slender.

Stout-bodied, somewhat humpbacked flies with short antenna (no longer than head). Pedicel not enlarged.

Wing neither conspicuously broad nor with anterior veins thicker than posterior veins; costa continuing around wing although weaker along hind margin. Antenna tapered and slender. Both sexes with eyes contiguous. Rarely collected, non-biting flies found only on seeps. Seepage midges. **Thaumaleidae**

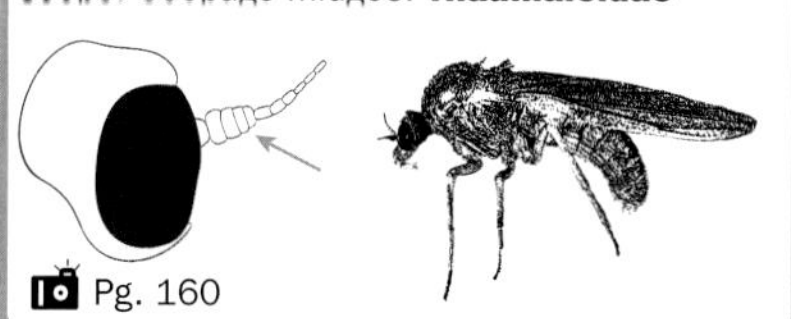

Pg. 160

Wing conspicuously broad, about twice as long as wide; anterior veins thicker than posterior veins, costa restricted to anterior margin. Antenna thick. Females with eyes separate. Black flies. **Simuliidae**

Pg. 158

First tarsomere at least as long as second, tarsus with 5 tarsomeres. Size variable but with at least 6 veins reaching wing margin; if delicate-bodied then pedicel conspicuously enlarged, often cuplike. Aquatic flies, infraorder Culicomorpha.

Appearance variable, but never with two anal veins, never with a distinct V-shaped suture on top of the thorax and R with fewer than 5 branches.

First tarsal segment reduced or absent, leaving only 4 long tarsomeres (tarsal segments). Usually tiny (1–3 mm) and delicate, with reduced wing venation. Antenna usually very long, second segment not enlarged. Terrestrial flies, part of the infraorder Bibionomorpha. Gall midges. **Cecidomyiidae** in part

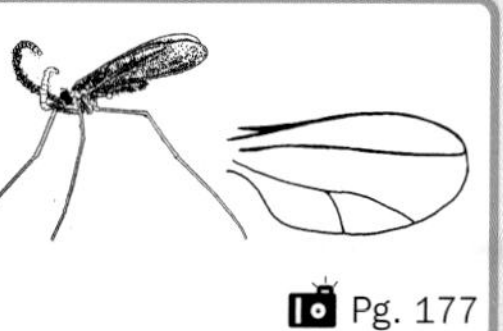

Pg. 177

KEY TWO

Lower Diptera, part two: Terrestrial families (plus Axymyiidae)

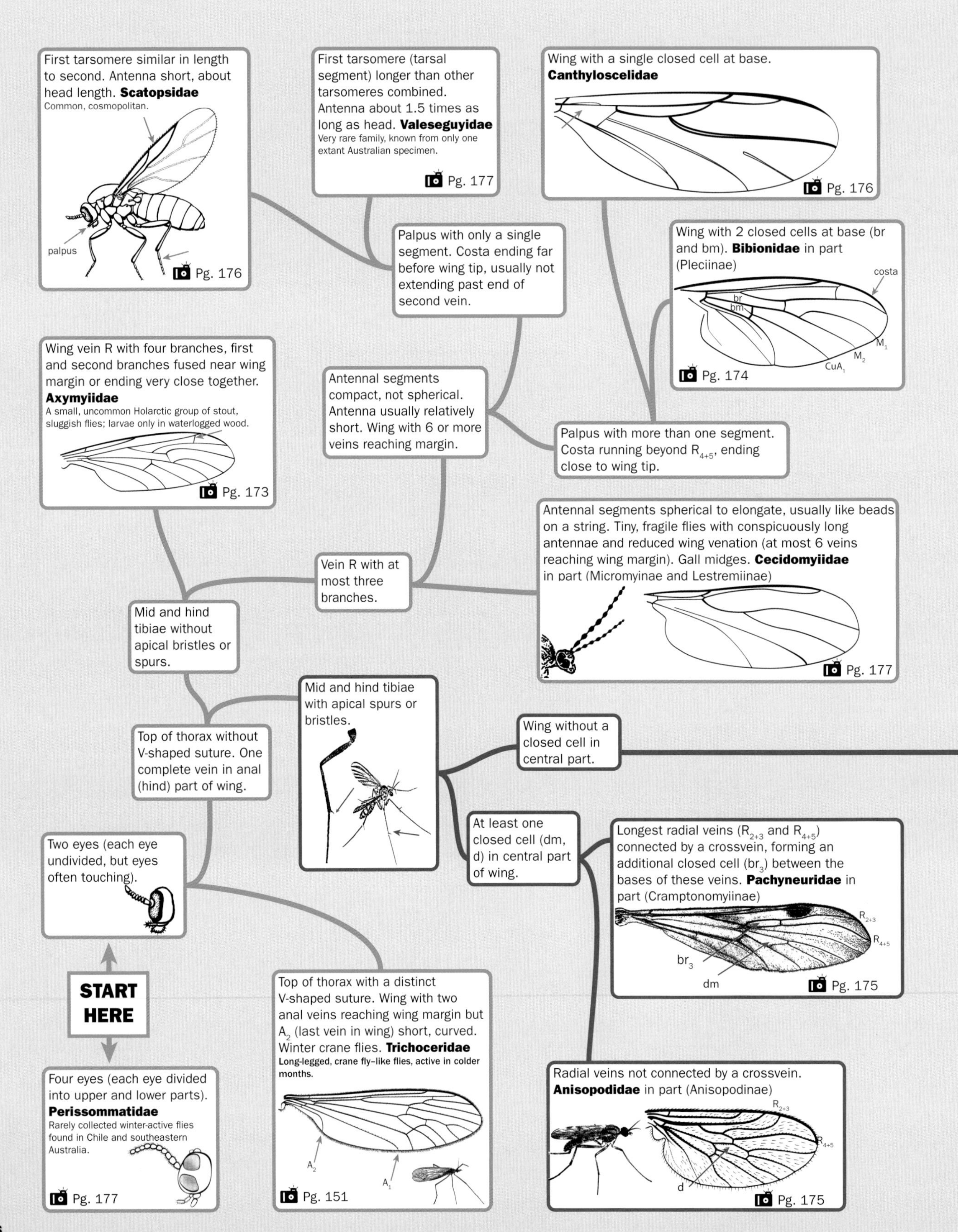

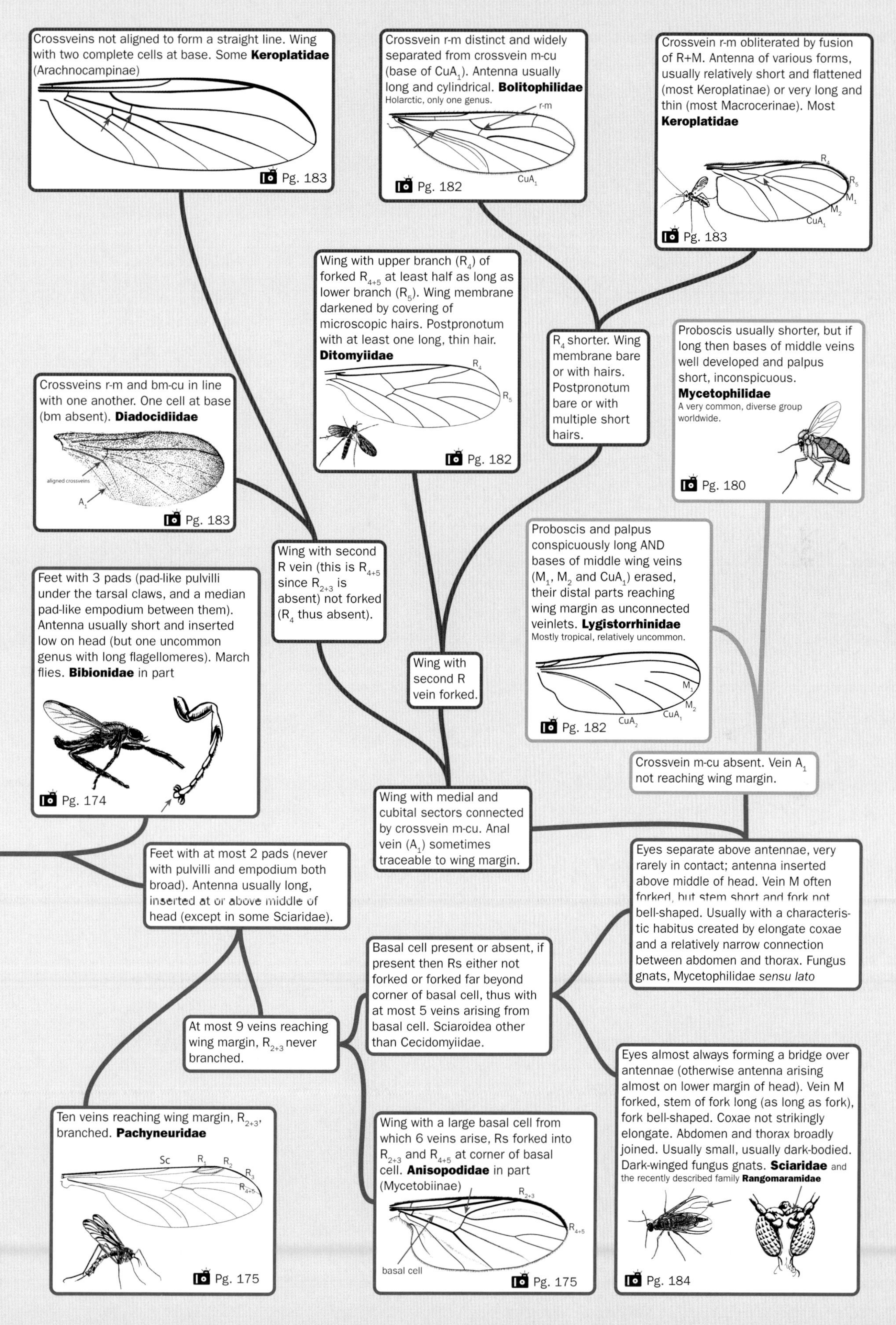
Crossveins not aligned to form a straight line. Wing with two complete cells at base. Some **Keroplatidae** (Arachnocampinae)
Pg. 183
Crossvein r-m distinct and widely separated from crossvein m-cu (base of CuA_1). Antenna usually long and cylindrical. **Bolitophilidae**
Holarctic, only one genus.
r-m
CuA_1
Pg. 182
Crossvein r-m obliterated by fusion of R+M. Antenna of various forms, usually relatively short and flattened (most Keroplatinae) or very long and thin (most Macrocerinae). Most **Keroplatidae**
R_4
R_5
M_1
M_2
CuA_1
Pg. 183
Wing with upper branch (R_4) of forked R_{4+5} at least half as long as lower branch (R_5). Wing membrane darkened by covering of microscopic hairs. Postpronotum with at least one long, thin hair. **Ditomyiidae**
R_4
R_5
Pg. 182
R_4 shorter. Wing membrane bare or with hairs. Postpronotum bare or with multiple short hairs.
Proboscis usually shorter, but if long then bases of middle veins well developed and palpus short, inconspicuous. **Mycetophilidae**
A very common, diverse group worldwide.
Pg. 180
Crossveins r-m and bm-cu in line with one another. One cell at base (bm absent). **Diadocidiidae**
aligned crossveins
A_1
Pg. 183
Wing with second R vein (this is R_{4+5} since R_{2+3} is absent) not forked (R_4 thus absent).
Proboscis and palpus conspicuously long AND bases of middle wing veins (M_1, M_2 and CuA_1) erased, their distal parts reaching wing margin as unconnected veinlets. **Lygistorrhinidae**
Mostly tropical, relatively uncommon.
M_1
M_2
CuA_1
CuA_2
Pg. 182
Feet with 3 pads (pad-like pulvilli under the tarsal claws, and a median pad-like empodium between them). Antenna usually short and inserted low on head (but one uncommon genus with long flagellomeres). March flies. **Bibionidae** in part
Pg. 174
Wing with second R vein forked.
Crossvein m-cu absent. Vein A_1 not reaching wing margin.
Wing with medial and cubital sectors connected by crossvein m-cu. Anal vein (A_1) sometimes traceable to wing margin.
Eyes separate above antennae, very rarely in contact; antenna inserted above middle of head. Vein M often forked, but stem short and fork not bell-shaped. Usually with a characteristic habitus created by elongate coxae and a relatively narrow connection between abdomen and thorax. Fungus gnats, Mycetophilidae *sensu lato*
Feet with at most 2 pads (never with pulvilli and empodium both broad). Antenna usually long, inserted at or above middle of head (except in some Sciaridae).
Basal cell present or absent, if present then Rs either not forked or forked far beyond corner of basal cell, thus with at most 5 veins arising from basal cell. Sciaroidea other than Cecidomyiidae.
At most 9 veins reaching wing margin, R_{2+3} never branched.
Eyes almost always forming a bridge over antennae (otherwise antenna arising almost on lower margin of head). Vein M forked, stem of fork long (as long as fork), fork bell-shaped. Coxae not strikingly elongate. Abdomen and thorax broadly joined. Usually small, usually dark-bodied. Dark-winged fungus gnats. **Sciaridae** and the recently described family **Rangomaramidae**
Pg. 184
Ten veins reaching wing margin, R_{2+3}, branched. **Pachyneuridae**
Sc
R_1
R_2
R_3
R_{4+5}
Pg. 175
Wing with a large basal cell from which 6 veins arise, Rs forked into R_{2+3} and R_{4+5} at corner of basal cell. **Anisopodidae** in part (Mycetobiinae)
R_{2+3}
R_{4+5}
basal cell
Pg. 175

KEY THREE

Brachycera, part one: Subgroups of Brachycera and families of Tabanomorpha

START HERE

Ptilinal fissure absent. R_{4+5} often forked. Antenna with or without arista.

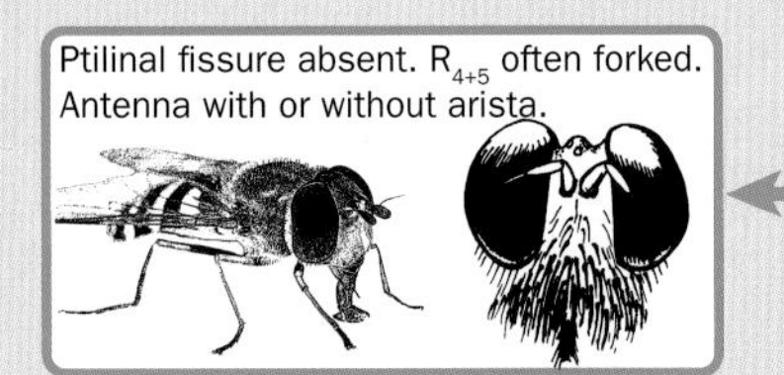

Ptilinal fissure present, arching over the antennae (weak in Conopidae and vestigial in *Sepedon*, a distinctive genus of Sciomyzidae). R_{4+5} not forked. Antenna with a hairlike or feather-like arista (lost in some rare acalyptrates). Schizophora. GO TO KEY FIVE

Feet with 2 pads (empodium bristlelike). If empodium slightly broadened (a few water-skating Empidoidea) then anal cell short, not pointed (CuA_2 joining A at an obtuse angle far from wing margin); otherwise anal cell variable. Asilomorpha, Empidoidea and Aschiza. GO TO KEY FOUR

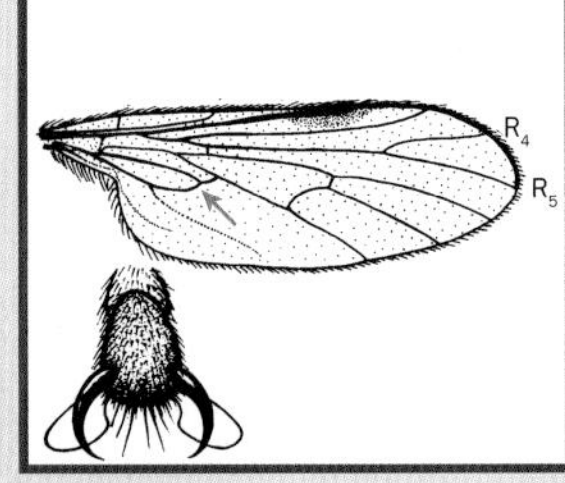

Feet with 3 pads (pulvilli under the claws and a pillow-like empodium in the middle). Anal cell either open or closed in a long point near wing margin (CuA_2 joining A at an acute angle near wing margin). Tabanomorpha (including Nemestrinidae and Acroceridae).

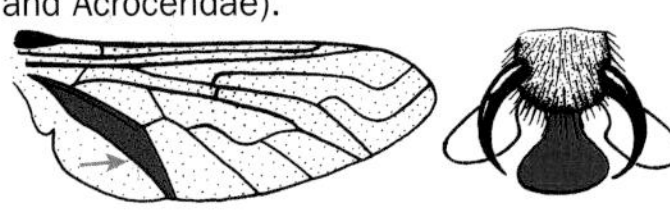

Head "normal," at least half as wide as thorax. Eyes of female distinctly separate. Lower lobe of calypter not enlarged.

Head small to very small in proportion to thorax; thorax often strikingly humpbacked. Both sexes with eyes meeting on top of head. Lower lobe of calypter very large. Small-headed flies. **Acroceridae**

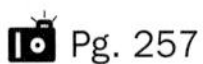
Pg. 257

Antenna not ending in a long, thin process; distal flagellomeres of various forms but never strikingly slender, hairlike or arista-like (ambiguous cases key both ways).

Antenna ending in a long, thin process, much thinner than first flagellomere (antennal segment 3).

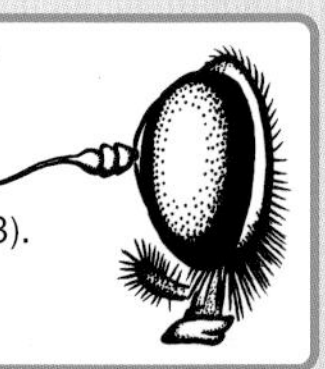

Veins appearing "tangled" as the branches of M curve forward and converge with R near apex; with a composite vein ("diagonal vein") straight from discal cell to outer margin of wing. Tangle-veined flies. **Nemestrinidae**

Often superficially similar to bee flies, Bombyliidae, which have a bristlelike empodium.

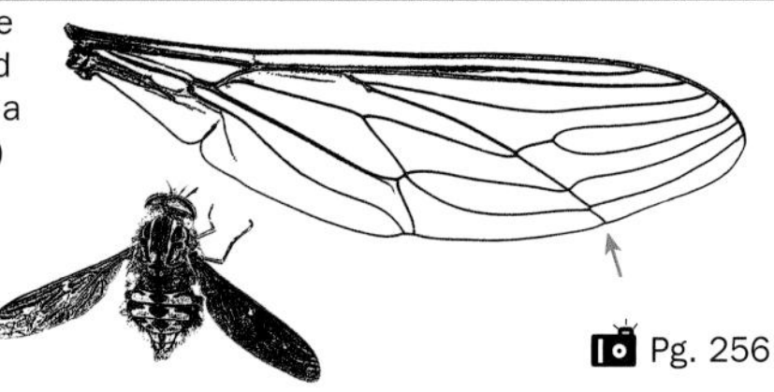

Pg. 256

Branches of M not curved forward, ending in wing margin beyond wing apex. Diagonal vein absent.

Radial veins grouped together, ending before wing tip; costa also usually ending well before wing apex. Cell d (discal cell) short, usually forming a distinct short, often squarish cell in middle of wing. Soldier flies. **Stratiomyidae** in part

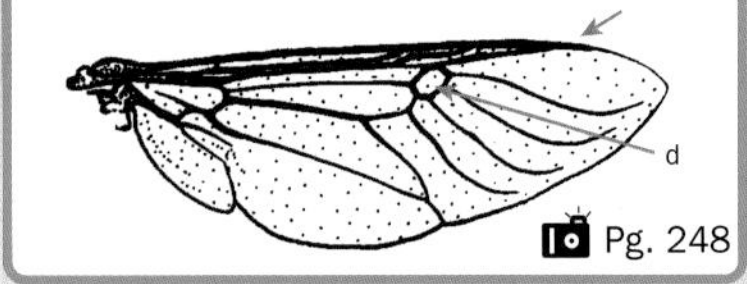

Pg. 248

Radial veins not crowded up before wing tip, costa ending at or beyond wing apex. Cell d (discal cell) much longer than wide.

Clypeus convex, bulbous. Foretibia usually without apical spur (spur present in some Athericidae).

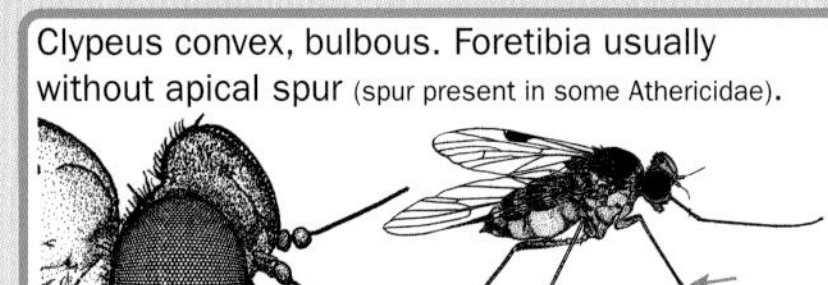

Clypeus flat or recessed. Foretibia with apical spur (spur sometimes small and inconspicuous).

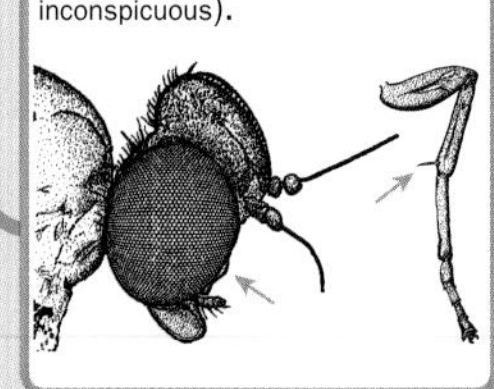

Metathorax without a postspiracular scale. Third antennal segment usually oval or conical with an apical stylus (but sometimes kidney-shaped with a dorsal stylus). Second and third veins (R_1, R_{2+3}) usually distinctly separated at apex. Snipe flies. **Rhagionidae** in part

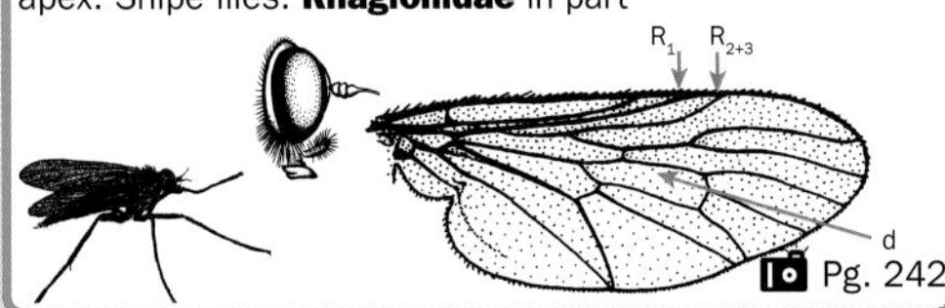

Pg. 242

Metathorax with a postspiracular scale (a lobe behind the posterior spiracle, just below the halter; also found in Tabanidae). Third antennal segment small but broad, stylus usually inserted dorsally. Veins R_1 and R_{2+3} close together or touching at apex. Water snipe flies. **Athericidae**

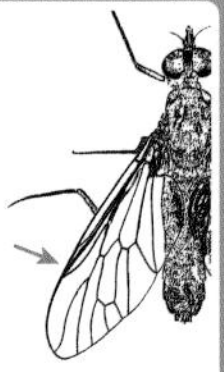

Pg. 244

Radial veins grouped together and ending before wing tip, costa also usually ending well before wing apex. Cell d (discal cell) short, forming a distinct squarish cell in middle of wing (rarely absent). Soldier flies. Most **Stratiomyidae**

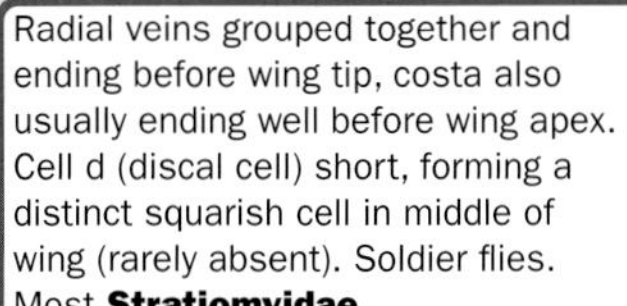

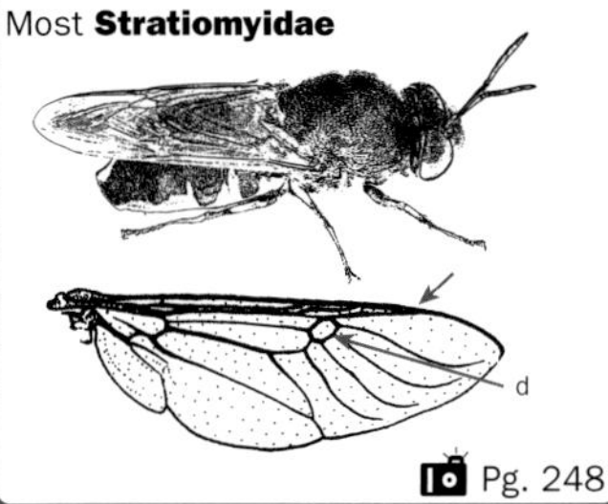

Pg. 248

Metathorax with a postspiracular scale or scale-like fold (a thin scale-like elevation behind the posterior spiracle, just below the halter). Subscutellum strongly developed, bulging.

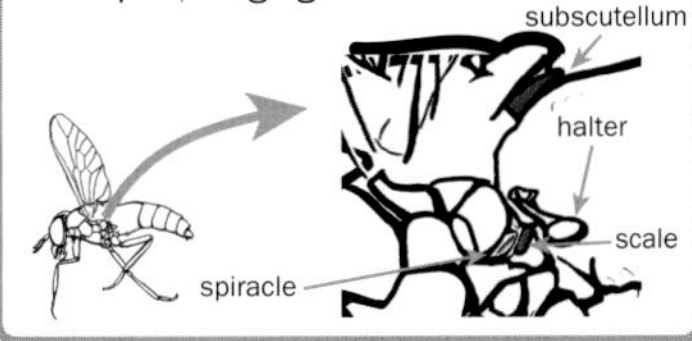

Lower calypter large, similar to upper calypter. First abdominal tergite deeply notched along hind margin, notch continuous with a median suture. Wing with broad Y-shaped R_{4+5} straddling wing tip. Females with biting mouthparts. Horse and deer flies. **Tabanidae**

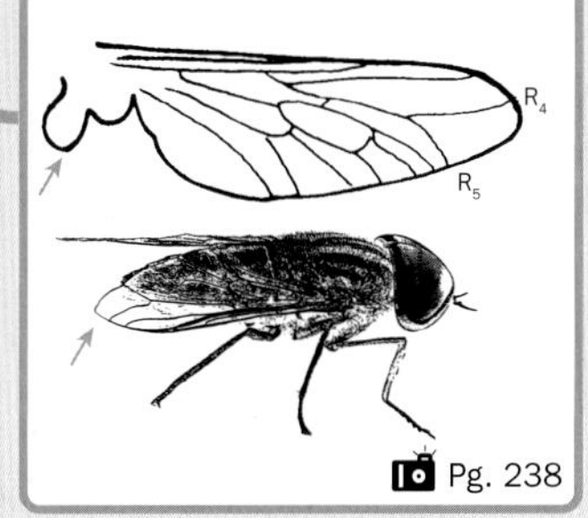

Pg. 238

Metathorax without a postspiracular scale or scale-like fold. Subscutellum not developed.

Lower calypter very small relative to upper calypter. First abdominal tergite without median notch or suture. Fork of R_{4+5} narrower. Neither sex with biting mouthparts.

Radial veins not crowded up before wing tip; costa ending at or beyond wing apex. Cell d (discal cell) much longer than wide.

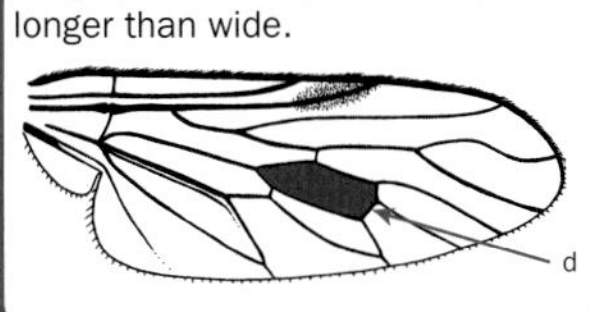

Anal cell closed far from wing margin. Prescutellum absent. Dull gray flies, 5–7 mm. **Oreoleptidae**
One rare western Nearctic species, known only from reared specimens.

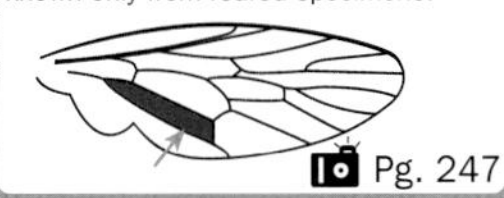

Pg. 247

Foretibia without apical spur. Alula well developed.

Foretibia with apical spur. Alula usually greatly reduced or absent. Most **Xylophagidae**

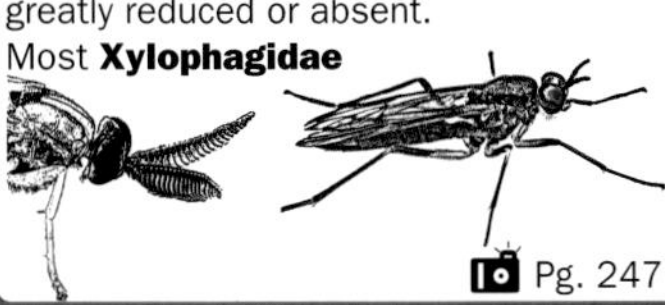

Pg. 247

A second closed cell (cell m_3) below discal cell.

d
alula
m_3

Wing without a closed cell immediately below discal cell.

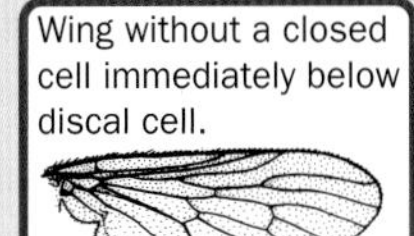

Anal cell open. Area between scutum and scutellum with a small swelling (prescutellum). Color and size variable. **Pelecorhynchidae**

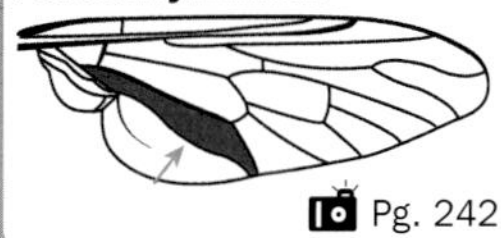

Pg. 242

Smaller (5–15 mm) flies often resembling ichneumonid wasps. No tuft of hairs below the posterior spiracles. Vein R_5 ending near wing tip. Widespread. **Xylomyidae**
Some specimens of the Australian Xylophagidae genus *Exeretonevra* will key out here.

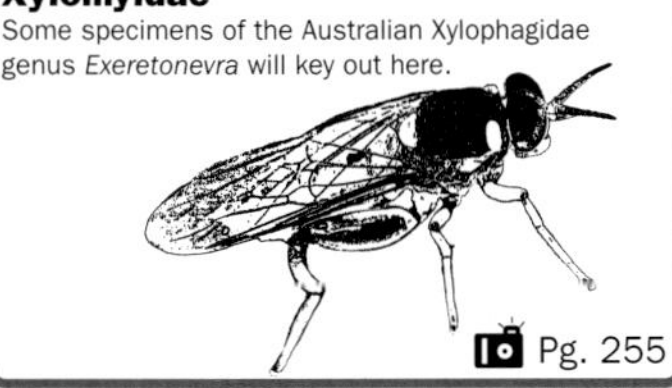

Pg. 255

Big (18–45 mm), broad, Neotropical flies resembling giant non-biting horse flies with a tuft of hairs below the posterior spiracle. R_5 ending far beyond wing tip. **Pantophthalmidae**

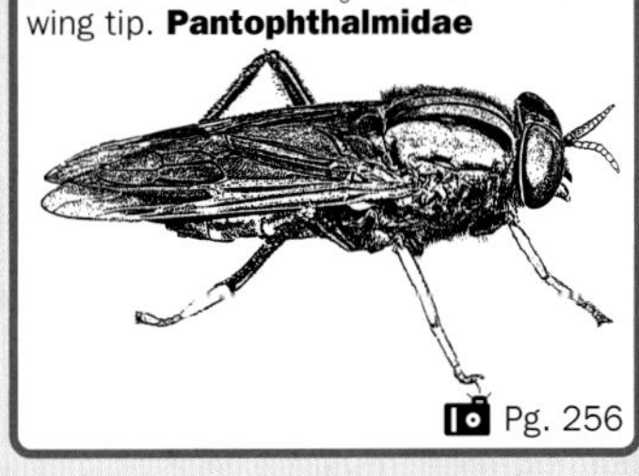

Pg. 256

Clypeus bulbous. **Rhagionidae** in part

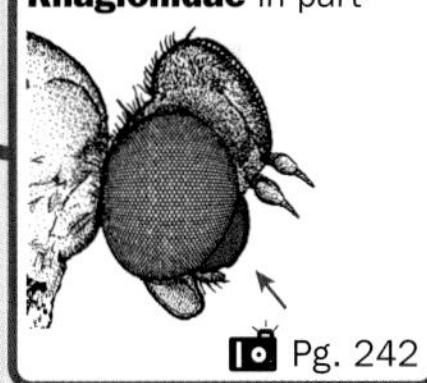

Pg. 242

Clypeus flat or recessed.

Base of wing narrow; alula absent, anal region not developed. Clypeus usually flat. Wormlions. **Vermileonidae**

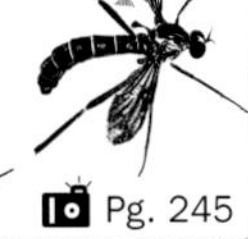

Pg. 245

Alula present. Clypeus deeply recessed. Some **Xylophagidae**

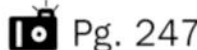

Pg. 247

Mid-tibial spur absent. 3–8 mm. Australia and South America. *Austroleptis*. **Austroleptidae**

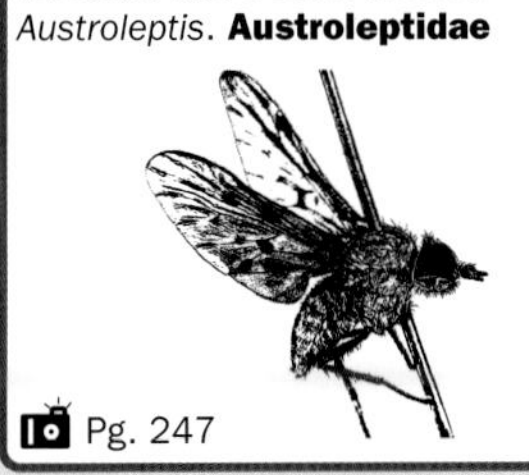

Pg. 247

Mid-tibial spur present. Less than 4 mm. Holarctic. *Bolbomyia* (**Rhagionidae** or Bolbomyiidae)

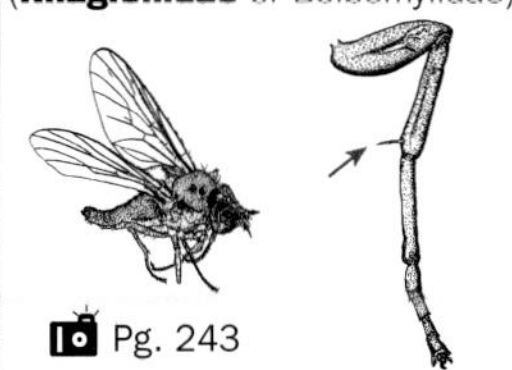

Pg. 243

KEY FOUR

Brachycera, part two: Asilomorpha, Empidoidea and Aschiza

START HERE

Predaceous flies with big, bulging eyes widely separated by a dorsal depression, and with a stout, shining beak. Face almost always with a distinctive row or tuft of bristles (this tuft, the mystax, is absent in some slender Neotropical species with stalked wings). Robber flies. **Asilidae**

Pg. 259

Mystax absent and eyes not separated by deep depression.

Wing with a "spurious vein" (a vein-like crease between radial and medial sectors) and/or first branch of M joining an unbranched R_{4+5} to make a closed cell r_{4+5}. Branches of M often running or almost parallel to wing margin. Hover flies or flower flies. **Syrphidae** A huge, cosmopolitan group of common, often brightly colored, small to large (4–25 mm) flies.

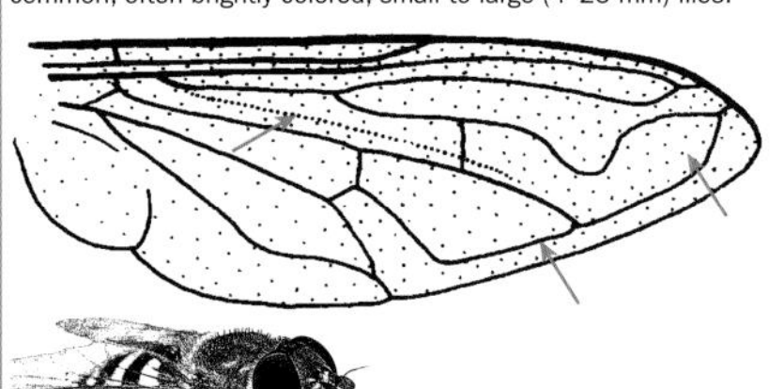

Pg. 413

Wing without a spurious vein, branches of M not turned up to run parallel to wing margin, and M_1 usually ending in wing margin.

Head hemispherical, very large in proportion to thorax, made up almost entirely of eyes. Antenna with a dorsal arista. Big-headed flies. **Pipunculidae** Small flies, syrphid-like in appearance and hovering habit.

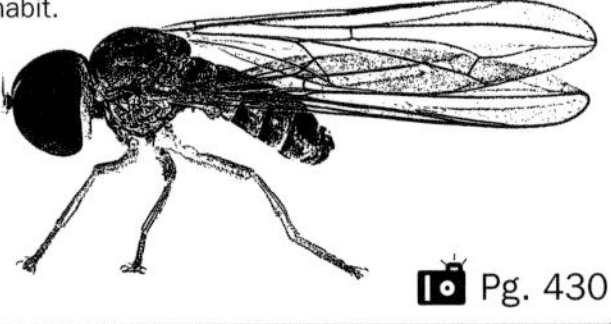

Pg. 430

Head not conspicuously large compared to thorax, rarely hemispherical (some Bombyliidae have a hemispherical head, but they have an apical style or arista).

First few wing veins (radial veins) usually short, thick, and crowded towards wing base; other veins weaker, not connected by crossveins. Pedicel of antenna usually hidden inside the next segment. Generally humpbacked, bristly flies with a distinctive habitus. Scuttle flies. **Phoridae** A huge, diverse, common, cosmopolitan group of small flies, ranging from less than 1 mm to over 5 mm.

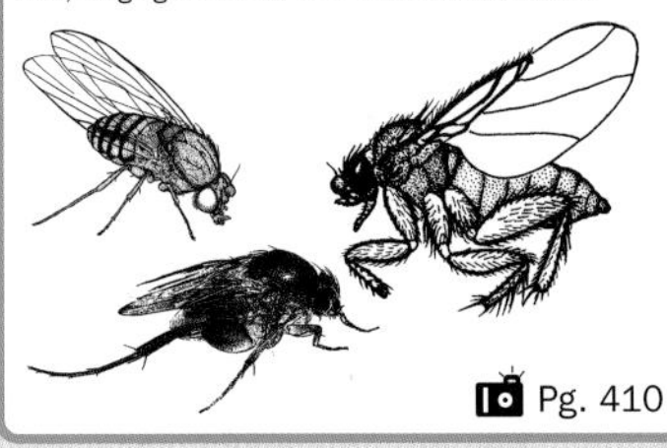

Pg. 410

Wing with radial veins not thickened and crowded towards wing base, crossveins present (usually) or absent (as in Lonchopteridae). Pedicel of antenna not hidden inside first flagellomere.

Anal cell, if present, usually shorter and separated from wing margin by more than its width (rare exceptions usually have R_{4+5} unforked and extending to the wing margin as a single vein, or are predaceous empidoids with unusual wing venation).

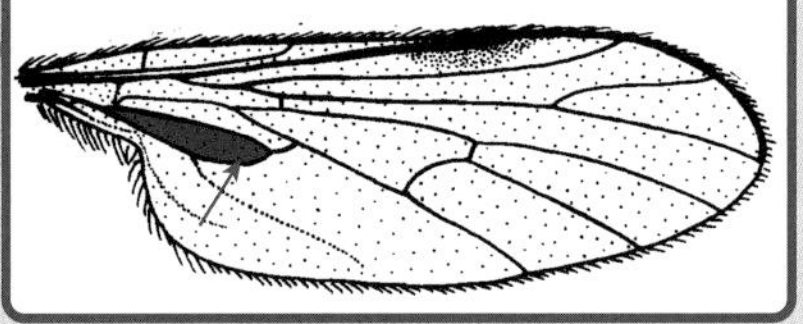

Wing with a long, usually pointed anal cell (cell cup), either closed very close to wing margin or open to wing margin (CuA_2 joining A near wing margin or ending in margin separate from end of A) and with R_{4+5} forked. Not predaceous.

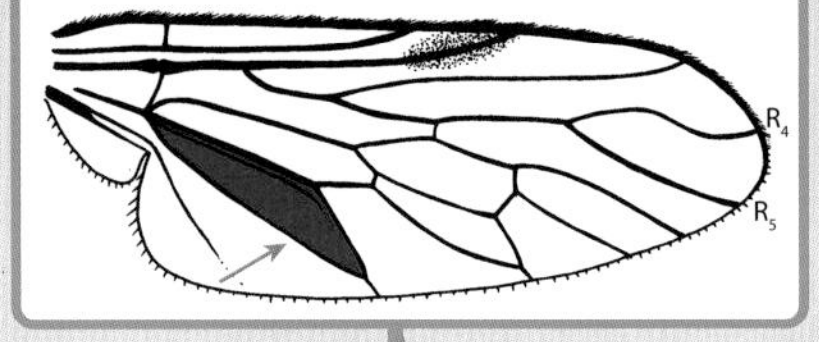

Wing veins usually not curved anteriorly; M_1 never running parallel to R_5, not ending in wing margin before wing apex.

Antennal flagellum (antenna beyond pedicel) a single, undivided piece with a bulbous base and a long, threadlike apical part. Anterior surface of hind coxa with strong, knoblike bulbous projection. Base of wing narrowed, with anal lobe greatly reduced. **Evocoidae** One distinctive species known only from central Chile.

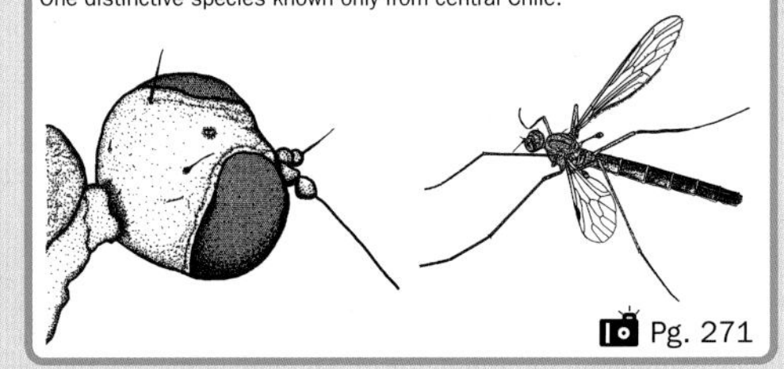

Pg. 271

Wing veins oddly curved anteriorly, distal parts of M_1 running parallel to R_5, reaching costa before wing apex. Large, robber fly–like flies.

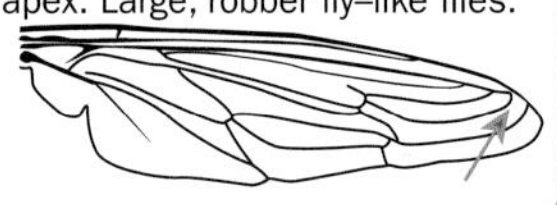

Antenna usually long, with a stalk and clubbed apex. M_2 curved forward, ending in eitheranother vein or in wing margin before wing apex. Form and color variable. Mydas flies. **Mydidae**

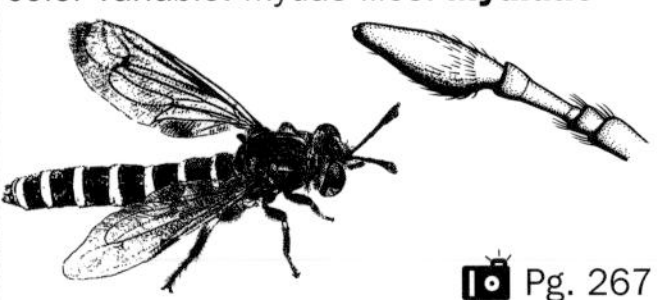

Pg. 267

Antenna usually shorter; flagellum shorter than head and similar in length to the combined scape and pedicel. M_2 ending in wing margin beyond apex. Usually black and gray. **Apioceridae**

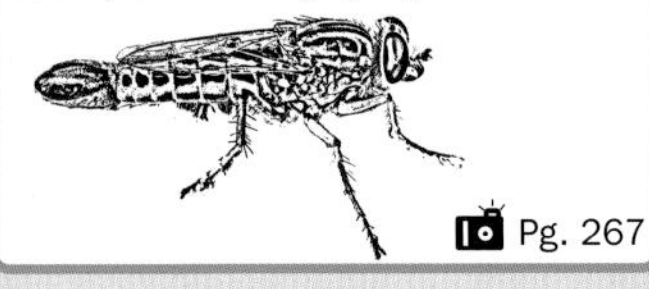

Pg. 267

Acrostichal setae reduced to a single row or absent. Arista with three segments. Tarsus of hind leg often greatly enlarged and flattened. Widespread. Flat-footed flies. **Platypezidae**

Pg. 408

Acrostichal setae in multiple rows. Arista with two segments. Tarsus of hind leg not enlarged and flattened. One European species. **Opetiidae**

Pg. 409

Base of foretibia without posteroventral gland. Apex of antenna usually without a thin, hairlike style. **Atelestidae**

Pg. 292

Base of foretibia with anteroroventral gland. Apex of antenna usually with a long, hairlike style. Hybotid dance flies. **Hybotidae**

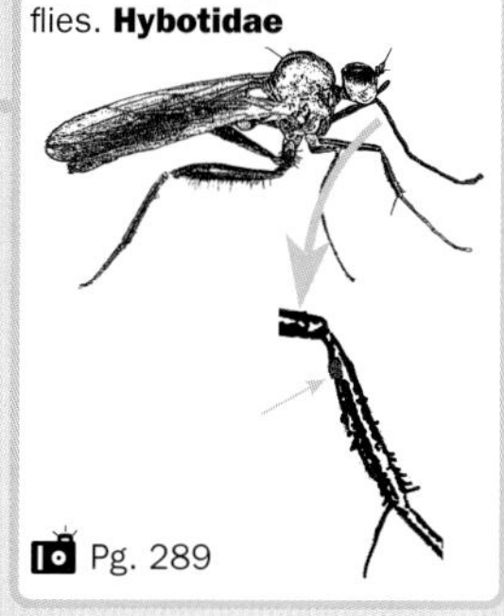

Pg. 289

Second antennal segment without lobes or apical processes on inner and outer surfaces. Subcosta free from R for most of its length. Male eye with upper facets conspicuously larger than lower facets. Widespread.

Rs originating far from base of wing, well beyond humeral crossvein. Male terminalia not conspicuously slung under body. Rarely metallic green.

Second antennal segment with lobes or apical processes on both inner and outer surfaces. Subcosta fused with R for most of its length but still ending free in costa. Male eye with uniformly sized facets. Australia. **Ironomyiidae**

Pg. 410

Subcosta complete and meeting wing margin (although sometimes dipped into R prior to apex) and anal vein usually complete and meeting wing margin (not quite reaching margin in male *Opetia*). Often velvety black phorid-like flies, at least males holoptic (with eyes meeting above head).

Wing with radial sector (Rs) originating near base of wing, at or near humeral crossvein (h). Male terminalia slung forward under body, often conspicuously so. Commonly metallic green. Long-legged flies. **Dolichopodidae**

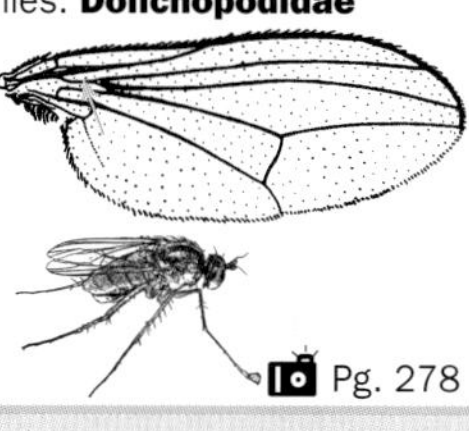

Pg. 278

Either subcosta incomplete and not meeting wing margin at apex or anal vein ending far short of wing margin. Eyes and vestiture variable.

Antenna apparently 3-segmented, third segment largest and with a minute stylus in an apical pit. Head without bristles. Crossvein r-m at or beyond middle of wing. Window flies. **Scenopinidae**

Pg. 270

R_{4+5} not forked

R_{4+5} forked. Dance flies. **Empididae**, Brachystomatidae, and unplaced genera sometimes treated as Homalocnemidae and Oreogetonidae

Pg. 283

End of antennae with an exposed stylus or arista. Head usually with bristles. Crossvein r-m (if present) usually before middle of wing.

Wing rounded at apex, with veins divergent and usually with crossveins. Veins bare above.

Neither R_{4+5} nor M_{1+2} branched. Micro bee flies. **Mythicomyiidae** Very small (1–3 mm) humpbacked flies with a conspicuous proboscis.

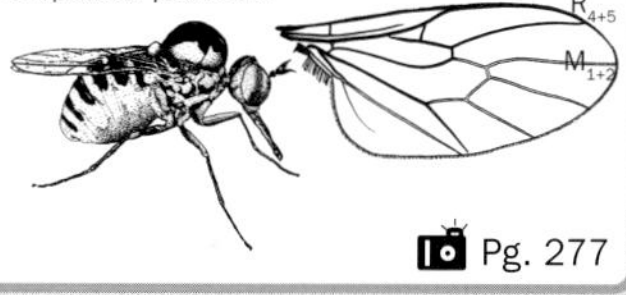

Pg. 277

Wing pointed, with parallel longitudinal veins and no crossveins beyond base; most veins with short black bristles on upper surface. Spear-winged flies. **Lonchopteridae** Cosmopolitan, few species but common.

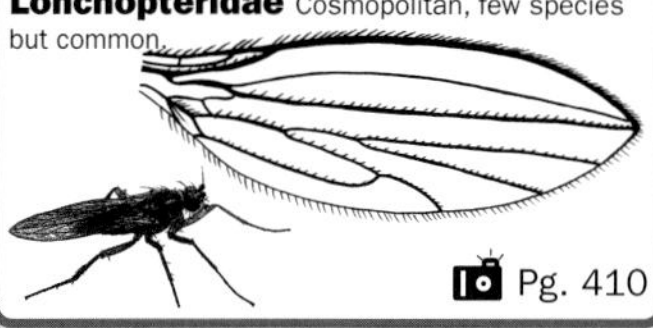

Pg. 410

Cell bm (the cell above the anal cell) with veins arising from three corners (or sometimes absent). Most commonly stout, short-bodied, often fuzzy flies with long wings held out from the body, but sometimes elongate and slender and sometimes tiny and empid-like. Discal cell present or absent, never followed by a tapered cell m_3 (m_3 absent or broadened towards wing tip).

At least R_{4+5} forked. If body length less than 5 mm then M_{1+2} usually branched as well.

Size usually larger (length 2–20 mm, usually 5–15 mm). M_{1+2} rarely forked but if so then fork conspicuously different from the forked R_{4+5}. Bee flies. **Bombyliidae** Widespread and common, usually fuzzy (not bristly), robust flies.

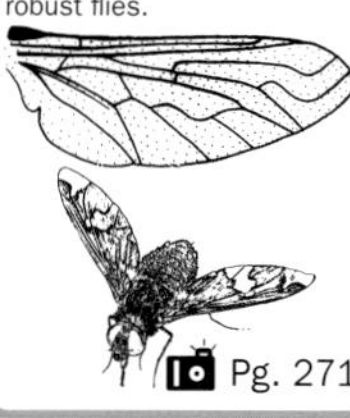

Pg. 271

Cell bm truncate (squared) distally, with separate veins arising from four separate corners. Discal cell present and elongate, followed by a tapered cell (m_3). Relatively elongate, slender flies, often superficially similar to Asilidae.

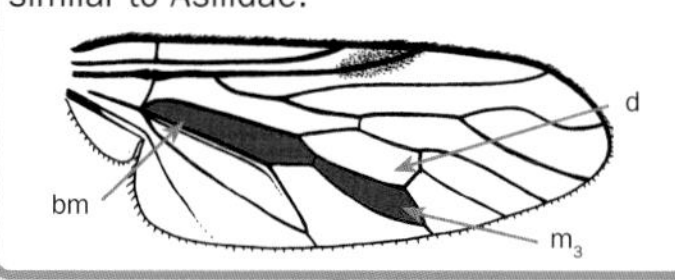

Antennal flagellum with more than one segment, apical part threadlike or not. Anterior surface of hind coxa without a bulbous projection. Base of wing with anal lobe developed.

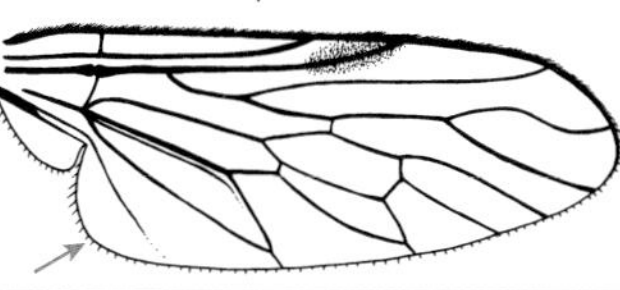

Small to very small (1–5 mm) empid-like flies. Wing with two similar forks (R_{4+5} and M_{1+2}) near tip.

Antenna usually ending in a short style much shorter than third antennal segment. R_{4+5} forked into a divergent Y encompassing wing tip (R_5 thus ending beyond wing tip). Widespread. Stiletto flies. **Therevidae**

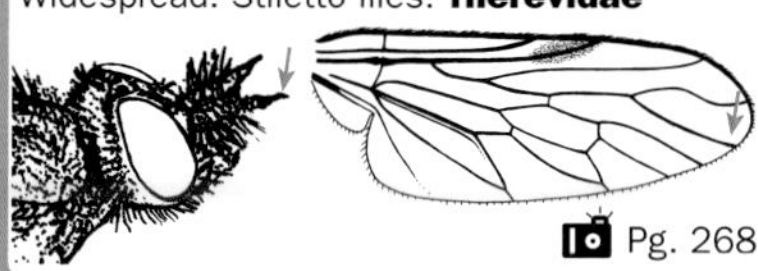

Pg. 268

Antenna ending in a long, thin style. R_{4+5} branched into a U with arms parallel, R_5 ending at wing tip. Tasmania, New Zealand and western Nearctic, uncommon. **Apsilocephalidae**

Pg. 271

Cell dm present. **Apystomyiidae** One very rare southwestern Nearctic species.

Pg. 292

Cell dm absent. **Hilarimorphidae** Uncommon, Holarctic-Oriental.

Pg. 292

KEY FIVE

Schizophora, part one, and the families of Calyptratae

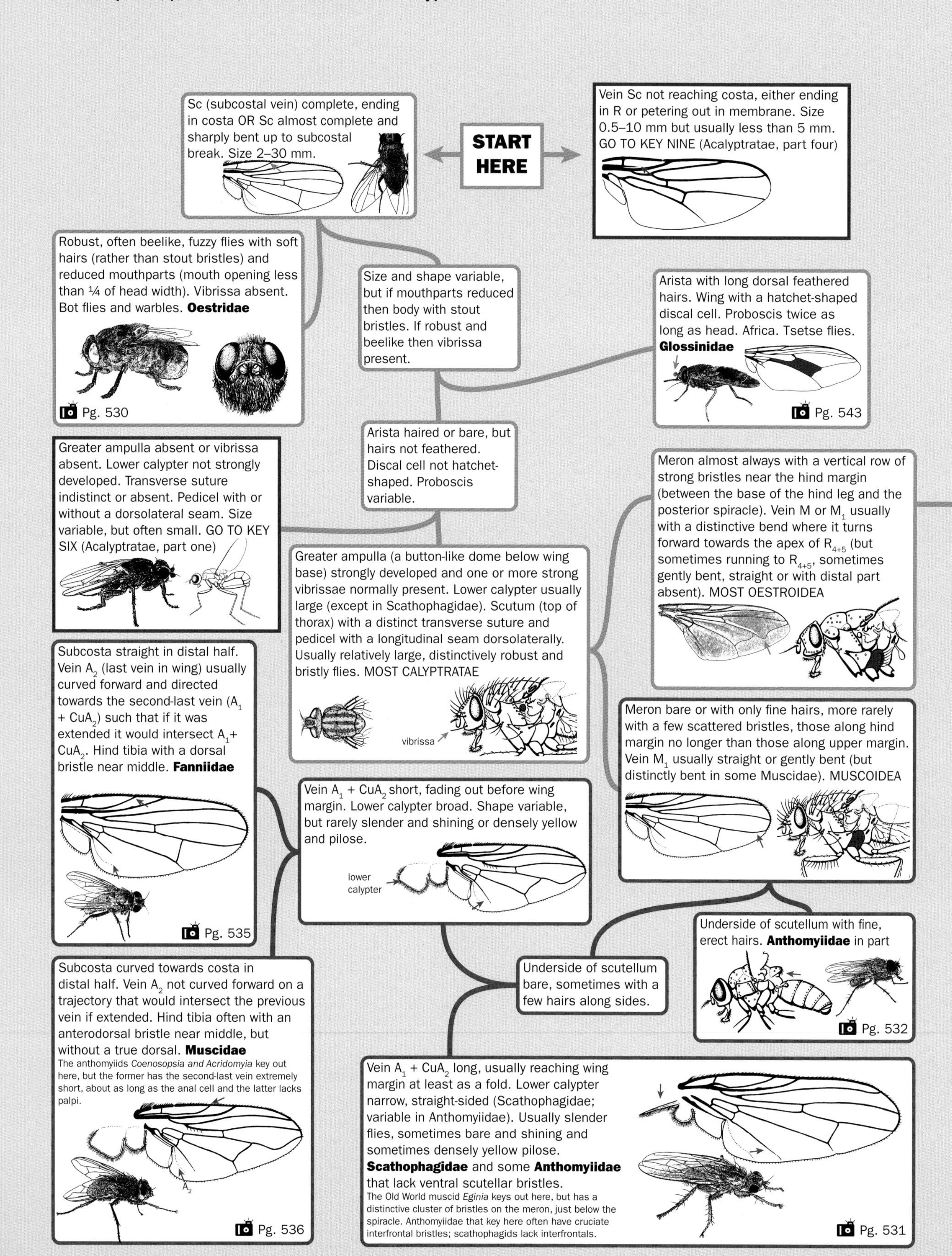

Underside of costa hairy only as far as humeral break (just beyond crossvein h), bare beyond this point. Occiput usually bare on upper half; lower facial margin usually distinctively protruding. Old World and Pacific. **Rhiniidae** (sometimes treated as the subfamily Rhiniinae in the family Calliphoridae)

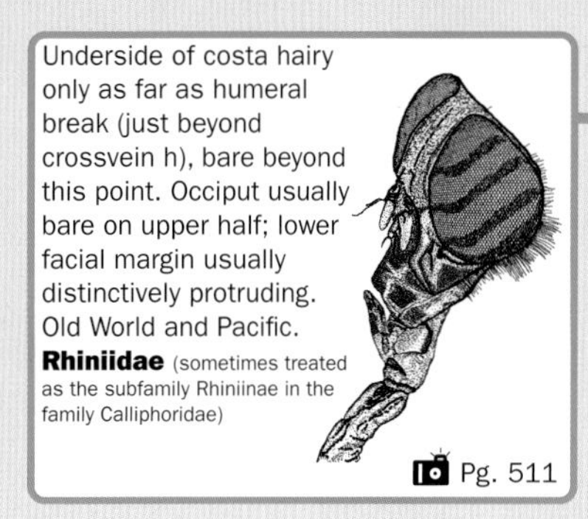

Pg. 511

Underside of costa hairy all the way to junction with subcostal vein. Occiput with setulae in upper part. Lower facial margin usually not protruding. Widespread, especially in warmer areas. **Calliphoridae** in part (subfamily Chrysomyiinae and a few unusual Luciliini)

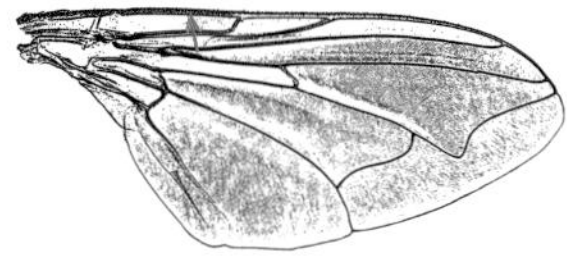

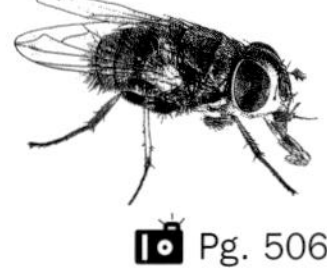

Pg. 506

Wing with setulae on posterodorsal surface of stem vein. Often partly or entirely shining metallic, rarely with gray and black abdominal pattern.

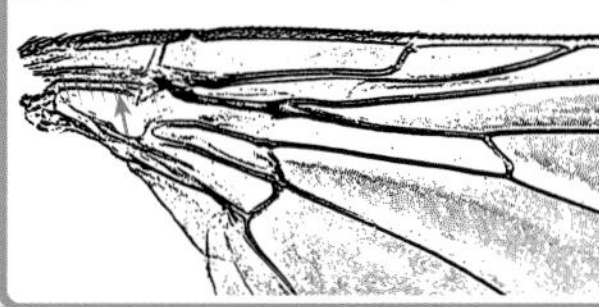

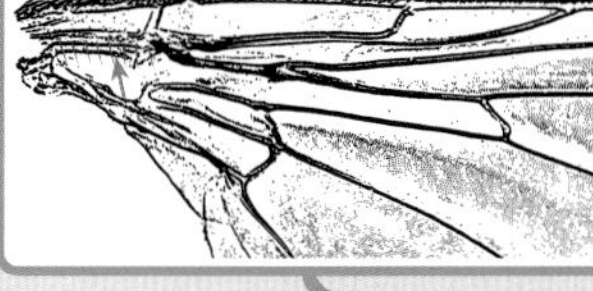

Wing without setulae on posterodorsal surface of stem vein. Color and pattern various, often shining metallic (most Calliphoridae) or gray and black (most Sarcophagidae).

Subscutellum forming a prominent convex lobe immediately under scutellum. Posterior spiracle usually flanked by two distinct flaps (lappets). Often (but not always) distinctively bristly, with long abdominal bristles. Wings in life usually held apart at about 45˚. Parasitic flies. **Tachinidae**

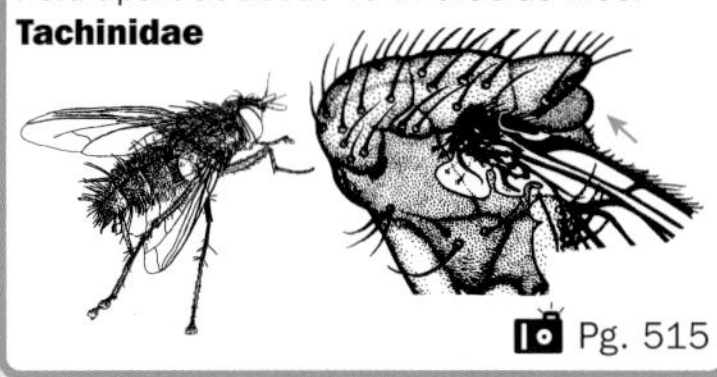

Pg. 515

Side of thorax with numerous and conspicuous long, crinkly golden hairs. Neither metallic nor conspicuously striped. Widespread. **Calliphoridae** in part (most Polleniinae, including common *Pollenia* species)

Pg. 506

Side of thorax without long golden hairs. Sometimes metallic.

Male antenna with segment 3 huge, often more or less axe-shaped. Female abdomen ending in long, curved process. Axe flies. **Axiniidae** (Australia, uncommon; sometimes included in Rhinophoridae)

Pg. 511

Posterior spiracle either flanked by distinct flaps (lappets) or fringe without overlapping long hairs. Bend of M usually (not always) more abrupt. Underside of scutellum variable. Size and appearance variable.

Antenna and abdomen not so modified.

Subscutellum absent or relatively weakly developed, with the membranous part between scutellum and subscutellum broader than the convex, sclerotized part of subscutellum. Posterior spiracle sometimes surrounded by fringe, sometimes by distinct lappets. Wings usually overlapping abdomen to a greater degree, diverging at a smaller angle. Other features variable.

Lower calypter oval or tongue-shaped, directed away from scutellum. Body usually uniformly dull, sometimes partly yellow but rarely shining metallic or distinctly striped or checkered. Relatively uncommon.

Lower calypter broader, not directed away from scutellum but instead running along scutellum for part of its length. Most either shining metallic or conspicuously striped or checkered. Common.

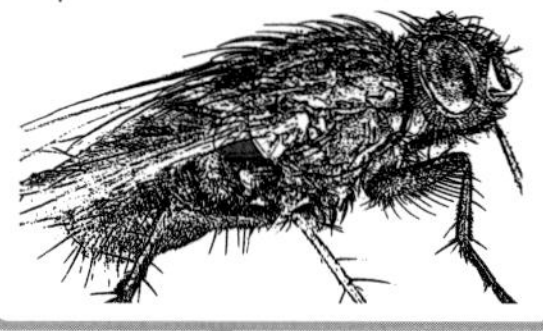

Posterior spiracle very large (as long as meron) with a fringe surrounding the rounded spiracular opening from below, in front and behind; fringe made of long, dense, usually pale hairs overlapped with a few longer dark ones. Bend of vein M gradual. Scutellum with fine hairs underneath. Large (9–15 mm), robust, abdomen often with green or purplish reflections. Wing membrane usually partly infuscated (smoky). **Mesembrinellidae** (Neotropical; often treated as the Calliphoridae subfamily Mesembrinellinae)

Pg. 510

Subscutellum at least slightly convex, but widely separated from scutellum. Posterior spiracle small, triangular, surrounded by a uniform and continuous fringe of short hairs or flanked by similar anterior and posterior tufts. M_1 sometimes turned abruptly up to meet R_{4+5} far before wing margin, creating a closed cell r_{4+5} (but M_1 sometimes gently bent to almost straight). Woodlouse flies. **Rhinophoridae**

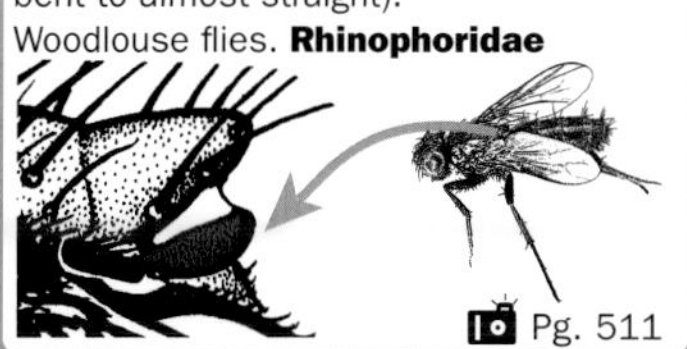

Pg. 511

Subscutellum flat. Posterior spiracle with unequal anterior and posterior flaps. M_1 curved towards wing tip but not meeting R before wing margin. Some **Calliphoridae**

Pg. 506

Usually metallic green, blue or black, at least abdomen usually with a metallic luster and lower calypter with black hairs on upper surface; sometimes not metallic, but never with a gray and black checkered abdomen. Sometimes uniformly yellow to brown. Notopleuron with 2 bristles. Arista usually plumose (long-haired) to tip. Most **Calliphoridae**

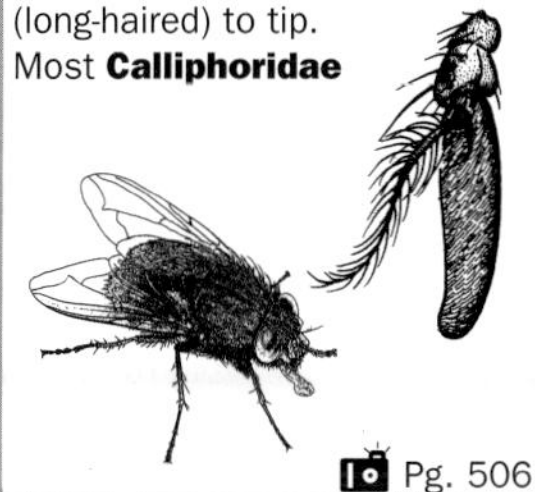

Pg. 506

Rarely metallic and shining but if so then lower calypter without hairs on upper surface and notopleuron with 4 bristles. Most species EITHER gray and black with top of thorax three-striped (and abdomen checkered, spotted or striped), antenna plumose in basal 2/3, and with 3-4 notopleural bristles (Sarcophaginae) OR more uniformly pigmented, tachinid-like flies with arista almost bare and 2 notopleural bristles (Miltogramminae). **Sarcophagidae**

The small Palaearctic calliphorid subfamily Helicoboscinae will key here but the area in front of the anterior spiracle is hairy rather than bare as in Sarcophagidae. The rare Australian McAlpine's fly (vein M straight, 3–4 short bristles on the meron) will also key here.

Pg. 512

KEY SIX

Schizophora, part two: Acalyptrates with a complete subcosta, part one.

Anal cell short. Slender, long-legged flies with distinct bristles and a strikingly elongate and geniculate proboscis. **Stylogastridae** Often treated as a subfamily of Conopidae

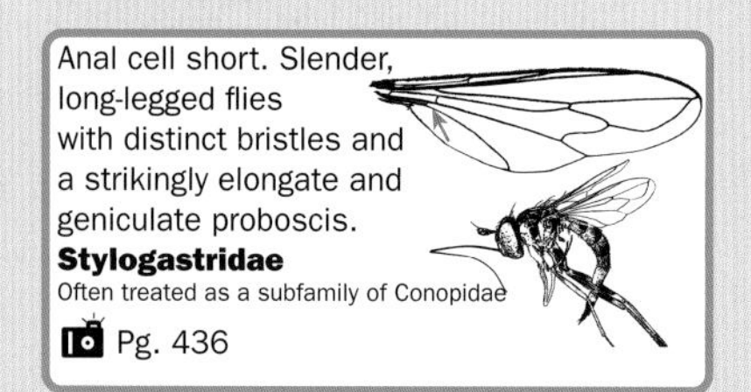

Pg. 436

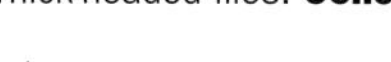

Anal cell long and pointed. Body relatively robust and without distinct bristles, often beelike or wasplike. Proboscis usually long and geniculate. Thick-headed flies. **Conopidae**

Pg. 437

Arista greatly reduced. Subcosta connected to R by a crossvein or narrow point of contact. Mouthparts usually long and sharply bent forward. Body often without distinct bristles.

Arista longer than third antennal segment; body with distinct bristles. Subcosta not connected to R by a crossvein. Mouthparts usually short (except in some small black Milichiidae).

Veins R_{4+5} and M almost always parallel or divergent (rare exceptions have a spinose costa or differ conspicuously from boxes below) . Ocellar bristles almost always present. Size variable.

Veins R_{4+5} and M converging towards wing tip. Ocellar bristles often absent. Usually over 7 mm, usually slender and long-legged.

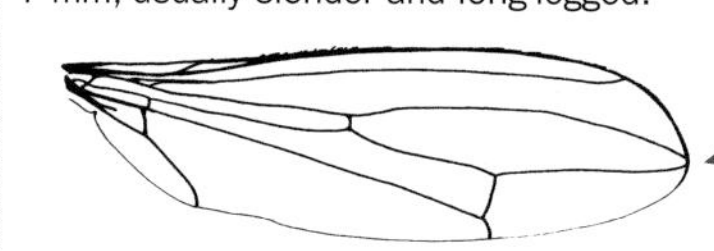

Sc not abruptly bent (rare exceptions either have R_1 bare, vibrissa present, subcostal break absent, ocelli absent, or frons without inclinate bristles).

START HERE

Robust flies with swollen femora and sunken vertex (top of the head). Face heavily sclerotized and prominent. Posterior thoracic spiracle with at least one bristle near top of hind margin. With ocellar bristles, without silver head patch. **Ropalomeridae** Neotropical and southern Nearctic.

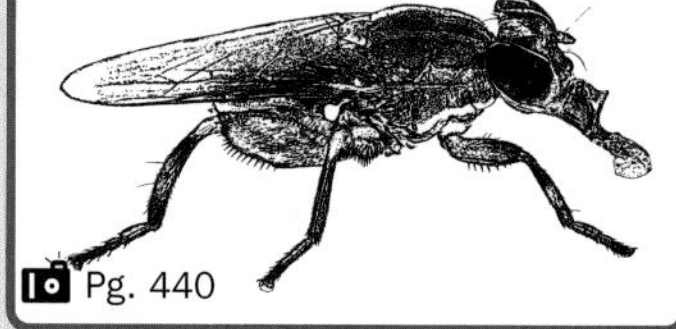

Pg. 440

Subcosta almost always abruptly bent up to costa (usually weak or transparent beyond bend) and vein R_1 setulose dorsally. Frons almost always with inclinate frontal bristles and reclinate orbital bristles. Costa broken at end of subcosta. Wings usually patterned, anal cell usually pointed. Ocelli present. Vibrissa absent. Fruit flies. **Tephritidae**

Pg. 450

Slender, conspicuously long-legged flies. Face soft, weakly sclerotized, and flat. Posterior spiracle without bristles. Ocellar bristles absent or top of head with a silver patch.

Arista apical or nearly so (arising near tip of third antennal segment). Front femur with ventral spines. Sometimes with a short, stout vibrissa. Cactus flies. **Neriidae**

Pg. 471

Scutellum distinctively inflated, usually larger than scutum and often large enough to cover most of the wings (creating a beetle-like appearance). Edge of scutellum unarmed.
Beetle flies. **Celyphidae**
Old World tropics.

Pg. 490

Scutellum with 2 long, tubular apical projections, usually ending in bristles. Head usually conspicuously broad (except in the small African subfamily Centrioncinae), eyes often on long stalks. Fore femur often thickened, armed with stout ventral spines.
Stalk-eyed flies. **Diopsidae**
Mostly Old World tropics, 1 genus in Holarctic region.

Pg. 473

Scutellum rarely greatly enlarged, but if so edge spinose rather than smooth.

If scutellum with long apical projections (rarely), then eyes not stalked and fore femur not thickened and armed.

Halter rarely black, but if so then either frontal bristles present, wing marked or anepisternum without a row of bristles. Other features variable.

Pronotum forming a distinct neck. Thorax and abdomen usually with yellow and black pattern. Abdomen short and broadly oval, wider than thorax. Arista with long hairs.
Somatiidae
Only 7 species, all similar, 4–5 mm. Neotropical.

Pg. 476

Pronotum short, inconspicuous. Rarely black and yellow patterned. Abdomen rarely wider than thorax (if so then arista without long hairs).

Halter black or dark brown. Head with no frontal bristles and a single pair of orbital bristles, curving backwards. Wing unmarked. Vibrissae absent. Anepisternum with bristles along hind margin. Usually small, shining dark flies. Females with a "lance-like" abdominal tip.
Lance flies. **Lonchaeidae**

Pg. 461

Gena without prominent upcurved bristles. Frons rarely with series of outcurved fronto-orbital bristles. GO TO KEY SEVEN

Cheek (gena) with one to several strong upcurved bristles below eye and frons with two or more outcurved fronto-orbital bristles. Female abdomen ending in upcurved spines.
Surf and beach flies.
Canacidae in part

Pg. 504

Anepisternal bristles absent. One or more katepisternal bristles present (often in a vertical row). Never with tomentose circular plate behind ocelli. R_1 with at most 2–3 dorsal setulae.
Micropezidae

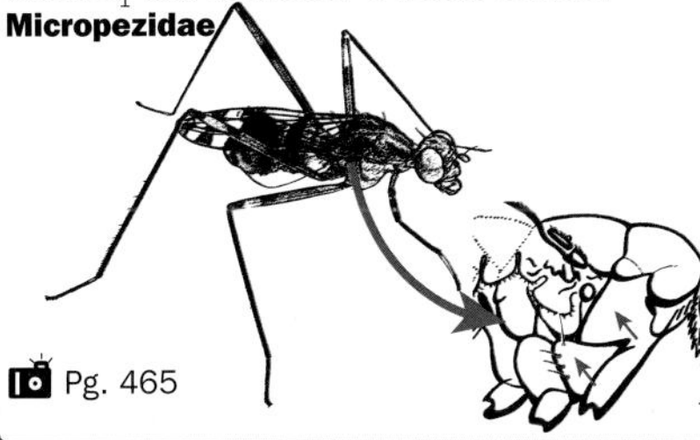

Pg. 465

Arista dorsobasal (arising near base of third antennal segment). Front femur without ventral spines. Vibrissa absent or slender and inconspicuous.

Anepisternum with one or more prominent bristles along posterior margin (in front of wing base). Katepisternum with only small hairs. Usually with silvery tomentose circular patch behind ocelli. Vein R_1 setulose dorsally.
Tanypezidae

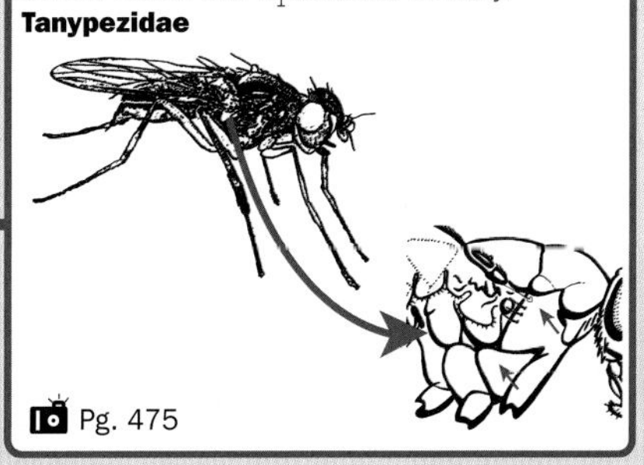

Pg. 475

KEY SEVEN

Acalyptratae, part two. Families keyed here have a vibrissa and a complete subcosta.

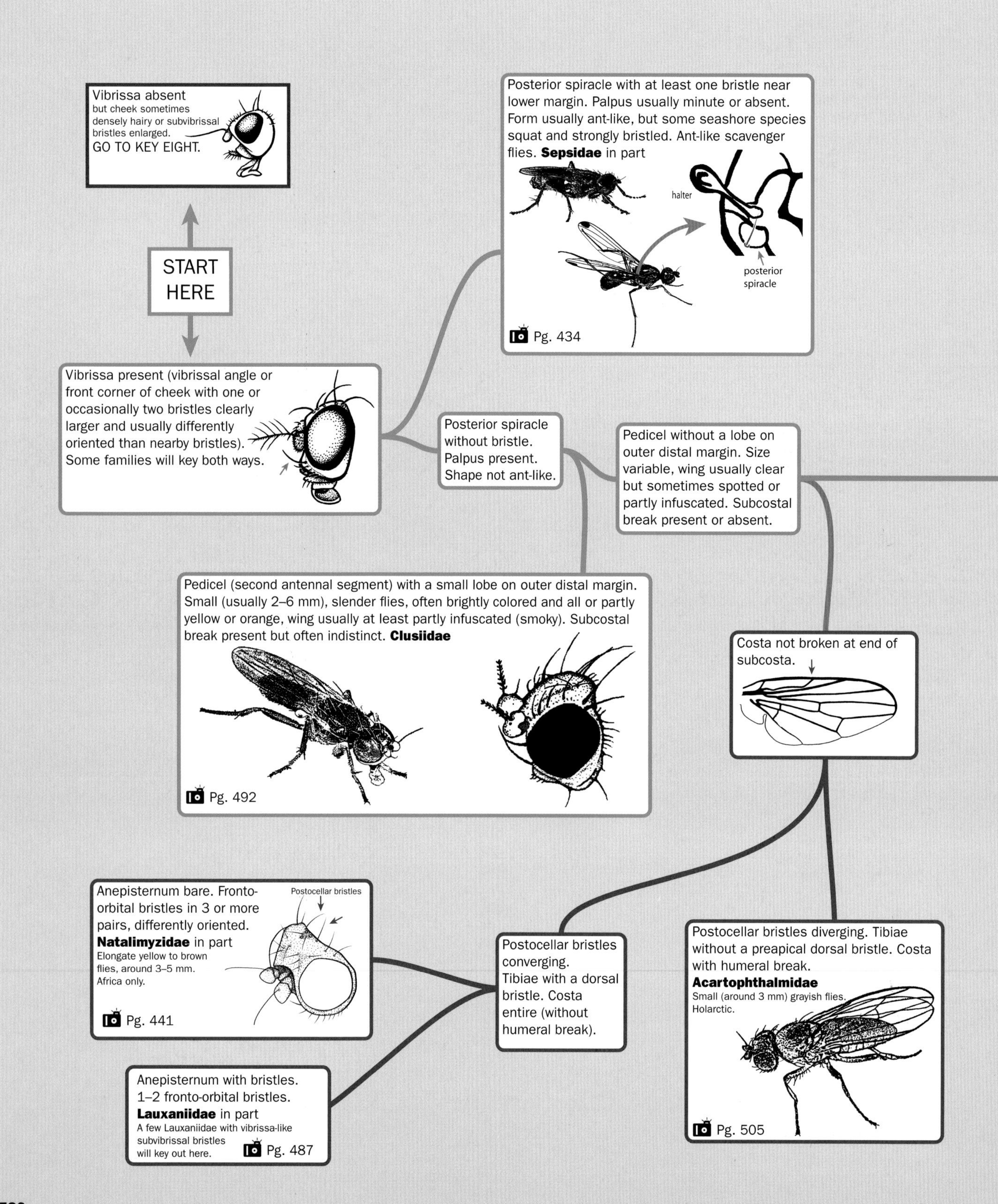

Top of thorax strongly arched, anterior half with only uniformly small bristles (no anterior dorsocentral bristles). Arista with long rays at least dorsally. **Curtonotidae**

Pg. 485

Subcosta weak apically or ending close to end of R_1. Frons with 2–8 frontal bristles, lower pairs curving inwards. Female abdomen with a conspicuously dark and hard conical oviscape. Usually 2–4 mm. Leaf-mining flies. **Agromyzidae** in part

Pg. 490

Top of thorax not strongly arched. Arista variable, but if plumose then anterior half of thorax with dorsocentral bristles. **Heleomyzidae** in part

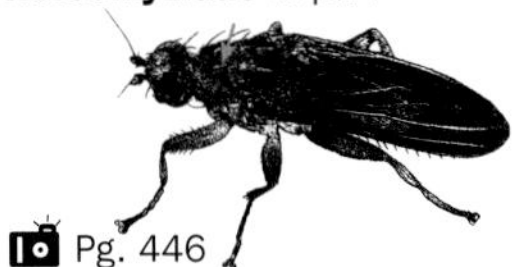

Pg. 446

Subcosta distinct, apex separate from R_1. Female abdomen with a telescoping apex, without a hard tubular oviscape. Frons with 0–4 frontal bristles, lower pairs rarely incurved. Usually 3–6 mm. **Piophilidae**

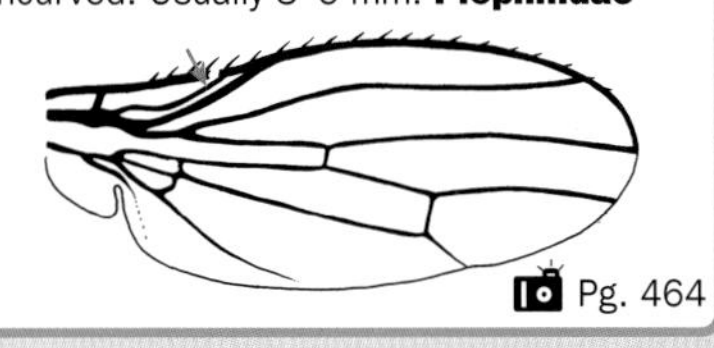

Pg. 464

At least one tibia with a strong preapical dorsal bristle. Costa often with spines.

Postocellar bristles divergent (very rarely parallel).

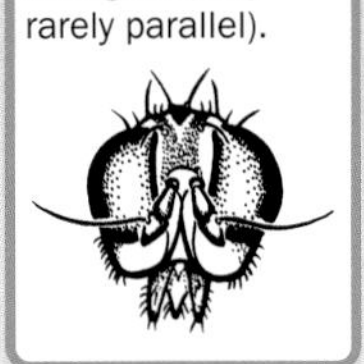

Costa broken at end of subcosta.

Tibiae without preapical dorsal bristles. Costa rarely spinose.

Postocellar bristles convergent, parallel or absent.

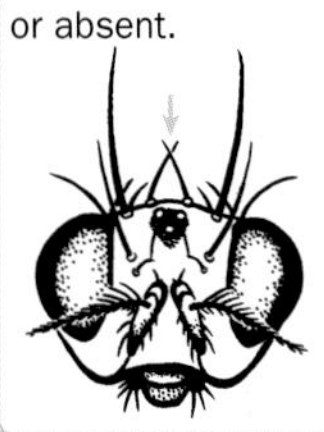

Mouthparts geniculate (bent at middle) OR anepisternum bare OR vibrissa well above lower margin of eye. Subcostal break sometimes very large and flanked by "lappet." Antennae not in deep depressions. Size 1–6 mm. **Milichiidae** in part.

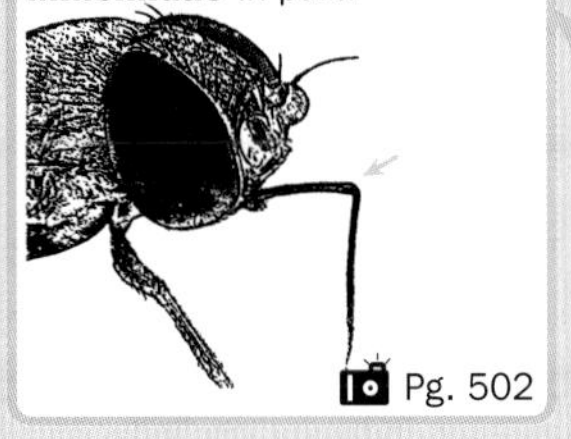

Pg. 502

Eyes not metallic (sometimes reddish). Size and wing variable. Vibrissa usually strong.

Small (usually 0.5–3.0 mm), all or partly pale yellowish flies with pale bristles and metallic green or red eyes (may be faded on dry specimens). Wing clear and subcosta distally weak. Vibrissa small and inconspicuous. **Chyromyidae** in part.

Some darker species differ from similar families in that R_{4+5} and M converge slightly towards the wing tip.

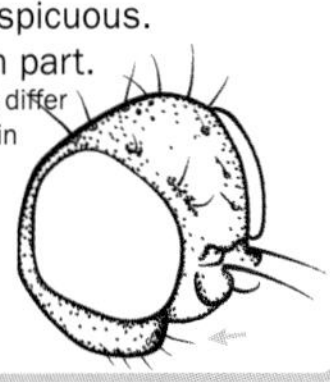

Pg. 449

Costa with a distinct humeral break. Lower fronto-orbital bristles incurved. Usually small and black, sometimes velvety, sometimes with a long, folded proboscis.

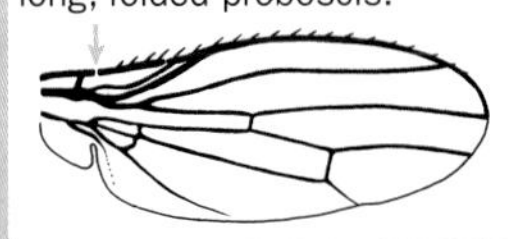

Costa without a humeral break. Lower frontal orbitals reclinate (backcurved). Size and color variable.

Mouthparts short, anepisternum with hairs or bristles, and vibrissa inserted below lower eye margin. Subcostal break without lappet. Antennae set in deep depressions. Size 1–3 mm. **Carnidae** in part

Pg. 505

Subcosta joining R_1 just before costa. Anepisternum with a row of bristles along posterior margin. Size usually 2–3 mm, usually silvery pruinose. Associated with saline environments. **Canacidae** in part (Tethininae)

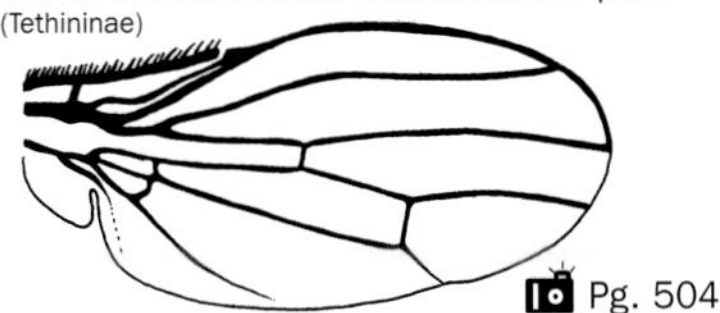

Pg. 504

Junction of subcosta and costa separate from R_1. Anepisternum with or without bristles. Size variable but usually over 4 mm, rarely silvery pruinose. Not associated with saline environments. Some **Heleomyzidae**

Heleomyzids will key out in several places but most have a dorsal bristle on at least one tibia.

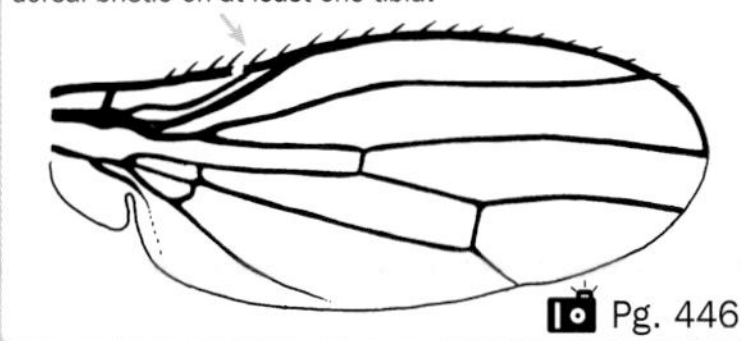

Pg. 446

KEY EIGHT

Acalyptratae, part three. Families keyed in this section have a complete subcosta but lack a distinct vibrissa.

Ocelli absent (almost always) and/or top of head with a distinct occipital slit. Wing usually patterned, never uniformly black. Uncommon crepuscular or nocturnal flies. **Pyrgotidae** in part

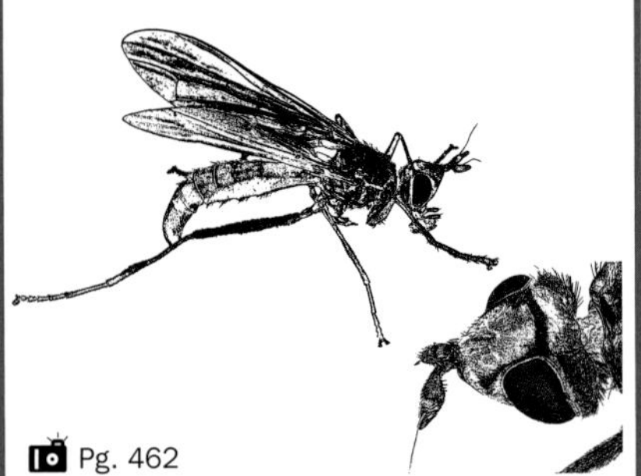

Pg. 462

START HERE

Ocelli almost always present (absent in an African platystomatid with black wings). Occipital slit absent. Normally diurnal flies.

Small (usually 0.5–3 mm), all or partly pale yellow flies with yellowish bristles and metallic greenish or reddish eyes (often faded on dead specimens). Wing clear and subcosta distally weak. Vibrissa present but thin, pale, easily overlooked. **Chyromyidae** in part. Pg. 449

Size and appearance variable, but never pale yellow with metallic eyes. Vibrissa absent, but cheek sometimes with multiple long hairs.

Posterior spiracle with at least one bristle along lower margin. Ant-shaped, with a spherical head and reduced palpus. **Sepsidae** in part Pg. 434

Posterior spiracle without a bristle along lower margin. Rarely ant-shaped, head not spherical, palpus usually well developed.

Costa unbroken (without a distinct gap or fracture near end of subcosta).

Last tarsal segment (terminal tarsomere) triangular, flattened and wider than other segments, with 2–3 long, curved apical dorsal bristles. Katepisternal bristle directed anteriorly. Seashore flies, usually dorsoventrally flattened with densely bristled cheeks and legs. **Coelopidae** Pg. 439

Last tarsal segment not widened and triangular. Katepisternal bristle directed upwards or posteriorly. Habitat and appearance variable.

Some or all tibiae with a preapical dorsal bristle.

Tibiae without a preapical dorsal bristle.

Costa broken or discontinuous near apex of subcosta (subcostal break; best viewed with transmitted light).

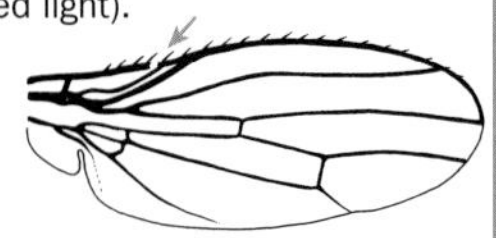

Thorax usually with at least one presutural dorsocentral bristle. Anal cell simple, without acute lobe. R_1 bare. Flutter flies. **Pallopteridae** Mostly Holarctic but also Chile, Argentina, New Zealand. Pg. 464

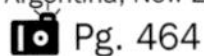

Presutural dorsocentral bristles absent (usually) and/or anal cell with an acute lobe at hind corner and/or R_1 setose dorsally.

Anal cell closed by a straight or convex crossvein, thus without a pointed hind corner. Head with one orbital bristle (sometimes only slightly larger than surrounding hairs). Some or all femora usually with stout ventral bristles. Anepimeron bare. Subcostal break distinct. New World, mostly Neotropical. **Richardiidae**

Pg. 462

Anal cell usually with a pointed hind corner, otherwise with at least two fronto-orbital bristles and/or anepimeron with hairs. Femora without stout ventral bristles. Subcostal break narrow and indistinct. Widespread. Picture-winged flies. **Ulidiidae** in part Pg. 458

Hind femur thicker than mid femur, usually with ventral rows of spines.

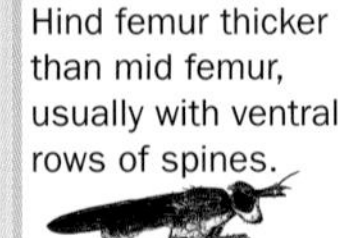

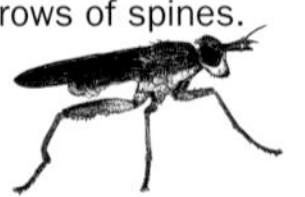

Hind femur not swollen, lacking ventral rows of spines

Postocellar bristles usually convergent or absent, sometimes parallel or slightly divergent. R_1 bare. Wing clear or patterned.

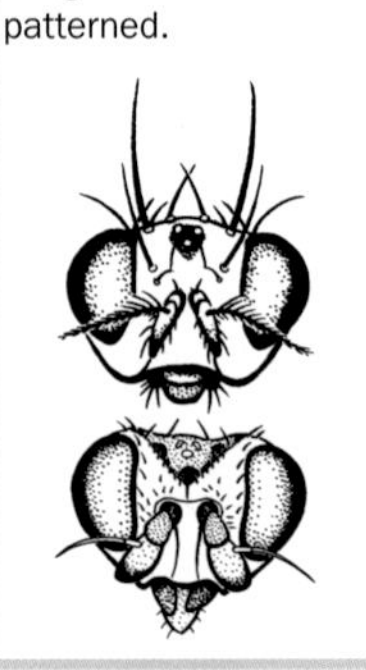

Postocellar bristles strongly divergent. Distal part of vein R_1 often with setae dorsally. Wing usually strongly patterned. Picture-winged flies. **Ulidiidae** in part. Pg. 458

Vein R_1 usually bare; if setose (some Ulidiidae) then anepimeron without bristles and costa with an indistinct subcostal break.

Vein R_1, and sometimes other veins, with dorsal bristles over entire length. Anepimeron with bristles. Costa not weakened at end of subcosta. Signal flies. **Platystomatidae** Most diverse in Old World tropics and Australasia. Pg. 455

Second antennal segment longer than third. Abdomen not narrowed or petiolate. Orbital bristles usually present. Widespread. **Sciomyzidae** in part
Pg. 431

Second antennal segment much shorter than third. Abdomen elongate and slender, sometimes petiolate. Orbital bristles absent. Oriental and Palaearctic.

Abdomen narrow but not petiolate. Third antennal segment (first flagellomere) not deflexed at an angle to basal segments, segment two without a narrow dorsal slit. Face reduced to medial fissure. Old World. **Megamerinidae**

Pg. 476

Abdomen petiolate (basally constricted). Third antennal segment deflexed at an angle to basal segments, antennal segment two with a narrow dorsal slit. Face and frons broad, flat, shining. Southeast Asia. **Gobryidae**

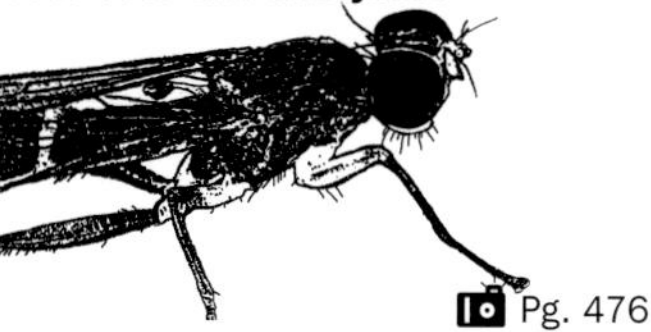

Pg. 476

Postscutellum not conspicuously enlarged. Shape and pigmentation variable. Widespread.

Hind tibia not expanded, first antennal segment small.

Shining black or brown. Anal vein almost reaching wing margin. **Paraleucopidae** in part
Pg. 490

Usually silvery dusted, sometimes black; if shining then anal vein short. Widespread. Aphid flies. **Chamaemyiidae**

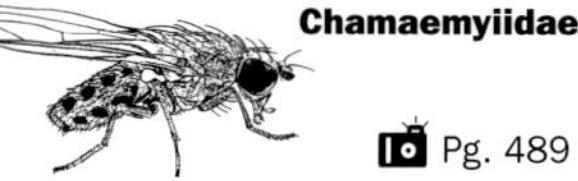

Pg. 489

Hind tibia expanded and flattened, first antennal segment longer than second. A few Neotropical **Lauxaniidae** (some Eurychoromyiinae)
Pg. 487

Long-legged, slender flies with large postscutellum projecting far behind scutellum. Prothorax elongate, with forelegs closer to mid legs than to head and with tarsus longer than tibia. Southeast Asia. **Nothybidae**

Pg. 476

Hind tibia with one (usually) or two preapical dorsal bristles. Widespread. Marsh flies. **Sciomyzidae** in part
Pg. 431

Hind tibia with multiple dorsal bristles (2 anterodorsal, 2–3 posterodorsal). Palaearctic. **Phaeomyiidae**
Pg. 441

Clypeus large, clearly visible below lower margin of face. Katepisternum with bristles (most genera) or long hairs (one seashore genus). Anepisternum without bristles. **Dryomyzidae**
Pg. 440

Clypeus small, usually concealed by lower margin of face. Katepisternum usually bare. Anepisternum with or without bristles.

Costa without spines.

Costa with stout spines evenly spaced along wing margin. **Helosciomyzidae**
Australia, New Zealand & South America

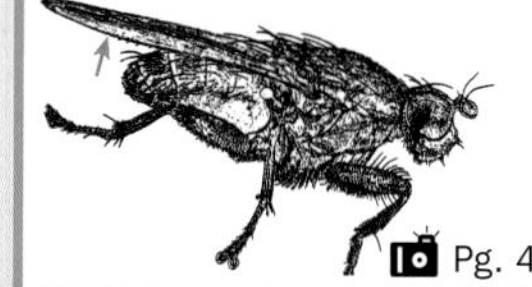

Pg. 441

Costa without stout bristles. Body and legs partly brown. **Heterocheilidae**

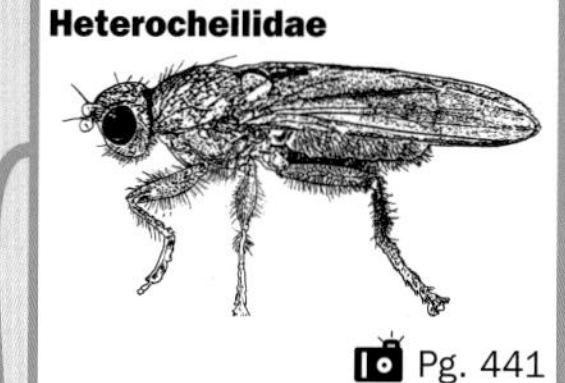

Pg. 441

Costa with stout, short bristles in addition to the usual small setulae. Color mainly gray. **Helcomyzidae**
Pg. 441

Cheek partly bare and/or cheek height much less than eye height. Underside of thorax without a sclerotized bridge anterior to forelegs (prosternum separated from propleuron by membrane). If clypeus conspicous then antennae contiguous at base. Sometimes coastal but rarely in seaweed; vestiture variable.

Eye small and cheek setulose. Thorax with a well-developed precoxal bridge (sclerotized portion under thorax in front of forelegs). Clypeus conspicuous and antennae widely separated at base. Pruinose flies found in seaweed on ocean beaches.

Vein A_1 + CuA_2 (from posterior corner of anal cell) usually extending to, or almost to, wing margin. Postocellar bristles divergent or parallel, rarely absent but if so then scutellum with one pair of bristles.

Vein A_1+CuA_2 extending as a short, sclerotized vein beyond apex of anal cell. Usually 1–2 fronto-orbital bristles; prosternum broad. Anepisternum usually with bristles (absent in some New Zealand species). Widespread.

Anepisternum with at least one bristle. Fronto-orbital bristles in 1 or 2 pairs. Abdominal sternite 1 present. Cosmopolitan. **Lauxaniidae**

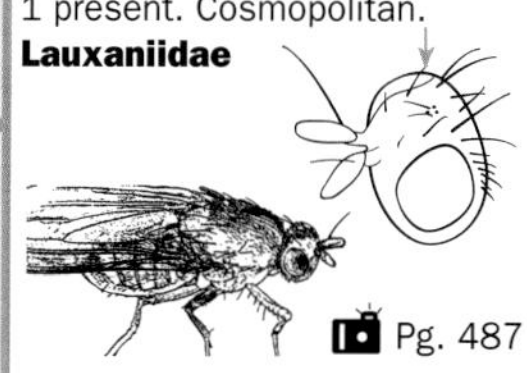

Pg. 487

Vein A_1 + CuA_2 (from corner of anal cell) short, ending far before wing margin. Postocellar bristles convergent, rarely absent or parallel but if so then scutellum with two pairs of bristles.

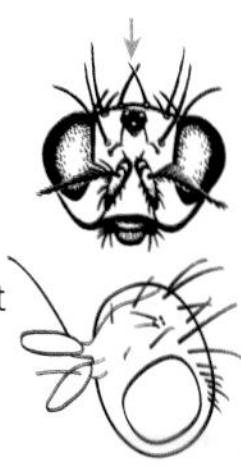

Vein A_1+CuA_2 not sclerotized beyond apex of anal cell. Usually more than 3 fronto-orbital bristles, bent in different directions. Prosternum narrow-oval, anepisternum without bristles. Africa. **Natalimyzidae** in part
Pg. 441

Anepisternum bare OR head with 3 fronto-orbital bristles. Abdominal sternite 1 absent. New Zealand. **Huttoninidae**

Pg. 441

KEY NINE

Acalyptratae, part four: Acalyptrates with an incomplete subcosta, part one.

START HERE

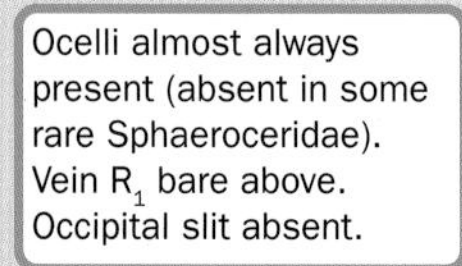

Ocelli almost always present (absent in some rare Sphaeroceridae). Vein R_1 bare above. Occipital slit absent.

Ocelli absent (almost always) and/or top of head with an occipital slit above eye. Vein R_1 usually setose above.

Head almost as large as thorax, proboscis small to absent. Female arista multi-branched, highly distinctive. Occipital slit absent. Rarely collected. **Ctenostylidae**

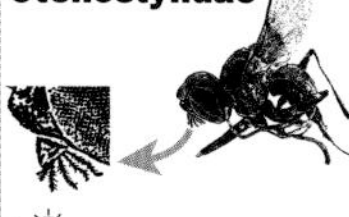

Pg. 465

Propleuron (sometimes hidden behind head) with a sharp raised vertical ridge. Ocellar triangle usually large, prominent and shining. Anal cell absent; vein CuA_1 usually with a characteristic kink. **Chloropidae**

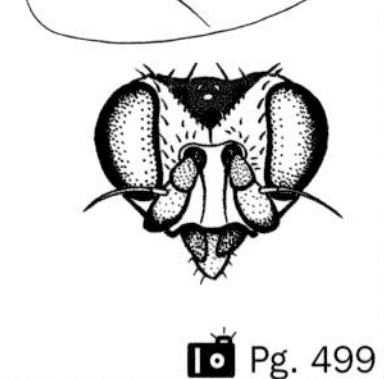

Pg. 499

Propleural carina absent (if apparently present then anal cell complete). Ocellar triangle usually smaller. Anal cell open or closed, but CuA_1 without a kink.

Head not unusually large, proboscis well developed. Arista not branched. Occipital slit present. **Pyrgotidae** in part

Pg. 462

Arista absent and first flagellomere (third antennal segment) wide. Minute and stout; black with blue metallic shine. **Cryptochetidae** in part

Pg. 487

First tarsal segment of hind leg not swollen, usually slender and longer than second tarsomere.

First tarsal segment (first tarsomere) of hind leg conspicuously short; first (and sometimes second) hind tarsomeres swollen and much larger than other segments (some rare species with all tarsomeres swollen and short). **Sphaeroceridae**

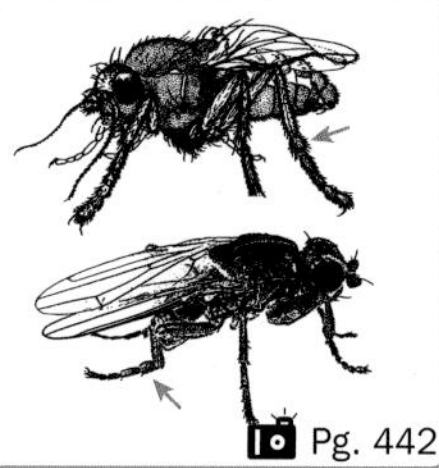

Pg. 442

Arista rarely absent or reduced, but if so then body neither stout nor metallic black.

Arista bare or with hairs both above and below; rarely with long dorsal hairs only, but if so then either anal cell complete or subcostal break absent. Face variable.

Arista with long dorsal hairs only, anal cell absent, and subcostal break present. Face often prominent. Shore flies. **Ephydridae** in part

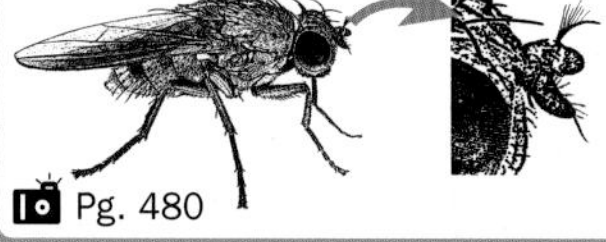

Pg. 480

R_{2+3} very short, ending near tip of R_1; anal cell absent or indistinct; veins R_{4+5} and M distally convergent. Arista sometimes with alternating dorsal and ventral rays, antenna zigzag on apical 2/3. **Asteiidae** in part

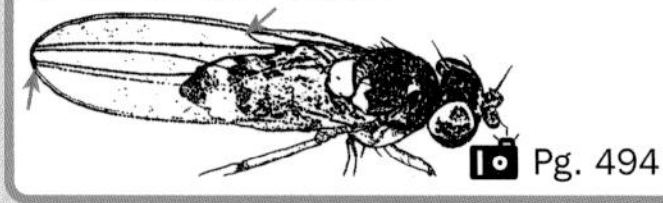

Pg. 494

R_{2+3} longer, not ending near tip of R_1. R_{4+5} and M rarely converging. Arista rarely zigzag. If arista zigag or R_{2+3} short then, anal cell complete.

Vibrissa or other enlarged bristles present at anteroventral corner of head, or lower face bulging and with bristles. If wing with an apical spot then more than one fronto-orbital bristle.

Vibrissa and facial bristles absent, although sometimes (*Geomyza*, with an apical wing spot and a single orbital bristle) with vibrissa-like subvibrissae.

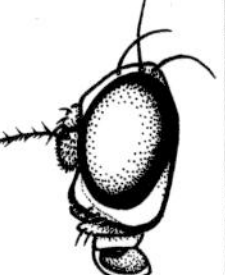

Wing usually clear, but if with apical spot then katepisternum without bristles. R_1 not kinked. Usually more than one fronto-orbital bristle.

Wing markings usually including an apical spot. R_1 with apical kink. Head rounded, with a single pair of fronto-orbital bristles. Katepisternum with a bristle. **Opomyzidae**

Pg. 497

Costa unbroken

Anal cell complete. Longest wing veins not converging.

Compact, shining metallic black. Pronotum short; hind femur without ventral spines. 2–3 mm. Australia and New World. **Paraleucopidae** in part

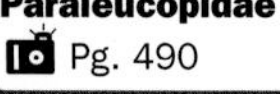

Pg. 490

Slender flies with a petiolate abdomen; pronotum forming a distinct neck. Hind femur with ventral spines. Neotropical, 4–6 mm. **Syringogastridae**

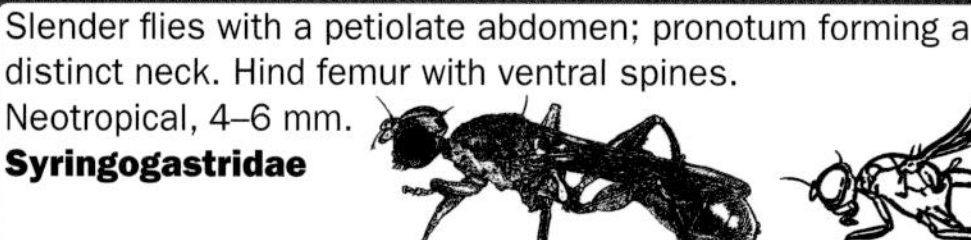

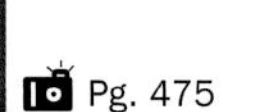

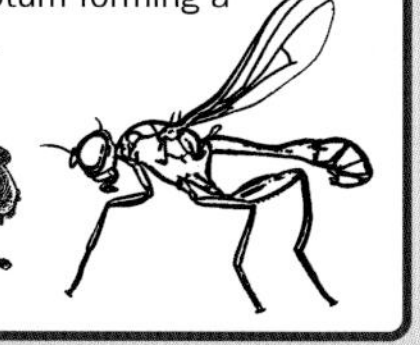

Pg. 475

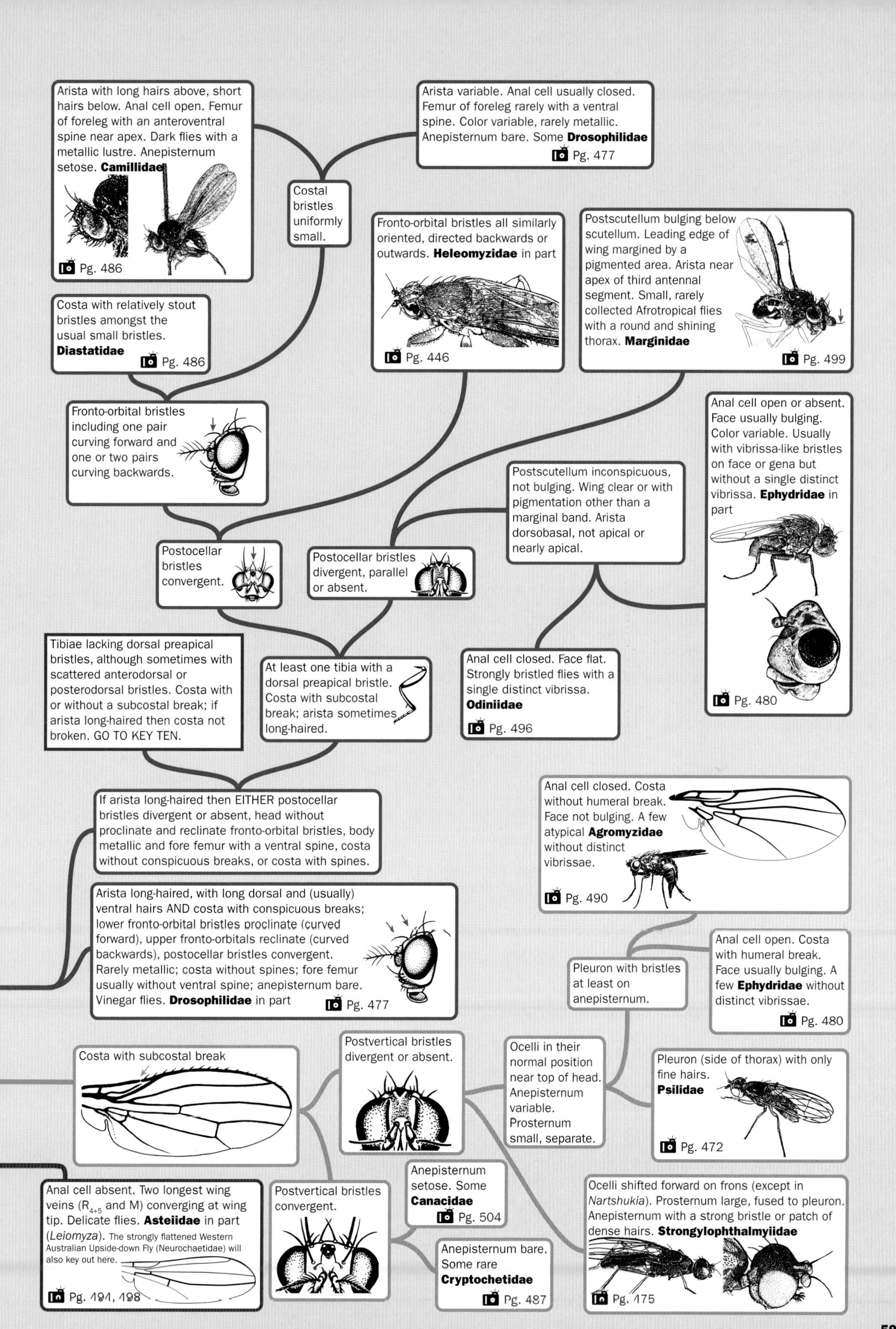
Arista with long hairs above, short hairs below. Anal cell open. Femur of foreleg with an anteroventral spine near apex. Dark flies with a metallic lustre. Anepisternum setose. Camillidae
Pg. 486
Arista variable. Anal cell usually closed. Femur of foreleg rarely with a ventral spine. Color variable, rarely metallic. Anepisternum bare. Some Drosophilidae
Pg. 477
Costal bristles uniformly small.
Fronto-orbital bristles all similarly oriented, directed backwards or outwards. Heleomyzidae in part
Pg. 446
Postscutellum bulging below scutellum. Leading edge of wing margined by a pigmented area. Arista near apex of third antennal segment. Small, rarely collected Afrotropical flies with a round and shining thorax. Marginidae
Pg. 499
Costa with relatively stout bristles amongst the usual small bristles. Diastatidae
Pg. 486
Fronto-orbital bristles including one pair curving forward and one or two pairs curving backwards.
Anal cell open or absent. Face usually bulging. Color variable. Usually with vibrissa-like bristles on face or gena but without a single distinct vibrissa. Ephydridae in part
Pg. 480
Postscutellum inconspicuous, not bulging. Wing clear or with pigmentation other than a marginal band. Arista dorsobasal, not apical or nearly apical.
Postocellar bristles convergent.
Postocellar bristles divergent, parallel or absent.
Tibiae lacking dorsal preapical bristles, although sometimes with scattered anterodorsal or posterodorsal bristles. Costa with or without a subcostal break; if arista long-haired then costa not broken. GO TO KEY TEN.
At least one tibia with a dorsal preapical bristle. Costa with subcostal break; arista sometimes long-haired.
Anal cell closed. Face flat. Strongly bristled flies with a single distinct vibrissa. Odiniidae
Pg. 496
If arista long-haired then EITHER postocellar bristles divergent or absent, head without proclinate and reclinate fronto-orbital bristles, body metallic and fore femur with a ventral spine, costa without conspicuous breaks, or costa with spines.
Anal cell closed. Costa without humeral break. Face not bulging. A few atypical Agromyzidae without distinct vibrissae.
Pg. 490
Arista long-haired, with long dorsal and (usually) ventral hairs AND costa with conspicuous breaks; lower fronto-orbital bristles proclinate (curved forward), upper fronto-orbitals reclinate (curved backwards), postocellar bristles convergent. Rarely metallic; costa without spines; fore femur usually without ventral spine; anepisternum bare. Vinegar flies. Drosophilidae in part
Pg. 477
Pleuron with bristles at least on anepisternum.
Anal cell open. Costa with humeral break. Face usually bulging. A few Ephydridae without distinct vibrissae.
Pg. 480
Costa with subcostal break
Postvertical bristles divergent or absent.
Ocelli in their normal position near top of head. Anepisternum variable. Prosternum small, separate.
Pleuron (side of thorax) with only fine hairs. Psilidae
Pg. 472
Anal cell absent. Two longest wing veins (R4+5 and M) converging at wing tip. Delicate flies. Asteiidae in part (Leiomyza). The strongly flattened Western Australian Upside-down Fly (Neurochaetidae) will also key out here.
Pg. 494, 498
Postvertical bristles convergent.
Anepisternum setose. Some Canacidae
Pg. 504
Anepisternum bare. Some rare Cryptochetidae
Pg. 487
Ocelli shifted forward on frons (except in Nartshukia). Prosternum large, fused to pleuron. Anepisternum with a strong bristle or patch of dense hairs. Strongylophthalmyiidae
Pg. 475

KEY TEN

Acalyptratae, part five: Acalyptrates with an incomplete subcosta, part two.

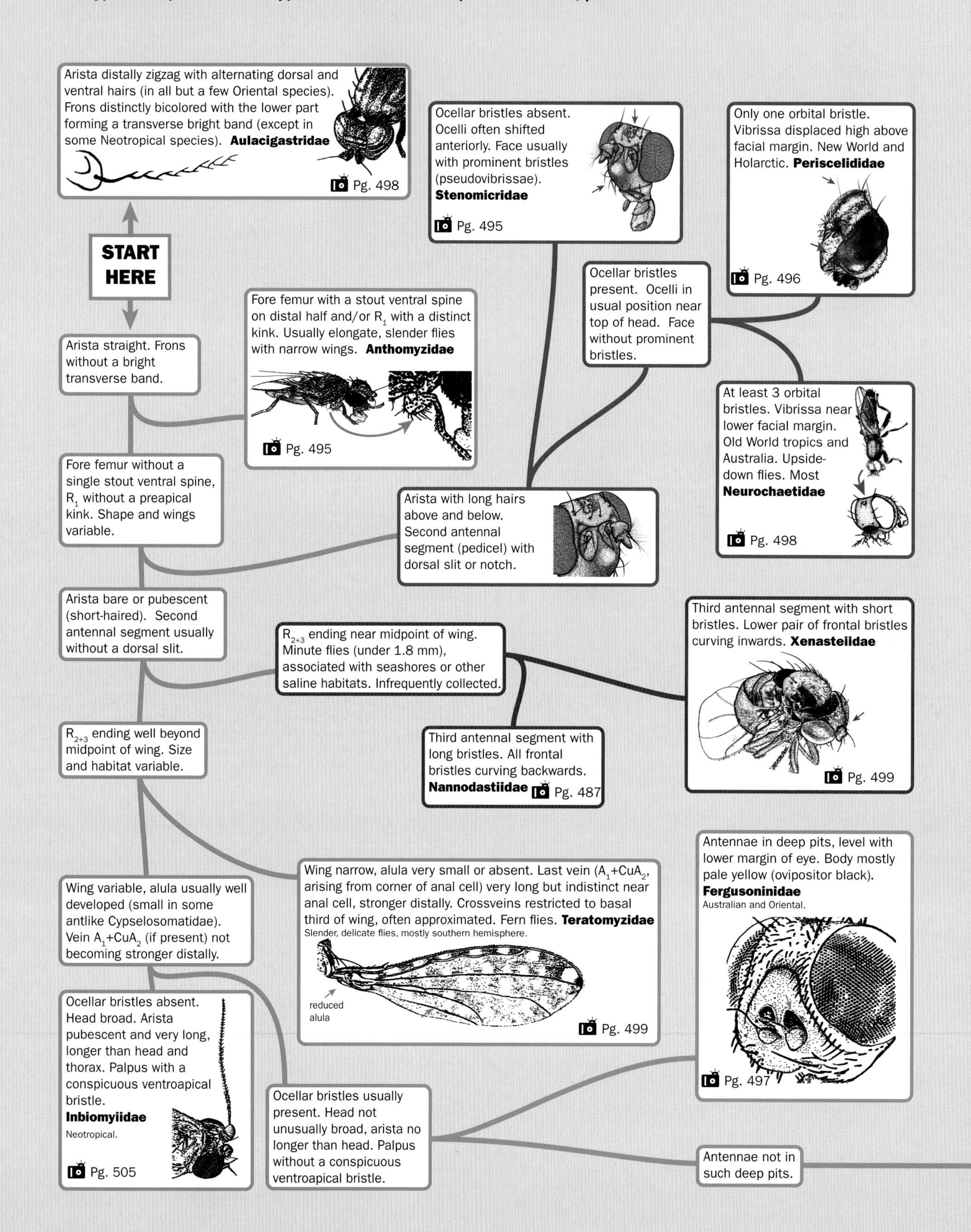

Vein CuA_1 extending far beyond cell dm. Anepisternum and/or katepisternum usually with bristles.

Gena with one to several upturned bristles below eye OR anepisternum setose and postocellar bristles convergent to absent. Fronto-orbital bristles usually outcurved. Surf flies and beach flies. **Canacidae** in part

Pg. 504

Gena without upturned bristles. Postocellar bristles diverging OR anepisternum bare.

Frons with 3–5 frontal bristles. Anepisternum with bristles. Female with a stout, tubular, non-retractible oviscape. Leaf-mining flies. **Agromyzidae** in part

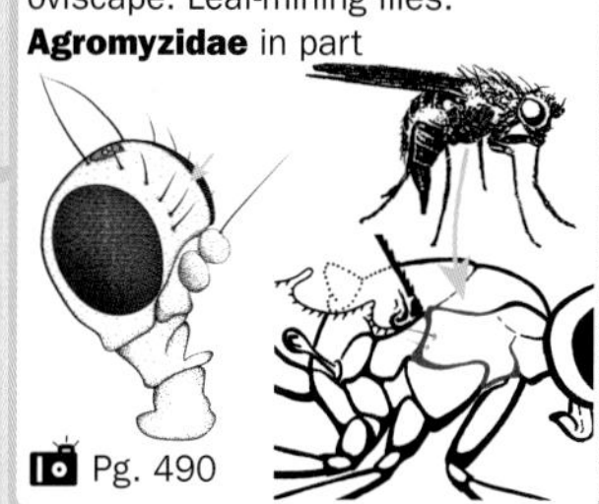

Pg. 490

Frons sometimes with small cruciate interfrontal bristles but without frontal bristles. Anepisternum bare. Female without a tubular oviscape. Small, slender, somewhat flattened. **Neminidae**
Australia and Old World tropics.

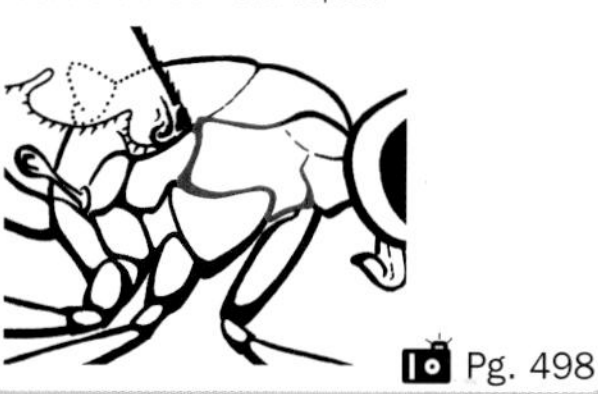

Pg. 498

Cell dm (longest closed cell in wing) with at most a short vein running off its distal corner (CuA_1 reduced or absent beyond cell dm). Anepisternal and katepisternal bristles absent. **Cypselosomatidae**
Australasian and Oriental.

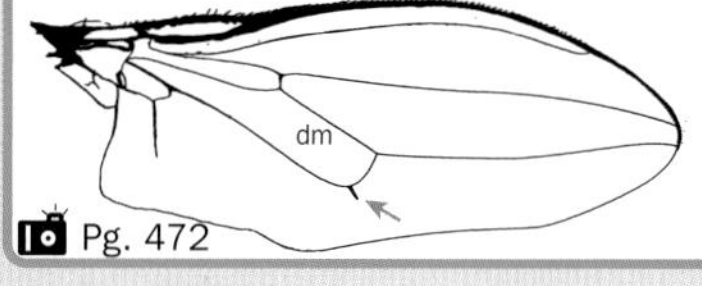
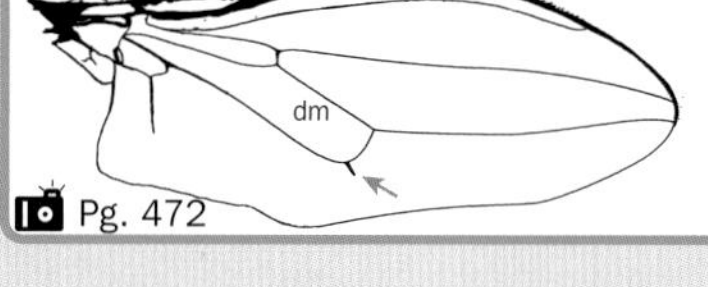
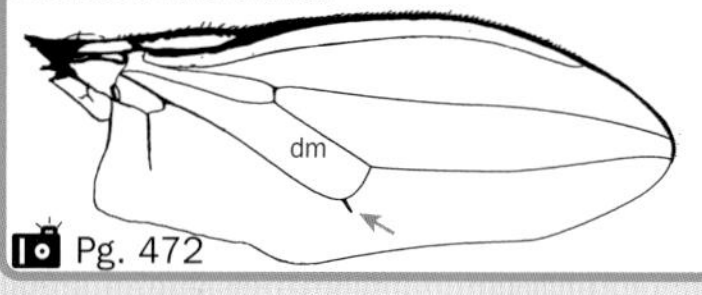

Pg. 472

Costa without a humeral break. Fronto-orbital bristles variable, but if numerous and inclinate then female abdomen with a conical, non-retractable oviscape.

Costa with humeral break (a gap or distinct weakening near crossvein h). Frons with at least 3 (usually 4 or more) fronto-orbital bristles, some usually inclinate. Conical oviscape absent.

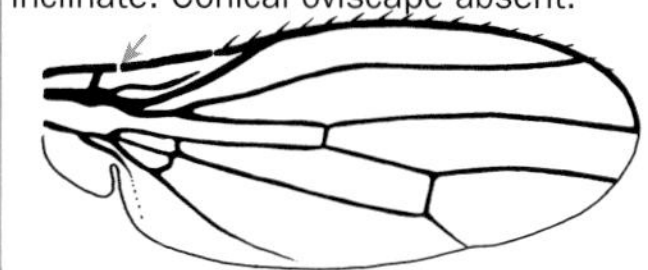

Wing with anal cell closed. Mouthparts sometimes long and geniculate.

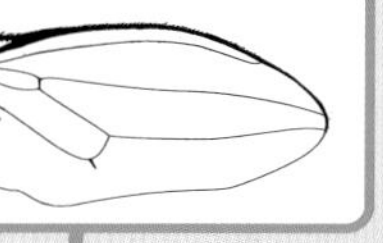

Mouthparts short and straight, anepisternum with hairs or bristles, AND vibrissa inserted at oral margin. 1–3 mm. Antennae often in depressions separated by sharp ridge. **Carnidae** in part

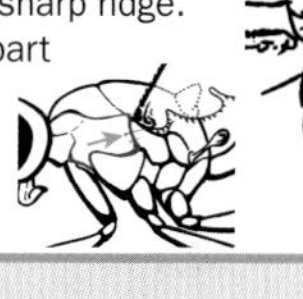

Pg. 505

Mouthparts geniculate (bent at middle) OR anepisternum bare OR vibrissa well above lower margin of eye. Subantennal depression absent or shallow. 1–6 mm.

Anal cell open or absent. Mouthparts short.

Fronto-orbital bristles variable, but lower 2 not inclinate. **Ephydridae** in part

Pg. 480

Frons with 4 fronto-orbital bristles, the lower 2 inclinate. **Carnidae** in part
Usually 1–2 mm, with pale halteres and crossveins near wing base.

Pg. 505

Fronto-orbital bristles usually at least partly inclinate or exclinate, if 2–4 strong reclinate pairs then postocellar bristles divergent. Antenna rarely porrect, vibrissa variable.

Frons with 2–4 (usually 3) pairs of reclinate orbital bristles and postocellar bristles strongly convergent (cruciate). Antenna porrect (directed forward). Vibrissa strong. **Pseudopomyzidae**

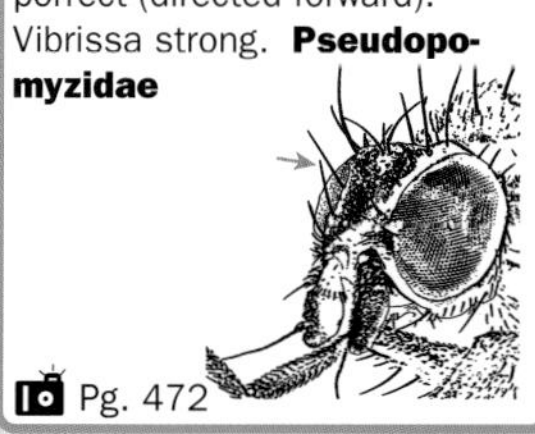

Pg. 472

More than 3 fronto-orbitals (almost always), or lower pair proclinate or exclinate. Katepisternum with 1–2 bristles. Mouthparts often geniculate (bent at middle). Vibrissa sometimes far above oral margin. Subcostal break sometimes very large and flanked by "lappet." Usually black. Widespread. **Milichiidae** in part

Pg. 502

Three fronto-orbital bristles, lower pair inclinate. Katepisternum with 2 bristles. Mouthparts short. Vibrissa at oral margin. Gray or brown microtomentose flies associated with seashores. Australia and New Zealand. **Australimyzidae**

Pg. 505

General Glossary and Commentary on Taxonomic Terminology

(Excluding most terms defined in the text and structures named on labeled diagrams)

Apomorphic (or apotypic) characters: "new" or "derived" characters found only in a single lineage. The ptilinum, for example, is an apomorphic character that arose in the common ancestor of the Schizophora and is unique to the Schizophora. The opposite of apomorphic is plesiomorphic; for example, the lack of a ptilinum is a plesiomorphic attribute of the Aschiza. Apomorphic characters shared by different taxa are called synapomorphies and provide evidence of close relationship between the taxa.

Basal: a relative term often used to refer to taxa or lineages that are related to, but lack the derived characters of, a taxon or lineage to which they are being compared. Thus, to say that lower Diptera are basal to Brachycera indicates that lower Diptera are flies that lack defining features (such as short antennae) of the Brachycera. Basal also serves as a simple and easily understood, if imprecise, shorthand for the oldest or earliest diverging lineages within a taxon. In this case, to say that lower Diptera are basal lineages of Diptera indicates that the lower dipteran lineages existed before the appearance of derived characters that define the rest of the Diptera.

Although the idea of basal lineages within a taxon is intuitive and convenient, it can be subjective, especially if used to refer to a single lineage as basal to another lineage; it is more precise to refer to such paired lineages as sister groups, as neither is really basal to the other in any absolute sense. Basal lineages are often inappropriately grouped together into paraphyletic taxa (such as Nematocera) that are defined by the absence of characters that define more derived lineages. In Figure 1.2 the lower Diptera comprise several basal lineages relative to the Brachycera (higher Diptera). In a taxonomic paper these lineages would be referred to in terms of their putative sister-group relationships: for example, Deuterophlebiidae is the sister group to all other Diptera, Nymphomyiidae is the sister group to all Diptera other than Deuterophlebiidae, Tipulomorpha is the sister group to all Diptera other than Deuterophlebiidae and Nymphomyiidae, and so forth.

Clade: a lineage or natural group that includes all the descendants of a single ancestor (and the ancestor itself). Think of it as a whole branch of a tree with all of its twigs. Clades are also referred to as monophyletic groups. In Figure 1.2, for example, compare the lower Diptera – which are not monophyletic, and thus a grade rather than a clade – to the higher Diptera, or Brachycera, which are monophyletic and thus a clade.

Common name: a local name for an organism, which often differs from place to place, unlike its scientific name. Formal common names that refer to a single species are capitalized (Black Blow Fly); those that refer to more than one species are not (blow fly). Common names containing the word fly are written as two words if they refer to Diptera and as one word if they refer to some other order such as dragonfly, mantisfly or mayfly.

Derived lineage: a lineage of relatively recent origin characterized by the "derived" or apomorphic states of the characters under consideration.

Described species: a species that has been given a scientific name (genus species) in a published description that meets the criteria outlined in the International Code of Zoological Nomenclature.

Disjunct distribution: a distribution with two or more widely separated, or disjunct, components.

Endemic: describes taxa that are found only in circumscribed areas, which are then called areas of endemism.

Exotic: describes taxa that are non-native to their environs. Usually they have been introduced from another continent by human activity.

Family: related genera that represent a distinctive, usually widespread, group that is generally recognizable in the field. See also Higher taxa.

Genus: a named group of related species (plural: genera). Some genera are defined very broadly and include hundreds of species, while others include only a single species. An ideal genus should be natural (monophyletic) and recognizable, but taxonomists often disagree about the limits of genera. See also Higher taxa.

Grade: a group of similar taxa that do not together form a monophyletic group (thus a grade is a paraphyletic group). The lower Diptera and the lower bee flies are examples of grades.

Groundplan: a term sometimes used for the characteristics of a hypothetical common ancestor of a group.

Higher: taxon names prefaced with "higher" are considered to be monophyletic or natural groups. The term distinguishes these lineages from related basal lineages, often collectively referred to as the "lower" group. Compare the lower Diptera (which are paraphyletic) to the higher Diptera or Brachycera (monophyletic) in Figure 1.2.

Higher taxa: taxa recognized at categorical levels above the species level. All species are organized into genera and families, genera are often divided into subgenera, families are almost always divided into subfamilies, and subfamilies are often divided into tribes. Groups of similar families are usually recognized as superfamilies. Although the International Code of Zoological Nomenclature regulates the names (not the definitions) of taxa up to the superfamily level, the limits of each taxon are subjective and taxonomists differ widely in how they define some taxa. The same lineage or clade might be split into multiple taxa, lumped into one taxon, or not recognized at all. Taxa above the superfamily level are not regulated by the Code.

Instar: the period between two larval molts or between hatching and the first molt. A first-instar larva is one that has hatched from the egg but not yet molted to the second instar.

Introduced: describes species that have been moved from one area to another, either accidentally or deliberately.

Kleptoparasite: an organism that steals food from other organisms.

Lower: "Lower Diptera" and "lower Brachycera" are widely used informal names that refer to groups lacking the derived characters that define their "higher" counterparts (Brachycera, Eremoneura). Groups defined on the lack of derived characters are usually artificial or paraphyletic groups and are often described as sets of "basal lineages";

this is almost certainly the case for the lower Diptera but less certain for the lower Brachycera. The Aschiza and Acalyptratae are also well-known groups defined (and named) for their lack of the derived characters that define their "higher" counterparts (Schizophora, Calyptratae); Aschiza and Acalyptratae could be referred to as the "lower Cyclorrhapha" and "lower Schizophora" respectively. A similar lower/higher distinction is used within some families, with "lower bee flies" and "lower Tephritidae" referring to the basal lineages in the Bombyliidae and Tephritidae respectively.

Larviparous (or viviparous): describes flies that do not lay eggs but instead deposit active larvae.

Molting: the act of casting the integument and forming a new, different or larger one to allow for growth or development. Higher Brachycera larvae molt three times as larvae (that is, they have three larval instars). The third molt is from larva to pupa and the fourth molt is from pupa to adult. The third and fourth molts are thus termed metamorphic molts.

Monophyletic: see Clade

Myrmecophilous: describes species associated with ants.

Native: describes species that occur naturally in a given area but are not restricted to that area (see also Endemic).

New species: see Undescribed species

Ovipositor: a general term referring to abdominal structures used for laying eggs. Diptera are often described as lacking a "true" ovipositor since they lack an appendicular ovipositor like that found in other orders. Some female flies have a prominent conical or tubular abdominal apex, the oviscape, formed by fusion of the tergites and sternites of segment 7. The oviscape can be a sheath into which the ovipositor retracts and it can play a role in piercing hosts or substrates.

Ovoviviparous: describes flies that lay eggs that hatch immediately, just after or even just before leaving the female's reproductive tract.

Paraphyletic, polyphyletic: describe groups that are artificial or non-monophyletic. Think of a paraphyletic group as a branch that is missing a few twigs, and a polyphyletic group as a bundle of two or more branches taken from different parts of the tree.

Parasites: organisms that feed on other animals without killing them.

Parasitoids: organisms that develop at the expense of one host animal, killing it in the process.

Parthenogenesis: asexual reproduction; without fertilization, and thus without males. Some flies are facultatively parthenogenetic and produce males only occasionally.

Phoretic: describes organisms that hitch rides on other organisms.

Phylogenetic analysis: research that seeks to discover the relationships between lineages and to identify the lineages or clades that we recognize and study as families and other taxa. The main structure of the Dipteran tree of life has been confidently recognized for a long time, but new taxonomic work and associated phylogenetic analyses continually refine our understanding of how the Diptera have diversified. Phylogenetic trees like figures 6.1 and 8.1 are only hypotheses and should be recognized as such. Some branches of the Dipteran tree of life – such as the Brachycera and the Schizophora – are strongly corroborated hypotheses that are reflected in stable and useful taxon names. Others are contentious or weakly supported, and some aspects of Diptera higher classification remain unstable as more and more conflicting phylogenetic analyses are published.

Phytophagous: describes flies that eat living plants.

Predaceous: describes flies that consume more than one host animal.

Saprophagous: describes flies that feed on decomposing material.

Sarcosaprophagous: describes flies that develop in the flesh of dead vertebrates.

Sclerites: the hardened plates that make up the fly's exoskeleton.

Sister groups: two taxa that are each other's closest relatives. Think of a branch with only two twigs – those two twigs are sister groups.

Speciation: the division of one species into two, usually following geographic separation. Speciation is the engine of diversity, and the hundreds of thousands of fly species out there are the products of speciation events.

Species: the working units of evolution, usually thought of as groups of populations that actually or potentially interbreed and that have some sort of specific mate-recognition system.

Sympatric: describes species that occur in the same areas (opposite: allopatric).

Synanthropic: describes species that are associated with humans or human-modified environments.

Taxon: a named group of organisms, such as a species, a genus, or a family (plural: taxa).

Undescribed species: species that have not yet been given a formal scientific name (also called unnamed or new species).

Vectors: arthropods that can transmit microorganisms that cause disease. For example, *Anopheles* species are the vectors of malaria.

Viviparous (or larviparous): describes flies that do not lay eggs, instead releasing active larvae.

Acknowledgments

The order Diptera includes more than 160,000 named species (about 10 percent of all living things), and it is not unreasonable to suggest that at least that many remain to be discovered. That adds up to far too many species for one author to reasonably overview, so I have relied on the kindness of many friends in the dipterological community for advice. In particular, I was helped a great deal by scientists at the important institutions that house crucial collections of both flies and fly specialists. This book simply would not have been possible without the cooperation of colleagues associated with the Canadian National Collection of Insects in Ottawa (Jeffrey Cumming, Scott Brooks, Bradley Sinclair, Jeffrey Skevington, Monty Wood, James O'Hara and Owen Lonsdale), the United States National Museum in Washington (Wayne Mathis, Allan Norrbom, Christian Thompson, Norman Woodley, Raymond Gagne and Lloyd Knutson), the Australian Museum in Sydney (David McAlpine, Daniel Bickel and Shane McEvey), the California Department of Food and Agriculture in Sacramento (Martin Hauser, Eric Fisher, Stephen Gaimari, Peter Kerr and Alessandra Rung) and of course the University of Guelph Insect Collection in Guelph (especially Joel Kits, Matthias Buck, Steven Paiero, John Klymko, Andrew Young, Gil Miranda, Stephen Luk and Morgan Jackson).

In addition to the help I received from these centers of dipterology, almost every dipterist I know provided me with some advice or help with identification of the images that enhance this book, and many of my good friends in the community were incredibly patient and generous with their specialized expertise. Some of the dipterists I turned to are Peter Adler, Dalton Amorim, Kevin Barber, Art Borkent, Brian Brown, Keith Bayless, Irina Brake, Claudio Carvalho, Pierfilippo Cerretti, Luis Guillermo Chaverri, Greg Courtney, Peter Cranston, Greg Curler, Doug Currie, Greg Dahlem, Marc DeMeyer, Carl Dick, Torsten Dikow, Martin Ebejer, Neil Evenhuis, Hans Feijen, Jon Gelhaus, Igor Grichanov, Joel Gibson, Christian Gonzalez, David Grimaldi, David Hancock, Ralph Harbach, John and Barbara Ismay, Mathias Jaschhof, Elena Kameneva, Peter Kerr, Ashley Kirk-Spriggs, Valery Korneyev, Jason Londt, Verner Michelsen, Kevin Moulton, Shelah Morita, Lorenzo Munari, Riley Nelson, Andrey Ozerov, Thomas Pape, Marc Pollet, Adrian Pont, Graham Rotheray, Menno Reemer, Knut Rognes, Jade Savage, Geir Soli, Aubrey Scarbrough, Jan Sevcik, Gunilla Stahls, James Sublette, Tony Thomas, Michael von Tschirnhaus, John Vargas, Pekka Vilkamaa, Andrew Whittington, Terry Whitworth, Shaun Winterton, Chen Young, Rüdiger Wagner, Richard Wilkerson, Andrew Woznica, Don Webb, David Yeates, Tadeusz Zatwarnicki and Manuel Zumbado.

I am particularly indebted to Dave Cheung for his help with graphics and laboratory photography, especially his work on the illustrated keys. Laura Johnson and Diana Shum also helped with graphics and specimen images for the keys, and some key character illustrations were provided by the University of Guelph Insect Collection image library. Sections of the book manuscript were kindly reviewed by Brad Sinclair and Jeff Cumming (Empidoidea), Greg Courtney (Nematocera), Christian Thompson (Aschiza), Monty Wood (Tachinidae), Thomas Pape (Calyptratae), Martin Hauser (Lower Brachycera) Matthias Buck, David McAlpine and Joel Kits (Acalyptratae). Daniel Bickel, Kevin Barber, Thomas Pape and Norman Woodley kindly provided comments on all or most of the manuscript. I've had the pleasure of field time with many of the above colleagues, but I am especially indebted to Thomas Pape for our adventures in Africa, to Greg Courtney for our Patagonian and west coast collecting and to Kevin Barber for 40 years of sharing the pursuit of flies. Last but not least I would like to acknowledge Christine Schisler for her proofreading and patience, and Alex and Stephen Marshall for assisting with field work in a dozen countries.

Although almost all the photographs in this book are my own, several friends and photographers kindly allowed me to use their images of taxa not represented in my collection. Each such image is credited in the captions.

Selected References

General References on Fly Biology and Diversity

Brown, B.V., A. Borkent, J.M. Cumming, D.M. Wood, N.E. Woodley and M. Zumbado, eds. 2009, 2010. *Manual of Central American Diptera.* 2 vols. Ottawa: NRC Research Press.

Chandler, P.J., ed. 2010. *A Dipterist's Handbook*, 2nd ed. The Amateur Entomologist Series 15. London: Amateur Entomologists' Society.

Colless, D.H., and D.K. McAlpine. 1991. "Diptera (Flies)." Chapter 39 in I.D. Naumann, P.B. Came, J.F. Lawrence, E.S. Nielsen, J.P. Spradbery, R.W. Taylor, M.J. Whitten and M.J. Littlejohn, eds., *The Insects of Australia: A Textbook for Students and Research Workers*, vol. 2. Melbourne, Australia: Melbourne University Press.

Crosskey, R.W., ed. 1980. *Catalogue of the Diptera of the Afrotropical Region.* London: British Museum (Natural History).

Delfinado, M.D., and D.E. Hardy, eds. 1973–77. *A Catalog of the Diptera of the Oriental Region.* Honolulu: University of Hawaii Press.

Evenhuis, N.L., ed. 1989. *Catalog of the Diptera of the Australasian and Oceanian Regions.* Bishop Museum Special Publication 86. Honolulu: Bishop Museum and E.J. Brill.

Ferrar, P. 1987. *A Guide to the Breeding Habits and Immature Stages of Diptera Cyclorrhapha.* 2 vols. Entomonograph 8. Leiden and Copenhagen: E.J. Brill and Scandinavian Science Press.

Grimaldi, D., and M.S. Engel. 2005. *Evolution of the Insects.* Cambridge: Cambridge University Press.

Hennig, W. 1973. "Diptera." In W. Kukenthal, ed., *Handbuch der Zoologie.* Vol. 4, *Arthropoda.* New York: de Gruyter.

McAlpine, J.F., B.V. Peterson, G.E. Shewell, H.J. Teskey, J.R. Vockeroth and D.M. Wood, eds. 1981, 1987. *Manual of Nearctic Diptera.* 2 vols. Monographs 27 and 28. Ottawa: Research Branch, Agriculture Canada.

McAlpine, J.F., and D.M. Wood, eds. 1989. *Manual of Nearctic Diptera,* vol. 3. Monograph 32. Ottawa: Research Branch, Agriculture Canada.

Oldroyd, H. 1964. *The Natural History of Flies.* Worthing, UK: Littlehampton Book Services.

Oosterbroek, P. 1998. *The Families of Diptera of the Malay Archipelago.* Fauna Melanesia Handbooks 1. Leiden: Brill.

Papavero, N., ed. 1966–78. *A Catalog of the Diptera of the Americas South of the United States.* São Paulo: Departamento de Zoologia, Secretaria da Agricultura.

Pape, T., D. Bickel and R. Meier, eds. 2009. *Diptera Diversity: Status, Challenges and Tools.* Leiden: Brill.

Pape, T., and F.C. Thompson, eds. 2012. *Systema Dipterorum.* http://www.diptera.org/ (accessed March 2012).

Papp, L., and B. Darvas, eds. 1997, 1998, 2000. *Contributions to a Manual of Palaearctic Diptera.* 3 vols. Budapest: Science Herald.

Soos, A., and L. Papp, eds. 1984–93. *Catalogue of Palaearctic Diptera.* Budapest: Akadémiai Kiadó.

Stone, A., C.W. Sabrosky, W.W. Wirth, R.H. Foote and J. Coulson, eds. 1965. *A Catalog of the Diptera of America North of Mexico.* Agricultural Handbook 276. Washington, DC: USDA.

PART ONE: NATURAL HISTORY

Chapter 1: Life Histories of Flies

Brown, B.V., S.A. Marshall and D.M. Wood. 2009. "Natural History." Chapter 3 in B.V. Brown, A. Borkent, J.M. Cumming, D.M. Wood, N.E. Woodley and M. Zumbado, eds., *Manual of Central American Diptera*, vol. 1. Ottawa: NRC Research Press.

Ibrahim, I.A., and A.M. Gad. 1975. "The occurrence of paedogenesis in *Eristalis* larvae (Diptera: Syrphidae)." *Journal of Medical Entomology* 12: 268.

Meier, R., M. Kotrba and P. Ferrar. 1999. "Ovoviviparity and viviparity in Diptera." *Biological Reviews* 74: 199–258.

Sinclair, B.J. 1992. "A phylogenetic interpretation of the Brachycera (Diptera) based on the larval mandible and associated mouthpart structures." *Systematic Entomology* 17: 233–52.

Sivinski, J.M. 1998. "Phototropism, bioluminescence and the Diptera." *Florida Entomologist* 81: 282–92.

Skevington, J.H., and P.T. Dang, eds. 2002. "Exploring the diversity of flies." *Biodiversity* 3: 3–27.

Chapter 2: Flies, Plants and Fungi

Gagné, R.J. 1989. *The Plant-Feeding Gall Midges of North America.* Ithaca, NY: Cornell University Press.

Goldblatt, P., P. Bernhardt, P. Vogan and J.C. Manning. 2004. "Pollination by fungus gnats (Diptera: Mycetophilidae) and self-recognition sites in *Tolmiea menziesii* (Saxifragaceae)." *Plant Systematics and Evolution* 244: 55–67.

Gorham, J.R. 1976. "Orchid pollination by *Aedes* mosquitoes in Alaska." *American Midland Naturalist* 95: 208–10.

Johnson, S.D., and K.E. Steiner. 1997. "Long-tongued fly pollination and evolution of floral spur length in the *Disa draconis* complex (Orchidaceae)." *Evolution* 51: 45–53.

Larson, B.M.H., P.G. Kevan and D.W. Inouye. 2001. "Flies and flowers: I. The taxonomic diversity of anthophilous pollinating flies." *Canadian Entomologist* 133: 439–65.

Sakai, S. 2002. "*Aristolochia* spp. (Aristolochiaceae) pollinated by flies breeding on decomposing flowers in Panama." *American Journal of Botany* 89: 527–34.

Sugiura, N. 1996. "Pollination of the orchid *Epipactis thunbergii* by syrphid flies (Diptera: Syrphidae)." *Ecological Research* 11: 249–55.

Chapter 3: Flies and Vertebrates

Dick, C.W., and B.D. Patterson. 2006. "Bat flies: Obligate ectoparasites of bats." Chapter 11 in S. Morand, B.R. Krasnov and R. Poulin, eds. *Micromammals and Macroparasites: From Evolutionary Ecology to Management.* Springer.

Greenberg, B. 1971. *Flies and Disease.* Vol. 1, *Ecology, Classification and Biotic Associations.* Princeton, NJ: Princeton University Press.

———. 1973. *Flies and Disease.* Vol. 2, *Biology and Disease Transmission.* Princeton, NJ: Princeton University Press.

Mullen, G.R., and L.A. Durden, eds. 2009. *Medical and Veterinary Entomology*, 2nd ed. Waltham, MA: Academic Press.

Nuertova, P. 1977. "Sarcosaprophagous Insects as Forensic Indicators." In G.C. Tedeshi, W.G. Eckert and L.G. Tedeshi, eds., *Forensic Medicine: A Study in Trauma and Environmental Hazards.* Philadelphia: Saunders.

Sherman, R.A., M.J. Hall and S. Thomas. 2000. "Medicinal maggots: An ancient remedy for some contemporary afflictions." *Annual Review of Entomology* 45: 55–81.

Smith, K.G.V. 1986. *A Manual of Forensic Entomology.* London: British Museum.

Chapter 4: Flies and Invertebrates

Grimaldi, D., and T. Nguyen. 1999. "Monograph on the spittlebug flies, genus *Cladochaeta* (Diptera: Drosophilidae: Cladochaetini)." *Bulletin of the American Museum of Natural History* 241: 1–326.

Matile, L. 1981. "Discovery in the Neotropical region of a parasitic genus of Keroplatidae, *Planivora* Hickman, and notes on its relationships (Diptera, Mycetophiloidea)." *Papaeis Avuloso Zoologie São Paulo* 34: 141–44.

Muratori, F.B., S. Borlee and R.H. Messing. 2010. "Induced niche shift as an anti-predator response for an endoparasitoid." *Proceedings of the Royal Society*, series B, 277 (1687): 1475–80.

Sivinski, J., S.A. Marshall and E. Petersson. 1999. "Kleptoparasitism and phoresy in the Diptera." *Florida Entomologist* 82: 179–97.

PART TWO: DIVERSITY

Chapter 5: Origins and Distribution of the Diptera

Blagoderov, V., D.A. Grimaldi and N.C. Fraser. 2007. "How time flies for flies: Diverse Diptera from the Triassic of Virginia and early radiation of the order." *American Museum Novitates* 3572: 1–39.

Buck, M., and S.A. Marshall. 2006. "Revision of the Neotropical family Inbiomyiidae (Diptera, Schizophora)." *Contributions in Science* 512: 1–30.

Cranston, P.S. 2005. "Biogeographical History." In D.K. Yeates and B.M. Wiegmann, eds. *The Evolutionary Biology of Flies*, 274–311. New York: Columbia University Press.

Evenhuis, N. *Fossil Diptera Catalog*. 1994. http://hbs.bishopmuseum.org/fossilcat/fossintro.html (accessed November 2010).

———. 2009. "Hawaii's Diptera Biodiversity." In T. Pape, D. Bickel and R. Meier, eds. *Diptera Diversity: Status, Challenges and Tools*, 47–68. Leiden: Brill .

Irwin, M.E., E.I. Schlinger and F.C. Thompson. 2003. "Diptera: True Flies." In S.M. Goodman and J.P. Benstead, eds. *The Natural History of Madagascar*, 692–702. Chicago: University of Chicago Press.

Magnacca, K.N., and D.K. Price. 2012. "New species of Hawaiian picture wing *Drosophila* (Diptera: Drosophilidae), with a key to species." *Zootaxa* 3188: 1–30.

Marshall, S.A. 1996. "The Sphaeroceridae of the Juan Fernandez Islands." *Studia Dipterologica* 4: 165–71.

———. 1997. "A revision of the *Sclerocoelus galapagensis* group (Diptera; Sphaeroceridae; Limosininae)." *Insecta Mundi* 11: 97–115.

Marshall, S.A., M. Buck, J.H. Skevington and D. Grimaldi. 2009. "A revision of the family Syringogastridae." *Zootaxa* 196: 1–80.

Mazzarolo, L.A., and D.S. Amorim. 2000. "*Cratomyia macrorrhyncha*, a Lower Cretaceous brachyceran fossil from the Santana Formation, Brazil, representing a new species, genus and family of the Stratiomyomorpha (Diptera)." *Insect Systematics and Evolution* 31: 91–102.

Oosterbroek, P. 1998. *The Families of Diptera of the Malay Archipelago.* Fauna Malesiana Handbooks 1. Leiden: Brill.

Sinclair, B.J. 2009. "Dipteran Biodiversity of the Galapagos." In T. Pape, D. Bickel and R. Meier, eds., *Diptera Diversity: Status, Challenges and Tools*, 98–117. Leiden: Brill.

Sinclair, B.J., and B.R. Stuckenberg. 1995. "Review of the Thaumaleidae (Diptera) of South Africa." *Annals of the Natal Museum* 36: 209–14.

Yeates, D.K., M.E. Irwin and B.M. Wiegmann. 2003. "Ocoidae, a new family of asiloid flies (Diptera: Brachycera: Asiloidea), based on *Ocoa chilensis* gen. and sp.n. from Chile, South America." *Systematic Entomology* 28: 417–31.

Chapter 6: The Lower Diptera

Adler, P.H., D.C. Currie and D.M. Wood. 2004. *The Black Flies (Simuliidae) of North America.* Ithaca, NY: Cornell University Press.

Amorim, D. de S. 2000. "A new phylogeny and phylogenetic classification for the Canthyloscelidae (Diptera: Psychodomorpha)." *Canadian Journal of Zoology* 78: 1067–77.

Amorim, D. de S., and D.A. Grimaldi. 2006. "Valeseguyidae, a new family of Diptera in the Scatopsoidea, with a new genus in Cretaceous amber from Myanmar." *Systematic Entomology* 31: 508–16.

Amorim, D. de S., and E. Rindal. 2007. "Phylogeny of the Mycetophiliformia, with proposal of the subfamilies Heterotrichinae, Ohakuneinae, and Chiletrichinae for the Rangomaramidae (Diptera, Bibionomorpha)." *Zootaxa* 1535: 1–92.

Amorim, D. de S., and D.K. Yeates. 2006. "Pesky gnats: Ridding dipteran classification of the Nematocera." *Studia Dipterologica* 13: 3–9.

Andersen, T., and O.A. Sæther. 1994. "*Usambaromyia nigrala* gen. n., sp.n., and Usambaromyiinae, a new subfamily among the Chironomidae (Diptera)." *Aquatic Insects* 16(1): 21–29.

Armitage, P.D., P.S. Cranston and L.C.V. Pinder, eds. 1995. *The Chironomidae: Biology and Ecology of Non-biting Midges.* Chapman and Hall.

Bertone, M.A., G.W. Courtney and B.M. Wiegmann. 2008. "Phylogenetics and temporal diversification of the earliest true flies (Insecta: Diptera) based on multiple nuclear genes." *Systematic Entomology* 33: 668–87.

Borkent, A. 2008. "The frog-biting midges of the world (Corethrellidae: Diptera)." *Zootaxa* 1804: 1–456.

Borkent, A., and W.W. Wirth. 1997. "World species of biting midges (Diptera: Ceratopogonidae)." *Bulletin of the American Museum of Natural History* 233: 1–257.

Colless, D.H. 1990. "*Valeseguya rieki*, a new genus and species of Diptera from Australia (Nematocera: Anisopodidae)." *Annales de la Societé Entomologique de France* 26(3): 351–53.

Courtney, G.W. 1991. "Phylogenetic analysis of the Blepharicerimorpha, with special reference to the mountain midges (Diptera: Deuterophlebiidae)." *Systematic Entomology* 16: 137–72.

———. 1994a. "Biosystematics of the Nymphomyiidae (Insecta: Diptera): Life history, morphology, and phylogenetic relationships." *Smithsonian Contributions to Zoology* 550: 1–41.

———. 1994b. "Revision of Palaearctic mountain midges (Diptera: Deuterophlebiidae), with phylogenetic and biogeographic analyses of world species." *Journal of Natural History* 24: 81–118.

Disney, R.H.L. 1971. "Association between black flies (Simuliidae) and prawns (Atyidae), with a discussion of this habit in simuliids." *Journal of Animal Ecology* 40: 39–51.

Grimaldi, D. 1991. "Mycetobiine woodgnats (Diptera: Anisopodidae) from the Oligo-Miocene amber of the Dominican Republic, and Old World affinities." *American Museum Novitates* 3014: 1–24.

Hennig, W. 1973. "Diptera." In W. Kukenthal, ed., *Handbuch der Zoologie*. Vol. 4, *Arthropoda.* New York: de Gruyter.

Jaschof, M. 2009. "*Ohakunea* Group." Chapter 18 in B.V. Brown, A. Borkent, J.M. Cumming, D.M. Wood, N.E. Woodley and M. Zumbado, eds., *Manual of Central American Diptera*, vol. 1. Ottawa: NRC Research Press.

Jaschhof, M., and R.K. Didham. 2002. "Rangomaramidae fam. nov. from New Zealand and its implications for the phylogeny of the Sciaroidea (Diptera: Bibionomorpha)." *Studia Dipterologica* Suppl. 11: 1–60.

Jaschhof, M., and C. Jaschhof. 2009. "The wood midges (Diptera: Cecidomyiidae: Lestremiinae) of Fennoscandia and Denmark." *Studia Dipterologica* Suppl. 18: 1–333.

Kjærandsen, J., and J.B. Jordal. 2007. "Fungus gnats (Diptera: Bolitophilidae, Diadocidiidae, Ditomyiidae, Keroplatidae and Mycetophilidae) from Møre og Romsdal." *Norwegian Journal of Entomology* 54: 147–71.

Madwar, S. 1935. "The biology and morphology of the immature stages of *Macrocera anglica* Edwards." *Psyche* 42: 25–34.

Matile, L. 1989. "Family Keroplatidae." In N. Evenhuis, ed., *Catalog of the Diptera of the Australasian and Oceanian Regions,* 28–133. Bishop Museum Special Publication 86. Honolulu and Leiden: Bishop Museum Press and E.J. Brill.

———. 1990. "Recherches sur la systématique et l'évolution des Keroplatidae (Diptera, Mycetophiloidea)." *Mémoires du Muséum National d'Histoire Naturelle* 148: 1–682.

Michelsen, V. 1996. "Neodiptera: New insights into the adult morphology and higher level phylogeny of Diptera (Insecta)." *Zoological Journal of the Linnaean Society* 117: 71–102.

Oosterbroek, P. 2005–12. *Catalogue of the Crane Flies of the World.* http://ip30.eti.uva.nl/ccw/ (accessed 2010).

Petersen, M.J., M.A. Bertone, B.M. Wiegmann and G.W. Courtney. 2010. "Phylogenetic synthesis of morphological and molecular data reveals new insights into the higher-level classification of Tipuloidea (Diptera)." *Systematic Entomology* 35: 526–45.

Reinert, J.F. 2000. "New classification for the composite genus *Aedes* (Diptera: Culicidae: Aedini), elevation of subgenus *Ochlerotatus* to generic rank, reclassification of the other subgenera, and notes on certain subgenera and species." *Journal of the American Mosquito Control Association* 16: 175–88.

Reinert, J.F., R.E. Harbach and I.J. Kitching. 2004. "Phylogeny and classification of Aedini (Diptera: Culicidae) based on morphological characters of all life stages." *Zoological Journal of the Linnean Society* 142: 289–368.

———. 2008. "Phylogeny and classification of *Ochlerotatus* and allied taxa (Diptera: Culicidae: Aedini) based on morphological data from all life stages." *Zoological Journal of the Linnean Society* 153: 29–114.

Rindal, E., Ø. Gammelmo and G. Søli. 2008. "On the family Keroplatidae in Norway (Diptera, Mycetophiliformia)." *Norwegian Journal of Entomology* 55: 81–85.

Rindal, E., G. Søli and L. Bachmann. 2009. "On the systematics of the fungus gnat subfamily Mycetophilinae (Diptera): A combined morphological and molecular approach." *Journal of Zoological Systematics and Evolutionary Research* 47(3): 227–33.

Vockeroth, J.R. 2009. "Mycetophilidae." Chapter 15 in B.V. Brown, A. Borkent, J.M. Cumming, D.M. Wood, N.E. Woodley and M. Zumbado, eds., *Manual of Central American Diptera*, vol. 1. Ottawa: NRC Research Press.

Wiegmann, B.M., M.D. Trautwein, I.S. Winkler, N.B. Barr, J.W. Kim, C. Lambkin, M.A. Bertone, B.K. Cassel, K.M. Bayless, A.M. Heimberg, B.M. Wheeler, K.J. Peterson, T. Pape, B.J. Sinclair, J.H. Skevington, V. Blagoderov, J. Caravas, S.N. Kutty, U. Schmidt-Ott, G.E. Kampmeier,

F.C. Thompson, D.A. Grimaldi, A.T. Beckenbach, G.W. Courtney, M. Friedrich, R. Meier and D.K. Yeates. 2011. "Episodic radiations in the fly tree of life." *Proceedings of the National Academy of Sciences* 108(21): 8731–36.

Wood, D.M., and A. Borkent. 1989. "Phylogeny and Classification of the Nematocera." In J.F. McAlpine and D.M. Wood, eds., *Manual of Nearctic Diptera*, vol. 3, 1333–70. Monograph 32. Ottawa: Research Branch, Agriculture Canada.

Woodley, N.E., A. Borkent and T.A. Wheeler. 2009. "Phylogeny of the Diptera." Chapter 5 in B.V. Brown, A. Borkent, J.M. Cumming, D.M. Wood, N.E. Woodley and M. Zumbado, eds., *Manual of Central American Diptera*, vol. 1. Ottawa: NRC Research Press.

Chapter 7: The Lower Brachycera and Empidoidea

Andersson, H. 1974. "Studies on the myrmecophilous fly *Glabellula arctica* (Zett.) (Diptera, Bombyliidae)." *Entomologica Scandinavica* 5: 29–38.

Bernardi, N. 1973. "The genera of the family Nemestrinidae (Diptera: Brachycera)." *Arquivos de Zoologia* (Brazil) 24: 211–318.

———. 1976. "Classificação de subfamilia Hirmoneurinae (Diptera, Nemestrinidae)." *Papéis Avulsos de Zoologia* 30: 25–33.

Bickel, D.J. 1994. "The Australian Sciapodinae (Diptera: Dolichopodidae), with a review of the Oriental and Australasian faunas, and a world conspectus of the subfamily." *Records of the Australian Museum* Suppl. 21: 1–394.

———. 2006. "*Papallacta* (Diptera: Dolichopodidae), a new stenopterous genus from the páramo of Ecuador." *Tijdschrift voor Entomologie* 149: 209–13.

———. 2009. "Why *Hilara* Is Not Amusing: The Problem of Open-Ended Taxa and the Limits of Taxonomic Knowledge." Chapter 10 in T. Pape, D. Bickel and R. Meier, eds., *Diptera Diversity :Status, Challenges and Tools.* Leiden: Brill.

Bickel, D., and M.C. Hernandez. 2004. "Neotropical *Thrypticus* (Diptera: Dolichopodidae) reared from Water Hyacinth, *Eichhornia crassipes*, and other Pontederiaceae." *Annals of the Entomological Society of America* 97: 437–49.

Brammer, C.A., and C.D. von Dohlen. 2007. "Evolutionary history of Stratiomyidae (Insecta: Diptera): the molecular phylogeny of a diverse family of flies." *Molecular Phylogenetics and Evolution* 43: 660–73.

Brooks, S. 2005. "Systematics and phylogeny of Dolichopodinae (Diptera: Dolichopodidae)." *Zootaxa* 857: 1–158.

Bybee, S.M, S.D. Taylor, C.R. Nelson and M.F. Whiting. 2004. "A phylogeny of robber flies (Diptera: Asilidae) at the subfamilial level: Molecular evidence." *Molecular Phylogenetics and Evolution* 30: 787–97.

Chvála, M. 1983. "The Empidoidea (Diptera) of Fennoscandia and Denmark. II. General Part. The families Hybotidae, Atelestidae and Microphoridae." *Fauna Entomologica Scandinavica* 12: 1–279.

Dikow, T. 2009a. "Phylogeny of Asilidae inferred from morphological characters of imagines (Insecta: Diptera: Brachycera: Asiloidea)." *Bulletin of the American Museum of Natural History* 319: 1–174.

———. 2009b. "A phylogenetic hypothesis for Asilidae based on a total evidence analysis of morphological and DNA sequence data (Insecta: Diptera: Brachycera: Asiloidea)." *Organisms, Diversity and Evolution* 9: 165–88.

Evenhuis, N.L. 1985. "New western North American homoeophthalmine Bombyliidae (Diptera)." *Polskie Pismo Entomologiczne* 55: 505–12.

———. 1990. "Systematics and evolution of the genera in the subfamilies Usiinae and Phthiriinae (Diptera: Bombyliidae) of the world." Entomonograph 11: 1–72.

———. 2002. "*Pieza*, a new genus of microbombyliids from the New World (Diptera: Mythicomyiidae)." *Zootaxa* 36: 1–28.

———. 2009. "Hawaii's Diptera Biodiversity." In T. Pape, D. Bickel and R. Meier, eds., *Diptera Diversity: Status, Challenges and Tools*, 47–68. Leiden: Brill.

Evenhuis, N.L., and D.J. Greathead. 1999. *World Catalog of Bee Flies (Diptera: Bombyliidae).* Leiden: Backhuys.

Fisher, E. 2009. "Asilidae." Chapter 45 in B.V. Brown, A. Borkent, J.M. Cumming, D.M. Wood, N.E. Woodley and M. Zumbado, eds., *Manual of Central American Diptera*, vol. 1. Ottawa: NRC Research Press.

Geller-Grimm, F. 1998. "Notes on the biology of *Dasypogon diadema* (Fabricius, 1781) (Diptera: Asilidae)." *Mitteilungen des Internationalen Entomologischen Vereins* 23(1/2): 17–32.

Griffiths, G.C.D. 1994. "Relationships among the major subgroups of Brachycera (Diptera): A critical review." *Canadian Entomologist* 126: 861–80.

Grimaldi, D., and J. Cumming. 1999. "Brachyceran Diptera in Cretaceous ambers and Mesozoic diversification of the Eremoneura." *Bulletin of the American Museum of Natural History* 239: 1–124.

Hull, F.M. 1973. "Bee flies of the world: The genera of the family Bombyliidae." *Bulletin of the United States National Museum* 286: 1–687.

Hurley, R.L., and J.B. Runyon. 2009. "A review of *Erebomyia* (Diptera: Dolichopodidae), with descriptions of three new species." *Zootaxa* 2054: 38–48.

Kerr, P.H. 2010. "Phylogeny and classification of Rhagionidae, with implications for Tabanomorpha (Diptera: Brachycera)." *Zootaxa* 2592: 1–133.

Mackerras, I.M., and M.E. Fuller. 1942. "The genus *Pelecorhynchus* (Diptera, Tabanoidea)." *Proceedings of the Linnean Society of New South Wales* 67: 9–76.

McAlpine, J.F., B.V. Peterson, G.E. Shewell, H.J. Teskey, J.R. Vockeroth and D.M. Wood, eds. 1981, 1987. *Manual of Nearctic Diptera.* Monographs 27 and 28. Ottawa: Research Branch, Agriculture Canada.

Moulton, J.K.. and B.M. Wiegmann. 2007. "The phylogenetic relationships of flies in the superfamily Empidoidea (Insecta: Diptera)." *Molecular Phylogenetics and Evolution* 43: 701–13.

Nagatomi, A. 1982. "The genera of Rhagionidae (Diptera)." *Journal of Natural History* 16: 31–70.

———. 1992. "Notes on the phylogeny of various taxa of the orthorrhaphous Brachycera (Insecta: Diptera)." *Zoological Science* 9: 843–57.

Nagatomi, A., and N. Liu. 1991. "Apystomyiidae, a new family of Asiloidea (Diptera)." *Acta Zoologica Academiae Scientiarum Hungaricae* 40(3): 203–18.

———. 1994. "Notes on the Proratinae (Diptera: Scenopinidae)." *South Pacific Study* 14: 137–222.

———. 1995. "Spermatheca and female terminalia of Pantophthalmidae and Xylophagidae s. lat." *Annals of the Entomological Society of America* 88: 603–26.

Nagatomi, A., T. Saigusa, H. Nagatomi and L. Lyneborg. 1991a. "Apsilocephalidae, a new family of the Orthorrhaphous Brachycera (Insecta, Diptera)." *Zoological Science* 8: 579–91.

———. 1991b. "The systematic position of the Apsilocephalidae, Rhagionempididae, Protempididae, Hilarimorphidae, Vermileonidae and some genera of Bombyliidae (Insecta, Diptera)." *Zoological Science* 8: 593–607.

Palmer, C.M., and D.K. Yeates. 2000. "The phylogenetic importance of immature stages: Solving the riddle of *Exeretonevra* Macquart (Diptera: Xylophagidae)." *Annals of the Entomological Society of America* 93: 15–27.

Pape, T., and F.C. Thompson, eds. 2012. *Systema Dipterorum.* http://www.diptera.org/ (accessed January 2012).

Paramonov, S.J. 1953. "A review of Australian Nemestrinidae (Diptera)." *Australian Journal of Zoology* 1(2): 242–90.

Pollet, M., and S. Brooks. 2008. "Long-legged Flies (Diptera: Dolichopodidae)." In J.L. Capinera, ed., *Encyclopedia of Entomology*, vol. 1, 2232–41. Springer.

Robinson, H. 1969. "A monographic study of the Mexican species of *Enlinia* (Diptera: Dolichopodidae)." *Smithsonian Contributions to Zoology* 25: 1–62.

———. 1975. "Bredin-Archbold-Smithsonian biological survey of Dominica: The family Dolichopodidae with some related Antillean and Panamanian species." *Smithsonian Contributions to Zoology* 185: 1–141.

Rozkošný, R. 1982. *A Biosystematic Study of the European Stratiomyidae (Diptera).* Vol. 1, *Introduction: Beridinae, Sarginae and Stratiomyinae.* Series Entomologica 21.

———. 1983. *A Biosystematic Study of the European Stratiomyidae (Diptera).* Vol. 2, *Clitellariinae, Hermetiinae, Pachygasterinae and Bibliography.* Series Entomologica 25.

Runyon, J.B., and R.L. Hurley. 2004. "A new genus of long-legged flies displaying remarkable wing directional asymmetry." *Proceedings of the Royal Society of London* Series B (Suppl.), 271: 114–16.

Schlinger, E.I. 1987. "The Biology of Acroceridae (Diptera): True Endoparasitoids of Spiders." In W. Nentwig, ed., *Ecophysiology of Spiders*, 319–27. Berlin: Springer.

Sinclair, B.J. 2010. "Revision and phylogenetic systematics of the Neotropical Ceratomerinae (Insecta: Diptera: Empidoidea: Brachystomatidae)." *Arthropod Systematics and Phylogeny* 68(2): 197–228.

Sinclair, B.J., and J.M. Cumming. 2006. "The morphology, higher-level phylogeny and classification of the Empidoidea (Diptera)." *Zootaxa* 1180: 1–172.

Sinclair, B.J., J.M. Cumming and D.M. Wood. 1994. "Homology and phylogenetic implications of male genitalia in Diptera–Lower Brachycera." *Entomologica Scandinavica* 24: 407–32.

Sinclair, B.J., and A.H. Kirk-Spriggs. 2010. "*Alavesia* Waters and Arillo: A Cretaceous-era genus discovered extant on the Brandberg Massif, Namibia (Diptera: Atelestidae)." *Systematic Entomology* 35: 268–76.

Sommerman, K.M. 1962. "Alaskan snipe fly immatures and their habitat (Rhagionidae: *Symphoromyia*)." *Mosquito News* 22: 116–23.

Stuckenberg, B.R. 1973. "The Athericidae, a new family in the lower Brachycera (Diptera)." *Annals of the Natal Museum* 21: 649–73.

———. 2000. "A new genus and species of Athericidae (Diptera: Tabanoidea) from Cape York Peninsula." *Records of the Australian Museum* 52: 151–59.

———. 2001. "Pruning the tree: A critical review of classifications of the Homeodactyla (Diptera, Brachycera), with new perspectives and an alternative classification." *Studia Dipterologica* 8: 1–41.

Stuckenberg, B.R., and F.C. Thompson. 2009. "Nearctic Diptera: Twenty Years Later." Chapter 1 in T. Pape, D. Bickel and R. Meier, eds., *Diptera Diversity: Status, Challenges and Tools.* Leiden: Brill.

Teskey, H.J. 1970. "The immature stages and phyletic position of *Glutops rossi* (Diptera: Pelecorhynchidae)." *The Canadian Entomologist* 102: 1171–79.

Trautwein, M.D., B.M. Wiegmann and D.K. Yeates. 2010. "A multigene phylogeny of the fly superfamily Asiloidea (Insecta): Taxon sampling and additional genes reveal the sister group to all higher flies (Cyclorrhapha)." *Molecular Phylogenetics and Evolution* 56: 918–30.

———. 2011. "Overcoming the effects of rogue taxa: Evolutionary relationships of the bee flies." *PLoS Currents: Tree of Life.* http://currents.plos.org/treeoflife/article/overcoming-the-effects-of-rogue-taxa-2cn7m3af919c4-1/.

Ulrich, H. 2003. "How recent are the Empidoidea of Baltic amber?" *Studia Dipterologica* 10: 321–27.

Val, F.C. 1976. "Systematics and evolution of the Pantophthalmidae (Diptera, Brachycera)." *Arquivos de Zoologia* 27: 51–164.

Webb, D.W. 1974. "A revision of the genus *Hilarimorpha* (Diptera: Hilarimorphidae)." *Journal of the Kansas Entomological Society* 47: 172–222.

———. 1994. "The immature stages of *Suragina concinna* (Williston) (Diptera: Athericidae)." *Journal of the Kansas Entomological Society* 67: 421–25.

Wheeler, W.M. 1930. *Demons of the Dust.* New York: W.W. Norton.

Wiegmann, B.M., D. K. Yeates, J.L. Thorne and H. Kishino. 2003. "Time flies: A new molecular time-scale for fly evolution without a clock." *Systematic Biology* 52: 745–56.

Wiegmann, B.M., M.D. Trautwein, I.S. Winkler, N.B. Barr, J.W. Kim, C. Lambkin, M.A. Bertone, B.K. Cassel, K.M. Bayless, A.M. Heimberg, B.M. Wheeler, K.J. Peterson, T. Pape, B.J. Sinclair, J.H. Skevington, V. Blagoderov, J. Caravas, S.N. Kutty, U. Schmidt-Ott, G.E. Kampmeier, F.C. Thompson, D.A. Grimaldi, A.T. Beckenbach, G.W. Courtney, M. Friedrich, R. Meier and D.K. Yeates. 2011. "Episodic radiations in the fly tree of life." *Proceedings of the National Academy of Sciences* 108(21): 8731–36.

Winterton, S.L., B.M. Wiegmann and E.I. Schlinger. 2007. "Phylogeny and Bayesian divergence time estimations of small-headed flies (Diptera: Acroceridae) using multiple molecular markers." *Molecular Phylogenetics and Evolution* 43: 808–32.

Woodley, N.E. 1986. "Parhadrestiinae, a new subfamily for *Parhadrestia* James and *Cretaceogaster* Teskey (Diptera: Stratiomyidae)." *Systematic Entomology* 11: 377–87.

———. 1989. "Phylogeny and Classification of the 'Orthorraphous' Brachycera." In J.F. McAlpine, ed., *Manual of Nearctic Diptera*, vol., 3, 1371–95. Monograph 32. Ottawa: Research Branch, Agriculture Canada.

———. 2001. *A World Catalog of the Stratiomyidae (Insecta: Diptera). Myia* 11: 1–473.

———. 2009. "Xylomyidae," "Stratiomyidae," and "Scenopinidae." Chapters 37, 38 and 47 in B.V. Brown, A. Borkent, J.M. Cumming, D.M. Wood, N.E. Woodley and M. Zumbado, eds., *Manual of Central American Diptera*, vol. 1. Ottawa: NRC Research Press.

Woodley, N.E., A. Borkent and T. Wheeler. 2009. "Phylogeny of the Diptera." Chapter 5 in B.V. Brown, A. Borkent, J.M. Cumming, D.M. Wood, N.E. Woodley and M. Zumbado, eds., *Manual of Central American Diptera*, vol. 1. Ottawa: NRC Research Press.

Yang, D., Y.J. Zhu, M.Q. Wang and L.L. Zhang. 2006. *World Catalog of Dolichopodidae (Insecta: Diptera).* Beijing: China Agricultural University Press.

Yeates, D.K. 1992. "Towards a monophyletic Bombyliidae (Diptera): The removal of the Proratinae (Diptera: Scenopinidae)." *American Museum Novitates* 3051: 1–30.

———. 1994. "Cladistics and classification of the Bombyliidae (Diptera: Asiloidea)." *Bulletin of the American Museum of Natural History* 219: 1–191.

———. 2002. "Relationships of extant lower Brachycera (Diptera): A quantitative synthesis of morphological characters." *Zoologica Scripta* 31: 105.

Yeates, D.K., and D. Greathead. 1997. "The evolutionary pattern of host use in the Bombyliidae: A diverse family of parasitoid flies." *Biological Journal of the Linnean Society* 60: 149–86.

Yeates, D.K., and M. E. Irwin. 1996. "Apioceridae (Insecta: Diptera): Cladistic reappraisal and biogeography." *Zoological Journal of the Linnean Society* 116: 247–301.

Yeates, D.K., M.E. Irwin and B.M. Wiegmann. 2003. "Ocoidae, a new family of asiloid flies (Diptera: Brachycera: Asiloidea), based on *Ocoa chilensis* gen. and sp.n. from Chile, South America." *Systematic Entomology* 28: 417–31.

———. 2006. "Evocoidae (Diptera: Asiloidea), a new family name for Ocoidae, based on *Evocoa*, a replacement name for the Chilean genus *Ocoa* Yeates, Irwin, and Wiegmann 2003." *Systematic Entomology* 31: 373.

Yeates, D.K., and C.L. Lambkin. 2004. "Bombyliidae. Bee Flies." Tree of Life Web Project, September 6. http://tolweb.org/Bombyliidae/23894/2004.09.06 (accessed March 2012).

Yeates, D.K., and B.M. Wiegmann. 2005. "Phylogeny and Evolution of Diptera: Recent Insights and New Perspectives." Chapter 2 in D.K. Yeates and B.M. Wiegmann, eds., *Evolutionary Biology of Flies.* New York: Columbia University Press.

Young, J.H., and D.J. Merritt. 2003. "The ultrastructure and function of the silk-producing basitarsus in the Hilarini (Diptera: Empididae)." *Arthropod Structure and Development* 32: 157–65.

Zloty, J., B.J. Sinclair and G. Pritchard. 2005. "Discovered in our backyard: A new genus and species of a new family from the Rocky Mountains of North America (Diptera, Tabanomorpha)." *Systematic Entomology* 30: 248–66.

Zumbado, M. 2006. *Diptera of Costa Rica and the New World Tropics.* Costa Rica: INBio Press.

Chapter 8: The Cyclorrhapha

Alcock, J., and D.W. Pyle. 1979. "The complex courtship of *Physiphora demandata* (Diptera: Otitidae)." *Zeitschrift für Tierpsychologie* 49: 352–62.

Baily, P.T. 1989. "The millipede parasitoid *Pelidnoptera nigripennis* (F.) (Diptera: Sciomyzidae) for the biological control of the millipede *Ommatoiulus moreleti* (Lucas) (Diplopoda: Julida: Julidae) in Australia." *Bulletin of Entomological Research* 79: 381–91.

Barnes, J.K. 1981. "Revision of the Helosciomyzidae (Diptera)." *Journal of the Royal Society of New Zealand* 11: 45–72.

Barraclough, D.A. 1995. "An illustrated identification key to the acalyptrate families (Diptera: Schizophora) occurring in southern Africa." *Annals of the Natal Museum* 36: 97–133.

———. 1999. "A review of South African species of Opomyzidae (Diptera: Schizophora), with description of a new species of *Opomyza* Fallén." *Annals of the Natal Museum* 40: 23–30.

———. 2007. "The distribution, ecology and phenology of the South African Natalimyzidae (Diptera: Schizophora: Sciomyzoidea)." *African Invertebrates* 48(2): 253–57.

Barraclough, D.A., and D.K. McAlpine. 2006. "Natalimyzidae, a new African family of acalyptrate flies." *African Invertebrates* 47: 117–34.

Bonduriansky, R. 1995. "A new Nearctic species of *Protopiophila* (Duda) (Diptera: Piophilidae), with notes on its behaviour and comparison with *P. latipes* (Meigen)." *Canadian Entomologist* 127: 859–63.

Brake, I. 2000. "Phylogenetic systematics of the Milichiidae (Diptera, Schizophora)." *Entomologica Scandinavica* Suppl. 57: 1–120.

Brake, I., and W. Mathis. 2007. "Revision of the genus *Australimyza* Harrison (Diptera: Australimyzidae)." *Systematic Entomology* 32: 252–75.

Brown, B.V. 1992. "Generic revision of Phoridae of the Nearctic region and phylogenetic classification of Phoridae, Sciadoceridae and Ironomyiidae (Diptera: Phoridea)." *Memoirs of the Entomological Society of Canada* 164: 1–144.

———. 1993. "Taxonomy and preliminary phylogeny of the parasitic genus *Apocephalus*, subgenus *Mesophora* (Diptera: Phoridae)." *Systematic Entomology* 18: 191–230.

———. 2005. "Revision of the *Melaloncha furcata*-group of bee-killing flies (Diptera: Phoridae)." *Insect Systematics and Evolution* 36: 241–58.

———. 2007. "A further new genus of primitive phorid fly (Diptera: Phoridae) from Baltic amber and its phylogenetic implications." *Contributions in Science* 513: 1–14.

Brown, B.V., and D.H. Feener, Jr. 1993. "Life history and immature stages of *Rhyncophoromyia maculineura*, an ant-parasitizing phorid fly (Diptera: Phoridae) from Peru." *Journal of Natural History* 27: 429–34.

Brown, B.V., and G. Kung. 2006. "Revision of the *Melaloncha ungulata*-group of bee-killing flies (Diptera: Phoridae)." *Contributions in Science* 507: 1–31.

Buck, M. 2006. "A new family and genus of acalypterate flies from the Neotropical region, with a phylogenetic analysis of Carnoidea family relationships (Diptera, Schizophora)." *Systematic Entomology* 31: 377–404.

Buck, M., and S.A. Marshall. 2006. "Revision of New World *Loxocera* (Diptera: Psilidae), with phylogenetic redefinition of Holarctic subgenera and species groups." *European Journal of Entomology* 103: 193–219.

Carles-Tolrá, M., P.C. Rodríguez and J. Verdú. 2010. "*Thyreophora cynophila* (Panzer, 1794): Collected in Spain 160 years after it was thought to be extinct (Diptera: Piophilidae: Thyreophorini)." *Boletín de la Sociedad Entomológica Aragonesa* 46: 1–7.

Chandler, P.J. 2001. "The flat-footed flies (Opetiidae and Platypezidae) of Europe." *Fauna Entomologica Scandinavica* 36: 1–27.

Chillcott, J.G. 1961. "A revision of the Nearctic species of Fanniinae (Diptera: Muscidae)." *Canadian Entomologist* Suppl. 14: 1–295.

Colless, D.H. 1994. "A new family of muscoid Diptera from Australasia, with sixteen new species in four new genera (Diptera: Axiniidae)." *Invertebrate Taxonomy* 8: 471–534.

Condon, M.A., D.C. Adams, D. Bann, K. Flaherty, J. Gammons, J. Johnson, M.L. Lewis, S. Marsteller, S.J. Scheffer, F. Serna and S. Swensen. 2008. "Uncovering tropical diversity: Six sympatric cryptic species of *Blepharoneura* (Diptera: Tephritidae) in flowers of *Gurania spinulosa* (Cucurbitaceae) in eastern Ecuador." *Biological Journal of the Linnean Society* 93: 779–97.

Crean, C.S., D.W. Dunn, T.H. Dat and A.S. Gilburn. 2000. "Female mate choice for large males in several species of seaweed fly (Diptera: Coelopidae)." *Animal Behavior* 59: 121–26.

Disney, R.H.L. 1986. "Two remarkable new species of scuttle-fly (Diptera: Phoridae) that parasitize termites (Isoptera) in Sulawesi." *Systematic Entomology* 11: 413–22.

———. 1991. "*Aenigmatistes* (Diptera: Phoridae), Aschiza with a ptilinum!" *Bonner Zoologische Beiträge* 42: 353–68.

———. 1994. *Scuttle Flies: The Phoridae.* Chapman and Hall.

Dobson, J.R. 1992. "Are adult Lonchaeidae (Diptera) specialized kleptoparasites of spiders' prey?" *British Journal of Entomology and Natural History* 5: 33–34.

Dupont, S., and T. Pape. 2009. "Termitophile and termite-associated scuttle flies (Diptera: Phoridae)." *Terrestrial Arthropod Reviews* 2: 3–40.

Ebejer, M.J. 2009. "A revision of Afrotropical Chyromyidae (excluding *Gymnochiromyia* Hendel) (Diptera: Schizophora), with the recognition of two subfamilies and the description of new genera." *African Invertebrates* 50: 321–434.

Eberhard, W.G. 1998. "Reproductive behavior of *Glyphidops flavifrons* and *Nerius plurivittatus* (Diptera: Neriidae)." *Journal of the Kansas Entomological Society* 71: 89–107.

Emden, F.I. van. 1950. "*Mormotomyia hirsuta* Austen (Diptera) and its systematic position." *Proceedings of the Royal Entomological Society of London* (B) 19: 121–28.

Feijen, H.R. 1983. "Systematics and phylogeny of Centrioncidae, a new afromontane family of Diptera (Schizophora)." *Zoologische Verhandelingen* 202: 1–137.

Feng-Yi Su, K., S.N. Kutty, S. Narayanan and R. Meier. 2008. "Morphology versus molecules: The phylogenetic relationships of Sepsidae (Diptera: Cyclorrhapha) based on morphology and DNA sequence data from ten genes." *Cladistics* 24: 902–16.

Ferrar, P. 1987. *A Guide to the Breeding Habits and Immature Stages of Diptera Cyclorrhapha.* 2 vols. Entomonograph 8. Leiden and Copenhagen: E.J. Brill and Scandinavian Science Press.

Foote, B.A. 1977. "Utilization of blue-green algae by larvae of shore flies." *Environmental Entomology* 6: 812–14.

Freidberg, A. 1981. "Taxonomy, natural history and immature stages of the bone skipper, *Centrophlebomyia furcata* (Fabricius) (Diptera: Piophilidae, Thyreophorina)." *Entomologica Scandinavica* 12: 320–26.

———. 1984. "The mating behavior of *Asteia elegantula*, with biological notes on some other Asteiidae (Diptera)." *Entomologia Generalis* 9(4): 217–-24.

———. 1994. "*Nemula*, a new genus of Neminidae (Diptera) from Madagascar." *Proceedings of the Entomological Society of Washington* 96: 471–82.

Gaimari, S.D., and V.C. Silva. 2010. "Revision of the Neotropical subfamily Eurychoromyiinae (Diptera: Lauxaniidae)." *Zootaxa* 2342: 1–64.

Gonzalez, L., and B.V. Brown. 2004. "New species and records of *Melaloncha* (*Udamochiras*) bee-killing flies (Diptera: Phoridae)." *Zootaxa* 730: 1–14.

Griffiths, G.C.D. 1972. *The Phylogenetic Classification of Diptera Cyclorrhapha, with Special Reference to the Structure of the Male Postabdomen.* The Hague: W. Junk.

Grimaldi, D.A. 2009. "The Asteioinea of Fiji (Insecta: Diptera: Periscelididae, Asteiidae, Xenasteiidae)." *American Museum Novitates* 3671: 1–59.

Guimarães, J.H. 1977. "A systematic revision of the Mesembrinellidae, stat. nov. (Diptera, Cyclorrhapha)." *Arquivos de Zoologia* 29: 1–109.

Han, H.Y. 2006. "Redescription of *Sinolochmostylia sinica* Yang, the first Palaearctic member of the little-known family Ctenostylidae (Diptera: Acalyptratae)." *Zoological Studies* 45: 3357–62.

Hardy, D.E. 1980. "Xenasteiidae, a new family of Schizophora (Diptera) from the Pacific and Indian Oceans." *Proceedings of the Hawaiian Entomological Society* 23: 205–25.

Hennig, W. 1969. "Neue Gattungen und Arten der Acalypteratae." *The Canadian Entomologist* 101: 589–633.

———. 1971. "Neue Untersuchungen über die Familien der Diptera Schizophora (Diptera: Cyclorrhapha)." *Stuttgarter Beiträge für Naturkunde* 226: 1–76.

Holloway, B.A. 1976. "A new bat-fly family from New Zealand (Diptera: Mystacinobiidae)." *New Zealand Journal of Zoology* 3: 279–301.

Klymko, J., and S.A. Marshall. 2008. "A review of the Nearctic Lonchopteridae." *Canadian Entomologist* 140: 649–73.

———. 2011. "Systematics of New World *Curtonotum* Macquart (Diptera: Curtonotidae)." *Zootaxa* 3079: 1–110.

Knutson, L., J.W. Stephenson and C.O. Berg. 1970. "Biosystematic studies of *Salticella fasciata* (Meigen), a snail-killing fly (Diptera: Sciomyzidae)." *Transactions of the Royal Entomological Society of London* 122: 81–100.

Knutson, L.V., and J-C. Vala. 2011. *Biology of Snail-Killing Sciomyzidae Flies.* Cambridge: Cambridge University Press.

Koenig, D., and C. Young. 2007. "First observation of parasitic relations between big-headed flies, *Nephrocerus* Zetterstedt (Diptera: Pipunculidae) and crane flies, *Tipula* Linnaeus (Diptera: Tipulidae Tipulinae), with larval and puparial descriptions for the genus *Nephrocerus*." *Proceedings of the Entomological Society of Washington* 109: 52–65.

Korneyev, V.A. 2000. "Phylogenetic Relationships among the Families of the Superfamily Tephritoidea." In M. Aluja and A.L. Norrbom, eds., *Fruit flies (Tephritidae): Phylogeny and Evolution of Behavior*, 3–22. Boca Raton, FL: CRC.

———. 2004. "Genera of Palaearctic Pyrgotidae (Diptera, Acalyptrata), with nomenclatural notes and a key." *Vestnik Zoologii* 38: 19–46.

Kutty, S.N., T. Pape, A.C. Pont, B.M. Wiegmann and R. Meier. 2008. "The Muscoidea (Diptera: Calyptratae) are paraphyletic: Evidence from four mitochondrial and four nuclear genes." *Molecular Phylogenetics and Evolution* 49: 639–52.

Kutty, S., T. Pape, B.M. Wiegmann and R. Meier. 2010. "Molecular phylogeny of the Calyptratae (Diptera: Cyclorrhapha), with an emphasis on the superfamily Oestroidea and the position of Mystacinobiidae and McAlpine's fly." *Systematic Entomology* 35: 614–35.

Lehrer, A.Z. 2005. *Bengaliidae du Monde (Insecta, Diptera).* Sofia and Moscow: Pensoft.

Lonsdale, O., and S.A. Marshall. 2005. "Family Clusiidae." Tree of Life Web Project. http://tolweb.org/tree?group=Clusiidae.

———. 2012. "*Sobarocephala* (Diptera: Clusiidae: Sobarocephalinae): Subgeneric classification and revision of the New World species." *Zootaxa* [in press].

Maa, T.C. 1963. "Genera and species of Hippoboscidae (Diptera): Types, synonymy, habitats and natural groupings." *Pacific Insects* Monograph 6: 1–186.

Malloch, J.R. 1932. "A new genus of diopsid-like Diptera (Periscelidae)." *Stylops* l: 266–68.

Mangan, R.L. 1979. "Reproductive behavior of the cactus fly, *Odontoloxozus longicornis*, male territoriality and female guarding as adaptive strategies." *Behavioral Ecology and Sociobiology* 4: 265–78.

Mangan, R.L., and D. Baldwin. 1986. "A new cryptic species of *Odontoloxozus* (Neriidae: Diptera) from the cape region of Baja California Sur (Mexico)." *Proceedings of the Entomological Society of Washington* 88: 110–21.

Marques, A.P., and R. Ale-Rocha. 2005. "Revisão do gênero *Willistoniella* Mik, 1895 (Diptera, Ropalomeridae) da Região Neotropical." *Revista Brasileira de Entomologia* 49(2): 210–27.

Marshall, S.A., and M. Buck. 2010. "Sphaeroceridae." Chapter 96 in B.V. Brown, A. Borkent, J.M. Cumming, D.M. Wood, N.E. Woodley and M. Zumbado, eds., *Manual of Central American Diptera*, vol. 2. Ottawa: NRC Research Press.

Marshall, S.A., M. Buck, J.H. Skevington and D. Grimaldi. 2009. "A revision of the family Syringogastridae." *Zootaxa* 196: 1–80.

Martin, O.Y., and D.J. Hosken. 2004. "Copulation reduces male but not female longevity in *Saltella sphondylli* (Diptera: Sepsidae)." *Journal of Evolutionary Biology* 17: 357–62.

Martín-Vega, D., A. Baz and V. Michelsen. 2010. "Back from the dead: *Thyreophora cynophila* (Panzer, 1798) (Diptera: Piophilidae) 'globally extinct' fugitive in Spain." *Systematic Entomology* 35: 607–13.

Mathis, W., and L. Papp, 1998. "Periscelididae." In L. Papp and B. Darvas, eds., *Contributions to a Manual of Palaearctic Diptera*, vol. 2. Budapest: Science Herald.

McAlpine, D.K. 1978. "Description and biology of a new genus of flies related to *Anthoclusia* and representing a new family (Diptera, Schizophora, Neurochaetidae)." *Annals of the Natal Museum* 23: 273–95.

———. 1983. "A new subfamily of Aulacigastridae (Diptera: Schizophora), with a discussion of aulacigastrid classification." *Australian Journal of Zoology* 31: 55–78.

———. 1985. "The Australian genera of Heleomyzidae (Diptera: Schizophora) and a reclassification of the family into tribes." *Records of the Australian Museum* 36: 203–51.

———. 1988. "Studies in upside-down flies (Diptera: Neurochaetidae). Part 2: Biology, adaptations and specific mating mechanisms." *Proceedings of the Linnaean Society of New South Wales* 110: 59–82.

———. 1990. "The taxonomic position of the Ctenostylidae (= Lochmostyliinae; Diptera: Schizophora)." *Memórias do Instituto Oswaldo Cruz* 84: 365–71.

———. 1991a. "Relationships of the genus *Heterocheila* (Diptera: Sciomyzoidea) with description of a new family." *Tijdschrift voor Entomologie* 134: 193–99.

———. 1991b. "Review of the Australian kelp flies (Diptera: Coelopidae)." *Systematic Entomology* 16: 29–84.

———. 1997a. Gobryidae, a new family of acalyptrate flies (Diptera: Diopsoidea), and a discussion of relationships of the diopsoid families. *Records of the Australian Museum* 49: 167-194.

———. 1997b. "Relationships of the Megamerinidae (Diptera: Nerioidea)." *Beiträge zur Entomologie* 47: 465–75.

———. 1998. "Neminidae." In P. Oosterbroek, *The Families of Diptera of the Malay Archipelago*. Fauna Melanesia Handbook 1. Leiden: Brill.

———. 2001. "Review of the Australasian genera of signal flies (Diptera: Platystomatidae)." *Records of the Australian Museum* 53: 113–99.

———. 2007. "Review of the Borboroidini or wombat flies (Diptera: Heteromyzidae), with reconsideration of the status of families Heleomyzidae and Sphaeroceridae, and descriptions of femoral gland-baskets." *Records of the Australian Museum* 59: 143–219.

———. 2008. "New extant species of ironic flies (Diptera: Ironomyiidae), with notes on ironomyiid morphology and relationships." *Proceedings of the Linnean Society of New South Wales* 129: 17–38.

McAlpine, D.K., and R.G. de Keyzer. 1994. "Generic classification of the fern flies (Diptera: Teratomyzidae), with a larval description." *Systematic Entomology* 19: 305–26.

McAlpine, J.F. 1967. "A detailed study of Ironomyiidae (Diptera: Phoridea)." *Canadian Entomologist* 99: 225–36.

———. 1989. "A revised classification of the Piophilidae, including 'Neottiophilinae' and 'Thyreophoridae' (Diptera: Schizophora)." *Memoirs of the Canadian Entomologist* 109: 1–66.

McAlpine, J.F., and G.C. Steyskal. 1982. "A revision of *Neosilba* McAlpine with a key to the world genera of Lonchaeidae (Diptera)." *Canadian Entomologist* 54: 504–5.

Meier, R. 1995. "Cladistic analysis of the Sepsidae (Cyclorrhapha: Diptera) based on a comparative scanning electron microscopic study of larvae." *Systematic Entomology* 20: 99–128.

———. 1996. "Larval morphology of the Sepsidae (Diptera: Sciomyzoidea), with a cladistic analysis using adult and larval characters." *Bulletin of the American Museum of Natural History* 228: 1–147.

Mello, R.L., and C.J.E. Lamas. 2010. "Revision of the South American species of Teretrurinae (Diptera: Pyrgotidae)." *Abstracts of the 7th International Congress of Dipterology*: 161.

Michelsen, V. 1983. "*Thyreophora anthropophaga* Robineau-Desvoidy, an 'extinct' bone-skipper rediscovered in Kashmir (Diptera: Piophilidae, Thyreophorina)." *Insect Systematics and Evolution* 14: 411–14.

———. 1991. "Revision of the aberrant New World genus *Coenosopsia* Diptera: Anthomyiidae, with a discussion of anthomyiid relationships." *Systematic Entomology* 16: 85–104.

———. 2007. "*Eginia ocypterata* (Meigen) (Diptera: Muscidae), an overlooked West Palaearctic parasitoid of Diplopoda, with an update of its known occurrence in Europe." *Studia Dipterologica* 13: 361–76.

Miller, R.M. 1984. "A new acalyptrate fly from southern Africa, possibly representing a new family." Abstracts of the 17th International Congress of Entomology, 20–26 August, Hamburg, Federal Republic of Germany: 32.

O'Grady, P., and R. DeSalle. 2008. "Out of Hawaii: The origin and biogeography of the genus *Scaptomyza* (Diptera: Drosophilidae)." *Biology Letters* 4: 195–99.

Papp, L. 1980. "New taxa of the acalyptrate flies (Diptera: Tunisimyiidae fam. n., Risidae, Ephydridae: Nannodastiinae subfam. n.)." *Acta Zoologica Academiae Scientiarum Hungaricae* 26: 415–31.

———. 1998. "Families of Heleomyzoidea." In L. Papp and B. Darvas, eds., *Contributions to a Manual of Palaearctic Diptera*, vol. 3. Budapest: Science Herald.

———. 2005. "Some acalyptrate flies from Taiwan." *Acta Zoologica Academiae Scientarum Hungaricae* 51: 187–213.

———. 2008. "New genera of the Old World Limosininae (Diptera, Sphaeroceridae)." *Acta Zoologica Academiae Scientiarum Hungaricae* 54 (Suppl. 2): 47–209.

———. 2011. "Description of a new genus and a new family, Circumphallidae fam. nov., of the acalyptrate flies (Diptera)." *Acta Zoologica Academiae Scientiarum Hungaricae* 57(4): 315–41.

Papp, L., and W.N. Mathis. 2001. "A review of the family Nannodastiidae (Diptera)." *Proceedings of the Entomological Society of Washington* 103(2): 337–48.

Papp, L., B. Merz and M. Foldvari. 2006. "Diptera of Thailand: A summary of the families and genera with reference to the species representations." *Acta Zoologica Academiae Scientiarum Hungaricae* 52: 97–269.

Pont, A.C. 1987. "The mysterious swarms of sepsid flies: An enigma solved?" *Journal of Natural History* 21: 305–17.

Preston-Mafham, K. 2001. "Resource defense mating system in two flies from Sulawesi: *Gymnonerius fuscus* Wiedemann and *Telostylinus* sp. near *duplicatus* Wiedemann (Diptera: Neriidae)." *Journal of Natural History* 35: 149–56.

Puniamoorthy, N., M.R.B. Ismail, D.S.H. Tan and R. Meier. 2009. From kissing to belly stridulation: Comparative analysis reveals surprising diversity, rapid evolution, and much homoplasy in the mating behaviour of 27 species of sepsid flies (Diptera: Sepsidae)." *Journal of Evolutionary Biology* 22: 2146–56.

Reemer, M. 2012. "Unravelling a hotchpotch: Phylogeny and classification of the Microdontinae (Diptera: Syrphidae)." PhD thesis, Leiden University.

Richards, O.W. 1973. "The Sphaeroceridae (= Borboridae or Cypselidae; Diptera Cyclorrhapha) of the Australian region." *Australian Journal of Zoology*, Suppl. ser. 22: 297–401.

Rognes, K. 1998. "Calliphoridae." In L. Papp and B. Darvas, eds., *Contributions to a Manual of Palaearctic Diptera*, vol. 3, 617–48. Budapest: Science Herald.

Rognes, K. 2011. A review of the monophyly and composition of the Bengaliinae with the description of a new genus and species, and new evidence for the presence of Melanomyinae in the Afrotropical Region (Diptera, Calliphoridae) *Zootaxa* 2964: 1-60.

Roháček, J. 2006. "A monograph of Palaearctic Anthomyzidae (Diptera), Part 1." *Časopis Slezského Zemského Muzea* 55: 1–328.

———. 2009. "A monograph of Palaearctic Anthomyzidae (Diptera), Part 2." *Časopis Slezského Zemského Muzea* 58: 1–180.

Roháček, J., and K.N. Barber. 2008. "New reduced-winged species of *Mumetopia*, with analysis of the relationships of this genus, *Chamaebosca* and allied genera (Diptera: Anthomyzidae)." *Acta Societas Zoologicae Bohemicae* 72: 191–215.

Rohácek, J., S.A. Marshall, A.L. Norrbom, M. Buck, D.I. Quiros and I. Smith. 2001. *World Catalog of Sphaeroceridae.* Opava, Czech Republic: Slezského Zemského Muzea.

Rotheray, G.E., M. Zumbado, E.G. Hancock and F.C. Thompson. 2000. "Remarkable aquatic predators in the genus *Ocyptamus* (Diptera, Syrphidae)." *Studia Dipterolologica* 7: 385–98.

Rozkosny, R. 1998. "Sciomyzidae." In L. Papp and B. Darvas, eds., *Contributions to a Manual of Palaearctic Diptera*, vol. 3. Budapest: Science Herald.

Rung, A. 2011. "A revision of the genus *Aulacigaster* Macquart (Diptera: Aulacigastridae)." *Smithsonian Contributions to Zoology* 633: 1–132.

Rung, A., W.N. Mathis and L. Papp. 2005. "*Curiosimusca*, gen. nov., and three new species in the family Aulacigastridae from the Oriental Region (Diptera: Opomyzoidea)." *Zootaxa* 1009: 21–36.

Sabrosky, C.W. 1957. "Synopsis of the New World species of the dipterous family Asteiidae." *Annals of the Entomological Society of America* 50: 43–61.

———. 1999. "Family-group names in Diptera: An annotated catalog." *Myia* 10: 1–360.

Scheffer, S.J., I.S. Winkler and B.M. Wiegmann. 2007. "Phylogenetic relationships within the leaf-mining flies (Diptera: Agromyzidae) inferred from sequence data from multiple genes." *Molecular Phylogenetics and Evolution* 42: 756–75.

Schneider, M.A. 2010. "A taxonomic revision of Australian Conopidae (Insecta: Diptera)." *Zootaxa* 2581: 1–246.

Schuehli, G.S. e, C.J.B. de Carvalho and B.M. Wiegmann. 2007. "Molecular phylogenetics of the Muscidae (Diptera: Calyptratae): New ideas in a congruence context." *Invertebrate Systematics* 21: 263–78.

Skevington, J.H. 2005. "Revision of Nearctic *Nephrocerus* Zetterstedt (Diptera: Pipunculidae)." *Zootaxa* 977: 1–36.

Skidmore, P. 1984. "*The biology of the Muscidae of the world.*" Series entomologica 29. Dordrecht, Netherlands: Junk.

Smith, K.G.V. 1967. "The biology and taxonomy of the genus *Stylogaster* Macquart, 1835 (Diptera: Conopidae, Stylogasterinae) in the Ethiopian and Malagasy regions." *Transactions of the Royal Entomological Society of London* 199: 47–69.

Smith, M.A., D.M. Wood, D.H. Janzen, W. Hallwachs and P.D.N. Hebert. 2007. "DNA barcodes affirm that 16 species of apparently generalist tropical parasitoid flies (Diptera: Tachinidae) are not all generalists." *Proceedings of the National Academy of Sciences USA* 104: 4967–72.

Smith, R.L. 1981. "The trouble with 'bobos,' *Paraleucopis mexicana* Steyskal, at Kino Bay, Sonora, Mexico (Diptera: Chamaemyiidae)." *Proceedings of the Entomological Society of Washington* 83: 406–12.

Ståhls, G., H. Hippa, G. Rotheray, J. Muona, and F. Gilbert. 2003. "Phylogeny of Syrphidae (Diptera) inferred from combined analysis of molecular and morphological characters." *Systematic Entomology* 28: 433–50.

Steffan, W.A. 1975. "Morphological and behavioral polymorphism in *Plastosciara perniciosa* (Diptera: Sciaridae)." *Proceedings of the Entomological Society of Washington* 77: 1–14.

Steyskal, G. 1981. "A new 'bobo' fly from the Gulf of California (Diptera: Chamaemyiidae: *Paraleucopis mexicana*)." *Proceedings of the Entomological Society of Washington* 83: 403–5.

Su, K.F.Y, S.N. Kutty, S. Narayanan and R. Meier. 2008. "Morphology versus molecules: The phylogenetic relationships of Sepsidae (Diptera: Cyclorrhapha) based on morphology and DNA sequence data from ten genes." *Cladistics* 24: 902–16.

Sueyoshi, M., L. Knutson and K. Ghorpade. 2009. "A taxonomic review of *Pelidnoptera* Rondani (Diptera: Sciomyzoidea), with discovery of a related new genus and species from Asia." *Insect Systematics and Evolution* 40: 389–409.

Sze, W.T., T. Pape and D.K O'Toole. 2008. "The first blow fly parasitoid takes a head start in its termite host (Diptera: Calliphoridae, Bengaliinae; Isoptera: Macrotermitidae)." *Systematics and Biodiversity* 6: 25–30.

Tenorio, J.M. 1972. "A revision of the Celyphidae (Diptera) of the Oriental region." *Transactions of the Royal Entomological Society of London* 123: 359–453.

Tschirnhaus, M. von. 2008. "Diptera Stelviana 4.3.01: Acartophthalmidae, Borboropsidae, Chyromyidae, Micropezidae, Odiniidae, Opetiidae, Periscelididae, Pseudopomyzidae, and Tanypezidae." *Studia Dipterologica* Suppl. 16: 65–97.

Ureña, O., and P. Hanson. 2010. "A fly larva (Syrphidae: *Ocyptamus*) that preys on adult flies." *Revista de Biologia Tropica* 58: 1157–63.

Vala, J., G. Gbedjissi, L. Knutson and C. Dossou. 2000. "Extraordinary feeding behavior in Diptera Sciomyzidae, snail-killing flies." *Comptes rendus de l'Académie des Sciences.* Series 3, *Sciences de la vie* 323: 299–304.

Valley, K., J.A. Novak and B.A. Foote. 1969. "Biology and immature stages of *Eumetopiella rufipes*." *Annals of the Entomological Society of America* 62: 227–34.

Wheeler, T.A. 2010. "Paraleucopidae." Chapter 73 in B.V. Brown, A. Borkent, J.M. Cumming, D.M. Wood, N.E. Woodley and M. Zumbado, eds., *Manual of Central American Diptera*, vol. 2. Ottawa: NRC Research Press.

White, A. 1916. "The Diptera-Brachycera of Tasmania. Part III: Families Asilidae, Bombyliidae, Empididae, Dolichopodidae and Phoridae." *Papers and Proceedings of the Royal Society of Tasmania* 217: 148–266.

Wiegmann, B.M., M.D. Trautwein, I.S. Winkler, N.B. Barr, J.W. Kim, C. Lambkin, M.A. Bertone, B.K. Cassel, K.M. Bayless, A.M. Heimberg, B.M. Wheeler, K.J. Peterson, T. Pape, B.J. Sinclair, J.H. Skevington, V. Blagoderov, J. Caravas, S.N. Kutty, U. Schmidt-Ott, G.E. Kampmeier, F.C. Thompson, D.A. Grimaldi, A.T. Beckenbach, G.W. Courtney, M. Friedrich, R. Meier and D.K. Yeates. 2011. "Episodic radiations in the fly tree of life." *Proceedings of the National Academy of Sciences* 108(21): 8731–36.

Zatwarnicki, T. 1992. "A new classification of Ephydridae based on phylogenetic reconstruction (Diptera: Cyclorrhapha)." *Genus* 3: 65–119.

Zumpt, F. 1965. *Myiasis in Man and Animals in the Old World.* London: Butterworths.

PART THREE: IDENTIFYING AND STUDYING FLIES

Chapters 9 and 10

Hamilton, J.R., D.K. Yeates, A. Hastings, D.H. Colless, D.K. McAlpine, D. Bickel, P.S. Cranston, M.A. Schneider, G. Daniels and S. Marshall. 2006. *On The Fly: The Interactive Atlas and Key to Australia Fly Families.* Compact disc. Queensland: Centre for Biological Information Technology.

Lonsdale, O., and S.A. Marshall. 2007. "Revision of the New World *Heteromeringia* (Diptera: Clusiidae: Clusiodinae)." *Beiträge zur Entomologie* 57: 37–80.

Lonsdale, O., D.K.B. Cheung and S.A. Marshall. 2011. "Key to the world genera and North American species of Clusiidae (Diptera: Schizophora)." *Canadian Journal of Arthropod Identification* 14 (3 May). http://www.biology.ualberta.ca/bsc/ejournal/lcm_14/lcm_14.html. doi:10.3752/cjai.2011.14. 2011.

Index

Note: Subgroups of the order Diptera (in **BOLD UPPERCASE** type) include all levels of classification down to the superfamily level. Families are listed separately (in **bold** type) and include subfamilies (in SMALL CAPITALS). Page numbers in bold type indicate a photograph in the text section.

A

D